AF719421

Human Diseases and Bacterial Pathogens

Human Diseases and Bacterial Pathogens

Editor: Abraham Bennett

AMERICAN
MEDICAL PUBLISHERS
www.americanmedicalpublishers.com

AMERICAN
MEDICAL PUBLISHERS
www.americanmedicalpublishers.com

Cataloging-in-Publication Data

Human diseases and bacterial pathogens / edited by Abraham Bennett.
p. cm.
Includes bibliographical references and index.
ISBN 978-1-63927-909-8
1. Bacterial diseases. 2. Pathogenic bacteria. 3. Medical bacteriology 4. Diseases. I. Bennett, Abraham.
QR201.B34 H86 2023
616.014--dc23

© American Medical Publishers, 2023

American Medical Publishers,
41 Flatbush Avenue,
1st Floor, New York,
NY 11217, USA

ISBN 978-1-63927-909-8 (Hardback)

This book contains information obtained from authentic and highly regarded sources. Copyright for all individual chapters remain with the respective authors as indicated. All chapters are published with permission under the Creative Commons Attribution License or equivalent. A wide variety of references are listed. Permission and sources are indicated; for detailed attributions, please refer to the permissions page and list of contributors. Reasonable efforts have been made to publish reliable data and information, but the authors, editors and publisher cannot assume any responsibility for the validity of all materials or the consequences of their use.

Trademark Notice: Registered trademark of products or corporate names are used only for explanation and identification without intent to infringe.

Contents

Preface

Bacteria can be found in almost every ecological niche on the planet. While most of them are harmless to humans, some of them can cause diseases. These bacteria are known as pathogenic bacteria. They have evolved sophisticated multi-protein secretion systems to interact with their environment. The proteins secreted by pathogenic bacteria into the cells of their host are termed as bacterial effectors. The secretion systems play a significant role in bacterial competitiveness and the pathogenesis of a variety of clinically relevant opportunistic and professional human pathogens. A wide range of intracellular processes including phagocytosis, endocytic trafficking, secretory trafficking, secretory pathway, apoptosis, etc. are regulated by bacterial effectors. The topics included in this book on bacterial pathogens and their relantionship with human diseases, are of utmost significance and bound to provide incredible insights to readers. It consists of contributions made by international experts. Those in search of information to further their knowledge will be greatly assisted by this book.

The information shared in this book is based on empirical researches made by veterans in this field of study. The elaborative information provided in this book will help the readers further their scope of knowledge leading to advancements in this field.

Finally, I would like to thank my fellow researchers who gave constructive feedback and my family members who supported me at every step of my research.

Editor

1

From Welfare to Warfare: The Arbitration of Host-Microbiota Interplay by the Type VI Secretion System

*Thomas E. Wood [1,2], Ezra Aksoy [3] and Abderrahman Hachani [4]**

[1] Division of Infectious Diseases, Department of Medicine, Massachusetts General Hospital, Boston, MA, United States, [2] Department of Microbiology, Harvard Medical School, Boston, MA, United States, [3] Centre for Biochemical Pharmacology, William Harvey Research Institute, Queen Mary University of London, London, United Kingdom, [4] Department of Microbiology and Immunology, University of Melbourne at the Peter Doherty Institute for Infection and Immunity, Melbourne, VIC, Australia

***Correspondence:**
Abderrahman Hachani
abderrahman.hachani@unimelb.edu.au

The health of mammals depends on a complex interplay with their microbial ecosystems. Compartments exposed to external environments such as the mucosal surfaces of the gastrointestinal tract accommodate the gut microbiota, composed by a wide range of bacteria. The gut microbiome confers benefits to the host, including expansion of metabolic potential and the development of an immune system that can robustly protect from external and internal insults. The cooperation between gut microbiome and host is enabled in part by the formation of partitioned niches that harbor diverse bacterial phyla. Bacterial secretion systems are commonly employed to manipulate the composition of these local environments. Here, we explore the roles of the bacterial type VI secretion system (T6SS), present in ~25% of gram-negative bacteria, including many symbionts, in the establishment and perturbation of bacterial commensalism, and symbiosis in host mucosal sites. This versatile apparatus drives bacterial competition, although in some cases can also interfere directly with host cells and facilitate nutrient acquisition. In addition, some bacterial pathogens cause disease when their T6SS leads to dysbiosis and subverts host immune responses in defined animal models. This review explores our knowledge of the T6SS in the context of the "host-microbiota-pathogen" triumvirate and examines contexts in which the importance of this secretion system may be underappreciated.

Keywords: gut microbiome, type six secretion system, commensal, symbiosis, dysbiosis, mucosal immunity, tolerance, MAMPs

INTRODUCTION

The gut tissue is composed of hundreds of millions of cells whilst providing a home for a microbiota containing trillions of bacteria (Sender et al., 2016). The association of the microbiota with our tissues is central for homeostatic and developmental mechanisms and thus governs many aspects of human health (Belkaid and Harrison, 2017). Due to this relationship, mammals in general may be considered as holobionts from an ecological perspective, in which the microbiota assists host metabolism and acts as an environmental training system for the associated tissues (Bäckhed et al., 2005; Al Nabhani et al., 2019; Tsolis and Bäumler, 2020). Microorganisms associate with the skin and mucosal surfaces such as the oral-nasal and vaginal cavities, respiratory and gastrointestinal tracts; with the gut microbiota constituting the best characterized community. We note that although we focus on the gut microbial ecosystem, the concepts may apply to all mucosal surfaces and potentially to other complex symbiotic communities.

The composition and community structure of the gut microbiota is complex and heterogenous. The distribution of microbial species within the large intestine is to be accounted with the diversity of residing immune cells, together forming a biodynamic ecosystem (Human Microbiome Project Consortium, 2012; James et al., 2020). Indeed, bacterial communities and immune cell populations exhibit great diversity in a niche-dependent fashion, with the latter displaying a wide range of transcriptional profiles within T and B cells of the adaptive immune system. The niches of gut commensals are determined by their metabolic activities and ability to stably associate with their local tissue environment (Lee et al., 2013; Ost and Round, 2018; Vonaesch et al., 2018). For example, some of the *Bacteroides* species are present in the intestinal lumen while others tightly associate with the mucus layering the epithelial surface of colonic crypts (Johansson et al., 2011). Yet, niche residency is not solely determined through dialogue with the host and critically depends on interactions with other microbes sharing nutritional niches (García-Bayona and Comstock, 2018). Here, bacteria vie for dominance, deploying a range of antibacterial toxins, some of which are delivered *via* membrane-embedded secretion systems.

The T6SS is prevalent in gram-negative bacteria, particularly in the phyla Proteobacteria and Bacteroidetes (Bingle et al., 2008; Russell et al., 2014b). This secretion apparatus is evolutionarily related to the bacteriophage tail, wherein contraction of a sheath propels a spiked-tube structure out of the bacterial cell, piercing the cell membrane of their targets to inject effector proteins (Pukatzki et al., 2007; Coulthurst, 2019). The cytoplasmic T6SS sheath, composed of a polymeric helix of TssB-TssC binds to a baseplate-like multi-protein platform, which itself associates with an envelope-spanning membrane complex of TssJ, TssL, and TssM (Durand et al., 2015; Nazarov et al., 2017). Phylogenetic analysis of TssC proteins found that type VI secretion systems cluster into three main groups, where subtypes I and II are proteobacterial, while subtype III is restricted to Bacteroidetes (Russell et al., 2014b). The inner tube is a stack of hexameric Hcp rings capped with a spike complex of a VgrG trimer, further sharpened with a PAAR protein tip; designed for effector and toxin delivery (Leiman et al., 2009; Shneider et al., 2013). T6SSs can directly target both prokaryotic and eukaryotic cells, as well as delivering effector proteins into the extracellular milieu in a contact-independent manner (Pukatzki et al., 2006; Hood et al., 2010; Si et al., 2017b). These effectors display a vast range of activities, including hydrolysis of peptidoglycan of peptidoglycan, nucleic acids, nucleotides, proteins, and lipids; membrane pore formation and metal ion binding, thus conferring a competition advantage to the T6SS-wielding bacterium and promoting its survival (Russell et al., 2014a; Wang et al., 2015; Ahmad et al., 2019). This review examines the relationship between the type VI secretion system and the microbiome in the context of both symbiosis and dysbiosis.

THE T6SS CONTRIBUTES TO THE FITNESS OF THE MICROBIOTA

The majority of the mammalian microbiome is acquired at birth, with the prevailing species seeded from the mother during delivery and influenced by breastfeeding and environmental exposure (Round et al., 2010). During the first year of life, the composition of the gut microbiome is highly dynamic, in part due to the weaning process, before stabilizing, and remaining consistent through adulthood (Faith et al., 2013; Verster et al., 2017; Al Nabhani et al., 2019). The major constituents of the gut community belong to the phyla Bacteroidetes, Firmicutes, Actinobacteria, and Proteobacteria, with members of the *Bacteroides* genus dominating the large intestine (Human Microbiome Project Consortium, 2012). Subtype III of T6SS (hereafter referred as $T6SS^{iii}$) is restricted to the Bacteroidetes phylum and has been shown to deliver antibacterial effectors resulting in microbial antagonism (Russell et al., 2014b).

Bioinformatic analyses of T6SS loci within the order Bacteroidales has classified them into three distinct "genetic architectures," designated GA1–3 (Coyne et al., 2016). GA1 and GA2 are found on integrative conjugative elements. Genomic analysis of the co-resident *Bacteroides* spp. isolated from human gut provided evidence of transfer of these elements between species *in situ*, implying that T6SS loci are under positive selection in the microbiome (Coyne et al., 2014). GA3 T6SSs are confined to *Bacteroides fragilis*, an obligate anaerobe, while GA1 and GA2 loci are more widespread within the phylum (Coyne et al., 2016). GA1–3 display distinct repertoires of effector-immunity pairs, possibly driving the incompatibility of these T6SSs within a single niche of an individual (Coyne and Comstock, 2019). One strain of *B. fragilis* tends to dominate the microbiota of an individual due to strain exclusion as the composition of the community stabilizes (Kostic et al., 2015; Yassour et al., 2016; Verster et al., 2017). Indeed, metagenomic analyses revealed that the abundance of GA3 T6SS loci is higher in infants, suggesting that competition between *B. fragilis* strains leads to stability of the microbial community in adulthood (Coyte et al., 2015; Verster et al., 2017). These observations should also be considered in light of the weaning process, wherein dietary changes lead to the influx of new bacterial competitors and dietary metabolites required for the host immune ontogeny (Al Nabhani et al., 2019). Co-existence of strains with different $T6SS^{iii}$ "genetic architectures" does arise but solely when bacterial species with overlapping nutritional niches become spatially segregated in the presence of a dense and diverse microbiota (Zitomersky et al., 2011; Hecht et al., 2016).

The use of gnotobiotic mouse models provided the empirical evidence supporting the roles of the T6SS in Bacteroidetes as ecological determinants, wherein T6SS expression and activity have been directly detected *in vivo* (Russell et al., 2014b; Chatzidaki-Livanis et al., 2016). *In vivo* competition assays have demonstrated that *B. fragilis* employs the T6SS to displace competitors from their niche in a contact-dependent manner, with several effector proteins supporting this elimination (**Table 1**) (Chatzidaki-Livanis et al., 2016; Hecht et al., 2016; Wexler et al., 2016; Ross et al., 2019). Furthermore, *in vitro* competition assays have found that T6SS-mediated antagonism of *Bacteroides* spp. targeted a narrow range of species, with most prey strains resistant to intoxication (Chatzidaki-Livanis et al., 2016; Wexler et al., 2016). Thus, the susceptibility to T6SS-dependent antagonism depends as much on the belligerent's identity as on the population distribution across topological niches.

Horizontal gene transfer facilitates the evolution of bacterial species in polymicrobial environments by enabling the positive selection of genes conferring a competitive advantage, a phenomenon also observed for T6SS loci (Unterweger et al., 2014). The existence of "orphan" T6SS immunity genes (conferring resistance to deleterious T6SS effector proteins; bearing no connection to the host immune system) in the absence of cognate effector genes was discovered in *Vibrio cholerae* isolates, leading to the hypothesis that their acquisition would subsequently protect the bearer against T6SS attacks from non-kin opponents (Kirchberger et al., 2017). The functionality of these orphan immunity genes was elegantly shown by Ross and colleagues in a recent study of members of the microbiome exhibiting extensive arrays called acquired interbacterial defense (AID) clusters (Ross et al., 2019). Here, many members of Bacteroidales were immune to T6SS antagonism by other species and may even possess immunity genes conferring resistance to anti-bacterial effectors associated to strategies beyond the T6SS (Zhang et al., 2012; Ross et al., 2019). However, immunity proteins are not the only way to mitigate the impact of antagonistic effectors. Several studies showed the inability of certain T6SS effectors to intoxicate prey cells lacking the cognate immunity proteins (Altindis et al., 2015; Ringel et al., 2017; Wood et al., 2019), and synergistic effector activities have also been described (LaCourse et al., 2018). Further protection strategies from T6SS-mediated killing, such as upregulation of envelope stress responses and production of extracellular polysaccharides, underscore the complexity of T6SS antagonism (Toska et al., 2018; Hersch et al., 2020).

T6SS-mediated bacterial antagonism targets specific competitors in the gut, helping to dictate niche occupancy. However, when considered in the broader ecological context of the microbiota and symbiosis with the host, the T6SS may also promote the symbiotic relationship with the host by enabling metabolic cooperation (Hooper et al., 2012; Vonaesch et al., 2018). Additionally, the presence of a stable microbiota provides resistance to dysbiosis and outcompetes invading microbial pathogens for nutrients. In terms of direct antibacterial warfare, the T6SS should be considered as a major armament of the microbiota in limiting infection (Kamada et al., 2013; Ducarmon et al., 2019). Indeed, mouse models have shown that the priority benefit of *B. fragilis* colonization may be protection against infection by enterotoxigenic *B. fragilis* strains, in a manner that depends on T6SS effector-immunity genotype (Hecht et al., 2016).

PROMOTION OF IMMUNE HOMEOSTASIS BY THE MICROBIOTA: A POTENTIAL ROLE FOR THE T6SS?

The intestinal microbiota is also crucial for the development of our immune system, as its absence leads to low antibody titer, poor glycosylation of mucosal surfaces, overt T_H2 responses and defective development of gut-associated lymphoid tissue in germ-free mice (Smith et al., 2007). The resident microbiota is proposed to train our immune system to actively tolerate the presence of distinct commensals whilst providing robust resistance against invading bacterial pathogens; presenting the intriguing teleological argument that commensal bacteria co-opt the host immune system to defend their niche (Round and Mazmanian, 2009). Evidence now strongly supports the idea

TABLE 1 | T6SS[iii] effectors of human symbionts.

Commensal bacterium	Antibacterial T6SS effector	Effector locus tag	Immunity protein(s) tag(s)	Immunity locus	References
Bacteroides dorei DSM 17855	"GA2_e14"	BACDOR_RS22955	"GA2_i14"	BACDOR_RS17020	Ross et al., 2019
Bacteroides fragilis 638R	Bfe1	BF638R_1988	Bfi1	BF638R_1987	Chatzidaki-Livanis et al., 2016
Bacteroides fragilis 638R	Bfe2	BF638R_1979	Bfi2	BF638R_1978	Chatzidaki-Livanis et al., 2016
Bacteroides fragilis 638R	–	–	Orphan Bti1 ("GA3_i6")	BF638R_2042	Ross et al., 2019
Bacteroides fragilis 638R	–	–	Orphan Bti2a,b ("GA3_i7ab")	BF638R_2053-4	Ross et al., 2019
Bacteroides fragilis 638R	–	–	Orphan "GA2_i11"	BF638R_1388	Ross et al., 2019
Bacteroides fragilis CL03T00C23	"GA2_e2"	HMPREF1079_RS08215	"GA2_i2"	HMPREF1079_RS08220	Ross et al., 2019
Bacteroides fragilis NCTC 9343	Bte1 ("GA3_e6")	BF9343_1937	Bti1 ("GA3_i6")	BF9343_1936	Wexler et al., 2016
Bacteroides fragilis NCTC 9343	Bte2	BF9343_1928	Bti2a,b ("GA3_i7ab")	BF9343_1927-6	Hecht et al., 2016; Wexler et al., 2016
Bacteroides fragilis NCTC 9343	–	–	Orphan "GA1_i5"	BF9343_1657	Ross et al., 2019
Bacteroides fragilis YCH46	"GA1_e5"	BF2850	"GA1_i5"	BF2851	Ross et al., 2019

of tolerogenic immune responses to commensal flora, rather than specifically ignoring these residents (Round et al., 2010). Tolerance is fostered through the detection of microbe-associated molecular patterns (MAMPs) and microbial metabolites, and extends beyond the local environment of the gut to promote appropriate systemic immune responses (**Figure 1**) (Clarke et al., 2010; Chu and Mazmanian, 2013). The detection of conserved MAMPs by pattern recognition receptors (PRRs) is one of the foundations of the innate immune system. Innate immune cells, particularly antigen-presenting dendritic cells (DCs) sense their environment in peripheral organs through continuous uptake and sampling of exogenously acquired antigens (Iwasaki and Medzhitov, 2015). Upon microbial encounter, the engagement of PRRs by MAMPs elicits an inflammatory genes program, enhances antigen processing and presentation processes in DCs; all critical for T cell mediated immune responses against pathogens (Medzhitov and Janeway, 1999; Takeuchi and Akira, 2010; Iwasaki and Medzhitov, 2015). MAMPs include lipopolysaccharide (LPS), peptidoglycan, lipoproteins and nucleic acids that trigger MAP kinase and NF-κB signaling leading to pro-inflammatory responses (Fitzgerald and Kagan, 2020). Yet, there is precedent for MAMPs to assist in the development of tolerogenic signals. Mucosal DCs interacting with commensal bacterial components directly or through indirect acquisition of secreted outer membrane vesicles (OMVs) prime host regulatory T cells (Tregs), a subset of T cells promoting tolerance to both food and microbial antigens, thus dampening immune responses to the resident bacterial communities.

The homeostasis of the host-microbiota axis is maintained by continuous immune system monitoring (Belkaid and Harrison, 2017). The best characterized example of immune modulation is the production of polysaccharide A (PSA) by *B. fragilis*, which signals *via* Toll-like receptor 2 (TLR2) on dendritic cells. This stimulates the differentiation of Tregs, producing an immunosuppressive environment through the secretion of the cytokine IL-10 (Mazmanian et al., 2005; O'Mahony et al., 2008; Round et al., 2011). In IL-10 deficient mice, commensal bacterium *Helicobacter hepaticus* exhibits colitogenic potential in the presence of gut microbiota, which has been reported to be suppressed by the T6SS of this ε-proteobacterium (Mazmanian et al., 2008; Chow and Mazmanian, 2010; Bartonickova et al., 2013; Jochum and Stecher, 2020). This highlights the interplay of tolerogenic signaling and the T6SS of resident members of the microbiota; however, the mechanistic details of this interaction are yet to be explored.

Tolerogenic immune signaling is also stimulated by commensal metabolites including the short chain fatty acids (SCFAs) acetate, propionate, and butyrate (Parada Venegas et al., 2019); intermediates of vitamin B2 and B9; and amino acid

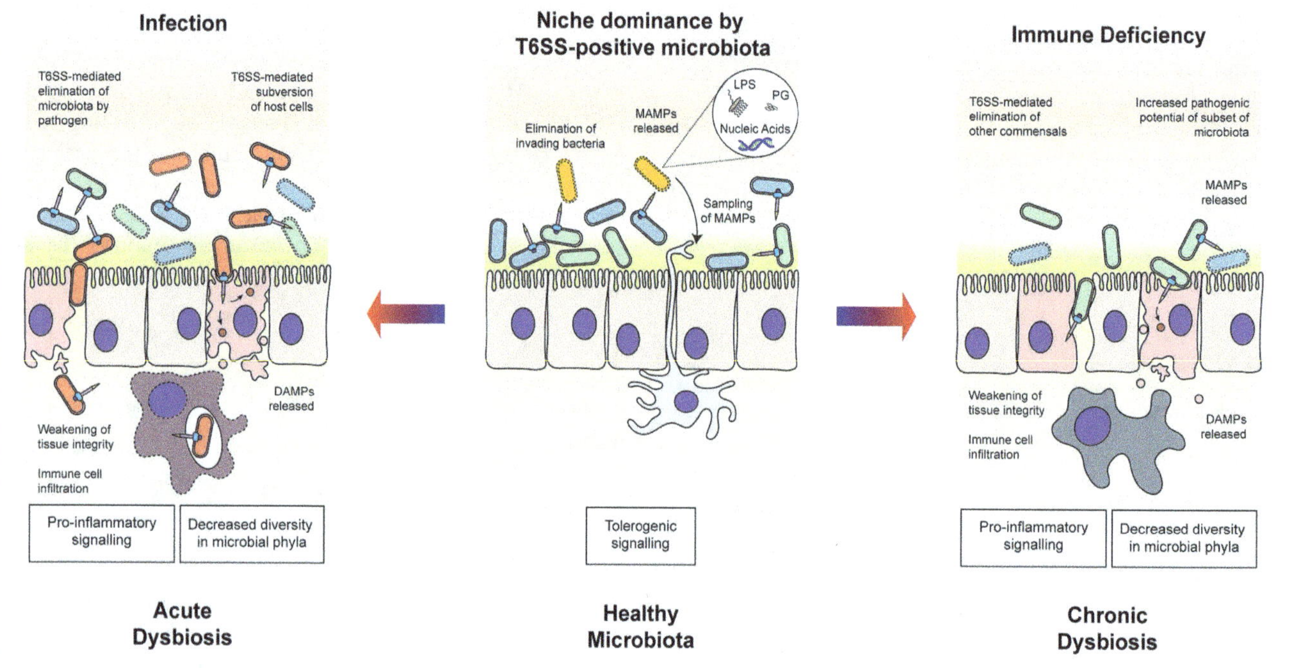

FIGURE 1 | Roles of the T6SS in host-microbiota-pathogen interactions. In healthy steady state conditions (middle panel), commensal bacteria use the T6SS to establish and maintain their niche in the host. The release of MAMPs through T6SS warfare can contribute to the establishment of immune tolerance, enhancing the symbiotic relationship. In the case of host immune deficiency (right panel), for example due to a genetic polymorphism in the host, cross-talk with the microbiota is compromised and the balance within the microbial community may be disrupted, resulting in chronic dysbiosis. The T6SS is likely to play a role in the modulation of competing commensal populations and subsequent decrease in diversity of bacteria phyla, as well as potentially directly manipulating host cells. In the case of infection by pathogenic bacteria wielding a T6SS (left panel), commensal bacteria are eliminated through both direct delivery of antibacterial effectors and indirect mechanisms such as host manipulation and nutrient competition. The state of dysbiosis that follows is acute but may be resolved through elimination of the pathogen by the host immune system. In both states of dysbiosis, the T6SS may play a determining role in eliciting the release of DAMPs, which influences the host immune response.

metabolism (Kjer-Nielsen et al., 2012; Venkatesh et al., 2014; Sasabe et al., 2016). Recent work has started to shed light on the numerous benefits that production of SCFAs by commensal bacteria confer to the host. One consequence is the upregulation of oxidative host metabolism by utilization of SCFAs as a carbon source, which bolsters the hypoxic microenvironment at the colonocyte surface, favoring the growth of obligate anaerobes (e.g., *Bacteroides* spp.) and limiting propagation of facultative aerobes, such as invasive pathogens like *Escherichia coli* (Litvak et al., 2018; Zhang et al., 2019). In addition, SCFAs act directly *via* immune cell receptors to modulate T cell subset expansion and macrophage polarization (Schulthess et al., 2019). These compounds commonly promote IL-10 production and suppress inflammation; however, they may also contribute toward effector T cell differentiation, depending on the overall local immunological context (Zhang et al., 2019). On the other hand, microbial metabolites in the intestine may stimulate virulence programs of invading bacteria, with several two-component signal transduction systems in T6SS-positive pathogens having been shown to respond to SCFAs and other metabolites produced by the microbiome (Lawhon et al., 2002; Gonzalez-Chavez et al., 2010; Kohli et al., 2018; Goodman et al., 2020). It is likely that T6SS-mediated turbulent population dynamics occurring during the microbiome development results in variation in the levels of these metabolites. Indeed, bacteria activate diverse antimicrobial programs upon non-kin recognition or danger sensing, including an as-yet uncharacterized diffusible signal from lysed *Pseudomonas aeruginosa* bacteria that heightens the antibacterial T6SS activity in kin (LeRoux et al., 2015). This antibacterial warfare would further alter levels of microbial products in the local milieu, tipping the ecological balance toward dysbiosis. Moreover, one could hypothesize that bacterial products resulting from the aftermath of T6SS-mediated bacterial antagonism may provide the ligands supporting the development of tolerogenic immune responses. Several lines of evidence from various models lend support to this hypothesis. T6SS-dependent exclusion of *Aliivibrio fischeri* non-kin strains has been reported during their colonization of the light organs of the Hawaiian bobtail squid *Euprymna scolopes* (Speare et al., 2018, 2020). The ensuing symbiosis results in morphogenesis of the organs, a process that a combination of *A. fischeri* LPS and specific monomeric peptidoglycan fragments, issued from cell wall remodeling occurring during bacterial growth and considered as a sign of bacterial viability (referred to as tracheal cytotoxin; TCT), are sufficient to stimulate (Koropatnick et al., 2004). In this case, the peptidoglycan fragments are actively released during *A. fischeri* growth. In the fruit fly *Drosophila melanogaster*, recognition of peptidoglycan by the peptidoglycan recognition protein (PGRP) scavenger receptors stimulates the Immune Deficiency (IMD) pathway, similar to that of tumor necrosis factor (TNF) in mammals (Kleino and Silverman, 2014). Alternative isoforms of PGRP can determine bacterial viability: recognition of TCT activates the pathway; whereas recognition of polymeric peptidoglycan fragments (issued from bacterial killing) by a splice variant exerts an inhibitory effect of signal transduction (Neyen et al., 2016). This effectively results in a dampened immune response as reduced pathogen viability could represent a reduced threat. Such interplay also occurs in the intestinal lymphoid tissues, where the generation of IgA-producing B cells is induced following the recognition of gram-negative bacterial peptidoglycan by NOD1 in epithelial cells (Bouskra et al., 2008). Other ligands provide additional cues for microbial viability in host cytosol, such as cyclic dinucleotides sensed by the cGAS-STING and RECON pathways (Moretti and Blander, 2018; Whiteley et al., 2019); and bacterial RNA sensing by TLR8 in the endosome of mammalian epithelial cells (Ugolini et al., 2018).

Equally, it is reasonable to envision T6SS machineries and their effectors as direct inducers of immune tolerance at mucosal sites. In agreement with such possibility, host cells of the innate immune system may forge tolerance by acquiring antigens through OMVs (Shen et al., 2012; Kaparakis-Liaskos and Ferrero, 2015; Chu et al., 2016; Durant et al., 2020). The association of TseF, an iron-acquiring T6SS effector of *Pseudomonas aeruginosa*, with OMVs may represent an underappreciated role for T6SS effectors in host-microbe interplay (Lin et al., 2017). A better understanding of the activities of T6SS effectors deployed by bacterial species at the interface of mucosal surfaces will illuminate the innate immune sensing and response mechanisms to bacterial molecules released into the host milieu, during homeostasis or under stress conditions.

T6SS DEPLOYMENT BY BACTERIAL PATHOGENS: UPSETTING THE APPLECART

By its sheer density, the microbiota offers high resistance to colonization by pathogens. Indeed, pathogens are vastly outnumbered at the start of infection and must compete with the host microbiota for space and nutrients, notwithstanding the contact-dependent and -independent mechanisms of bacterial warfare. Although the T6SS was initially associated with bacterial virulence, the precise role of this apparatus in host infection has not always been clear (Hachani et al., 2016). Recently, studies have highlighted the role of the T6SS in bacterial antagonism during infection, rather than through a direct interaction with host cells. Early evidence for T6SS-mediated competition *in vivo* emerged from a transposon library screen of *Vibrio cholerae* strains for impaired colonization of the infant rabbit intestine (Fu et al., 2013). The authors found that *tsiV3*, encoding the immunity protein to the specialized peptidoglycan hydrolase effector VgrG3, is necessary to alleviate a colonization bottleneck in this model of intestinal infection. Further analysis of T6SS dynamics during *V. cholerae* colonization found that its role in commensal elimination is largely confined to the jejunum, suggesting that this antibacterial activity may be targeted toward specific microbial residents of this niche (Fu et al., 2018). The T6SS of gastrointestinal pathogens *Salmonella enterica* serovar Typhimurium and *Shigella sonnei* are also required for complete virulence, with evidence supporting a role in antagonism of members of the microbiota (Sana et al., 2016; Anderson et al., 2017). Yet, similar to *V. cholerae*, S. Typhimurium exhibited a limited target range in bacterial competition assays against gram-negative members of the microbiota, again hinting at specific targeting during infection (Sana et al., 2016). Although the abundance of proteobacterial commensals is low in comparison to members of the Bacteroidetes and Firmicutes phyla, they are enriched in many niches, for example *Acinetobacter* spp. in colonic crypts, and *Escherichia* and *Shigella* species in the sigmoid colon (Pédron et al., 2012; James et al., 2020). Due to the clash of nutritional niches between many proteobacterial gut residents and their pathogenic proteobacterial counterparts, T6SS-mediated antagonism is likely to unfold between them. Moreover, metagenomic analyses indicate the presence of species possessing T6SS^{i} components, which are absent from the Bacteroidetes subgroup, thereby supporting the notion of T6SS^{i}-mediated warfare waged by commensal bacteria (Coyne and Comstock, 2019).

The induction of inflammatory host responses is a common mechanism of mass disruption by bacterial competitors, which promotes elimination of the microbial community and dysbiosis (Ackermann et al., 2008). For example, by triggering macrophage pyroptosis, an invasive subpopulation of S. Typhimurium can elicit a large inflammatory response leading to the release of pro-inflammatory cytokines from epithelial cells (Thiennimitr et al., 2012). Although this tissue-invasive S. Typhimurium subpopulation is eliminated by the subsequent infiltration of immune cells, the ensuing inflammatory response (notably the IL-22 signaling axis) reduces iron availability in the lumen. Due to its numerous metal ion acquisition systems, the luminal S. Typhimurium subpopulation is able to outcompete the commensal inhabitants and replicate in the lumen (Behnsen et al., 2014). Similarly, the secretion of cholera toxin by *V. cholerae* results in iron depletion to favor the pathogen's proliferation at the detriment of the microbiota (Rivera-Chávez and Mekalanos, 2019). The antibacterial activity of the T6SS itself can also stimulate host inflammation. Bacterial lysis mediated by the *V. cholerae* T6SS in mice mono-colonized with a commensal *E. coli* strain elicits a host transcriptional response, elevating expression of antimicrobial peptides and NF-κB signaling components (Zhao et al., 2018). NF-κB induction could be recapitulated *in vitro* using supernatants from T6SS-dependent killing assays, suggesting that MAMPs released from T6SS-mediated bacterial lysis may be the factors supporting the induction of this host response. Furthermore, El Tor pandemic strains of *V. cholerae* display higher levels of T6SS gene expression than reference clinical isolates, thus underpinning the association of T6SS antibacterial activity with pathology (Zhao et al., 2018). In the TRUC murine model for ulcerative colitis, the presence of a commensal bacterial population promotes spontaneous disease onset in this susceptible host (Garrett et al., 2007). Here, the presence of *Proteus mirabilis* and *Klebsiella pneumoniae* in this commensal community correlated with colitogenic potential (Garrett et al., 2010). Both of these species possess T6SSs that display antibacterial activity (Alteri et al., 2013; Hsieh et al., 2019), while this secretion system has also been shown to contribute to the fitness of the pathogens *in vivo* (Lery et al., 2014; Debnath et al., 2018). One can therefore contemplate a role for this secretion system in the TRUC model whereby T6SS-mediated elimination of commensal bacteria promotes an inflammatory response that cannot be restrained due to the immune genes deficiency of the host, resulting in colitis.

Non-mammalian models also support the notion of T6SS-dependent dysbiosis as a driving force for disease symptoms and pathology. A recent study found that *Pseudomonas protegens* uses antibacterial effectors to antagonize the gut microbiota of butterfly larvae, enabling tissue invasion and disease onset (Vacheron et al., 2019). Infection of *D. melanogaster* with *V. cholerae* results in diarrheal symptoms and gut inflammation (Blow et al., 2005), and the T6SS of the pathogen was found to promote mortality in a manner dependent on the presence of constituents of the microbiota (Fast et al., 2018). The IMD pathway also contributes to this pathology, suggesting that elimination of the fly gut commensal bacteria can be lethal due to exacerbated host inflammatory response (Ryu et al., 2008; Fast et al., 2018). T6SS-mediated depletion of the polymicrobial community impacts tissue repair during fly infection, mirroring the pioneering work establishing the role of the human gut microbiota in tissue homeostasis (Rakoff-Nahoum et al., 2004; Fast et al., 2020).

The competition for nutrients is a key aspect of colonization resistance in the host environment. As discussed above, microbiota niche occupancy is partly dictated by the ability to use specific carbon and nitrogen sources. Around one fifth of the genome of *Bacteroides* spp. encodes proteins involved in polysaccharide catabolism, conferring great metabolic versatility (Sonnenburg et al., 2005; Schwalm and Groisman, 2017). Besides, the host accentuates the state of nutritional immunity by sequestering metal ions upon infection to limit the replication of pathogens. Recent work by the Shen laboratory and others has revealed a role for the T6SS in nutrient acquisition, whereby the secretion of metal ion-binding proteins facilitates the uptake of zinc, iron, copper or manganese (Wang et al., 2015; Lin et al., 2017; Si et al., 2017a,b; Han et al., 2019). A *T6SS-4* mutant of *Yersinia pestis* exhibited reduced pathogenicity in an orogastric mouse model, indicating the role of this virulence factor in overcoming nutritional immunity during infection (Wang et al., 2015). It is likely that members of the microbiota utilize the T6SS for nutrient acquisition too; however, no T6SS effectors have been described to date. The role of the T6SS of bacterial pathogens in disrupting the steady state of microbiota-host ecosystems is becoming increasingly clear and underscores the importance of the microbiota in colonization resistance alongside the versatility of this secretion system.

DIRECT HOST CELL CONTACT: T6SS ENCOUNTERS OF THE THIRD KIND

The T6SS versatility extends beyond its prominent antibacterial role in many gram-negative bacteria. As the most evolved member of the contractile injection systems, it delivers effectors into the extracellular milieu or directly into neighboring bacteria and/or eukaryotic targets. Many anti-eukaryotic activities of the T6SS have been described, including the manipulation of biochemical processes governing the physiology of phagocytes and epithelial cells (reviewed in Hachani et al., 2016). Furthermore, several studies have found that the T6SS can target fungal cells, and whereas the human microbiota also harbors fungi such as *Candida albicans*, these interactions within a host remain unexplored (Haapalainen et al., 2012; Trunk et al., 2018; Storey et al., 2020). A summary of T6SS effector proteins with roles distinct from direct bacterial antagonism are listed in **Table 2**.

Once pathogens gain a foothold by ousting the residing microbiota in their desired niche, they must contend with the microbial clearance mechanisms of the host. After phagocytosis by immune cells, phagosomal bacteria are subjected to the oxidative burst, where the nicotinamide adenine dinucleotide phosphate (NADPH) oxidase membrane complex produces superoxide radicals in the vacuole to destroy the engulfed microbe. During oxidative stress, such as after uptake by macrophages, enterohaemorrhagic *E. coli* (EHEC) secretes the T6SS catalase effector KatN to detoxify the local environment (Wan et al., 2017). Intriguingly, despite being important in the survival of EHEC in macrophages, the absence of KatN did not impact virulence in a streptomycin-treated mouse model. However, the T6SS itself was required for complete virulence suggesting the presence of other host cell-targeted effectors or undescribed compensatory host cell mechanisms. The T6SSs secreting metal binding effectors are upregulated upon oxidative stress, suggesting they likely play a role in defense against reactive oxygen species produced by immune cells (Wang et al., 2015; Lin et al., 2017; Si et al., 2017a,b). Indeed, the zinc-binding effector YezP of *Y. pestis* is required for intracellular survival in macrophages (Wang et al., 2015).

Burkholderia cenocepacia, an opportunistic pathogen of cystic fibrosis patients, resides primarily in alveolar macrophages where it resists killing (Schwab et al., 2014). The delivery of the T6SS effector TecA into the macrophage cytosol leads to the deamidation of Rho GTPases, which hampers the activity of the NADPH complex (Rosales-Reyes et al., 2012; Aubert et al., 2016). Yet, this inactivation of Rho GTPases is detected by the pyrin inflammasome, leading to caspase-1 activation, pyroptosis and inflammation (Xu et al., 2014). Inflammasomes are vital for enacting cell-autonomous immunity. Thus, they are frequently targeted by invasive bacterial pathogens (Sanchez-Garrido et al., 2020). The NLRC4 and NLRP3 inflammasomes are activated by the type III secretion system of *Edwardsiella piscicida* after phagocytosis. However, this bacterium is also able to impair the activation of caspase-1 using its T6SS effector EvpP (Chen et al., 2017). The mode of action of this effector remains elusive but appears to prevent the induction of ASC-mediated canonical inflammasome seeding by inhibition of calcium-dependent JNK activation.

The facultative intracellular pathogen *Francisella tularensis* avoids destruction by macrophages through the action of its T6SS (Nano et al., 2004). Proteomics analysis identified several T6SS effector proteins that are required for escape from the phagosome, and cytosolic replication (Eshraghi et al., 2016). One of these is the phosphatidylinositol 3-kinase (PI3K)-like effector OpiA, which remodels the phospholipid content of

TABLE 2 | T6SS effectors with roles beyond bacterial antagonism.

T6SS Effector	Bacterium	Function	References
VgrG1[AD]	*Aeromonas dhakensis*	Cytoskeletal disruption	Suarez et al., 2010
TecA	*Burkholderia cenocepacia*	Inhibition of Rho GTPases	Rosales-Reyes et al., 2012; Aubert et al., 2016
TseZ	*Burkholderia thailandensis*	Acquisition of Zn^{2+}	Si et al., 2017a
TseM	*B. thailandensis*	Acquisition of Mn^{2+}	Si et al., 2017b
VgrG5	*Burkholderia pseudomallei*; *B. thailandensis*	Formation of multi-nucleated giant cells	Schwarz et al., 2014; Toesca et al., 2014
EvpP	*Edwardsiella piscicida*	Inhibition of inflammasome formation	Chen et al., 2017
KatN	Enterohaemorraghic *Escherichia coli*	Protection against oxidative stress	Wan et al., 2017
OpiA	*Francisella tularensis*	Phagosomal escape	Eshraghi et al., 2016; Ledvina et al., 2018
Azu	*Pseudomonas aeruginosa*	Acquisition of Cu^{2+}	Han et al., 2019
TseF	*P. aeruginosa*	Acquisition of Fe^{3+}	Lin et al., 2017
PldA	*P. aeruginosa*	Internalization into non-phagocytic cells	Jiang et al., 2014
PldB	*P. aeruginosa*	Internalization into non-phagocytic cells	Jiang et al., 2014
Tle4[PA]	*P. aeruginosa*	Disruption of ER homeostasis	Jiang et al., 2016
VgrG2b	*P. aeruginosa*	Cytoskeletal manipulation	Sana et al., 2012
Tfe1	*Serratia marcescens*	Membrane depolarization	Trunk et al., 2018
Tfe2	*S. marcescens*	Metabolic dysregulation	Trunk et al., 2018
VgrG1[VC]	*Vibrio cholerae*	Cytoskeletal disruption	Pukatzki et al., 2007; Ma et al., 2009
VasX	*V. cholerae*	Formation of membrane pores	Miyata et al., 2011
YezP	*Yersinia pestis*	Acquisition of Zn^{2+}	Wang et al., 2015

the phagosomal membrane to delay its maturation in the endosomal compartment, thereby facilitating pathogen escape prior to lysosomal fusion (Ledvina et al., 2018). The H2-T6SS of *Pseudomonas aeruginosa* also delivers membrane targeting effector proteins into host cells, namely the phospholipases PldA, PldB, and Tle4 (Jiang et al., 2014, 2016; Wettstadt et al., 2019). While PldA and PldB promote internalization of *P. aeruginosa* by manipulating the PI3K-Akt signaling axis, Tle4 fragments the endoplasmic reticulum, activating the unfolded protein response and autophagy. However, the benefits of these cellular modifications for the bacterium remain unclear. The T6SS yielded by bacterial pathogens targeting host cells presents a further risk to the microbiota, since the ensuing subversion of host processes affects their ecological niche. Indeed, such indirect impact has been demonstrated in the zebrafish model of cholera, where the actin-crosslinking domain of VgrG1 of *V. cholerae* stimulates peristalsis, resulting in the collapse of the resident microbial community (Logan et al., 2018). The gradual repopulation by the commensal microbiota may evict the invading pathogen despite the reversal of the numerical advantage; yet the niche must still be conducive for this repopulation to occur.

CONCLUSIONS

The extended versatility of the T6SS enriches both the panel of virulence factors of bacterial pathogens, and the mutualism toolkit of symbiotic bacteria. The T6SS plays an underappreciated role in the maintenance of this synergistic steady state in the microbiota. Notwithstanding its original designation as a virulence factor, the T6SS is clearly beneficial to the host in facilitating stable colonization of the microbiota. Further investigation into the genetic architecture of the T6SS[iii] of Bacteroidales, its target range, and effector-immunity repertoire will provide deeper insight into the ecology of the microbiota. Contact-dependent signaling has been described for CDI toxin delivery into immune prey (Garcia et al., 2016) and analogous processes may also be operated by T6SS effectors targeting both bacteria and eukaryotic cells. Exploring the interactions between the T6SS of commensal bacterial and host cells may illuminate the factors commandeering a homeostatic and balanced tolerogenic signaling; with broader implications in infection, diet, autoimmune and autoinflammatory disorders. In all, we describe the underappreciated roles of the T6SS at the nexus of the microbiota, host and the defense against incoming pathogens; and propose further avenues of investigation to dissect the role of this versatile secretion machine in the establishment and homeostasis of holobionts.

AUTHOR CONTRIBUTIONS

All authors have intellectually revised this work together and approved it for publication.

ACKNOWLEDGMENTS

We apologize to all colleagues whose work has not been mentioned in this manuscript due to space limitations.

REFERENCES

Ackermann, M., Stecher, B., Freed, N. E., Songhet, P., Hardt, W.-D., and Doebeli, M. (2008). Self-destructive cooperation mediated by phenotypic noise. *Nature* 454, 987–990. doi: 10.1038/nature07067

Ahmad, S., Wang, B., Walker, M. D., Tran, H.-K. R., Stogios, P. J., Savchenko, A., et al. (2019). An interbacterial toxin inhibits target cell growth by synthesizing (p)ppApp. *Nature* 575, 674–678. doi: 10.1038/s41586-019-1735-9

Al Nabhani, Z., Dulauroy, S., Marques, R., Cousu, C., Al Bounny, S., Déjardin, F., et al. (2019). A weaning reaction to microbiota is required for resistance to immunopathologies in the adult. *Immunity* 50, 1276–1288.e5. doi: 10.1016/j.immuni.2019.02.014

Alteri, C. J., Himpsl, S. D., Pickens, S. R., Lindner, J. R., Zora, J. S., Miller, J. E., et al. (2013). Multicellular bacteria deploy the type VI secretion system to preemptively strike neighboring cells. *PLoS Pathog.* 9:e1003608. doi: 10.1371/journal.ppat.1003608

Altindis, E., Dong, T. G., Catalano, C., and Mekalanos, J. J. (2015). Secretome analysis of *Vibrio cholerae* type VI secretion system reveals a new effector-immunity pair. *MBio* 6:e00075-15. doi: 10.1128/mBio.00075-15

Anderson, M. C., Vonaesch, P., Saffarian, A., Marteyn, B. S., and Sansonetti, P. J. (2017). *Shigella sonnei* encodes a functional T6SS used for interbacterial competition and niche occupancy. *Cell Host Microbe* 21, 769–776.e3. doi: 10.1016/j.chom.2017.05.004

Aubert, D. F., Xu, H., Yang, J., Shi, X., Gao, W., Li, L., et al. (2016). A burkholderia type VI effector deamidates Rho GTPases to activate the pyrin inflammasome and trigger inflammation. *Cell Host Microbe* 19, 664–674. doi: 10.1016/j.chom.2016.04.004

Bäckhed, F., Ley, R. E., Sonnenburg, J. L., Peterson, D. A., and Gordon, J. I. (2005). Host-bacterial mutualism in the human intestine. *Science* 307, 1915–1920. doi: 10.1126/science.1104816

Bartonickova, L., Sterzenbach, T., Nell, S., Kops, F., Schulze, J., Venzke, A., et al. (2013). Hcp and VgrG1 are secreted components of the *Helicobacter hepaticus* type VI secretion system and VgrG1 increases the bacterial colitogenic potential. *Cell. Microbiol.* 15, 992–1011. doi: 10.1111/cmi.12094

Behnsen, J., Jellbauer, S., Wong, C. P., Edwards, R. A., George, M. D., Ouyang, W., et al. (2014). The cytokine IL-22 promotes pathogen colonization by suppressing related commensal bacteria. *Immunity* 40, 262–273. doi: 10.1016/j.immuni.2014.01.003

Belkaid, Y., and Harrison, O. J. (2017). Homeostatic immunity and the microbiota. *Immunity* 46, 562–576. doi: 10.1016/j.immuni.2017.04.008

Bingle, L. E. H., Bailey, C. M., and Pallen, M. J. (2008). Type VI secretion: a beginner's guide. *Curr. Opin. Microbiol.* 11, 3–8. doi: 10.1016/j.mib.2008.01.006

Blow, N. S., Salomon, R. N., Garrity, K., Reveillaud, I., Kopin, A., Jackson, F. R., et al. (2005). *Vibrio cholerae* infection of *Drosophila melanogaster* mimics the human disease cholera. *PLoS Pathog.* 1:e8. doi: 10.1371/journal.ppat.0010008

Bouskra, D., Brézillon, C., Bérard, M., Werts, C., Varona, R., Boneca, I. G., et al. (2008). Lymphoid tissue genesis induced by commensals through NOD1 regulates intestinal homeostasis. *Nature* 456, 507–510. doi: 10.1038/nature07450

Chatzidaki-Livanis, M., Geva-Zatorsky, N., and Comstock, L. E. (2016). *Bacteroides* fragilis type VI secretion systems use novel effector and immunity proteins to antagonize human gut *Bacteroidales* species. *Proc. Natl. Acad. Sci. U.S.A.* 113, 3627–3632. doi: 10.1073/pnas.1522510113

Chen, H., Yang, D., Han, F., Tan, J., Zhang, L., Xiao, J., et al. (2017). The bacterial T6SS effector EvpP prevents NLRP3 inflammasome activation by inhibiting the Ca^{2+}-dependent MAPK-Jnk pathway. *Cell Host Microbe* 21, 47–58. doi: 10.1016/j.chom.2016.12.004

Chow, J., and Mazmanian, S. K. (2010). A pathobiont of the microbiota balances host colonization and intestinal inflammation. *Cell Host Microbe* 7, 265–276. doi: 10.1016/j.chom.2010.03.004

Chu, H., Khosravi, A., Kusumawardhani, I. P., Kwon, A. H. K., Vasconcelos, A. C., Cunha, L. D., et al. (2016). Gene-microbiota interactions contribute to the pathogenesis of inflammatory bowel disease. *Science* 352, 1116–1120. doi: 10.1126/science.aad9948

Chu, H., and Mazmanian, S. K. (2013). Innate immune recognition of the microbiota promotes host-microbial symbiosis. *Nat. Immunol.* 14, 668–675. doi: 10.1038/ni.2635

Clarke, T. B., Davis, K. M., Lysenko, E. S., Zhou, A. Y., Yu, Y., and Weiser, J. N. (2010). Recognition of peptidoglycan from the microbiota by Nod1 enhances systemic innate immunity. *Nat. Med.* 16, 228–231. doi: 10.1038/nm.2087

Coulthurst, S. J. (2019). The type VI secretion system: a versatile bacterial weapon. *Microbiology* 165, 503–515. doi: 10.1099/mic.0.000789

Coyne, M. J., Roelofs, K. G., and Comstock, L. E. (2016). Type VI secretion systems of human gut *Bacteroidales* segregate into three genetic architectures , two of which are contained on mobile genetic elements. *BMC Genomics* 17:58. doi: 10.1186/s12864-016-2377-z

Coyne, M. J., Zitomersky, N. L., McGuire, A. M., Earl, A. M., and Comstock, L. E. (2014). Evidence of extensive DNA transfer between *Bacteroidales* species within the human gut. *MBio* 5:e01305-14. doi: 10.1128/mBio.01305-14

Coyne, M. J., and Comstock, L. E. (2019). Type VI secretion systems and the gut microbiota. *Microbiol. Spectr.* 7, 343–350. doi: 10.1128/microbiolspec.PSIB-0009-2018

Coyte, K. Z., Schluter, J., and Foster, K. R. (2015). The ecology of the microbiome: Networks, competition, and stability. *Science* 350, 663–666. doi: 10.1126/science.aad2602

Debnath, I., Stringer, A. M., Smith, S. N., Bae, E., Mobley, H. L. T., Wade, J. T., et al. (2018). MrpJ directly regulates *Proteus mirabilis* virulence factors, including fimbriae and type VI secretion, during urinary tract infection. *Infect. Immun.* 86:e00388-18. doi: 10.1128/IAI.00388-18

Ducarmon, Q. R., Zwittink, R. D., Hornung, B. V. H., van Schaik, W., Young, V. B., and Kuijper, E. J. (2019). Gut microbiota and colonization resistance against bacterial enteric infection. *Microbiol. Mol. Biol. Rev.* 83:e00007-19. doi: 10.1128/MMBR.00007-19

Durand, E., Nguyen, V. S., Zoued, A., Logger, L., Péhau-Arnaudet, G., Aschtgen, M.-S., et al. (2015). Biogenesis and structure of a type VI secretion membrane core complex. *Nature* 523, 555–560. doi: 10.1038/nature 14667

Durant, L., Stentz, R., Noble, A., Brooks, J., Gicheva, N., Reddi, D., et al. (2020). *Bacteroides thetaiotaomicron*-derived outer membrane vesicles promote regulatory dendritic cell responses in health but not in inflammatory bowel disease. *Microbiome* 8:88. doi: 10.1186/s40168-020-00868-z

Eshraghi, A., Kim, J., Walls, A. C., Ledvina, H. E., Miller, C. N., Ramsey, K. M., et al. (2016). Secreted effectors encoded within and outside of the *Francisella* pathogenicity island promote intramacrophage growth. *Cell Host Microbe* 20, 573–583. doi: 10.1016/j.chom.2016.10.008

Faith, J. J., Guruge, J. L., Charbonneau, M., Subramanian, S., Seedorf, H., Goodman, A. L., et al. (2013). The long-term stability of the human gut microbiota. *Science* 34:1237439. doi: 10.1126/science.1237439

Fast, D., Kostiuk, B., Foley, E., and Pukatzki, S. (2018). Commensal pathogen competition impacts host viability. *Proc. Natl. Acad. Sci. U.S.A.* 115, 7099–7104. doi: 10.1073/pnas.1802165115

Fast, D., Petkau, K., Ferguson, M., Shin, M., Galenza, A., Kostiuk, B., et al. (2020). *Vibrio cholerae*-symbiont interactions inhibit intestinal repair in *Drosophila*. *Cell Rep.* 30, 1088–1100.e5. doi: 10.1016/j.celrep.2019.12.094

Fitzgerald, K. A., and Kagan, J. C. (2020). Toll-like receptors and the control of immunity. *Cell* 18, 1044–1066. doi: 10.1016/j.cell.2020.02.041

Fu, Y., Ho, B. T., and Mekalanos, J. J. (2018). Tracking *Vibrio cholerae* cell-cell interactions during infection reveals bacterial population dynamics within intestinal microenvironments. *Cell Host Microbe* 23, 274–281.e2. doi: 10.1016/j.chom.2017.12.006

Fu, Y., Waldor, M. K., and Mekalanos, J. J. (2013). Tn-seq analysis of *Vibrio cholerae* intestinal colonization reveals a role for T6SS-mediated antibacterial activity in the host. *Cell Host and Microbe* 14, 652–663. doi: 10.1016/j.chom.2013.11.001

Garcia, E. C., Perault, A. I., Marlatt, S. A., and Cotter, P. A. (2016). Interbacterial signaling via *Burkholderia* contact-dependent growth inhibition system proteins. *Proc. Natl. Acad. Sci. U.S.A.* 113, 8296–8301. doi: 10.1073/pnas.1606323113

García-Bayona, L., and Comstock, L. E. (2018). Bacterial antagonism in host-associated microbial communities. *Science* 361:eaat2456. doi: 10.1126/science.aat2456

Garrett, W. S., Gallini, C. A., Yatsunenko, T., Michaud, M., DuBois, A., Delaney, M. L., et al. (2010). *Enterobacteriaceae* act in concert with the gut microbiota to induce spontaneous and maternally transmitted colitis. *Cell Host Microbe* 8, 292–300. doi: 10.1016/j.chom.2010.08.004

Garrett, W. S., Lord, G. M., Punit, S., Lugo-Villarino, G., Mazmanian, S. K., Ito, S., et al. (2007). Communicable ulcerative colitis induced by T-bet deficiency in the innate immune system. *Cell* 131, 33–45. doi: 10.1016/j.cell.2007.08.017

Gonzalez-Chavez, R., Alvarez, A. F., Romeo, T., and Georgellis, D. (2010). The physiological stimulus for the BarA sensor kinase. *J. Bacteriol.* 192, 2009–2012. doi: 10.1128/JB.01685-09

Goodman, K. N., Powers, M. J., Crofts, A. A., Trent, M. S., and Hendrixson, D. R. (2020). Campylobacter jejuni BumSR directs a response to butyrate via sensor phosphatase activity to impact transcription and colonization. *Proc. Natl. Acad. Sci. U.S.A.* 117, 11715–11726. doi: 10.1073/pnas.1922719117

Haapalainen, M., Mosorin, H., Dorati, F., Wu, R. F., Roine, E., Taira, S., et al. (2012). Hcp2, a secreted protein of the phytopathogen Pseudomonas syringae pv. tomato DC3000, is required for fitness for competition against bacteria and yeasts. *J. Bacteriol.* 194, 4810–4822. doi: 10.1128/JB.00611-12

Hachani, A., Wood, T. E., and Filloux, A. (2016). Type VI secretion and anti-host effectors. *Curr. Opin. Microbiol.* 29, 81–93. doi: 10.1016/j.mib.2015.11.006

Han, Y., Wang, T., Chen, G., Pu, Q., Liu, Q., Zhang, Y., et al. (2019). A *Pseudomonas aeruginosa* type VI secretion system regulated by CueR facilitates copper acquisition. *PLoS Pathog.* 15:e1008198. doi: 10.1371/journal.ppat.1008198

Hecht, A. L., Casterline, B. W., Earley, Z. M., Goo, Y. A., Goodlett, D. R., and Bubeck Wardenburg, J. (2016). Strain competition restricts colonization of an enteric pathogen and prevents colitis. *EMBO Rep.*17, 1281–1291. doi: 10.15252/embr.201642282

Hersch, S. J., Watanabe, N., Stietz, M. S., Manera, K., Kamal, F., Burkinshaw, B., et al. (2020). Envelope stress responses defend against type six secretion system attacks independently of immunity proteins. *Nat. Microbiol.* 5, 706–714. doi: 10.1038/s41564-020-0672-6

Hood, R. D., Singh, P., Hsu, F., Güvener, T., Carl, M. A., Trinidad, R. R. S., et al. (2010). A type VI secretion system of *Pseudomonas aeruginosa* targets a toxin to bacteria. *Cell Host Microbe* 7, 25–37. doi: 10.1016/j.chom.2009.12.007

Hooper, L. V., Littman, D. R., and Macpherson, A. J. (2012). Interactions between the microbiota and the immune system. *Science* 336, 1268–1273. doi: 10.1126/science.1223490

Hsieh, P.-F., Lu, Y.-R., Lin, T.-L., Lai, L.-Y., and Wang, J.-T. (2019). *Klebsiella pneumoniae* Type VI secretion system contributes to bacterial competition, cell invasion, type-1 fimbriae expression, and *in vivo* colonization. *J. Infect. Dis.* 219, 637–647. doi: 10.1093/infdis/jiy534

Human Microbiome Project Consortium (2012). Structure, function and diversity of the healthy human microbiome. *Nature* 486, 207–214. doi: 10.1038/nature11234

Iwasaki, A., and Medzhitov, R. (2015). Control of adaptive immunity by the innate immune system. *Nat. Immunol.* 16, 343–353. doi: 10.1038/ni.3123

James, K. R., Gomes, T., Elmentaite, R., Kumar, N., Gulliver, E. L., King, H. W., et al. (2020). Distinct microbial and immune niches of the human colon. *Nat. Immunol.* 21, 343–353. doi: 10.1038/s41590-020-0602-z

Jiang, F., Wang, X., Wang, B., Chen, L., Zhao, Z., Waterfield, N. R., et al. (2016). The *Pseudomonas aeruginosa* type VI secretion PGAP1-like effector induces host autophagy by activating endoplasmic reticulum stress. *Cell Rep.* 16, 1502–1509. doi: 10.1016/j.celrep.2016.07.012

Jiang, F., Waterfield, N. R., Yang, J., Yang, G., and Jin, Q. (2014). A *Pseudomonas aeruginosa* type VI secretion phospholipase D effector targets both prokaryotic and eukaryotic cells. *Cell Host Microbe* 15, 600–610. doi: 10.1016/j.chom.2014.04.010

Jochum, L., and Stecher, B. (2020). Label or concept – what is a pathobiont? *Trends Microbiol.*, 28, 789–792. doi: 10.1016/j.tim.2020.04.011

Johansson, M. E. V., Larsson, J. M. H., and Hansson, G. C. (2011). The two mucus layers of colon are organized by the MUC2 mucin, whereas the outer layer is a legislator of host-microbial interactions. *Proc. Natl. Acad. Sci. U.S.A.* 108(Suppl.), 4659–4665. doi: 10.1073/pnas.1006451107

Kamada, N., Seo, S.-U., Chen, G. Y., and Núñez, G. (2013). Role of the gut microbiota in immunity and inflammatory disease. *Nat. Rev. Immunol.* 13, 321–35. doi: 10.1038/nri3430

Kaparakis-Liaskos, M., and Ferrero, R. L. (2015). Immune modulation by bacterial outer membrane vesicles. *Nat. Rev. Immunol.* 15, 375–387. doi: 10.1038/nri3837

Kirchberger, P. C., Unterweger, D., Provenzano, D., Pukatzki, S., and Boucher, Y. (2017). Sequential displacement of type VI secretion system effector genes leads to evolution of diverse immunity gene arrays in *Vibrio cholerae*. *Sci. Rep.* 7:45133. doi: 10.1038/srep45133

Kjer-Nielsen, L., Patel, O., Corbett, A. J., Le Nours, J., Meehan, B., Liu, L., et al. (2012). MR1 presents microbial vitamin B metabolites to MAIT cells. *Nature* 491, 717–723. doi: 10.1038/nature11605

Kleino, A., and Silverman, N. (2014). The *Drosophila* IMD pathway in the activation of the humoral immune response. *Dev. Comp. Immunol.* 42, 25–35. doi: 10.1016/j.dci.2013.05.014

Kohli, N., Crisp, Z., Riordan, R., Li, M., Alaniz, R. C., and Jayaraman, A. (2018). The microbiota metabolite indole inhibits *Salmonella* virulence: involvement of the PhoPQ two-component system. *PLoS ONE* 13:e0190613. doi: 10.1371/journal.pone.0190613

Koropatnick, T. A., Engle, J. T., Apicella, M. A., Stabb, E. V., Goldman, W. E., and McFall-Ngai, M. J. (2004). Microbial factor-mediated development in a host-bacterial mutualism. *Science* 306, 1186–1188. doi: 10.1126/science.1102218

Kostic, A. D., Gevers, D., Siljander, H., Vatanen, T., Hyötyläinen, T., Hämäläinen, A.-M., et al. (2015). The dynamics of the human infant gut microbiome in development and in progression toward type 1 diabetes. *Cell Host Microbe* 17, 260–273. doi: 10.1016/j.chom.2015.01.001

LaCourse, K. D., Peterson, S. B., Kulasekara, H. D., Radey, M. C., Kim, J., and Mougous, J. D. (2018). Conditional toxicity and synergy drive diversity among antibacterial effectors. *Nat. Microbiol.* 3, 440–446. doi: 10.1038/s41564-018-0113-y

Lawhon, S. D., Maurer, R., Suyemoto, M., and Altier, C. (2002). Intestinal short-chain fatty acids alter *Salmonella typhimurium* invasion gene expression and virulence through BarA/SirA. *Mol. Microbiol.* 46, 1451–1464. doi: 10.1046/j.1365-2958.2002.03268.x

Ledvina, H. E., Kelly, K. A., Eshraghi, A., Plemel, R. L., Peterson, S. B., Lee, B., et al. (2018). A phosphatidylinositol 3-kinase effector alters phagosomal maturation to promote intracellular growth of *Francisella*. *Cell Host Microbe* 24, 285–295.e8. doi: 10.1016/j.chom.2018.07.003

Lee, S. M., Donaldson, G. P., Mikulski, Z., Boyajian, S., Ley, K., and Mazmanian, S. K. (2013). Bacterial colonization factors control specificity and stability of the gut microbiota. *Nature* 501, 426–429. doi: 10.1038/nature12447

Leiman, P. G., Basler, M., Ramagopal, U. A., Bonanno, J. B., Sauder, J. M., Pukatzki, S., et al. (2009). Type VI secretion apparatus and phage tail-associated protein complexes share a common evolutionary origin. *Proc. Natl. Acad. Sci. U.S.A.* 106, 4154–4159. doi: 10.1073/pnas.0813360106

LeRoux, M., Kirkpatrick, R. L., Montauti, E. I., Tran, B. Q., Peterson, S. B., Harding, B. N., et al. (2015). Kin cell lysis is a danger signal that activates antibacterial pathways of *Pseudomonas aeruginosa*. *eLife* 4:e05701. doi: 10.7554/eLife.05701

Lery, L. M. S., Frangeul, L., Tomas, A., Passet, V., Almeida, A. S., Bialek-Davenet, S., et al. (2014). Comparative analysis of *Klebsiella pneumoniae* genomes identifies a phospholipase D family protein as a novel virulence factor. *BMC Biol.* 12:41. doi: 10.1186/1741-7007-12-41

Lin, J., Zhang, W., Cheng, J., Yang, X., Zhu, K., Wang, Y., et al. (2017). A *Pseudomonas* T6SS effector recruits PQS-containing outer membrane vesicles for iron acquisition. *Nat. Commun.* 8:14888. doi: 10.1038/ncomms 14888

Litvak, Y., Byndloss, M. X., and Bäumler, A. J. (2018). Colonocyte metabolism shapes the gut microbiota. *Science* 362:eaat9076. doi: 10.1126/science.aat9076

Logan, S. L., Thomas, J., Yan, J., Baker, R. P., Shields, D. S., Xavier, J. B., et al. (2018). The *Vibrio cholerae* type VI secretion system can modulate host intestinal mechanics to displace gut bacterial symbionts. *Proc. Natl. Acad. Sci. U.S.A.* 115, E3779–E3787. doi: 10.1101/226472

Ma, A. T., McAuley, S. B., Pukatzki, S., and Mekalanos, J. J. (2009). Translocation of a *Vibrio cholerae* type VI secretion effector requires bacterial endocytosis by host cells. *Cell Host Microbe* 5, 234–243. doi: 10.1016/j.chom.2009.02.005

Mazmanian, S. K., Liu, C. H., Tzianabos, A. O., and Kasper, D. L. (2005). An immunomodulatory molecule of symbiotic bacteria directs maturation of the host immune system. *Cell* 122, 107–18. doi: 10.1016/j.cell.2005.05.007

Mazmanian, S. K., Round, J. L., and Kasper, D. L. (2008). A microbial symbiosis factor prevents intestinal inflammatory disease. *Nature* 453, 620–625. doi: 10.1038/nature07008

Medzhitov, R., and Janeway, C. A. (1999). Innate immune induction of the adaptive immune response. *Cold Spring Harb. Symp. Quant. Biol.* 64, 429–436. doi: 10.1101/sqb.1999.64.429

Miyata, S. T., Kitaoka, M., Brooks, T. M., McAuley, S. B., and Pukatzki, S. (2011). *Vibrio cholerae* requires the type VI secretion system virulence factor Vasx to kill *Dictyostelium discoideum*. *Infect. Immun.* 79, 2941–2949. doi: 10.1128/IAI.01266-10

Moretti, J., and Blander, J. M. (2018). Detection of a vita-PAMP STINGs cells into reticulophagy. *Autophagy* 14, 1102–1104. doi: 10.1080/15548627.2018.1441471

Nano, F. E., Zhang, N., Cowley, S. C., Klose, K. E., Cheung, K. K. M., Roberts, M. J., et al. (2004). A *Francisella tularensis* pathogenicity island required for intramacrophage growth. *J. Bacteriol.* 186, 6430–6436. doi: 10.1128/JB.186.19.6430-6436.2004

Nazarov, S., Schneider, J. P., Brackmann, M., Goldie, K. N., Stahlberg, H., and Basler, M. (2017). Cryo-EM reconstruction of type VI secretion system baseplate and sheath distal end. *EMBO J.* 37:e97103. doi: 10.15252/embj.201797103

Neyen, C., Runchel, C., Schüpfer, F., Meier, P., and Lemaitre, B. (2016). The regulatory isoform rPGRP-LC induces immune resolution *via* endosomal degradation of receptors. *Nat. Immunol.* 17, 1150–1158. doi: 10.1038/ni.3536

O'Mahony, C., Scully, P., O'Mahony, D., Murphy, S., O'Brien, F., Lyons, A., et al. (2008). Commensal-induced regulatory T cells mediate protection against pathogen-stimulated NF-κB activation. *PLoS Pathog.* 4:e1000112. doi: 10.1371/journal.ppat.1000112

Ost, K. S., and Round, J. L. (2018). Communication between the microbiota and mammalian immunity. *Annu. Rev. Microbiol.* 72, 399–422. doi: 10.1146/annurev-micro-090817-062307

Parada Venegas, D., de la Fuente, M. K., Landskron, G., González, M. J., Quera, R., Dijkstra, G., et al. (2019). Short chain fatty acids (SCFAs)-mediated gut epithelial and immune regulation and its relevance for inflammatory bowel diseases. *Front. Immunol.* 10:277. doi: 10.3389/fimmu.2019.01486

Pédron, T., Mulet, C., Dauga, C., Frangeul, L., Chervaux, C., Grompone, G., et al. (2012). A crypt-specific core microbiota resides in the mouse colon. *MBio* 3:e00116-12. doi: 10.1128/mBio.00116-12

Pukatzki, S., Ma, A. T., Revel, A. T., Sturtevant, D., and Mekalanos, J. J. (2007). Type VI secretion system translocates a phage tail spike-like protein into target cells where it cross-links actin. *Proc. Natl. Acad. Sci. U.S.A.* 104, 15508–15513. doi: 10.1073/pnas.0706532104

Pukatzki, S., Ma, A. T., Sturtevant, D., Krastins, B., Sarracino, D., Nelson, W. C., et al. (2006). Identification of a conserved bacterial protein secretion system in *Vibrio cholerae* using the *Dictyostelium* host model system. *Proc. Natl. Acad. Sci. U.S.A.* 103, 1528–1533. doi: 10.1073/pnas.0510322103

Rakoff-Nahoum, S., Paglino, J., Eslami-Varzaneh, F., Edberg, S., and Medzhitov, R. (2004). Recognition of commensal microflora by toll-like receptors is required for intestinal homeostasis. *Cell* 118, 229–241. doi: 10.1016/j.cell.2004.07.002

Ringel, P. D., Hu, D., and Basler, M. (2017). The role of type VI secretion system effectors in target cell lysis and subsequent horizontal gene transfer. *Cell Rep.* 21, 3927–3940. doi: 10.1016/j.celrep.2017.12.020

Rivera-Chávez, F., and Mekalanos, J. J. (2019). Cholera toxin promotes pathogen acquisition of host-derived nutrients. *Nature* 572, 244–248. doi: 10.1038/s41586-019-1453-3

Rosales-Reyes, R., Skeldon, A. M., Aubert, D. F., and Valvano, M. A. (2012). The type VI secretion system of *Burkholderia cenocepacia* affects multiple

Rho family GTPases disrupting the actin cytoskeleton and the assembly of NADPH oxidase complex in macrophages. *Cell Microbiol.* 14, 255–273. doi: 10.1111/j.1462-5822.2011.01716.x

Ross, B. D., Verster, A. J., Radey, M. C., Schmidtke, D. T., Pope, C. E., Hoffman, L. R., et al. (2019). Human gut bacteria contain acquired interbacterial defence systems. *Nature* 575, 224–228. doi: 10.1038/s41586-019-1708-z

Round, J. L., Lee, S. M., Li, J., Tran, G., Jabri, B., Chatila, T. A., et al. (2011). The Toll-like receptor 2 pathway establishes colonization by a commensal of the human microbiota. *Science* 332, 974–977. doi: 10.1126/science.1206095

Round, J. L., and Mazmanian, S. K. (2009). The gut microbiota shapes intestinal immune responses during health and disease. *Nat. Rev. Immunol.* 9, 313–323. doi: 10.1038/nri2515

Round, J. L., O'Connell, R. M., and Mazmanian, S. K. (2010). Coordination of tolerogenic immune responses by the commensal microbiota. *J. Autoimmun.* 34, J220–J225. doi: 10.1016/j.jaut.2009.11.007

Russell, A. B., Peterson, S. B., and Mougous, J. D. (2014a). Type VI secretion system effectors: poisons with a purpose. *Nat. Rev. Microbiol.* 12, 137–148. doi: 10.1038/nrmicro3185

Russell, A. B., Wexler, A. G., Harding, B. N., Whitney, J. C., Bohn, A. J., Goo, Y. A., et al. (2014b). A type VI secretion-related pathway in *Bacteroidetes* mediates interbacterial antagonism. *Cell Host Microbe* 16, 227–236. doi: 10.1016/j.chom.2014.07.007

Ryu, J.-H., Kim, S.-H., Lee, H.-Y., Bai, J. Y., Nam, Y.-D., Bae, J.-W., et al. (2008). Innate immune homeostasis by the homeobox gene *Caudal* and commensal-gut mutualism in *Drosophila*. *Science* 319, 777–782. doi: 10.1126/science.1149357

Sana, T. G., Flaugnatti, N., Lugo, K. A., Lam, L. H., Jacobson, A., Baylot, V., et al. (2016). *Salmonella Typhimurium* utilizes a T6SS-mediated antibacterial weapon to establish in the host gut. *Proc. Natl. Acad. Sci. U.S.A.* 113, E5044–E5051. doi: 10.1073/pnas.1608858113

Sana, T. G., Hachani, A., Bucior, I., Soscia, C., Garvis, S., Termine, E., et al. (2012). The second type VI secretion system of *Pseudomonas aeruginosa* strain PAO1 is regulated by quorum sensing and *fur* and modulates internalization in epithelial cells. *J. Biol. Chem.* 287, 27095–27105. doi: 10.1074/jbc.M112.376368

Sanchez-Garrido, J., Slater, S. L., Clements, A., Shenoy, A. R., and Frankel, G. (2020). Vying for the control of inflammasomes: the cytosolic frontier of enteric bacterial pathogen-host interactions. *Cell.Microbiol.* 22:e13184. doi: 10.1111/cmi.13184

Sasabe, J., Miyoshi, Y., Rakoff-Nahoum, S., Zhang, T., Mita, M., Davis, B. M., et al. (2016). Interplay between microbial D-amino acids and host D-amino acid oxidase modifies murine mucosal defence and gut microbiota. *Nat. Microbiol.* 1:16125. doi: 10.1038/nmicrobiol.2016.125

Schulthess, J., Pandey, S., Capitani, M., Rue-Albrecht, K. C., Arnold, I., Franchini, F., et al. (2019). The short chain fatty acid butyrate imprints an antimicrobial program in macrophages. *Immunity* 50, 432–445.e7. doi: 10.1016/j.immuni.2018.12.018

Schwab, U., Abdullah, L. H., Perlmutt, O. S., Albert, D., Davis, C. W., Arnold, R. R., et al. (2014). Localization of *Burkholderia cepacia* complex bacteria in cystic fibrosis lungs and interactions with *Pseudomonas aeruginosa* in hypoxic mucus. *Infect. Immun.* 82, 4729–4745. doi: 10.1128/IAI.01876-14

Schwalm, N. D., and Groisman, E. A. (2017). Navigating the gut buffet: control of polysaccharide utilization in Bacteroides spp. *Trends Microbiol.* 25, 1005–1015. doi: 10.1016/j.tim.2017.06.009

Schwarz, S., Singh, P., Robertson, J. D., LeRoux, M., Skerrett, S. J., Goodlett, D. R., et al. (2014). VgrG-5 is a *Burkholderia* type VI secretion system-exported protein required for multinucleated giant cell formation and virulence. *Infect. Immun.* 82, 1445–1452. doi: 10.1128/IAI.01368-13

Sender, R., Fuchs, S., and Milo, R. (2016). Revised estimates for the number of human and bacteria cells in the body. *PLoS Biol.* 14:e1002533. doi: 10.1371/journal.pbio.1002533

Shen, Y., Giardino Torchia, M. L., Lawson, G. W., Karp, C. L., Ashwell, J. D., and Mazmanian, S. K. (2012). Outer membrane vesicles of a human commensal mediate immune regulation and disease protection. *Cell Host Microbe* 12, 509–520. doi: 10.1016/j.chom.2012.08.004

Shneider, M. M., Buth, S. A., Ho, B. T., Basler, M., Mekalanos, J. J., and Leiman, P. G. (2013). PAAR-repeat proteins sharpen and diversify the type VI secretion system spike. *Nature* 500, 350–353. doi: 10.1038/nature12453

Si, M., Wang, Y., Zhang, B., Zhao, C., Kang, Y., Bai, H., et al. (2017a). The type VI secretion system engages a redox-regulated dual-functional heme transporter for zinc acquisition. *Cell Rep.* 20, 949–959. doi: 10.1016/j.celrep.2017.06.081

Si, M., Zhao, C., Burkinshaw, B. J., Zhang, B., Wei, D., Wang, Y., et al. (2017b). Manganese scavenging and oxidative stress response mediated by type VI secretion system in *Burkholderia thailandensis*. *Proc. Natl. Acad. Sci. U.S.A.* 114, E2233–E2242. doi: 10.1073/pnas.1614902114

Smith, K., McCoy, K. D., and Macpherson, A. J. (2007). Use of axenic animals in studying the adaptation of mammals to their commensal intestinal microbiota. *Semin. Immunol.* 19, 59–69. doi: 10.1016/j.smim.2006.10.002

Sonnenburg, J. L., Xu, J., Leip, D. D., Chen, C.-H., Westover, B. P., Weatherford, J., et al. (2005). Glycan foraging *in vivo* by an intestine-adapted bacterial symbiont. *Science* 307, 1955–1959. doi: 10.1126/science.1109051

Speare, L., Cecere, A. G., Guckes, K. R., Smith, S., Wollenberg, M. S., Mandel, M. J., et al. (2018). Bacterial symbionts use a type VI secretion system to eliminate competitors in their natural host. *Proc. Natl. Acad. Sci. U.S.A.* 115, E8528–E8537. doi: 10.1073/pnas.1808302115

Speare, L., Smith, S., Salvato, F., Kleiner, M., and Septer, A. N. (2020). Environmental viscosity modulates interbacterial killing during habitat transition. *MBio* 11, 1–14. doi: 10.1128/mBio.03060-19

Storey, D., McNally, A., Åstrand, M., sa-Pessoa Graca Santos, J., Rodriguez-Escudero, I., Elmore, B., et al. (2020). *Klebsiella pneumoniae* type VI secretion system-mediated microbial competition is PhoPQ controlled and reactive oxygen species dependent. *PLoS Pathog.*16:e1007969. doi: 10.1371/journal.ppat.1007969

Suarez, G., Sierra, J. C., Erova, T. E., Sha, J., Horneman, A. J., and Chopra, A. K. (2010). A type VI secretion system effector protein, VgrG1, from *Aeromonas hydrophila* that induces host cell toxicity by ADP ribosylation of actin. *J. Bacteriol.* 192, 155–168. doi: 10.1128/JB.01260-09

Takeuchi, O., and Akira, S. (2010). Pattern recognition receptors and inflammation. *Cell* 140, 805–820. doi: 10.1016/j.cell.2010.01.022

Thiennimitr, P., Winter, S. E., and Bäumler, A. J. (2012). *Salmonella*, the host and its microbiota. *Curr. Opin. Microbiol.* 15, 108–114. doi: 10.1016/j.mib.2011.10.002

Toesca, I. J., French, C. T., and Miller, J. F. (2014). The type VI secretion system spike protein VgrG5 mediates membrane fusion during intercellular spread by *pseudomallei* group *Burkholderia* species. *Infect. Immun.* 82, 1436–1444. doi: 10.1128/IAI.01367-13

Toska, J., Ho, B. T., and Mekalanos, J. J. (2018). Exopolysaccharide protects *Vibrio cholerae* from exogenous attacks by the type 6 secretion system. *Proc. Natl. Acad. Sci. U.S.A.* 115, 7997–8002. doi: 10.1073/pnas.1808469115

Trunk, K., Peltier, J., Liu, Y.-C., Dill, B. D., Walker, L., Gow, N. A. R., et al. (2018). The type VI secretion system deploys antifungal effectors against microbial competitors. *Nat. Microbiol.* 3, 920–931. doi: 10.1038/s41564-018-0191-x

Tsolis, R. M., and Bäumler, A. J. (2020). Gastrointestinal host-pathogen interaction in the age of microbiome research. *Curr. Opin. Microbiol.* 53, 78–89. doi: 10.1016/j.mib.2020.03.002

Ugolini, M., Gerhard, J., Burkert, S., Jensen, K. J., Georg, P., Ebner, F., et al. (2018). Recognition of microbial viability *via* TLR8 drives T_{FH} cell differentiation and vaccine responses. *Nat. Immunol.* 19, 386–396. doi: 10.1038/s41590-018-0068-4

Unterweger, D., Miyata, S. T., Bachmann, V., Brooks, T. M., Mullins, T., Kostiuk, B., et al. (2014). The *Vibrio cholerae* type VI secretion system employs diverse effector modules for intraspecific competition. *Nat. Commun.* 5:3549. doi: 10.1038/ncomms4549

Vacheron, J., Péchy-Tarr, M., Brochet, S., Heiman, C. M., Stojiljkovic, M., Maurhofer, M., et al. (2019). T6SS contributes to gut microbiome invasion and killing of an herbivorous pest insect by plant-beneficial *Pseudomonas protegens*. *ISME J.* 13, 1318–1329. doi: 10.1038/s41396-019-0353-8

Venkatesh, M., Mukherjee, S., Wang, H., Li, H., Sun, K., Benechet, A. P., et al. (2014). Symbiotic bacterial metabolites regulate gastrointestinal barrier function via the xenobiotic sensor PXR and toll-like receptor 4. *Immunity* 41, 296–310. doi: 10.1016/j.immuni.2014.06.014

Verster, A. J., Ross, B. D., Radey, M. C., Bao, Y., Goodman, A. L., Mougous, J. D., et al. (2017). The landscape of type VI secretion across human gut microbiomes reveals its role in community composition. *Cell Host Microbe* 22, 411–419.e4. doi: 10.1016/j.chom.2017.08.010

Vonaesch, P., Anderson, M., and Sansonetti, P. J. (2018). Pathogens, microbiome and the host: emergence of the ecological Koch's postulates. *FEMS Microbiol. Rev.* 42, 273–292. doi: 10.1093/femsre/fuy003

Wan, B., Zhang, Q., Ni, J., Li, S., Wen, D., Li, J., et al. (2017). Type VI secretion system contributes to enterohemorrhagic *Escherichia coli* virulence by secreting catalase against host reactive oxygen species (ROS). *PLoS Pathog.* 13:e1006246. doi: 10.1371/journal.ppat.1006246

Wang, T., Si, M., Song, Y., Zhu, W., Gao, F., Wang, Y., et al. (2015). Type VI secretion system transports Zn^{2+} to combat multiple stresses and host immunity. *PLoS Pathog.* 11:e1005020. doi: 10.1371/journal.ppat.1005020

Wettstadt, S., Wood, T. E., Fecht, S., and Filloux, A. (2019). Delivery of the *Pseudomonas aeruginosa* phospholipase effectors PldA and PldB in a VgrG- and H2-T6SS-dependent manner. *Front. Microbiol.* 10:1718. doi: 10.3389/fmicb.2019.01718

Wexler, A. G., Bao, Y., Whitney, J. C., Bobay, L.-M., Xavier, J. B., Schofield, W. B., et al. (2016). Human symbionts inject and neutralize antibacterial toxins to persist in the gut. *Proc. Natl. Acad. Sci. U.S.A.*113, 3639–3644. doi: 10.1073/pnas.1525637113

Whiteley, A. T., Eaglesham, J. B., de Oliveira Mann, C. C., Morehouse, B. R., Lowey, B., Nieminen, E. A., et al. (2019). Bacterial cGAS-like enzymes synthesize diverse nucleotide signals. *Nature* 567, 194–199. doi: 10.1038/s41586-019-0953-5

Wood, T. E., Howard, S. A., Förster, A., Nolan, L. M., Manoli, E., Bullen, N. P., et al. (2019). The *Pseudomonas aeruginosa* T6SS delivers a periplasmic toxin that disrupts bacterial cell morphology. *Cell Rep.* 29, 187–201.e7. doi: 10.1016/j.celrep.2019.08.094

Xu, H., Yang, J., Gao, W., Li, L., Li, P., Zhang, L., et al. (2014). Innate immune sensing of bacterial modifications of Rho GTPases by the Pyrin inflammasome. *Nature* 513, 237–241. doi: 10.1038/nature13449

Yassour, M., Vatanen, T., Siljander, H., Hämäläinen, A.-M., Härkönen, T., Ryhänen, S. J., et al. (2016). Natural history of the infant gut microbiome and impact of antibiotic treatment on bacterial strain diversity and stability. *Sci. Transl. Med.* 8:343ra81. doi: 10.1126/scitranslmed.aad0917

Zhang, D., de Souza, R. F., Anantharaman, V., Iyer, L. M., and Aravind, L. (2012). Polymorphic toxin systems: comprehensive characterization of trafficking modes, processing, mechanisms of action, immunity and ecology using comparative genomics. *Biol. Direct* 7:18. doi: 10.1186/1745-6150-7-18

Zhang, Z., Tang, H., Chen, P., Xie, H., and Tao, Y. (2019). Demystifying the manipulation of host immunity, metabolism, and extraintestinal tumors by the gut microbiome. *Signal Transduct. Target. Ther.* 4:41. doi: 10.1038/s41392-019-0074-5

Zhao, W., Caro, F., Robins, W., and Mekalanos, J. J. (2018). Antagonism toward the intestinal microbiota and its effect on *Vibrio cholerae* virulence. *Science* 359, 210–213. doi: 10.1126/science.aap8775

Zitomersky, N. L., Coyne, M. J., and Comstock, L. E. (2011). Longitudinal analysis of the prevalence, maintenance, and IgA response to species of the order *Bacteroidales* in the human gut. *Infect. Immun.* 79, 2012–2020. doi: 10.1128/IAI.01348-10

2

The *Coxiella burnetii* T4SS Effector AnkF is Important for Intracellular Replication

Julian Pechstein[1], Jan Schulze-Luehrmann[1], Stephanie Bisle[1], Franck Cantet[2], Paul A. Beare[3], Martha Ölke[1], Matteo Bonazzi[2], Christian Berens[4] and Anja Lührmann[1]*

[1] *Mikrobiologisches Institut-Klinische Mikrobiologie, Immunologie und Hygiene, Universitätsklinikum Erlangen, Friedrich-Alexander-Universität Erlangen-Nürnberg, Erlangen, Germany,* [2] *Institut de Recherche en Infectiologie de Montpellier (IRIM), Centre National de la Recherche Scientifique (CNRS), Université de Montpellier, Montpellier, France,* [3] *Coxiella Pathogenesis Section, Laboratory of Bacteriology, Rocky Mountain Laboratories, National Institute of Allergy and Infectious Diseases, National Institutes of Health, Hamilton, MT, United States,* [4] *Friedrich-Loeffler-Institut, Institut für Molekulare Pathogenese, Jena, Germany*

Correspondence:
Anja Lührmann
anja.luehrmann@uk-erlangen.de

Coxiella burnetii is an obligate intracellular pathogen and the causative agent of the zoonotic disease Q fever. Following uptake by alveolar macrophages, the pathogen replicates in an acidic phagolysosomal vacuole, the *C. burnetii*-containing vacuole (CCV). Effector proteins translocated into the host cell by the type IV secretion system (T4SS) are important for the establishment of the CCV. Here we focus on the effector protein AnkF and its role in establishing the CCV. The *C. burnetii* AnkF knock out mutant invades host cells as efficiently as wild-type *C. burnetii*, but this mutant is hampered in its ability to replicate intracellularly, indicating that AnkF might be involved in the development of a replicative CCV. To unravel the underlying reason(s), we searched for AnkF interactors in host cells and identified vimentin through a yeast two-hybrid approach. While AnkF does not alter vimentin expression at the mRNA or protein levels, the presence of AnkF results in structural reorganization and vesicular co-localization with recombinant vimentin. Ectopically expressed AnkF partially accumulates around the established CCV and endogenous vimentin is recruited to the CCV in a time-dependent manner, suggesting that AnkF might attract vimentin to the CCV. However, knocking-down endogenous vimentin does not affect intracellular replication of *C. burnetii*. Other cytoskeletal components are recruited to the CCV and might compensate for the lack of vimentin. Taken together, AnkF is essential for the establishment of the replicative CCV, however, its mode of action is still elusive.

Keywords: Q fever, type IV secretion system, effector protein, vimentin, phagosome, *Coxiella burnetii*

INTRODUCTION

The obligate intracellular Gram-negative pathogen *Coxiella burnetii* is the causative agent of the zoonotic disease Q fever (Maurin and Raoult, 1999). With the exception of New Zealand, *C. burnetii* is distributed worldwide. The bacterium can infect a vast range of species (Maurin and Raoult, 1999; Gonzalez-Barrio and Ruiz-Fons, 2019), but live-stock animals, such as cattle, sheep, and goats, are the most important natural reservoirs and also the main source of human infections (Rodolakis, 2009). An acute infection might be symptom-free or cause a flu-like illness (Maurin and Raoult, 1999). The development of pneumonia or granulomatous hepatitis are also common symptoms of acute Q fever (Raoult et al., 2005). Immunocompromised people with preceding cardiac valve pathology and pregnant women are mainly at risk of developing chronic Q fever. Its typical symptoms include endocarditis and vascular infections (Maurin and Raoult, 1999). While good treatment options are available for acute Q fever, they are missing for chronic Q fever. Thus, chronic Q fever is treated by a combination of doxycycline and hydroxychloroquine for at least 18 months (Kersh, 2013). This lengthy treatment comes with severe side effects and, as a consequence, limited compliance.

Primary infection in humans occurs in alveolar macrophages after inhalation of *C. burnetii*-contaminated dust particles or aerosols (Khavkin and Tabibzadeh, 1988). However, non-phagocytic cells are also susceptible to infection (Voth and Heinzen, 2007). Host cell invasion of professional phagocytes by *C. burnetii* involves $\alpha_v\beta_3$ integrin receptors and actin-dependent membrane ruffling (Baca et al., 1993; Capo et al., 1999; Dellacasagrande et al., 2000; Aguilera et al., 2009). In non-professional phagocytes, the bacterial invasin OmpA and cortactin are involved (Rosales et al., 2012; Martinez et al., 2014). Following internalization, the bacteria reside within the *C. burnetii*-containing vacuole (CCV), which develops into a phagolysosomal-like compartment with an acidic pH (van Schaik et al., 2013). Most bacteria are killed under these conditions, but for *C. burnetii* they are optimal for proliferation (Hackstadt and Williams, 1981). Moreover, expression and activity of the type IV secretion system (T4SS) is enabled under acidic conditions (Coleman et al., 2004; Newton et al., 2013). The fact that bacteria lacking the T4SS are unable to replicate intracellularly (Carey et al., 2011) demonstrates that the T4SS is a major virulence determinant. It is used to inject virulence factors, so-called effector proteins, which allows reprograming of the host cell for the benefit of the pathogen (Lührmann et al., 2017). Translocation of effector proteins starts around 8 h post-infection and translocation rates increase in a time-dependent manner (Newton et al., 2013). Several of the ~150 identified effector proteins interfere with vesicular trafficking or localize to the CCV membrane. The activity of T4SS effector proteins allows the massive expansion of the CCV, which can occupy the majority of the host cell's volume (Lührmann et al., 2017). How *C. burnetii* ensures the stability of this huge compartment is not understood, but galectins (Mansilla Pareja et al., 2017) and actin (Colonne et al., 2016; Miller et al., 2018) might be involved.

Here we report that the T4SS effector protein AnkF (CBU0447) is important for optimal intracellular replication of *C. burnetii*. Ectopically expressed AnkF localized throughout the cytosol, the nucleus and partially around the CCV. It also interacted with the intermediate filament vimentin, which is recruited to the CCV in a time-dependent manner. However, siRNA-mediated knock-down of vimentin did not reduce *C. burnetii* replication.

MATERIALS AND METHODS

Reagents and Antibodies

Unless stated otherwise, reagents were purchased from Carl Roth, Sigma-Aldrich or Thermo Fisher. The following primary antibodies were used: anti-*Coxiella*, anti-AnkF (Davids Biotechnologie), anti-vimentin (Cell Signaling #5741S, or Sigma-Aldrich, #V6630), anti-tubulin (Cell Signaling, #3873) and anti-cytokeratin 18 (Thermo Fisher, #MA5-12104). Actin was stained with the Phalloidin-Alexa Fluor-647 conjugate (A2066, Sigma-Aldrich). The LAMP2 (ABL-93) specific primary antibody was developed by J.T. August and obtained from the Developmental Studies Hybridoma Bank, University of Iowa, Department of Biology, Iowa City, IA, USA. Secondary antibodies conjugated with Alexa Fluor-488 or -594 were purchased from Dianova. For STED confocal microscopy, Alexa Fluor-580-STAR antibodies (#2-0012-005-8, Abberior GmbH, Göttingen, Germany) were kindly provided by Dr. Ralph Palmisano, Optical Imaging Center Erlangen (OICE), Germany.

Bacterial Strains, Yeast Strains, and Cell Lines

Escherichia coli DH10β were cultivated in Luria Bertani (1% bacto tryptone, 0.5% yeast extract and 1% NaCl) broth supplemented with 100 µg/ml ampicillin or 50 µg/ml kanamycin where appropriate. *Coxiella burnetii* Nine Mile II (NMII) RSA439 clone 4 were grown in acidified citrate-cysteine medium (ACCM-2) at 37°C, 2.5% O_2, and 5% CO_2. Axenic media were supplemented with 3 µg/ml chloramphenicol where appropriate for selection. The leucine- and tryptophan-auxotrophic *Saccharomyces cerivisiae* strains Y187, AH109, and Y190 were grown in YPAD (1% yeast extract, 2% caseine peptone, 2% glucose, and 0.01% adenine hemisulfate) or SCAD (2% glucose, 0.6% yeast nitrogen base, 0.06% amino acid mix, pH 5.8) with medium shaking or on agar plates (media supplemented with 1.5% agar) at 30°C.

CHO-FcR cells (Chinese hamster fibroblasts endogenously expressing the macrophage-lymphocyte Fc receptor) were maintained in Dulbecco's Modified Eagle's Medium (DMEM, Thermo Fisher). HeLa (human cervical carcinomal epithelial cells), U2OS and U2OS-vimentin-rsEGFP (recombinant human bone osteosarcoma cells endogenously expressing vimentin-rsEGFP (Ratz et al., 2015)) were maintained in DMEM. HeLa cells stably transfected with pWHE644/655-AnkF were cultured in DMEM supplemented with 1% Penicillin/Streptomycin (Thermo Fisher), 0.3 mg/ml Geneticin (G418) and 0.25 µg/ml puromycin (Berens et al., 2015; Bisle et al., 2016). All media were supplemented with 5% heat-inactivated fetal bovine serum (FBS, Biochrom, Berlin, Germany) during infection with *C. burnetii* or 10% FCS when cells were cultured in the absence of bacteria.

Analysis of *ankF* in 52 *C. burnetii* Strains

Genome assemblies of strains, which had been uploaded at the complete genome, chromosome, scaffold and contig levels and for which information on their genome group classification was known (Hemsley et al., 2019), were identified using the search term "*Coxiella burnetii*" in NCBI Genome (https://www.ncbi.nlm.nih.gov/genome/), downloaded and used to create a Coxiella-WGS database in Geneious Prime 2019.2.3 (Biomatters, New Zealand). Only a single sequence was taken from strains with multiple entries or passage variants. The respective *ankF* sequences were identified by BLAST analysis of Coxiella-WGS using the *ankF* coding sequence from the RSA493 Nine Mile strain as reference and the Geneious default parameters. Sequences from two strains were discarded due to sequence ambiguity (Cb171_QLYMPHOMA; CDBG01000000) and a partial sequence at the end of a contig (Q321; AAYJ01000000), so that 52 sequences remained for the mutational analysis.

Analysis of *C. burnetii ankF*::Tn

C. burnetii and the transposon mutant *C. burnetii ankF*::Tn (*ankF*::Tn) were analyzed by PCR for the presence of *ankF* and the insertion of the transposon with the primers 608 and 609 amplifying the *ankF* codon sequence.

Infection With *C. burnetii*

Infection of cell lines with *C. burnetii* was performed as described elsewhere (Schulze-Luehrmann et al., 2016). Briefly, *C. burnetii* were cultured axenically for 3 days at 37°C, 2.5 % O_2 and 5 % CO_2 and 1 day at RT and normal atmosphere. To infect cell lines, bacteria were pelleted and adjusted spectrophotometrically in PBS to yield respective MOIs. Bacteria were then added to 300 µl or 600 µl of cell culture media used for cells seeded 24 h earlier in 24-well or 6-well plates, respectively.

Immunofluorescence

Cells seeded in 24-well plates were washed three times with 1 ml PBS and fixed with 4% paraformaldehyde (Alfa Aesar, Karlsruhe, Germany) in PBS for 15 min at RT. Following three wash steps with PBS, cells were permeabilized with 0.1% Triton X-100 (Sigma-Aldrich) in PBS for 3 min, washed again, and blocked and quenched with PBS containing 50 mM NH_4Cl (Roth, Karlsruhe, Germany) and 5% Goat serum (Thermo Fisher) for 30 min at RT. Cells were subsequently incubated with primary antibodies in 5% Goat serum in PBS for 45 min at RT. Primary antibodies were washed off three times with PBS, and cells were incubated for 30 min at RT with fluorophore-coupled secondary antibodies in 5% Goat serum in PBS. Secondary antibodies were washed off three times with PBS, and stained cells were mounted on glass slides with ProLong Diamond containing DAPI to stain nuclei or bacterial DNA.

Human Peripheral Blood-Derived Macrophages

From three healthy donors (Ethical Committee Erlangen approval number 111-12B) peripheral blood were obtained, and macrophages were derived as previously described (Hayek et al., 2019).

Colony-Forming Units (CFU)

The infected primary macrophages were washed with PBS, incubated for 40 min in ice-cold H_2O and pipetted repeatedly with a syringe carrying a 25G needle (25 G 1" 0.5 mm × 25 mm, BD Microlance 3, Spain) to lyse the cells. The lysates were centrifuged (10 min, 1000 rpm, 4°C) and the supernatant was pelleted (1 min, 14000 rpm, 4°C) and resuspended in 200 µl PBS pH 7.4. A serial dilution was performed and pipetted in triplicates on ACCM-D/0.5% agarose plates. The plates were incubated for 2 weeks at 37°C, 5% CO_2 and 2.5% O_2 and the CFUs were counted.

Phenotypic Screening

For phenotypic screening, samples were imaged with an ArrayScan VTI Live epifluorescence automated microscope (Cellomics) equipped with an ORCA-ER CCD camera (Hamamatsu). Twenty-five fields per well were acquired for image analysis. Phenotypic profiles (expressed as z-scores) were calculated using CellProfiler, from triplicate experiments as previously described (Martinez et al., 2015) following median based normalization of 96-well plates. Plate effects were corrected by the median value across wells that are annotated as control.

Labeling of DQ-Red BSA–Positive Proteolytic Organelles

HeLa cells were infected with *C. burnetii* or the transposon mutant *ankF*::Tn at an MOI of 50. Following infection, cells were incubated with 10 µg/ml DQ-Red BSA (Life technologies) for 16 h. Cells were subjected to immunofluorescence staining as described.

Generation of the *C. burnetii* Δ*ankF* Mutant and the Complemented Strain

C. burnetii Nine Mile phase II was electroporated with 10 µg pJC-CAT::*ankF*-5′3′-*lysCA* as previously described (Beare and Heinzen, 2014). Co-integrants were selected by culturing the bacteria in ACCM-D media lacking lysine, but containing 2% sucrose for 5 days as previously described (Beare et al., 2018). Surviving transformants were expanded in ACCM-D media lacking lysine for 7 days. After spreading the diluted culture on 0.25% ACCM-D agarose without lysine, clonal Δ*ankF* mutants were picked after 10 days of culture. The picked clones were expanded in ACCM-D media without lysine.

Complementation of Δ*ankF* was achieved by electroporation of the mutant strain with 10 µg pMiniTn7T-*ankF*::AnkF. Integrants were selected by culturing the bacteria in ACCM-D media lacking lysine and lacking arginine, but containing citrulline for 5 days as previously described (Beare et al., 2018). The diluted culture was spread on 0.25% ACCM-D agarose without lysine and arginine, but containing citrulline for 10 days. Individual clones were picked and expanded in ACCM-D medium lacking lysine and arginine, but with citrulline.

Preparation of a HeLa cDNA Library for the Yeast Two-Hybrid Assay

A cDNA library (Clontech, #HL4000AA) from the HeLa S3 cell line was kindly provided by Prof. Dr. Hashemolhosseini (Institute for Biochemistry, University of Erlangen-Nuremberg). The cDNA library was restricted with EcoRI and XhoI and ligated into the de-phosphorylated and likewise-digested vector pGAD-GH.

Transformation of *S. cerevisiae* Y187 With Plasmid DNA

To transform the tryptophan- and leucine-auxotrophic *S. cerevisiae* Y187 with pGBT9-AnkF (prey-construct), a single yeast colony was grown on complete YPAD agar at 30°C. The next day, a colony was used for inoculation of YPAD medium at 30°C and 170 rpm shaking. The culture was grown to mid-logarithmic growth phase and sequentially washed in sterile ddH_2O and twice in Lithium-acetate (LiAc). After another centrifugation step the pellet was re-suspended in 50% PEG 3350, 1 M LiAc, 5 µl of 10 mg/ml salmon sperm DNA (Invitrogen) and 3 µg of pGBT9-AnkF plasmid DNA. The transformation was performed via the heat-shock method at 42°C for 20 min. Next, the culture was pelleted, re-suspended in ddH_2O, plated dropwise onto SCAD agar plates lacking tryptophan ($SCAD^{-T}$) for selection and incubated 2 to 3 days at 30°C.

Transformation of *S. cerevisiae* AH109 With a HeLa Genomic Library

S. cerevisiae AH109 were grown as described above. The next day, the mid-log phase culture was washed sequentially in sterile ddH_2O and TE/LiAc-buffer (10 mM Tris pH 7.4, 1 mM EDTA, 100 mM LiAc) including 10 mg salmon sperm DNA and 3 mg of purified HeLa genomic Library. The yeast suspension was subsequently incubated for 30 min at 30°C and another 15 min at 42°C supplemented with 10% dimethyl sulfoxide (Sigma-Aldrich). Following a 2-min incubation on ice, the suspension was re-suspended in ddH_2O, plated on SCAD plates lacking tryptophan ($SCAD^{-T}$) and incubated for 3 days at 30 °C. The colonies were pooled in YPAD medium with 20% w/v glycerol and further cultured at 30°C for 2.5 h. Respective cultures were aliquoted and stored at −80°C.

Yeast Two-Hybrid Screen

The Yeast Two Hybrid screen was performed by mating the yeast strain *S. cerevisiae* Y187-pGBT9-AnkF with the strain *S. cerevisiae* AH109 containing the HeLa genomic library. For this purpose, a pre-culture of *S. cerevisiae* Y187-pGBT9-AnkF was inoculated and cultivated in $SCAD^{-T}$-medium at 30°C and 170 rpm shaking. Next, a $SCAD^{-T}$ over-night (ON) culture was inoculated and incubated at 30°C and 170 rpm shaking until reaching an OD_{600} of approximately 0.8 to 1.2. Cultures were pelleted at 4,500*g* at RT. Meanwhile, two aliquots of the *S. cerevisiae* AH109 strain containing the HeLa genomic library were thawed in a 30°C water bath and subsequently added to the pelleted *S. cerevisiae* Y187-pGBT9-AnkF culture. The culture mix was vortexed, pelleted again, re-suspended in residual supernatant and subsequently plated onto YPAD plates. Following a 4.5-h incubation at 30°C, clones were washed off the YPAD plates twice with medium and pelleted. Next, the pellet was re-suspended in water, plated onto selective SCAD plates lacking histidine, leucine, and tryptophan ($SCAD^{-HLT}$) and incubated for five days at 30°C for clonal isolation.

LacZ Filter-Lift Assay

Clones of the mated yeast strains *S. cerevisiae* Y187-pGBT9-AnkF and *S. cerevisiae* AH109 grown on selective $SCAD^{-HLT}$ plates were considered to harbor both bait- and prey plasmids and to express the GAL4 transcription factor and the HIS3 gene. GAL4 initiates expression of the *lacZ* reporter gene serving as a read out for bait (AnkF) and prey (HeLa genomic library) interaction. HIS3 serves to complement the histidine auxotrophy.

Mated yeasts were plated onto nitrocellulose membranes placed on $SCAD^{-HLT}$ plates and incubated at 30°C for 3 days. To read out the LacZ reporter activity, nitrocellulose membranes were drowned in liquid nitrogen and then placed onto filter paper soaked in 2 ml LacZ buffer (60 mM Na_2HPO_4, 40 mM NaH_2PO_4, 10 mM KCl, 1 mM $MgSO_4$, pH 7.0), supplemented with 66.2 µl 2 % X-Gal (Biomol) in dimethylformamide and 5.4 µl beta-mercaptoethanol). Following another incubation step of 3 to 5 h at 30°C, respective clones were checked for blue dye precipitation on nitrocellulose membranes.

Reverse Transcription-Quantitative Real-Time PCR (RT-qPCR) of Vimentin in Presence of AnkF

For isolation of RNA, 1.2×10^6 stably-transfected HeLa-pWHE644/655-AnkF cells were seeded in 6-well plates in 3 ml per well and induced for AnkF expression with 1 µg/ml doxycycline. Cells were washed with PBS and lysed with RLT buffer (RNeasy® Plus Mini Kit, Qiagen, Germany) supplemented with β-mercaptoethanol. Total RNA was isolated using the RNeasy® Plus Mini Kit (Qiagen) and the QIAshredder™ kit according to manufacturer's instructions. RNA was eluted from columns in RNAse-free ddH_2O. The reverse transcription of RNA into cDNA was performed with the SuperScript™ II Reverse Transkriptase kit (Thermo Fisher) with random $Oligo(dT)_{12-18}$ primers (Thermo Fisher). QPCR was performed to quantify the amount of vimentin cDNA using primers 936 and 937 and quantification of actin cDNA using primers 827 and 828 (**Table 1**) with the QuantiFast SYBR Green PCR kit (Qiagen, Germany). Amplification of cDNA was performed in 384-well optical plates in an ABI Prism 7900HT. Relative amounts of vimentin cDNA were calculated by the ΔΔCT method using the housekeeping gene GAPDH for normalization.

Immunoblot Analysis

Proteins were separated by SDS-PAGE Bolt™ Bis-Tris Plus gradient 4% to 12% polyacrylamide gels (Thermo Fisher) and transferred to 0.45 µm pore size Immobilon-P PVDF membranes (Merck Millipore) by the semi-dry transfer method. Blotted membranes were incubated with respective primary antibodies and HRP-conjugated secondary antibodies. Immunodetection was performed with the Pierce ECL Western Blotting Substrate (Thermo Fisher) according to the manufacturer's instructions and exposure of X-ray films (GE Healthcare) to chemiluminescence.

Transient Transfection of CHO or HeLa Cells

2×10^4 *C. burnetii*-infected or uninfected cells were seeded on coverslips in a 24-well plate in 1 ml medium and incubated at 37°C, 5% CO_2 for 24 h. Cells were transfected with 250 to 500 ng of plasmid DNA using the X-tremeGENE 9 DNA

TABLE 1 | Primers used in this study.

No.	Sequence (5′->3′)*	Restriction site
40	AAGGATCCCTACCGCTGGAAGCCGC	BamHI
53	CCGGATCCATGAGACAGCGTGAAATTAATG	PstI
79	CCGGTACCCTACCGCTGGAAGCCGC	KpnI
608	ATGCGCCAGCGTGAAATTAATGATGAAGCTAT	
609	CTACCGCTGGAAGCCGCGATTATTGTGTTTTT	
650	CCGGATCCTTAGACAGCGTGAAATTAATGAT	BamHI
672	AATTTCATCGTTCCCGGCAG	
673	GCCGCGTTTACTAATCCCCA	
710	GCGAATTCATATGTCCACCAGGTCCGTGT	EcoRI
711	GCCGGTACCTTATTCAAGGTCATCGTGATGC	KpnI
827	CCAACCGCGAGAAGATGA	
828	CCAGAGGCGTACAGGGAT	
936	TCCAGCAGCTTCCTGTAGGT	
937	CCC TCACCTGTGAAGTGGAT	
1063	CAGGAAACAGAATTCATGGTGTCAAAAGGAGAAGAAG	EcoRI
1065	AGAGGTACCGAGCTCTTATTTATAAAGTTCATCCATGCC	SacI

Restriction sites are underlined.

Transfection Reagent (Roche, Switzerland) following the manufacturer´s protocol.

Confocal Microscopy

Fixed and stained cells on cover slips were mounted on glass slides and visualized with a Zeiss LSM 700 confocal laser scanning microscope. Image acquisition was performed with Zeiss Zen software (Carl Zeiss, Oberkochen, Germany).

STED Microscopy

For high-resolution STED microscopy, fixed and stained cells on cover slips (12 mm radial cover slips, 0.17± 0.005 mm) were visualized with use of an Abberior 3D STED 2-Channel Super Resolution- and resolft microscope (Abberior Instruments GmbH, Göttingen, Germany). Images were acquired with the Imspector software (Abberior GmbH, Göttingen, Germany).

Live Cell Imaging

U2-OS-vimentin-rsEGFP cells infected with *C. burnetii*-tdTomato were cultivated in µ-slides (Ibidi, Planegg, Germany) and visualized with a Zeiss Spinning Disc Axio Observer Z1 (Carl Zeiss, Oberkochen, Germany).

siRNA-Mediated Knock-Down of Vimentin in HeLa Cells and Quantification of Intracellular *C. burnetii*

Transfection of 1×10^5 HeLa cells in 12-well plates was performed by transfection of a 50 nM On-Targetplus human siRNA pool (Dharmafect) specific for human vimentin (vim) or a non-targeting siRNA pool as a control with the DharmaFECT 1 transfection reagent (Thermo Fisher) according to the manufacturer's instructions. Following a 24-h incubation at 37°C and 5% CO_2, cells were infected with *C. burnetii*. At the indicated time-points post-infection, *C. burnetii* was isolated from HeLa cells by osmotic lysis in sterile ddH_2O. In detail, *C. burnetii* infected HeLa cells were washed and subsequently lysed with 2 ml sterile ddH_2O for 30 min at 37°C and 5% CO_2. Lysed cells were re-suspended thoroughly, and bacteria were isolated by differential centrifugation at 300*g* for 10 min at RT and afterward at 20.000*g* for 2 min at RT. For isolation of *C. burnetii* genomic DNA (gDNA), pelleted *C. burnetii* were processed using the Illustra Bacteria Genomic Prep Mini Spin Kit (GE Healthcare) according to the manufacturer's instructions. Isolated gDNA from axenically-grown *C. burnetii* was used as a genomic equivalent (GE) standard ranging between 10^3 and 10^7 copies for absolute quantification. Calculation of GEs for the standard was performed as described elsewhere (Schulze-Luehrmann et al., 2016). Amplification of the *IS1111* insertion sequence was performed in 384-well optical plates in an ABI Prism 7900HT using primers 672 and 673 (**Table 1**).

Plasmid Construction

Restriction enzymes were purchased from NEB or Thermo Scientific. The antarctic phosphatase and T4-ligase were purchased from Thermo Fisher. Primer sequences and constructed plasmids are listed in **Table 2** and **Table 3**, respectively. For creation of pCMV-HA-vimentin, the vimentin coding sequence was amplified for cloning from isolated HeLa genomic DNA with the primers 710 and 711 by PCR with Q5-Phusion polymerase (NEB), purified, restricted with EcoRI and KpnI and ligated into the likewise-restricted and de-phosphorylated vector pCMV-HA. To create pEGFP-C2-AnkF, the AnkF coding sequence was amplified from *C. burnetii* NMII genomic DNA with primers 53 and 79, restricted with PstI and KpnI and ligated into the likewise-restricted and de-phosphorylated vector pEGFP-C2. For construction of pKM244mod.-tdT_{cc}, a *Coxiella*-codon optimized coding sequence of tandem-di-tomato (tdT_{cc}), synthesized and

TABLE 2 | Plasmids constructed in this study.

Plasmid	Primers	Reference
pCMV-HA-vimentin	No. 53 and No. 79	This study
pEGFP-C2-AnkF	No. 710 and No. 711	This study
pKM244mod.-tdT_{cc}	No. 1063 and No. 1065	This study
pJC-CAT::*ankF-5′3′-lysCA*		This study
pMini-Tn7T-*ankF*::AnkF		This study

TABLE 3 | Analysis and grouping of the *ankF* gene in 52 *C. burnetii* strains.

Genome Group	Strain	AnkF group	Accession #	Sequence Reference
I	RSA493 (NM-I)	1	AE016828	(Seshadri et al., 2003)
	RSA315 (Turkey)	1	NOLO00000000	(Beare et al., 2017b)
	RSA435 (Dyer)	1	NOLQ00000000	(Beare et al., 2017b)
	RSA270 (Ohio314)	1	NOLT00000000	(Beare et al., 2017c)
	RSA329 (California33)	1	NOLV00000000	(Beare et al., 2017c)
	RSA350 (California16)	1	NOLU00000000	(Beare et al., 2017c)
	RSA514 (NM-Crazy)	1	NOVG00000000	(Beare et al., 2018)
	Cb_C2	1	CCAJ010000000	(Sidi-Boumedine et al., 2014)
(I)	Cb175_Guyana	1	HG825990	(D'Amato et al., 2015)
IIa	RSA331 (Henzerling)	1	CP000890, CP014559	(Beare et al., 2006; Kuley et al., 2017)
	Heizberg	1	CP014561	(Kuley et al., 2017)
	RSA461 (M44_Clone1)	1	NOVI00000000	(Beare et al., 2018)
	Cb185	1	CBTH01000000	(Million et al., 2014)
IIb	CbCVIC1	1	CP014549	(Kuley et al., 2017)
	Z3055	1	LK937696	(D'Amato et al., 2014b)
	NL-Limburg	1	JZWL01	(Hammerl et al., 2015)
	NL3262	1	CP013667	(Kuley et al., 2016)
	NLhu3345937	1	CP014354	(Kuley et al., 2016)
	602 (14160-002)	1	CP014836	(Kuley et al., 2017)
	42785537	1	CP014548	(Kuley et al., 2017)
	EV-Cb_C13	1	CCAM010000000	(Sidi-Boumedine et al., 2014)
	Q540	1	PPFP01000000	(Hemsley et al., 2019)
	Cb_D2 (DSTL_2)	1	RQJT01000000	(Hemsley et al., 2019)
	Cb_D8 (DSTL_8)	1	RQJS01000000	(Hemsley et al., 2019)
	Cb_D10 (DSTL_10)	1	RQJR01000000	(Hemsley et al., 2019)
	Cb109	1	AKYP01000000	(Rouli et al., 2012)
III	Idaho Goat_Q195	1	NOLR00000000	(Beare et al., 2017c)
	2574	1	CP014555	(Kuley et al., 2017)
	601 (14160-001)	1	CP014551	(Kuley et al., 2017)
	18430	1	CP014557	(Kuley et al., 2017)
	701CbB1	1	CP014553	(Kuley et al., 2017)
	Cb_B1	1	CCAH010000000	(Sidi-Boumedine et al., 2014)
	Cb_B18	1	CCAI010000000	(Sidi-Boumedine et al., 2014)
	EV-Cb_BK18	1	CCAL010000000	(Sidi-Boumedine et al., 2014)
	Q532	1	PPFQ01000000	(Hemsley et al., 2019)
	Q545	1	PPFO01000000	(Hemsley et al., 2019)
	Cb_D1 (DSTL_1R)	1	RQJU01000000	(Hemsley et al., 2019)
	Q556	1	PPFN01000000	(Hemsley et al., 2019)
	Q559	1	PPFM01000000	(Hemsley et al., 2019)
IV	Schperling	4	CP014563	(Kuley et al., 2017)
	Cbu_K154	4	CP001020	(Beare et al., 2009)
	'MSU Goat Q177 (Priscilla)	4	CP018150	(Walter et al., 2016) Unpublished
	Leningrad-2	4	PDLP00000000	(Freylikhman et al., 2017) Unpublished
	Namibia	5	CP007555	(Walter et al., 2014)
	AuQ01 (Arandale)	4	JPVV01000000	(Walter et al., 2014)
	Cb196_SaudiArabia	4	CCXO01000000	(D'Amato et al., 2014b)
	Cb_O184	4	CCAK010000000	(Sidi-Boumedine et al., 2014)
V	Cbu G_Q212	2	CP001019	(Beare et al., 2009)
	Scurry S_Q217	2	CP014565	(Kuley et al., 2017)
	Ko_Q229	2	NOLP00000000	(Beare et al., 2017b)
	Dog Utad	2	CCNR01000000; CCYB01000000	(D'Amato et al., 2014a)
VI	Dugway 5J108-111, 7D77-80, 7E65-68	3	CP000733; NOLN01000000, NOLM01000000	(Beare et al., 2009; Beare et al., 2017a)

cloned into pEX-K4 (pEX-K4-tdT$_{cc}$), was ordered from Eurofins (Luxemburg). The tdT$_{cc}$- coding sequence was amplified by PCR from pEX-K4-tdT$_{cc}$ with the primers 1063 and 1065 and cloned into the EcoRI-digested vector pKM244mod with use of the GeneArt® Seamless Cloning and Assembly Kit (Thermo Fisher) according to the manufacturer's instructions to create pKM244mod.-tdT$_{cc}$. The plasmid pGBT9-AnkF was cloned by PCR from the AnkF coding sequence of *C. burnetii* NMII genomic DNA with the primers 650 and 40 followed by fragment purification, restriction with BamHI and ligation into the likewise-restricted and de-phosphorylated vector pGBT9.

For construction of pJC-CAT::*ankF-5′3′-lysCA*, the 5′ and 3′ regions of *ankF* were amplified by PCR from NMII genomic DNA using the specific primer sets (5′-CGGTACCC GGGGAT CCCATATCGATAATGTGTTGATGG and 3′-CACCCATATGCGACGCGAGCGTCGA GTTCTTTCTCTA CCTAATTAAACTTTATG) and (5′-CGTCGCATATGGGTG CGCATG TACGTCTCCGCTAAGTAGCCCGTATG and 3′-GAA

CCTGTTTGTCGACGCTTGAGA TTCAGCGGGTGG), respectively. The 5′ and 3′ PCR products were cloned into BamHI/SalI-digested pJC-CAT by In-Fusion, resulting in formation of an internal NdeI site between the 5′ and 3′ fragments and creation of pJC-CAT::*ankF*-5′3′. The *1169*^{P}*-lysCA* cassette was amplified from pJC-CAT::*1169*^{P}*-lysCA* (Beare et al., 2018) by PCR with specific primers a450 and a451 and cloned by In-Fusion into NdeI-digested pJC-CAT::*ankF*-5′3′ to create pJC-CAT::*ankF*-5′3′*-lysCA*.

For construction of pMini-Tn7T-*ankF*::AnkF the *ankF* gene and its native promotor were amplified by PCR from NMII genomic DNA using the specific primers (5′-GATGAATT CGACGAGCAAAGGAGCCCT and 3′-GTAGAATT CTTCGCCATCTTC TTAGCGCAC) followed by fragment purification, restriction with EcoRI and ligation into the likewise-digested and de-phosphorylated vector pMini-Tn7T-ArgGH (Sandoz et al., 2016; Beare et al., 2018).

Statistical Analysis

Statistical analysis was conducted with Prism 8 (GraphPad software). Bar graphs depict mean data ± standard deviation from three independent experiments. An unpaired Student's t-test was performed to determine significance of each data point. A p-value of < 0.05 was considered significant.

RESULTS

AnkF Is a Highly Conserved Effector Protein

The T4SS effector protein AnkF was one of the first *C. burnetii* T4SS effector proteins identified (Pan et al., 2008). However, its function has not been studied in detail so far. In a previous publication, the *ankF* sequences from four different *C. burnetii* isolates were analyzed. As the *ankF* gene appeared to be highly conserved (Voth et al., 2009), it was suggested that AnkF might be an important virulence factor. In the meantime, additional *C. burnetii* isolates have been sequenced. Thus, we compared the *ankF* sequences from 52 *C. burnetii* strains (**Table 3**). Our analysis demonstrates that these strains encode five different alleles of the *ankF* gene (**Figure 1**). Isolates assigned to the

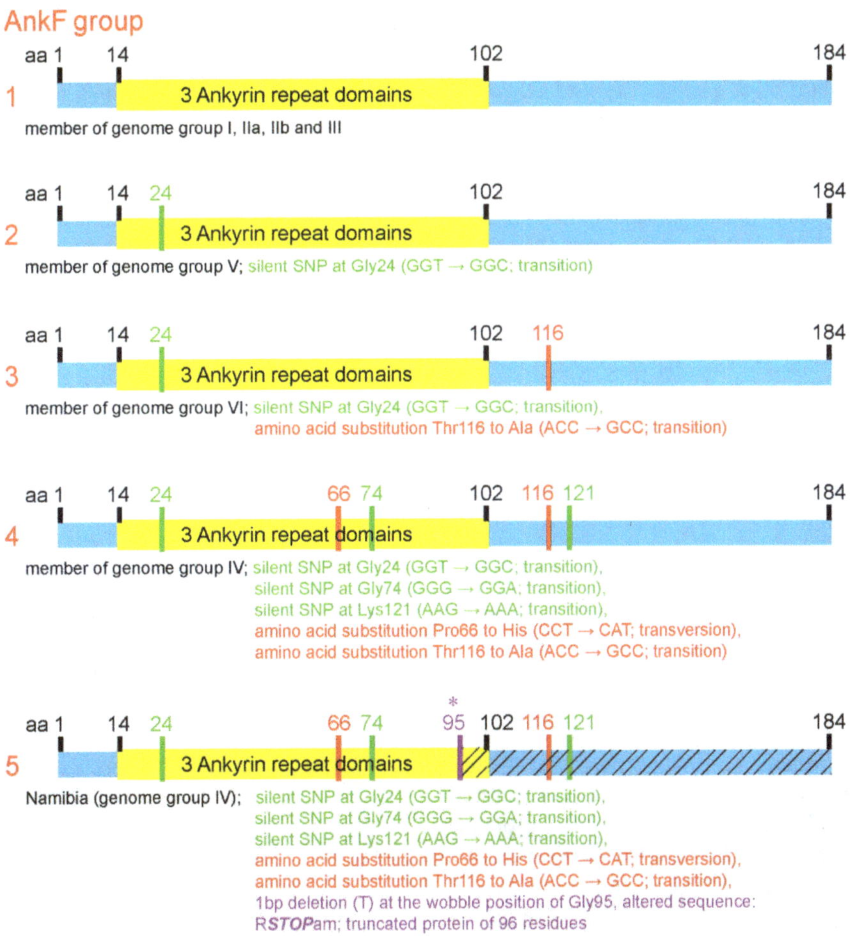

FIGURE 1 | Schematic diagram of five alleles of the *ankF* gene. The sequences of the respective *ankF* gene from 52 *C. burnetii* strains (**Table 3**) were analyzed. Depicted are the three ankyrin repeat domains (in yellow), silent nucleotide polymorphisms (in green), nucleotide exchanges (in red) and a frameshift mutation leading to a premature stop (in purple; marked with *). Beneath the diagram of the alleles, the corresponding *C. burnetii* genome group and the nucleotide substitutions are given.

genome groups I, IIa, IIb, and III (Hemsley et al., 2019; Long et al., 2019) express the wild-type sequence of the Nine Mile reference strain. All genome group V members contain a silent single-nucleotide polymorphism (SNP) at Gly24, thus, also encode a wild-type protein. The Dugway strain, representing genome group VI, encodes a full-length protein with the Gly24 SNP and an additional amino acid substitution from threonine to alanine at residue 116. The *ankF* genes from strains classified as belonging to genome group IV contain the latter two mutations. In addition, they also contain silent SNPs in the codons for Gly74 and Lys121 as well as a mutation leading to a proline to histidine exchange at position 66. There is only one exception to this sequence/genome group correlation in genome group IV. Here, the Namibia strain has an additional frameshift mutation at residue Gly95, resulting in a protein of 96 residues. Overall, our analysis suggests that *ankF* is highly conserved between the different isolates, which supports the assumption that AnkF might be an essential virulence factor.

AnkF Transposon Mutants Are Infectious but Fail to Establish a Replicative CCV

In order to determine whether AnkF is involved in *C. burnetii* pathogenesis, we analyzed AnkF mutants in their ability to infect cells and to replicate intracellularly. To this end we used an *ankF* transposon mutant of *C. burnetii* (*ankF*::Tn), which harbors a transposable element integrated between bps 507 and 508 of the *ankF* coding sequence (**Figure 2A**). First, we confirmed clonality by performing a PCR of wild-type and *ankF*::Tn *C. burnetii* (**Figure 2B**). Next, we analyzed the capability of the transposon mutant to grow in axenic culture. As shown in **Figure 2C**, axenic growth of the mutant is comparable to that of the wild-type strain over seven days. In order to elucidate the role of AnkF during infection, the transposon mutant was characterized regarding internalization and intracellular replication. Thus, HeLa cells were infected with the *ankF* mutant and wild-type *C. burnetii* and immunofluorescence was performed at 4 and 48 h post-infection. At 4 h post-infection, roughly 40% to 50% of the cells contained intracellular bacteria for each condition, implying that the *ankF*::Tn and wild-type strains are equally infectious (**Figure 3A**). However, *ankF*::Tn failed to replicate intracellularly in HeLa cells (**Figure 3B**), primary human monocyte-derived macrophages (**Figure 3C**), and in U2OS cells (**Figure 3D**), demonstrating that this replication defect is not cell-type specific. Next, to further investigate the replication defect of *ankF*::Tn mutants, we performed multi-parametric phenotypic profiling. For this purpose, we resorted to U2OS cells, as their morphology simplifies phenotypic characterization. In addition to their replication defect within infected cells, the *AnkF*::Tn mutants were impacted in their ability to develop CCVs (**Figure 3E**). Accordingly, CCVs formed by the *ankF*::Tn mutants harbored less bacteria as compared to wild-type. Other parameters were largely unaffected, with the exception of the area occupied by lysosomes in infected cells, which is consistent with a defect in CCV biogenesis. Thus, the *ankF*::Tn mutant has a defect in intracellular replication, which might be caused by disturbed CCV development.

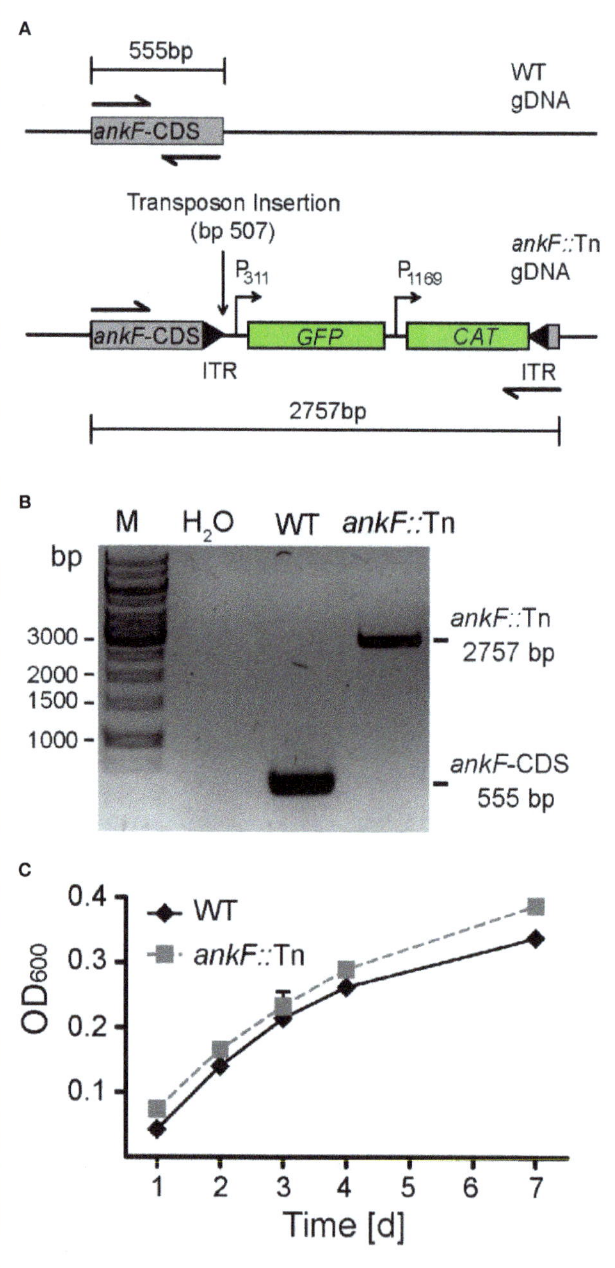

FIGURE 2 | The *ankF*::Tn mutant has no replication defect in axenic culture. **(A)** Scheme of the genomic region of wild-type *C. burnetii* and *C. burnetii-ankF*::Tn at the *ankF* locus. The transposable element is inserted between bp 507 and 508 of the *ankF* coding sequence and contains a GFP reporter under the control of the *C. burnetii* promotor P_{311} and a chloramphenicol resistance cassette under the control of the *C. burnetii* promotor P_{1169}. **(B)** Agarose gel of PCR products from wild-type *C. burnetii* and *ankF*::Tn generated with primers specific for the *ankF* coding sequence (shown as halved arrows in A). **(C)** Wild-type *C. burnetii* and *ankF*::Tn were inoculated at an OD_{600} of 0.01 in ACCM-2 medium and incubated at 37°C, 2.5% O_2 and 5% CO_2. OD_{600} was determined by spectrophotometric analysis at the indicated time-points post-inoculation. Error bars represent the mean standard deviation of three independent experiments.

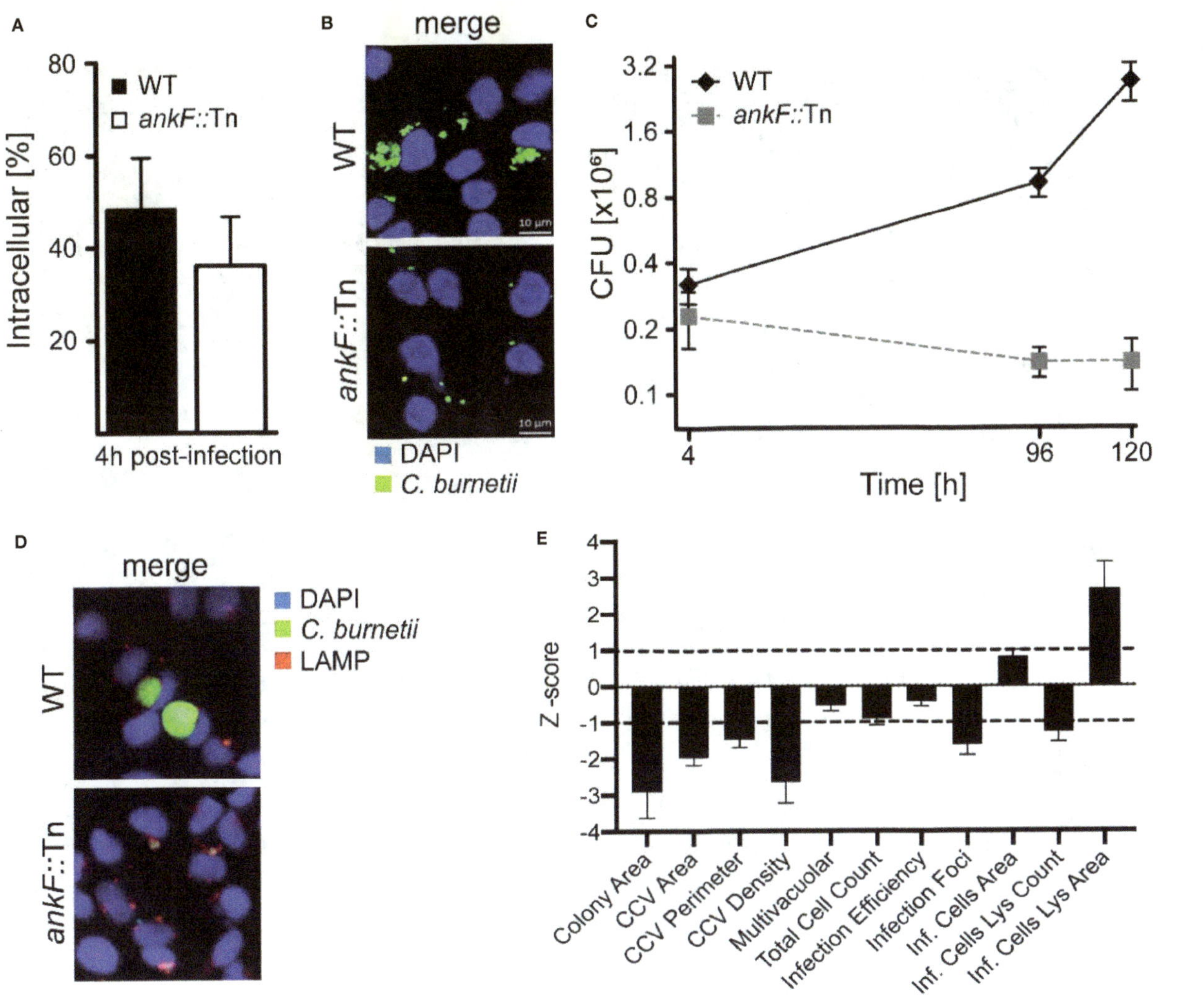

FIGURE 3 | The *C. burnetii* T4SS effector AnkF is essential for intracellular replication. **(A, B)** HeLa cells were infected with wild-type *C. burnetii* (WT) and *ankF*::Tn (*ankF*::Tn) at an MOI of 50. **(A)** 4 h post-infection the number of intracellular bacteria was determined from 100 infected cells. Error bars represent mean standard deviations of three independent experiments. **(B)** 48 h post-infection, the cells were fixed and stained with antisera against *C. burnetii* (green). Nuclei and bacterial DNA were stained with DAPI (blue). Representative LSM images are shown. **(C)** Human monocyte-derived macrophages were infected with wild-type *C. burnetii* (WT) and *ankF*::Tn (*ankF*::Tn) at an MOI of 10 for 4, 96, and 120 h. The bacterial numbers were determined by colony-forming unit (CFU) counts. Error bars represent the mean standard deviation of three independent experiments. **(D, E)** U2OS cells were challenged either with wild-type *C. burnetii* or the *ankF*::Tn mutant strain, both expressing GFP, at an MOI of 100. Six days post-infection, cells were fixed and labeled with an anti-LAMP1 antibody (red) and Hoechst (blue) to visualize CCVs and host cell nuclei, respectively. **(D)** Images were acquired with an Arrayscan VTI Live epifluorescence automated microscope equipped with a 20x objective and an ORCA ER CCD camera. Representative images are shown. **(E)** An average of 50,000 cells were then automatically imaged and analyzed from triplicate experiments for each condition and the phenotypic profile of the *ankF*::Tn mutant was compared to that of wild-type *C. burnetii* and expressed as z-scores over 11 independent features.

AnkF Is Dispensable for CCV Maturation, but Might Act on CCV Characteristics

C. burnetii requires the maturation of the CCV into a lysosomal-like compartment with an acidic pH (van Schaik et al., 2013; Lührmann et al., 2017; Newton et al., 2020). Thus, immunofluorescence was performed with infected HeLa cells to monitor the presence of the lysosomal marker LAMP2 on the CCV. Additionally, the degradative activity of the CCV lumen was visualized by fluorescent fluorophore-coupled BSA (DQ-Red BSA) using LSM. DQ-Red is a self-quenched substrate that emits fluorescence after cleavage by proteases. The presence of LAMP2 around the CCV and fluorescent DQ-Red BSA inside the CCV was prominent at 24 h post-infection in HeLa cells infected with wild-type *C. burnetii* (**Figures 4A–D**). Importantly, the small CCVs formed by *ankF*::Tn were also decorated with LAMP2 and possessed a degradative lumen (**Figures 4A–D**). These data

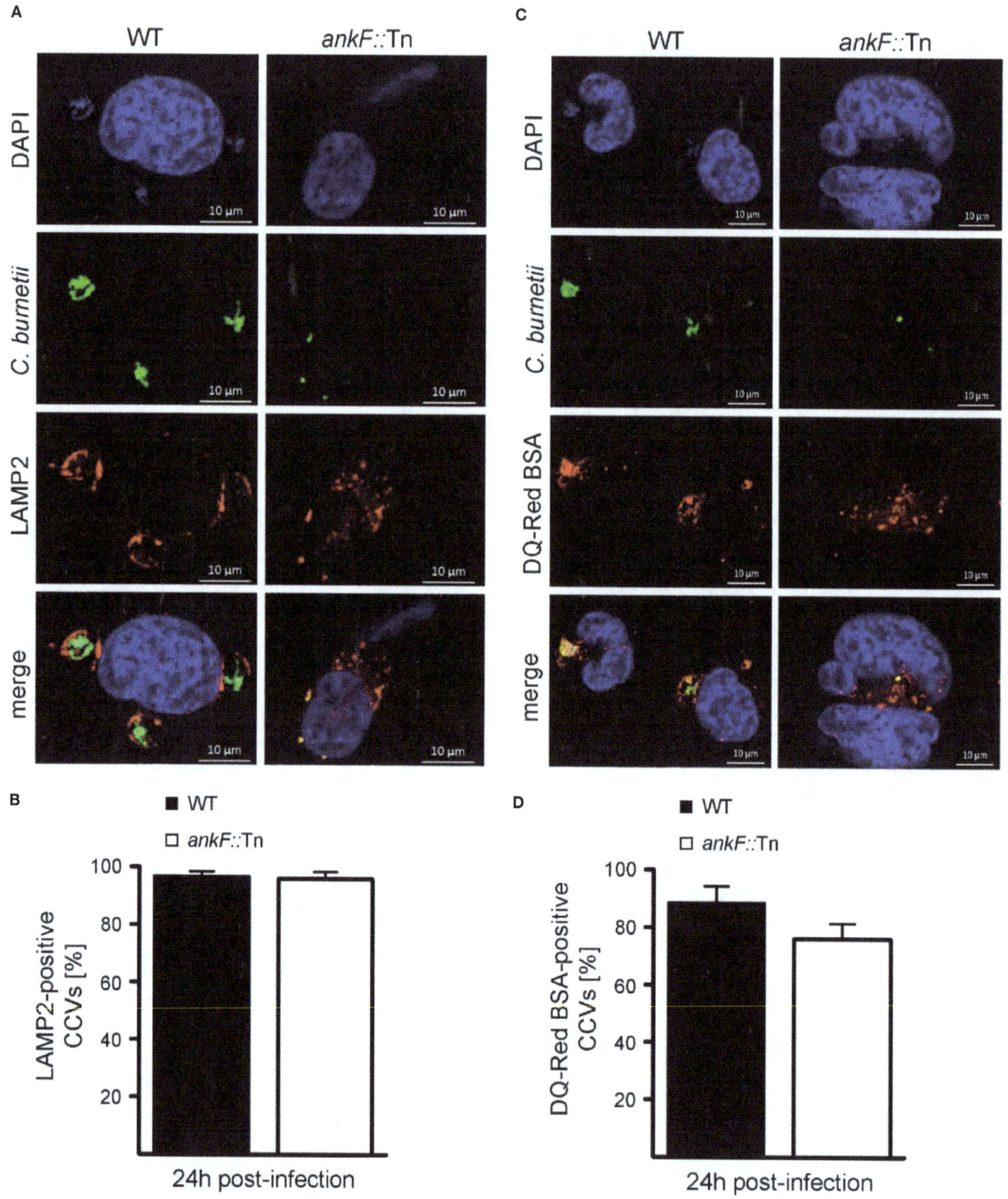

FIGURE 4 | The *C. burnetii* T4SS effector AnkF is dispensable for CCV maturation. **(A, B)** HeLa cells were infected with wild-type *C. burnetii* (WT) and the *ankF*::Tn (*ankF*::Tn) mutant at an MOI of 50. At 24 h post-infection the cells were fixed and stained with an anti-LAMP2 antibody (red) and with antisera against *C. burnetii* (green). Nuclei and bacterial DNA were stained with DAPI (blue). Cells were visualized using LSM. **(A)** Representative LSM images are shown. **(B)** 100 infected cells were analyzed for the association of LAMP2 with the CCV. Error bars represent mean standard deviations of three independent experiments. **(C, D)** HeLa cells were infected with wild-type *C. burnetii* (WT) and *ankF*::Tn (*ankF*::Tn) at an MOI of 50. Cells were incubated with DQ-Red BSA for 16 h. At 24 h post-infection, the cells were fixed and stained with antisera against *C. burnetii* (green). Nuclei and bacterial DNA was stained with DAPI (blue). Cleaved DQ-Red BSA emits fluorescence (red). Cells were visualized using LSM. **(C)** Representative LSM images are shown. **(D)** 100 CCVs were analyzed for red fluorescence using an epifluorescence microscope. Error bars represent mean standard deviation of three independent experiments.

suggest that the inability of *ankF*::Tn to replicate intracellularly was not mediated by altered maturation of the CCV.

The Replication Defect of AnkF Mutants Can Be Partially Complemented

Confirmation of loss-of-function studies by transposon mutagenesis requires phenotypic complementation. For this purpose, we infected HeLa cells with an equal MOI of both tdTomato-expressing wild-type *C. burnetii* and *ankF*::Tn mutants expressing GFP and analyzed CCV formation and bacterial replication by immunofluorescence. In cells infected with only a single bacterium we observed at 72 h post-infection either a spacious CCV in case of wild-type bacteria or small CCVs in case of the transposon mutant (**Figure 5**).

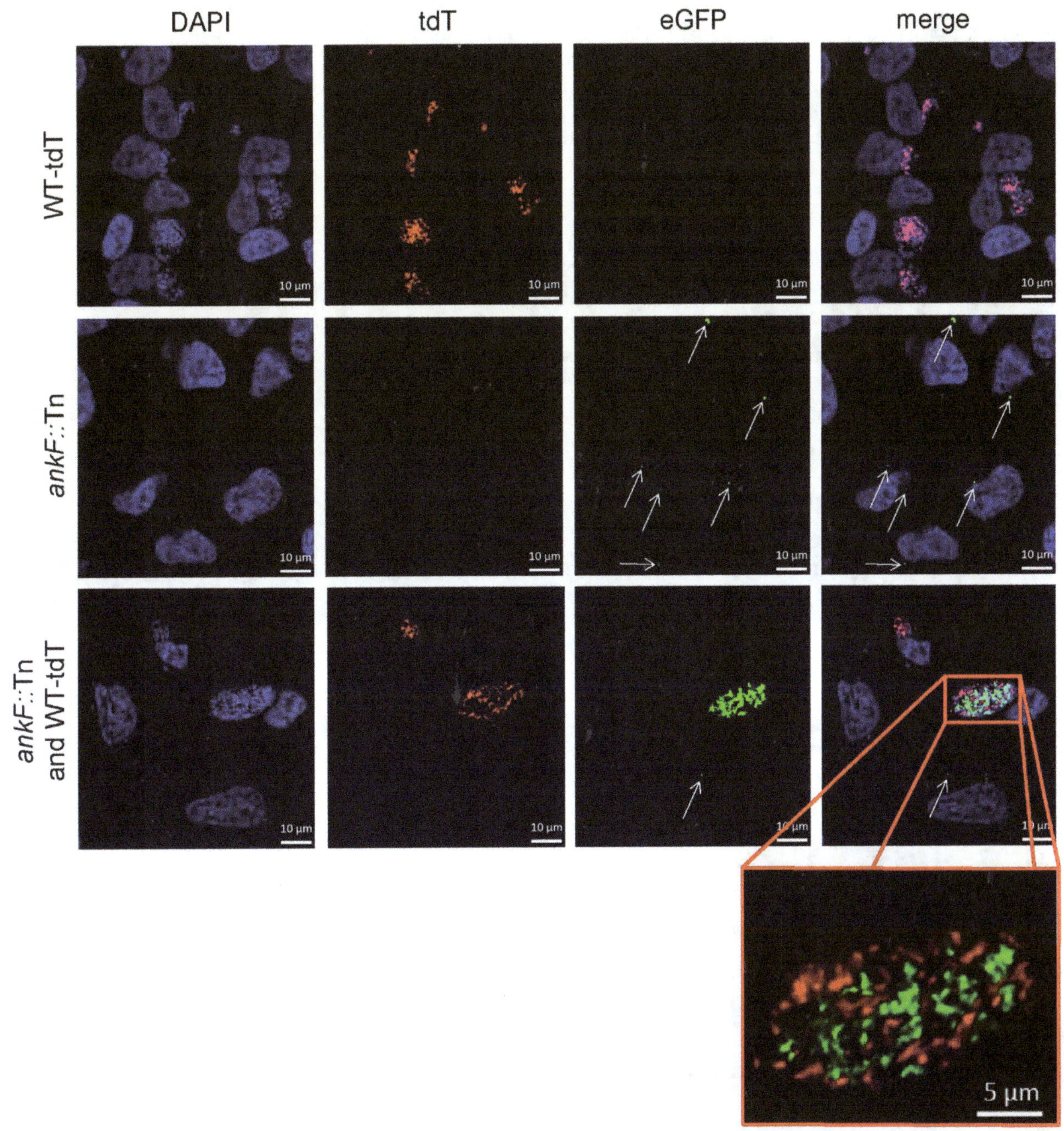

FIGURE 5 | The AnkF transposon mutant replicates inside a CCV harboring wild-type bacteria. HeLa cells were infected with *C. burnetii* expressing IPTG-inducible tdTomato (WT-tdT) in the presence of 2 mM IPTG and/or *ankF*::Tn (*ankF*::Tn) endogenously expressing GFP at an MOI of 50 each as indicated. At 72 h post-infection, the cells were fixed and nuclei and bacterial DNA was stained with DAPI (blue). Representative LSM images of two independent experiments are shown. A red box represents a close-up without DAPI staining of the merged image. White arrows indicate single *ankF*::Tn cells.

However, if wild-type and *ankF*::Tn bacteria share the same vacuole, we observed high numbers of wild-type and *ankF*::Tn bacteria (**Figure 5**). Importantly, we never detected such high numbers of *ankF*::Tn bacteria if the wild-type *C. burnetii* were in a different CCV, but still in the same cell. Thus, we reasoned that AnkF might be altering the CCV in a way, which allows bacterial replication.

Next, we generated a clean AnkF deletion mutant (Δ*ankF*) and the respective complemented strain to prove the importance of AnkF for intracellular replication. Thus, we infected HeLa cells with wild-type, Δ*ankF* and Δ*ankF*::AnkF and analyzed the size of the CCVs and the infection rates at 14 and 60 h post-infection by immunofluorescence microscopy. As shown in **Figure 6A** at 60 h post-infection, wild-type *C. burnetii* establishes large CCVs, while the size of the CCVs harboring *C. burnetii* lacking AnkF were significantly smaller and the CCVs containing the complemented strain showed an intermediate vacuole size. In addition, while the infection rate at 14 h post-infection was 80% to 90% and did not differ between the three different strains (data not shown), at 60 h post-infection the infection rate of cells infected with the AnkF deletion mutant were reduced by ~40% in comparison to cells infected with wild-type *C. burnetii*. The complemented strains showed an intermediate phenotype (**Figure 6B**). These data supports the assumption that the T4SS effector protein AnkF is involved in intracellular replication of *C. burnetii*.

Taken together, our results demonstrate that AnkF is also important for *C. burnetii* replication in macrophages and, thus, in the primary host cell of *C. burnetii*. In addition, our results suggest, that AnkF might influence CCV properties.

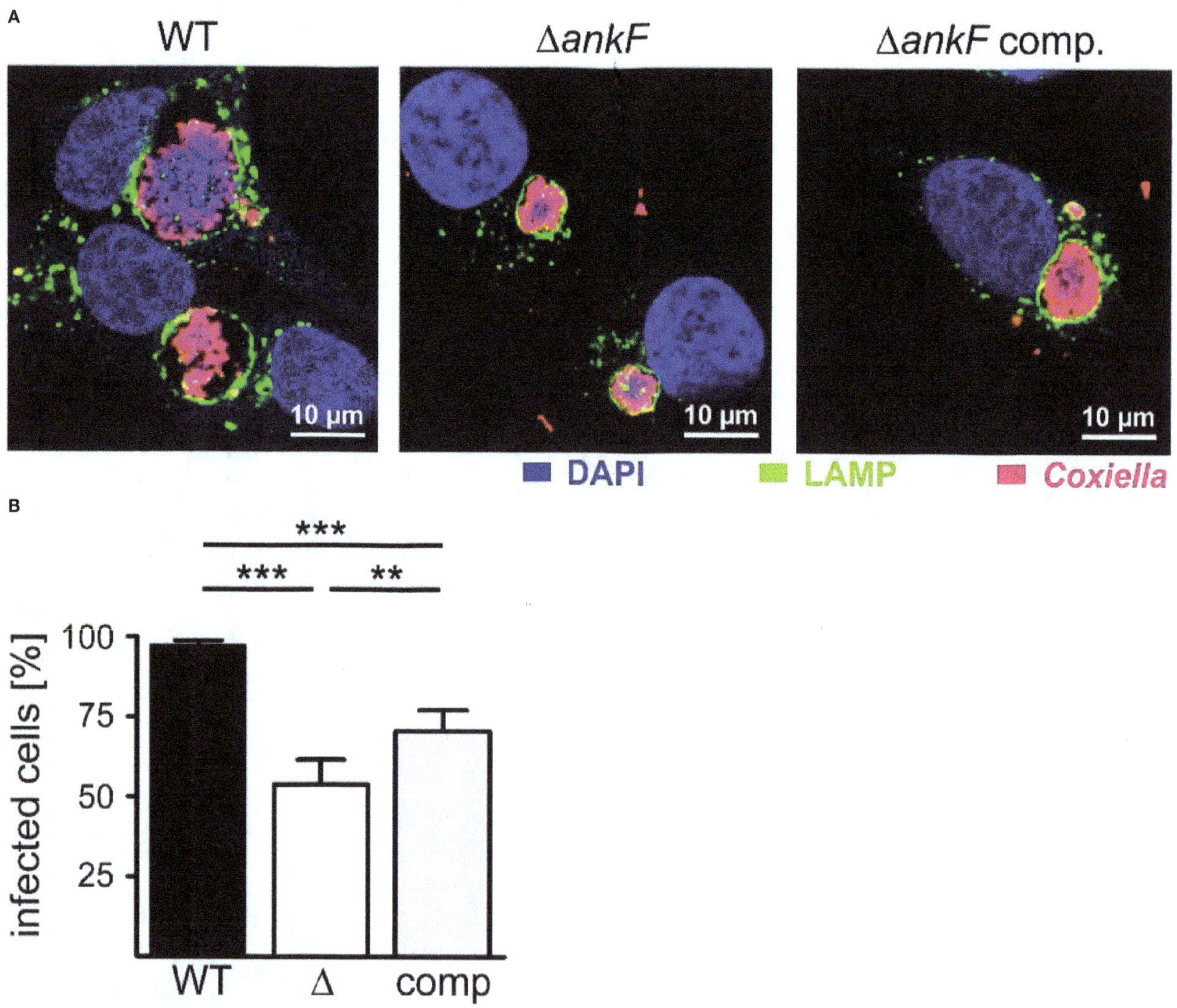

FIGURE 6 | The replication defect of Δ*ankF C. burnetii* can be complemented. HeLa cells were infected with *C. burnetii* wild-type (WT), Δ*ankF* or Δ*ankF*::AnkF at an MOI of 100. At 60 h post-infection, the cells were fixed and stained with an anti-LAMP2 antibody (green) and with antisera against *C. burnetii* (red). Nuclei and bacterial DNA were stained with DAPI (blue). **(A)** Representative LSM images of three independent experiments are shown. **(B)** The percentage of infected cells were determined by analyzing 100 cells per experiment in three independent experiments. Error bars indicate ± SD. $^{**}p < 0.01$, $^{***}p < 0.001$.

AnkF Binds Vimentin and Alters Its Localization

To learn how AnkF might alter the CCV to allow bacterial replication, we performed a yeast two-hybrid assay to identify potential host cell interaction partners. Using a HeLa genomic library, the intermediate filament vimentin was found to bind AnkF. Performing a LacZ filter-lift assay, X-Gal dye precipitation by interaction of AnkF and vimentin exceeded the signal obtained with the positive control (p53 and t-antigen (**Figure 7A**). Next, we determined how the AnkF-vimentin interaction might influence vimentin function. Modification of vimentin by bacterial products has been reported before. Thus, *Mycobacterium tuberculosis* down-regulates vimentin expression (Garg et al., 2006; Mahesh et al., 2016). In order to test whether vimentin was modified at the level of gene expression or steady-state protein in the presence of ectopic AnkF, quantification of vimentin expression at both mRNA and protein levels was performed. Stable HeLa-*ankF* cell lines harboring a doxycycline-inducible AnkF expression system (Berens et al., 2015) were used to quantify vimentin mRNA expression by RT-PCR and protein synthesis by immunoblot analysis. While we detected increasing protein levels of AnkF starting from 4 h post-induction, we did not observe an influence of AnkF on the vimentin protein level or stability (**Figure 7B**). No change in mRNA expression occurred in the presence of expressed AnkF (**Figure 7C**). Thus, AnkF does not influence vimentin transcriptionally or translationally.

Next, we performed co-localization studies. Over-expression of GFP-tagged AnkF and HA-tagged vimentin in CHO cells showed a clear co-localization (**Figure 7D**). Moreover the co-expression alters the filamentous appearance of HA-vimentin into a punctate-like localization (**Figure 7D**), suggesting that AnkF might alter vimentin assembly and localization within the cell.

Vimentin Is Recruited to the CCV

Physiologically, vimentin is involved in intracellular trafficking events (Styers et al., 2005). Depending on the intracellular conditions and the stresses involved, vimentin can either be flexible, but also of stabilizing nature (Janmey et al., 1991). Thus, vimentin provides a stabilizing scaffold for *C. trachomatis* inclusions during infection (Kumar and Valdivia, 2008). Additionally, siRNA-mediated knock-down of vimentin was shown to reduce the number of CCVs in host cells (McDonough et al., 2013). This led us to investigate whether vimentin is recruited to the CCV during infection. Thus, HeLa cells were infected with *C. burnetii* and localization of endogenous vimentin and LAMP2 was visualized by immunofluorescence. At 24 h post-infection, vimentin recruitment to the CCV was visible, as judged by co-localization with LAMP2 (**Figure 8A**). In order to underline the close association of vimentin filaments around the CCV, STED-LSM was performed, demonstrating the close localization of vimentin around the CCV (**Figure 8B**). We quantified the acquisition of vimentin at the CCV and demonstrate that vimentin was recruited to the CCV in a time-dependent manner (**Figure 8C**). The recruitment of vimentin to the CCV correlated with CCV growth, suggesting that association of vimentin with the CCV depends on bacterial replication. However, vimentin filaments are highly dynamic. Thus, analyzing vimentin in fixed cells might be prone to artifacts. Therefore, we infected U2OS-rseGFP-vimentin cells (Ratz et al., 2015) with tdTomato-*C burnetii* and tracked vimentin recruitment to the CCV by live-cell-imaging. **Figure 8D** depicts polymerization of vimentin fibers around the surface of the CCV within a timeframe of 1 h, confirming recruitment of vimentin to the CCV in living cells.

AnkF Partially Accumulates Around the CCV

In the next step, we investigated whether AnkF, like vimentin, is localized at the CCV. In uninfected cells, GFP- and HA-tagged AnkF exhibit cytoplasmic and nuclear localization when ectopically expressed in HeLa cells (Rodriguez-Escudero et al., 2016). Here, we determined the subcellular localization of AnkF in *C. burnetii* infected cells. In agreement with previous reports (Rodriguez-Escudero et al., 2016), ectopically expressed AnkF localized in the cytosol and within the nucleus. However, we also detected ectopically expressed AnkF in proximity to the CCV (**Figure 8E**), which led us to hypothesize that AnkF might be involved in the recruitment of vimentin to the CCV.

Vimentin Is Dispensable for *C. burnetii* Replication

While we have demonstrated that AnkF is important for efficient intracellular replication, the role of vimentin for bacterial progeny was uncertain. Vimentin is involved in host cell invasion of pathogenic bacteria and supports bacterial replication by providing stability for the bacteria-containing vacuole (Janmey, 1991; Mak and Bruggemann, 2016). To determine the role of vimentin in the host-pathogen-interaction, a siRNA-mediated knock-down of vimentin was conducted in *C. burnetii*-infected HeLa cells. An efficient knock-down of vimentin was detected by immunoblot analysis (**Figure 9A**). Our immunofluorescence analysis, in contrast, revealed the presence of smaller vimentin fragments (**Figure 9B**). These fragments might represent soluble tetrameric vimentin (Soellner et al., 1985), which lack the stabilizing activity of filamentous vimentin. Thus, siRNA-mediated knock-down of vimentin allows to determine the role of stabilizing vimentin for *C. burnetii* replication. The knock-down of vimentin reduced the number of *C. burnetii* at 4 h post-infection (**Figure 9C**). At 24 and 48 h post-infection, the absence of vimentin did not seem to influence bacterial replication. These results suggest that the stabilizing function of vimentin is dispensable for intracellular replication of *C. burnetii*.

Other CCV-Associated Structural Components Are Not Influenced by AnkF Expression

However, it might be possible that other intermediate filaments, microfilaments or microtubules compensate for the lack of

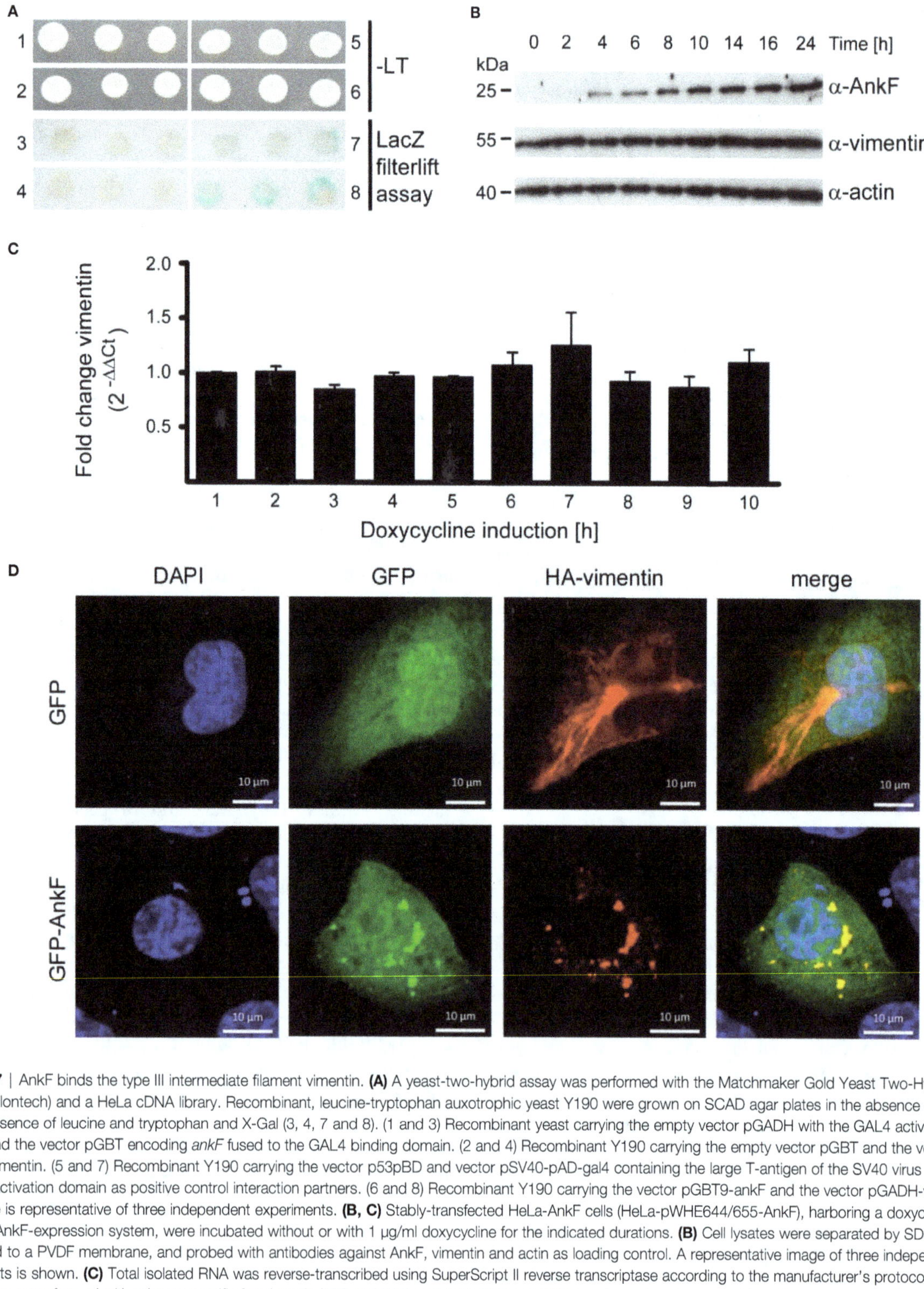

FIGURE 7 | AnkF binds the type III intermediate filament vimentin. **(A)** A yeast-two-hybrid assay was performed with the Matchmaker Gold Yeast Two-Hybrid System (Clontech) and a HeLa cDNA library. Recombinant, leucine-tryptophan auxotrophic yeast Y190 were grown on SCAD agar plates in the absence (1, 2, 5 and 6) and presence of leucine and tryptophan and X-Gal (3, 4, 7 and 8). (1 and 3) Recombinant yeast carrying the empty vector pGADH with the GAL4 activation domain and the vector pGBT encoding *ankF* fused to the GAL4 binding domain. (2 and 4) Recombinant Y190 carrying the empty vector pGBT and the vector pGADH-vimentin. (5 and 7) Recombinant Y190 carrying the vector p53pBD and vector pSV40-pAD-gal4 containing the large T-antigen of the SV40 virus fused to the Gal4 activation domain as positive control interaction partners. (6 and 8) Recombinant Y190 carrying the vector pGBT9-ankF and the vector pGADH-vimentin. The image is representative of three independent experiments. **(B, C)** Stably-transfected HeLa-AnkF cells (HeLa-pWHE644/655-AnkF), harboring a doxycycline-inducible AnkF-expression system, were incubated without or with 1 µg/ml doxycycline for the indicated durations. **(B)** Cell lysates were separated by SDS-PAGE, transferred to a PVDF membrane, and probed with antibodies against AnkF, vimentin and actin as loading control. A representative image of three independent experiments is shown. **(C)** Total isolated RNA was reverse-transcribed using SuperScript II reverse transcriptase according to the manufacturer's protocol and a qRT-PCR was performed with primers specific for *vimentin* (936 and 937) and *actin* (827 and 828) as a housekeeping gene. Vimentin was normalized to actin and is depicted as fold change compared to non-induced cells. **(D)** CHO-FcR cells were transiently transfected with plasmids pCMV-HA-vimentin (red) together with pGFP (green) alone (upper panel) or together with pEGFP-AnkF (lower panel). 24 h post-transfection, cells were fixed and HA-vimentin was stained with specific antibodies by indirect immunofluorescence (red). Nuclei were stained with DAPI (blue). Transient expression of HA- and GFP-containing proteins was visualized by epifluorescence microscopy. Images are representative of three independent experiments.

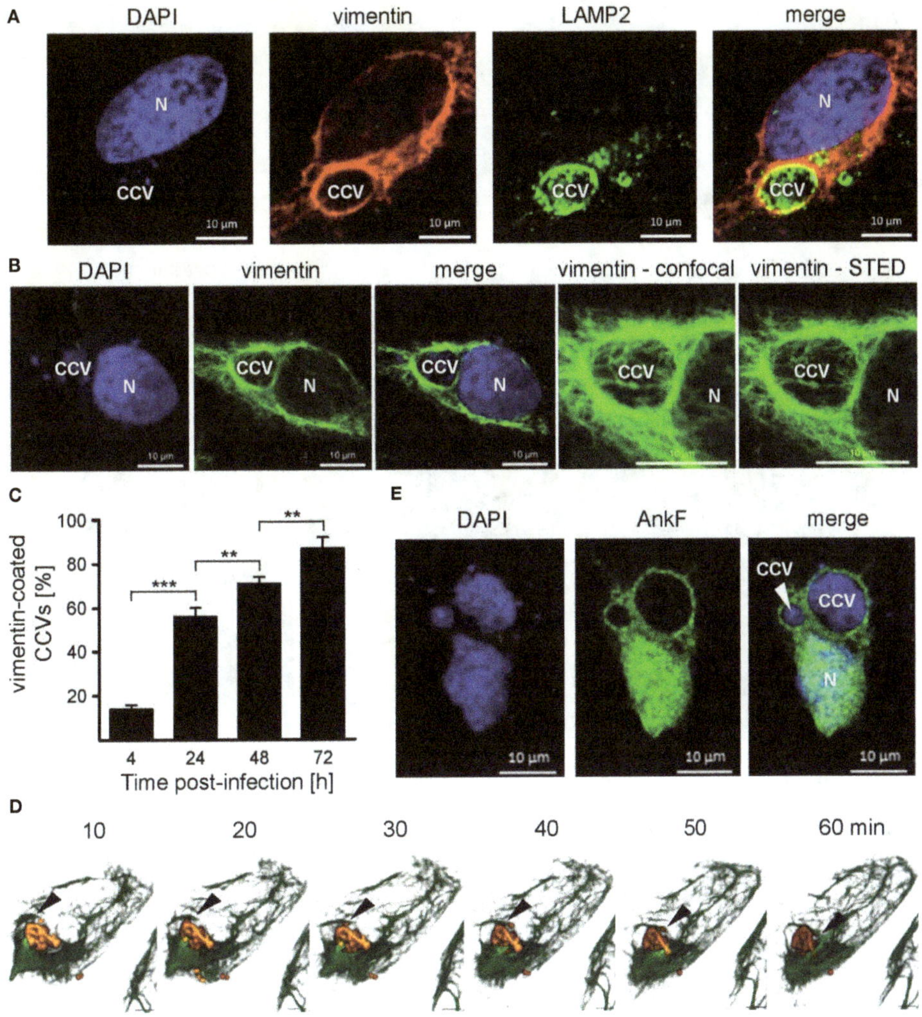

FIGURE 8 | Vimentin and AnkF associate with the CCV. **(A)** HeLa cells were infected with *C. burnetii* at an MOI of 50 for 24 h. Cells were fixed and stained with vimentin- and LAMP2-specific antibodies by indirect immunofluorescence (red and green, respectively). Nuclei and bacterial DNA were stained with DAPI (blue). Cells were visualized using LSM. **(B)** Infected cells were fixed and stained with a vimentin-specific antibody by indirect immunofluorescence (green). Nuclei and bacterial DNA were stained with DAPI (blue). Cells were visualized using LSM or Stimulated Emission Depletion (STED) high-resolution laser scanning fluorescence microscopy to highlight the close association of vimentin with the CCV. **(C)** HeLa cells were infected with *C. burnetii* at an MOI of 50 for up to 72 h. At the time points indicated, cells were fixed and vimentin and *C. burnetii* were stained with specific antibodies by indirect immunofluorescence. Using LSM, 100 cells were counted per each of three independent experiments for association of vimentin with the CCV. Error bars indicated ± SD. $^{**}p < 0.01$, $^{***}p < 0.001$. **(D)** Stable U2-OS-rseGFP-vimentin cells were infected with recombinant, inducible td-Tomato-expressing *C. burnetii* in the presence of 2 mM IPTG at an MOI of 100. At 72 h post infection, live-cell imaging was performed using Spinning Disc Confocal Microscopy. Z-stack images were acquired in a 15-µm range in 0.2-µm intervals every 10 min during a time period of 1 h. 3D re-construction was performed using ZEN software (Carl Zeiss AG) to visualize rseGFP-vimentin (green) and *C. burnetii*-tdTcc (red). Black arrows indicate the growing tip of vimentin filaments on the CCV. Scale bar: 5 µm. **(E)** HeLa cells were infected with *C. burnetii* at an MOI of 50. After 6 days, cells were transfected with pcDNA-AnkF. 24 h post-transfection, the cells were fixed and stained with an anti-AnkF-serum (green). Nuclei and bacterial DNA were stained with DAPI (blue). A representative LSM image from three independent experiments is shown.

vimentin. Indeed, cytokeratin 8 and 18 as well as actin act in concert with vimentin to ensure stability of the *C. trachomatis* inclusion (Kumar and Valdivia, 2008). Furthermore, actin was shown to associate with the CCV, and network disruption might result in smaller CCVs (Aguilera et al., 2009; Miller et al., 2018). It was proposed that actin stabilizes the CCV (Colonne et al., 2016) and participates in vesicular trafficking events (Aguilera et al., 2009; Miller et al., 2018). In order to elucidate the association of microfilaments, microtubules, and cytokeratins with bacterial compartment, immunofluorescence of *C. burnetii*-infected HeLa cells was performed. At 72 h post-infection, actin forms a punctate-like pattern around the bacterial compartment, whereas tubulin and cytokeratin 18 associate around the bacterial compartment in a filamentous pattern (**Figure 10**). Thus, these cytoskeletal components might be involved in stabilizing the CCV and might compensate for vimentin deficiency.

In order to elucidate whether AnkF expression might also influence the subcellular localization of these cytoskeletal

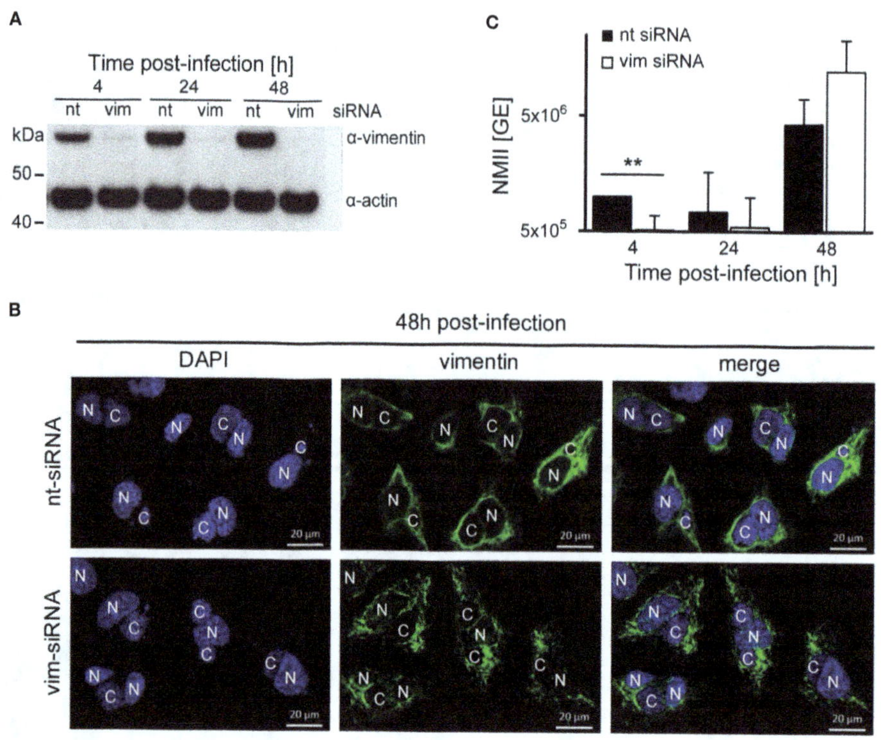

FIGURE 9 | Vimentin is dispensable for *C. burnetii* replication. **(A–C)** HeLa cells were transfected with 50 nM non-targeting (nt) or vimentin-targeting (vim) siRNA. After 24 h post-transfection, cells were infected with *C. burnetii* (WT) at a MOI of 50. **(A)** At the time points post-infection indicated, total cell lysates were separated by SDS-PAGE and analyzed by immunoblot analysis using antibodies against vimentin and actin as loading control. The immune blot is representative of four independent experiments. **(B)** 48 h post-infection, cells were fixed and stained with an anti-vimentin antibody (green). Nuclei and bacterial DNA were stained with DAPI (blue). Images are representative of two independent experiments. N: nucleus, C: *C. burnetii*-containing vacuole. **(C)** At the time-points indicated, cell lysates were prepared by osmotic lysis. Bacteria, isolated by differential centrifugation, were used for preparation of genomic DNA (gDNA). Absolute quantification of bacterial genomic equivalents was performed by quantitative real-time PCR with primers specific for genomic *IS1111* sequences. gDNA, prepared from axenically-grown *C. burnetii* cultures served as standard for genomic equivalents. Error bars represent mean standard deviations of four independent experiments. $^{**}p < 0.01$.

components, we ectopically expressed GFP-AnkF or GFP as control in HeLa cells and analyzed the structure of actin, tubulin, and cytokeratin by immunofluorescence. As demonstrated in **Figure 11**, the expression of GFP-AnkF did not disturb the localization of these cytoskeletal components, suggesting that AnkF might specifically modify vimentin localization.

DISCUSSION

The *C. burnetii* protein AnkF was shown to be injected into the host cell in a T4SS-dependent manner (Pan et al., 2008). The chaperone IcmS, which is required for the translocation of a subset of *C. burnetii* T4SS effector proteins (Larson et al., 2019), however, is dispensable for AnkF translocation (Voth et al., 2009). Where AnkF localizes within the host cell is still unclear. Ectopically expressed AnkF was found in the host cell cytoplasm, the nucleus (Rodriguez-Escudero et al., 2016) and in the vicinity of the CCV (**Figure 8E**). This multiple and diverse localization may explain why our attempts to localize translocated AnkF have failed so far. Another reason might be the short intracellular half-life of AnkF (Pan et al., 2008).

By comparing five different isolates, AnkF was identified as a highly conserved effector protein (Voth et al., 2009; Larson et al., 2016). Our analysis of 52 isolates confirmed this assumption (**Figure 1**). This is an interesting observation, as the majority of genes encoding for effector proteins were characterized by considerable heterogeneity between different *C. burnetii* isolates (Bisle et al., 2016; Larson et al., 2016). Conserved effector proteins might therefore be important virulence determinants. Indeed, the insertion of a transposon in the *ankF* gene at position 507 results in a markedly reduced ability to replicate intracellularly (**Figures 3B–E, 4A, C**). In a previous publication, a different transposon mutant carrying a transposon insertion at base pair 291 in *ankF* showed no effect on bacterial intracellular replication (Martinez et al., 2014). However, our experiments indicate that this transposon mutant is not clonal and contains wild-type AnkF (data not shown), which might explain the lack of phenotype. Importantly, while we observed an inability of the *ankF* transposon mutant to replicate intracellularly, we only noted a moderate reduction in

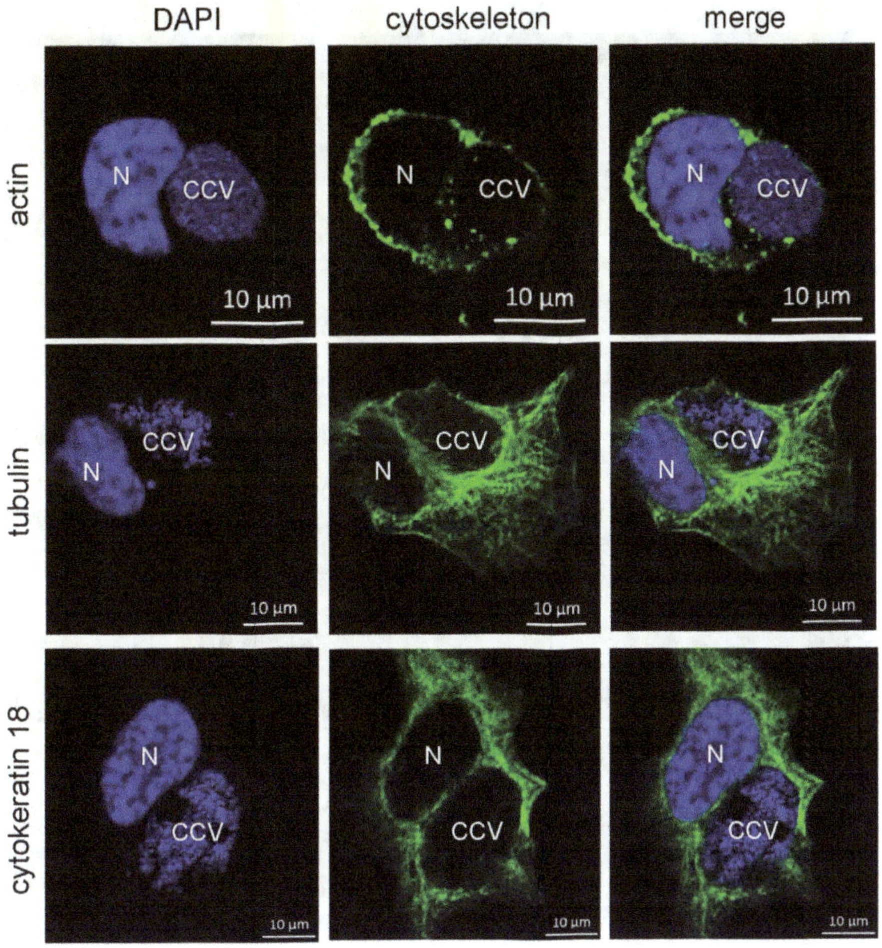

FIGURE 10 | Cytoskeletal filaments decorate the CCV. HeLa cells were infected with *C. burnetii* at an MOI of 50. At 72 h post-infection, cells were fixed and stained with Phallotoxin-647 for actin, or with tubulin- and cytokeratin 18-specific antibodies by indirect immunofluorescence (green). Nuclei and bacterial DNA were stained with DAPI (blue). Cells were visualized using LSM. N: Nucleus. CCV: *C. burnetii*-containing vacuole.

intracellular replication of the *ankF* deletion mutant (**Figures 6A, B**). The underlying reason for this difference might be transposon-mediated off-target effects in the genomic neighborhood of *ankF*. Indeed, a transposon insertion within *cbu0446*, a gene flanking *ankF* (*cbu0447*), was shown to cause a strong replication defect (Martinez et al., 2014). Further experiments are required to elucidate whether AnkF and CBU0446 are influencing each other.

While the activity of AnkF is still unclear, we showed that it binds vimentin (**Figure 7A**) and modulates vimentin filamentous assembly (**Figure 7D**). We propose that AnkF recruits vimentin to the CCV, as AnkF partially localizes to the CCV (**Figure 8E**) and the timing of recruitment of vimentin to the CCV (**Figure 8C**) correlates with effector protein translocation, which starts not earlier than 8 h and peaks around 24 h post-infection (Newton et al., 2013). While the domain important for binding to vimentin was not identified in this study, it is tempting to speculate that the ankyrin repeat domains in AnkF (Voth et al., 2009) might be involved. Ankyrin repeats are important protein-protein interaction domains (Mosavi et al., 2004), and the protein Ankyrin 1 was shown to bind to vimentin and mediate the association of vimentin with erythrocyte membranes (Georgatos and Marchesi, 1985).

Vimentin belongs to the class of intermediate filaments (IF), which are major elements of the cytoskeleton (Herrmann et al., 2009). Assembly and disassembly of vimentin filaments is mediated by post-translational modifications (Eriksson et al., 2004). In comparison to actin and tubulin, vimentin fibers are less rigid but more resistant to tensile forces (Janmey, 1991). Vimentin plays an important role in stabilizing cellular organelles and in organelle-positioning within the cytoplasm (Minin and Moldaver, 2008). In addition to its role as a cytosolic protein, vimentin has been reported to be cell surface-located and extracellularly localized (Mor-Vaknin et al., 2003). Vimentin is also involved in bacterial infections. Thus, vimentin and keratin 18 bind to the *Shigella flexneri* type III secretion system

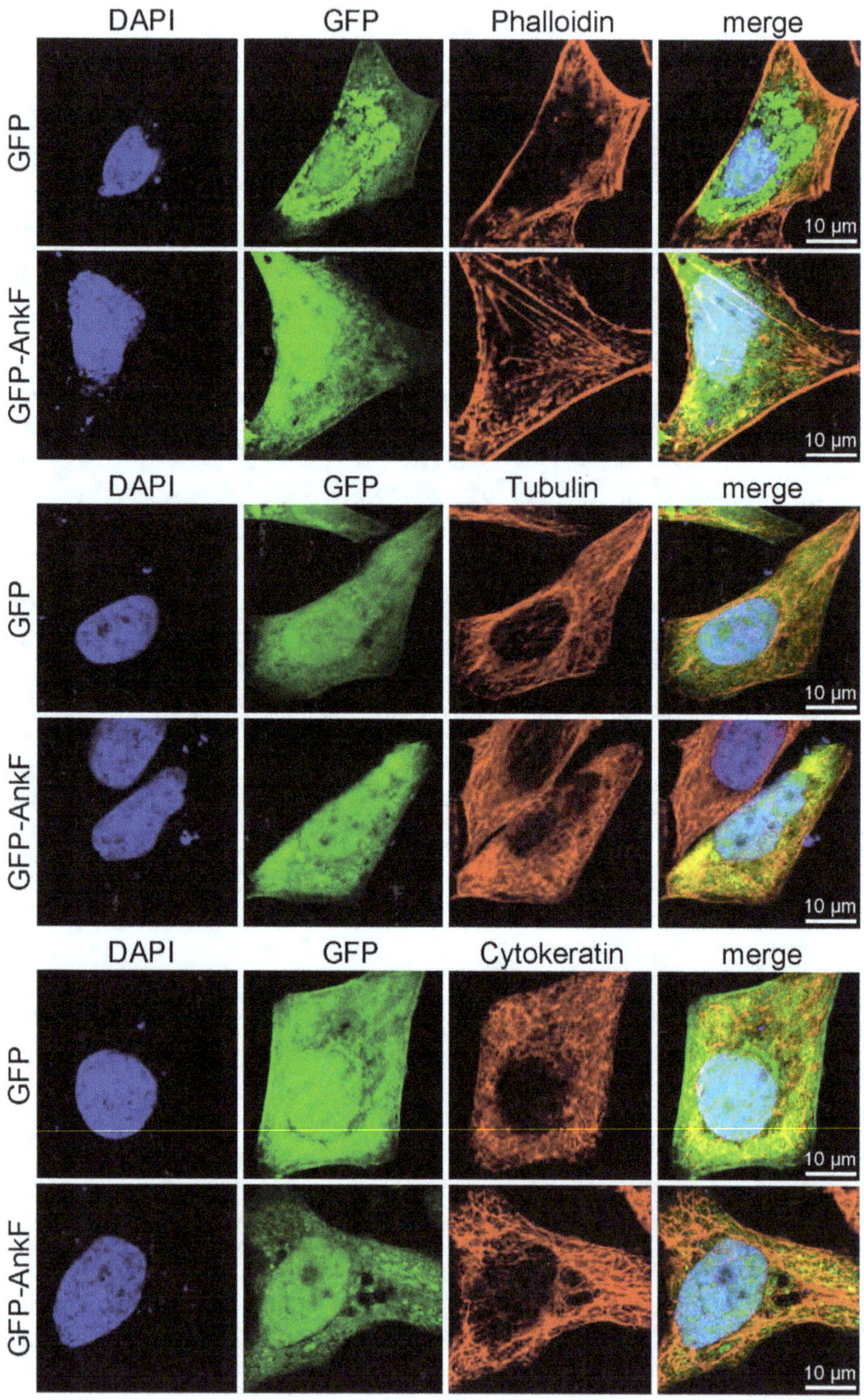

FIGURE 11 | Cytoskeletal filaments are not modified by AnkF expression. HeLa cells were transiently transfected with pGFP alone or with pEGFP-AnkF (green). 24 h post-transfection, cells were fixed and stained with Phalloidin-Alexa647 for actin, or with tubulin- and cytokeratin 18-specific antibodies by indirect immunofluorescence (red). Nuclei and bacterial DNA were stained with DAPI (blue). Cells were visualized using LSM. Representative images from two independent experiments are shown.

(T3SS) translocon pore protein IpaC. This interaction facilitates the docking of the bacteria to the host cell and enables effector protein translocation, which is crucial for virulence (Russo et al., 2016). Vimentin not only influences the translocation of effector proteins, it also influences bacterial invasion, the stability of the replicative niche and innate immune responses (Mak and Bruggemann, 2016). Thus, surface-located vimentin mediates invasion of Group B *streptococci*, *Listeria monocytogenes*,

Escherichia coli K1 and *Propionibacterium acnes* (Mak et al., 2012; Huang et al., 2016; Bastounis et al., 2018; Ghosh et al., 2018; Deng et al., 2019). Vimentin is also important for cell entry of several viruses, including SARS-CoV (Mak and Bruggemann, 2016; Yu et al., 2016). Our data suggest that vimentin might also be involved in the entry of *C. burnetii* in non-phagocytic cells (**Figure 9C**). Whether vimentin plays a role in *C. burnetii* invasion in phagocytic cells requires testing. Complement receptor 3 and $\alpha_v\beta_3$ integrin are involved in the phagocytosis of *C. burnetii* by phagocytic cells (Capo et al., 1999). Vimentin interacts with β_3 integrin, which is proposed to increase β_3 integrin clustering at the plasma membrane and to support β_3 integrin-ligand binding (Kim et al., 2016). This makes it likely that vimentin participates in *C. burnetii* uptake into phagocytic cells.

Moreover, expression of vimentin influences immune signaling, including activation of NF-κB (Mor-Vaknin et al., 2013) and the MAP kinase ERK1/2. Thus, vimentin promotes ERK1/2 signaling during *S. enterica* infection (Murli et al., 2001). Similarly, vimentin-induced ERK1/2 activation facilitates an *A. phagocytophilum* infection (Sukumaran et al., 2011). Likewise, a *C. burnetii* infection leads to the activation of the MAP kinase ERK1/2 (Boucherit et al., 2012; Graham et al., 2013), which is required for the anti-apoptotic activity of *C. burnetii* (Voth and Heinzen, 2009). Whether ERK1/2 activation during *C. burnetii* infection is mediated by vimentin is unknown. It is possible that vimentin is not only required as scaffolding for the CCV, but also as an inducer of immune signaling.

In addition, vimentin stabilizes or positions bacteria-containing vacuoles (Guignot and Servin, 2008; Kumar and Valdivia, 2008; Mak and Bruggemann, 2016; Truchan et al., 2016). Thus, vimentin contributes to the stability of the *C. trachomatis* inclusion. However, while the absence of vimentin influences the stability and morphology of the inclusion, it does not seem to affect bacterial replication (Kumar and Valdivia, 2008). The *Anaplasma phagocytophilum*-containing vacuole is also encased by vimentin (Truchan et al., 2016). In contrast to infection with *C. trachomatis* (Kumar and Valdivia, 2008) and *C. burnetii* (**Figures 7B, C**), the infection with *A. phagocytophilum* resulted in increased vimentin expression (Truchan et al., 2016). Pharmacologic inhibition of soluble vimentin did not reduce bacterial replication when administered to *A. phagocytophilum*-infected cells (Truchan et al., 2016). Similarly, siRNA-mediated knock-down of vimentin did not influence *Salmonella* Typhimurium replication (Guignot and Servin, 2008). Thus, our observation that the lack of vimentin does not influence bacterial replication (**Figure 9C**) is in line with previous publications (Guignot and Servin, 2008; Kumar and Valdivia, 2008; Truchan et al., 2016). The reason why vimentin seems to be dispensable for bacterial replication is elusive. One explanation might be functional redundancy. Thus, other intermediate filaments or microfilaments might be able to compensate for the lack of vimentin. Of note, the recruitment of vimentin to the *C. trachomatis* inclusion is dependent on actin microfilaments, which also decorate the inclusion. In addition, the intermediate filaments cytokeratin 8 and 18 were also recruited to the inclusion providing stability (Kumar and Valdivia, 2008). Our data demonstrate that microtubules and the intermediate filament cytokeratin 18 associate around the bacteria (**Figure 10**), making it possible that microtubules and other intermediate filaments compensate for the vimentin knock-down. The functional redundancy might explain the lack of a replication-defect in the absence of vimentin (**Figure 9C**) (Guignot and Servin, 2008; Kumar and Valdivia, 2008; Truchan et al., 2016). Furthermore, actin patches surrounded the CCV (**Figure 9**), as reported previously (Miller et al., 2018). Miller and colleagues showed that the lack of actin patches did not affect *C. burnetii* replication. However, actin filaments produced in an Arp2/3-dependent manner were required for vesicular trafficking events, and, thus, for CCV generation (Miller et al., 2018). Based on these data, we suggest that microfilaments and different intermediate filaments, including vimentin and cytokeratin 18, are recruited to the CCV to provide stability; at the same time they provide a platform for fusion and fission events, which allows the replicative CCV to form.

AUTHOR CONTRIBUTIONS

JP, JS-L, SB, FC, and MÖ performed the experiments and analyzed the data. MB and PAB provided resources, CB analyzed data, and AL conceived the study, obtained funding, supervised the study and drafted the manuscript. All authors contributed to the article and approved the submitted version.

ACKNOWLEDGMENTS

This work was supported by the Deutsche Forschungsgemeinschaft (DFG) through the SPP1580 to AL, by the Bundesministerium für Bildung und Forschung (BMBF) under the project number 01KI1726A/01KI1726C of "Q-GAPS" as part of the research network zoonotic infectious diseases to AL and CB, and by the French National Research Agency (ANR; ANR-14-CE14-0012-01) to MB. We thank Rita A. Eckart, Nadine Übelmesser, Christian Söldner for experimental help, Uwe Appelt (Core Facility for Cell Sorting and Immunomonitoring of the Universitätsklinikum Erlangen) for cell sorting and Philipp Tripal (Optical Imaging Center Erlangen (OICE)) for support with confocal and STED microscopy and with live cell imaging. We acknowledge the Montpellier imaging facility MRI, member of the national infrastructure France-BioImaging supported by the French National Research Agency (ANR-10-INBS-04).

REFERENCES

Aguilera, M., Salinas, R., Rosales, E., Carminati, S., Colombo, M. I., and Beron, W. (2009). Actin dynamics and Rho GTPases regulate the size and formation of parasitophorous vacuoles containing *Coxiella burnetii*. *Infect. Immun.* 77, 4609–4620. doi: 10.1128/IAI.00301-09

Baca, O. G., Klassen, D. A., and Aragon, A. S. (1993). Entry of *Coxiella burnetii* into host cells. *Acta Virol.* 37, 143–155.

Bastounis, E. E., Yeh, Y. T., and Theriot, J. A. (2018). Matrix stiffness modulates infection of endothelial cells by *Listeria monocytogenes* via expression of cell surface vimentin. *Mol. Biol. Cell* 29, 1571–1589. doi: 10.1091/mbc.E18-04-0228

Beare, P. A., and Heinzen, R. A. (2014). Gene inactivation in *Coxiella burnetii*. *Methods Mol. Biol.* 1197, 329–345. doi: 10.1007/978-1-4939-1261-2_19

Beare, P. A., Samuel, J. E., Howe, D., Virtaneva, K., Porcella, S. F., and Heinzen, R. A. (2006). Genetic diversity of the Q fever agent, *Coxiella burnetii*, assessed by microarray-based whole-genome comparisons. *J. Bacteriol.* 188, 2309–2324. doi: 10.1128/JB.188.7.2309-2324.2006

Beare, P. A., Unsworth, N., Andoh, M., Voth, D. E., Omsland, A., Gilk, S. D., et al. (2009). Comparative Genomics Reveal Extensive Transposon-Mediated Genomic Plasticity and Diversity among Potential Effector Proteins within the Genus *Coxiella*. *Infect. Immun.* 77, 642–656. doi: 10.1128/IAI.01141-08

Beare, P. A., Jeffrey, B. M., Martens, C. A., and Heinzen, R. A. (2017a). Draft Genome Sequences of the Avirulent *Coxiella burnetii* Dugway 7D77-80 and Dugway 7E65-68 Strains Isolated from Rodents in Dugway, Utah. *Genome Announcements* 5, e00984-17. doi: 10.1128/genomeA.00984-17

Beare, P. A., Jeffrey, B. M., Martens, C. A., and Heinzen, R. A. (2017b). Draft Genome Sequences of Three *Coxiella burnetii* Strains Isolated from Q Fever Patients. *Genome Announcements* 5, e00986-17. doi: 10.1128/genomeA.00986-17

Beare, P. A., Jeffrey, B. M., Martens, C. A., Pearson, T., and Heinzen, R. A. (2017c). Draft Genome Sequences of Historical Strains of *Coxiella burnetii* Isolated from Cow's Milk and a Goat Placenta. *Genome Announcements* 5, e00985-17. doi: 10.1128/genomeA.00985-17

Beare, P. A., Jeffrey, B. M., Long, C. M., Martens, C. M., and Heinzen, R. A. (2018). Genetic mechanisms of *Coxiella burnetii* lipopolysaccharide phase variation. *PloS Pathog.* 14, e1006922. doi: 10.1371/journal.ppat.1006922

Berens, C., Bisle, S., Klingenbeck, L., and Lührmann, A. (2015). Applying an Inducible Expression System to Study Interference of Bacterial Virulence Factors with Intracellular Signaling. *J. Visualized Experiments JoVE* 100, e52903. doi: 10.3791/52903

Bisle, S., Klingenbeck, L., Borges, V., Sobotta, K., Schulze-Luehrmann, J., Menge, C., et al. (2016). The Inhibition of the Apoptosis Pathway by the *Coxiella burnetii* Effector Protein CaeA requires the EK Repetition Motif, but is Independent of Survivin. *Virulence* 7, 400–412. doi: 10.1080/21505594.2016.1139280

Boucherit, N., Barry, A. O., Mottola, G., Trouplin, V., Capo, C., Mege, J. L., et al. (2012). Effects of *Coxiella burnetii* on MAPKinases phosphorylation. *FEMS Immunol. Med. Microbiol.* 64, 101–103. doi: 10.1111/j.1574-695X.2011.00852.x

Capo, C., Lindberg, F. P., Meconi, S., Zaffran, Y., Tardei, G., Brown, E. J., et al. (1999). Subversion of monocyte functions by *Coxiella burnetii*: impairment of the cross-talk between alphavbeta3 integrin and CR3. *J. Immunol.* 163, 6078–6085.

Carey, K. L., Newton, H. J., Lührmann, A., and Roy, C. R. (2011). The *Coxiella burnetii* Dot/Icm system delivers a unique repertoire of type IV effectors into host cells and is required for intracellular replication. *PloS Pathog.* 7, e1002056. doi: 10.1371/journal.ppat.1002056

Coleman, S. A., Fischer, E. R., Howe, D., Mead, D. J., and Heinzen, R. A. (2004). Temporal analysis of *Coxiella burnetii* morphological differentiation. *J. Bacteriol.* 186, 7344–7352. doi: 10.1128/JB.186.21.7344-7352.2004

Colonne, P. M., Winchell, C. G., Graham, J. G., Onyilagha, F. I., MacDonald, L. J., Doeppler, H. R., et al. (2016). Vasodilator-Stimulated Phosphoprotein Activity Is Required for *Coxiella burnetii* Growth in Human Macrophages. *PloS Pathog.* 12, e1005915. doi: 10.1371/journal.ppat.1005915

Dellacasagrande, J., Ghigo, E., Hammami, S. M., Toman, R., Raoult, D., Capo, C., et al. (2000). alpha(v)beta(3) integrin and bacterial lipopolysaccharide are involved in *Coxiella burnetii*-stimulated production of tumor necrosis factor by human monocytes. *Infect. Immun.* 68, 5673–5678. doi: 10.1128/IAI.68.10.5673-5678.2000

Deng, L., Spencer, B. L., Holmes, J. A., Mu, R., Rego, S., Weston, T. A., et al. (2019). The Group B Streptococcal surface antigen I/II protein, BspC, interacts with host vimentin to promote adherence to brain endothelium and inflammation during the pathogenesis of meningitis. *PloS Pathog.* 15, e1007848. doi: 10.1371/journal.ppat.1007848

D'Amato, F., Million, M., Edouard, S., Delerce, J., Robert, C., Marrie, T., et al. (2014a). Draft genome sequence of *Coxiella burnetii* Dog Utad, a strain isolated from a dog-related outbreak of Q fever. *New Microbes New Infect.* 2, 136–137. doi: 10.1002/nmi2.55

D'Amato, F., Rouli, L., Edouard, S., Tyczka, J., Million, M., Robert, C., et al. (2014b). The genome of *Coxiella burnetii* Z3055, a clone linked to the Netherlands Q fever outbreaks, provides evidence for the role of drift in the emergence of epidemic clones. *Comp. Immunol. Microbiol. Infect. Dis.* 37, 281–288. doi: 10.1016/j.cimid.2014.08.003

D'Amato, F., Eldin, C., Georgiades, K., Edouard, S., Delerce, J., Labas, N., et al. (2015). Loss of TSS1 in hypervirulent *Coxiella burnetii* 175, the causative agent of Q fever in French Guiana. *Comp. Immunol. Microbiol. Infect. Dis.* 41, 35–41. doi: 10.1016/j.cimid.2015.04.003

Eriksson, J. E., He, T., Trejo-Skalli, A. V., Harmala-Brasken, A. S., Hellman, J., Chou, Y. H., et al. (2004). Specific in vivo phosphorylation sites determine the assembly dynamics of vimentin intermediate filaments. *J. Cell Sci.* 117, 919–932. doi: 10.1242/jcs.00906

Garg, A., Barnes, P. F., Porgador, A., Roy, S., Wu, S., Nanda, J. S., et al. (2006). Vimentin expressed on *Mycobacterium tuberculosis*-infected human monocytes is involved in binding to the NKp46 receptor. *J. Immunol.* 177, 6192–6198. doi: 10.4049/jimmunol.177.9.6192

Georgatos, S. D., and Marchesi, V. T. (1985). The binding of vimentin to human erythrocyte membranes: a model system for the study of intermediate filament-membrane interactions. *J. Cell Biol.* 100, 1955–1961. doi: 10.1083/jcb.100.6.1955

Ghosh, P., Halvorsen, E. M., Ammendolia, D. A., Mor-Vaknin, N., O'Riordan, M. X. D., Brumell, J. H., et al. (2018). Invasion of the Brain by *Listeria monocytogenes* Is Mediated by InlF and Host Cell Vimentin. *mBio* 9, e00160-18. doi: 10.1128/mBio.00160-18

Gonzalez-Barrio, D., and Ruiz-Fons, F. (2019). *Coxiella burnetii* in wild mammals: A systematic review. *Transboundary Emerg. Dis.* 66, 662–671. doi: 10.1111/tbed.13085

Graham, J. G., MacDonald, L. J., Hussain, S. K., Sharma, U. M., Kurten, R. C., and Voth, D. E. (2013). Virulent *Coxiella burnetii* pathotypes productively infect primary human alveolar macrophages. *Cell Microbiol.* 15, 1012–1025. doi: 10.1111/cmi.12096

Guignot, J., and Servin, A. L. (2008). Maintenance of the *Salmonella*-containing vacuole in the juxtanuclear area: a role for intermediate filaments. *Microbial. Pathogen.* 45, 415–422. doi: 10.1016/j.micpath.2008.09.007

Hackstadt, T., and Williams, J. C. (1981). Biochemical stratagem for obligate parasitism of eukaryotic cells by *Coxiella burnetii*. *Proc. Natl. Acad. Sci. U. S. A.* 78, 3240–3244. doi: 10.1073/pnas.78.5.3240

Hammerl, J. A., Mertens, K., Sprague, L. D., Hackert, V. H., Buijs, J., Hoebe, C. J., et al. (2015). First Draft Genome Sequence of a Human *Coxiella burnetii* Isolate, Originating from the Largest Q Fever Outbreak Ever Reported, the Netherlands 2007 to 2010. *Genome Announcements* 3, e00445-15. doi: 10.1128/genomeA.00445-15

Hayek, I., Fischer, F., Schulze-Luehrmann, J., Dettmer, K., Sobotta, K., Schatz, V., et al. (2019). Limitation of TCA Cycle Intermediates Represents an Oxygen-Independent Nutritional Antibacterial Effector Mechanism of Macrophages. *Cell Rep.* 26, 3502–3510. doi: 10.1016/j.celrep.2019.02.103

Hemsley, C. M., O'Neill, P. A., Essex-Lopresti, A., Norville, I. H., Atkins, T. P., and Titball, R. W. (2019). Extensive genome analysis of *Coxiella burnetii* reveals limited evolution within genomic groups. *BMC Genomics* 20, 441. doi: 10.1186/s12864-019-5833-8

Herrmann, H., Strelkov, S. V., Burkhard, P., and Aebi, U. (2009). Intermediate filaments: primary determinants of cell architecture and plasticity. *J. Clin. Invest.* 119, 1772–1783. doi: 10.1172/JCI38214

Huang, S. H., Chi, F., Peng, L., Bo, T., Zhang, B., Liu, L. Q., et al. (2016). Vimentin, a Novel NF-kappaB Regulator, Is Required for Meningitic *Escherichia coli* K1-Induced Pathogen Invasion and PMN Transmigration across the Blood-Brain Barrier. *PloS One* 11, e0162641. doi: 10.1371/journal.pone.0162641

Janmey, P. A., Euteneuer, U., Traub, P., and Schliwa, M. (1991). Viscoelastic properties of vimentin compared with other filamentous biopolymer networks. *J. Cell Biol.* 113, 155–160. doi: 10.1083/jcb.113.1.155

Janmey, P. A. (1991). Mechanical properties of cytoskeletal polymers. *Curr. Opin. Cell Biol.* 3, 4–11. doi: 10.1016/0955-0674(91)90159-V

Kersh, G. J. (2013). Antimicrobial therapies for Q fever. *Expert Rev. Anti-infective Ther.* 11, 1207–1214. doi: 10.1586/14787210.2013.840534

Khavkin, T., and Tabibzadeh, S. S. (1988). Histologic, immunofluorescence, and electron microscopic study of infectious process in mouse lung after intranasal challenge with *Coxiella burnetii*. *Infect. Immun.* 56, 1792–1799. doi: 10.1128/IAI.56.7.1792-1799.1988

Kim, J., Yang, C., Kim, E. J., Jang, J., Kim, S. J., Kang, S. M., et al. (2016). Vimentin filaments regulate integrin-ligand interactions by binding to the cytoplasmic tail of integrin beta3. *J. Cell Sci.* 129, 2030–2042. doi: 10.1242/jcs.180315

Kuley, R., Smith, H. E., Janse, I., Harders, F. L., Baas, F., Schijlen, E., et al. (2016). First Complete Genome Sequence of the Dutch Veterinary *Coxiella burnetii* Strain NL3262, Originating from the Largest Global Q Fever Outbreak, and Draft Genome Sequence of Its Epidemiologically Linked Chronic Human Isolate NLhu3345937. *Genome Announcements* 4, e00245-16. doi: 10.1128/genomeA.00245-16

Kuley, R., Kuijt, E., Smits, M. A., Roest, H. I. J., Smith, H. E., and Bossers, A. (2017). Genome Plasticity and Polymorphisms in Critical Genes Correlate with Increased Virulence of Dutch Outbreak-Related *Coxiella burnetii* Strains. *Front. Microbiol.* 8, 1526. doi: 10.3389/fmicb.2017.01526

Kumar, Y., and Valdivia, R. H. (2008). Actin and intermediate filaments stabilize the *Chlamydia trachomatis* vacuole by forming dynamic structural scaffolds. *Cell Host Microbe* 4, 159–169. doi: 10.1016/j.chom.2008.05.018

Larson, C. L., Martinez, E., Beare, P. A., Jeffrey, B., Heinzen, R. A., and Bonazzi, M. (2016). Right on Q: genetics begin to unravel *Coxiella burnetii* host cell interactions. *Future Microbiol.* 11, 919–939. doi: 10.2217/fmb-2016-0044

Larson, C. L., Beare, P. A., and Heinzen, R. A. (2019). Dependency of *Coxiella burnetii* Type 4B Secretion on the Chaperone IcmS. *J. Bacteriol.* 201, e00431-19. doi: 10.1128/JB.00431-19

Long, C. M., Beare, P. A., Cockrell, D. C., Larson, C. L., and Heinzen, R. A. (2019). Comparative virulence of diverse *Coxiella burnetii* strains. *Virulence* 10, 133–150. doi: 10.1080/21505594.2019.1575715

Lührmann, A., Newton, H. J., and Bonazzi, M. (2017). Beginning to Understand the Role of the Type IV Secretion System Effector Proteins in *Coxiella burnetii* Pathogenesis. *Curr. Topics Microbiol. Immunol.* 413, 243–268. doi: 10.1007/978-3-319-75241-9_10

Mahesh, P. P., Retnakumar, R. J., and Mundayoor, S. (2016). Downregulation of vimentin in macrophages infected with live *Mycobacterium tuberculosis* is mediated by Reactive Oxygen Species. *Sci. Rep.* 6, 21526. doi: 10.1038/srep21526

Mak, T. N., and Bruggemann, H. (2016). Vimentin in Bacterial Infections. *Cells* 5, 18. doi: 10.3390/cells5020018

Mak, T. N., Fischer, N., Laube, B., Brinkmann, V., Metruccio, M. M., Sfanos, K. S., et al. (2012). *Propionibacterium acnes* host cell tropism contributes to vimentin-mediated invasion and induction of inflammation. *Cell Microbiol.* 14, 1720–1733. doi: 10.1111/j.1462-5822.2012.01833.x

Mansilla Pareja, M. E., Bongiovanni, A., Lafont, F., and Colombo, M. I. (2017). Alterations of the *Coxiella burnetii* Replicative Vacuole Membrane Integrity and Interplay with the Autophagy Pathway. *Front. Cell Infect. Microbiol.* 7, 112. doi: 10.3389/fcimb.2017.00112

Martinez, E., Cantet, F., Fava, L., Norville, I., and Bonazzi, M. (2014). Identification of OmpA, a *Coxiella burnetii* protein involved in host cell invasion, by multi-phenotypic high-content screening. *PloS Pathog.* 10, e1004013. doi: 10.1371/journal.ppat.1004013

Martinez, E., Cantet, F., and Bonazzi, M. (2015). Generation and multi-phenotypic high-content screening of *Coxiella burnetii* transposon mutants. *J. Visualized Experiments JoVE* 99, e52851. doi: 10.3791/52851

Maurin, M., and Raoult, D. (1999). Q fever. *Clin. Microbiol. Rev.* 12, 518–553. doi: 10.1128/CMR.12.4.518

McDonough, J. A., Newton, H. J., Klum, S., Swiss, R., Agaisse, H., and Roy, C. R. (2013). Host pathways important for *Coxiella burnetii* infection revealed by genome-wide RNA interference screening. *mBio* 4, e00606–e00612. doi: 10.1128/mBio.00606-12

Miller, H. E., Larson, C. L., and Heinzen, R. A. (2018). Actin polymerization in the endosomal pathway, but not on the *Coxiella*-containing vacuole, is essential for pathogen growth. *PloS Pathog.* 14, e1007005. doi: 10.1371/journal.ppat.1007005

Million, M., Roblot, F., Carles, D., D'Amato, F., Protopopescu, C., Carrieri, M. P., et al. (2014). Reevaluation of the risk of fetal death and malformation after Q Fever. *Clin. Infect. Dis.* 59, 256–260. doi: 10.1093/cid/ciu259

Minin, A. A., and Moldaver, M. V. (2008). Intermediate vimentin filaments and their role in intracellular organelle distribution. *Biochem. Biokhimiia* 73, 1453–1466. doi: 10.1134/S0006297908130063

Mor-Vaknin, N., Punturieri, A., Sitwala, K., and Markovitz, D. M. (2003). Vimentin is secreted by activated macrophages. *Nat. Cell Biol.* 5, 59–63. doi: 10.1038/ncb898

Mor-Vaknin, N., Legendre, M., Yu, Y., Serezani, C. H., Garg, S. K., Jatzek, A., et al. (2013). Murine colitis is mediated by vimentin. *Sci. Rep.* 3, 1045. doi: 10.1038/srep01045

Mosavi, L. K., Cammett, T. J., Desrosiers, D. C., and Peng, Z. Y. (2004). The ankyrin repeat as molecular architecture for protein recognition. *Protein Sci. Publ. Protein Soc.* 13, 1435–1448. doi: 10.1110/ps.03554604

Murli, S., Watson, R. O., and Galan, J. E. (2001). Role of tyrosine kinases and the tyrosine phosphatase SptP in the interaction of *Salmonella* with host cells. *Cell Microbiol.* 3, 795–810. doi: 10.1046/j.1462-5822.2001.00158.x

Newton, H. J., McDonough, J. A., and Roy, C. R. (2013). Effector protein translocation by the *Coxiella burnetii* Dot/Icm type IV secretion system requires endocytic maturation of the pathogen-occupied vacuole. *PloS One* 8, e54566. doi: 10.1371/journal.pone.0054566

Newton, P., Thomas, D. R., Reed, S. C. O., Lau, N., Xu, B., Ong, S. Y., et al. (2020). Lysosomal degradation products induce *Coxiella burnetii* virulence. *Proc. Natl. Acad. Sci. U. S. A.* 117, 6801–6810. doi: 10.1073/pnas.1921344117

Pan, X., Lührmann, A., Satoh, A., Laskowski-Arce, M. A., and Roy, C. R. (2008). Ankyrin repeat proteins comprise a diverse family of bacterial type IV effectors. *Science* 320, 1651–1654. doi: 10.1126/science.1158160

Raoult, D., Marrie, T., and Mege, J. (2005). Natural history and pathophysiology of Q fever. *Lancet Infect. Dis.* 5, 219–226. doi: 10.1016/S1473-3099(05)70052-9

Ratz, M., Testa, I., Hell, S. W., and Jakobs, S. (2015). CRISPR/Cas9-mediated endogenous protein tagging for RESOLFT super-resolution microscopy of living human cells. *Sci. Rep.* 5, 9592. doi: 10.1038/srep09592

Rodolakis, A. (2009). Q Fever in dairy animals. *Ann. New Y. Acad. Sci.* 1166, 90–93. doi: 10.1111/j.1749-6632.2009.04511.x

Rodriguez-Escudero, M., Cid, V. J., Molina, M., Schulze-Luehrmann, J., Lührmann, A., and Rodriguez-Escudero, I. (2016). Studying *Coxiella burnetii* Type IV Substrates in the Yeast Saccharomyces cerevisiae: Focus on Subcellular Localization and Protein Aggregation. *PloS One* 11, e0148032. doi: 10.1371/journal.pone.0148032

Rosales, E. M., Aguilera, M. O., Salinas, R. P., Carminati, S. A., Colombo, M. I., Martinez-Quiles, N., et al. (2012). Cortactin is involved in the entry of *Coxiella burnetii* into non-phagocytic cells. *PloS One* 7, e39348. doi: 10.1371/journal.pone.0039348

Rouli, L., Rolain, J. M., El Filali, A., Robert, C., and Raoult, D. (2012). Genome sequence of *Coxiella burnetii* 109, a doxycycline-resistant clinical isolate. *J. Bacteriol.* 194, 6939. doi: 10.1128/JB.01856-12

Russo, B. C., Stamm, L. M., Raaben, M., Kim, C. M., Kahoud, E., Robinson, L. R., et al. (2016). Intermediate filaments enable pathogen docking to trigger type 3 effector translocation. *Nat. Microbiol.* 1, 16025. doi: 10.1038/nmicrobiol.2016.25

Sandoz, K. M., Beare, P. A., Cockrell, D. C., and Heinzen, R. A. (2016). Complementation of Arginine Auxotrophy for Genetic Transformation of *Coxiella burnetii* by Use of a Defined Axenic Medium. *Appl. Environ. Microbiol.* 82, 3042–3051. doi: 10.1128/AEM.00261-16

Schulze-Luehrmann, J., Eckart, R. A., Olke, M., Saftig, P., Liebler-Tenorio, E., and Lührmann, A. (2016). LAMP proteins account for the maturation delay during the establishment of the *Coxiella burnetii*-containing vacuole. *Cell Microbiol.* 18, 181–194. doi: 10.1111/cmi.12494

Seshadri, R., Paulsen, I. T., Eisen, J. A., Read, T. D., Nelson, K. E., Nelson, W. C., et al. (2003). Complete genome sequence of the Q fever pathogen *Coxiella burnetii*. *Proc. Natl. Acad. Sci. U. S. A.* 100, 5455–5460. doi: 10.1073/pnas.0931379100

Sidi-Boumedine, K., Ellis, R. J., Adam, G., Prigent, M., Angen, O., Aspan, A., et al. (2014). Draft Genome Sequences of Six Ruminant *Coxiella burnetii* Isolates of European Origin. *Genome Announcements* 2, e00285-14. doi: 10.1128/genomeA.00285-14

Soellner, P., Quinlan, R. A., and Franke, W. W. (1985). Identification of a distinct soluble subunit of an intermediate filament protein: tetrameric vimentin from living cells. *Proc. Natl. Acad. Sci. U. S. A.* 82, 7929–7933. doi: 10.1073/pnas.82.23.7929

Styers, M. L., Kowalczyk, A. P., and Faundez, V. (2005). Intermediate filaments and vesicular membrane traffic: the odd couple's first dance? *Traffic* 6, 359–365. doi: 10.1111/j.1600-0854.2005.00286.x

Sukumaran, B., Mastronunzio, J. E., Narasimhan, S., Fankhauser, S., Uchil, P. D., Levy, R., et al. (2011). *Anaplasma phagocytophilum* AptA modulates Erk1/2 signalling. *Cell Microbiol.* 13, 47–61. doi: 10.1111/j.1462-5822.2010.01516.x

Truchan, H. K., Cockburn, C. L., May, L. J., VieBrock, L., and Carlyon, J. A. (2016). *Anaplasma phagocytophilum*-Occupied Vacuole Interactions with the Host Cell Cytoskeleton. *Veterinary Sci.* 3, 25. doi: 10.3390/vetsci3030025

van Schaik, E. J., Chen, C., Mertens, K., Weber, M. M., and Samuel, J. E. (2013). Molecular pathogenesis of the obligate intracellular bacterium *Coxiella burnetii*. *Nat. Rev. Microbiol.* 11, 561–573. doi: 10.1038/nrmicro3049

Voth, D. E., and Heinzen, R. A. (2007). Lounging in a lysosome: the intracellular lifestyle of *Coxiella burnetii*. *Cell Microbiol.* 9, 829–840. doi: 10.1111/j.1462-5822.2007.00901.x

Voth, D. E., and Heinzen, R. A. (2009). Sustained activation of Akt and Erk1/2 is required for *Coxiella burnetii* antiapoptotic activity. *Infect. Immun.* 77, 205–213. doi: 10.1128/IAI.01124-08

Voth, D. E., Howe, D., Beare, P. A., Vogel, J. P., Unsworth, N., Samuel, J. E., et al. (2009). The *Coxiella burnetii* ankyrin repeat domain-containing protein family is heterogeneous, with C-terminal truncations that influence Dot/Icm-mediated secretion. *J. Bacteriol.* 191, 4232–4242. doi: 10.1128/JB.01656-08

Walter, M. C., Ohrman, C., Myrtennas, K., Sjodin, A., Bystrom, M., Larsson, P., et al. (2014). Genome sequence of *Coxiella burnetii* strain Namibia. *Standards Genom. Sci.* 9, 22. doi: 10.1186/1944-3277-9-22

Yu, Y. T., Chien, S. C., Chen, I. Y., Lai, C. T., Tsay, Y. G., Chang, S. C., et al. (2016). Surface vimentin is critical for the cell entry of SARS-CoV. *J. Biomed. Sci.* 23, 14. doi: 10.1186/s12929-016-0234-7

3

Host Cell Targets of Released Lipid and Secreted Protein Effectors of *Mycobacterium tuberculosis*

*Jacques Augenstreich and Volker Briken**

Department of Cell Biology and Molecular Genetics, University of Maryland, College Park, MD, United States

***Correspondence:**
Volker Briken
vbriken@umd.edu

Mycobacterium tuberculosis (Mtb) is a very successful pathogen, strictly adapted to humans and the cause of tuberculosis. Its success is associated with its ability to inhibit host cell intrinsic immune responses by using an arsenal of virulence factors of different nature. It has evolved to synthesize a series of complex lipids which form an outer membrane and may also be released to enter host cell membranes. In addition, secreted protein effectors of Mtb are entering the host cell cytosol to interact with host cell proteins. We briefly discuss the current model, involving the ESX-1 type seven secretion system and the Mtb lipid phthiocerol dimycoserosate (PDIM), of how Mtb creates pores in the phagosomal membrane to allow Mtb proteins to access to the host cell cytosol. We provide an exhaustive list of Mtb secreted proteins that have effector functions. They modify (mostly inhibit but sometimes activate) host cell pathways such as: phagosome maturation, cell death, cytokine response, xenophagy, reactive oxygen species (ROS) response via NADPH oxidase 2 (NOX2), nitric oxide (NO) response via NO Synthase 2 (NOS2) and antigen presentation via MHC class I and class II molecules. We discuss the host cell targets for each lipid and protein effector and the importance of the Mtb effector for virulence of the bacterium.

Keywords: *Mycobacterium tuberculosis*, effector, cell death, lipids, cytokines, phagosome maturation, NOX2, ESX-1

INTRODUCTION

Mtb is the bacterium responsible for the highest number of deaths annually caused by an infection disease. Its success as human pathogen is at least partially due to the ability of Mtb to evade the host immune response (Liu et al., 2017; Queval et al., 2017; Upadhyay et al., 2018; Bussi and Gutierrez, 2019; Sia and Rengarajan, 2019). On the other side there was intense selective pressure on humans to develop immune responses that could drive active tuberculosis infections into latent ones in order for the host to survive (Moreira-Teixeira et al., 2018; Olive and Sassetti, 2018; Simmons et al., 2018; Correa-Macedo et al., 2019). Ever since the first observation that intracellular Mtb inhibits the normal progression of phagosome maturations in the 1970s (Armstrong and Hart, 1971), the "How" and "What" of Mtb-mediated host cell manipulation have been under intense investigation. The translational research potential to exploit this knowledge, either for directly targeting Mtb effectors or for developing host targeted therapeutics, is important due to the urgent need of novel drugs to treat tuberculosis. Nevertheless, progress was limited for a long time due to the absence of genetic tools to modify Mtb. This changed in the early 1990s when pioneering work established the first tools to allow foreign gene expression in mycobacteria (Jacobs et al., 1987; Ranes et al., 1990) and to create a specific gene deletion mutant (Bardarov et al., 2002). Recently, ORBIT and CRISPRi technologies have been developed to simplify gene deletion/suppression approaches in Mtb (Rock et al., 2017; Murphy et al., 2018). The genetic toolbox was quickly expanded to include high-throughput gene disruption capacity via transposon (Tn)-mediated approaches which allowed for loss-of-function genetic screens (Camacho et al., 1999; Cox et al., 1999). The transposon approach was improved and combined with advances in gene sequencing technology to generate the TnSeq approach which allows for loss-of-function screens on a population basis (Long et al., 2015). Furthermore, shuttle cosmids were developed to be able to express large regions of the Mtb genome in non-tuberculoid mycobacteria in order to execute gain-of-function genetic screens (Bange et al., 1999; Velmurugan et al., 2007). Similarly, the genetic modification of host cells via siRNA and CRISPR/Cas9 allows for genome-wide genetic screens that can be combined with high-throughput readouts in order to identify host cell genes involved in specific host responses to Mtb infections (Kumar et al., 2010). The Collaborative Cross collection of mouse strains was used to identify host susceptibility and resistance genes following Mtb infections (Smith et al., 2019). These breakthroughs in experimental approaches have already led to seminal findings on Mtb lipid and protein effectors and their host cell targets. The goal of this review is to provide an overview of the current knowledge of secreted Mtb proteins and lipids, their mechanisms of action and their importance for virulence of Mtb.

MTB EFFECTOR PROTEINS AND THEIR TARGETS IN THE HOST CELL

What are some of the important characteristics of Mtb proteins that interact with the host cell? Many effectors need to be secreted by the bacterium and hence the secretion system for which they are a substrate is an important feature of the effector. The Mtb genome encodes for five type-seven secretion systems (T7SS), ESX-1 to ESX-5. They mediate the secretion of two family of proteins, EsxA-like proteins which are part of the Trp-X-Gly (WXG) family, and proteins harboring a Pro-Glu (PE) or Pro-Pro-Glu (PPE) N-terminal motif. The main substrates of ESX-1 are EsxA and EsxB. The ESX-3 system transports the EsxA/B paralogues EsxG/H. There is very little information on the substrates of the ESX-2 and ESX-4 systems. Finally, ESX-5 is responsible for secretion of a large number of proteins many of which are part of the PE/PPE superfamily (Gröschel et al., 2016; Bosserman and Champion, 2017). The heterologous expression of PE/PPE proteins in *M. smegmatis* (Msme) is a commonly used approach to characterize these proteins since they are absent in this mycobacterial species. Nevertheless, Msme also does not express an ESX-5 system, so the transport of these heterologously-expressed Mtb proteins is unknown. In order to focus the scope of this review we have decided to not include a discussion of Mtb PE/PPE effectors expressed in Msme, unless there is additional experimental evidence derived from Mtb deletion mutant or ectopic expression of the Mtb protein in a host cell. Furthermore, we also decided to limit the review to proteins that are secreted and released by Mtb which excludes many PE/PPE family proteins because they remain associated with the Mtb cell wall (e.g., PE-PGRS33). We point the reader to a current and exhaustive review on PE/PPE Mtb proteins (De Maio et al., 2020). In the absence of experimental data, we used SignalP 5.0 (Almagro Armenteros et al., 2019) to predict the presence of a signal peptide which indicates secretion via the SecA1/2 or Tat secretion systems.

Another important characteristic is the function of the effector. Protein tyrosine phosphatases A and -B (PtpA, PtpB) and the secreted acid phosphatase of Mtb (SapM) are the only three phosphatases known to be secreted by Mtb (Koul et al., 2000; Saleh and Belisle, 2000) (for more detailed review see Wong et al., 2013). The Mtb genome encodes 11 serine/threonine protein kinases of which 9 have a single transmembrane domain, an extracellular sensor domain and the intracellular kinase domain (Av-Gay and Everett, 2000; Prisic and Husson, 2014). PknG and PknK have no transmembrane domain but only PknG has been shown to be secreted (Prisic and Husson, 2014). In addition, Mtb encodes for the tyrosine kinase PtkA which phosphorylates and activates the phosphatase PtpA (Zhou et al., 2015; Jaiswal et al., 2019). In addition to kinases and phosphatases there is a wide range of other enzymatic activities reported for effectors (**Table 1**) but these are not always associated with their function in host cell manipulation but rather with their primary function in Mtb homeostasis (e.g., echA2 Truong and Penn, 2020). Finally, for many effectors no specific activity has been determined yet.

One of the most interesting feature of a secreted effector is its host cell target (**Table 1**). The identification of a target can be fairly straight forward if there is a strong binding affinity which allows for co-immunoprecipitation or column-based enrichment approaches. Interactions that are of low affinity are much harder to characterize which is probably one reason that for many Mtb effectors their host cell targets have not been identified yet (**Table 1**). Related to the targeted host cell protein is the impact of the Mtb effector on the host cell signaling pathways. The manipulation of proinflammatory cytokine responses, phagosome maturation, autophagy and host cell death are the major pathways targeted by Mtb that have been identified today (**Table 1**). Since these host cell pathways are interconnected (for example, proinflammatory cytokine signaling may affect host cell death) a specific Mtb effector may affect more than one host cell signaling pathway.

Finally, what is the contribution of a given Mtb protein effector to the virulence of the bacterium? The answer to that question is complicated by many factors. One of them being the redundancy of Mtb effector proteins for a given host cell signaling pathway. For example, many proteins and some lipids target the maturation of the Mtb phagosome (**Tables 1, 2**). So, studying their individual effect on phagosome maturation will show meaningful differences but when taken into context of an infected mouse lung these differences observed during *ex vivo* infections may not be important enough to affect the overall survival of Mtb in the mouse lung. An approach to overcoming this hurdle would be by generating double or even triple Mtb gene deletion mutants. Another complicating factor is how to be sure if the observed *in vivo* effect of a given Mtb mutant

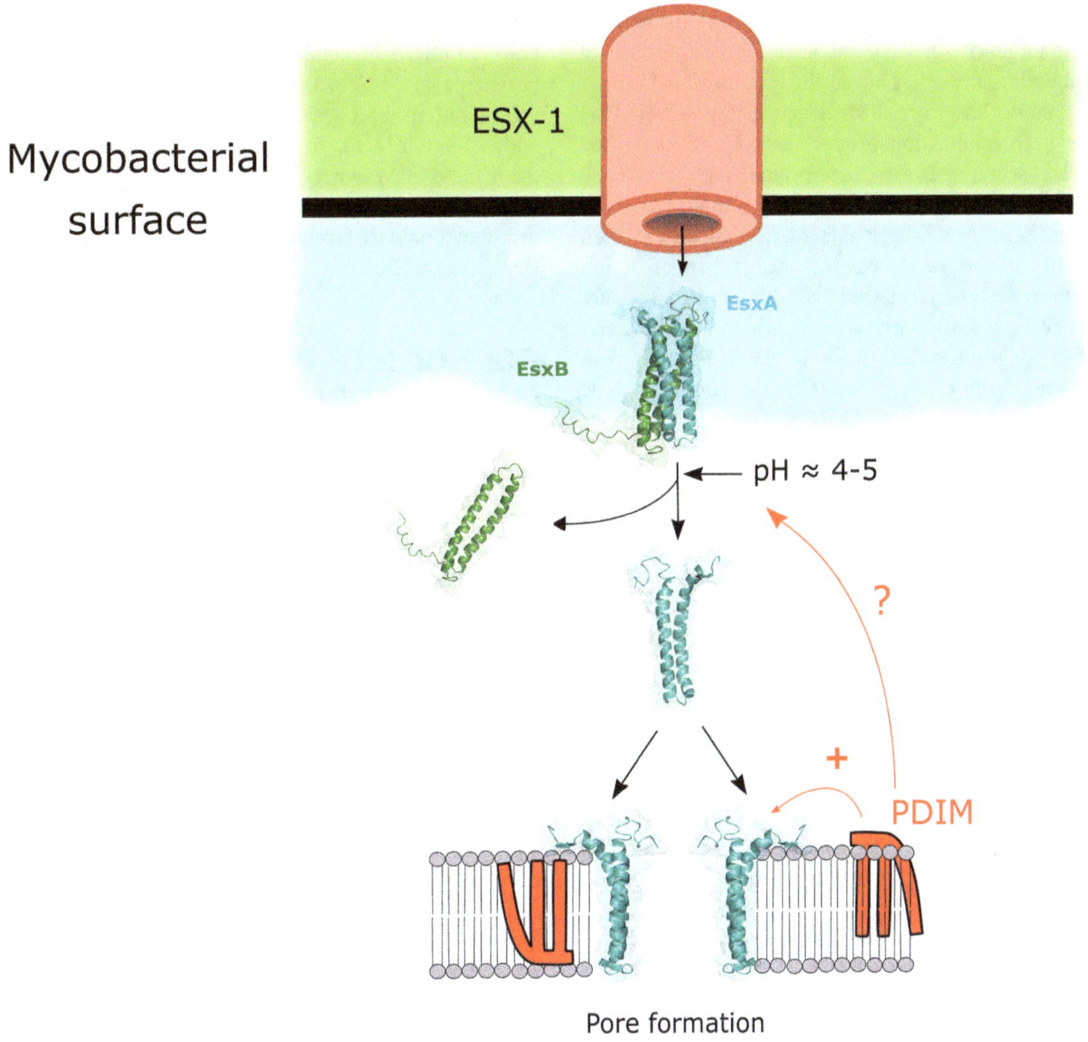

FIGURE 1 | Model of the mechanism of membrane lysis by the cooperation of ESX-1 and PDIM. ESX-1 secretes EsxA and EsxB in a 1:1 heterodimer. This dimer is separated upon pH decrease and / or post-translational modifications, and free EsxA can induce pore formation. PDIM can potentiate EsxA membranolytic activity by either synergizing the pore forming activity of EsxA or by acting on the EsxA/B heterodimer complex separation.

is actually due to the lack of manipulation of the specific host cell signaling pathway and not instead due to some secondary function of the Mtb protein during *in vivo* infections. In the most extreme cases of moonlighting, Mtb effector proteins such as echA2 (Truong and Penn, 2020) this is obvious because of the known function of echA2 in bacterial homeostasis, but it is actually a concern for any Mtb deletion mutant used during *in vivo* infections. One way to address this question is to use a genetic approach by infecting wild type mice but also a knock-out mouse strain that is deficient in the host cell signaling pathway that the Mtb effector is targeting. If the knock-out mouse strain rescues the virulence of the Mtb mutant strain when compared to wild-type Mtb, it is fair to assume that the *in vivo* attenuation (or hypovirulence for that matter) is due to the specific host cell signaling pathway that the Mtb effector protein targets.

HOW DO MTB PROTEINS GAIN ACCESS TO THE HOST CELL CYTOSOL?

In order for secreted Mtb proteins to reach host cell targets they have to overcome at least two barriers: the first one being the double membrane of Mtb (Hoffmann et al., 2008; Zuber et al., 2008) and the second one being the phagosomal membrane (**Figure 1**). The protein secretion systems of Mtb have been extensively reviewed elsewhere (Ligon et al., 2012; Majlessi et al., 2015). ESX-1 is arguably the most extensively studied member of T7SS in Mtb as its crucial importance in pathogenesis was described in numerous studies and reviewed extensively (Gröschel et al., 2016; Bosserman and Champion, 2017; Vaziri and Brosch, 2019). It is likely that the tremendous importance of ESX-1 for the virulence of Mtb is due to the fact that it allows Mtb effectors to gain access to the host cell cytosol by permeabilization of the phagosomal membrane (**Figure 1**). Gaining access to the cytosol via phagosomal membrane permeabilization in order to target host cell targets with bacterial effectors is a successful strategy shared with many other intracellular pathogens (Kumar and Valdivia, 2009). The crucial virulence factor secreted by ESX-1 for the permeabilization process is EsxA (also named ESAT-6), which is secreted as a 1:1 heterodimer with EsxB (also named CFP-10) (Renshaw et al., 2005). EsxA membranolytic activity was first observed by Hsu et al. (2003). Since then numerous studies aimed to dissect the mechanism of action leading to membrane lysis (reviewed elsewhere Peng and Sun, 2016). But recently a study cast a doubt on published EsxA *in vitro* studies suggesting that traces of detergent left from the purification process were responsible for the membranolytic activity observed (Conrad et al., 2017). Nevertheless, some studies were performed with detergent-free purification process and showed an EsxA pore-forming activity on membranes (de Jonge et al., 2007; Ma et al., 2015; Zhang et al., 2016; Ray et al., 2019; Aguilera et al., 2020; Augenstreich et al., 2020). Regardless of this confounding issue, there is good agreement that one crucial step for the lytic activity of EsxA is the separation of EsxA from EsxB after the secretion of the heterodimer through the ESX-1 system. That process can be mediated through an acidification that can lead to EsxA release and membrane binding (de Jonge et al., 2007). In contrast, other reports show that the membrane lysis process seem to happen at a mildly acidic pH and Mtb could rupture the phagosomal membrane even after a bafilomycin treatment which inhibits phagosome acidification (Simeone et al., 2015; Augenstreich et al., 2017). Hemolysis studies also suggested that a RD1-mediated lysis can happen at pH7 (Smith et al., 2008; Conrad et al., 2017; Augenstreich et al., 2020). Thus, the separation of the EsxA/EsxB complex seems to take place at neutral pH. A new study started to unravel this mechanism by showing EsxA undergo acetylation through secretion that increases EsxA/EsxB separation by decreasing the complex stability and contributes to *M. marinum* (Mm) phagosome escape (Aguilera et al., 2020). The lipidic virulence factor PDIM also showed to be essential for phagosomal escape during macrophages infection by Mtb (Augenstreich et al., 2017; Barczak et al., 2017; Quigley et al., 2017; Lerner et al., 2018). It is tempting to hypothesize that PDIM might also play a part in this process, that would explain the membrane lysis observed at neutral pH and that the lysis occurs at the contact point between the bacteria and the target membrane. But this will require additional studies to finally resolve the exact molecular mechanism of EsxA-mediated membrane lysis. Finally, even if EsxA is the main factor responsible for the lysis, other factors may be participating in the process, like the sphingomyelinase SpmT of Mtb (Speer et al., 2015) or uncharacterized ESX-1 associated factor(s) in Mm (Lienard et al., 2020).

It was long thought that the EsxA/ESX1-dependent phagosomal rupture was restricted to Mtb and Mm, since Msme despite encoding an ESX-1 system fails to escape the phagosome in macrophages. But recently it was described that *M. abscessus* was also able to rupture the phagosome, but through the use of ESX-4 (Laencina et al., 2018). Interestingly, *eccB4* deficient *M. abscessus* strains failed to inhibit phagosome acidification and to induce phagosomal rupture. It is linked to a secretion defect of the EsxA/EsxB-like complex EsxT/EsxU and one might speculate they act like their ESX-1 counterparts (Laencina et al., 2018). Thus, for *M. abscessus* that lacks an ESX-1 system, ESX-4 seems to perform the functions associated with ESX-1 in Mtb. In a potentially analogous mechanism to the EsxA/PDIM synergy described for Mtb, a new glycolipid was characterized in *M. abscessus* and its transport to the mycobacterial surface is required for phagosomal rupture (Dubois et al., 2018).

ROLE OF MTB PROTEINS IN HOST MANIPULATION (TABLE 1)

Phagosome Maturation

The uptake of Mtb by phagocytes generates a phagosome, the Mtb-containing vacuole (MCV). The normal maturation process of a phagosome is to fuse with early endosomes, then late endosomes and finally lysosomes to gradually acidify its lumen, acquire acidic protease and hydrolase in order to destroy the phagosomal bacterium (Upadhyay et al., 2018). One of the first immune evasion mechanisms assigned to Mtb was its capacity to prevent the fusion of lysosomes with the MCV (Armstrong and Hart, 1971). These excellent recent reviews provide additional information of Mtb-mediated inhibition of phagosome maturation (Upadhyay et al., 2018; Bussi and Gutierrez, 2019).

EsxH: The ESX-3 secretion system has a limited set of substrates comprised by EsxG-EsxH, PE5-PPE4, and the PE15-PPE20 heterodimers (Tufariello et al., 2016). EsxG and EsxH are EsxA and EsxB-like proteins respectively, which resolved in a 1:1 heterodimer structure is very similar to the EsxA/B complex but with a different function (Ilghari et al., 2011). A yeast two-hybrid (Y2H) screen identified the interaction of Mtb but not Msme EsxH with HRS (Hepatocyte Growth Factor-Regulated Tyrosine Kinase Substrate) (Mehra et al., 2013). HRS is important in the initial assembly of the ESCRT machinery which functions in transport of endosome to lysosomes for certain receptors and cargo (Szymanska et al., 2018). This interaction could be confirmed by co-immunoprecipitation experiments in HEK293 cells expressing EsxH and EsxG (Mehra et al., 2013). Importantly, the overexpression of EsxG/EsxH in Mtb increased the capacity of the bacteria to inhibit phagosome maturation (Mehra et al., 2013). The knock-down of HRS as well as downstream effector Tsg101 (one component of the ESCRT machinery) both resulted in a decreased MCV maturation showing that the host cell target of EsxH is functionally important (Mehra et al., 2013). It was subsequently shown that the inhibition of ESCRT by EsxH also reduced the capacity of macrophages and dendritic cells to present Mtb antigens and prime T-cells *ex vivo* and *in vivo* (Portal Celhay et al., 2016). Finally, ESCRT is recruited to the site of minor membrane damage and since Mtb, via ESX-1 and PDIM, permeabilizes the phagosomal membrane it was consistent with an ESX-1 dependent recruitment of ESCRT to the MCV (Mittal et al., 2018). Mtb is required to permeabilize the phagosomal membrane in order to manipulate the host cell.

TABLE 1 | Summary overview of Mtb protein effectors.

Name	Gene ID	Secretion pathway	Function	Host target	Host cell process	Impact of gene deletion on Mtb virulence
PtpB	Rv0153c	?	Phosphatase	?	?	Attenuated *ex vivo* and *in vivo* (guinea pig)
HBHA	Rv0475	?	?		Apoptosis (A)	No effect *in vivo*
SodA	Rv3846	SecA2	Superoxide dismutase	Phagosomal Superoxides	Apoptosis (I)	Attenuated *in vivo*
Rv3654c	Rv3654c	Predicted TAT	?	PSF	Apoptosis (I)	Attenuated *ex vivo*
Rv3033	Rv3033	Predicted SecA1/2 SP	?	?	Apoptosis (I)	Attenuated *ex vivo* but no data for *in vivo*
GroEL2/HSP65	Rv0440	?	Chaperone	Mortalin	Apoptosis (I)	?
Eis	Rv2416c	?	lysine Nε-acetyltransferase activity	JNK	Apoptosis (I), Xenophagy (I), Cytokine response (I)	Not attenuated *in vivo*
MPT53/DsbE	Rv2878c	Predicted SecA1/2	Disulfide oxidoreductase	Tak1	Cytokine response (A)	Hypervirulent *in vivo*
PPE13		ESX-5	?	NLRP3	Cytokine response (A)	?
EchA1	Rv0222	?	Probable enoyl-CoA hydratase	SHP1, TRAF6	Cytokine response (I)	Attenuated *in vivo*
EsxA	Rv3875	ESX-1	?	TLR-2, SR-B1, B2M	Cytokine response (I), Antigen presentation(I), Invasion(A), Pore formation (A)	Attenuated *ex vivo* and *in vivo*
Hip1	Rv2224c	Predicted SecA1/2	Esterase and Protease activity	GroEL2 (a secreted Mtb protein!)	Cytokine response (I), Apoptosis (I)	Attenuated *ex vivo* and *in vivo*
LpqN	Rv0583c	Predicted SecA1/2	?	CBL	Cytokine response (M)	Attenuated *ex vivo* and *in vivo*
CpnT/TNT	Rv3903c	?	hydrolyses NAD+	NAD+	Necrosis (A)	Not attenuated *in vivo*
PPE2	Rv0256c	ESX-5?	Transcriptional repressor	inos gene promotor, p67phox	NO and ROS production (I)	Attenuated *ex vivo*
SapM	Rv3310	SecA2	Phosphatase	Phosphatidyl-inositol-3-phosphate	Phagosome maturation (I)	Attenuated *ex vivo* and in *vivo* (guinea pig)
PknG	Rv0410c	SecA2	Serine/Threonine kinase	Rab7L1/Rab29	Phagosome maturation (I)	Attenuated *ex vivo* and *in vivo*
CpsA	Rv3484	?	?: contains LCP and LytR domains	?: inhibits NOX2 activation	Phagosome maturation (I)	Attenuated *ex vivo* and *in vivo* (mouse and zebrafish model)
TlyA	Rv1694	?	rRNA methylase, hemolysin	?	Phagosome maturation (I)	Attenuated *ex vivo* and *in vivo*
LpdC	Rv0462	SecA2	Lipoamide dehydrogenase	Coronin-1	Phagosome maturation (I)	Attenuated *in vivo* (probably due to role on metabolism)
EsxH	Rv0288	ESX-3	?	HRS	Phagosome maturation (I)	Attenuated *ex vivo* and *in vivo*
PE_PGRS30	Rv1651c	ESX-5	?	?	Phagosome maturation (I)	Attenuated *ex vivo* and *in vivo*
NdkA	Rv2445c	SecA2	GTPase Activation Protein (GAP)	Rab5, Rab7, Rac1	Phagosome maturation (I), Apoptosis (I)	Attenuated *ex vivo* and *in vivo* (SCID mouse model only)
PtpA	Rv2234	?	Phosphatase	VPS33B, Subunit H of V-ATPase, ubiquitin, GSK3	Phagosome maturation (I), Cytokine Response (I), Apoptosis (I)	Attenuated *ex vivo* not *in vivo*
Rv3364c	Rv3364c	?	?	Cathepsin G	Pyroptosis (I)	?
PE_PGRS47	Rv2741	ESX-5 ?, Predicted SecA1/2	?	?	Xenophagy (I)	Attenuated *ex vivo* and *in vivo*

All listed proteins have been shown to be secreted. If data is available, the secretion mechanisms is indicated. If the secretion pathway is not yet determined, we used SignalP-5.0 to check for signal peptide prediction. The parentheses behind the host cell process indicate Activation (A), Inhibition (I) or Modulation (M). If not stated explicitly the virulence impact refers to mouse studies.

This damage can be recognized by the host cell ESCRT machinery which could result in increased phagosome maturation and antigen presentation. Hence, the bacterial adaptation to inhibit ESCRT-recruitment via secretion of its own effector EsxH. The importance of this system for Mtb virulence was demonstrated by the strong attenuation of an *esxH* deletion mutant in the mouse model with a 3–4 log reduction of CFU in the lungs (Portal Celhay et al., 2016; Tufariello et al., 2016).

PE_PGRS30: The ortholog of this Mtb gene in Mm (*mag* 24-1) is highly upregulated in bacteria present in the granuloma and its deletion results in loss of bacterial virulence *ex vivo* and *in vivo* (Ramakrishnan et al., 2000). Mag 24-1 is important for the capacity of the bacteria to inhibit phagosome maturation and exclusion of the vacuolar H^+-ATPase from the MCV is a likely mechanism (Hagedorn and Soldati, 2007). The investigation of Mtb PE_PGRS30 led to very similar findings in regard to virulence and the importance of this gene for the capacity to inhibit phagosome maturation (Iantomasi et al., 2012).

TlyA: The TlyA protein has been shown to have rRNA methylase activity which functions in the methylation of 16S and 23R rRNA nucleotides (Johansen et al., 2006). Nevertheless, TlyA of Mtb also showed hemolysin activity when expressed in Msme (King et al., 1993; Wren et al., 1998). Consistently, purified TlyA can form oligomers on purified phagosomes and red blood cells leading to lysis (Rahman et al., 2010). TlyA peptides were identified in the culture filtrate of Mtb via mass spec analysis (Kelkar et al., 2011). The secretion of TlyA to the cell membrane does not require Tat or SecA2 secretion systems and TlyA is also included in membrane vesicles secreted by Msme (Kumar et al., 2015). In two gain-of-function approaches by expressing Mtb TlyA in Msme or coating latex beads with purified TlyA it could be demonstrated that TlyA mediates the inhibition of phagosome maturation by reducing Rab5, EEA1, and Rab7 recruitment to the phagosome but increasing Rab14 (Mittal et al., 2014). What the actual target of TlyA in the host cell is or how the phagosomal lytic activity connects to phagosome maturation inhibition remains to be determined. The Mtb TlyA mutant is attenuated in *ex vivo* infected macrophages and *in vivo* in BALB/c mice after aerosol infection but this study did not include a complemented mutant strain (Rahman et al., 2015).

LdpC: Is a lipoamide dehydrogenase which has an important function in the metabolization of branched-chain amino acids (Venugopal et al., 2011). Nevertheless, besides its cytosolic location and function, LdpC is also secreted via the SecA2 pathway (Zulauf et al., 2018). As such it was identified in the host cell cytosol to bind to host cell coronin-1 protein (Deghmane et al., 2007). This binding resulted in the retention of coronin-1 on the MCV membrane which inhibits phagosome maturation (Deghmane et al., 2007). The *ldpC* Mtb mutant is attenuated in the mouse model but this might be due to its function in Mtb metabolism which affects bacterial growth (Venugopal et al., 2011).

PtpA: PtpA is one of the three secreted phosphatases of Mtb (Koul et al., 2000; Saleh and Belisle, 2000). The tyrosine kinase PtkA of Mtb phosphorylates and activates cytosolic PtpA (Zhou et al., 2015; Jaiswal et al., 2019). Secreted PtpA affect three different host cell response pathways: (1) It binds to host cell ubiquitin which activates its phosphatase activity and leads to dephosphorylation of the host cell kinases JNK and p38 which reduces the pro-inflammatory cytokine response of the host cell (Wang et al., 2015a). (2) PtpA binds to and dephosphorylates host cell GSK3α which leads to less active caspase-3 and thus reduced host cell apoptosis (Poirier et al., 2014). (3) Mtb is able to inhibit recruitment of the host cell vacuolar-H^+-ATPase (V-ATPase) to the MCV (Sturgill-Koszycki et al., 1994). PtpA binds subunit H of the V-ATPase (Wong et al., 2011) and it also binds to and dephosphorylates VPS33B (Bach et al., 2008), a protein enabling endosome to lysosome trafficking (Galmes et al., 2015). How these two capacities of PtpA are linked is not known but binding of PtpA to V-ATPase is a prerequisite for the dephosphorylation of VPS33B (Wong et al., 2011). The *PtpA* Mtb mutant is deficient in its capacity to inhibit the acidification of the MCV, inhibit phagosome maturation and growth in *ex vivo* infected macrophages (Bach et al., 2008; Wong et al., 2011). Nevertheless, the absence of PtpA does not alter growth of Mtb in the mouse model (Grundner et al., 2008). PtpA is phosphorylated on two tyrosines by the tyrosine kinase PtkA of Mtb which leads to activation of the phosphatase activity of PtpA (Zhou et al., 2015; Jaiswal et al., 2019). Consistently, the deletion of PtkA in Mtb leads to growth reduction in infected macrophages (Wong et al., 2018).

SapM: This phosphatase was identified via an elegant screen of Mtb culture filtrate fractions for acid phosphatase activity (Saleh and Belisle, 2000). It is secreted via the SecA2 pathway (Zulauf et al., 2018). SapM, unlike PtpA, has phosphatase activity on the lipid phosphatidylinositol-3-phosphate (PI3P) in addition to its tyrosine phosphatase activity (Vergne et al., 2005). The PI3P phosphatase activity of SapM is required to reduce the accumulation of PI3P on the MCV membrane which is key to inhibiting phagosome maturation (Vergne et al., 2005; Koliwer-Brandl et al., 2019). A point mutation abolishes phosphatase activity of SapM and the mutated protein is unable to mediate inhibition of phagosome maturation (Zulauf et al., 2018). A SapM deletion mutant is attenuated for *ex vivo* growth in macrophages (Saikolappan et al., 2012; Puri et al., 2013; Koliwer-Brandl et al., 2019) and *in vivo* growth in guinea pigs (Puri et al., 2013).

PknG: Mtb genome encodes 11 serine/threonine protein kinases of which 9 have a single transmembrane domain, an extracellular sensor domain and the intracellular kinase domain (Av-Gay and Everett, 2000; Prisic and Husson, 2014). PknG and PknK have no transmembrane domain but only PknG has been shown to be secreted (Prisic and Husson, 2014). The cytosolic location of PknG reflects the location of its many intracellular substrates (Baros et al., 2020) but PknG is also secreted via the SecA2 pathway (van der Woude et al., 2014; Zulauf et al., 2018). After infection of macrophages PknG can be found in the host cell cytosol where it mediates the inhibition of phagosome maturation and survival of bacteria (Walburger et al., 2004; van der Woude et al., 2014; Zulauf et al., 2018). The human RabGTPase protein Rab7L1 (Rab29 in mouse) is the host cell target of PknG (Pradhan et al., 2018). At the Golgi apparatus to PknG blocks the transition of inactive GDP-associated Rab7L1 to the active GTP-associated Rab7L1. This leads to the lack of recruitment of Rab7L1-GTP to the MCV which helps to inhibit phagosome maturation (Pradhan et al., 2018). The *pknG* Mtb mutant is attenuated in immunodeficient SCID mice and immunocompetent BALB/c mice after intra venous injections but not in the lungs of CD-1 mice after aerosol delivery (Cowley et al., 2004).

NdkA: The nucleoside diphosphate kinase A is secreted (Chopra et al., 2003; Målen et al., 2007) via the SecA2 pathway (Zulauf et al., 2018) and it seems to be an essential gene since it was not possible to generate deletion mutants neither in Mtb nor BCG (Sun et al., 2010, 2013). NdkA has GTPase activation protein activity which means that it accelerates the transition of small GTPase proteins from the GTP-bound (active) state to their GDP-bound (inactive) state (Chopra et al., 2004; Sun et al., 2010, 2013). Rab5 and Rab7 are small GTPase involved in vesicle trafficking and therefore also in the maturation process from early phagosome to late phagosome (Rab5) and late phagosome to phagolysosome (Rab7). Interestingly, NdkA isolated from Mtb but not NdkA from Msme can bind to Rab5 and Rab7 and dephosphorylate associated GTP. The *NdkA* knock-down strain of BCG is unable to inhibit the maturation process of the MCV (Sun et al., 2010). Another target of NdkA is the small GTPase Racl (Chopra et al., 2004; Sun et al., 2013). The inactivation of Racl results in a deficiency of the macrophages to assemble the functional NADPH oxidase (NOX2) complex on the MCV (Sun et al., 2013). The *NdkA* knock-down Mtb strain is inducing higher levels host cell ROS and host cell apoptosis. The strain is also attenuated for growth in *ex vivo* infected macrophages and in immunodeficient SCID mice (Sun et al., 2013).

Cell Death

There are many different ways that a cell may die but for the purpose of this review we want to focus on apoptosis, necroptosis, and pyroptosis (Galluzzi et al., 2018). In the case of Mtb the current working model is that host cell apoptosis is detrimental for the virulence of Mtb, whereas host cell necrosis favors the pathogen (Behar et al., 2010; Srinivasan et al., 2014). It is thus not surprising that Mtb inhibits extrinsic and intrinsic apoptosis signaling pathways and has developed effectors to induces host cell necrosis (Moraco and Kornfeld, 2014; Srinivasan et al., 2014; Mohareer et al., 2018). The overall role of pyroptosis during the course of Mtb infections is not yet determined.

Pyroptosis

Rv3364c: The *Rv3361c-Rv3365c* operon is upregulated in Mtb after infection of macrophages and then the secreted Rv3364c can enter the host cell cytosol. Rv3364c binds to and inhibits expression and activity of Cathepsin G which leads to a reduced activation of the inflammatory caspase-1 and consequently less pyroptosis (Danelishvili et al., 2011).

Apoptosis

Rv3654c: This gene is expressed in an operon (*Rv3654c-Rv3660c*) which is upregulated after Mtb infection of macrophages. Rv3654c can reach the host cell cytosol where it binds to PSF (protein-associated splicing factor) (Danelishvili et al., 2010). The binding of Rv3654c to PSF leads to cleavage of the protein and reduced expression which results in reduced expression of caspase-8, a protease essential for signaling in the extrinsic apoptosis pathway (Danelishvili et al., 2010). Consequently, the *Rv3654c* Mtb mutant induced higher levels of TNF-mediated apoptosis and showed less survival during *ex vivo* infection of macrophages (Danelishvili et al., 2010).

Rv3033: A genome-wide screen for Mtb transposon mutants that have reduced survival in macrophages identified Rv3033 as important for survival in resting macrophages and macrophages treated with IFN-γ after infection (Rengarajan et al., 2005). Overexpression of Rv3033 in Msme conferred a reduction in host cell apoptosis (Zhang et al., 2018). The anti-apoptotic capacity of Rv3033 could be confirmed by expressing Rv3033 in a macrophage cell line and challenging with Mtb H37Ra infection (Zhang et al., 2018). Mtb Rv3033 targets the intrinsic, caspase-9 dependent, apoptosis pathway (Zhang et al., 2018). The deletion of Rv3033 reduces viability in *ex vivo* infected macrophages but no *in vivo* data is available (Rengarajan et al., 2005).

SodA: Superoxide Dismutase A is secreted via the SecA2-system and might be involved in the neutralization of superoxides generated in the MCV by the NOX2 phagocyte oxidase (Braunstein et al., 2003). The *SodA* gene is essential but an antisense inhibition of *SodA* expression in Mtb led to an important attenuation of the strain *in vivo*, marked by a high induction of apoptosis in the lungs of mice (Edwards et al., 2001). Later it was demonstrated that the absence of secreted SodA and the increase in host cell apoptosis *in vivo* leads to an increase in the presentation of Mtb-derived antigens (Hinchey et al., 2007).

Eis: The Enhanced Intracellular Survival protein was first identified in a gain-of-function screen of Mtb genes expressed in Msme that would increase survival of the bacteria after infection of macrophages (Wei et al., 2000). Eis is secreted by Mtb but it is unclear via which secretion pathway (Dahl et al., 2001) and it is able to reach the host cell cytosol (Samuel et al., 2007). The deletion of *eis* in Mtb causes an increase in host cell JNK kinase activation which leads to increased NOX2-mediated ROS generation, causing increased pro-inflammatory cytokine secretion, autophagy and host cell death (Samuel et al., 2007; Shin et al., 2010) The Mtb and Msme Eis proteins both have aminoglycoside N-acetyltransferase activity which confer resistance to antibiotics but on Mtb Eis also has lysine N^{ε}-acetyltransferase activity (Kim et al., 2012). The latter activity of Mtb Eis targets acetylation of the host cell phosphatase DUSP16/MKP-7 which potentially increases its binding to JNK and inhibits its activation (Kim et al., 2012). The deletion of *eis* in Mtb does not reduce its virulence in mice (Samuel et al., 2007; Shin et al., 2010).

Hipl and GroEL2: he Hipl protein is likely cell membrane associated since it has a predicted lipoprotein signal peptide but one transmembrane domain predicted (Krogh et al., 2001; Almagro Armenteros et al., 2019) but was found in membrane and culture filtrate fractions (Målen et al., 2010; de Souza et al., 2011). Hipl has esterase and protease activity (Naffin-Olivos et al., 2014) and GroEL2 is an Mtb chaperon that interacts among others with Mtb DnaK protein. GroEL2 is found in the cytosol and cell wall of Mtb and is one substrate for proteolytic cleavage by Hipl which results in the release of a shorter protein into the supernatant (Rengarajan et al., 2008; Naffin-Olivos et al., 2014). New data shows that after Mtb infection GroEL2 gets released from the cell wall and can actually enter the host cell cytosol where ultimately it binds to the mitochondrial protein mortalin which has homology to Mtb DnaK protein (Joseph et al., 2017). Surprisingly, GroEL2 can access the host cell cytosol even in cells infected with the *M. bovis* BCG vaccine strain which lacks a functional ESX-1 system (Joseph et al., 2017). The interaction of GroEL2 with host cell mortalin mediates the inhibition of host cell apoptosis. There is no data available on the impact of GroEL2 deficiency on Mtb virulence, which is difficult to assess because GroEL2 is essential for *in vitro* growth of Mtb (Dejesus et al., 2017).

HBHA: eparin-binding hemagglutinin (HBHA) of Mtb is found in the cell wall and culture filtrate and is important of bacterial adhesion to epithelial cells but not macrophages (Menozzi et al., 1996). The deletion of *hbha* in Mtb results in a mutant strain that has similar virulence in the lungs of mice after intranasal infection but strongly reduced the ability for extrapulmonary dissemination as measured via CFU in the spleen (Pethe et al., 2001). Either heterologous expression of HBHA in Msme or the *hbha* Mtb deletion mutant demonstrated that this protein may lead to the increase in host cell apoptosis (Sohn et al., 2011). HBHA localizes to host cell mitochondria where it leads to increased activation of the pro-apoptosis protein BAX and increased levels of mitochondrial reactive superoxide generation (Sohn et al., 2011). The impact of increased host cell apoptosis during the context of *in vivo* infection in the lung is not significant (Pethe et al., 2001).

Necrosis

Mtb needs to escape its intracellular niche at some point in order to disseminate and infect other cells. Host cell necrosis favors pathogenesis of Mtb as has been shown by modulating host factors that tip the cell death modality toward necrosis instead of apoptosis (Behar et al., 2010). Host cell necrosis can actually stimulate the growth of Mtb (Dallenga et al., 2017; Lerner et al., 2017). In addition, the deletion of a transcriptional repressor (*Rv3167c*) resulted in an Mtb mutant strain that induced higher levels of host cell necrosis due to increased PDIM expression and was also hypervirulent in mice (Srinivasan et al., 2016). Despite its importance for virulence very little is known about Mtb effectors that induce host cell necrosis.

CpnT/TNT: The N-terminal domain of CpnT has pore forming capacity which is involved in uptake of small molecules through the mycomembrane (Danilchanka et al., 2014). The C-terminal domain (Tuberculosis necrotizing toxin, TNT) can be release after proteolytic cleavage and will target host cell coenzyme NAD^+ for hydrolysis (Sun et al., 2015). The host cell depletion of NAD^+ leads to necroptosis via the RIPK3/MLKL pathway but without activation of upstream signaling components such as RIPK1 (Pajuelo et al., 2018). The *cpnT* Mtb deletion mutant is not attenuated in mice (Danilchanka et al., 2014), suggesting that Mtb has redundant pathways for inducing host necrosis and most likely additional secreted effectors.

Xenophagy

Xenophagy is a specialized form of canonical autophagy which results in the encapsulation pathogens by a double membrane autophagosome (Upadhyay and Philips, 2019). Xenophagy is a cell intrinsic defense mechanism against Mtb infection (Gutierrez et al., 2004). The ubiquitination of Mtb is dependent on the ESX-1 system and extracellular Mtb DNA (Watson et al., 2012). Ubiquitinated Mtb gets recognized by cytosolic autophagy receptors p62 and NDP52 which initiates autophagosome formation (Watson et al., 2012). The ubiquitin ligase Parkin 2 and Smurfl are of critical importance for the ubiquitination of Mtb and host resistance to Mtb in the mouse (Manzanillo et al., 2013; Franco et al., 2016). Interestingly, Mtb expresses surface protein (Rv1468c) containing a eukaryotic-like ubiquitin-associated domain that binds ubiquitin and recruits p62 facilitating the xenophagic clearance of Mtb (Chai et al., 2019). Importantly, Mtb has yet to be defined mechanisms to inhibit host cell clearance via xenophagy as described for an ESX-1 dependent inhibition of autophagic flux (Romagnoli et al., 2012; Chandra et al., 2015; Cardenal-Muñoz et al., 2017). Please refer to following review for more background information (Khaminets et al., 2016; Upadhyay and Philips, 2019).

PE_PGRS47

This protein was identified in a gain-of-function screen using Msme for Mtb genes that mediated the inhibition of antigen presentation (Saini et al., 2016). Expression of PE_PGRS47 in Msme demonstrated that transport of the protein to the cell wall fraction (Saini et al., 2016) although EM studies in infected cells showed a location of PE_PGRS47 in the host cell cytosol (Saini et al., 2016). Msme does not express an ESX-5 secretion system and hence if it secretes PE_PGRS47 it has to be via a different secretion system which is somewhat surprising since ESX-5 is the major system for secretion of PE_PGRS proteins (Abdallah et al., 2009). The ability of PE_PGRS47 to limit antigen presentation is most likely indirect via its capacity to inhibit xenophagy and associated phagosome-lysosome fusion which results in the generation of Mtb peptides that can be presented at the cell surface (Saini et al., 2016). The PE_PGRS47 Mtb mutant is attenuated in immunodeficient and immunocompetent mice (Saini et al., 2016). The molecular mechanism and host target of PE_PGRS47 remain to be characterized.

Cytokine Response

Host cell cytokines have important functions in host defense as demonstrated by the increased susceptibility of $tnf^{-/-}$ and $ifn\text{-}\gamma^{-/-}$ mice to Mtb infections (Flynn et al., 1993, 1995). In contrast, IFN-β, a cytokine associated with anti-viral immunity, is exacerbating Mtb infections in mice and humans (Antonelli et al., 2010; Berry et al., 2010). Extracellular pattern recognition receptors such as TLRs and intracellular PRR such as NLRs are the sensors for pathogen associated molecular patterns (PAMPs) and after binding of a PAMP initiate a signaling cascade that leads to the production of cytokines. The importance of cytokines for host immunity to Mtb is reviewed in more detail in these excellent reviews (Mayer-Barber and Sher, 2015; Domingo-Gonzalez et al., 2016; Sia and Rengarajan, 2019).

Hip1and GroEL2

The interaction of Hip1 and GroEL2 have already been described in a previous section. The *hip1* Mtb deletion mutant induces increased pro-inflammatory cytokine production (TNF, IL-1β, IL-18, IL-6) in macrophages and dendritic cells via a TLR2/MyD88 signaling pathway (Madan-Lala et al., 2011, 2014). Importantly, just the overexpression of the cleaved and secreted GroEL2 fragment in the *hip1* Mtb mutant reverts the phenotype of the mutant back to wild-type Mtb, suggesting that the GroEL2 fragment is blocking the host cell TLR2/MyD88 signaling pathway (Naffin-Olivos et al., 2014). The *hip1* Mtb mutant is attenuated in *ex vivo* infected macrophages and *in vivo* infected mice (Rengarajan et al., 2008; Vandal et al., 2009) it is unclear to date though, if the attenuation *in vivo* is only due to the observed effect on GroEL2 or also due to the general susceptibility of the *hip1* mutant to low pH and ROS (Vandal et al., 2009).

PPE13

This member of the PE/PPE protein family contains a NxGxNxG motif which is characteristic for the major polymorphic tandem repeat (MPTR) subfamily of PPE proteins (Hermans et al., 1992). PPE13 does not contain a signal peptide and is secreted via the ESX-5 secretion system (Abdallah et al., 2009). Heterologous expression of PPE13 in Msme or ectopic expression in eukaryotic cells demonstrate that PPE13 activates the NLRP3 inflammasome leading to increased IL-1β secretion (Yang et al., 2020). Furthermore, PPE13 binds to NLRP3's NACHT and LRR domains via its MPTR domain (Yang et al., 2020). The PPE13-NLRP3 interaction facilitates homodimerization of NLRP3 and recruitment of the NLRP3 activator protein NEK7 (Yang et al., 2020). There is no data available on the impact of Mtb PPE13 on bacterial virulence.

LpqN

The lipoprotein has a signal peptide suggesting secretion via the SecA1/2 pathway and was found in the culture filtrate of Mtb (Målen et al., 2007). It was identified to bind CBL during a screening of 105 secreted Mtb proteins for host cell binding partners (Penn et al., 2018). CBL is a ubiquitin ligase that is increasingly phosphorylated after Mtb infections (Penn et al., 2018). The Mtb *lpqN* deletion mutant is growing less efficiently in *ex vivo* infected macrophages and *in vivo* (Penn et al., 2018). Importantly, the growth deficiency of the mutant in macrophages could be rescued by the deletion of host cell *Cbl* gene (Penn et al., 2018). The study shows data in support of CBL being a regulatorwhich suppresses anti-viral but supports anti-bacterial host cell intrinsic defense pathways; for example, $cbl^{-/-}$ derived BMDMs are intrinsically more resistant to viral infection when compared to wild-type BMDMs. It is proposed that Mtb, by inhibiting CBL, induces an anti-viral host response which favors its own survival because anti-bacterial defense mechanisms are not induced (Penn et al., 2018).

EchA1

The enoyl-CoA hydratase A1 is involved in the lipid metabolism of Mtb but is also secreted via an unknown mechanism (no predicted signal peptide) and reaches the host cell cytosol (Wang et al., 2020). After ubiquitination of EchA1 by host cell ubiquitin ligase ANAPC2, EchA1 binds TRAF6 and SHP1 which prevents activation of TRAF6 and thus reduces the production of pro-inflammatory cytokines (Wang et al., 2020). The echA1 deletion mutant of Mtb is attenuated in the mouse model after aerosol infection (Wang et al., 2020).

PtpB

This is a broad-spectrum phosphatase that dephosphorylates phosphotyrosine, -serine and -threonine substrates in addition to various phosphoinositides (Beresford et al., 2007). Ectopic expression of PtpB in RAW264.7 murine macrophages conveyed inhibition of IFN-γ-mediated activation of the ERK1/2 and p38 signaling pathway toward increased IL-6 production and inhibition host cell apoptosis (Zhou et al., 2010). Nevertheless, these findings need to be confirmed via infection of cells with Mtb and a specific *PtpB* Mtb mutant. The deletion of *PtpB* resulted in a mutant that was less virulent in *ex vivo* macrophages infection models (Singh et al., 2003; Beresford et al., 2009; Koliwer-Brandl et al., 2019). Furthermore, the Mtb deletion mutant had an approximative 100 fold reduction in lung CFUs in the guinea pig model when compared to Mtb (Singh et al., 2003).

MPT53/DsbE

The protein is found in the culture filtrate (Målen et al., 2007) and has a predicted signal peptide (Almagro Armenteros et al., 2019) and is a disulfide bond-forming (Dsb)-like protein. In a screen of 208 secreted Mtb effectors expressed in HEK293T cells that changed NF-κB activation, DsbE was found to activate the NF-κB reporter gene (Wang et al., 2019). DsbE was found to bind to TGF-β-activated kinase 1 (TAK1) which is an important signaling molecule downstream of the TLR/TRAF6/TAB2 or TAB3 signaling pathway (Wang et al., 2019). TAK1 may activate NF-κB and the kinases JNKs and p38 which leads to the biosynthesis of pro-inflammatory cytokines (TNF, IL-6, IL-12). The binding of DsbE to TAK1 increased its phosphorylation which is required for activation (Wang et al., 2019). The enzymatic activity of DsbE is required for binding since a disulfide-oxidoreductase inactive mutant of DsbE fails to activate TAK1 (Wang et al., 2019). Consistent with this data the *dsbE* Mtb deletion mutant induced less TNF and IL-6 production in *ex vivo* infected macrophages and in the lungs of aerosol infected mice. The mutant was also hypervirulent in the mice with a 10 to 100-fold increase in lung CFUs after 21 d of infection (Wang et al., 2019). The fact that the secreted DsbE actually activates protective host responses suggest that its recognition by TAK1 is actually a host defense mechanism and the data that other Mtb and *E. coli* proteins with disulfide-oxidoreductase activity may also activate TAK1 supports this model (Wang et al., 2019). Thus, the sensing of bacterial disulfide-oxidoreductase activity in the host cell cytosol maybe a case of effector-triggered immunity (Lopes Fischer et al., 2020).

NOX2/NOS2

The production of phagosomal ROS by the activated NOX2 and cellular NO by NOS2 are associated with cell intrinsic host defense (Bedard and Krause, 2007; Bogdan, 2015). Mtb is relative resistant to direct killing by ROS but the increase in phagosomal ROS observed after infection with *nuoG* and *secA2* Mtb mutants leads to an increase in host cell apoptosis which attenuates these mutants and leads to increased host cell antigen presentation (Hinchey et al., 2007; Velmurugan et al., 2007; Miller et al., 2010). The proteasome of Mtb is important for resistance of Mtb to killing via NO-derived reactive nitrogen intermediates (Darwin et al., 2003). Nevertheless, a more complex role for NOX2 and NOS2 during *in vivo* infections has emerged which associates them with a role in host immune tolerance (Olive and Sassetti, 2018).

CpsA

The protein was found in the culture filtrate (Målen et al., 2007) but has no predicted signal peptide sequence (Almagro Armenteros et al., 2019). A Y2H screen showed that CpsA binds to TAX1BP1 and NDP53 which are two proteins involved in xenophagy (Mehra et al., 2013), and SMCO1 (Penn et al., 2018). The use of knock-out host cells deficient in xenophagy, LC3-associated phagocytosis (LAP; see this reference for review Upadhyay and Philips, 2019) or both demonstrated that Mtb CpsA is involved in the inhibition of LAP (Koster et al., 2017). The precise host cell target of CpsA has not been determined but the exclusion of activated NOX2 from the nascent MCV is clearly an important aspect of the molecular mechanism of CpsA-mediated host cell manipulation (Koster et al., 2017). Overall, the end result is that a *CpsA* deletion mutant ends up in an MCV that fuses with lysosomes which results in decreased intracellular survival (Koster et al., 2017). The Mtb *CpsA* mutant is also attenuated in the mouse model (Koster et al., 2017; Malm et al., 2018) and deletion of the Mm homolog attenuated this pathogen in the zebrafish model (Wang et al., 2015b). The *CpsA* mutant induces increased ROS due to the activated NOX2 and it is thus likely that host cell apoptosis levels are also increased as this was shown before for several Mtb mutants that results in increased phagosomal ROS (Hinchey et al., 2007; Miller et al., 2010; Sun et al., 2013).

PPE2

This protein is secreted (Bhat et al., 2013) and since it is a member of the PE/PPE family it most likely is a substrate of the ESX-5 secretion system. PPE2 is targeted to the host cell nucleus via a nuclear location signal and binds to the promotor of the *Nos2* gene (Bhat et al., 2017). Consequently, infection of macrophages with the Mtb *Ppe2* deletion mutant and Msme overexpressing Mtb PPE2 result in increased or decreased NOS2 protein expression (Bhat et al., 2013, 2017). This activity resulted in increased survival of the Msme-PPE2 strain compared to Msme in *ex vivo* infected macrophages or *in vivo* infected mice (Bhat et al., 2017). PPE2 contains an SH3-like domain that allows for binding of the host cell p67phox NOX2 subunit. The binding inhibits p67phox recruitment to the phagosomal membrane and subsequent NOX2 assembly and activation (Srivastava et al., 2019). The overall result is that less ROS will be produced in the MCV which helps survival of bacteria during *ex vivo* macrophage infections (Srivastava et al., 2019).

Multiple Host Cell Targets of Mtb EsxA

EsxA

Due to the already discussed importance of EsxA on phagosomal membrane permeabilization it is challenging to experimentally dissociate phenotypes of the Mtb *esxA* mutant that are mediated by a direct effector activity of EsxA and those due to a lack of phagosomal membrane permeabilization and thus lack of access of other effectors to their host cell targets. Consequently, we focused this discussion on pathways in which a direct binding of EsxA to an effector protein could be shown (**Figure 2**), while acknowledging that other observed phenotypes of the Mtb *esxA* mutant might still be due to direct activity of EsxA. At the macrophages surface, EsxA inhibits TLR2 signaling by antagonistic binding to the receptor (Pathak et al., 2007). The inhibition of the downstream NF-κB pathway was also observed after incubation of cells with EsxA protein alone or in complex with EsxB (Ganguly et al., 2008). Both of these inhibitions lead to a decreased cytokine response by the infected host cell (**Figure 2**). EsxA can also interfere with antigen presentation to cytolytic T-cells by binding to the β2-macroglobuline which is associated with MHC class I. The binding of EsxA decrease the capacity of the MHC class I to present peptides due to decreased cell surface expression (Sreejit et al., 2014). A recent study found that EsxA can physically bind to the scavenger receptor B1 (SR-B1) and allow Mtb to cross the pulmonary epithelium through M cells (Khan et al., 2020). An Mtb-human protein-protein interactome screen identified several additional potential host cell binding proteins for EsxA but they require further validation (Penn et al., 2018). There have been many studies showing an impact of deletion of Mtb *esxA* on the host cell death response (apoptosis, necrosis, and pyroptosis) but in these studies it is difficult to discriminate between a direct or an indirect effect of EsxA.

MTB LIPIDS AS EFFECTOR MOLECULES

Mtb produces a wide variety of unique lipids which are important for host cell manipulation and virulence of Mtb (Neyrolles and Guilhot, 2011; Arbues et al., 2014; Gago et al., 2017; Queiroz and Riley, 2017). These lipids are localized in the mycobacterial envelope and have a very unique structure and role for pathogenesis (reviewed in Vincent et al., 2018; Dulberger et al., 2020, **Figure 3**). Briefly, the envelope consists of: (1) a plasma membrane which is mainly composed of phospholipids, (2) a superstructure made up of a layer of peptidoglycan covalently linked to arabinogalactan, and (3) a mycomembrane which as its inner leaflet has mycolic acids that are esterified with the underlying arabinogalactan (**Figure 3**). The outer leaflet of the mycomembrane is made up of a wide variety of lipids and almost all of them are involved in the host immune response manipulation by Mtb (Vincent et al., 2018; Daffe and Marrakchi, 2019; Dulberger et al., 2020) (**Figure 3**).

HOW DO MTB LIPIDS GET INTO THE HOST CELL?

Mtb lipids are able to reach other organelles and the plasma membrane during infection providing evidence that some of

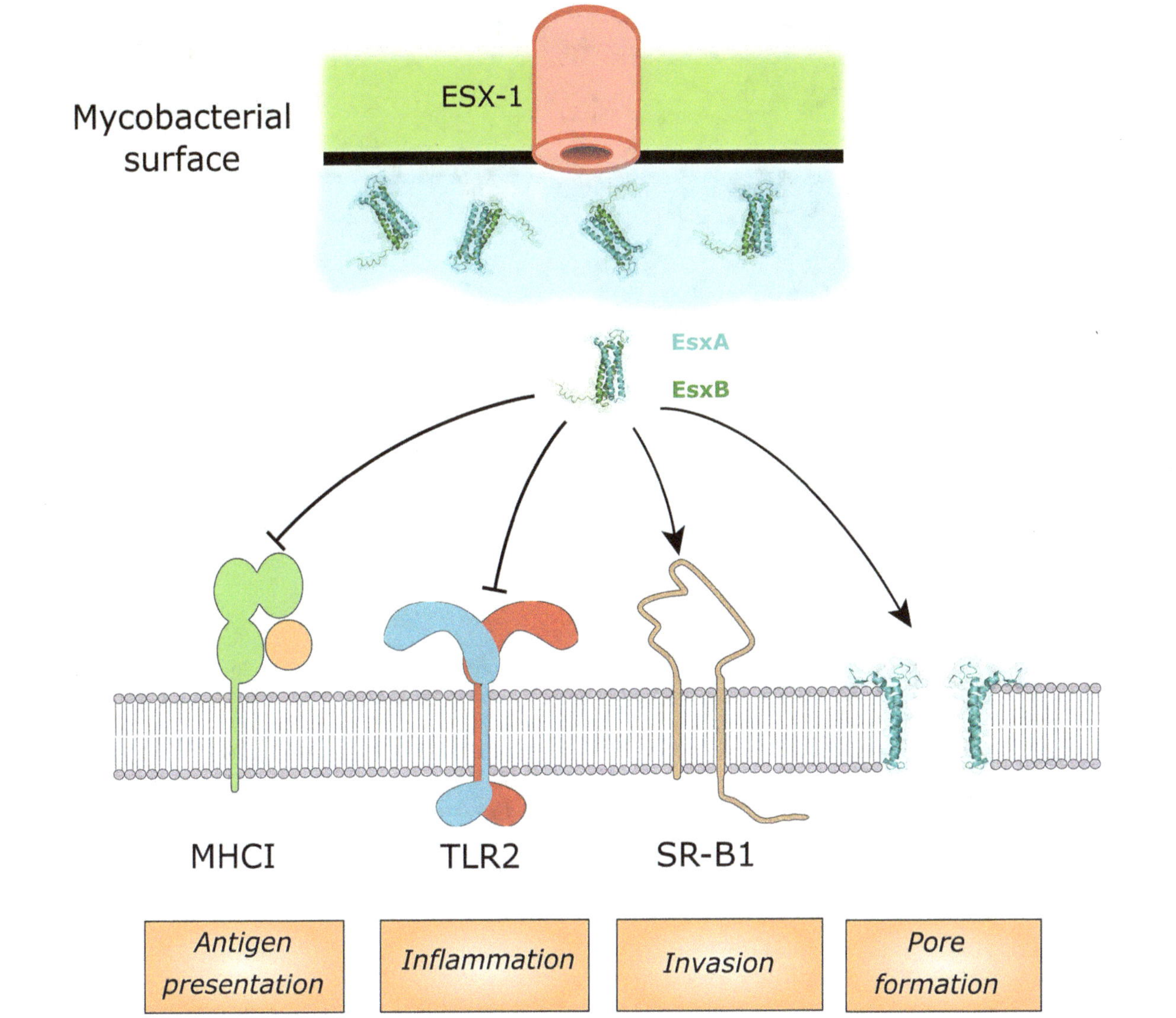

FIGURE 2 | Host cell targets of EsxA and/or the EsxA/B complex. EsxA can antagonistically bind MHCI and TLR2, while it can bind to SR-B1 to enhance invading the lung epithelium. EsxA targets the phagosomal membrane for pore formation.

the Mtb surface lipids could be released (Beatty et al., 2000; Rhoades et al., 2003). Mtb lipids are also found incorporated into exosomes and can thus reach uninfected bystander cells (Beatty et al., 2000; Athman et al., 2015). Thus, studying the way lipids are released by Mtb and their impact on the host is crucial to understand the pathogeny of Mtb. The section below summarizes the advances in the study of release of lipids by Mtb (**Figure 3**).

Passive Release of Mtb Lipids

The first potential release mechanism is through direct contact of the bacteria with the host membranes, or by mild shedding of the capsular layer to which many lipids are loosely associated (Ortalo-Magné et al., 1996; Chiaradia et al., 2017). Indeed, in some phagosomes, mycobacteria are tightly surrounded by the phagosomal membrane and this feature was associated with the ability of *M. avium* to inhibit phagosome lysosome fusion (de Chastellier et al., 2009). Recent work showed an IFN-γ and Rab20 dependent increase of MCV volume which correlated with increased phagosome maturation (Schnettger et al., 2017). Moreover gold immunolabeling of Lipoarabinomannan (LAM) and phosphatidylinositol mannosides (PIM) suggested a transfer of these lipids from Mtb to the macrophages membranes through contact and/or release from the bacterial surface as the cryo sections did not show any vesicles emission (Beatty et al., 2000). Supporting these results, another structural study of Mtb found that the capsular layer can be removed through agitation of Mtb (Sani et al., 2010). A passive secretion followed by an insertion and diffusion supports their major impact on the immune response by also affecting bystander cells. PDIM, another major virulence lipid of Mtb, was found in

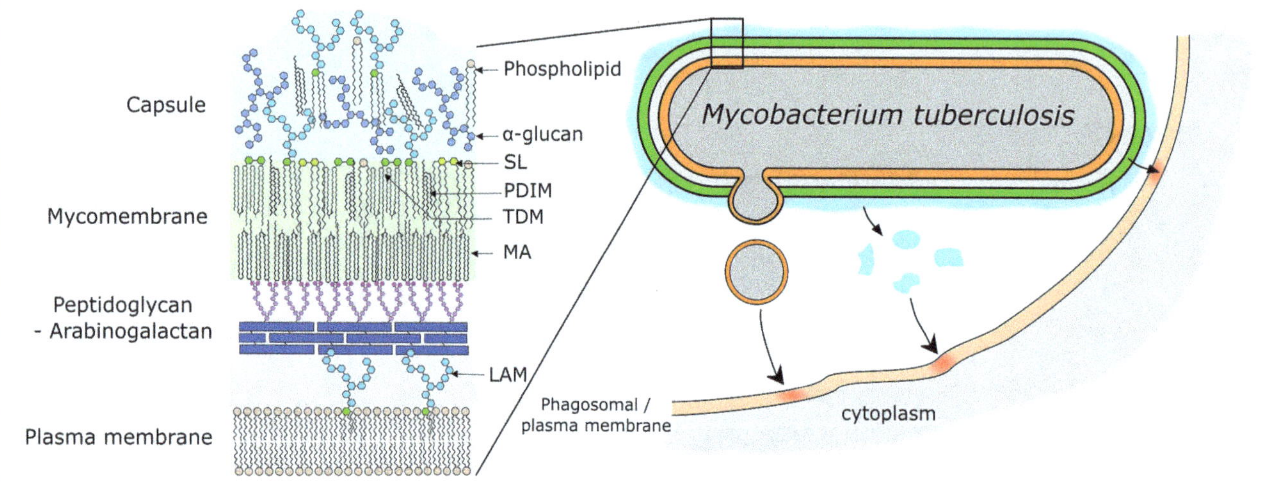

FIGURE 3 | Models of mycobacterial envelope and lipid transfer. **(Left)** Simplified model of the organization of the mycobacterial envelope with the main virulence lipids highlighted. **(Right)** Representation of the different potential mechanisms of lipid release and transfer to the host cell membrane. Lipids can be release by emission of membrane vesicles, or by shedding of the capsular layer into the phagosome lumen or due to close physical contact directly into the phagosome membrane. SL, Sulfolipids; PDIM, Phthiocerol Dimycerosates; TDM, Trehalose Dimycolates; MA, Mycolic Acid; LAM, Lipoarabinomannan.

macrophages membranes during infection (Augenstreich et al., 2019). Preliminary results on Mm coating with fluorescent modified PDIM indicated a transfer and a diffusion of these lipids upon contact with macrophages (Cambier et al., 2020). These results also support a passive transfer of PDIM upon contact of the bacteria with host cells. Such transfer by contact has also been observed for another bacterial species, *Borrelia burgdorferi*, where bacterial cholesterol-glycolipids were observed being transfer from the bacterial surface to the host cells plasma membrane (Crowley et al., 2013). Interestingly, the contact-mediated transfer accounted for 50% of lipid transfer, while the other part happened through membrane vesicles (MV) secretion by the bacteria (Crowley et al., 2013). Therefore, it is conceivable that for Mtb both contact dependent and independent transfers of lipids occur during infection (**Figure 3**).

Active Release of Mtb Lipids

The second mechanism of lipid release that has been recently described is through the release of MV (Prados Rosales et al., 2011) (for extensive review on this topic Brown et al., 2015; Layre, 2020) (**Figure 3**). These vesicles of $\approx$ 150 nm of diameter are apparently emitted by budding from the bacterial surface mostly observed in culture (Prados Rosales et al., 2011, 2014b; Athman et al., 2015), but production is also observed by bacteria localized in the host cell phagosome (Prados Rosales et al., 2011). Proteomics analysis indicated that some characterized antigenic factors were associated to the MVs also known for their inflammatory properties (Prados Rosales et al., 2011) and subsequent immunization of animals with Mtb MVs induced a protective immune response (Prados Rosales et al., 2014a). The lipid composition of the MVs is still only partially characterized, but phosphatidylinositol mannosides (PIM), lipoarabinomannan (LAM), poly-acyltrehaloses (PAT), and phenol glycolipids (PGL) (Prados Rosales et al., 2011). In addition, MVs include hundreds of Mtb proteins as determined by the host antibody responses to injected MVs (Prados Rosales et al., 2014a) and proteomics analyses (Prados Rosales et al., 2011). The PIM and LAM in MVs were associated to the lipid induced dampening of the acquired immune response by inhibiting T-cell activation (Athman et al., 2017). In contrast, some of these lipids were originally described as pro-inflammatory by activating TLR2 signaling in the case of lipomannan (Quesniaux et al., 2004). PGL are also present in MVs and they are also known as antagonist inhibition of TLR2 signaling (Arbues et al., 2016). The abundance of virulence lipids mainly present at the surface of the bacteria might indicate the other lipids like PDIM and trehalose dimycolate (TDM) that can diffuse out from the bacteria, may be present in the MVs but further investigations will be needed to attest of that. Also, the suggested lipid transfer from MV emission by Mtb to host cell membranes (Athman et al., 2015), still need to be formerly determined. The Mtb-produced MVs could potentially mediate immune modulatory effects beyond the site of infection and this might be of particular importance during the extracellular replication of Mtb within necrotic regions of human granulomas.

Role of Mtb Lipids in Host Manipulation (Table 2)

The main effect described so far of the lipid panoply produced by Mtb is as inflammatory regulators (**Table 2**). An interesting "lipidic immunostat" model was proposed in which lipids are separated into pro- and anti-inflammatory (Queiroz and Riley, 2017). Indeed, the majority of the lipids produced are glycolipids whose sugar moieties are recognized by a wide range of pattern recognition receptors (PRR) (Ishikawa et al., 2017; Queiroz and Riley, 2017).

TABLE 2 | Summary overview of Mtb lipid effectors.

Name	Known release mechanism	Host target	Host cell process	Impact of gene deletion on Mtb virulence
PIM, LM, LAM, ManLAM	Extracellular vesicles, shedding	TLR4, TLR2-TLR1-6, MR, DC-SIGN, Dectin-2	Cytokine response (A/I), Phagocytosis (A), phagosome maturation (I), cell death (A)	Essential, variability in acylations and LAM capping sugar moiety linked to the degree of virulence
TMM / TDM	Shedding	Mincle	Cytokine response (I), phagosome maturation (I)	Essential, but inability to form "cords" which is dependent on TDM attenuates the strains.
DAT/PAT	Extracellular vesicles, shedding	?	Cytokine response (I), phagosome maturation (I), cell death (A)	Decreased persistence *in vivo* in PDIM-deficient strains, decreased growth in human macrophages.
PGL	Extracellular vesicles, shedding	TLR2	Cytokine response (I)	Attenuation *in vivo*
PDIM	Shedding	?	Cytokine responses (I) phagocytosis (A), phagosome maturation (I), phagosome escape (A), autophagy (A) apoptosis (A), necrosis (A)	Attenuation *ex vivo* and *in vivo*
SL-1	Shedding	TLR-2	Cytokine reponse (I), Phagosome maturation (I), autophagy (I), cell death (A)	No attenuation *in vivo*, lack of coughing and transmission in guinea pigs
1-TbAd	Shedding	?	Phagosome acidification (I)	Attenuation in human macrophages

The parentheses behind the host cell process indicate Activation (A), Inhibition (I) or Modulation (M). If not stated explicitly the virulence impact refers to mouse studies.

PIM, LM, LAM

This group of lipids is composed of lipids synthesized from phosphatidylinositol to generate the phosphatidylinositol mannosides (PIM) intermediates followed by the lipomannan (LM) and finally lipoarabinomannan (LAM) (Briken et al., 2004; Guerin et al., 2010; Sancho-Vaello et al., 2017). PIM (Gilleron et al., 2003) and to a lesser extent LAM/ManLAM (Nigou et al., 2008) are TLR2 ligands with PIM species acting as agonists, and LAM species acting as anti-inflammatory molecules (Quesniaux et al., 2004). LM is also a TLR2 agonist leading to cell signaling that induces IL-12 production and apoptosis (Dao et al., 2004). These lipids can therefore regulate TLR2/MyD88/NF-κB dependent production and secretion of numerous inflammatory cytokines such as TNF, IL12p40 or IL-8. A mannose-capped LAM (ManLAM) dampens the immune response through binding and inhibition of DC-SIGN signaling (Maeda et al., 2003). Alternatively, ManLAM can bind to Dectin-2 receptor, inducing an inflammatory response that appeared to be detrimental for mycobacteria in mice (Yonekawa et al., 2014). The mannose moieties of LAM can bind to the mannose receptor and stimulate the phagocytosis of Mtb (Maeda et al., 2003; Torrelles et al., 2006). Once in the bacteria are internalized, PIM (Vergne et al., 2004) and LAM (Fratti et al., 2001, 2003) contribute to the capacity of Mtb to inhibit phagosome maturation. PIM actually stimulates the fusion of the MCV with early endosomes which helps to avoid fusion with late endosomes (Vergne et al., 2004). The recruitment of the early endosome autoantigen (EEA1) protein to early phagosomes is an essential step in phagosome maturation that Mtb LAM is able to inhibit (Fratti et al., 2001). These excellent reviews provide a more in-depth overview on the activity of PIM/LM/LAM Mtb glycolipids (Vergne et al., 2014; Garcia-Vilanova et al., 2019).

TDM, DAT/PAT, SL-1

This group of Mtb lipids is composed of the trehalose-containing lipids (Garcia-Vilanova et al., 2019). Probably the best known lipid within this group is the essential lipid trehalose dimycolate (TDM), also called "cord factor" because it is required for the cording phenotype of Mtb which was originally described by Robert Koch in 1882 (Glickman et al., 2000). TDM binds to the host cell receptor Mincle and leads to macrophage and dendritic cell activation (Ishikawa et al., 2009; Ostrop et al., 2015) and also to the TDM-induced granuloma formation in the lungs of mice injected with TDM (Ishikawa et al., 2009). Inside macrophages, TDM contributes to enhance the survival of Mtb, as they are involved in phagosome maturation inhibition (Indrigo et al., 2003) and intracellular cording was recently associated to an inhibition of cytosolic detection of Mtb in endothelial cells, thus favoring persistence in lymphoid tissues (Lerner et al., 2020). Accordingly, Mtb mutants which have a deficiency of cycloprane modification in the mycolic acid chains of TDM show a defect in cording and a decrease in the granulomatous response *in vivo*, as well as the persistence in the host (Glickman et al., 2000; Rao et al., 2005).

A second group of compounds are the di- / poly-acyltrehaloses (DAT/PAT) (Garcia-Vilanova et al., 2019). They have no effect on virulence of Mtb in the mouse and guinea pig model (Rousseau et al., 2003a; Chesne-Seck et al., 2008; Passemar et al., 2014) but in the absence of PDIM a role in virulence could be detected for DAT/PAT in the mouse model (Passemar et al., 2014). At the cellular level DAT/PAT stimulate binding and entry of Mtb into macrophages and epithelial cells (Rousseau et al., 2003a). Mtb DAT/PAT are also important for the Mtb-mediated phagosome maturation inhibition (Brodin et al., 2010; Passemar et al., 2014).

The last member of this group is the Sulfoglycolipid-1 (SL-1) is only synthesized by Mtb and *M. canetti* and characteristically contains sulfated trehalose. Several studies demonstrated a role of SL-1 in the phagosome-lysosome fusion inhibition by Mtb (Goren et al., 1976; Brodin et al., 2010; Passemar et al., 2014). A Tn-mutagenesis genetic screen in Mtb identified Mtb mutants in genes involved in SL-1 biosynthesis to show increased activation of NF-κB after THP-1 cell infection when compared to Mtb (Blanc et al., 2017). Purified and synthetic SL-1 has antagonistic binding activity to TLR-2 which mediates decreased NF-κB activation, reduced pro-inflammatory cytokine production and costimulatory molecule expression (Blanc et al., 2017). In the same fashion, a report found that SL can inhibit autophagy through MyD88 signaling (Bah et al., 2020). SL-1 have no effect on virulence of Mtb in the mouse and guinea pig model (Rousseau et al., 2003b; Chesne-Seck et al., 2008). Recently, a study found a crucial role of SL-1 in the transmission process of Mtb by stimulating cough in guinea pigs (Ruhl et al., 2020). Indeed, SL-1 activates nociceptive neurons which triggers the coughing reflex and consistently, guinea pigs infected with a SL-1-deficient Mtb strain do not cough and do not transmit bacteria to uninfected guinea pigs (Ruhl et al., 2020).

PDIM, PGL

Another group is composed of the lipid DIM/PDIM and its glycosylated form the phenol glycolipids (PGLs). PDIM are only produced by pathogenic mycobacteria in the MTB complex (Goren et al., 1974; Vincent et al., 2018). PDIM-deficient strains are attenuated in the Guinea pig model (Goren et al., 1974), and these findings could be confirmed, two decades later, after a screening of a transposon mutant library of H37Rv for loss of *in vivo* virulence in the mouse model (Camacho et al., 1999; Cox et al., 1999). The attenuation of Mtb PDIM-deficient strains is more remarkable during the first weeks after infection suggesting a function in defense against the innate immune response (Rousseau et al., 2004; Murry et al., 2009; Kirksey et al., 2011; Day et al., 2014). At the cellular level, PDIM are anti-inflammatory lipids since a PDIM-deficient Mtb strain causes increased proinflammatory cytokines responses such as TNF and IL-6 in macrophages and dendritic cells (Rousseau et al., 2004). Consistent with this anti-inflammatory effect a PDIM-deficient Mm strain shows an increase in MyD88-dependent recruitment of macrophages to the granuloma in the zebrafish model (Cambier et al., 2014). Their presence also stimulates Mtb phagocytosis mediated by CR3 (Astarie-Dequeker et al., 2009), contributes to the phagosome maturation inhibition (Astarie-Dequeker et al., 2009; Passemar et al., 2014), modulates autophagic response (Bah et al., 2020), is required for phagosomal escape and cell death induction (Augenstreich et al., 2017; Barczak et al., 2017; Quigley et al., 2017). Studies of infection of human endothelial cells also showed that PDIMs are required for phagosomal escape (Lerner et al., 2018) and intracellular cording (Lerner et al., 2020). Presently, no host cell receptor for PDIM was identified, as the effect PDIM on CR3-mediated phagocytosis did not reveal any binding (Arbues et al., 2016).

PGLs are a glycosylated form of PDIM, with sugar moieties varying depending on the mycobacterial strain that produces the PGL (Arbues et al., 2014). The Mtb Beijing strains that are highly prevalent in Asia (Huet et al., 2009), *M.leprae* (Hunter and Brennan, 1981), Mm and *M. ulcerans* produce PGL lipid species (Daffé and Lanéelle, 1988). Mtb- or Mm-derived PGL have mostly an anti-inflammatory role by inhibiting inflammatory cytokines secretion (Reed et al., 2004; Robinson et al., 2008) and also contribute to phagosome maturation inhibition (Robinson et al., 2008). More recently it was found that Mtb PGL or its trisaccharide domain can bind to TLR2 and inhibit the NF-κB pathway (Arbues et al., 2016). Interestingly, in the zebrafish model the PGL of Mm are associated with a CCR2-mediated recruitment of permissive macrophages in order to increase virulence of Mm (Cambier et al., 2014).

1-TbAd

The 1-tuberculosinyladenosine (1-TbAd) was discovered using an HPLC–mass spectrometry (MS)-based lipidomics approach and by comparing lipid profile of H37Rv and the *M. bovis*-derived vaccine strain BCG (Layre et al., 2014). This is a di-terpene linked adenosine lipid which was detected both associated on the bacteria and in the culture supernatant, suggesting a release of the lipid by shedding (Layre et al., 2014; Buter et al., 2019). This lipid was found in the vast majority of clinical isolates tested and appeared highly abundant (Buter et al., 2019). More interestingly, this lipid can act as an antacid when Mtb resides in a phagosome, so it can counter the decrease in pH due to phagosomal maturation (Buter et al., 2019). 1-TbAd can also diffuse out of the phagosome and induce swelling in lysosomes, thus inhibiting their fusion to the mycobacterial phagosome (Buter et al., 2019).

Mtb Lipids as Modifiers of Host Membrane Biophysical Properties

In general, the role of lipids was mainly described as pathogen associated molecular pattern (PAMP). Nevertheless, their ability to transfer from the bacteria mycomembrane to the host cell macrophages membranes during the infection led to study their potential impact on host cell membrane structure and organization. For example, Mtb-derived ManLAM can disrupt microdomain (often called Raft) in artificial membranes and vesicles fusion (Hayakawa et al., 2007). They were also observed in these domains in ManLAM-treated cells and associated with a defect in phagosomes-lysosomes fusion (Welin et al., 2008). This property was also observed during Mtb infection of macrophages (Fratti et al., 2003). More recently, ManLAM was also found to bind to lactosylceramide in rafts at the plasma membrane and the phagosomal membrane, to induce phagocytosis and to inhibit phagosome-lysosome fusion, respectively (Nakayama et al., 2016).

An alteration of the membrane biophysical properties were also found with TDM inserted into artificial and isolated mitochondria (Sut et al., 1990; Harland et al., 2008) and they were able to decrease membrane fusion in liposomes model (Spargo et al., 1991). SL-1 increases membrane polarity on THP-1 treated with the purified lipid and that was associated with its autophagy inhibitory properties (Mishra et al., 2019; Dadhich et al., 2020).

PDIM lipids may not have a host cell receptor that mediates their effects because of their purely lipidic nature and it is thus conceivable their broad effects on macrophage responses are due to alterations in the membrane organization. Indeed, PDIM were found to decrease membrane polarity in macrophages infected with BCG (Astarie-Dequeker et al., 2009). Moreover, it was recently characterized that PDIM can adopt a conical shape in membranes that is responsible for an increase curvature of artificial membranes (Augenstreich et al., 2019). PDIM treatment of macrophages prior of infection can also rescue the phagocytosis level of PDIM deficient strain of Mtb. Interestingly, treating macrophages with the conical lipid Palmitoyl-Oleoyl Phosphatidylethanolamine can restore the phagocytosis of a PDIM-deficient strain of Mtb at the same level as a PDIM treatment (Augenstreich et al., 2019). This strongly suggested a tight link between PDIM conical shape and its effect on macrophages responses. All these observations on PDIM reveal that the biophysical impact of the insertion of Mtb virulence lipids into the host cells membranes is potentially underestimated for the other virulence lipids such as TDM, PGL, DAT/PAT or SL-1, and could explain part of the crucial importance the lipids play in Mtb virulence.

DISCUSSION

The knowledge of how Mtb effectors interact with the host cell has increased tremendously over the last decade and the successful application of system biology approaches to identify effector-host cell interactions has already revealed many new potential interactions that will certainly generate compelling new findings in the years to come (Mehra et al., 2013; Penn et al., 2018; Wang et al., 2020). We focused this review on proteins and lipids of Mtb that are secreted and released by the bacterium and affect host cell defense pathways. We certainly did not intend to diminish the importance of the other strategies of Mtb to manipulate the host cell: (1) Interactions of cell wall-anchored Mtb proteins and host cell membrane receptors (e.g., PE_PGRS33), (2) Mtb secretes nucleotides [c-di-AMP (Dey et al., 2015, 2017), Mtb DNA (Watson et al., 2012, 2015; Collins et al., 2015; Wassermann et al., 2015) and RNA (Cheng and Schorey, 2018)] that clearly interact with the host cell to; for example, induce IFN-β production in the case of secreted Mtb DNA (Collins et al., 2015; Wassermann et al., 2015; Watson et al., 2015) and (3) Mtb secretes membrane vesicles that contain cargo and include membrane bound lipids and proteins that will interact with the host cell (Prados Rosales et al., 2011; Brown et al., 2015; Lee et al., 2015).

AUTHOR CONTRIBUTIONS

JA and VB wrote and edited the manuscript and tables. JA created the figures. All authors contributed to the article and approved the submitted version.

REFERENCES

Abdallah, A. M., Verboom, T., Weerdenburg, E. M., Gey van Pittius, N. C., Mahasha, P. W., Jiménez, C., et al. (2009). PPE and PE_PGRS proteins of Mycobacterium marinumare transported via the type VII secretion system ESX-5. *Mol. Microbiol.* 73, 329–340. doi: 10.1111/j.1365-2958.2009.06783.x

Aguilera, J., Karki, C. B., Li, L., Vazquez Reyes, S., Estevao, I., Grajeda, B. I., et al. (2020). Nα-Acetylation of the virulence factor EsxA is required for mycobacterial cytosolic translocation and virulence. *J. Biol. Chem.* 295, 5785–5794. doi: 10.1074/jbc.RA119.012497

Almagro Armenteros, J. J., Tsirigos, K. D., Sønderby, C. K., Petersen, T. N., Winther, O., Brunak, S., et al. (2019). SignalP 5.0 improves signal peptide predictions using deep neural networks. *Nat. Biotechnol.* 37, 420–423. doi: 10.1038/s41587-019-0036-z

Antonelli, L. R. V., Gigliotti Rothfuchs, A., Gonçalves, R., Roffê, E., Cheever, A. W., Bafica, A., et al. (2010). Intranasal Poly-IC treatment exacerbates tuberculosis in mice through the pulmonary recruitment of a pathogen-permissive monocyte/macrophage population. *J. Clin. Invest.* 120, 1674–1682. doi: 10.1172/JCI40817

Arbues, A., Lugo-Villarino, G., Neyrolles, O., Guilhot, C., and Astarie-Dequeker, C. (2014). Playing hide-and-seek with host macrophages through the use of mycobacterial cell envelope phthiocerol dimycocerosates and phenolic glycolipids. *Front. Cell. Infect. Microbiol.* 4:173. doi: 10.3389/fcimb.2014.00173

Arbues, A., Malaga, W., Constant, P., Guilhot, C., Prandi, J., and Astarie-Dequeker, C. (2016). Trisaccharides of phenolic glycolipids confer advantages to pathogenic mycobacteria through manipulation of host-cell pattern-recognition receptors. *ACS Chem. Biol.* 11, 2865–2875. doi: 10.1021/acschembio.6b00568

Armstrong, J. A., and Hart, P. D. (1971). Response of cultured macrophages to *Mycobacterium tuberculosis*, with observations on fusion of lysosomes with phagosomes. *J. Exp. Med.* 134, 713–740. doi: 10.1084/jem.134.3.713

Astarie-Dequeker, C., Le Guyader, L., Malaga, W., Seaphanh, F.-K., Chalut, C., Lopez, A., et al. (2009). Phthiocerol dimycocerosates of *M. tuberculosis* participate in macrophage invasion by inducing changes in the organization of plasma membrane lipids. *PLoS Pathog* 5:e1000289. doi: 10.1371/journal.ppat.1000289

Athman, J. J., Sande, O. J., Groft, S. G., Reba, S. M., Nagy, N., Wearsch, P. A., et al. (2017). *Mycobacterium tuberculosis* membrane vesicles inhibit T Cell activation. *J. Immunol.* 198, 2028–2037. doi: 10.4049/jimmunol.1601199

Athman, J. J., Wang, Y., McDonald, D. J., Boom, W. H., Harding, C. V., and Wearsch, P. A. (2015). bacterial membrane vesicles mediate the release of *Mycobacterium tuberculosis* lipoglycans and lipoproteins from infected macrophages. *J. Immunol.* 195, 1044–1053. doi: 10.4049/jimmunol.1402894

Augenstreich, J., Arbues, A., Simeone, R., Haanappel, E., Wegener, A., Sayes, F., et al. (2017). ESX-1 and phthiocerol dimycocerosates of *Mycobacterium tuberculosis* act in concert to cause phagosomal rupture and host cell apoptosis. *Cell. Microbiol.* 19, 1–19. doi: 10.1111/cmi.12726

Augenstreich, J., Haanappel, E., Ferré, G., Czaplicki, G., Jolibois, F., Destainville, N., et al. (2019). The conical shape of DIM lipids promotes *Mycobacterium tuberculosis* infection of macrophages. *Proc. Natl. Acad. Sci. U.S.A.* 116, 25649–25658. doi: 10.1073/pnas.1910368116

Augenstreich, J., Haanappel, E., Sayes, F., Simeone, R., Guillet, V., Mazères, S., et al. (2020). phthiocerol dimycocerosates from *Mycobacterium tuberculosis* increase the membrane activity of bacterial effectors and host receptors. *Front. Cell. Infect. Microbiol.* 10:420. doi: 10.3389/fcimb.2020.00420

Av-Gay, Y., and Everett, M. (2000). The eukaryotic-like Ser/Thr protein kinases of *Mycobacterium tuberculosis*. *Trends Microbiol.* 8, 238–244. doi: 10.1016/S0966-842X(00)01734-0

Bach, H., Papavinasasundaram, K. G., Wong, D., Hmama, Z., and Av-Gay, Y. (2008). *Mycobacterium tuberculosis* virulence is mediated by PtpA dephosphorylation of human vacuolar protein sorting 33B. *Cell Host Microbe* 3, 316–322. doi: 10.1016/j.chom.2008.03.008

Bah, A., Sanicas, M., Nigou, J., Guilhot, C., Astarie-Dequeker, C., and Vergne, I. (2020). The Lipid Virulence Factors of *Mycobacterium tuberculosis* exert multilayered control over autophagy-related pathways in infected human macrophages. *Cells* 9, 666. doi: 10.3390/cells9030666

Bange, F. C., Collins, F. M., and Jacobs, W. R. (1999). Survival of mice infected with *Mycobacterium smegmatis* containing large DNA fragments from *Mycobacterium tuberculosis*. *Tuber. Lung Dis.* 79, 171–180. doi: 10.1054/tuld.1998.0201

Barczak, A. K., Avraham, R., Singh, S., Luo, S. S., Zhang, W. R., Bray, M.-A., et al. (2017). Systematic, multiparametric analysis of *Mycobacterium tuberculosis* intracellular infection offers insight into coordinated virulence. *PLoS Pathog* 13:e1006363. doi: 10.1371/journal.ppat.1006363

Bardarov, S., Bardarov, S. J. Jr., Pavelka, M. S. J. Jr., Sambandamurthy, V., Larsen, M., Tufariello, J., et al. (2002). Specialized transduction: an efficient method for generating marked and unmarked targeted gene disruptions in *Mycobacterium tuberculosis, M. bovis* BCG and *M. smegmatis. Microbiology* 148, 3007–3017. doi: 10.1099/00221287-148-10-3007

Baros, S. S., Blackburn, J. M., and Soares, N. C. (2020). phosphoproteomic approaches to discover novel substrates of mycobacterial ser/thr protein kinases. *Mol. Cell Proteomics* 19, 233–244. doi: 10.1074/mcp.R119.001668

Beatty, W. L., Rhoades, E. R., Ullrich, H. J., Chatterjee, D., Heuser, J. E., and Russell, D. G. (2000). Trafficking and release of mycobacterial lipids from infected macrophages. *Traffic* 1, 235–247. doi: 10.1034/j.1600-0854.2000.010306.x

Bedard, K., and Krause, K.-H. (2007). The NOX family of ROS-generating NADPH oxidases: physiology and pathophysiology. *Physiol. Rev* 87, 245–313. doi: 10.1152/physrev.00044.2005

Behar, S. M., Divangahi, M., and Remold, H. G. (2010). Evasion of innate immunity by *Mycobacterium tuberculosis*: is death an exit strategy? *Nat. Rev. Microbiol.* 8, 668–674. doi: 10.1038/nrmicro2387

Beresford, N., Patel, S., Armstrong, J., Szöor, B., Fordham-Skelton, A. P., and Tabernero, L. (2007). MptpB, a virulence factor from *Mycobacterium tuberculosis*, exhibits triple-specificity phosphatase activity. *Biochem. J.* 406, 13–18. doi: 10.1042/BJ20070670

Beresford, N. J., Mulhearn, D., Szczepankiewicz, B., Liu, G., Johnson, M. E., Fordham-Skelton, A., et al. (2009). Inhibition of MptpB phosphatase from *Mycobacterium tuberculosis* impairs mycobacterial survival in macrophages. *J. Antimicrob. Chemother.* 63, 928–936. doi: 10.1093/jac/dkp031

Berry, M. P. R., Graham, C. M., McNab, F. W., Xu, Z., Bloch, S. A. A., Oni, T., et al. (2010). An interferon-inducible neutrophil-driven blood transcriptional signature in human tuberculosis. *Nature* 466, 973–977. doi: 10.1038/nature09247

Bhat, K. H., Das, A., Srikantam, A., and Mukhopadhyay, S. (2013). PPE2 protein of *Mycobacterium tuberculosis* may inhibit nitric oxide in activated macrophages. *Ann. N. Y. Acad. Sci.* 1283, 97–101. doi: 10.1111/nyas.12070

Bhat, K. H., Srivastava, S., Kotturu, S. K., Ghosh, S., and Mukhopadhyay, S. (2017). The PPE2 protein of *Mycobacterium tuberculosis* translocates to host nucleus and inhibits nitric oxide production. *Sci. Rep.* 7, 39706–39713. doi: 10.1038/srep39706

Blanc, L., Gilleron, M., Prandi, J., Song, O.-R., Jang, M.-S., Gicquel, B., et al. (2017). *Mycobacterium tuberculosis* inhibits human innate immune responses via the production of TLR2 antagonist glycolipids. *Proc. Natl Acad Sci. U.S.A.* 114, 11205–11210. doi: 10.1073/pnas.1707840114

Bogdan, C. (2015). Nitric oxide synthase in innate and adaptive immunity: an update. *Trends Immunol.* 36, 161–178. doi: 10.1016/j.it.2015.01.003

Bosserman, R. E., and Champion, P. A. (2017). Esx Systems and the mycobacterial cell envelope: what's the connection? *J. Bacteriol.* 199, e00131–e00117. doi: 10.1128/JB.00131-17

Braunstein, M., Espinosa, B. J., Chan, J., Belisle, J. T., and Jacobs, W. R. (2003). SecA2 functions in the secretion of superoxide dismutase A and in the virulence of *Mycobacterium tuberculosis*. *Mol. Microbiol.* 48, 453–464. doi: 10.1046/j.1365-2958.2003.03438.x

Briken, V., Porcelli, S. A., Besra, G. S., and Kremer, L. (2004). Mycobacterial lipoarabinomannan and related lipoglycans: from biogenesis to modulation of the immune response. *Mol. Microbiol.* 53, 391–403. doi: 10.1111/j.1365-2958.2004.04183.x

Brodin, P., Poquet, Y., Levillain, F., Peguillet, I., Larrouy-Maumus, G., Gilleron, M., et al. (2010). High content phenotypic cell-based visual screen identifies *Mycobacterium tuberculosis* acyltrehalose-containing glycolipids involved in phagosome remodeling. *PLoS Pathog.* 6:e1001100. doi: 10.1371/journal.ppat.1001100

Brown, L., Wolf, J. M., Prados Rosales, R., and Casadevall, A. (2015). Through the wall: extracellular vesicles in gram-positive bacteria, mycobacteria and fungi. *Nat. Rev. Microbiol.* 13, 620–630. doi: 10.1038/nrmicro3480

Bussi, C., and Gutierrez, M. G. (2019). *Mycobacterium tuberculosis* infection of host cells in space and time. *FEMS Microbiol. Rev.* 43, 341–361. doi: 10.1093/femsre/fuz006

Buter, J., Cheng, T.-Y., Ghanem, M., Grootemaat, A. E., Raman, S., Feng, X., et al. (2019). *Mycobacterium tuberculosis* releases an antacid that remodels phagosomes. *Nat. Chem. Biol.* 15, 889–899. doi: 10.1038/s41589-019-0336-0

Camacho, L. R., Ensergueix, D., Perez, E., Gicquel, B., and Guilhot, C. (1999). Identification of a virulence gene cluster of *Mycobacterium tuberculosis* by signature-tagged transposon mutagenesis. *Mol. Microbiol.* 34, 257–267. doi: 10.1046/j.1365-2958.1999.01593.x

Cambier, C. J., Banik, S. M., Buonomo, J. A., and Bertozzi, C. R. (2020). Spreading of a mycobacterial cell surface lipid into host epithelial membranes promotes infectivity. *bioRxiv* 83, 505–554. doi: 10.1101/845081

Cambier, C. J., Takaki, K. K., Larson, R. P., Hernandez, R. E., Tobin, D. M., Urdahl, K. B., et al. (2014). Mycobacteria manipulate macrophage recruitment through coordinated use of membrane lipids. *Nature* 505, 218–222. doi: 10.1038/nature12799

Cardenal-Muñoz, E., Arafah, S., López-Jiménez, A. T., Kicka, S., Falaise, A., Bach, F., et al. (2017). *Mycobacterium marinum* antagonistically induces an autophagic response while repressing the autophagic flux in a TORC1- and ESX-1-dependent manner. *PLoS Pathog.* 13:e1006344. doi: 10.1371/journal.ppat.1006344

Chai, Q., Wang, X., Qiang, L., Zhang, Y., Ge, P., Lu, Z., et al. (2019). A *Mycobacterium tuberculosis* surface protein recruits ubiquitin to trigger host xenophagy. *Nat. Commun.* 10:973. doi: 10.1038/s41467-019-09955-8

Chandra, P., Ghanwat, S., Matta, S. K., Yadav, S. S., Mehta, M., Siddiqui, Z., et al. (2015). *Mycobacterium tuberculosis* Inhibits rab7 recruitment to selectively modulate autophagy flux in macrophages. *Sci. Rep.* 5:16320. doi: 10.1038/srep16320

Cheng, Y., and Schorey, J. S. (2018). *Mycobacterium tuberculosis*-induced IFN-β production requires cytosolic DNA and RNA sensing pathways. *J. Exp. Med.* 215, 2919–2935. doi: 10.1084/jem.20180508

Chesne-Seck, M.-L., Barilone, N., Boudou, F., Gonzalo-Asensio, J., Kolattukudy, P. E., Martin, C., et al. (2008). A point mutation in the two-component regulator PhoP-PhoR accounts for the absence of polyketide-derived acyltrehaloses but not that of phthiocerol dimycocerosates in *Mycobacterium tuberculosis* H37Ra. *J. Bacteriol.* 190, 1329–1334. doi: 10.1128/JB.01465-07

Chiaradia, L., Lefebvre, C., Parra, J., Marcoux, J., Burlet-Schiltz, O., Etienne, G., et al. (2017). Dissecting the mycobacterial cell envelope and defining the composition of the native mycomembrane. *Sci. Rep.* 7, 12807–12812. doi: 10.1038/s41598-017-12718-4

Chopra, P., Koduri, H., Singh, R., Koul, A., Ghildiyal, M., Sharma, K., et al. (2004). Nucleoside diphosphate kinase of *Mycobacterium tuberculosis* acts as GTPase-activating protein for Rho-GTPases. *FEBS Lett.* 571, 212–216. doi: 10.1016/j.febslet.2004.06.073

Chopra, P., Singh, A., Koul, A., Ramachandran, S., Drlica, K., Tyagi, A. K., et al. (2003). Cytotoxic activity of nucleoside diphosphate kinase secreted from *Mycobacterium tuberculosis*. *Eur. J. Biochem.* 270, 625–634. doi: 10.1046/j.1432-1033.2003.03402.x

Collins, A. C., Cai, H., Li, T., Franco, L. H., Li, X.-D., Nair, V. R., et al. (2015). Cyclic GMP-AMP synthase is an innate immune DNA sensor for *Mycobacterium tuberculosis*. *Cell Host Microbe* 17, 820–828. doi: 10.1016/j.chom.2015.05.005

Conrad, W. H., Osman, M. M., Shanahan, J. K., Chu, F., Takaki, K. K., Cameron, J., et al. (2017). Mycobacterial ESX-1 secretion system mediates host cell lysis through bacterium contact-dependent gross membrane disruptions. *Proc. Natl. Acad. Sci. U.S.A.* 114, 1371–1376. doi: 10.1073/pnas.1620133114

Correa-Macedo, W., Cambri, G., and Schurr, E. (2019). The interplay of human and *Mycobacterium tuberculosis* genomic variability. *Front. Genet.* 10:865. doi: 10.3389/fgene.2019.00865

Cowley, S., Ko, M., Pick, N., Chow, R., Downing, K. J., Gordhan, B. G., et al. (2004). The *Mycobacterium tuberculosis* protein serine/threonine kinase PknG is linked to cellular glutamate/glutamine levels and is important for growth *in vivo*. *Mol. Microbiol.* 52, 1691–1702. doi: 10.1111/j.1365-2958.2004.04085.x

Cox, J. S., Chen, B., McNeil, M., and Jacobs, W. R. (1999). Complex lipid determines tissue-specific replication of *Mycobacterium tuberculosis* in mice. *Nature* 402, 79–83. doi: 10.1038/47042

Crowley, J. T., Toledo, A. M., LaRocca, T. J., Coleman, J. L., London, E., and Benach, J. L. (2013). Lipid exchange between Borrelia burgdorferi and host cells. *PLoS Pathog.* 9:e1003109. doi: 10.1371/journal.ppat.1003109

Dadhich, R., Mishra, M., Ning, S., Jana, S., Sarpe, V. A., Mahato, J., et al. (2020). A Virulence-associated glycolipid with distinct conformational attributes: impact on lateral organization of host plasma membrane, autophagy and signaling. *ACS Chem. Biol.* 15, 740–750. doi: 10.1021/acschembio.9b00991

Daffé, M., and Lanéelle, M. A. (1988). Distribution of phthiocerol diester, phenolic mycosides and related compounds in mycobacteria. *J. Gen. Microbiol.* 134, 2049–2055. doi: 10.1099/00221287-134-7-2049

Daffe, M., and Marrakchi, H. (2019). Unraveling the structure of the mycobacterial envelope. *Microbiol. Spect.* 7, 1–11. doi: 10.1128/microbiolspec.GPP3-0027-2018

Dahl, J. L., Wei, J., Moulder, J. W., Laal, S., and Friedman, R. L. (2001). Subcellular localization of the Iitracellular survival-enhancing eis protein of *Mycobacterium tuberculosis*. *Infect. Immun.* 69, 4295–4302. doi: 10.1128/IAI.69.7.4295-4302.2001

Dallenga, T., Repnik, U., Corleis, B., Eich, J., Reimer, R., Griffiths, G. W., et al. (2017). *M. tuberculosis*-induced necrosis of infected neutrophils promotes bacterial growth following phagocytosis by macrophages. *Cell Host Microbe* 22, 519–530.e3. doi: 10.1016/j.chom.2017.09.003

Danelishvili, L., Everman, J. L., McNamara, M. J., and Bermudez, L. E. (2011). Inhibition of the plasma-membrane-associated serine protease cathepsin G by *Mycobacterium tuberculosis* Rv3364c suppresses caspase-1 and pyroptosis in macrophages. *Front. Microbiol.* 2:281. doi: 10.3389/fmicb.2011.00281

Danelishvili, L., Yamazaki, Y., Selker, J., and Bermudez, L. E. (2010). Secreted *Mycobacterium tuberculosis* Rv3654c and Rv3655c proteins participate in the suppression of macrophage apoptosis. *PLoS ONE* 5:e10474. doi: 10.1371/journal.pone.0010474

Danilchanka, O., Sun, J., Pavlenok, M., Maueröder, C., Speer, A., Siroy, A., et al. (2014). An outer membrane channel protein of *Mycobacterium tuberculosis* with exotoxin activity. *Proc. Natl. Acad. Sci. U.S.A.* 111, 6750–6755. doi: 10.1073/pnas.1400136111

Dao, D. N., Kremer, L., Guérardel, Y., Molano, A., Jacobs, W. R., Porcelli, S. A., et al. (2004). *Mycobacterium tuberculosis* lipomannan induces apoptosis and interleukin-12 production in macrophages. *Infect. Immun.* 72, 2067–2074. doi: 10.1128/IAI.72.4.2067-2074.2004

Darwin, K. H., Ehrt, S., Gutierrez-Ramos, J.-C., Weich, N., and Nathan, C. F. (2003). The proteasome of *Mycobacterium tuberculosis* is required for resistance to nitric oxide. *Science* 302, 1963–1966. doi: 10.1126/science.1091176

Day, T. A., Mittler, J. E., Nixon, M. R., Thompson, C., Miner, M. D., Hickey, M. J., et al. (2014). *Mycobacterium tuberculosis* strains lacking surface lipid phthiocerol dimycocerosate are susceptible to killing by an early innate host response. *Infect. Immun.* 82, 5214–5222. doi: 10.1128/IAI.01340-13

de Chastellier, C., Forquet, F., Gordon, A., and Thilo, L. (2009). Mycobacterium requires an all-around closely apposing phagosome membrane to maintain the maturation block and this apposition is re-established when it rescues itself from phagolysosomes. *Cell. Microbiol.* 11, 1190–1207. doi: 10.1111/j.1462-5822.2009.01324.x

de Jonge, M. I., Pehau-Arnaudet, G., Fretz, M. M., Romain, F., Bottai, D., Brodin, P., et al. (2007). ESAT-6 from *Mycobacterium tuberculosis* dissociates from its putative chaperone CFP-10 under acidic conditions and exhibits membrane-lysing activity. *J. Bacteriol.* 189, 6028–6034. doi: 10.1128/JB.00469-07

De Maio, F., Berisio, R., Manganelli, R., and Delogu, G. (2020). PE_PGRS proteins of *Mycobacterium tuberculosis*: a specialized molecular task force at the forefront of host-pathogen interaction. *Virulence* 11, 898–915. doi: 10.1080/21505594.2020.1785815

de Souza, G. A., Leversen, N. A., Målen, H., and Wiker, H. G. (2011). Bacterial proteins with cleaved or uncleaved signal peptides of the general secretory pathway. *J. Proteomics* 75, 502–510. doi: 10.1016/j.jprot.2011.08.016

Deghmane, A.-E., Soualhine, H., Soulhine, H., Bach, H., Sendide, K., Itoh, S., et al. (2007). Lipoamide dehydrogenase mediates retention of coronin-1 on BCG vacuoles, leading to arrest in phagosome maturation. *J. Cell Sci.* 120, 2796–2806. doi: 10.1242/jcs.006221

Dejesus, M. A., Gerrick, E. R., Xu, W., Park, S. W., Long, J. E., Boutte, C. C., et al. (2017). comprehensive essentiality analysis of the *Mycobacterium tuberculosis* genome via saturating transposon mutagenesis. *MBio* 8:e02133-e02116. doi: 10.1128/mBio.02133-16

Dey, B., Dey, R. J., Cheung, L. S., Pokkali, S., Guo, H., Lee, J.-H., et al. (2015). A bacterial cyclic dinucleotide activates the cytosolic surveillance pathway and mediates innate resistance to tuberculosis. *Nat. Med.* 21, 401–406. doi: 10.1038/nm.3813

Dey, R. J., Dey, B., Zheng, Y., Cheung, L. S., Zhou, J., Sayre, D., et al. (2017). Inhibition of innate immune cytosolic surveillance by an *M. tuberculosis* phosphodiesterase. *Nat. Chem. Biol.* 13, 210–217. doi: 10.1038/nchembio.2254

Domingo-Gonzalez, R., Prince, O., Cooper, A., and Khader, S. A. (2016). Cytokines and Chemokines in *Mycobacterium tuberculosis* Infection. *Microbiol. Spectr.* 4, 1–58. doi: 10.1128/microbiolspec.TBTB2-0018-2016

Dubois, V., Viljoen, A., Laencina, L., Le Moigne, V., Bernut, A., Dubar, F., et al. (2018). MmpL8MAB controls *Mycobacterium abscessus* virulence and production of a previously unknown glycolipid family. *Proc. Natl. Acad. Sci. U.S.A.* 115, E10147–E10156. doi: 10.1073/pnas.1812984115

Dulberger, C. L., Rubin, E. J., and Boutte, C. C. (2020). The mycobacterial cell envelope - a moving target. *Nat. Rev. Microbiol.* 18, 47–59. doi: 10.1038/s41579-019-0273-7

Edwards, K. M., Cynamon, M. H., Voladri, R. K., Hager, C. C., DeStefano, M. S., Tham, K. T., et al. (2001). Iron-cofactored superoxide dismutase inhibits host responses to *Mycobacterium tuberculosis*. *Am. J. Respir. Crit. Care Med.* 164, 2213–2219. doi: 10.1164/ajrccm.164.12.2106093

Flynn, J. L., Chan, J., Triebold, K. J., Dalton, D. K., Stewart, T. A., and Bloom, B. R. (1993). An essential role for interferon gamma in resistance to *Mycobacterium tuberculosis* infection. *J. Exp. Med.* 178, 2249–2254. doi: 10.1084/jem.178.6.2249

Flynn, J. L., Goldstein, M. M., Chan, J., Triebold, K. J., Pfeffer, K., Lowenstein, C. J., et al. (1995). Tumor necrosis factor-alpha is required in the protective immune response against *Mycobacterium tuberculosis* in mice. *Immunity* 2, 561–572. doi: 10.1016/1074-7613(95)90001-2

Franco, L. H., Nair, V. R., Scharn, C. R., Xavier, R. J., Torrealba, J. R., Shiloh, M. U., et al. (2016). The ubiquitin ligase smurf1 functions in selective autophagy of *Mycobacterium tuberculosis* and anti-tuberculous host defense. *Cell Host Microbe* 21, 59–72. doi: 10.1016/j.chom.2017.08.005

Fratti, R. A., Backer, J. M., Gruenberg, J., Corvera, S., and Deretic, V. (2001). Role of phosphatidylinositol 3-kinase and Rab5 effectors in phagosomal biogenesis and mycobacterial phagosome maturation arrest. *J. Cell Biol.* 154, 631–644. doi: 10.1083/jcb.200106049

Fratti, R. A., Chua, J., Vergne, I., and Deretic, V. (2003). *Mycobacterium tuberculosis* glycosylated phosphatidylinositol causes phagosome maturation arrest. *Proc. Natl. Acad. Sci. U.S.A.* 100, 5437–5442. doi: 10.1073/pnas.0737613100

Gago, G., Diacovich, L., and Gramajo, H. (2017). Lipid metabolism and its implication in mycobacteria-host interaction. *Curr. Opin. Microbiol.* 41, 36–42. doi: 10.1016/j.mib.2017.11.020

Galluzzi, L., Vitale, I., Aaronson, S. A., Abrams, J. M., Adam, D., Agostinis, P., et al. (2018). molecular mechanisms of cell death: recommendations of the nomenclature committee on cell death 2018. *Cell Death Differ.* 25, 486–541. doi: 10.1038/s41418-017-0012-4

Galmes, R., ten Brink, C., Oorschot, V., Veenendaal, T., Jonker, C., van der Sluijs, P., et al. (2015). Vps33B is required for delivery of endocytosed cargo to lysosomes. *Traffic* 16, 1288–1305. doi: 10.1111/tra.12334

Ganguly, N., Giang, P. H., Gupta, C., Basu, S. K., Siddiqui, I., Salunke, D. M., et al. (2008). *Mycobacterium tuberculosis* secretory proteins CFP-10, ESAT-6 and the CFP10:ESAT6 complex inhibit lipopolysaccharide-induced NF-kappaB transactivation by downregulation of reactive oxidative species (ROS) production. *Immunol. Cell Biol.* 86, 98–106. doi: 10.1038/sj.icb.7100117

Garcia-Vilanova, A., Chan, J., and Torrelles, J. B. (2019). underestimated manipulative roles of *Mycobacterium tuberculosis* cell envelope glycolipids during infection. *Front. Immunol.* 10:2909. doi: 10.3389/fimmu.2019.02909

Gilleron, M., Quesniaux, V. F. J., and Puzo, G. (2003). Acylation state of the phosphatidylinositol hexamannosides from *Mycobacterium bovis* bacillus Calmette Guerin and *Mycobacterium tuberculosis* H37Rv and its implication in toll-like receptor response. *J. Biol. Chem.* 278, 29880–29889. doi: 10.1074/jbc.M303446200

Glickman, M. S., Cox, J. S., and Jacobs, W. R. (2000). A novel mycolic acid cyclopropane synthetase is required for cording, persistence, and virulence of *Mycobacterium tuberculosis*. *Mol. Cell* 5, 717–727. doi: 10.1016/S1097-2765(00)80250-6

Goren, M. B., Brokl, O., and Schaefer, W. B. (1974). Lipids of putative relevance to virulence in *Mycobacterium tuberculosis*: phthiocerol dimycocerosate and the attenuation indicator lipid. *Infect. Immun.* 9, 150–158. doi: 10.1128/IAI.9.1.150-158.1974

Goren, M. B., D'Arcy Hart, P., Young, M. R., and Armstrong, J. A. (1976). Prevention of phagosome-lysosome fusion in cultured macrophages by sulfatides of *Mycobacterium tuberculosis*. *Proc. Natl. Acad. Sci. U.S.A.* 73, 2510–2514. doi: 10.1073/pnas.73.7.2510

Gröschel, M. I., Sayes, F., Simeone, R., Majlessi, L., and Brosch, R. (2016). ESX secretion systems: mycobacterial evolution to counter host immunity. *Nat Rev. Microbiol.* 14, 677–691. doi: 10.1038/nrmicro.2016.131

Grundner, C., Cox, J. S., and Alber, T. (2008). Protein tyrosine phosphatase PtpA is not required for *Mycobacterium tuberculosis* growth in mice. *FEMS Microbiol. Lett.* 287, 181–184. doi: 10.1111/j.1574-6968.2008.01309.x

Guerin, M. E., Kordulákóvá, J., Alzari, P. M., Brennan, P. J., and Jackson, M. (2010). Molecular basis of phosphatidyl-myo-inositol mannoside biosynthesis and regulation in mycobacteria. *J. Biol. Chem.* 285, 33577–33583. doi: 10.1074/jbc.R110.168328

Gutierrez, M. G., Master, S. S., Singh, S. B., Taylor, G. A., Colombo, M. I., and Deretic, V. (2004). Autophagy is a defense mechanism inhibiting BCG and *Mycobacterium tuberculosis* survival in infected macrophages. *Cell* 119, 753–766. doi: 10.1016/j.cell.2004.11.038

Hagedorn, M., and Soldati, T. (2007). Flotillin and RacH modulate the intracellular immunity of dictyostelium to *Mycobacterium marinum* infection. *Cell. Microbiol.* 9, 2716–2733. doi: 10.1111/j.1462-5822.2007.00993.x

Harland, C. W., Rabuka, D., Bertozzi, C. R., and Parthasarathy, R. (2008). The *Mycobacterium tuberculosis* virulence factor trehalose dimycolate imparts desiccation resistance to model mycobacterial membranes. *Biophys. J.* 94, 4718–4724. doi: 10.1529/biophysj.107.125542

Hayakawa, E., Tokumasu, F., Nardone, G. A., Jin, A. J., Hackley, V. A., and Dvorak, J. A. (2007). A *Mycobacterium tuberculosis*-derived lipid inhibits membrane fusion by modulating lipid membrane domains. *Biophys. J.* 93, 4018–4030. doi: 10.1529/biophysj.107.104075

Hermans, P. W., van Soolingen, D., and van Embden, J. D. (1992). Characterization of a major polymorphic tandem repeat in *Mycobacterium tuberculosis* and its potential use in the epidemiology of *Mycobacterium kansasii* and *Mycobacterium gordonae*. *J. Bacteriol.* 174, 4157–4165. doi: 10.1128/JB.174.12.4157-4165.1992

Hinchey, J., Lee, S., Jeon, B. Y., Basaraba, R. J., Venkataswamy, M. M., Chen, B., et al. (2007). Enhanced priming of adaptive immunity by a proapoptotic mutant of *Mycobacterium tuberculosis*. *J. Clin. Invest.* 117, 2279–2288. doi: 10.1172/JCI31947

Hoffmann, C., Leis, A., Niederweis, M., Plitzko, J. M., and Engelhardt, H. (2008). Disclosure of the mycobacterial outer membrane: cryo-electron tomography and vitreous sections reveal the lipid bilayer structure. *Proc. Natl. Acad. Sci. U.S.A.* 105, 3963–3967. doi: 10.1073/pnas.0709530105

Hsu, T., Hingley-Wilson, S. M., Chen, B., Chen, M., Dai, A. Z., Morin, P. M., et al. (2003). The primary mechanism of attenuation of bacillus Calmette-Guerin is a loss of secreted lytic function required for invasion of lung interstitial tissue. *Proc. Natl. Acad. Sci. U.S.A.* 100, 12420–12425. doi: 10.1073/pnas.1635213100

Huet, G., Constant, P., Malaga, W., Laneelle, M.-A., Kremer, K., van Soolingen, D., et al. (2009). A lipid profile typifies the beijing strains of *Mycobacterium tuberculosis* identification of a mutation responsible for a modification of the structures of phthiocerol dimycocerosates and phenolic glycolipids. *J. Biol. Chem.* 284, 27101–27113. doi: 10.1074/jbc.M109.041939

Hunter, S. W., and Brennan, P. J. (1981). A novel phenolic glycolipid from *Mycobacterium leprae* possibly involved in immunogenicity and pathogenicity. *J. Bacteriol.* 147, 728–735. doi: 10.1128/JB.147.3.728-735.1981

Iantomasi, R., Sali, M., Cascioferro, A., Palucci, I., Zumbo, A., Soldini, S., et al. (2012). PE_PGRS30 is required for the full virulence of *Mycobacterium tuberculosis*. *Cell. Microbiol.* 14, 356–367. doi: 10.1111/j.1462-5822.2011.01721.x

Ilghari, D., Lightbody, K. L., Veverka, V., Waters, L. C., Muskett, F. W., Renshaw, P. S., et al. (2011). Solution structure of the *Mycobacterium tuberculosis* EsxG·EsxH complex: functional implications and comparisons with other *M. tuberculosis* Esx family complexes. *J. Biol. Chem.* 286, 29993–30002. doi: 10.1074/jbc.M111.248732

Indrigo, J., Hunter, R. L., and Actor, J. K. (2003). Cord factor trehalose 6,6'-dimycolate (TDM) mediates trafficking events during mycobacterial infection of murine macrophages. *Microbiology* 149, 2049–2059. doi: 10.1099/mic.0.26226-0

Ishikawa, E., Ishikawa, T., Morita, Y. S., Toyonaga, K., Yamada, H., Takeuchi, O., et al. (2009). Direct recognition of the mycobacterial glycolipid, trehalose dimycolate, by C-type lectin Mincle. *J. Exp. Med.* 206, 2879–2888. doi: 10.1084/jem.20091750

Ishikawa, E., Mori, D., and Yamasaki, S. (2017). Recognition of mycobacterial lipids by immune receptors. *Trends Immunol.* 38, 66–76. doi: 10.1016/j.it.2016.10.009

Jacobs, W. R., Tuckman, M., and Bloom, B. R. (1987). Introduction of foreign DNA into mycobacteria using a shuttle phasmid. *Nature* 327, 532–535. doi: 10.1038/327532a0

Jaiswal, S., Chatterjee, A., Pandey, S., Lata, K., Gadi, R. K., Manda, R., et al. (2019). Mycobacterial protein tyrosine kinase, PtkA phosphorylates PtpA at tyrosine residues and the mechanism is stalled by the novel series of inhibitors. *J. Drug Target* 27, 51–59. doi: 10.1080/1061186X.2018.1473407

Johansen, S. K., Maus, C. E., Plikaytis, B. B., and Douthwaite, S. (2006). Capreomycin binds across the ribosomal subunit interface using tlyA-encoded 2'-O-methylations in 16S and 23S rRNAs. *Mol. Cell* 23, 173–182. doi: 10.1016/j.molcel.2006.05.044

Joseph, S., Yuen, A., Singh, V., and Hmama, Z. (2017). *Mycobacterium tuberculosis* Cpn60.2 (GroEL2) blocks macrophage apoptosis via interaction with mitochondrial mortalin. *Biol. Open* 6, 481–488. doi: 10.1242/bio.023119

Kelkar, D. S., Kumar, D., Kumar, P., Balakrishnan, L., Muthusamy, B., Yadav, A. K., et al. (2011). Proteogenomic analysis of *Mycobacterium tuberculosis* by high resolution mass spectrometry. *Mol. Cell. Proteomics* 10:M111.011627. doi: 10.1074/mcp.M111.011627

Khaminets, A., Behl, C., and Dikic, I. (2016). Ubiquitin-dependent and independent signals in selective autophagy. *Trends Cell. Biol.* 26, 6–16. doi: 10.1016/j.tcb.2015.08.010

Khan, H. S., Nair, V. R., Ruhl, C. R., Alvarez-Arguedas, S., Galvan Rendiz, J. L., Franco, L. H., et al. (2020). Identification of scavenger receptor B1 as the airway microfold cell receptor for *Mycobacterium tuberculosis*. *Elife* 9:883. doi: 10.7554/eLife.52551.sa2

Kim, K. H., An, D. R., Song, J., Yoon, J. Y., Kim, H. S., Yoon, H. J., et al. (2012). *Mycobacterium tuberculosis* Eis protein initiates suppression of host immune responses by acetylation of DUSP16/MKP-7. *Proc. Natl. Acad. Sci. U.S.A.* 109, 7729–7734. doi: 10.1073/pnas.1120251109

King, C. H., Mundayoor, S., Crawford, J. T., and Shinnick, T. M. (1993). Expression of contact-dependent cytolytic activity by *Mycobacterium tuberculosis* and isolation of the genomic locus that encodes the activity. *Infect. Immun.* 61, 2708–2712. doi: 10.1128/IAI.61.6.2708-2712.1993

Kirksey, M. A., Tischler, A. D., Simeone, R., Hisert, K. B., Uplekar, S., Guilhot, C., et al. (2011). Spontaneous phthiocerol dimycocerosate-deficient variants of *Mycobacterium tuberculosis* are susceptible to gamma interferon-mediated immunity. *Infect. Immun.* 79, 2829–2838. doi: 10.1128/IAI.00097-11

Koliwer-Brandl, H., Knobloch, P., Barisch, C., Welin, A., Hanna, N., Soldati, T., et al. (2019). Distinct *Mycobacterium marinum* phosphatases determine pathogen vacuole phosphoinositide pattern, phagosome maturation, and escape to the cytosol. *Cell. Microbiol.* 21:e13008. doi: 10.1111/cmi.13008

Koster, S., Upadhyay, S., Chandra, P., Papavinasasundaram, K., Yang, G., Hassan, A., et al. (2017). *Mycobacterium tuberculosis* is protected from NADPH oxidase and LC3-associated phagocytosis by the LCP protein CpsA. *Proc. Natl. Acad. Sci. U.S.A.* 114, E8711–E8720. doi: 10.1073/pnas.1707792114

Koul, A., Choidas, A., Treder, M., Tyagi, A. K., Drlica, K., Singh, Y., et al. (2000). Cloning and characterization of secretory tyrosine phosphatases of *Mycobacterium tuberculosis. J. Bacteriol.* 182, 5425–5432. doi: 10.1128/JB.182.19.5425-5432.2000

Krogh, A., Larsson, B., von Heijne, G., and Sonnhammer, E. L. (2001). Predicting transmembrane protein topology with a hidden markov model: application to complete genomes. *J. Mol. Biol.* 305, 567–580. doi: 10.1006/jmbi.2000.4315

Kumar, D., Nath, L., Kamal, M. A., Varshney, A., Jain, A., Singh, S., et al. (2010). Genome-wide analysis of the host intracellular network that regulates survival of *Mycobacterium tuberculosis. Cell* 140, 731–743. doi: 10.1016/j.cell.2010.02.012

Kumar, S., Mittal, E., Deore, S., Kumar, A., Rahman, A., and Krishnasastry, M. V. (2015). Mycobacterial tlyA gene product is localized to the cell-wall without signal sequence. *Front. Cell. Infect. Microbiol.* 5:60. doi: 10.3389/fcimb.2015.00060

Kumar, Y., and Valdivia, R. H. (2009). Leading a sheltered life: intracellular pathogens and maintenance of vacuolar compartments. *Cell Host Microbe* 5, 593–601. doi: 10.1016/j.chom.2009.05.014

Laencina, L., Dubois, V., Le Moigne, V., Viljoen, A., Majlessi, L., Pritchard, J., et al. (2018). Identification of genes required for *Mycobacterium abscessus* growth *in vivo* with a prominent role of the ESX-4 locus. *Proc. Natl. Acad. Sci. U.S.A.* 115, E1002–E1011. doi: 10.1073/pnas.1713195115

Layre, E. (2020). Trafficking of *Mycobacterium tuberculosis* envelope components and release within extracellular vesicles: host-pathogen interactions beyond the wall. *Front. Immunol.* 11:1230. doi: 10.3389/fimmu.2020.01230

Layre, E., Lee, H. J., Young, D. C., Martinot, A. J., Buter, J., Minnaard, A. J., et al. (2014). Molecular profiling of *Mycobacterium tuberculosis* identifies tuberculosinyl nucleoside products of the virulence-associated enzyme Rv3378c. *Proc. Natl. Acad. Sci. U.S.A.* 111, 2978–2983. doi: 10.1073/pnas.1315883111

Lee, J., Choi, D.-S., Lee, J. S., Kim, D.-K., Go, G., Park, S.-M., et al. (2015). Proteomic analysis of extracellular vesicles derived from *Mycobacterium tuberculosis. Proteomics* 15, 3331–3337. doi: 10.1002/pmic.201500037

Lerner, T. R., Borel, S., Greenwood, D. J., Repnik, U., Russell, M. R. G., Herbst, S., et al. (2017). *Mycobacterium tuberculosis* replicates within necrotic human macrophages. *J. Cell Biol.* 216, 583–594. doi: 10.1083/jcb.2016 03040

Lerner, T. R., Queval, C. J., Fearns, A., Repnik, U., Griffiths, G., and Gutierrez, M. G. (2018). Phthiocerol dimycocerosates promote access to the cytosol and intracellular burden of *Mycobacterium tuberculosis* in lymphatic endothelial cells. *BMC Biol.* 16:1. doi: 10.1186/s12915-017-0471-6

Lerner, T. R., Queval, C. J., Lai, R. P., Russell, M. R., Fearns, A., Greenwood, D. J., et al. (2020). *Mycobacterium tuberculosis* cords within lymphatic endothelial cells to evade host immunity. *JCI Insight* 5:221. doi: 10.1172/jci.insight. 136937

Lienard, J., Nobs, E., Lovins, V., Movert, E., Valfridsson, C., and Carlsson, F. (2020). The *Mycobacterium marinum* ESX-1 system mediates phagosomal permeabilization and type I interferon production via separable mechanisms. *Proc. Natl. Acad. Sci. U.S.A.* 117, 1160–1166. doi: 10.1073/pnas.1911646117

Ligon, L. S., Hayden, J. D., and Braunstein, M. (2012). The ins and outs of *Mycobacterium tuberculosis* protein export. *Tuberculosis* 92, 121–132. doi: 10.1016/j.tube.2011.11.005

Liu, C. H., Liu, H., and Ge, B. (2017). Innate immunity in tuberculosis: host defense vs pathogen evasion. *Nat. Publ. Group* 14, 963–975. doi: 10.1038/cmi.2017.88

Long, J. E., DeJesus, M., Ward, D., Baker, R. E., Ioerger, T., and Sassetti, C. M. (2015). Identifying essential genes in *Mycobacterium tuberculosis* by global phenotypic profiling. *Methods Mol. Biol.* 1279, 79–95. doi: 10.1007/978-1-4939-2398-4_6

Lopes Fischer, N., Naseer, N., Shin, S., and Brodsky, I. E. (2020). Effector-triggered immunity and pathogen sensing in metazoans. *Nat. Microbiol.* 5, 14–26. doi: 10.1038/s41564-019-0623-2

Målen, H., Berven, F. S., Fladmark, K. E., and Wiker, H. G. (2007). Comprehensive analysis of exported proteins from *Mycobacterium tuberculosis* H37Rv. *Proteomics* 7, 1702–1718. doi: 10.1002/pmic.200600853

Målen, H., Pathak, S., Søfteland, T., de Souza, G. A., and Wiker, H. G. (2010). Definition of novel cell envelope associated proteins in Triton X-114 extracts of *Mycobacterium tuberculosis* H37Rv. *BMC Microbiol.* 10:132. doi: 10.1186/1471-2180-10-132

Ma, Y., Keil, V., and Sun, J. (2015). Characterization of *Mycobacterium tuberculosis* EsxA membrane insertion: roles of N- and C-terminal flexible arms and central helix-turn-helix motif. *J. Biol. Chem.* 290, 7314–7322. doi: 10.1074/jbc.M114.622076

Madan-Lala, R., Peixoto, K. V., Re, F., and Rengarajan, J. (2011). *Mycobacterium tuberculosis* Hip1 dampens macrophage proinflammatory responses by limiting toll-like receptor 2 activation. *Infect. Immun.* 79, 4828–4838. doi: 10.1128/IAI.05574-11

Madan-Lala, R., Sia, J. K., King, R., Adekambi, T., Monin, L., Khader, S. A., et al. (2014). *Mycobacterium tuberculosis* impairs dendritic cell functions through the serine hydrolase Hip1. *J. Immunol.* 192, 4263–4272. doi: 10.4049/jimmunol.1303185

Maeda, N., Nigou, J., Herrmann, J. L., Jackson, M., Amara, A., Lagrange, P. H., et al. (2003). The cell surface receptor DC-SIGN discriminates between Mycobacterium species through selective recognition of the mannose caps on lipoarabinomannan. *J. Biol. Chem.* 278, 5513–5516. doi: 10.1074/jbc.C200586200

Majlessi, L., Prados Rosales, R., Casadevall, A., and Brosch, R. (2015). Release of mycobacterial antigens. *Immunol. Rev.* 264, 25–45. doi: 10.1111/imr.12251

Malm, S., Maaß, S., Schaible, U. E., Ehlers, S., and Niemann, S. (2018). *In vivo* virulence of *Mycobacterium tuberculosis* depends on a single homologue of the LytR-CpsA-Psr proteins. *Sci. Rep.* 8, 3936–3910. doi: 10.1038/s41598-018-22012-6

Manzanillo, P. S., Ayres, J. S., Watson, R. O., Collins, A. C., Souza, G., Rae, C. S., et al. (2013). The ubiquitin ligase parkin mediates resistance to intracellular pathogens. *Nature* 501, 512–516. doi: 10.1038/nature12566

Mayer-Barber, K. D., and Sher, A. (2015). Cytokine and lipid mediator networks in tuberculosis. *Immunol. Rev.* 264, 264–275. doi: 10.1111/imr.12249

Mehra, A., Zahra, A., Thompson, V., Sirisaengtaksin, N., Wells, A., Porto, M., et al. (2013). *Mycobacterium tuberculosis* type VII secreted effector EsxH targets host ESCRT to impair trafficking. *PLoS Pathog.* 9:e1003734. doi: 10.1371/journal.ppat.1003734

Menozzi, F. D., Rouse, J. H., Alavi, M., Laude-Sharp, M., Muller, J., Bischoff, R., et al. (1996). Identification of a heparin-binding hemagglutinin present in mycobacteria. *J. Exp. Med.* 184, 993–1001. doi: 10.1084/jem.184.3.993

Miller, J. L., Velmurugan, K., Cowan, M. J., and Briken, V. (2010). The type I NADH dehydrogenase of *Mycobacterium tuberculosis* counters phagosomal NOX2 activity to inhibit TNF-alpha-mediated host cell apoptosis. *PLoS Pathog.* 6:e1000864. doi: 10.1371/journal.ppat.1000864

Mishra, M., Adhyapak, P., Dadhich, R., and Kapoor, S. (2019). Dynamic remodeling of the host cell membrane by virulent mycobacterial sulfoglycolipid-1. *Sci. Rep.* 9, 12844–12813. doi: 10.1038/s41598-019-49343-2

Mittal, E., Kumar, S., Rahman, A., and Krishnasastry, M. V. (2014). Modulation of phagolysosome maturation by bacterial tlyA gene product. *J. Biosci.* 39, 821–834. doi: 10.1007/s12038-014-9472-6

Mittal, E., Skowyra, M. L., Uwase, G., Tinaztepe, E., Mehra, A., Koster, S., et al. (2018). *Mycobacterium tuberculosis* Type VII secretion system effectors differentially impact the ESCRT endomembrane damage response. *MBio* 9:2763. doi: 10.1128/mBio.01765-18

Mohareer, K., Asalla, S., and Banerjee, S. (2018). Cell death at the cross roads of host-pathogen interaction in *Mycobacterium tuberculosis* infection. *Tuberculosis* 113, 99–121. doi: 10.1016/j.tube.2018.09.007

Moraco, A. H., and Kornfeld, H. (2014). Cell death and autophagy in tuberculosis. *Semin. Immunol.* 26, 497–511. doi: 10.1016/j.smim.2014.10.001

Moreira-Teixeira, L., Mayer-Barber, K., Sher, A., and O'Garra, A. (2018). Type I interferons in tuberculosis: foe and occasionally friend. *J. Exp. Med.* 215, 1273–1285. doi: 10.1084/jem.20180325

Murphy, K. C., Nelson, S. J., Nambi, S., Papavinasasundaram, K., Baer, C. E., and Sassetti, C. M. (2018). ORBIT: a new paradigm for genetic engineering of mycobacterial chromosomes. *MBio* 9:6836. doi: 10.1128/mBio.01467-18

Murry, J. P., Pandey, A. K., Sassetti, C. M., and Rubin, E. J. (2009). Phthiocerol dimycocerosate transport is required for resisting interferon-gamma-independent immunity. *J. Infect. Dis.* 200, 774–782. doi: 10.1086/605128

Naffin-Olivos, J. L., Georgieva, M., Goldfarb, N., Madan-Lala, R., Dong, L., Bizzell, E., et al. (2014). *Mycobacterium tuberculosis* Hip1 modulates macrophage responses through proteolysis of GroEL2. *PLoS Pathog.* 10:e1004132. doi: 10.1371/journal.ppat.1004132

Nakayama, H., Kurihara, H., Morita, Y. S., Kinoshita, T., Mauri, L., Prinetti, A., et al. (2016). Lipoarabinomannan binding to lactosylceramide in lipid rafts is essential for the phagocytosis of mycobacteria by human neutrophils. *Sci. Signal.* 9:ra101. doi: 10.1126/scisignal.aaf1585

Neyrolles, O., and Guilhot, C. (2011). Recent advances in deciphering the contribution of *Mycobacterium tuberculosis* lipids to pathogenesis. *Tuberculosis* 91, 187–195. doi: 10.1016/j.tube.2011.01.002

Nigou, J., Vasselon, T., Ray, A., Constant, P., Gilleron, M., Besra, G. S., et al. (2008). Mannan chain length controls lipoglycans signaling via and binding to TLR2. *J. Immunol.* 180, 6696–6702. doi: 10.4049/jimmunol.180.10.6696

Olive, A. J., and Sassetti, C. M. (2018). Tolerating the unwelcome guest; how the host withstands persistent *Mycobacterium tuberculosis*. *Front. Immunol.* 9:2094. doi: 10.3389/fimmu.2018.02094

Ortalo-Magné, A., Lemassu, A., Lanéelle, M. A., Bardou, F., Silve, G., Gounon, P., et al. (1996). Identification of the surface-exposed lipids on the cell envelopes of *Mycobacterium tuberculosis* and other mycobacterial species. *J. Bacteriol.* 178, 456–461. doi: 10.1128/JB.178.2.456-461.1996

Ostrop, J., Jozefowski, K., Zimmermann, S., Hofmann, K., Strasser, E., Lepenies, B., et al. (2015). Contribution of MINCLE-SYK signaling to activation of primary human apcs by mycobacterial cord factor and the novel adjuvant TDB. *J. Immunol.* 195, 2417–2428. doi: 10.4049/jimmunol.1500102

Pajuelo, D., González-Juarbe, N., Tak, U., Sun, J., Orihuela, C. J., and Niederweis, M. (2018). NAD+ depletion triggers macrophage necroptosis, a cell death pathway exploited by *Mycobacterium tuberculosis*. *Cell Rep.* 24, 429–440. doi: 10.1016/j.celrep.2018.06.042

Passemar, C., Arbues, A., Malaga, W., Mercier, I., Moreau, F., Lepourry, L., et al. (2014). Multiple deletions in the polyketide synthase gene repertoire of *Mycobacterium tuberculosis* reveal functional overlap of cell envelope lipids in host-pathogen interactions. *Cell. Microbiol.* 16, 195–213. doi: 10.1111/cmi.12214

Pathak, S. K., Basu, S., Basu, K. K., Banerjee, A., Pathak, S., Bhattacharyya, A., et al. (2007). Direct extracellular interaction between the early secreted antigen ESAT-6 of *Mycobacterium tuberculosis* and TLR2 inhibits TLR signaling in macrophages. *Nat. Immunol.* 8, 610–618. doi: 10.1038/ni1468

Peng, X., and Sun, J. (2016). Mechanism of ESAT-6 membrane interaction and its roles in pathogenesis of *Mycobacterium tuberculosis*. *Toxicon* 116, 29–34. doi: 10.1016/j.toxicon.2015.10.003

Penn, B. H., Netter, Z., Johnson, J. R., Von Dollen, J., Jang, G. M., Johnson, T., et al. (2018). An Mtb-human protein-protein interaction map identifies a switch between host antiviral and antibacterial responses. *Mol. Cell* 71, 637–648.e5. doi: 10.1016/j.molcel.2018.07.010

Pethe, K., Alonso, S., Biet, F., Delogu, G., Brennan, M. J., Locht, C., et al. (2001). The heparin-binding haemagglutinin of *M. tuberculosis* is required for extrapulmonary dissemination. *Nature* 412, 190–194. doi: 10.1038/35084083

Poirier, V., Bach, H., and Av-Gay, Y. (2014). *Mycobacterium tuberculosis* promotes anti-apoptotic activity of the macrophage by PtpA protein-dependent dephosphorylation of host GSK3α. *J. Biol. Chem.* 289, 29376–29385. doi: 10.1074/jbc.M114.582502

Portal Celhay, C., Tufariello, J. M., Srivastava, S., Zahra, A., Klevorn, T., Grace, P. S., et al. (2016). *Mycobacterium tuberculosis* EsxH inhibits ESCRT-dependent CD4(+) T-cell activation. *Nat. Microbiol.* 2:16232. doi: 10.1038/nmicrobiol.2016.232

Pradhan, G., Shrivastva, R., and Mukhopadhyay, S. (2018). Mycobacterial PknG targets the rab7l1 signaling pathway to inhibit phagosome-lysosome fusion. *J. Immunol.* 201, 1421–1433. doi: 10.4049/jimmunol.1800530

Prados Rosales, R., Baena, A., Martinez, L. R., Luque-Garcia, J., Kalscheuer, R., Veeraraghavan, U., et al. (2011). Mycobacteria release active membrane vesicles that modulate immune responses in a TLR2-dependent manner in mice. *J. Clin. Invest.* 121, 1471–1483. doi: 10.1172/JCI44261

Prados Rosales, R., Carreño, L. J., Batista-Gonzalez, A., Baena, A., Venkataswamy, M. M., Xu, J., et al. (2014a). Mycobacterial membrane vesicles administered systemically in mice induce a protective immune response to surface compartments of *Mycobacterium tuberculosis*. *MBio* 5:e01921–e01914. doi: 10.1128/mBio.01921-14

Prados Rosales, R., Weinrick, B. C., Piqué, D. G., Jacobs, W. R., Casadevall, A., and Rodriguez, G. M. (2014b). Role for *Mycobacterium tuberculosis* membrane vesicles in iron acquisition. *J. Bacteriol.* 196, 1250–1256. doi: 10.1128/JB.01090-13

Prisic, S., and Husson, R. N. (2014). *Mycobacterium tuberculosis* serine/threonine protein kinases. *Microbiol. Spectr.* 2, 1–42. doi: 10.1128/microbiolspec.MGM2-0006-2013

Puri, R. V., Reddy, P. V., and Tyagi, A. K. (2013). Secreted acid phosphatase (sapm) of *Mycobacterium tuberculosis* is indispensable for arresting phagosomal maturation and growth of the pathogen in guinea pig tissues. *PLoS ONE* 8:e70514. doi: 10.1371/journal.pone.0070514

Queiroz, A., and Riley, L. W. (2017). Bacterial immunostat: *Mycobacterium tuberculosis* lipids and their role in the host immune response. *Rev. Soc. Bras. Med. Trop.* 50, 9–18. doi: 10.1590/0037-8682-0230-2016

Quesniaux, V. J., Nicolle, D. M., Torres, D., Kremer, L., Guérardel, Y., Nigou, J., et al. (2004). Toll-like receptor 2 (TLR2)-dependent-positive and TLR2-independent-negative regulation of proinflammatory cytokines by mycobacterial lipomannans. *J. Immunol.* 172, 4425–4434. doi: 10.4049/jimmunol.172.7.4425

Queval, C. J., Brosch, R., and Simeone, R. (2017). The macrophage: a disputed fortress in the battle against *Mycobacterium tuberculosis*. *Front. Microbiol.* 8:883. doi: 10.3389/fmicb.2017.02284

Quigley, J., Hughitt, V. K., Velikovsky, C. A., Mariuzza, R. A., El-Sayed, N. M., and Briken, V. (2017). The cell wall lipid PDIM contributes to phagosomal escape and host cell exit of *Mycobacterium tuberculosis*. *MBio* 8, e00148–e00117. doi: 10.1128/mBio.00148-17

Rahman, A., Srivastava, S. S., Sneh, A., Ahmed, N., and Krishnasastry, M. V. (2010). Molecular characterization of tlyA gene product, Rv1694 of *Mycobacterium tuberculosis*: a non-conventional hemolysin and a ribosomal RNA methyl transferase. *BMC Biochem.* 11:35. doi: 10.1186/1471-2091-11-35

Rahman, M. A., Sobia, P., Dwivedi, V. P., Bhawsar, A., Singh, D. K., Sharma, P., et al. (2015). *Mycobacterium tuberculosis* TlyA protein negatively regulates T helper (Th) 1 and Th17 differentiation and promotes tuberculosis pathogenesis. *J. Biol. Chem.* 290, 14407–14417. doi: 10.1074/jbc.M115.653600

Ramakrishnan, L., Federspiel, N. A., and Falkow, S. (2000). Granuloma-specific expression of Mycobacterium virulence proteins from the glycine-rich PE-PGRS family. *Science* 288, 1436–1439. doi: 10.1126/science.288.5470.1436

Ranes, M. G., Rauzier, J., Lagranderie, M., Gheorghiu, M., and Gicquel, B. (1990). Functional analysis of pAL5000, a plasmid from mycobacterium fortuitum: construction of a "mini" mycobacterium-*Escherichia coli* shuttle vector. *J. Bacteriol.* 172, 2793–2797. doi: 10.1128/JB.172.5.2793-2797.1990

Rao, V., Fujiwara, N., Porcelli, S. A., and Glickman, M. S. (2005). *Mycobacterium tuberculosis* controls host innate immune activation through cyclopropane modification of a glycolipid effector molecule. *J. Exp. Med.* 201, 535–543. doi: 10.1084/jem.20041668

Ray, S., Vazquez Reyes, S., Xiao, C., and Sun, J. (2019). Effects of membrane lipid composition on *Mycobacterium tuberculosis* EsxA membrane insertion: a dual play of fluidity and charge. *Tuberculosis* 118:101854. doi: 10.1016/j.tube.2019.07.005

Reed, M. B., Domenech, P., Manca, C., Su, H., Barczak, A. K., Kreiswirth, B. N., et al. (2004). A glycolipid of hypervirulent tuberculosis strains that inhibits the innate immune response. *Nature* 431, 84–87. doi: 10.1038/nature 02837

Rengarajan, J., Bloom, B. R., and Rubin, E. J. (2005). Genome-wide requirements for *Mycobacterium tuberculosis* adaptation and survival in macrophages. *Proc. Natl. Acad. Sci. U.S.A.* 102, 8327–8332. doi: 10.1073/pnas.0503272102

Rengarajan, J., Murphy, E., Park, A., Krone, C. L., Hett, E. C., Bloom, B. R., et al. (2008). *Mycobacterium tuberculosis* Rv2224c modulates innate immune responses. *Proc. Natl. Acad. Sci. U.S.A.* 105, 264–269. doi: 10.1073/pnas.0710601105

Renshaw, P. S., Lightbody, K. L., Veverka, V., Muskett, F. W., Kelly, G., Frenkiel, T. A., et al. (2005). Structure and function of the complex formed by the tuberculosis virulence factors CFP-10 and ESAT-6. *EMBO J.* 24, 2491–2498. doi: 10.1038/sj.emboj.7600732

Rhoades, E., Hsu, F. F., Torrelles, J. B., Turk, J., Chatterjee, D., and Russell, D. G. (2003). Identification and macrophage-activating activity of glycolipids released from intracellular *Mycobacterium bovis* BCG. *Mol. Microbiol.* 48, 875–888. doi: 10.1046/j.1365-2958.2003.03473.x

Robinson, N., Kolter, T., Wolke, M., Rybniker, J., Hartmann, P., and Plum, G. (2008). Mycobacterial phenolic glycolipid inhibits phagosome maturation and subverts the pro-inflammatory cytokine response. *Traffic* 9, 1936–1947. doi: 10.1111/j.1600-0854.2008.00804.x

Rock, J. M., Hopkins, F. F., Chavez, A., Diallo, M., Chase, M. R., Gerrick, E. R., et al. (2017). Programmable transcriptional repression in mycobacteria using an orthogonal CRISPR interference platform. *Nat. Microbiol.* 2:16274. doi: 10.1038/nmicrobiol.2016.274

Romagnoli, A., Etna, M. P., Giacomini, E., Pardini, M., Remoli, M. E., Corazzari, M., et al. (2012). ESX-1 dependent impairment of autophagic flux by *Mycobacterium tuberculosis* in human dendritic cells. *Autophagy* 8, 1357–1370. doi: 10.4161/auto.20881

Rousseau, C., Neyrolles, O., Bordat, Y., Giroux, S., Sirakova, T. D., Prevost, M.-C., et al. (2003a). Deficiency in mycolipenate- and mycosanoate-derived acyltrehaloses enhances early interactions of *Mycobacterium tuberculosis* with host cells. *Cell. Microbiol.* 5, 405–415. doi: 10.1046/j.1462-5822.2003.00289.x

Rousseau, C., Turner, O. C., Rush, E., Bordat, Y., Sirakova, T. D., Kolattukudy, P. E., et al. (2003b). Sulfolipid deficiency does not affect the virulence of *Mycobacterium tuberculosis* H37Rv in mice and guinea pigs. *Infect. Immun.* 71, 4684–4690. doi: 10.1128/IAI.71.8.4684-4690.2003

Rousseau, C., Winter, N., Pivert, E., Bordat, Y., Neyrolles, O., Avé, P., et al. (2004). Production of phthiocerol dimycocerosates protects *Mycobacterium tuberculosis* from the cidal activity of reactive nitrogen intermediates produced by macrophages and modulates the early immune response to infection. *Cell. Microbiol.* 6, 277–287. doi: 10.1046/j.1462-5822.2004.00368.x

Ruhl, C. R., Pasko, B. L., Khan, H. S., Kindt, L. M., Stamm, C. E., Franco, L. H., et al. (2020). *Mycobacterium tuberculosis* sulfolipid-1 activates nociceptive neurons and induces cough. *Cell* 181, 293–305.e11. doi: 10.1016/j.cell.2020.02.026

Saikolappan, S., Estrella, J., Sasindran, S. J., Khan, A., Armitige, L. Y., Jagannath, C., et al. (2012). The fbpA/sapM double knock out strain of *Mycobacterium tuberculosis* is highly attenuated and immunogenic in macrophages. *PLoS ONE* 7:e36198. doi: 10.1371/journal.pone.0036198

Saini, N. K., Baena, A., Ng, T. W., Venkataswamy, M. M., Kennedy, S. C., Kunnath-Velayudhan, S., et al. (2016). Suppression of autophagy and antigen presentation by *Mycobacterium tuberculosis* PE_PGRS47. *Nat. Microbiol.* 1:16133. doi: 10.1038/nmicrobiol.2016.133

Saleh, M. T., and Belisle, J. T. (2000). Secretion of an acid phosphatase (SapM) by *Mycobacterium tuberculosis* that is similar to eukaryotic acid phosphatases. *J. Bacteriol.* 182, 6850–6853. doi: 10.1128/JB.182.23.6850-6853.2000

Samuel, L. P., Song, C.-H., Wei, J., Roberts, E. A., Dahl, J. L., Barry, C. E., et al. (2007). Expression, production and release of the eis protein by *Mycobacterium tuberculosis* during infection of macrophages and its effect on cytokine secretion. *Microbiology* 153, 529–540. doi: 10.1099/mic.0.2006/002642-0

Sancho-Vaello, E., Albesa-Jové, D., Rodrigo-Unzueta, A., and Guerin, M. E. (2017). Structural basis of phosphatidyl-myo-inositol mannosides biosynthesis in mycobacteria. *BBA Mol. Cell Biol. Lipids* 1862, 1355–1367. doi: 10.1016/j.bbalip.2016.11.002

Sani, M., Houben, E. N. G., Geurtsen, J., Pierson, J., de Punder, K., van Zon, M., et al. (2010). Direct visualization by cryo-EM of the mycobacterial capsular layer: a labile structure containing ESX-1-secreted proteins. *PLoS Pathog.* 6:e1000794. doi: 10.1371/journal.ppat.1000794

Schnettger, L., Rodgers, A., Repnik, U., Lai, R. P., Pei, G., Verdoes, M., et al. (2017). A Rab20-Dependent membrane trafficking pathway controls *M. tuberculosis* replication by regulating phagosome spaciousness and integrity. *Cell Host Microbe* 21, 619–628.e5. doi: 10.1016/j.chom.2017.04.004

Shin, D.-M., Jeon, B.-Y., Lee, H.-M., Jin, H. S., Yuk, J.-M., Song, C.-H., et al. (2010). *Mycobacterium tuberculosis* eis regulates autophagy, inflammation, and cell death through redox-dependent signaling. *PLoS Pathog.* 6:e1001230. doi: 10.1371/journal.ppat.1001230

Sia, J. K., and Rengarajan, J. (2019). Immunology of *Mycobacterium tuberculosis* infections. *Microbiol. Spectr.* 7, 1–37. doi: 10.1128/microbiolspec.GPP3-0022-2018

Simeone, R., Sayes, F., Song, O., Gröschel, M. I., Brodin, P., Brosch, R., et al. (2015). Cytosolic access of *Mycobacterium tuberculosis*: critical impact of phagosomal acidification control and demonstration of occurrence *in vivo*. *PLoS Pathog.* 11:e1004650. doi: 10.1371/journal.ppat.1004650

Simmons, J. D., Stein, C. M., Seshadri, C., Campo, M., Alter, G., Fortune, S., et al. (2018). Immunological mechanisms of human resistance to persistent *Mycobacterium tuberculosis* infection. *Nat. Rev. Immunol.* 18, 575–589. doi: 10.1038/s41577-018-0025-3

Singh, R., Rao, V., Shakila, H., Gupta, R., Khera, A., Dhar, N., et al. (2003). Disruption of mptpB impairs the ability of *Mycobacterium tuberculosis* to survive in guinea pigs. *Mol. Microbiol.* 50, 751–762. doi: 10.1046/j.1365-2958.2003.03712.x

Smith, C. M., Proulx, M. K., Lai, R., Kiritsy, M. C., Bell, T. A., Hock, P., et al. (2019). functionally overlapping variants control tuberculosis susceptibility in collaborative cross mice. *MBio* 10, 621. doi: 10.1128/mBio.02791-19

Smith, J., Manoranjan, J., Pan, M., Bohsali, A., Xu, J., Liu, J., et al. (2008). Evidence for pore formation in host cell membranes by ESX-1-secreted ESAT-6 and its role in *Mycobacterium marinum* escape from the vacuole. *Infect. Immun.* 76, 5478–5487. doi: 10.1128/IAI.00614-08

Sohn, H., Kim, J.-S., Shin, S. J., Kim, K., Won, C.-J., Kim, W. S., et al. (2011). Targeting of *Mycobacterium tuberculosis* heparin-binding hemagglutinin to mitochondria in macrophages. *PLoS Pathog.* 7:e1002435. doi: 10.1371/journal.ppat.1002435

Spargo, B. J., Crowe, L. M., Ioneda, T., Beaman, B. L., and Crowe, J. H. (1991). Cord factor (alpha,alpha-trehalose 6,6'-dimycolate) inhibits fusion between phospholipid vesicles. *Proc. Natl. Acad. Sci. U.S.A.* 88, 737–740. doi: 10.1073/pnas.88.3.737

Speer, A., Sun, J., Danilchanka, O., Meikle, V., Rowland, J. L., Walter, K., et al. (2015). Surface hydrolysis of sphingomyelin by the outer membrane protein Rv0888 supports replication of *Mycobacterium tuberculosis* in macrophages. *Mol. Microbiol.* 97, 881–897. doi: 10.1111/mmi.13073

Sreejit, G., Ahmed, A., Parveen, N., Jha, V., Valluri, V. L., Ghosh, S., et al. (2014). The ESAT-6 protein of *Mycobacterium tuberculosis* interacts with beta-2-microglobulin (β2m) affecting antigen presentation function of macrophage. *PLoS Pathog.* 10:e1004446. doi: 10.1371/journal.ppat.1004446

Srinivasan, L., Ahlbrand, S., and Briken, V. (2014). Interaction of *Mycobacterium tuberculosis* with host cell death pathways. *Cold Spring Harb. Perspect. Med.* 4:a022459. doi: 10.1101/cshperspect.a022459

Srinivasan, L., Gurses, S. A., Hurley, B. E., Miller, J. L., Karakousis, P. C., and Briken, V. (2016). Identification of a transcription factor that regulates host cell exit and virulence of *Mycobacterium tuberculosis*. *PLoS Pathog.* 12:e1005652. doi: 10.1371/journal.ppat.1005652

Srivastava, S., Battu, M. B., Khan, M. Z., Nandicoori, V. K., and Mukhopadhyay, S. (2019). *Mycobacterium tuberculosis* PPE2 protein interacts with p67phox and inhibits reactive oxygen species production. *J. Immunol.* 2019:ji1801143. doi: 10.4049/jimmunol.1801143

Sturgill-Koszycki, S., Schlesinger, P. H., Chakraborty, P., Haddix, P. L., Collins, H. L., Fok, A. K., et al. (1994). Lack of acidification in Mycobacterium phagosomes produced by exclusion of the vesicular proton-ATPase. *Science* 263, 678–681. doi: 10.1126/science.8303277

Sun, J., Singh, V., Lau, A., Stokes, R. W., Obregón-Henao, A., Orme, I. M., et al. (2013). *Mycobacterium tuberculosis* nucleoside diphosphate kinase inactivates small GTPases leading to evasion of innate immunity. *PLoS Pathog.* 9:e1003499. doi: 10.1371/journal.ppat.1003499

Sun, J., Siroy, A., Lokareddy, R. K., Speer, A., Doornbos, K. S., Cingolani, G., et al. (2015). The tuberculosis necrotizing toxin kills macrophages by hydrolyzing NAD. *Nat. Struct. Mol. Biol* 22, 672–678. doi: 10.1038/nsmb.3064

Sun, J., Wang, X., Lau, A., Liao, T.-Y. A., Bucci, C., and Hmama, Z. (2010). Mycobacterial nucleoside diphosphate kinase blocks phagosome maturation in murine RAW 264.7 macrophages. *PLoS ONE* 5:e8769. doi: 10.1371/journal.pone.0008769

Sut, A., Sirugue, S., Sixou, S., Lakhdar-Ghazal, F., Tocanne, J. F., and Lanéelle, G. (1990). Mycobacteria glycolipids as potential pathogenicity effectors: alteration of model and natural membranes. *Biochemistry* 29, 8498–8502. doi: 10.1021/bi00488a042

Szymanska, E., Budick-Harmelin, N., and Miaczynska, M. (2018). Endosomal "sort" of signaling control: the role of ESCRT machinery in regulation of receptor-mediated signaling pathways. *Semin. Cell Dev. Biol.* 74, 11–20. doi: 10.1016/j.semcdb.2017.08.012

Torrelles, J. B., Azad, A. K., and Schlesinger, L. S. (2006). Fine discrimination in the recognition of individual species of phosphatidyl-myo-inositol mannosides from *Mycobacterium tuberculosis* by C-type lectin pattern recognition receptors. *J. Immunol.* 177, 1805–1816. doi: 10.4049/jimmunol.177.3.1805

Truong, T., and Penn, B. H. (2020). An *M. tuberculosis* metabolic enzyme moonlights as an anti-inflammatory effector protein. *Cell Host Microbe* 27, 310–312. doi: 10.1016/j.chom.2020.02.012

Tufariello, J. M., Chapman, J. R., Kerantzas, C. A., Wong, K.-W., Vilchèze, C., Jones, C. M., et al. (2016). Separable roles for *Mycobacterium tuberculosis* ESX-3 effectors in iron acquisition and virulence. *Proc. Natl. Acad. Sci. U.S.A.* 113, E348–E357. doi: 10.1073/pnas.1523321113

Upadhyay, S., Mittal, E., and Philips, J. A. (2018). Tuberculosis and the art of macrophage manipulation. *Pathog. Dis* 76:37. doi: 10.1093/femspd/fty037

Upadhyay, S., and Philips, J. A. (2019). LC3-associated phagocytosis: host defense and microbial response. *Curr. Opin. Immunol.* 60, 81–90. doi: 10.1016/j.coi.2019.04.012

van der Woude, A. D., Stoop, E. J. M., Stiess, M., Wang, S., Ummels, R., van Stempvoort, G., et al. (2014). Analysis of SecA2-dependent substrates in *Mycobacterium marinum* identifies protein kinase G (PknG) as a virulence effector. *Cell. Microbiol.* 16, 280–295. doi: 10.1111/cmi.12221

Vandal, O. H., Roberts, J. A., Odaira, T., Schnappinger, D., Nathan, C. F., and Ehrt, S. (2009). Acid-susceptible mutants of *Mycobacterium tuberculosis* share hypersusceptibility to cell wall and oxidative stress and to the host environment. *J. Bacteriol.* 191, 625–631. doi: 10.1128/JB.00932-08

Vaziri, F., and Brosch, R. (2019). ESX/Type VII secretion systems-an important way out for mycobacterial proteins. *Microbiol. Spectr.* 7, 351–362. doi: 10.1128/9781683670285.ch28

Velmurugan, K., Chen, B., Miller, J. L., Azogue, S., Gurses, S., Hsu, T., et al. (2007). *Mycobacterium tuberculosis* nuoG is a virulence gene that inhibits apoptosis of infected host cells. *PLoS Pathog.* 3:e110. doi: 10.1371/journal.ppat.0030110

Venugopal, A., Bryk, R., Shi, S., Rhee, K., Rath, P., Schnappinger, D., et al. (2011). Virulence of *Mycobacterium tuberculosis* depends on lipoamide dehydrogenase, a member of three multienzyme complexes. *Cell Host Microbe* 9, 21–31. doi: 10.1016/j.chom.2010.12.004

Vergne, I., Chua, J., Lee, H.-H., Lucas, M., Belisle, J., and Deretic, V. (2005). Mechanism of phagolysosome biogenesis block by viable *Mycobacterium tuberculosis. Proc. Natl. Acad. Sci. U.S.A.* 102, 4033–4038. doi: 10.1073/pnas.0409716102

Vergne, I., Fratti, R. A., Hill, P. J., Chua, J., Belisle, J., and Deretic, V. (2004). *Mycobacterium tuberculosis* phagosome maturation arrest: mycobacterial phosphatidylinositol analog phosphatidylinositol mannoside stimulates early endosomal fusion. *Mol. Biol. Cell.* 15, 751–760. doi: 10.1091/mbc.e03-05-0307

Vergne, I., Gilleron, M., and Nigou, J. (2014). Manipulation of the endocytic pathway and phagocyte functions by *Mycobacterium tuberculosis* lipoarabinomannan. *Front. Cell. Infect. Microbiol.* 4:187. doi: 10.3389/fcimb.2014.00187

Vincent, A. T., Nyongesa, S., Morneau, I., Reed, M. B., Tocheva, E. I., and Veyrier, F. J. (2018). The mycobacterial cell envelope: a relict from the past or the result of recent evolution? *Front. Microbiol.* 9:2341. doi: 10.3389/fmicb.2018.02341

Walburger, A., Koul, A., Ferrari, G., Nguyen, L., Prescianotto-Baschong, C., Huygen, K., et al. (2004). Protein kinase G from pathogenic mycobacteria promotes survival within macrophages. *Science* 304, 1800–1804. doi: 10.1126/science.1099384

Wang, J., Li, B.-X., Ge, P.-P., Li, J., Wang, Q., Gao, G. F., et al. (2015a). *Mycobacterium tuberculosis* suppresses innate immunity by coopting the host ubiquitin system. *Nat. Immunol.* 16, 237–245. doi: 10.1038/ni.3096

Wang, L., Liu, Z., Wang, J., Liu, H., Wu, J., Tang, T., et al. (2019). Oxidization of TGFβ-activated kinase by MPT53 is required for immunity to *Mycobacterium tuberculosis. Nat. Microbiol.* 4, 1378–1388. doi: 10.1038/s41564-019-0436-3

Wang, L., Wu, J., Li, J., Yang, H., Tang, T., Liang, H., et al. (2020). Host-mediated ubiquitination of a mycobacterial protein suppresses immunity. *Nature* 577, 682–688. doi: 10.1038/s41586-019-1915-7

Wang, Q., Zhu, L., Jones, V., Wang, C., Hua, Y., Shi, X., et al. (2015b). CpsA, a LytR-CpsA-Psr family protein in mycobacterium marinum, is required for cell wall integrity and virulence. *Infect. Immun.* 83, 2844–2854. doi: 10.1128/IAI.03081-14

Wassermann, R., Gulen, M. F., Sala, C., Perin, S. G., Lou, Y., Rybniker, J., et al. (2015). *Mycobacterium tuberculosis* differentially activates cgas- and inflammasome-dependent intracellular immune responses through ESX-1. *Cell Host Microbe* 17, 799–810. doi: 10.1016/j.chom.2015.05.003

Watson, R. O., Bell, S. L., MacDuff, D. A., Kimmey, J. M., Diner, E. J., Olivas, J., et al. (2015). The Cytosolic sensor cgas detects *Mycobacterium tuberculosis* DNA to induce type i interferons and activate autophagy. *Cell Host Microbe* 17, 811–819. doi: 10.1016/j.chom.2015.05.004

Watson, R. O., Manzanillo, P. S., and Cox, J. S. (2012). Extracellular *M. tuberculosis* DNA targets bacteria for autophagy by activating the host DNA-sensing pathway. *Cell* 150, 803–815. doi: 10.1016/j.cell.2012.06.040

Wei, J., Dahl, J. L., Moulder, J. W., Roberts, E. A., O'Gaora, P., Young, D. B., et al. (2000). Identification of a *Mycobacterium tuberculosis* gene that enhances mycobacterial survival in macrophages. *J. Bacteriol.* 182, 377–384. doi: 10.1128/JB.182.2.377-384.2000

Welin, A., Winberg, M. E., Abdalla, H., Sarndahl, E., Rasmusson, B., Stendahl, O., et al. (2008). Incorporation of *Mycobacterium tuberculosis* lipoarabinomannan into macrophage membrane rafts is a prerequisite for the phagosomal maturation block. *Infect. Immun.* 76, 2882–2887. doi: 10.1128/IAI.01549-07

Wong, D., Bach, H., Sun, J., Hmama, Z., and Av-Gay, Y. (2011). *Mycobacterium tuberculosis* protein tyrosine phosphatase (PtpA) excludes host vacuolar-H+-ATPase to inhibit phagosome acidification. *Proc. Natl. Acad. Sci. U.S.A.* 108, 19371–19376. doi: 10.1073/pnas.1109201108

Wong, D., Chao, J. D., and Av-Gay, Y. (2013). *Mycobacterium tuberculosis*-secreted phosphatases: from pathogenesis to targets for TB drug development. *Trends Microbiol.* 21, 100–109. doi: 10.1016/j.tim.2012.09.002

Wong, D., Li, W., Chao, J. D., Zhou, P., Narula, G., Tsui, C., et al. (2018). Protein tyrosine kinase, PtkA, is required for *Mycobacterium tuberculosis* growth in macrophages. *Sci. Rep.* 8, 155–112. doi: 10.1038/s41598-017-18547-9

Wren, B. W., Stabler, R. A., Das, S. S., Butcher, P. D., Mangan, J. A., Clarke, J. D., et al. (1998). Characterization of a haemolysin from *Mycobacterium tuberculosis* with homology to a virulence factor of *Serpulina hyodysenteriae. Microbiology* 144, 1205–1211. doi: 10.1099/00221287-144-5-1205

Yang, Y., Xu, P., He, P., Shi, F., Tang, Y., Guan, C., et al. (2020). Mycobacterial PPE13 activates inflammasome by interacting with the NATCH and LRR domains of NLRP3. *FASEB J.* 182:902. doi: 10.1096/fj.202000200RR

Yonekawa, A., Saijo, S., Hoshino, Y., Miyake, Y., Ishikawa, E., Suzukawa, M., et al. (2014). Dectin-2 is a direct receptor for mannose-capped lipoarabinomannan of mycobacteria. *Immunity* 41, 402–413. doi: 10.1016/j.immuni.2014.08.005

Zhang, Q., Wang, D., Jiang, G., Liu, W., Deng, Q., Li, X., et al. (2016). EsxA membrane-permeabilizing activity plays a key role in mycobacterial cytosolic translocation and virulence: effects of single-residue mutations at glutamine 5. *Sci. Rep.* 6, 32618–32615. doi: 10.1038/srep32618

Zhang, W., Lu, Q., Dong, Y., Yue, Y., and Xiong, S. (2018). Rv3033, as an emerging anti-apoptosis factor, facilitates mycobacteria survival via inhibiting macrophage intrinsic apoptosis. *Front. Immunol.* 9:2136. doi: 10.3389/fimmu.2018.02136

Zhou, B., He, Y., Zhang, X., Xu, J., Luo, Y., Wang, Y., et al. (2010). Targeting mycobacterium protein tyrosine phosphatase B for antituberculosis agents. *Proc. Natl Acad. Sci. U.S.A.* 107, 4573–4578. doi: 10.1073/pnas.0909133107

Zhou, P., Li, W., Wong, D., Xie, J., and Av-Gay, Y. (2015). Phosphorylation control of protein tyrosine phosphatase A activity in *Mycobacterium tuberculosis. FEBS Lett.* 589, 326–331. doi: 10.1016/j.febslet.2014.12.015

Zuber, B., Chami, M., Houssin, C., Dubochet, J., Griffiths, G., and Daffe, M. (2008). Direct visualization of the outer membrane of mycobacteria and corynebacteria in their native state. *J. Bacteriol.* 190, 5672–5680. doi: 10.1128/JB.01919-07

Zulauf, K. E., Sullivan, J. T., and Braunstein, M. (2018). The SecA2 pathway of *Mycobacterium tuberculosis* exports effectors that work in concert to arrest phagosome and autophagosome maturation. *PLoS Pathog.* 14:e1007011. doi: 10.1371/journal.ppat.1007011

4

Campylobacter jejuni Serine Protease HtrA Cleaves the Tight Junction Component Claudin-8

Irshad Sharafutdinov[1†], Delara Soltan Esmaeili[1†], Aileen Harrer[1†], Nicole Tegtmeyer[1], Heinrich Sticht[2] and Steffen Backert[1]*

[1] Department of Biology, Division of Microbiology, University of Erlangen-Nuremberg, Erlangen, Germany, [2] Division of Bioinformatics, Institute of Biochemistry, University of Erlangen-Nuremberg, Erlangen, Germany

***Correspondence:**
Steffen Backert
Steffen.Backert@fau.de

[†]These authors have contributed equally to this work

Campylobacter jejuni express the high temperature requirement protein A (HtrA), a secreted serine protease, which is implicated in virulence properties of the pathogen. Previous studies have shown that *C. jejuni* HtrA can cleave the epithelial transmembrane proteins occludin and E-cadherin in the tight and adherens junctions, respectively. In the present report, we studied the interaction of HtrA with another human tight junction protein, claudin-8. Confocal immunofluorescence experiments have shown that *C. jejuni* infection of the intestinal polarized epithelial cells *in vitro* leads to a relocation of claudin-8. Wild-type *C. jejuni* induced the downregulation of claudin-8 signals in the tight junctions and an accumulation of claudin-8 agglomerates in the cytoplasm, which were not seen during infection with isogenic Δ*htrA* knockout deletion or protease-inactive S197A point mutants. Western blotting of protein samples from infected *vs*. uninfected cells revealed that an 18-kDa carboxy-terminal fragment is cleaved-off from the 26-kDa full-length claudin-8 protein, but not during infection with the isogenic Δ*htrA* mutant. These results were confirmed by *in vitro* cleavage assays using the purified recombinant *C. jejuni* HtrA and human claudin-8 proteins. Recombinant HtrA cleaved purified claudin-8 *in vitro* giving rise to the same 18-kDa sized carboxy-terminal cleavage product. Mapping studies revealed that HtrA cleavage occurs in the first extracellular loop of claudin-8. Three-dimensional modeling of the claudin-8 structure identified an exposed HtrA cleavage site between the amino acids alanine 58 and asparagine 59, which is in well agreement with the mapping studies. Taken together, HtrA operates as a secreted virulence factor targeting multiple proteins both in the tight and adherens junctions. This strategy may help the bacteria to open the cell-to-cell junctions, and to transmigrate across the intestinal epithelium by a paracellular mechanism and establish an acute infection.

Keywords: claudin-8, occludin, tight junction, E-cadherin, *Campylobacter*, protease, HtrA

INTRODUCTION

Campylobacter jejuni is a Gram-negative, microaerophilic, spiral-shaped, flagellated proteobacterium that commensally colonizes the mucus layer in the lower intestinal tract of many birds and various mammals (Burnham and Hendrixson, 2018). The genus *Campylobacter* is evolutionarily related to the gastric pathogen *Helicobacter pylori* and considered to be the most common bacterial cause of human gastroenteritis in the world (Marder et al., 2017). *C. jejuni* represents a major zoonotic pathogen, which can be transmitted *via* a fecal > oral route through the consumption of contaminated poultry meats as well as other animal-derived food products (Hale et al., 2012). The rates of gastroenteric infections caused by *C. jejuni* are very high both in developed and developing countries (Kaakoush et al., 2015). Upon infection, *C. jejuni* appears to present lipooligosaccharides (LOS) at the bacterial cell surface, which in some patients can cross-react with human gangliosides resulting in autoimmune disorders such as the Guillian-Barré or Miller-Fisher syndromes (Ang et al., 2002; Yuki et al., 2004; Charles et al., 2017). Furthermore, in some cases this pathogen can promote colorectal cancer tumorigenesis through the action of microbial cytolethal distending toxin (CDT) (Brauner et al., 2010; He et al., 2019). Interestingly, CDT from *C. jejuni* was also extensively studied as an anti-cancer therapeutic agent for potential clinical applications (Lai et al., 2016). The pathogenic process generated by *C. jejuni* in the human intestine develops upon reaching the gut, where the bacteria attach to and then invade into epithelial cells, resulting in host tissue damage (Van Spreeuwel et al., 1985; Wooldridge and Ketley, 1997). Various bacterial adhesion proteins (adhesins) provide stable attachment through specific interaction with host cell receptors, which is a necessary requirement for subsequent host cell entry (Hermans et al., 2011; Backert et al., 2013). Several major adhesins were identified and have been reported to provide effective *C. jejuni* adhesion to a host cell, with the CadF (*Campylobacter* adhesin to fibronectin) protein being central in this process (Konkel et al., 1997; Schmidt et al., 2019). However, other adhesins also play important roles in cell attachment and include FlpA (fibronectin like protein A), JlpA (*jejuni* lipoprotein A), PEB1 (periplasmic binding protein 1), MOMP (major outer membrane protein), and some others (O Cróinín and Backert, 2012). Upon attachment, *C. jejuni* can enter host target cells *via* a signaling process involving the small Rho GTPases Rac1 and Cdc42 (Krause-Gruszczynska et al., 2007a; Krause-Gruszczynska et al., 2007b; Boehm et al., 2011; Krause-Gruszczynska et al., 2011; Eucker and Konkel, 2012). In addition, *C. jejuni* utilizes its flagellum as a type III secretion system (fT3SS) for the secretion or injection of effector proteins that interfere with host cell functions (Young et al., 1999; Christensen et al., 2009; Barrero-Tobon and Hendrixson, 2012). Furthermore, *C. jejuni* can transmigrate to the basolateral site of the intestinal epithelium by disrupting the host cellular junctions, and serine protease HtrA appears to play a driving role in this process (Boehm et al., 2018; Harrer et al., 2019).

HtrA proteins represent ATP-independent serine type proteases and are widely distributed both in prokaryotic and eukaryotic organisms (Clausen et al., 2002; Neddermann and Backert, 2019). Many bacteria can encode either one or more HtrA homologs (Li et al., 1996; Humphreys et al., 1999; Cortes et al., 2002; Purdy et al., 2002; Mo et al., 2006; Wilson et al., 2006; Ye et al., 2016). HtrA proteins due to their structure combine both protease and chaperone functions (Clausen et al., 2011). Bacterial HtrAs consist of an amino-terminal signal peptide, a trypsin-like serine protease domain and one or two PDZ-domains at the carboxy-terminus that regulate interactions with itself or other proteins (Kim and Kim, 2005; Skorko-Glonek et al., 2013). *Escherichia coli* is considered to be the best studied model organism concerning HtrA. This species encodes as many as three HtrA homologs, namely, DegP, DegQ, and DegS (Kim and Kim, 2005; Clausen et al., 2011). Their main function is to protect the bacterium against heat and other stresses, as well as to remove misfolded proteins. For example, inactivation of the *htrA* gene in *Streptococcus mutans* has been shown to affect its resistance to low and high temperatures, low pH as well as oxidative and DNA damaging agents (Biswas and Biswas, 2005). HtrAs can also play a major role in the pathogenesis of other Gram-positive and Gram-negative microbes (Backert et al., 2018). For example, the human pathogen *Streptococcus pyogenes* with impaired HtrA function expressed reduced amounts of mature streptococcal pyrogenic exotoxin B (SpeB), as it was shown by Western blot analysis and protease assays (Cole et al., 2007). In addition, *H. pylori* was the first bacterium shown to secrete HtrA into the extracellular environment, which was associated with the paracellular transmigration of the pathogen through cleavage of the host adherens junction protein E-cadherin (Harrer et al., 2018).

Campylobacter jejuni encodes one HtrA homolog, whose function was analysed by biochemical assays *in vitro* and in the bacteria *in vivo* (Brondsted et al., 2005; Zarzecka et al., 2020). Cryo-electron microscopy revealed the architecture of *C. jejuni* HtrA defined as a dodecamer, assembled by four trimers (Zarzecka et al., 2020). However, HtrA of *C. jejuni* can be also secreted into the extracellular space, where it has been shown to cleave the extracellular domain of E-cadherin at various positions (Boehm et al., 2012; Hoy et al., 2012). As a result, this helps *C. jejuni* to transmigrate between neighbouring host cells to the basal side of polarized gut epithelium. However, prior to the reach adherence junctions, *C. jejuni* faces the tight junction barrier, consisting of several proteins including tricellulin, occludin, claudins, and junction adhesion molecules (JAMs) (Guttman and Finlay, 2009; Gunzel and Yu, 2013; Van Itallie and Anderson, 2014; Slifer and Blikslager, 2020). This multiprotein junctional complex maintains "fence" tasks providing cell polarity and "gate" function, which provides selective transport of small molecules through the apical-basal barrier (Zihni et al., 2016). In addition, tight junction transmembrane proteins bind to intracellular scaffold proteins such as zonula occludens (ZO) -1, -2, and -3 forming tight connection with the actin cytoskeleton (Zihni et al., 2016). More recently, we have shown that *C. jejuni* can disrupt the tight junction protein occludin in an HtrA-dependent manner

(Harrer et al., 2019). However, it is still unclear, whether other tight junction proteins such as the claudins may also be affected during *C. jejuni* paracellular transmigration.

Claudins represent tight JAMs responsible for the paracellular barrier function and account for at least 27 members in mammals (Tsukita et al., 2019). Based on their sequence homology, claudin family members consist of four putative transmembrane segments, a large extracellular loop (ECL1) containing a consensus sequence motif, and a second shorter extracellular loop (ECL2) also known as extracellular segments 1 and 2, respectively (Gunzel and Yu, 2013). Interestingly, *C. jejuni* has been shown to disrupt tight junctions during bacterial invasion of non-tumorigenic canine intestinal epithelial cells through claudin-4 cleavage (Lamb-Rosteski et al., 2008). Fluorescence microscopy revealed that infection of the epithelial monolayer with *C. jejuni* results in disruption of pericellular claudin-4, while Western blotting showed significantly less total claudin-4 (Lamb-Rosteski et al., 2008). However, the molecular background of claudin degradation during *C. jejuni* infection remained unclear. *C. jejuni* has been already shown to disrupt the major junction proteins occludin and E-cadherin during paracellular migration in an HtrA-dependent manner (Boehm et al., 2012; Harrer et al., 2019), which may be associated with shedded outer-membrane vesicles (Elmi et al., 2018). The results presented here suggest that claudin-8 is a major novel cleavage target for *C. jejuni* HtrA, and besides occludin, the second target protein in the tight junctions, which may help the pathogen to disrupt the epithelial barrier during infection.

MATERIALS AND METHODS

Campylobacter Strains and Infection Assays

The *C. jejuni* wild-type (wt) strain 81-176 and its isogenic knockout mutant *C. jejuni ΔhtrA*, the complemented mutant *ΔhtrA/htrA* and protease-inactive S197A point mutation in the *htrA* gene were used throughout this study (Boehm et al., 2012; Boehm et al., 2015). Bacterial cells were cultured using *Campylobacter* blood-free selective agar base including *Campylobacter* growth supplement provided by Oxoid (Wesel, Germany). Alternatively, the bacteria were grown on Mueller-Hinton (MH) agar supplemented with chloramphenicol (20 µg/ml) or kanamycin (30 µg/ml), respectively. Incubation was for 48 h at 37°C in jars using microaerobic conditions provided by the CampyGen™ system from Oxoid. All *C. jejuni* strains were harvested using sterile cotton swabs and resuspended in liquid BHI medium. The optical density (OD) was measured at 600 nm in an Eppendorf spectrophotometer to calculate the number of bacterial cells followed by host cell infection of *C. jejuni* using a multiplicity of infection (MOI) of 100.

Cell Culture and Immunofluorescence Staining

The human intestinal cell lines Caco-2 (ATCC HTB-37) and T84 (ATCC CCL-248) were seeded into 75 cm^2 cell culture flasks and finally in 12-well plates using DMEM medium including 10% FCS (Invitrogen) and 4 mM glutamine (Invitrogen, Karlsruhe, Germany). The cells formed proper monolayers and were incubated for 14 days to allow proper cell polarization. After a 12-h co-incubation with *C. jejuni*, the infected cells were washed twice with sterile PBS buffer followed immunofluorescence staining according to a previous protocol (Krause-Gruszczynska et al., 2007a). In brief, the cells have been fixed for 10 min in 4% PFA (paraformaldehyde) at 20°C. Afterwards, the cells were permeabilized for 1 min using 0.25% Triton-X100 and then blocked for 1 h in PBS buffer containing 5% BSA. Immunostaining of the cells was performed with the following antibodies: α-claudin-8 (#710222 and #40-0700Z, Invitrogen) and α-occludin (#sc-133256, Santa Cruz Biotechnology). The *C. jejuni* bacteria were visualized by α-*Campylobacter* antibody (Dako, Glostrup, Denmark). The DNA in the nucleus was stained by DAPI (4'-6-diamidino-2-phenylindole dihydrochloride) (Thermo Fisher Scientific). TRITC (tetramethylrhodamine isothiocynate)-conjugated α-rabbit, TRITC-conjugated α-mouse, FITC (fluorescein isothiocyanate)-conjugated α-rabbit and Alexa-633-conjugated α-rabbit (Thermo Fisher Scientific) were utilized as secondary antibodies. All samples were investigated by confocal fluorescence microscopy using a Leica SP5 (Leica Microsystems, Wetzlar, Germany). Excitation/emission of the fluorescence from DAPI, FITC and TRITC was processed at 405/413–460 nm, 488/496–550 nm, and 561/571–630 nm wavelengths, respectively. The obtained data were visualized using LAS AF computer software (Leica Microsystems). All microscopic experiments were performed at the Optical Imaging Centre Erlangen (OICE, Erlangen, Germany).

Quantification of the Fluorescence Intensities of Claudin-8 and Occludin

The images of T84 cells after confocal immunofluoresence microscopy were subjected for further analysis in the Fiji platform (Schindelin et al., 2012). Cellular localization of claudin-8 and occludin in the T84 cell monolayers was analyzed by segmentation of the tight junctions area and the cytoplasm area into regions of interest (ROIs). In the defined ROIs (tight junctions or cytoplasm), we quantified the relative fluorescence units (RFUs) of claudin-8 and occludin, respectively. The mean value of RFUs for each condition (claudin-8 in tight junctions vs. claudin-8 in the cytoplasm, occludin in tight junctions vs. occludin in the cytoplasm) was calculated from 10 cells in the T84 monolayer and presented as mean ± standard deviation. In addition, the fluorescence intensity of cells was assessed in single representative T84 cells (mock control cells or cells infected with wt *C. jejuni*). The RFUs were counted within a straight line passing through a cell including tight junctions and the cytoplasm areas, as marked, between the yellow arrows of **Figure 3B**.

Cloning, Expression, and Purification of *C. jejuni* HtrA

Recombinant *C. jejuni* HtrA was purified under native conditions as described previously (Zarzecka et al., 2018).

For this purpose, *C. jejuni* HtrA of strain 81-176 without the signal peptide (amino acids 17-472) was amplified from genomic DNA and cloned in the expression plasmid pGEX-6P-1 (GE Healthcare Life Sciences, Munich, Germany) as a GST-fusion protein using the restriction sites *Bam*HI and *Xma*I. The expression was performed in *Escherichia coli* BL21 and the purification protocol was described previously in detail (Löwer et al., 2008). *E. coli* LPS has been removed by incubating 100 µg/ml of HtrA for 1 h using 10 µg/ml of polymyxin B (Sigma Aldrich) at 20°C (Brisslert et al., 2005). The final purity of HtrA was determined to be more than 95% by SDS-PAGE electrophoresis and Coomassie staining (Moese et al., 2001).

In Vitro HtrA Cleavage Assays

Cleavage assays were performed with recombinant *C. jejuni* HtrA and purified human full-length Claudin-8 coupled to GST (# H00009073-P01, Abnova, Taipei City, Taiwan). For this purpose, 100 ng of GST-claudin-8 were incubated with 30 ng of purified *C. jejuni* HtrA in 25 µl 50 mM HEPES buffer (pH 7.4) for 16 h at 37°C. The resulting cleavage products were analysed by SDS-PAGE and Western blotting using α-claudin-8, α-GST, and α-HtrA antibodies as described below.

Western Blotting Studies

Proteins derived from infected cells and *in vitro* HtrA cleavage assays were loaded on 8% SDS-PAGE gels and blotted on PVDF membranes as described (Hartung et al., 2015). Afterwards, the membranes were prepared for blocking for 1 h using either 3% BSA or 5% skim milk in TBST (25 mM Tris-HCl pH 7.4, 140 mM NaCl, 0.1% Tween-20) buffer at 20°C. The following antibodies were used: rabbit α-claudin-8 (#710222, Invitrogen), mouse α-GAPDH (Sigma Aldrich, Taufkrichen, Germany), or rabbit α-HtrA (Brondsted et al., 2005). These antibodies were incubated overnight at 4°C using the manufacturer's or published protocols. Horseradish peroxidase-conjugated α-rabbit or polyvalent α-mouse immunoglobulin were utilized as secondary antibodies (Life Technologies, Darmstadt, Germany). The detection of bound antibodies was accomplished using the ECL Plus chemiluminescence Western Blot kit (GE Healthcare) (Hirsch et al., 2012).

Bioinformatics

The sequence logo describing the HtrA cleavage site was generated using the "frequency plot" option of WebLogo (Crooks et al., 2004). The search for HtrA cleavage sites in claudin-8 was done with ScanProsite (de Castro et al., 2006). Modeling of the claudin-8 structure was performed with HHpred (Zimmermann et al., 2018) and Modeller (Webb and Sali, 2017) using the structure of Claudin-9 (Vecchio and Stroud, 2019) as a template. RasMol (Sayle and Milnerwhite, 1995) was used for structure analysis and visualization.

Statistics

All data were evaluated *via* two-tailed Mann-Whitney test with GraphPad Prism 6 (Version 6.01). The obtained p-values $p < 0.001$ (***) and $p < 0.0001$ (****) were defined as statistically significant; ns, non-significant.

RESULTS

Rearrangement of Claudin-8 in the Tight Junctions of Polarized Intestinal Epithelial Cell Lines by *C. jejuni* HtrA

To evaluate whether *C. jejuni* HtrA affects the claudin-8 distribution in the cellular tight junctions, immunofluorescence microscopy has been applied. For this purpose, confluent Caco-2 epithelial cells were grown in monolayers for 14 days to achieve proper polarization and detectable expression of claudin-8. The cells were then infected either with *C. jejuni* wt strain 81-176 or its isogenic Δ*htrA* knockout mutant. As control experiments, Caco-2 cells were infected with *C. jejuni* 81-176 Δ*htrA* complemented either with wt *htrA* ($\Delta htrA/htrA^{wt}$) or a protease-inactive S197A point mutant ($\Delta htrA/htrA^{SA}$). After 12 h of infection, Caco-2 cells were fixed in PFA and immunostained with antibodies against claudin-8 and *C. jejuni* as labelled with green and red fluorophores, respectively. Immunofluorescence microscopy revealed that non-infected Caco-2 monolayers are characterized by the uniform distribution of claudin-8 in the tight junction areas (**Figure 1A**), while infection with wt *C. jejuni* or the wt complemented strain ($\Delta htrA/htrA^{wt}$) led to mislocalization of claudin-8 out of the tight junctions to form agglomerates in the cytoplasm (**Figures 1B, E**, white arrows). Interestingly, when Caco-2 monolayers were infected with *C. jejuni* carrying deleted *htrA* or the protease-inactive $htrA^{S197A}$ point-mutated gene, claudin-8 localization within cellular tight junctions was not or only slightly affected (**Figures 1C, D**).

To corroborate the above findings with claudin-8, we utilized a second intestinal epithelial cell line, T84. To this end, T84 cell monolayers were grown for 14 days and then infected with the above described *C. jejuni* strains under identical conditions. Besides claudin-8, we additionally counterstained the cells with the tight junction protein occludin, which we have shown recently to be cleaved by *C. jejuni* HtrA (Harrer et al., 2019). As expected, in non-infected T84 control cells or cells infected with *C. jejuni* carrying a defective *htrA* gene, both proteins (occludin and claudin-8) revealed similar staining patterns in the cellular tight junctions (**Figures 2A, C, D**). In contrast, infection with wt *C. jejuni* or the wt complemented strain ($\Delta htrA/htrA^{wt}$) led to the disruption of both occludin and claudin-8 in tight junctions along with their appearance in the cytoplasmic area (**Figures 2B, E**, white arrows). The inlays on top of each panel show enlarged sections of tight junctions from corresponding areas marked with smaller white boxes.

Quantification of Claudin-8 Signals in the Tight Junctions and Cytoplasm During *C. jejuni* Infection

To further analyze the distribution of occludin and claudin-8 in the host cells, we quantified the mean fluorescence intensity of the proteins in the areas of membrane-associated tight junctions and cytoplasm within the overall cell population. In mock control cells, both occludin and claudin-8 revealed a higher fluorescence intensity within the tight junctions compared to

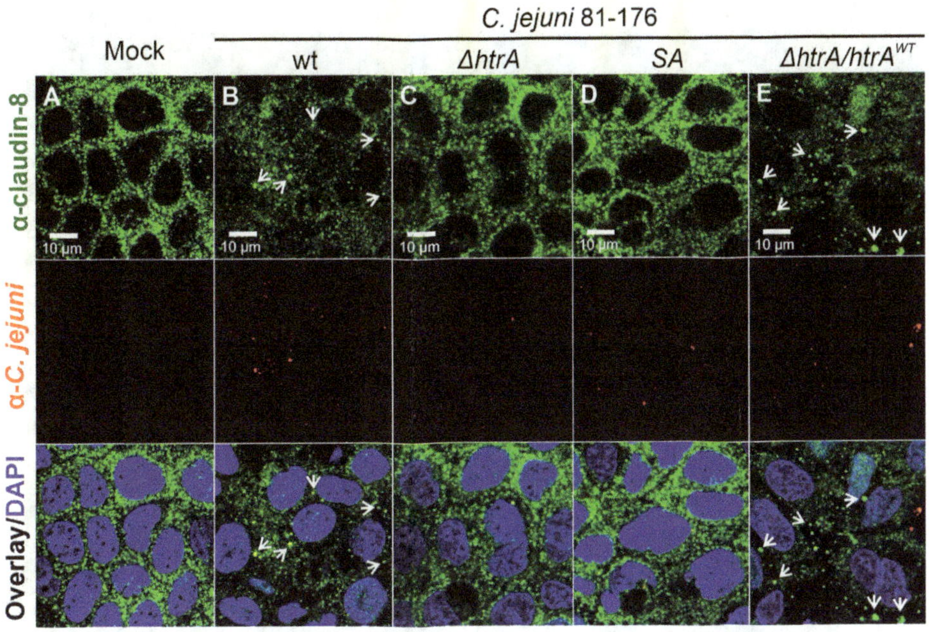

FIGURE 1 | Infection of polarized Caco-2 epithelial cell monolayers by *C. jejuni* 81-176 results in the disruption of claudin-8 patterns. Epithelial cells were immunostained with α-claudin-8 (green) and α-*C. jejuni* (red) without infection **(A)**, or after infection with *C. jejuni* wild-type (wt) strain **(B)**, Δ*htrA* knockout mutant **(C)**, Δ*htrA* complemented with protease inactive SA mutant **(D)** or wt *htrA* **(E)**. Infection was performed for 12 h at an MOI of 100. DAPI staining (blue) was used for the nuclear DNA counterstaining. White arrows indicate claudin-8 redistribution to agglomerates.

the cytoplasmic proteins (**Figure 3A**). Similar patterns of occludin and claudin-8 in the tight junctions were found in T84 cells infected with *C. jejuni* carrying an impaired *htrA* gene. In contrast, infection of T84 cells with wt *C. jejuni* or the wt complemented strain led to a significant drop of fluorescence intensity of occludin and claudin-8 in the cellular tight junctions. In particular, the mean fluorescence of cytosolic occludin slightly increased when infected with wt *C. jejuni* ($p < 0.0001$) or with the wt complemented strain ($p < 0.01$), which is in agreement with our previous studies (Harrer et al., 2019). The mean fluorescence of cytosolic claudin-8 upon infection did not change significantly within overall cell populations. However, the bright fluorescent aggregates of the cytosolic claudin-8 were only detected in T84 cells infected with wt *C. jejuni* carrying an intact *htrA* gene, confirming the re-distribution of the protein from the cell membrane (**Figure 2**, white arrows and **Figure 3B**). The fluorescence intensity of cells was further assessed in representative single cells between the yellow arrows as marked in **Figure 3B**. In non-infected T84 cells the peaks of fluorescence for occludin and claudin-8 appeared in the cell periphery, suggesting their prevailing membrane localization (**Figure 3C**). After infection of T84 cells with wt *C. jejuni*, the fluorescence intensity of occludin and claudin-8 at the cell periphery was significantly reduced. Moreover, in infected T84 cells the strong fluorescence signals of claudin-8 (comparable to the fluorescence of the protein in the membrane of non-infected cells) were also detectable in the cytoplasm (**Figure 2**, white arrows, and **Figure 3C**). Based on these immunofluorescence experiments, we can conclude that HtrA is involved in the disturbance of claudin-8 and its translocation from tight junctions into the cytoplasm during infection with *C. jejuni*.

HtrA Induces Claudin-8 Cleavage During *C. jejuni* Infection of Polarized Intestinal Epithelial Cells

Since infection by *C. jejuni* expressing intact HtrA leads to disorganisation of claudin-8 in Caco-2 cells, we came into assumption that HtrA might have a specific proteolytic activity against claudin-8. To test whether *C. jejuni* HtrA is inherently involved in claudin-8 proteolytic cleavage, immunoblotting assays were used. Polarized Caco-2 monolayers were infected with either wt *C. jejuni*, expressing intact HtrA or Δ*htrA* deletion mutant. After 12-h infection, immunoblotting was applied by using α-claudin-8, α-HtrA and α-GAPDH antibodies (**Figure 4A**). When infected with wt *C. jejuni* we observed full-length claudin-8 at about 26 kDa, and an additional band of lower intensity that appeared at approximately 18 kDa. In contrast, Caco-2 cells that were exposed to *htrA*-deficient *C. jejuni*, an 18-kDa band was not detected, assuming the absence of cleavage. The immunogenicity of α-claudin-8 antibody is directed against amino acids 206–225 at the carboxy-terminus of the protein, which corresponds to the QKSYHTGKKSPSVYSRSQYV sequence. Since the antibody recognized an 18-kDa claudin-8 fragment, we proposed that this fragment corresponds to the carboxy-terminus and cleavage of claudin-8 by *C. jejuni* HtrA takes place at the amino-terminus of the protein, presumably located in the first extracellular loop (**Figures 4B, C**). We were not able to detect the remaining very small amino-terminal

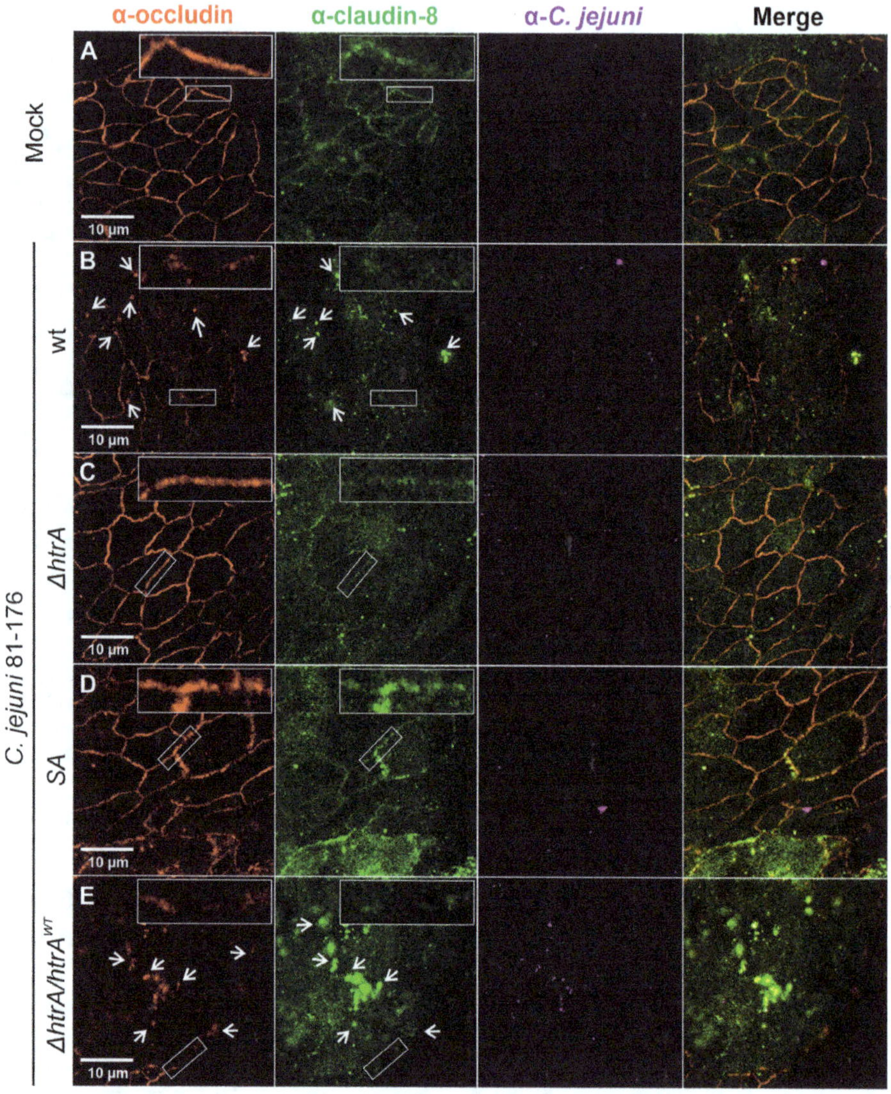

FIGURE 2 | Infection of polarized T84 epithelial cell monolayers by *C. jejuni* 81-176 leads to disruption of claudin-8 and occludin in the tight junctions. Epithelial cells were immunostained with α-occludin (red), α-claudin-8 (green) and α-*C. jejuni* (magenta) without infection **(A)**, or after infection with *C. jejuni* wild-type (wt) strain **(B)**, Δ*htrA* knockout mutant **(C)**, protease inactive SA mutant **(D)** or Δ*htrA* complemented with wt *htrA* **(E)**. White arrows indicate occludin and claudin-8 redistribution to protein agglomerates. White boxes contain enlarged parts of tight junctions from corresponding smaller white boxes.

cleaved peptide of claudin-8 due to lack of antibodies against this part of the amino-terminal tail.

C. jejuni HtrA Cleaves Claudin-8 by an *In Vitro* Cleavage Assay

To confirm whether the carboxy-terminal 18-kDa fragment of claudin-8 cleavage is the result of direct HtrA proteolytic activity, and not by any other bacterial or cellular protease, an *in vitro* cleavage assay using the recombinant proteins was conducted. For this purpose, we utilized recombinant amino-terminally GST-tagged claudin-8 (rGST-Claudin-8, about 51 kDa), which was incubated with purified *C. jejuni* HtrA for 12 h at 37°C. The *in vitro* cleavage reactions were subjected to immunoblotting using α-claudin-8, α-GST and α-HtrA antibodies. The immunostaining with α-claudin-8 revealed the appearance of the same sized carboxy-terminal 18-kDa fragment of claudin-8 through cleavage by HtrA (**Figure 5A**). GST is a protein of 25 kDa. The remaining N-terminal fragment was detected after staining with α-GST at 33 kDa, which corresponds to the GST protein fused to the cleaved 8-kDa amino-terminus of claudin-8 (**Figures 5A, B**).

Mapping of the *C. jejuni* HtrA Cleavage Site for Claudin-8 Proteolysis

The sequence preferences for *C. jejuni* HtrA cleavage were identified recently based on the cleavage sites detected in the β-casein and lysozyme substrates by mass spectrometry (Zarzecka et al., 2020). This analysis revealed that the cleavage

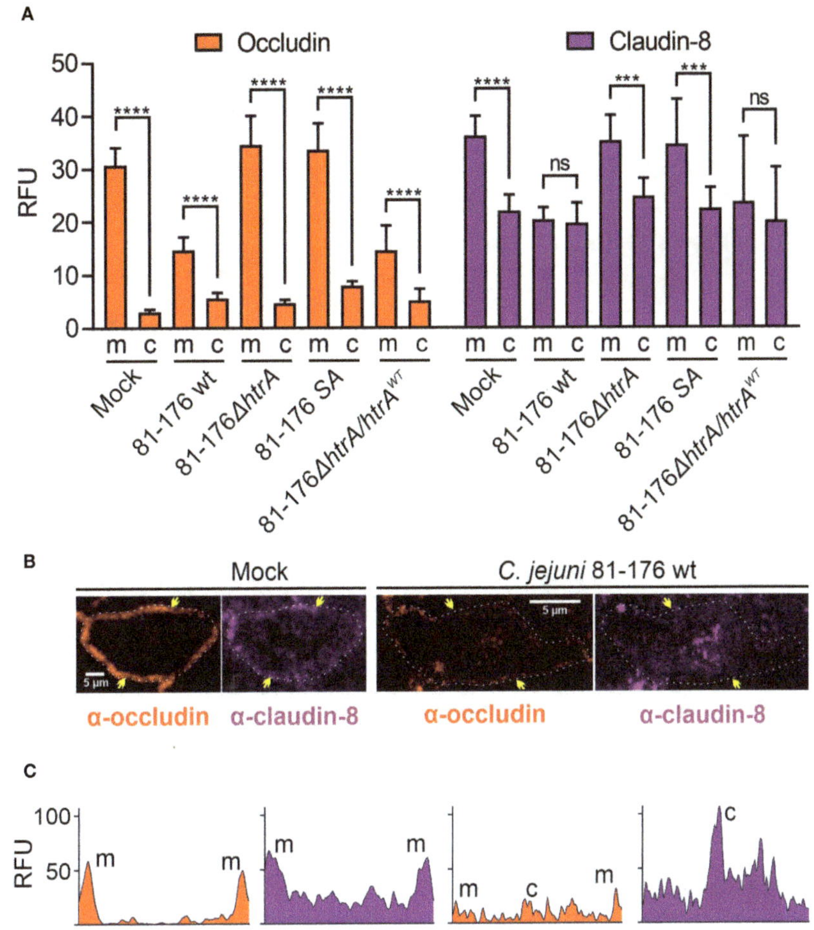

FIGURE 3 | Localization of occludin and claudin-8 proteins in tight junctions and cytoplasm of T84 cells. The mean florescence intensity of the proteins was assessed separately in the membrane (m)-associated tight junctions and cytoplasm (c), and was calculated from 10 T84 cells per sample **(A)**. Single cell analysis of representative T84 cells without or with infection by wt *C. jejuni*. Infection of T84 cells leads to the drop of fluorescence intensity of claudin-8 in tight junctions along with its appearance in the cytoplasm **(B)**. Corresponding fluorescence intensity plots for occludin and claudin-8 within single cells between two yellow arrows, as marked in panel **(B)**. When infected with wt *C. jejuni*, a strong fluorescence intensity can be found in the cell cytoplasm **(C)**. RFU, relative fluorescence units of membrane and cytoplasmic localization. ***$p < 0.001$, ****$p < 0.0001$, ns, non-significant.

site exhibits a rather large sequence heterogeneity and that the highest conservation is observed for the P1 and P1' sites. Based on the three predominant residues observed at the P1 and P1' site, we defined the pattern [VAL]-[SKN] to search for HtrA cleavage sites in the extracellular domain 1 (ECD1) of claudin-8. A frequency plot of the HtrA cleavage sites was established (**Figure 6A**). The height of each character is proportional to the frequency of the amino acid residue at the individual position of the cleaved peptide. This analysis revealed that there are three instances of this pattern in ECD1, namely, V32-S33, A58-N59, and L73-S74. The location of these sites in the claudin-8 structure is shown in **Figures 6B, C**. The V32-S33 site is buried in a β-sheet and therefore only hardly accessible to HtrA cleavage. The L73-S74 site is located in the immediate vicinity to a transmembrane helix; therefore, cleavage is sterically hindered by the presence of the membrane. In contrast, A58-N59 is located in an exposed turn connecting two β-strands and therefore is accessible. Thus, the A58-N59 position represents the most likely site for *C. jejuni* HtrA cleavage in claudin-8.

DISCUSSION

The tight junctions in the intestine play a major role both in epithelial cell monolayer integrity and in selective transport of molecules across neighboring epithelial cells. The key proteins building the tight junction net include claudin family members, occludin and tricellulin that localize at the upper tricellular contacts facing the lumen. During infection, some microbial pathogens can target the tight junction complex leading to protein dysregulation, which results in the disruption of intestinal barrier homeostasis in order to support microbial survival, spread and sometimes persistence (Vogelmann et al., 2004). For example, the human pathogen *Vibrio cholerae* secretes

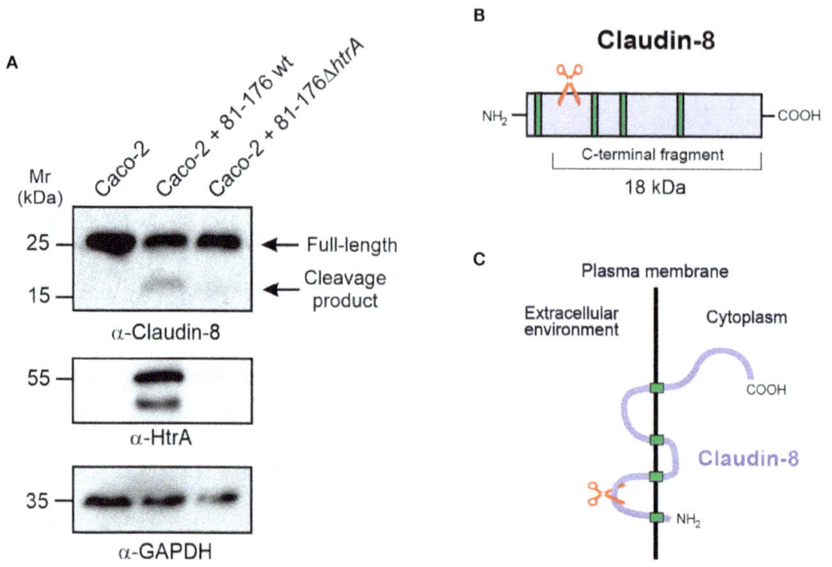

FIGURE 4 | HtrA cleaves claudin-8 during *C. jejuni* infection *in vivo*. **(A)** Immunoblotting of protein extracts from polarized Caco-2 cell monolayers infected with *C. jejuni* wild-type (wt) or *ΔhtrA* mutant. The blots were stained with polyclonal α-claudin-8 antibodies recognizing the carboxy-terminus of the protein. The α-HtrA and α-GAPDH blots served as controls. Besides the full-length protein (~26 kDa), a cleaved carboxy-terminal fragment (~18 kDa) was visualized with α- claudin-8 antibodies upon infection with *C. jejuni* wt. **(B)** Proposed cleavage site position in the amino-terminus of claudin-8 for HtrA protease according to the size of protein cleavage products determined by SDS-PAGE marker proteins. **(C)** A model for claudin-8 cleavage in the first extracellular loop (ECL1) by HtrA protease.

a metalloprotease, haemagglutinin/protease (HA/P), which degrades the occludin extracellular domain and subsequently affects the host cell actin cytoskeleton due to impaired interactions with the scaffold protein ZO-1 (Schubert-Unkmeir et al., 2010). *Neisseria meningitidis* was shown to activate matrix metalloproteinase 8 in human brain microvascular endothelial cells, resulting in cleavage of occludin. This resulted in disappearance of occludin from the cell periphery and cleavage to a lower-sized 50-kDa protein (Schubert-Unkmeir et al., 2010). Finally, multiple Gram-negative bacteria including *C. jejuni*, *H. pylori*, *Escherichia coli*, and *Shigella flexneri* secrete the HtrA protease towards adherence junctions, where it cleaves the adherens junction receptor protein E-cadherin (Schmidt et al., 2016). Recently, *C. jejuni* HtrA was shown to cleave E-cadherin and occludin upon infection of intestinal polarized Caco-2 cells (Harrer et al., 2019). In our previous analysis, we have identified the cleavage sites of *H. pylori* HtrA in the E-cadherin protein with a consensus cleavage sequence occurring at the [VITA]-[VITA]-x-x-D-[DN] motif (Schmidt et al., 2016). We have then identified a simplified pattern [VITA]-[VITA]-x (2,4)-[DN] of an HtrA cleavage site in occludin, which occurred in the second extracellular loop and was characterized by lack of secondary structure (Harrer et al., 2019). The present results shown here demonstrate that HtrA from *C. jejuni* can cleave another major tight junction protein, claudin-8, as this has been shown both *in vitro* with purified proteins and upon infection of cultured polarized Caco-2 and T84 cells *in vivo*.

Claudins represent highly conserved 20- to 27-kDa proteins that are differentially expressed along the various epithelial compartments of the gastrointestinal tract (Kim et al., 2019). Claudins are composed of four transmembrane regions including a short intracellular amino-terminal sequence (~1–7 residues), a large first extracellular loop (~52 residues), a shorter second extracellular loop (16–33 residues), and a cytoplasmic carboxy-terminal domain that varies considerably in length between different isoforms (21–63 residues) (Kim et al., 2019). The function of these proteins is mainly determined by their extracellular loops ECL1 and ECL2 (Krause et al., 2008). When the larger ECL1 is supposed to provide paracellular tightness and the selective ion permeability, the smaller ECL2 might contribute in narrowing of the paracellular cleft and hold neighboring cell membranes (Wen et al., 2004; Piontek et al., 2008). The decreased barrier function due to disruption of tight junctions leads to alterations in levels of pro-inflammatory signaling that apparently results in variety of pathologies (Harhaj and Antonetti, 2004; Turner, 2006). Thus, dysregulation of claudins in the gastrointestinal tract can lead to various illnesses such as inflammatory bowel disease, celiac disease and gastroesophageal reflux disease (Zeissig et al., 2007; Szakal et al., 2010; Monkemuller et al., 2012). Claudin-8, in particular, has been shown to be expressed in the small and large intestines, liver and gallbladder and to be involved in the tight junction barrier function (Jeansonne et al., 2003; Kim et al., 2019). Downregulation and redistribution of claudin-8 along with claudin-5 lead to alterations in tight junction's structure and pronounced barrier dysfunction both in mild and moderately active Crohn's disease (Zeissig et al., 2007).

Since claudin-8 plays an important role in barrier function of intestinal epithelial cell monolayers, we aimed to elucidate if this tight junction protein could be a target for *C. jejuni* HtrA.

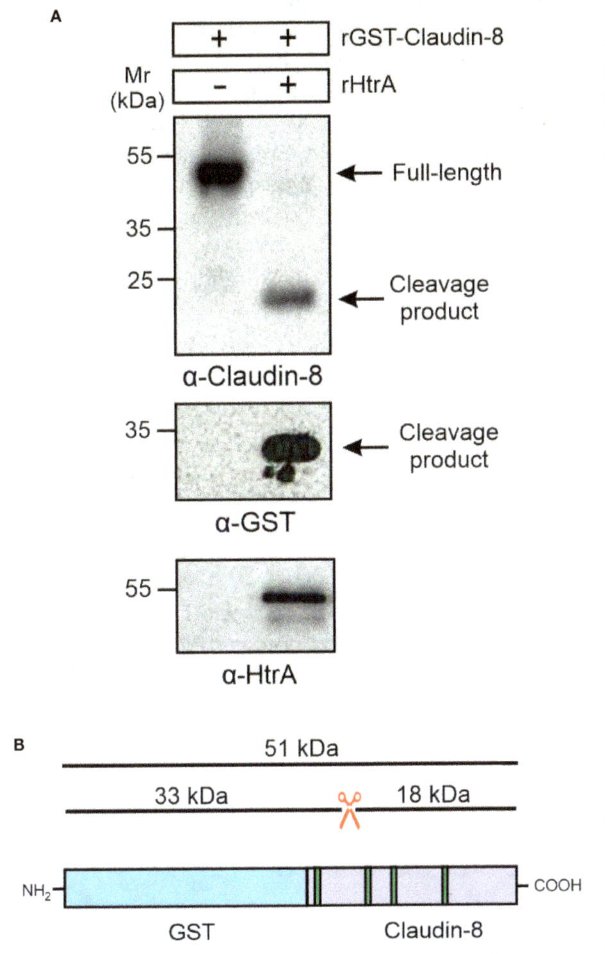

FIGURE 5 | HtrA cleaves claudin-8 in a cleavage assay *in vitro*. **(A)** Recombinant GST-tagged human claudin-8 was incubated with either purified rHtrA *in vitro*. Immunoblotting against α-claudin-8 showed cleavage of the full-length GST-tagged claudin-8 protein, when incubated with recombinant HtrA resulted in the appearance of an 18-kDa claudin-8 fragment. In addition, a cleaved 33-kDa fragment was visualized in the blot with α-GST antibodies. **(B)** Mapping of the GST-claudin-8 cleavage fragments. The full-length GST-tagged claudin-8 protein (51 kDa) is cleaved into a carboxy-terminal 18-kDa fragment of claudin-8 and a 33 kDa amino-terminal fragment containing the GST-tag and the remaining 8-kDa amino-terminal fragment of claudin-8.

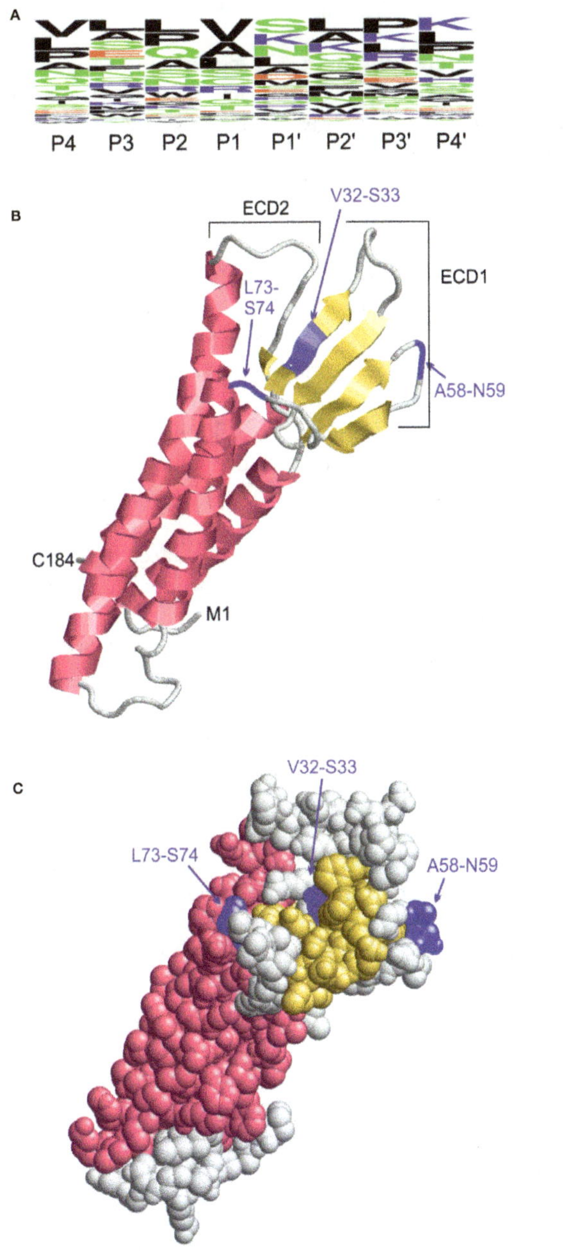

FIGURE 6 | Identification of candidate HtrA cleavage sites in human claudin-8. **(A)** Sequence logo derived from the HtrA cleavage sites in β-casein and lysozyme. The height of the letters is proportional to the relative frequency of the amino acid at the respective sequence position (P4 to P4') denote the individual positions of the HtrA cleavage site. **(B)** Model of the claudin-8 structure indicating the position of candidate cleavage sites (blue) in the extracellular domain 1 that confer the pattern [VAL]-[SKN]. **(C)** The space-filled-presentation of claudin-8 shows that only the A58-N59 site is fully accessible for recognition by a protease.

We have previously shown that upon infection of Caco-2 wt cells, *C. jejuni* exploits the secreted serine protease HtrA to cleave the adherens junction protein E-cadherin (Boehm et al., 2012). Then, we have demonstrated that both the apical tight junction proteins occludin (Harrer et al., 2019) and claudin-8 (this work) are also disrupted by HtrA during *C. jejuni* infection facilitating pathogen entry into the intercellular space between neighboring cells of the gut epithelium. This approach may help the bacteria to transmigrate across the intestinal epithelium by a paracellular mechanism and reach basal surfaces and the fibronectin-integrin complex that connect epithelial cells with underlying tissue (Backert et al., 2013; Backert et al., 2018). In particular, *C. jejuni* uses the fibronectin and integrin receptors to enter host cells in a CadF/FlpA-dependent manner (Boehm et al., 2011; Krause-Gruszczynska et al., 2011; Eucker and Konkel, 2012). While some major junctional proteins such as claudin-8, occludin and E-cadherin are targeted by HtrA, *htrA*-deficient bacteria do not and are strongly diminished in transmigration, adhesion and invasion of polarized Caco-2 cells (Harrer et al., 2019).

While tight junctions are essential in regulating the permeability across the epithelia, tight junctions can also mediate signaling pathways in response to other factors, for instance, through phosphorylation (Nunbhakdi-Craig et al., 2002; Fujibe et al., 2004). This can directly regulate the permeability of the cell monolayer, by promoting the barrier function or increasing Mg^{2+} transport (Ishizaki et al., 2003; Ikari et al., 2008). For example, a mutant protein kinase WNK4 present in patients with the pseudohypoaldosteronism type II disorder was found to phosphorylate claudins 1-4 resulting in elevated paracellular permeability (Yamauchi et al., 2004). In general, phosphorylation of claudins appeared as a vital process required for the maintenance of cell homeostasis but this also makes it an attractive target for the pathogens. Though we did not determine the claudin-8 phosphorylation status in the present report, due to lack of corresponding antibodies, *C. jejuni* HtrA could either potentially affect phosphorylation-mediated downstream signaling *via* claudin-8 or directly cleave target proteins. This idea should be studied in more detail in future experiments. Finally, tight junction proteins such as the claudins can be targeted not only by *C. jejuni*, but also other microbial pathogens. For instance, *H. pylori* HtrA can also cleave claudin-8 resulting in the same 18-kDa carboxy-terminal fragment (Tegtmeyer et al., 2017). This suggests that *H. pylori* and *C. jejuni* HtrAs most likely cleave the same sequence between the A58-N59 position in claudin-8. In addition, *Clostridium perfringens* has been shown to destroy the epithelial cell layer through the interaction of an enterotoxin with the claudin-4 protein by using it as a host cell receptor (Eichner et al., 2017). In a similar way, *C. perfringens* enterotoxin has been found to interact with other claudin members including claudin-5, -6, -8 and -14, confirming that they share a similar structural topology (Lal-Nag et al., 2012; Shrestha and McClane, 2013; Liao et al., 2016). Thus, the claudins in the tight junctions represent preferred targets by multiple microbial pathogens.

Taken together, we found that *C. jejuni*, in addition to occludin and E-cadherin, is capable to cleave claudin-8, which results in the disruption of major junction proteins in an HtrA-dependent manner. Thereby, this microbial pathogen can reach the basal side of polarized epithelial cells by transmigration through the tight and adherens junctions disrupted by secreted HtrA. However, cleavage of occludin and claudin-8 by *C. jejuni* HtrA might be just one option for the paracellular transmigration, while another intriguing mechanism could be the control of potential phosphorylation of claudins triggered by bacterium. *C. jejuni* is widely known to hijack host molecular signaling for its own benefit, for instance, by phosphorylation of host cell receptors such as EGFR, PDGFR and other signaling proteins (Tegtmeyer et al., 2020), and further investigation of this strategy can provide new insights in the pathogenesis of these important bacteria.

AUTHOR CONTRIBUTIONS

IS, DS, and AH performed the infection studies. IS performed the immunofluorescence experiments. DS, AH, and NT did the *in vitro* cleavage experiments. HS performed the bioinformatics analysis and cleavage site identification. IS analyzed the data. SB conceptualized the study, analyzed the data, and wrote the paper together with IS. All authors contributed to the article and approved the submitted version.

ACKNOWLEDGMENTS

We thank Wilhelm Brill and Nina Rottmann for excellent technical assistance.

REFERENCES

Ang, C. W., Laman, J. D., Willison, H. J., Wagner, E. R., Endtz, H. P., De Klerk, M. A., et al. (2002). Structure of *Campylobacter jejuni* lipopolysaccharides determines antiganglioside specificity and clinical features of Guillain-Barre, and Miller Fisher patients. *Infect. Immun.* 70, 1202–1208. doi: 10.1128/iai.70.3.1202-1208.2002

Backert, S., Boehm, M., Wessler, S., and Tegtmeyer, N. (2013). Transmigration route of *Campylobacter jejuni* across polarized intestinal epithelial cells: paracellular, transcellular or both? *Cell. Commun. Signal.* 11, 72. doi: 10.1186/1478-811x-11-72

Backert, S., Bernegger, S., Skorko-Glonek, J., and Wessler, S. (2018). Extracellular HtrA serine proteases: An emerging new strategy in bacterial pathogenesis. *Cell. Microbiol.* 20, e12845. doi: 10.1111/cmi.12845

Barrero-Tobon, A. M., and Hendrixson, D. R. (2012). Identification and analysis of flagellar coexpressed determinants (Feds) of *Campylobacter jejuni* involved in colonization. *Mol. Microbiol.* 84, 352–369. doi: 10.1111/j.1365-2958.2012.08027.x

Biswas, S., and Biswas, I. (2005). Role of HtrA in surface protein expression and biofilm formation by *Streptococcus mutans*. *Infect. Imum.* 73, 6923–6934. doi: 10.1128/iai.73.10.6923-6934.2005

Boehm, M., Krause-Gruszczynska, M., Rohde, M., Tegtmeyer, N., Takahashi, S., Oyarzabal, O. A., et al. (2011). Major host factors involved in epithelial cell invasion of *Campylobacter jejuni*: role of fibronectin, integrin beta1, FAK, Tiam-1, and DOCK180 in activating Rho GTPase Rac1. *Front. Cell Infect. Microbiol.* 1, 17. doi: 10.3389/fcimb.2011.00017

Boehm, M., Hoy, B., Rohde, M., Tegtmeyer, N., Baek, K. T., Oyarzabal, O. A., et al. (2012). Rapid paracellular transmigration of *Campylobacter jejuni* across polarized epithelial cells without affecting TER: role of proteolytic-active HtrA cleaving E-cadherin but not fibronectin. *Gut Pathog.* 4, 3. doi: 10.1186/1757-4749-4-3

Boehm, M., Lind, J., Backert, S., and Tegtmeyer, N. (2015). *Campylobacter jejuni* serine protease HtrA plays an important role in heat tolerance, oxygen resistance, host cell adhesion, invasion, and transmigration. *Eur. J. Microbiol. Immunol. (Bp)* 5, 68–80. doi: 10.1556/eujmi-d-15-00003

Boehm, M., Simson, D., Escher, U., Schmidt, A. M., Bereswill, S., Tegtmeyer, N., et al. (2018). Function of Serine Protease HtrA in the Lifecycle of the Foodborne Pathogen *Campylobacter jejuni*. *Eur. J. Microbiol. Immunol. (Bp)* 8, 70–77. doi: 10.1556/1886.2018.00011

Brauner, A., Brandt, L., Frisan, T., Thelestam, M., and Ekbom, A. (2010). Is there a risk of cancer development after *Campylobacter* infection? *Scand. J. Gastroenterol.* 45, 893–897. doi: 10.3109/00365521003734133

Brisslert, M., Enarsson, K., Lundin, S., Karlsson, A., Kusters, J. G., Svennerholm, A. M., et al. (2005). *Helicobacter pylori* induce neutrophil transendothelial migration: role of the bacterial HP-NAP. *FEMS Microbiol. Lett.* 249, 95–103. doi: 10.1016/j.femsle.2005.06.008

Brondsted, L., Andersen, M. T., Parker, M., Jorgensen, K., and Ingmer, H. (2005). The HtrA protease of *Campylobacter jejuni* is required for heat and oxygen tolerance and for optimal interaction with human epithelial cells. *Appl. Environ. Microbiol.* 71, 3205–3212. doi: 10.1128/aem.71.6.3205-3212.2005

Burnham, P. M., and Hendrixson, D. R. (2018). *Campylobacter jejuni*: collective components promoting a successful enteric lifestyle. *Nat. Rev. Microbiol.* 16, 551–565. doi: 10.1038/s41579-018-0037-9

Charles, J. L., Bell, J. A., Gadsden, B. J., Malik, A., Cooke, H., Van de Grift, L. K., et al. (2017). Guillain Barre Syndrome is induced in Non-Obese Diabetic (NOD) mice following *Campylobacter jejuni* infection and is exacerbated by antibiotics. *J. Autoimmun.* 77, 11–38. doi: 10.1016/j.jaut.2016.09.003

Christensen, J. E., Pacheco, S. A., and Konkel, M. E. (2009). Identification of a *Campylobacter jejuni*-secreted protein required for maximal invasion of host cells. *Mol. Microbiol.* 73, 650–662. doi: 10.1111/j.1365-2958.2009.06797.x

Clausen, T., Southan, C., and Ehrmann, M. (2002). The HtrA family of proteases: Implications for protein composition and cell fate. *Mol. Cell.* 10, 443–455. doi: 10.1016/s1097-2765(02)00658-5

Clausen, T., Kaiser, M., Huber, R., and Ehrmann, M. (2011). HTRA proteases: regulated proteolysis in protein quality control. *Nat. Rev. Mol. Cell. Biol.* 12, 152–162. doi: 10.1038/nrm3065

Cole, J. N., Aquilina, J. A., Hains, P. G., Henningham, A., Sriprakash, K. S., Caparon, M. G., et al. (2007). Role of group A *Streptococcus* HtrA in the maturation of SpeB protease. *Proteomics* 7, 4488–4498. doi: 10.1002/pmic.200700626

Cortes, G., de Astorza, B., Benedi, V. J., and Alberti, S. (2002). Role of the *htrA* gene in *Klebsiella pneumoniae* virulence. *Infect. Immun.* 70, 4772–4776. doi: 10.1128/IAI.70.9.4772-4776.2002

Cróinín, O., and Backert, S. (2012). Host epithelial cell invasion by *Campylobacter jejuni*: trigger or zipper mechanism? *Front. Cell. Infect. Microbiol.* 2, 25. doi: 10.3389/fcimb.2012.00025

Crooks, G. E., Hon, G., Chandonia, J. M., and Brenner, S. E. (2004). WebLogo: A sequence logo generator. *Genome Res.* 14, 1188–1190. doi: 10.1101/gr.849004

de Castro, E., Sigrist, C. J. A., Gattiker, A., Bulliard, V., Langendijk-Genevaux, P. S., Gasteiger, E., et al. (2006). ScanProsite: detection of PROSITE signature matches and ProRule-associated functional and structural residues in proteins. *Nucleic Acids Res.* 34, W362–W365. doi: 10.1093/nar/gkl124

Eichner, M., Augustin, C., Fromm, A., Piontek, A., Walther, W., Bucker, R., et al. (2017). In Colon Epithelia, *Clostridium perfringens* Enterotoxin Causes Focal Leaks by Targeting Claudins Which are Apically Accessible Due to Tight Junction Derangement. *J. Infect. Dis.* 217, 147–157. doi: 10.1093/infdis/jix485

Elmi, A., Dorey, A., Watson, E., Jagatia, H., Inglis, N. F., Gundogdu, O., et al. (2018). The bile salt sodium taurocholate induces *Campylobacter jejuni* outer membrane vesicle production and increases OMV-associated proteolytic activity. *Cell Microbiol.* 20 (3), 12814. doi: 10.1111/cmi.12814

Eucker, T. P., and Konkel, M. E. (2012). The cooperative action of bacterial fibronectin-binding proteins and secreted proteins promote maximal *Campylobacter jejuni* invasion of host cells by stimulating membrane ruffling. *Cell. Microbiol.* 14, 226–238. doi: 10.1111/j.1462-5822.2011.01714.x

Fujibe, M., Chiba, H., Kojima, T., Soma, T., Wada, T., Yamashita, T., et al. (2004). Thr(203) of claudin-1, a putative phosphorylation site for MAP kinase, is required to promote the barrier function of tight junctions. *Exp. Cell. Res.* 295, 36–47. doi: 10.1016/j.yexcr.2003.12.014

Gunzel, D., and Yu, A. S. (2013). Claudins and the modulation of tight junction permeability. *Physiol. Rev.* 93, 525–569. doi: 10.1152/physrev.00019.2012

Guttman, J. A., and Finlay, B. B. (2009). Tight junctions as targets of infectious agents. *Biochim. Biophys. Acta* 1788, 832–841. doi: 10.1016/j.bbamem.2008.10.028

Hale, C. R., Scallan, E., Cronquist, A. B., Dunn, J., Smith, K., Robinson, T., et al. (2012). Estimates of Enteric Illness Attributable to Contact With Animals and Their Environments in the United States. *Clin. Infect. Dis.* 54, S472–S479. doi: 10.1093/cid/cis051

Harhaj, N. S., and Antonetti, D. A. (2004). Regulation of tight junctions and loss of barrier function in pathophysiology. *Int. J. Biochem. Cell Biol.* 36, 1206–1237. doi: 10.1016/j.biocel.2003.08.007

Harrer, A., Tegtmeyer, N., Boehm, M., and Backert, S. (2018). Overexpression of serine protease HtrA in *Helicobacter pylori* enhances disruption of adherens junctions, paracellular transmigration of the bacteria and type IV secretion of CagA. *Helicobacter* 9, 40. doi: 10.1186/s13099-017-0189-6

Harrer, A., Bucker, R., Boehm, M., Zarzecka, U., Tegtmeyer, N., Sticht, H., et al. (2019). *Campylobacter jejuni* enters gut epithelial cells and impairs intestinal barrier function through cleavage of occludin by serine protease HtrA. *Gut Pathog.* 11, 4. doi: 10.1186/s13099-019-0283-z

Hartung, M. L., Gruber, D. C., Koch, K. N., Gruter, L., Rehrauer, H., Tegtmeyer, N., et al. (2015). *H. pylori*-Induced DNA Strand Breaks Are Introduced by Nucleotide Excision Repair Endonucleases and Promote NF-kappaB Target Gene Expression. *Cell Rep.* 13, 70–79. doi: 10.1016/j.celrep.2015.08.074

He, Z., Gharaibeh, R. Z., Newsome, R. C., Pope, J. L., Dougherty, M. W., Tomkovich, S., et al. (2019). *Campylobacter jejuni* promotes colorectal tumorigenesis through the action of cytolethal distending toxin. *Gut* 68, 289–300. doi: 10.1136/gutjnl-2018-317200

Hermans, D., Van Deun, K., Martel, A., Van Immerseel, F., Messens, W., Heyndrickx, M., et al. (2011). Colonization factors of *Campylobacter jejuni* in the chicken gut. *Vet. Res.* 42, 82. doi: 10.1186/1297-9716-42-82

Hirsch, C., Tegtmeyer, N., Rohde, M., Rowland, M., Oyarzabal, O. A., and Backert, S. (2012). Live *Helicobacter pylori* in the root canal of endodontic-infected deciduous teeth. *J. Gastroenterol.* 47, 936–940. doi: 10.1007/s00535-012-0618-8

Hoy, B., Geppert, T., Boehm, M., Reisen, F., Plattner, P., Gadermaier, G., et al. (2012). Distinct roles of secreted HtrA proteases from gram-negative pathogens in cleaving the junctional protein and tumor suppressor E-cadherin. *J. Biol. Chem.* 287, 10115–10120. doi: 10.1074/jbc.C111.333419

Humphreys, S., Stevenson, A., Bacon, A., Weinhardt, A. B., and Roberts, M. (1999). The alternative sigma factor, sigmaE, is critically important for the virulence of *Salmonella typhimurium*. *Infect. Immun.* 67, 1560–1568. doi: 10.1128/IAI.67.4.1560-1568.1999

Ikari, A., Ito, M., Okude, C., Sawada, H., Harada, H., Degawa, M., et al. (2008). Claudin-16 is directly phosphorylated by protein kinase A independently of a vasodilator-stimulated phosphoprotein-mediated pathway. *J. Cell. Physiol.* 214, 221–229. doi: 10.1002/jcp.21178

Ishizaki, T., Chiba, H., Kojima, T., Fujibe, M., Soma, T., Miyajima, H., et al. (2003). Cyclic AMP induces phosphorylation of claudin-5 immunoprecipitates and expression of claudin-5 gene in blood-brain-barrier endothelial cells via protein kinase A-dependent and -independent pathways. *Exp. Cell Res.* 290, 275–288. doi: 10.1016/s0014-4827(03)00354-9

Jeansonne, B., Lu, Q., Goodenough, D. A., and Chen, Y. H. (2003). Claudin-8 interacts with multi-PDZ domain protein 1 (MUPP1) and reduces paracellular conductance in epithelial cells. *Cell. Mol. Biol.* 49, 13–21.

Kaakoush, N. O., Castano-Rodriguez, N., Mitchell, H. M., and Man, S.II (2015). Global Epidemiology of *Campylobacter* Infection. *Clin. Microbiol. Rev.* 28, 687–720. doi: 10.1128/cmr.00006-15

Kim, D. Y., and Kim, K. K. (2005). Structure and function of HtrA family proteins, the key players in protein quality control. *J. Biochem. Mol. Biol.* 38, 266–274. doi: 10.5483/bmbrep.2005.38.3.266

Kim, D. Y., Furuta, G. T., Nguyen, N., Inage, E., and Masterson, J. C. (2019). Epithelial Claudin Proteins and Their Role in Gastrointestinal Diseases. *J. Pediatr. Gastroenterol. Nutr.* 68, 611–614. doi: 10.1097/mpg.0000000000002301

Konkel, M. E., Garvis, S. G., Tipton, S. L., Anderson, D. E., and Cieplak, W. (1997). Identification and molecular cloning of a gene encoding a fibronectin-binding protein (CadF) from *Campylobacter jejuni*. *Mol. Microbiol.* 24, 953–963. doi: 10.1046/j.1365-2958.1997.4031771.x

Krause, G., Winkler, L., Mueller, S. L., Haseloff, R. F., Piontek, J., and Blasig, I. E. (2008). Structure and function of claudins. *Biochim. Biophys. Acta* 1778, 631–645. doi: 10.1016/j.bbamem.2007.10.018

Krause-Gruszczynska, M., Rohde, M., Hartig, R., Genth, H., Schmidt, G., Keo, T., et al. (2007a). Role of the small Rho GTPases Rac1 and Cdc42 in host cell invasion of *Campylobacter jejuni*. *Cell. Microbiol.* 9, 2431–2444. doi: 10.1111/j.1462-5822.2007.00971.x

Krause-Gruszczynska, M., van Alphen, L. B., Oyarzabal, O. A., Alter, T., Hanel, I., Schliephake, A., et al. (2007b). Expression patterns and role of the CadF protein in *Campylobacter jejuni* and *Campylobacter coli*. *FEMS Microbiol. Lett.* 274, 9–16. doi: 10.1111/j.1574-6968.2007.00802.x

Krause-Gruszczynska, M., Boehm, M., Rohde, M., Tegtmeyer, N., Takahashi, S., Buday, L., et al. (2011). The signaling pathway of *Campylobacter jejuni*-induced Cdc42 activation: Role of fibronectin, integrin beta1, tyrosine kinases and guanine exchange factor Vav2. *Cell Commun. Signal.* 9 (1), 32. doi: 10.1186/1478-811x-9-32

Lai, C. K., Chen, Y. A., Lin, C. J., Lin, H. J., Kao, M. C., Huang, M. Z., et al. (2016). Molecular Mechanisms and Potential Clinical Applications of *Campylobacter jejuni* Cytolethal Distending Toxin. *Front. Cell. Infect. Microbiol.* 6, 9. doi: 10.3389/fcimb.2016.00009

Lal-Nag, M., Battis, M., Santin, A. D., and Morin, P. J. (2012). Claudin-6: a novel receptor for CPE-mediated cytotoxicity in ovarian cancer. *Oncogenesis* 1 (11), 33. doi: 10.1038/oncsis.2012.32

Lamb-Rosteski, J. M., Kalischuk, L. D., Inglis, G. D., and Buret, A. G. (2008). Epidermal growth factor inhibits *Campylobacter jejuni*-induced claudin-4 disruption, loss of epithelial barrier function, and *Escherichia coli* translocation. *Infect. Immun.* 76, 3390–3398. doi: 10.1128/iai.01698-07

Li, S. R., Dorrell, N., Everest, P. H., Dougan, G., and Wren, B. W. (1996). Construction and characterization of a *Yersinia enterocolitica* O:8 high-temperature requirement (*htrA*) isogenic mutant. *Infect. Immun.* 64, 2088–2094. doi: 10.1128/IAI.64.6.2088-2094.1996

Liao, Z. B., Yang, Z. G., Piontek, A., Eichner, M., Krause, G., Li, L. X., et al. (2016). Specific binding of a mutated fragment of *Clostridium perfringens* enterotoxin to endothelial claudin-5 and its modulation of cerebral vascular permeability. *Neuroscience* 327, 53–63. doi: 10.1016/j.neuroscience.2016.04.013

Löwer, M., Weydig, C., Metzler, D., Reuter, A., Starzinski-Powitz, A., Wessler, S., et al. (2008). Prediction of extracellular proteases of the human pathogen *Helicobacter pylori* reveals proteolytic activity of the Hp1018/19 protein HtrA. *PLoS ONE* 3, 3510–3563. doi: 10.1371/journal.pone.0003510

Marder, E. P., Cieslak, P. R., Cronquist, A. B., Dunn, J., Lathrop, S., Rabatsky-Ehr, T., et al. (2017). Incidence and Trends of Infections with Pathogens Transmitted Commonly Through Food and the Effect of Increasing Use of Culture-Independent Diagnostic Tests on Surveillance - Foodborne Diseases Active Surveillance Network, 10 US Sites 2013-2016. *MMWR Morb. Mortal Wkly. Rep.* 66, 397–403. doi: 10.15585/mmwr.mm6615a1

Mo, E., Peters, S. E., Willers, C., Maskell, D. J., and Charles, I. G. (2006). Single, double and triple mutants of *Salmonella enterica* serovar Typhimurium degP (*htrA*), degQ (*hhoA*) and degS (*hhoB*) have diverse phenotypes on exposure to elevated temperature and their growth in vivo is attenuated to different extents. *Microb. Pathog.* 41, 174–182. doi: 10.1016/j.micpath.2006.07.004

Moese, S., Selbach, M., Zimny-Arndt, U., Jungblut, P. R., Meyer, T. F., and Backert, S. (2001). Identification of a tyrosine-phosphorylated 35 kDa carboxy-terminal fragment (p35CagA) of the *Helicobacter pylori* CagA protein in phagocytic cells: processing or breakage? *Proteomics* 1, 618–629. doi: 10.1002/1615-9861(200104)1:4<618::AID-PROT618>3.0.CO;2-C

Monkemuller, K., Wex, T., Kuester, D., Fry, L. C., Kandulski, A., Kropf, S., et al. (2012). Role of tight junction proteins in gastroesophageal reflux disease. *BMC Gastroenterol.* 12:128. doi: 10.1186/1471-230x-12-128

Neddermann, M., and Backert, S. (2019). Quantification of serine protease HtrA molecules secreted by the foodborne pathogen *Campylobacter jejuni*. *Gut Pathog.* 11, 14. doi: 10.1186/s13099-019-0295-8

Nunbhakdi-Craig, V., Machleidt, T., Ogris, E., Bellotto, D., White, C. L., and Sontag, E. (2002). Protein phosphatase 2A associates with and regulates atypical PKC and the epithelial tight junction complex. *J. Cell Biol.* 158, 967–978. doi: 10.1083/jcb.200206114

Piontek, J., Winkler, L., Wolburg, H., Muller, S. L., Zuleger, N., Piehl, C., et al. (2008). Formation of tight junction: determinants of homophilic interaction between classic claudins. *FASEB J.* 22, 146–158. doi: 10.1096/fj.07-8319com

Purdy, G. E., Hong, M., and Payne, S. M. (2002). *Shigella flexneri* DegP facilitates IcsA surface expression and is required for efficient intercellular spread. *Infect. Immun.* 70, 6355–6364. doi: 10.1128/iai.70.11.6355-6364.2002

Sayle, R. A., and Milnerwhite, E. J. (1995). RASMOL - biomolecular graphics for all. *Trends Biochem. Sci.* 20, 374–376. doi: 10.1016/s0968-0004(00)89080-5

Schindelin, J., Arganda-Carreras, I., Frise, E., Kaynig, V., Longair, M., Pietzsch, T., et al. (2012). Fiji: an open-source platform for biological-image analysis. *Nat. Methods* 9, 676–682. doi: 10.1038/nmeth.2019

Schmidt, T. P., Perna, A. M., Fugmann, T., Bohm, M., Jan, H., Haller, S., et al. (2016). Identification of E-cadherin signature motifs functioning as cleavage sites for *Helicobacter pylori* HtrA. *Sci. Rep.* 6, 23264. doi: 10.1038/srep23264

Schmidt, A. M., Escher, U., Mousavi, S., Tegtmeyer, N., Boehm, M., Backert, S., et al. (2019). Immunopathological properties of the *Campylobacter jejuni* flagellins and the adhesin CadF as assessed in a clinical murine infection model. *Gut Pathog.* 11, 24. doi: 10.1186/s13099-019-0306-9

Schubert-Unkmeir, A., Konrad, C., Slanina, H., Czapek, F., Hebling, S., and Frosch, M. (2010). *Neisseria meningitidis* Induces Brain Microvascular Endothelial Cell Detachment from the Matrix and Cleavage of Occludin: A Role for MMP-8. *PLoS Pathog.* 6 (4), 1000874. doi: 10.1371/journal.ppat.1000874

Shrestha, A., and McClane, B. A. (2013). Human Claudin-8 and-14 Are Receptors Capable of Conveying the Cytotoxic Effects of *Clostridium perfringens* Enterotoxin. *mBio* 4, e00594–e00512. doi: 10.1128/mBio.00594-12

Skorko-Glonek, J., Zurawa-Janicka, D., Koper, T., Jarzab, M., Figaj, D., Glaza, P., et al. (2013). HtrA Protease Family as Therapeutic Targets. *Curr. Pharm. Des.* 19, 977–1009. doi: 10.2174/1381612811319060003

Slifer, Z. M., and Blikslager, A. T. (2020). The Integral Role of Tight Junction Proteins in the Repair of Injured Intestinal Epithelium. *Int. J. Mol. Sci.* 21, 972. doi: 10.3390/ijms21030972

Szakal, D. N., Gyorffy, H., Arato, A., Cseh, A., Molnar, K., Papp, M., et al. (2010). Mucosal expression of claudins 2, 3 and 4 in proximal and distal part of duodenum in children with coeliac disease. *Virchows Arch.* 456, 245–250. doi: 10.1007/s00428-009-0879-7

Tegtmeyer, N., Wessler, S., Necchi, V., Rohde, M., Harrer, A., Rau, T. T., et al. (2017). *Helicobacter pylori* Employs a Unique Basolateral Type IV Secretion Mechanism for CagA Delivery. *Cell Host Microbe* 22, 552–560. doi: 10.1016/j.chom.2017.09.005

Tegtmeyer, N., Sharafutdinov, I., Harrer, A., Esmaeili, D. S., Linz, B., and Backert, S. (2020). *Campylobacter* virulence factors and molecular host-pathogen interactions. *Curr. Top. Microbiol. Immunol.* (in press).

Tsukita, S., Tanaka, H., and Tamura, A. (2019). The Claudins: From Tight Junctions to Biological Systems. *Trends Biochem. Sci.* 44, 141–152. doi: 10.1016/j.tibs.2018.09.008

Turner, J. R. (2006). Molecular basis of epithelial barrier regulation - From basic mechanisms to clinical application. *Am. J. Pathol.* 169, 1901–1909. doi: 10.2353/ajpath.2006.060681

Van Itallie, C. M., and Anderson, J. M. (2014). Architecture of tight junctions and principles of molecular composition. *Semin. Cell. Dev. Biol.* 36, 157–165. doi: 10.1016/j.semcdb.2014.08.011

Van Spreeuwel, J. P., Duursma, G. C., Meijer, C. J., Bax, R., Rosekrans, P. C., and Lindeman, J. (1985). Campylobacter colitis: histological immunohistochemical and ultrastructural findings. *Gut* 26, 945–951. doi: 10.1136/gut.26.9.945

Vecchio, A. J., and Stroud, R. M. (2019). Claudin-9 structures reveal mechanism for toxin-induced gut barrier breakdown. *Proc. Natl. Acad. Sci. U. S. A.* 116, 17817–17824. doi: 10.1073/pnas.1908929116

Vogelmann, R., Amieva, M. R., Falkow, S., and Nelson, W. J. (2004). Breaking into the epithelial apical-junctional complex–news from pathogen hackers. *Curr. Opin. Cell. Biol.* 16, 86–93. doi: 10.1016/j.ceb.2003.12.002

Webb, B., and Sali, A. (2017). Protein Structure Modeling with MODELLER. *Methods Mol. Biol.* 1654, 39–54. doi: 10.1007/978-1-4939-7231-9_4

Wen, H. J., Watry, D. D., Marcondes, M. C. G., and Fox, H. S. (2004). Selective decrease in paracellular conductance of tight junctions: role of the first extracellular domain of claudin-5. *Mol. Cell Biol.* 24, 8408–8417. doi: 10.1128/mcb.24.19.8408-8417.2004

Wilson, R. L., Brown, L. L., Kirkwood-Watts, D., Warren, T. K., Lund, S. A., King, D. S., et al. (2006). *Listeria monocytogenes* 10403S HtrA is necessary for resistance to cellular stress and virulence. *Infect. Immun.* 74, 765–768. doi: 10.1128/iai.74.1.765-768.2006

Wooldridge, K. G., and Ketley, J. M. (1997). Campylobacter-host cell interactions. *Trends Microbiol.* 5, 96–102. doi: 10.1016/s0966-842x(97)01004-4

Yamauchi, K., Rai, T., Kobayashi, K., Sohara, E., Suzuki, T., Itoh, T., et al. (2004). Disease-causing mutant WNK4 increases paracellular chloride permeability and phosphorylates claudins. *Proc. Natl. Acad. Sci. U. S. A.* 101, 4690–4694. doi: 10.1073/pnas.0306924101

Ye, M., Sharma, K., Thakur, M., Smith, A. A., Buyuktanir, O., Xiang, X., et al. (2016). HtrA, a Temperature- and Stationary Phase-Activated Protease Involved in Maturation of a Key Microbial Virulence Determinant, Facilitates Borrelia burgdorferi Infection in Mammalian Hosts. *Infect. Immun.* 84, 2372–2381. doi: 10.1128/iai.00360-16

Young, G. M., Schmiel, D. H., and Miller, V. L. (1999). A new pathway for the secretion of virulence factors by bacteria: The flagellar export apparatus functions as a protein-secretion system. *Proc. Natl. Acad. Sci. U. S. A.* 96, 6456–6461. doi: 10.1073/pnas.96.11.6456

Yuki, N., Susuki, K., Koga, M., Nishimoto, Y., Odaka, M., Hirata, K., et al. (2004). Carbohydrate mimicry between human ganglioside GM1 and *Campylobacter jejuni* lipooligosaccharide causes Guillain-Barre syndrome. *Proc. Natl. Acad. Sci. U. S. A.* 101, 11404–11409. doi: 10.1073/pnas.0402391101

Zarzecka, U., Modrak-Wojcik, A., Bayassi, M., Szewczyk, M., Gieldon, A., Lesner, A., et al. (2018). Biochemical properties of the HtrA homolog from bacterium *Stenotrophomonas maltophilia*. *Int. J. Biol. Macromol.* 109, 992–1005. doi: 10.1016/j.ijbiomac.2017.11.086

Zarzecka, U., Grinzato, A., Kandiah, E., Cysewski, D., Berto, P., Skorko-Glonek, J., et al. (2020). Functional analysis and cryo-electron microscopy of *Campylobacter jejuni* serine protease HtrA. *Gut Microbes* 12, 1–16. doi: 10.1080/19490976.2020.1810532

Zeissig, S., Burgel, N., Gunzel, D., Richter, J., Mankertz, J., Wahnschaffe, U., et al. (2007). Changes in expression and distribution of claudin 2, 5 and 8 lead to discontinuous tight junctions and barrier dysfunction in active Crohn's disease. *Gut* 56, 61–72. doi: 10.1136/gut.2006.094375

Zihni, C., Mills, C., Matter, K., and Balda, M. S. (2016). Tight junctions: from simple barriers to multifunctional molecular gates. *Nat. Rev. Mol. Cell Biol.* 17, 564–580. doi: 10.1038/nrm.2016.80

Zimmermann, L., Stephens, A., Nam, S. Z., Rau, D., Kubler, J., Lozajic, M., et al. (2018). A Completely Reimplemented MPI Bioinformatics Toolkit with a New HHpred Server at its Core. *J. Mol. Biol.* 430, 2237–2243. doi: 10.1016/j.jmb.2017.12.007

5

Solving the Puzzle: Connecting a Heterologous *Agrobacterium tumefaciens* T6SS Effector to a *Pseudomonas aeruginosa* Spike Complex

Sarah Wettstadt[1†]*, Erh-Min Lai*[2] *and Alain Filloux*[1*]

[1] *MRC Centre for Molecular Bacteriology and Infection, Department of Life Sciences, Imperial College London, London, United Kingdom,* [2] *Institute of Plant and Microbial Biology, Academia Sinica, Taipei, Taiwan*

***Correspondence:**
Alain Filloux
a.filloux@imperial.ac.uk

†Present address:
Sarah Wettstadt, Department of Environmental Protection, Estación Experimental de Zaidín – Consejo Superior de Investigaciones Científicas, Granada, Spain

The type VI secretion system (T6SS) is a contractile injection apparatus that translocates a spike loaded with various effectors directly into eukaryotic and prokaryotic target cells. Such T6SS spike consists of a needle-shaped trimer of VgrG proteins topped by a conical and sharp PAAR protein that facilitates puncturing of the target membrane. T6SS-delivered effector proteins can be either fused to one of the two spike proteins or interact with either in a highly specific manner. In *Agrobacterium tumefaciens* the T6SS effector Tde1 is targeted to its cognate VgrG1 protein. Here, we attempted to use a VgrG shuttle to deliver a heterologous T6SS effector by directing Tde1 onto a T6SS spike in *Pseudomonas aeruginosa*. For this, we designed chimeras between VgrG1 from *A. tumefaciens* and VgrG1a from *P. aeruginosa* and showed that modification of the spike protein hampered T6SS functionality in the presence of the Tde1 effector complex. We provide evidence suggesting that Tde1 specifically binds to the VgrG spike in the heterologous environment and propose that there are additional requirements to allow proper effector delivery and translocation. Our work sheds light on complex aspects of the molecular mechanisms of T6SS delivery and highlights some limitations on how effectors can be translocated using this nanomachine.

Keywords: type VI secretion system, bacterial toxin, VgrG, PAAR, Tde1

INTRODUCTION

The T6SS is a versatile secretion system, injecting effector proteins into target cells which equips bacteria with the ability to establish a niche in any given polymicrobial environment or modulate host cell responses. The T6SS is anchored to the bacterial cytoplasmic membrane *via* a so-called membrane complex (Durand et al., 2012, 2015) which is connected to a cytosolic membrane bound baseplate (Brunet et al., 2015; Planamente et al., 2016). The cytosolic tubular sheath attaches to the baseplate at the inner membrane and encompasses a tube composed of Hcp hexamers that is propelled out of the cell upon sheath contraction (Pukatzki et al., 2006; Leiman et al., 2009; Brunet et al., 2014). On top of the Hcp tube and residing within the baseplate complex sits the so-called T6SS spike consisting of a needle-shaped trimer of VgrG proteins and a conically-shaped PAAR protein (Shneider et al., 2013). The VgrG protein consists of a gp5- and a gp27-like domain that, when assembled to a trimer, form a rigid structure due to the intertwining of the C-terminal hydrophobic β-sheets (Kanamaru et al., 2002). Each last β-sheet binds to the hydrophobic surface of a cognate PAAR protein (Shneider et al., 2013). The VgrG-PAAR spike complex has two main functions: it facilitates puncturing of target membranes while it is also directly involved in carrying T6SS effectors into the target cell (Shneider et al., 2013).

T6SS effectors are classified into two groups: specialized effectors and cargo effectors (Durand et al., 2014). A specialized, or evolved, effector contains an N-terminal domain that is a structural component, like VgrG, PAAR, or Hcp, essential for T6SS assembly. The C-terminal domain, however, is an extension with an effector domain and is not required for delivery of the VgrG-PAAR spike complex (Ma et al., 2009; Wood et al., 2019b). In a different scenario, cargo effectors interact non-covalently with structural components, like Hcp, VgrG, or PAAR, and once the T6SS propels out the spike, the cargo effector is delivered in a "piggy-back ride" (Hachani et al., 2014). This so-called "*á la carte*" effector delivery concept describes that a VgrG recruits and delivers one specific effector (Hachani et al., 2014) and that this specific interaction is mediated by the C-terminal residues or even C-terminal domains of VgrG proteins. This interaction is the prerequisite for effector delivery and was shown for Tde1 binding VgrG1 in *Agrobacterium tumefaciens*, Tle1 binding VgrG1 in enteroaggregative *Escherichia coli* and PldA and PldB binding VgrG4b and VgrG5, respectively, in *P. aeruginosa*, amongst others (Flaugnatti et al., 2015; Unterweger et al., 2015; Bondage et al., 2016; Wettstadt et al., 2019).

Furthermore, some T6SS effectors were identified that require additional proteins for their delivery, which can be of the DUF1795, DUF2169, or DUF4123 family (Diniz and Coulthurst, 2015; Liang et al., 2015; Unterweger et al., 2015; Bondage et al., 2016; Quentin et al., 2018). *DUF4123* genes, also coined *tap* (Unterweger et al., 2015), or *tec* (Liang et al., 2015), can be found in the vicinity of a range of T6SS effector-encoding genes, together with a gene encoding a VgrG or PAAR, mediating delivery of the effector. Tap components are proven to be essential for the delivery of a range of effectors, like Tde1 from *A. tumefaciens*, TseL from *Vibrio cholerae*, or TseF from *P. aeruginosa* (Liang et al., 2015; Unterweger et al., 2015; Bondage et al., 2016). The current model suggests that Tap binds and stabilizes its cognate effector. Tap then facilitates binding of the effector to the C-terminus of the cognate VgrG or PAAR protein and subsequently dissociates from the tip. After dissociation of Tap, the effector remains bound to the VgrG or PAAR protein in a yet unknown mechanism but upon sheath contraction and by pushing the spike complex in the cell envelope, the effector is then transported (Bondage et al., 2016; Burkinshaw et al., 2018).

To broaden our knowledge on the molecular mechanisms of effector delivery, our aim was to achieve heterologous effector delivery. We used the nuclease effector Tde1 from *A. tumefaciens* and attempted to connect it to the VgrG1a spike in *P. aeruginosa*. For this, we constructed VgrG1a chimeras containing the C-terminal Tde1-binding extension from the *A. tumefaciens* VgrG1. We could show that these chimeras bind the cognate Tde1, however effector delivery could not be attained. This highlights the specificity of the T6SS spike for its effectors and outlines limitations for T6SS-mediated effector delivery.

MATERIALS AND METHODS

Bacterial Strains and Growth Conditions

Bacterial strains used in this study are described in **Table S1**. *P. aeruginosa* strains were grown in tryptone soy broth (TSB) or LB supplemented with antibiotics where appropriate (spectinomycin 2,000 $\mu g\ mL^{-1}$) at 37°C with agitation. *Escherichia coli* strains were grown in LB broth supplemented with antibiotics where appropriate (streptomycin 50 $\mu g\ mL^{-1}$, kanamycin 50 $\mu g\ mL^{-1}$). *A. tumefaciens* was grown at 28°C in minimal medium as described before (Lin et al., 2013).

DNA Manipulation

DNA purification was performed using the PureLink Genomic DNA minikit (Life Technologies) while plasmid DNA isolation using the QIAprep spin miniprep kit (Qiagen). Restriction endonucleases were used according to the manufacturer's specifications (New England Biolabs or Roche) and all used oligonucleotides are listed in **Table S2** and were purchased from Sigma. KOD Hot Start DNA Polymerase (Novagen) was used to amplify genes or DNA fragments used for the construction of mutator plasmids and deletion mutants as described by the manufacturer with the inclusion of 0.5 M betaine (Sigma). Colony PCR was performed with Taq polymerase (New England Biolabs) and DNA sequencing was performed by GATC Biotech.

Construction of *P. aeruginosa* Mutants

P. aeruginosa deletion mutants were constructed as described previously (Vasseur et al., 2005; Ventre et al., 2006) using the suicide plasmid pKNG101 (Herrero et al., 1990; Kaniga et al., 1991). Briefly, to create PAO1Δ*tse6tsi6*, 500-bp DNA fragments of the 5′ (up) and 3′ (down) ends of the *tse6-tsi6* gene pair were obtained by PCR using PAO1 chromosomal DNA as a template with the oligonucleotides Δ*tse6tsi6*_upF and Δ*tse6tsi6*_upR as well as with Δ*tse6tsi6*_dnF and Δ*tse6tsi6*_dnR (**Table S3**). A third PCR step using Δ*tse6tsi6*_upF and Δ*tse6tsi6*_dnR resulted in a DNA fragment with a clean deletion of the *tse6-tsi6* gene pair. To create the chimeric *vgrG1a* genes, splicing by overlap extension PCRs was performed initiated by three single PCR fragments. Gene fragments containing ~500 bp upstream and downstream of the splice junction were amplified using the overlapping primers *construct*_upR and *construct*_dnF as well as the upstream *vgrG1a*_F and downstream *vgrG1a*_R primers from the *P. aeruginosa* genome. A third gene fragment containing the fusion fragment of interest was obtained by using primers *construct*_upF and *construct*_dnR that are overlapping with *construct*_upR and *construct*_dnF, respectively. Subsequently, two overlap extension PCR steps were undertaken, employing an equimolar ratio of the upstream and downstream fragments as the DNA template. The gene fragments were cloned into pCR-BluntII-TOPO (Invitrogen), their sequences confirmed and sub-cloned into the pKNG101 suicide vector (**Table S2**). The pKNG-derivatives were maintained in *E. coli* strain CC118λpir and mobilized into *P. aeruginosa* PAK using *E. coli* 1,047 carrying the conjugative plasmid pRK2013 (Figurski and Helinski, 1979). Clones, in which double recombination events occurred, resulting in the deletion of the gene of interest *(GOI)* or fusion to *GOI*, were isolated using counterselection on sucrose plates as previously described (Vasseur et al., 2005). Gene deletions were verified by PCR using external primers and gene fusions confirmed by sequencing.

Secretion Assay

Secretion assays were performed as previously described (Hachani et al., 2011). Bacterial suspension was diluted from overnight cultures in TSB to OD_{600} of 0.1 and grown at 37°C to an OD_{600} of 5. A bacterial culture sample adjusted to OD_{600} of 1 was harvested by centrifugation and served as the whole cell sample. Thirteen milliliters of culture was centrifuged at 4,000 g for 20 min at 4°C to separate the bacterial cells and the culture supernatant. Then, 10 mL of the supernatant was transferred into falcon tubes and centrifuged again; 7 mL of the uppermost supernatant was transferred into new tubes and centrifuged. Two hundred microliters trichloroacetic acid were added to 1.8 mL supernatant fraction to precipitate proteins overnight at 4°C. The protein precipitate was obtained by centrifugation at 16,000 g for 30 min at 4°C, washed with cold 90% (v/v) acetone and centrifuged one more time. After removing the supernatant, the washed pellet was air-dried for 30 min and resuspended in 1x Laemmli buffer to an OD_{600} equivalent of 10.

Western Blot Analysis and SDS-PAGE

For SDS-PAGE analysis, cell extracts were loaded/migrated onto SDS polyacrylamide gels, and proteins transferred to a nitrocellulose membrane at 3 mA/cm^2. Following transfer, membranes were incubated overnight in blocking buffer (5% milk powder, 0.1% Tween 20 in Tris-buffered saline, pH 8.0). Polyclonal antibodies against VgrG1abc were used at a dilution of 1:1,000 (Hachani et al., 2011), Hcp1 at 1:1,000 (Hachani et al., 2011), Tde1 at 1:1,000 (Bondage et al., 2016), VgrG1^{Atu} at 1:1,000 (Bondage et al., 2016), and Tse3 at 1:500 (Hachani et al., 2011). Monoclonal antibodies against the β subunit of RNA polymerase (RpoB, NeoClone) were used at 1:5,000. Secondary antibodies conjugated to horseradish peroxidase were used at a dilution of 1:5,000. Western blots were developed using Super-Signal West Pico Chemiluminescent Substrate (Pierce) and visualized on a LAS3000 Fuji Imager.

Relative Protein Quantification

The intensities of the bands corresponding to the VgrG1a-chimeras were analyzed using the intensity measurement tool in the software ImageJ. Figures from three independent experiments were analyzed. The band corresponding to VgrG1a in wild-type (WT) cells was set as a standard for VgrG1a content in the cell. All measured intensities for other bands were divided by the intensity for VgrG1a under WT conditions. RpoB was used as protein loading control.

Interbacterial Competition Assays

Interbacterial competition assays were conducted on solid media due to the contact-dependent killing of the T6SS. Prey *P. aeruginosa* strains contained the Mini-CTX-*lacZ* integrated at the *att* site, consequently resulting in blue colonies on X-gal-containing plates. Overnight cultures in TSB were collected by centrifugation at 8,000 g for 3 min before washing twice in 1 mL sterile PBS and normalized to OD_{600} of 1.0. The OD_{600} was measured again for confirmation and 100 μL of attacker and 20 μL prey strains were mixed. This mixture was centrifuged at 8,000 g for 3 min and 20 μL supernatant was removed to give a competition mixture ratio of 5:1 of attacker and prey strains. Twenty microliters of each competition mix was spotted in duplicates onto LB-agar, the spots dried and the Petri dish lids were secured using parafilm M (Bemis). Competition plates were inverted and incubated at 37°C for 5 h for H1-T6SS-inducive killing.

The input competitions were serially diluted to 10^{-7}, plated on selective media for both attacker and prey (LB agar with 100 μg mL^{-1} X-gal for blue/white *P. aeruginosa* prey/attacker differentiation) and grown overnight at 37°C to confirm the input ratios. Competition spots were collected using 10 μL inoculation loops (VWR) and resuspended in 1 mL PBS. The competition output mixture was serially diluted to 10^{-7}, plated on selective media and grown overnight at 37°C similarly to the input. Both attacker and prey colony forming units were enumerated on both input and output dilution plates. All competition assays were repeated three times unless otherwise stated and the mean colony forming units (cfu) of survived prey strains obtained from all experiments with the standard deviation was plotted.

Bacterial-Two-Hybrid Assay

Genes expressing *tde1* or *tap1-tde1* were cloned into pUT18C and pKT25 and sequence confirmed. Variations of both vectors were introduced into *E. coli* DHM1 via heat shock and selected on LB plates containing both Kanamycin and Ampicillin. Resulting colonies were picked and grown in LB containing both antibiotics and overnight cultures were spotted onto LB containing Kanamycin (100 μg μL^{-1}), Ampicillin (100 μg μL^{-1}), IPTG (1 μM), and X-gal (100 μg μL^{-1}) and grown for 48 h at 30°C.

RESULTS

Heterologous Secretion of Tde1 From *A. tumefaciens*

In this study, we meant to investigate whether a heterologous T6SS cargo effector could be delivered by a T6SS solely by manipulating the VgrG tip. We chose Tde1 from *A. tumefaciens* as the heterologous effector, firstly because the Tde1 orthologs are only found in a-proteobacteria and secondly because, its basic delivery system has previously been studied in great details (Ma et al., 2014; Bondage et al., 2016; Wu et al., 2020). The current model states that VgrG1 from *A. tumefaciens* (VgrG1^A) is assembled to a functional trimer and capped by the cognate PAAR protein, with one PAAR protein binding the three last VgrG1^A β-strands. Concomitantly, Tap1 interacts with and stabilizes Tde1 within the cell and the Tap1-Tde1 complex is recruited to the C-terminal amino acids of VgrG1^A.

Since Tap1 was not detected in the supernatant fraction, it is believed to be released prior to secretion, while Tde1 remains bound to the VgrG1^{A} spike, which is propelled out of the cell (Bondage et al., 2016).

We aimed at directing the heterologous effector Tde1 toward VgrG1a from the H1-T6SS in *P. aeruginosa* (VgrG1a^{P}). To assess whether *P. aeruginosa* is naturally able to deliver Tde1 using its H1-T6SS, we introduced the plasmid pTrc200 containing the *A. tumefaciens tap1-tde1-tdi1-paar* (t-t-t-p) (**Figure 1A**) into *P. aeruginosa* PAKΔ*retS*, which has an active H1-T6SS, and performed secretion assays (**Figure 1B**). For all following experiments, we used a plasmid expressing a catalytic mutant of Tde1, as *P. aeruginosa* growth was inhibited when cells expressed WT Tde1 likely due to Tde1 toxicity (Ma et al., 2014). Tde1 (**Figure 1B**, top panel) is expressed in consistent amounts (lanes 2 and 3), and no secretion is observed in a T6SS-positive background (lane 9). Intriguingly, a weak Tde1 band is detectable in the supernatant fraction of T6SS-inactive strains (lane 10) but this also correlates with elevated detection of RpoB in this fraction (third panel) and suggests partial cell lysis. We then attempted to connect Tde1 onto the *P. aeruginosa* T6SS by co-expressing the cognate *A. tumefaciens vgrG1^{A}* (**Figure 1B**, second panel). In this case, we observed that the bands corresponding to Tde1 increased in intensity (lanes 6 and 7), which suggests that a specific interaction between Tde1 and VgrG1^{A} may occur in *P. aeruginosa* resulting in Tde1 stabilization. However, neither VgrG1^{A} nor Tde1 could be detected in the supernatant fractions (lanes 13) suggesting no efficient Tde1 secretion even in the presence of its cognate VgrG. Again, some traces of Tde1 are found in the supernatant of the T6SS-inactive background (lane 14) but it correlates with elevated RpoB levels. Interestingly, a decreased amount of Hcp1 in the supernatant fraction (lane 13) could clearly be seen which suggests that the H1-T6SS function is altered in the presence of VgrG1^{A}, possibly because the heterologous VgrG is able to partially connect to the *P. aeruginosa* T6SS but is then not further engaged in the secretion process.

Design of VgrG Chimeras to Connect Tde1 to the H1-T6SS

As an alternative to using the entire *A. tumefaciens* VgrG1^{A}, we decided to design chimeras to directly connect Tde1 to the *P. aeruginosa* VgrG1a. It was established that the 31 C-terminal amino acids of VgrG1^{A} is required for binding the Tap1-Tde1-complex and a prerequisite for Tde1 delivery in *A. tumefaciens* (Bondage et al., 2016). We designed three chimeras, which we named A, B, and C, between VgrG1a^{P} from *P. aeruginosa* (**Figure 2**, green) and VgrG1^{A} from *A. tumefaciens* (**Figure 2**, blue) to connect Tde1 to the *P. aeruginosa* H1-T6SS as shown in **Figure 2**, while any of the chimera would replace the WT *vgrG1a* gene on the chromosome. Construct A (VgrG1a^{P}-G1^{A31}) contains the full length VgrG1a^{P} extended by 31 C-terminal amino acids from VgrG1^{A}, thus including amino acids likely responsible for binding the cognate *A. tumefaciens* PAAR and

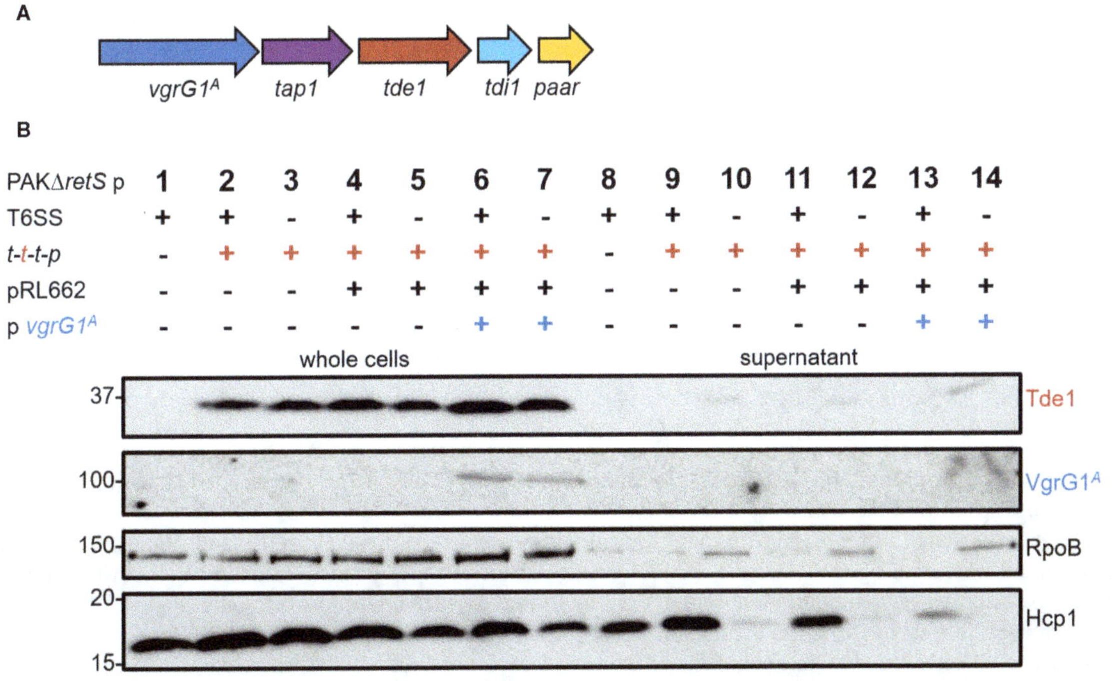

FIGURE 1 | *P. aeruginosa* cannot perform heterologous secretion of *A. tumefaciens* VgrG1^{A} and Tde1. **(A)** The set of genes *tap1-tde1-tdi1-paar* (purple-red-cyan-orange) were expressed from pTrc200, while *vgrG1^{A}* (blue) was expressed from pRL662 in *P. aeruginosa* (Bondage et al., 2016). **(B)** Representative figure of a western blot from a secretion assay of PAKΔ*retS* pTrc200 (p) with an active (+) or inactive (–) H1-T6SS expressing (+) *tap1-tde1-tdi1-paar* (*t-t-t-p*) or containing pRL662 (+) with *vgrG1^{A}* (+). Antibodies used (from top to bottom) are against Tde1, VgrG1^{A}, RpoB, and Hcp1 as indicated on the right.

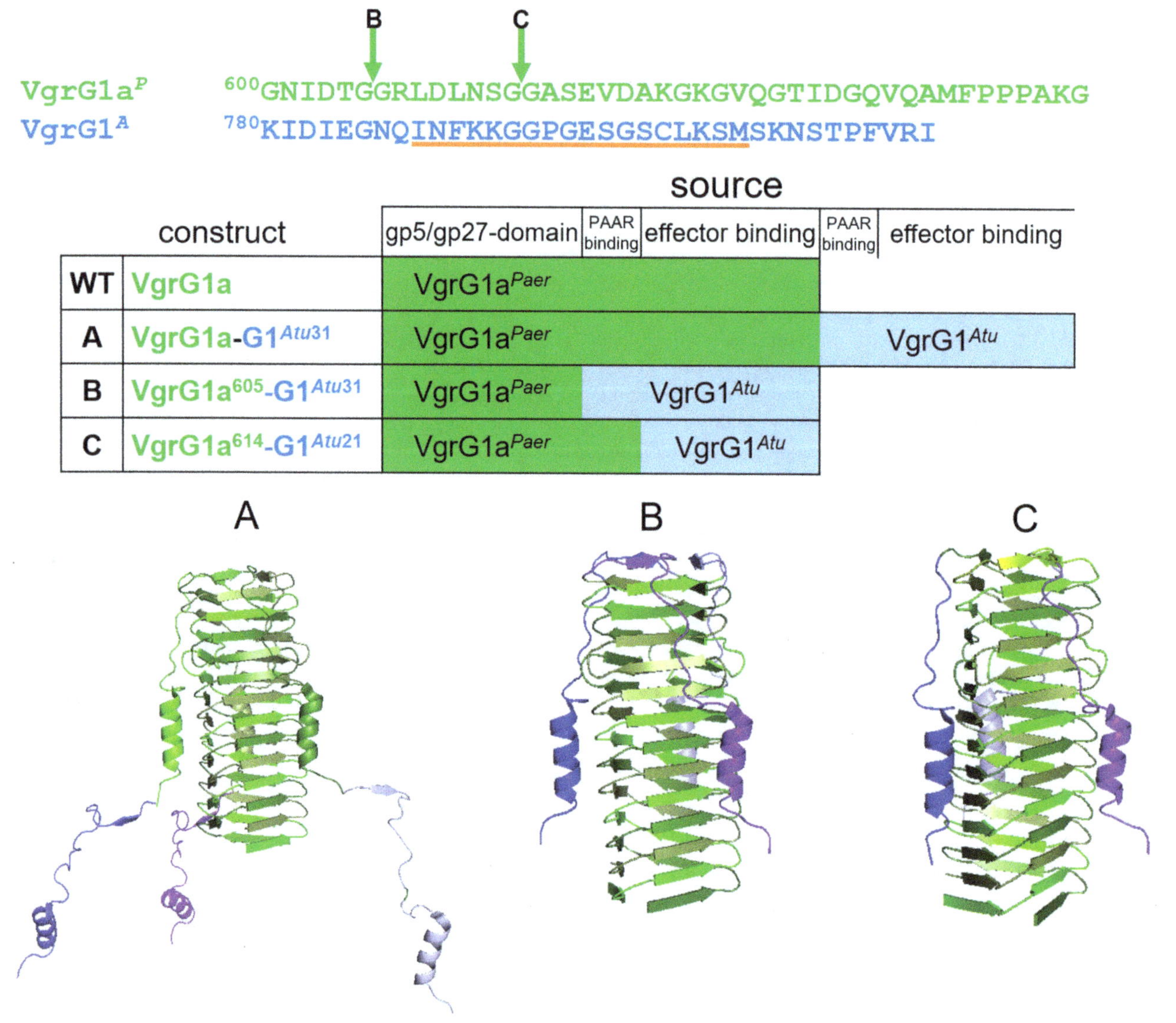

FIGURE 2 | Construction of VgrG chimeras. The top panel represents the amino acid sequences of the C-terminal extensions of VgrG1a^P (green) and VgrG1^A (blue). The arrows correspond to the swapping points after amino acid (aa) 605 in chimera B and aa 614 in chimera C. The orange line corresponds to the epitope, against which the VgrG1 antibody was raised. In the lower panel structural models (pdb:4mtk) of the needle parts of the three designed chimeras are shown. In construct A, the STOP codon of *vgrG1a^P* (green) was replaced by the gene portion corresponding to the last 31 aa of VgrG1^A (blue). Constructs B,C were designed by substituting the gene portions corresponding to the last 35 aa and 29 aa of *vgrG1a^P* with the gene portions corresponding to the last 31 aa and 21 aa, respectively, of VgrG1^A.

Tde1 (Bondage et al., 2016). Construct B (VgrG1a^{P605}-G1^{A31}) covers the first 605 amino acids from VgrG1a^P and 31 C-terminal amino acids from VgrG1^A, while construct C (VgrG1a^{P614}-G1^{A21}) comprises of the first 614 amino acids from VgrG1a^P and 21 C-terminal amino acids from VgrG1^A.

First, we investigated whether modification of the VgrG1a spike protein impacts its ability to form a functional T6SS spike and therefore its secretion. We integrated the chimeras into PAKΔ*retS* background and performed secretion assays (**Figure 3A**) and detected all three chimeras with an antibody against VgrG1abc (top panel, lanes 3–5). While chimera B seemed to be less stable within the whole cell fraction, we only detected chimera A in the supernatant fraction (lane 8) suggesting it to be secreted. We then asked whether the chimeras would also be stable in absence of the other two VgrG proteins VgrG1b and VgrG1c and whether their absence would impact secretion of any of the chimeras. Using PAKΔ*retS*Δ*vgrG1b*Δ*vgrG1c* cells (**Figure 3B**), thus lacking the two other VgrGs related to the H1-T6SS, we confirmed data from previous studies showing that VgrG1a on its own is able to form a functional T6SS spike promoting efficient Hcp1 secretion, as well as secretion of the Hcp1-dependent effector Tse3 (**Figure 3B**, lane 16) (Hachani et al., 2014). When testing any of the VgrG1a chimeras, only VgrG1a-construct A expressing full

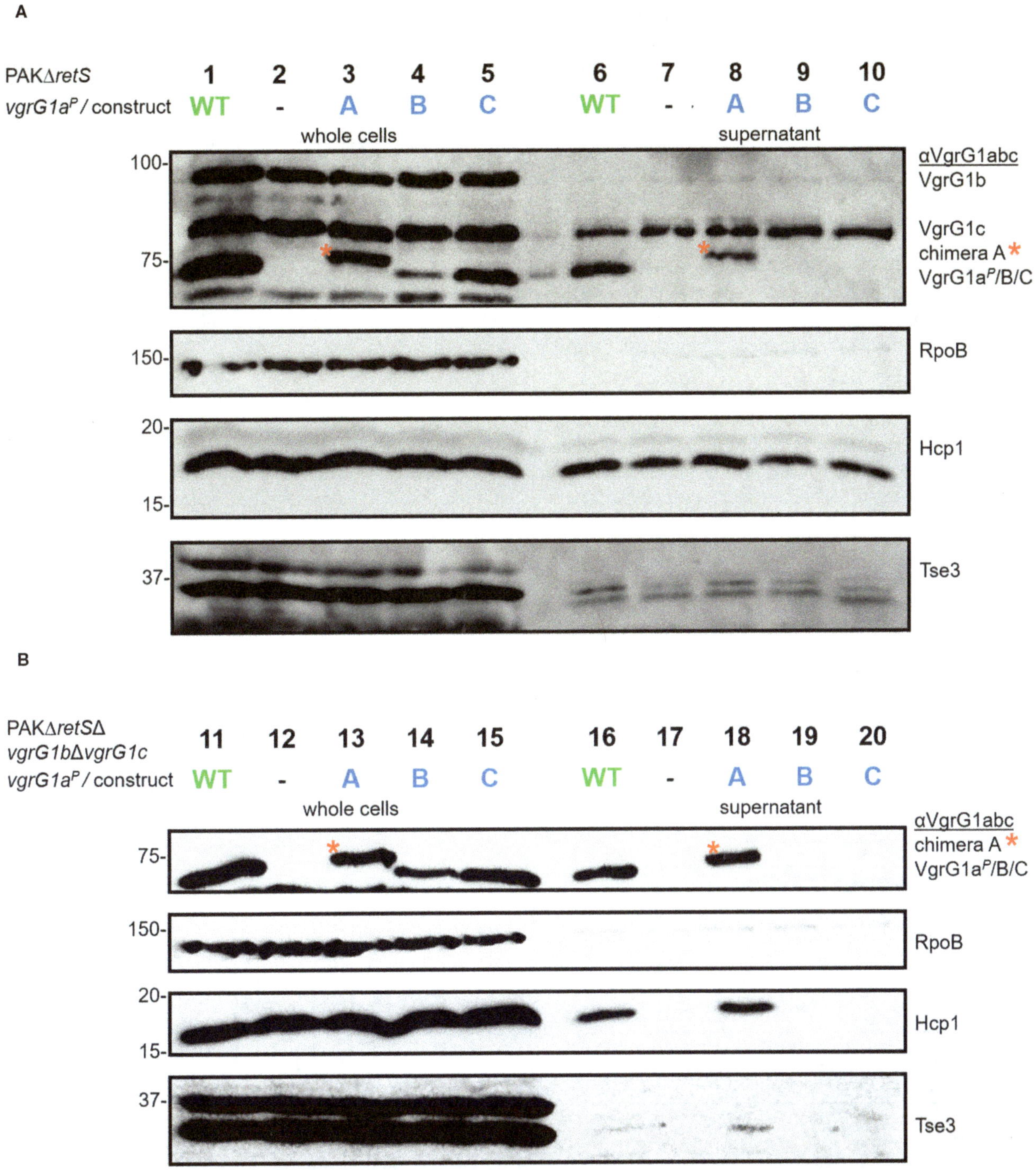

FIGURE 3 | Modifying the C-terminus of VgrG1a inhibits T6SS functionality. Representative figures of western blots from three independent secretion assays of **(A)** PAKΔ*retS* and **(B)** PAKΔ*retS*Δ*vgrG1b*Δ*vgrG1c* expressing Wild type VgrG1a^P (WT), no VgrG1a^P (-) or the chimeras A, B, C between VgrG1a^P and VgrG1^A. Antibodies used (from top to bottom) are against VgrG1abc, RpoB, Hcp1, and the Hcp1-dependent effector Tse3 as indicated on the right.

length VgrG1a^P fused to G1^{A31} is able to execute this function as seen by identifying Hcp1, VgrG1a^P-G1^{A31}, and Tse3 in the supernatant (**Figure 3B**, lane 18, red asterisk). None of the other two chimeras seems to be able to form a functional T6SS spike.

Connecting Tde1 to the VgrG1a Spike

We then assessed whether Tde1 could bind to the C-terminus of VgrG1^A, when the latter is fused to a heterologous VgrG vehicle as is the case in our three chimeric constructs. We used a bacterial-two-hybrid (BTH) assay, in which the three chimeras as well as a catalytic Tde1 mutant carry the T18- or T25-domains (T18/25) at their C-termini. When testing interactions between Tde1 and any of the VgrG-constructs, no blue colonies appeared indicating that the two proteins do not interact (**Figure 4A**, top panel). Since Tde1 only binds VgrG1^{A} in presence of Tap1 in *A. tumefaciens* (Bondage et al., 2016), we then cloned the *tap1* gene upstream of *tde1* and re-tested the interactions (bottom panel). When Tap1 is present, dark blue spots could be readily observed suggesting strong interactions between Tap1-Tde1 and any of the VgrG1a^{P}-G1^{A}-chimeras. The fact that construct C efficiently associates with Tap1-Tde1 suggests that as little as the 21 C-terminal amino acids of VgrG1^{A} are sufficient to mediate this interaction. Intriguingly, Tap1-Tde1 binds only very weakly to VgrG1^{A}, which might be due to the lack of the cognate PAAR (Bondage et al., 2016).

This result suggests that Tap1-Tde1 could recognize the C-terminus of VgrG1^{A} when it is plugged onto a heterologous VgrG core. We then investigated whether this interaction would also occur in *P. aeruginosa* cells, which would be the prerequisite for Tde1 secretion. For this, PAKΔ*retS* cells expressing WT VgrG1a^{P} or any of the chimeras (A/B/C) were grown for 5 h at 37°C in the presence (+) or absence (-) of a plasmid carrying the *tap1-tde1-tdi1-paar (t-t-t-p)* genes. Western blot analysis was performed on whole cell lysates probed for Tde1 and VgrG1abc (**Figure 4B**) as well as RpoB as a loading control. Images from three independent experiments were analyzed using ImageJ and the intensity of the VgrG1a^{P} band in absence of Tde1 was quantified (bottom plot) and served as a standard for VgrG1a^{P} level. The intensity of any bands corresponding to VgrG1a^{P}/constructABC was quantified and divided by the standard intensity for VgrG1a^{P} in Tde1 absence (**Figure 4B**, lane 1). According to this calculation, the closer a ratio is to 1, more of this protein is present in the cell. No difference in intensity between the VgrG1a^{P} bands in absence (**Figure 4B**, lane 1) or presence (**Figure 4B**, lane 2) of Tap1-Tde1-Tdi1-PAAR was observed. However, the intensity of the bands representing the VgrG1a^{P}-G1^{A} chimeras shows variability when in presence or absence of Tap1-Tde1-Tdi1-PAAR. Indeed, a drastic decrease in abundance is observed in absence of Tap1-Tde1-Tdi1-PAAR (**Figure 4B**, lanes 3, 5, and 7) while co-expression of Tap1-Tde1-Tdi1-PAAR led to a significant increase in intensity of the corresponding bands (**Figure 4B**, lanes 4, 6,

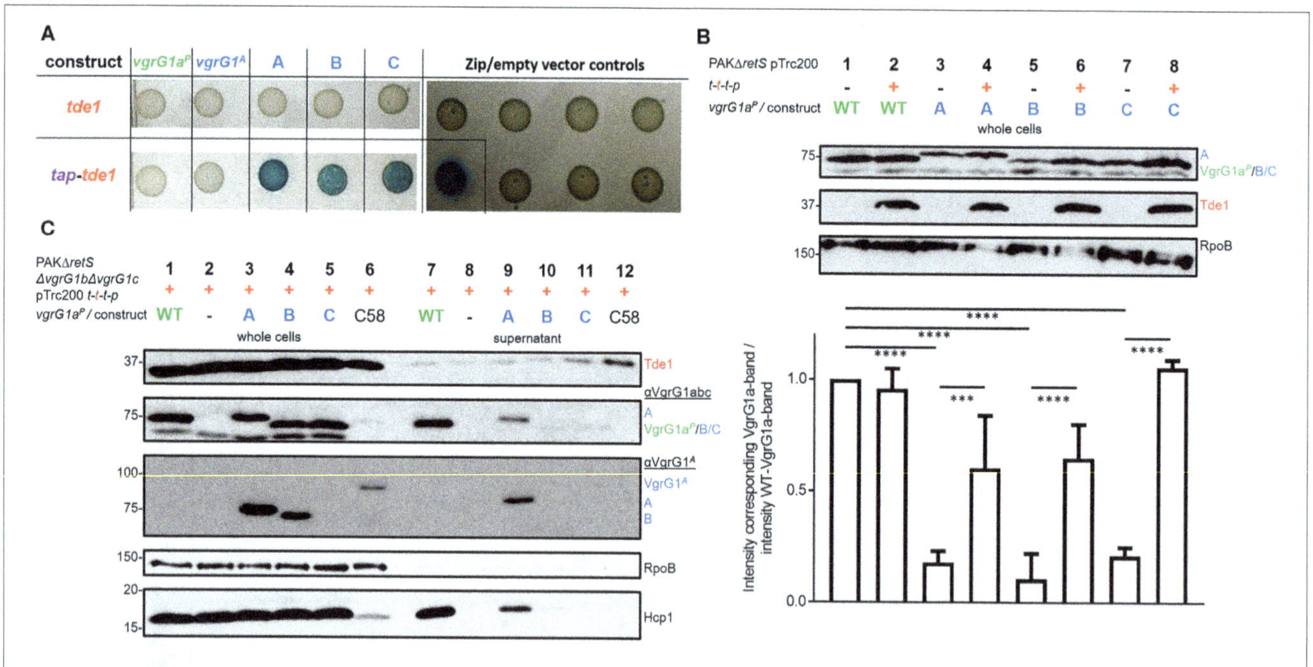

FIGURE 4 | All chimeras interact with Tde1 but are unable to facilitate effector secretion. **(A)** T25 and T18-domains were fused to the C-termini of either VgrG or a catalytic mutant of Tde1 (red) in presence of the cognate Tap1 (purple). Representative bacterial-two-hybrid plates tested for interactions between VgrG1a^{P}-G1^{A}-chimeras and Tde1 on its own (top panel) or in presence of the Tap1 protein (bottom panel) (Bondage et al., 2016). All interactions were tested with all gene sequences cloned into both pKT25 and pUT18C represented as two columns for each tested interaction. Empty vector controls as well as the leucine-*zipper* pUT18C-*Zip* and pKT25-*Zip* vectors are shown in the right panel. **(B)** Top panel: Representative western blot of whole cell lysates of PAKΔ*retS* containing pTrc200 and expressing chimera A, B, C between VgrG1a^{P} and VgrG1^{A} in absence (-) and presence (+) of *tap1-tde1-tdi1-paar* (*t-t-t-p*). As a control for VgrG1a^{P} stability, the parental strain was included and grown in absence and presence of *t-t-t-p* (lanes 1 and 2). Antibodies were used against VgrG1abc, Tde1, and RpoB as indicated on the right. Bottom panel: Intensity of the corresponding VgrG1a^{P} bands from three independent experiments was quantified using ImageJ (https://imagej.nih.gov/ij/index.html) and divided by the intensity of the band belonging to WT VgrG1a^{P} in absence of Tde1 (lane 1) as a standard for VgrG1a stability. One-Way ANOVA with Tukey's multiple comparison test was conducted between datasets as indicated with $^{****}p < 0.0001$. **(C)** Representative figures of western blots from three independent secretion assays of PAKΔ*retS*Δ*vgrG1b*Δ*vgrG1c* pTrc200 expressing chimeras A, B, C between VgrG1a^{P} and VgrG1^{A} in presence (+) of *tap1-tde1-tdi1-paar* (*t-t-t-p*). As a control for Tde1 secretion the supernatant fraction of *A. tumefaciens* C58 was included (lanes 6 and 12). Antibodies (from top to bottom) against Tde1, VgrG1abc, VgrG1^{A}, RpoB, and Hcp1 were used as indicated. $^{***}p < 0.001$.

and 8). This suggests a lack of stability of the chimeric VgrG in absence of the Tap1-Tde1-Tdi1-PAAR complex. Since PAAR proteins were shown to bind VgrG spike proteins (Shneider et al., 2013), one could suggest this to be the case here. However, chimera C does not contain the interaction motif for PAAR but its stability is the most increased amongst the chimeras (**Figure 4B**, lane 8). Hence, interactions between the chimeras and the PAAR protein can be ruled out. Furthermore, no immunity protein was ever shown to interact directly with a VgrG protein, so it is unlikely that the immunity Tdi1 would interact with any chimera. Since Tde1 was shown to interact with VgrG1^A in presence of Tap1 *in vivo* (Bondage et al., 2016), we propose that here a Tap1-Tde1 complex could bind to any of the tested chimeras.

Connecting Tde1 to the VgrG1a Spike Is Not Sufficient for T6SS-Mediated Secretion

Having established that chimera A is proficiently secreted and that the heterologous effector Tde1 seems to bind to it for stability, we then aimed at testing whether chimera A could be a secretion vehicle to deliver Tde1 from *P. aeruginosa*. In **Figure 3** we showed that stability of the chimeras is independent of the presence of the other two VgrG proteins and western blot detection of the chimeras is facilitated in the absence of VgrG1b and VgrG1c due to cross-recognition by the antibody. Furthermore, Hcp1 secretion in the PAKΔ*retS*Δ*vgrG1b*Δ*vgrG1c* background represents an adequate readout for T6SS functionality (**Figure 3B**), and we chose to test Tde1 secretion from *P. aeruginosa* in the absence of VgrG1b and VgrG1c. Hence, we performed standard secretion assays of *P. aeruginosa* strains that additionally expressed *tap1-tde1-tdi1-paar* genes (**Figure 4C**). Western blot assays using antibodies against both VgrG1a^P and an amino acid stretch from the C-terminus of VgrG1^A (**Figure 2**, orange line) (Bondage et al., 2016) revealed that VgrG1a^P and all VgrG1a^P-VgrG1^A-chimeras are produced in significant amounts (**Figure 4C**, second and third panels, lanes 1, 3–5). We did not detect construct C (lane 5) with the antibody against the C-terminal amino acids of VgrG1^A, which is surprising as this antibody was raised against an amino acid stretch that includes most of the 21 amino acids, hence one might have expected the antibody would recognize our chimera. However, since the antibody against VgrG1abc from *P. aeruginosa* did detect this chimera, we could assume production of the protein. Yet, in the supernatant fractions, we could only detect VgrG1a^P (lane 7) and construct A (lane 9). We also monitored Hcp1 as a readout for T6SS functionality (bottom panel) but detected this protein only in the supernatant fractions of strains expressing and secreting VgrG1a^P or construct A. Neither chimera B or C, nor Hcp1, were detected in the supernatant fractions when using strains expressed chimera B or C (lanes 10 and 11). This confirms that only construct A is able to form a functional T6SS tip.

Even though we observed faint bands for Tde1 in the supernatant fractions of all tested strains (**Figure 4C**, top panel) we suggest it is unlikely resulting from T6SS-dependent secretion since there is no particular increase in the intensity of the Tde1 band in presence of the functional secreted chimera A (lane 9) as compared to other non-secreted chimeras (lanes 10 and 11). As a control for Tde1 secretion, *A. tumefaciens* was grown under T6SS-inducing conditions (Lin et al., 2013) displaying an intense band for Tde1 in the supernatant (lane 12).

Tde1 Presence Interferes With T6SS Functionality and Effector Delivery

Despite our data suggesting that Tde1 may interact with and stabilize chimeric VgrG1a spikes in *P. aeruginosa*, and that at least one chimera (construct A) is secreted, we did not find conditions which resulted in effective Tde1 secretion. There are a few observations that may explain this controversy. First, we noticed an impact on the secretion efficiency of construct A in presence of Tap1-Tde1-Tdi1-PAAR (**Figures 5A,B**, compare lanes 6). We quantified these data by measuring the intensities of the corresponding bands from the supernatant fractions, and from three independent experiments, using ImageJ (**Figure 5C**). Approximately 60% of the produced VgrG1a^P and construct A are secreted into the supernatant in absence of Tap1-Tde1-Tdi1-PAAR (lanes 1 and 3). However, whereas in the presence of Tap1-Tde1-Tdi1-PAAR, VgrG1a^P secretion level is unaffected (lane 2), the secretion of construct A drops to 20% (lane 4). This is a remarkable finding which suggests that Tap1-Tde1-Tdi1-PAAR does interfere with construct A secretion but not with VgrG1a^P and this correlates with the evidence that Tap1-Tde1-Tdi1-PAAR interacts with construct A but not with VgrG1a^P.

Since construct A secretion is impacted by the presence of Tap1-Tde1-Tdi1-PAAR, we hypothesized that delivery of the VgrG1a^P-dependent PAAR effector Tse6 into prey cells would be affected as well. In WT cells, VgrG1a^P is able to drive killing of Tse6-sensitive cells as shown previously (Hachani et al., 2014). Since construct A secretion is similar to that of VgrG1a^P, one would expect a similar killing of Tse6-sensitive cells, except if the presence of Tap1-Tde1-Tdi1-PAAR interferes with construct A to bind and deliver Tse6. In a competition assay, we used a prey strain lacking *tsei6* which has no longer immunity to Tse6, while the attacker uses either VgrG1a^P or construct A in presence or absence of Tap1-Tde1-Tdi1-PAAR (**Figure 5D**). Note that a catalytic mutant of Tde1 was also used to rule out any toxic effect on the prey in case Tde1 was delivered even in small amounts.

The number of recovered prey cells (cfu) was lower when incubated with strains expressing VgrG1a^P (**Figure 5D**, lane 1) than with strains lacking VgrG1a^P (lane 5), confirming that Tse6 delivery depends on VgrG1a^P (Hachani et al., 2014). Neither of these outcomes was affected by Tde1 presence (lanes 2 and 6). When the attacking strains express construct A but no Tde1 (lane 3), Tse6-dependent killing is reflected by low prey survival similar to WT levels. This confirms that not only is construct A secreted, but it also drives Tse6 delivery into prey cells, corroborating its functionality despite the additional C-terminal residues from VgrG1^A. Remarkably, co-expression of Tap1-Tde1-Tdi1-PAAR (lane 4) leads to an intermediate phenotype suggesting that Tse6 delivery is affected but not completely abrogated. This correlates with the fact that construct A is still secreted but in smaller amounts (**Figure 5A**, lane 6), which would mean that secretion of the whole VgrG-PAAR complex is affected. Further, this suggests

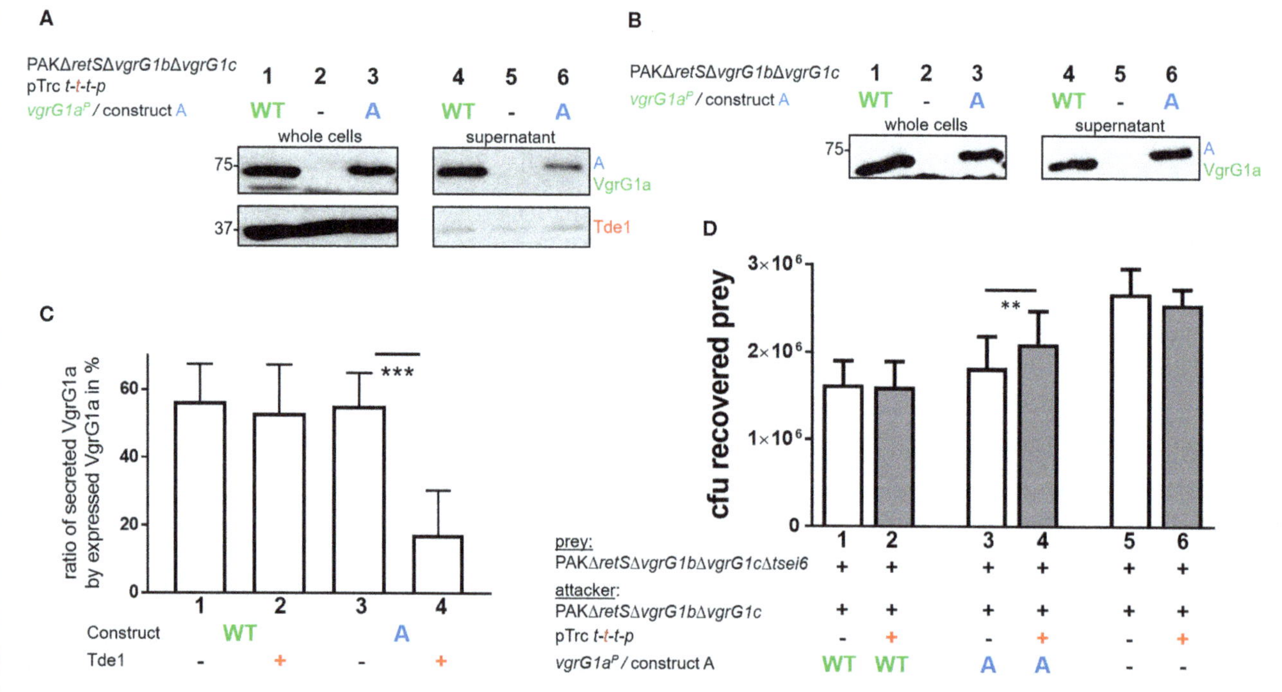

FIGURE 5 | Presence of Tde1 interferes with delivery of construct A and the VgrG1a-dependent effector Tse6. **(A,B)** Sections of western blots from secretion assays of PAK*ΔretSΔvgrG1bΔvgrG1c* expressing native VgrG1a^{P} (WT) or chimera A **(A)**. In A, *tap1-tde1-tdi1-paar* is co-expressed from pTrc200, but not in **(B)**. Antibodies were used against VgrG1abc and Tde1 as indicated on the right. **(C)** Percentage of secreted VgrG1a^{P} (WT) or construct A **(A)** in absence (-) or presence (+) of *tap1-tde1-tdi1-paar*. Intensity of bands corresponding to VgrG1a^{P} was measured using ImageJ and plotted. Dunnett's multiple comparisons test was conducted with $^{***}p < 0.001$. **(D)** Plot of recovered cfu of prey strain PAK*ΔretSΔvgrG1bΔvgrG1cΔtsei6::lacZ* after contact with attacker strain that expressed VgrG1a^{P} (WT) or construct A **(A)** in absence (-) or presence (+) of *tap1-tde1-tdi1-paar* (*t-t-t-p*) from pTrc200 from five independent experiments. Spots were incubated for 5 h at 37°C in a 5:1 ratio. One-Way ANOVA analysis with Sidak's multiple comparisons test was conducted between the datasets obtained from the two genetically same strain backgrounds with $^{**}p < 0.01$.

that either binding of Tap1-Tde1 to the tip composed of construct A does not entirely prevent Tse6 secretion, or that only a fraction of construct A is bound by Tap1-Tde1 and others which are associated with Tse6 could deliver Tse6.

DISCUSSION

Bacterial protein secretion systems are highly specific and proteins secreted by a given secretion type (T1SS-T7SS) usually fail to be recognized by another type due to the lack of appropriate secretion motifs (Voulhoux et al., 2000; Filloux, 2010, 2011). Even proteins secreted by the same type but by distinct systems within the same species would usually fail to be effectively released (Ball et al., 2002). Despite such strong selectivity several studies have successfully managed to reprogram secretion by engineering chimeric effectors or secretion machines (Nicolay et al., 2015; van Ulsen et al., 2018). The present study aimed at evaluating potential venues to redirect a heterologous cargo effector from *A. tumefaciens* to the T6SS of *P. aeruginosa*. We reasoned that a T6SS spike/VgrG could accommodate to the endogenous T6SS machine through its conserved domain, while the additional variable domain could be manipulated to adapt recognition of a heterologous effector. With this in mind we hypothesized that specific interactions between the effector Tde1 and the appropriate C terminal region of its cognate VgrG1^{A} might be sufficient to attach the effector to a chimeric but heterologous VgrG spike. We constructed three chimeras made of the *P. aeruginosa* VgrG1a^{P} and the Tde1 binding amino acids from VgrG1^{A}, but none of them was able to complete Tde1 delivery in *P. aeruginosa*.

Although highly complex in terms of interpretation, several findings from our study could be explained with a concept based on steric hindrance as sketched in **Figure 6**. In an original *P. aeruginosa* context (**Figure 6A**), VgrG1a binds and delivers its cognate Tse6 PAAR effector, while EagT6 (magenta) and the elongation factor 2 (Ef2, cyan) also tightly interact with and sterically occupy the space around the VgrG tip (Whitney et al., 2015; Quentin et al., 2018). However, when chimera A is expressed, the only chimera which *per se* is secreted by *P. aeruginosa*, the same Tse6-associated components are interacting with the tip, since none of the interaction surfaces between VgrG1a and the Tse6 PAAR domain was modified. Additionally, three Tap1-Tde1-complexes (purple-red and only two of them depicted in the figure) could also bind to each monomer of the chimera A spike complex, with such interaction having been confirmed by the bacterial-two-hybrid assay (**Figure 4A**). In all, we suggest that the steric hindrance due to a wealth of binding partners around the VgrG tip might be a limitation factor

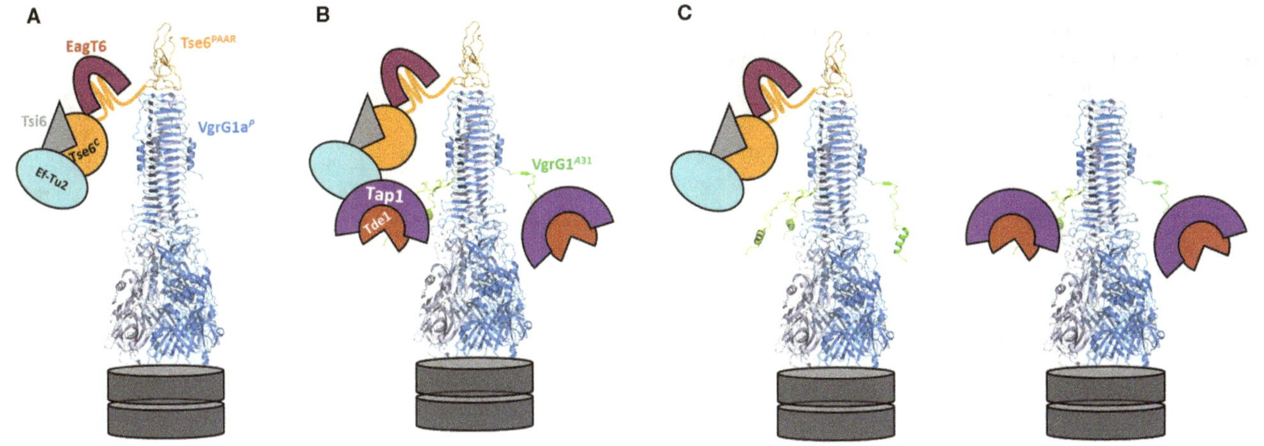

FIGURE 6 | Model of the decoration of the VgrG1a/chimera A-tip in absence and presence of Tap-Tde1. VgrG1a^{P} is represented as a blue trimer (pdb: 4mtk), topped by the PAAR domain of Tse6 (orange) with the C-terminal domain of Tse6 floating around the tip bound to its cognate immunity Tsi6 (gray). The chaperone EagT6 (magenta) and the elongation factor EfTu2 (cyan) additionally bind the C-terminal toxin domain of Tse6. **(A)** Configuration of the VgrG1a tip in the parental strain. **(B)** In cells expressing construct A, VgrG1a is extended by 31 aa originally from VgrG1^{A} (green) resulting in an extended VgrG spike protein. In the presence of Tap (purple) and Tde1 (red), three Tap1-Tde1-complexes bind to each of the C-terminal extensions of chimera A (only two are shown here) leading to a sterically crowded complex tip. **(C)** Two construct A spikes coexist in *P. aeruginosa* cells. One complex is comparable to the WT spike with Tse6 binding to the three last β-strands of construct A, resulting in a secretion-competent T6SS tip. A second complex interacts with the Tap1-Tde1-complex but not with Tse6 and this complex cannot be secreted resulting in less overall secretion of construct A in comparison to VgrG1a spikes.

(**Figure 6B**). Since we observed a hampered construct A secretion (**Figure 5A**, lane 6) and Tse6 delivery (**Figure 5D**, lane 4) in the presence of Tap1-Tde1-complexes, one might also suggest that two complexes coexist in *P. aeruginosa* cells (**Figure 6C**). A set of construct A proteins might form a functional T6SS spike capped by Tse6 resulting in secretion of the complex. A second set of construct A might be bound to the Tap1-Tde1-complex, which might form a VgrG trimer, but does not bind Tse6, thus not leading to a functional trimer that is being secreted. One might suggest the production of both WT VgrG1a and chimera A within one background to facilitate the production of a heterotrimer that could secrete both Tse6 as a PAAR and Tde1 as a cargo effector. However, production of such heterotrimer would be challenging as it would consist of three VgrG proteins with an unknown ratio of WT VgrG1a and chimera A. Furthermore, in a previous study we saw that in presence of two VgrG1a species, WT VgrG1a and modified VgrG1a (Wettstadt and Filloux, 2020), WT VgrG1a was secreted in higher amounts than the modified VgrG1a suggesting a preferred formation of WT VgrG1a spikes to be secreted.

Our results may emphasize a major limitation of T6SS-mediated effector delivery, which is the available space around the VgrG tip and probably within the T6SS membrane complex. Recent studies visualized the baseplate structure (Nazarov et al., 2018) and the membrane complex (Durand et al., 2015) of the T6SS and cavities around the spike complexes which were unoccupied and proposed to allow the accommodation of effector proteins. However, it is not clear whether there is enough space to accommodate additional and folded cargo and a wealth of decorations around the VgrG tip would prevent the fit into the membrane complex. It is also to take into consideration that the dimension from the T6SS machine originating from *V. cholerae* and *E. coli* might not be similar for the H1-T6SS from *P. aeruginosa*.

Our study highlights the fine balance between the functionality of the spike, e.g., its ability to perforate the cell envelope once embedded in the T6SS membrane complex and thus its delivery into the extracellular medium, and its capacity to bind a cargo effector and drive its secretion. Previous studies in *Serratia marcescens*, *Acinetobacter baylyi*, and *A. tumefaciens* demonstrated that at least one VgrG protein and its cognate PAAR need to assemble to form a functional T6SS tip (Shneider et al., 2013; Bondage et al., 2016; Cianfanelli et al., 2016; Wu et al., 2020). In the case of *S. marcescens*, three such assemblies can form: VgrG1 and its cognate PAAR; VgrG2 and Rhs1 or VgrG2 and Rhs2. Each of the two PAAR domains of the Rhs effectors can top VgrG2 and thus form a functional spike complex. Similarly, in *A. tumefaciens* two such assemblies were confirmed: VgrG1 with PAAR delivering Tde1 and VgrG2 with the PAAR effector Tde2 (Bondage et al., 2016). In *P. aeruginosa*, it was previously demonstrated, that each VgrG associated with the H1-T6SS, VgrG1a, VgrG1b, or VgrG1c, can associate with a cognate PAAR effector, Tse6, Tse7, and Tse5, respectively (Hachani et al., 2014). Here, we showed in *P. aeruginosa* that presence of the full length VgrG1a is required to form a functional H1-T6SS spike while any chimera lacking parts of the C-terminal residues of VgrG1a (construct B or C) was not able to perform the same function. This might suggest that the C-terminal residues of VgrG1a are required for specific interactions with its cognate PAAR effector Tse6, and ultimately the presence of the PAAR in the spike would be needed for effective secretion of the whole spike.

This explanation is consistent with recent findings that effector loading onto its cognate VgrG spike activates T6SS assembly (Liang et al., 2019; Wu et al., 2020). This is also supported by data showing that the three last β-sheets of VgrG1b specifically interact with the cognate PAAR effector Tse7 (Shneider et al., 2013; Pissaridou et al., 2018). Yet, the cargo effectors from *P. aeruginosa* and *E. coli*, PldA, PldB and Tle1, that do not contain N-terminal PAAR domains, were shown to specifically bind to the C-terminal domains of their cognate VgrGs, VgrG4b, VgrG5, and VgrG1, respectively (Flaugnatti et al., 2015; Wettstadt et al., 2019). In this case it is likely that the T6SS spike would be completed by a standalone PAAR domain (Wood et al., 2019a).

Our study confirmed that heterologous secretion cannot easily result from simple and straightforward genetic manipulations and the recognition of a secreted effector by its own machinery has likely resulted from a longstanding co-evolution which guarantees specificity. This way it would be hard to hijack the process and only intended effectors are released upon specific conditions, avoiding also any leakage of intracellular proteins. In the T6SS where C terminal motifs in the VgrG spike are proven to confer effector recognition specificity, search for motifs within the effector *per se* has not given what one could consider a universal T6SS motif. Yet, a conserved motif has been found in a subset of T6SS effectors, which has been called the MIX motif (Marker for type sIX effectors) (Salomon et al., 2014). In several other T6SS effectors, a different but conserved domain is also found at the N terminus, that has been called FIX (Jana et al., 2019). Whether these domains could be used to engineer chimeric T6SS effectors that will be retargeted to heterologous T6SS machine is yet to be fully investigated.

AUTHOR CONTRIBUTIONS

SW contributed to conceptualization, investigation, methodology, and writing the original draft. E-ML and AF were responsible for the conceptualization, funding acquisition, supervision, and reviewing and editing the manuscript. All authors contributed to the article and approved the submitted version.

REFERENCES

Ball, G., Durand, E., Lazdunski, A., and Filloux, A. (2002). A novel type II secretion system in *Pseudomonas aeruginosa*. *Mol. Microbiol.* 43, 475–485. doi: 10.1046/j.1365-2958.2002.02759.x

Bondage, D. D., Lin, J. S., Ma, L. S., Kuo, C. H., and Lai, E. M. (2016). VgrG C terminus confers the type VI effector transport specificity and is required for binding with PAAR and adaptor-effector complex. *Proc. Natl. Acad. Sci. U.S.A.* 113, E3931–3940. doi: 10.1073/pnas.1600428113

Brunet, Y. R., Henin, J., Celia, H., and Cascales, E. (2014). Type VI secretion and bacteriophage tail tubes share a common assembly pathway. *EMBO Rep.* 15, 315–321. doi: 10.1002/embr.201337936

Brunet, Y. R., Zoued, A., Boyer, F., Douzi, B., and Cascales, E. (2015). The type VI secretion TssEFGK-VgrG phage-like baseplate is recruited to the TssJLM membrane complex via multiple contacts and serves as assembly platform for tail tube/sheath polymerization. *PLoS Genet.* 11:e1005545. doi: 10.1371/journal.pgen.1005545

Burkinshaw, B. J., Liang, X., Wong, M., Le, A. N. H., Lam, L., and Dong, T., G. (2018). A type VI secretion system effector delivery mechanism dependent on PAAR and a chaperone-co-chaperone complex. *Nat. Microbiol.* 3, 632–640. doi: 10.1038/s41564-018-0144-4

Cianfanelli, F. R., Alcoforado Diniz, J., Guo, M., De Cesare, V., Trost, M., and Coulthurst, S., J. (2016). VgrG and PAAR proteins define distinct versions of a functional type VI secretion system. *PLoS Pathog.* 12:e1005735. doi: 10.1371/journal.ppat.1005735

Diniz, J. A., and Coulthurst, S. J. (2015). Intra-species competition in *Serratia marcescens* is mediated by type VI secretion Rhs effectors and a conserved effector-associated accessory protein. *J. Bacteriol.* 197, 2350–2360. doi: 10.1128/JB.00199-15

Durand, E., Cambillau, C., Cascales, E., and Journet, L. (2014). VgrG, Tae, Tle, and beyond: the versatile arsenal of type VI secretion effectors. *Trends Microbiol.* 22, 498–507. doi: 10.1016/j.tim.2014.06.004

Durand, E., Nguyen, V. S., Zoued, A., Logger, L., Pehau-Arnaudet, G., Aschtgen, M. S., et al. (2015). Biogenesis and structure of a type VI secretion membrane core complex. *Nature* 523, 555–560. doi: 10.1038/nature14667

Durand, E., Zoued, A., Spinelli, S., Watson, P. J., Aschtgen, M. S., Journet, L., et al. (2012). Structural characterization and oligomerization of the TssL protein, a component shared by bacterial type VI and type IVb secretion systems. *J. Biol. Chem.* 287, 14157–14168. doi: 10.1074/jbc.M111.338731

Figurski, D. H., and Helinski, D. R. (1979). Replication of an origin-containing derivative of plasmid RK2 dependent on a plasmid function provided in trans. *Proc. Natl. Acad. Sci. U.S.A.* 76, 1648–1652. doi: 10.1073/pnas.76.4.1648

Filloux, A. (2010). Secretion signal and protein targeting in bacteria: a biological puzzle. *J. Bacteriol.* 192, 3847–3849. doi: 10.1128/JB.00565-10

Filloux, A. (2011). Protein secretion systems in *Pseudomonas aeruginosa*: an essay on diversity, evolution, and function. *Front. Microbiol.* 2:155. doi: 10.3389/fmicb.2011.00155

Flaugnatti, N., Le, T. T., Canaan, S., Aschtgen, M. S., Nguyen, V. S., Blangy, S., et al. (2015). A phospholipase A anti-bacterial T6SS effector interacts directly with the C-terminal domain of the VgrG spike protein for delivery. *Mol. Microbiol.* 99, 1099–1118doi: 10.1111/mmi.13292

Hachani, A., Allsopp, L. P., Oduko, Y., and Filloux, A. (2014). The VgrG proteins are "a la carte" delivery systems for bacterial type VI effectors. *J. Biol. Chem.* 289, 17872–17884. doi: 10.1074/jbc.M114.563429

Hachani, A., Lossi, N. S., Hamilton, A., Jones, C., Bleves, S., Albesa-Jove, D., et al. (2011). Type VI secretion system in *Pseudomonas aeruginosa*: secretion and multimerization of VgrG proteins. *J. Biol. Chem.* 286, 12317–12327. doi: 10.1074/jbc.M110.193045

Herrero, M., de Lorenzo, V., and Timmis, K. N. (1990). Transposon vectors containing non-antibiotic resistance selection markers for cloning and stable chromosomal insertion of foreign genes in gram-negative bacteria. *J. Bacteriol.* 172, 6557–6567. doi: 10.1128/JB.172.11.6557-6567.1990

Jana, B., Fridman, C. M., Bosis, E., and Salomon, D. (2019). A modular effector with a DNase domain and a marker for T6SS substrates. *Nat. Commun.* 10:3595. doi: 10.1038/s41467-019-11546-6

Kanamaru, S., Leiman, P. G., Kostyuchenko, V. A., Chipman, P. R., Mesyanzhinov, V. V., Arisaka, F., et al. (2002). Structure of the cell-puncturing device of bacteriophage T4. *Nature* 415, 553–557. doi: 10.1038/415553a

Kaniga, K., Delor, I., and Cornelis, G. R. (1991). A wide-host-range suicide vector for improving reverse genetics in gram-negative bacteria: inactivation of the blaA gene of *Yersinia enterocolitica*. *Gene* 109, 137–141. doi: 10.1016/0378-1119(91)90599-7

Leiman, P. G., Basler, M., Ramagopal, U. A., Bonanno, J. B., Sauder, J. M., Pukatzki, S., et al. (2009). Type VI secretion apparatus and phage tail-associated protein complexes share a common evolutionary origin. *Proc. Natl. Acad. Sci. U.S.A.* 106, 4154–4159. doi: 10.1073/pnas.0813360106

Liang, X., Kamal, F., Pei, T. T., Xu, P., Mekalanos, J. J., and Dong, T. G. (2019). An onboard checking mechanism ensures effector delivery of the type VI secretion system in *Vibrio cholerae*. *Proc. Natl. Acad. Sci. U.S.A.* 116, 23292–23298. doi: 10.1073/pnas.1914202116

Liang, X., Moore, R., Wilton, M., Wong, M. J., Lam, L., and Dong, T. G. (2015). Identification of divergent type VI secretion effectors using a conserved chaperone domain. *Proc. Natl. Acad. Sci. U.S.A.* 112, 9106–9111. doi: 10.1073/pnas.1505317112

Lin, J. S., Ma, L. S., and Lai, E. M. (2013). Systematic dissection of the agrobacterium type VI secretion system reveals machinery and secreted components for subcomplex formation. *PLoS ONE* 8:e67647. doi: 10.1371/journal.pone.0067647

Ma, A. T., McAuley, S., Pukatzki, S., and Mekalanos, J. J. (2009). Translocation of a *Vibrio cholerae* type VI secretion effector requires bacterial endocytosis by host cells. *Cell Host Microbe* 5, 234–243. doi: 10.1016/j.chom.2009.02.005

Ma, L. S., Hachani, A., Lin, J. S., Filloux, A., and Lai, E. M. (2014). *Agrobacterium tumefaciens* deploys a superfamily of type VI secretion DNase effectors as weapons for interbacterial competition in planta. *Cell Host Microbe* 16, 94–104. doi: 10.1016/j.chom.2014.06.002

Nazarov, S., Schneider, J. P., Brackmann, M., Goldie, K. N., Stahlberg, H., and Basler, M. (2018). Cryo-EM reconstruction of type VI secretion system baseplate and sheath distal end. *EMBO J.* 37:e97103. doi: 10.15252/embj.201797103

Nicolay, T., Vanderleyden, J., and Spaepen, S. (2015). Autotransporter-based cell surface display in Gram-negative bacteria. *Crit. Rev. Microbiol.* 41, 109–123. doi: 10.3109/1040841X.2013.804032

Pissaridou, P., Allsopp, L. P., Wettstadt, S., Howard, S. A., Mavridou, D. A. I., and Filloux, A. (2018). The *Pseudomonas aeruginosa* T6SS-VgrG1b spike is topped by a PAAR protein eliciting DNA damage to bacterial competitors. *Proc. Natl. Acad. Sci. U.S.A.* 115, 12519–12524. doi: 10.1073/pnas.1814181115

Planamente, S., Salih, O., Manoli, E., Albesa-Jove, D., Freemont, P. S., and Filloux, A. (2016). TssA forms a gp6-like ring attached to the type VI secretion sheath. *EMBO J.* 35, 1613–1627. doi: 10.15252/embj.201694024

Pukatzki, S., Ma, A. T., Sturtevant, D., Krastins, B., Sarracino, D., Nelson, W. C., et al. (2006). Identification of a conserved bacterial protein secretion system in *Vibrio cholerae* using the Dictyostelium host model system. *Proc. Natl. Acad. Sci. U.S.A.* 103, 1528–1533. doi: 10.1073/pnas.0510322103

Quentin, D., Ahmad, S., Shanthamoorthy, P., Mougous, J. D., Whitney, J. C., and Raunser, S. (2018). Mechanism of loading and translocation of type VI secretion system effector Tse6. *Nat. Microbiol.* 3, 1142–1152. doi: 10.1038/s41564-018-0238-z

Salomon, D., Kinch, L. N., Trudgian, D. C., Guo, X., Klimko, J. A., Grishin, N. V., et al. (2014). Marker for type VI secretion system effectors. *Proc. Natl. Acad. Sci. U.S.A.* 111, 9271–9276. doi: 10.1073/pnas.1406110111

Shneider, M. M., Buth, S. A., Ho, B. T., Basler, M., Mekalanos, J. J., and Leiman, P. G. (2013). PAAR-repeat proteins sharpen and diversify the type VI secretion system spike. *Nature* 500, 350–353. doi: 10.1038/nature12453

Unterweger, D., Kostiuk, B., Otjengerdes, R., Wilton, A., Diaz-Satizabal, L., and Pukatzki, S. (2015). Chimeric adaptor proteins translocate diverse type VI secretion system effectors in *Vibrio cholerae*. *EMBO J.* 34, 2198–2210. doi: 10.15252/embj.201591163

van Ulsen, P., Zinner, K. M., Jong, W. S. P., and Luirink, J. (2018). On display: autotransporter secretion and application. *FEMS Microbiol. Lett.* 365:fny165. doi: 10.1093/femsle/fny165

Vasseur, P., Vallet-Gely, I., Soscia, C., Genin, S., and Filloux, A. (2005). The pel genes of the *Pseudomonas aeruginosa* PAK strain are involved at early and late stages of biofilm formation. *Microbiology* 151(Pt 3), 985–997. doi: 10.1099/mic.0.27410-0

Ventre, I., Goodman, A. L., Vallet-Gely, I., Vasseur, P., Soscia, C., Molin, S., et al. (2006). Multiple sensors control reciprocal expression of *Pseudomonas aeruginosa* regulatory RNA and virulence genes. *Proc. Natl. Acad. Sci. U.S.A.* 103, 171–176. doi: 10.1073/pnas.0507407103

Voulhoux, R., Taupiac, M. P., Czjzek, M., Beaumelle, B., and Filloux, A. (2000). Influence of deletions within domain II of exotoxin A on its extracellular secretion from *Pseudomonas aeruginosa*. *J. Bacteriol.* 182, 4051–4058. doi: 10.1128/JB.182.14.4051-4058.2000

Wettstadt, S., and Filloux, A. (2020). Manipulating the type VI secretion system spike to shuttle passenger proteins. *PLoS ONE* 15:e0228941. doi: 10.1371/journal.pone.0228941

Wettstadt, S., Wood, T., E., Fecht, S., and Filloux, A. (2019). Delivery of the *Pseudomonas aeruginosa* phospholipase effectors PldA and PldB in a VgrG- and H2-T6SS-dependent manner. *Front. Microbiol.* 10:1718. doi: 10.3389/fmicb.2019.01718

Whitney, J. C., Quentin, D., Sawai, S., LeRoux, M., Harding, B. N., Ledvina, H. E., et al. (2015). An interbacterial NAD(P)(+) glycohydrolase toxin requires elongation factor Tu for delivery to target cells. *Cell* 163, 607–619. doi: 10.1016/j.cell.2015.09.027

Wood, T. E., Howard, S., A., Wettstadt, S., and Filloux, A. (2019a). PAAR proteins act as the 'sorting hat' of the type VI secretion system. *Microbiology* 165, 1203–1218. doi: 10.1099/mic.0.000842

Wood, T. E., Howard, S. A., Forster, A., Nolan, L. M., and Manoli, E., Bullen, N. P., et al. (2019b). The *Pseudomonas aeruginosa* T6SS delivers a periplasmic toxin that disrupts bacterial cell morphology. *Cell Rep. 29,* 187–201.e7. doi: 10.1016/j.celrep.2019.08.094

Wu, C. F., Lien, Y. W., Bondage, D., Lin, J. S., and Pilhofer, M., Shih, Y. L., et al. (2020). Effector loading onto the VgrG carrier activates type VI secretion system assembly. *EMBO Rep.* 21:e47961. doi: 10.15252/embr.201947961

6

Interfering with Autophagy: The Opposing Strategies Deployed by *Legionella pneumophila* and *Coxiella burnetii* Effector Proteins

David R. Thomas, Patrice Newton, Nicole Lau and Hayley J. Newton*

Department of Microbiology and Immunology, Peter Doherty Institute for Infection and Immunity, The University of Melbourne, Melbourne, VIC, Australia

Correspondence:
Hayley J. Newton
hnewton@unimelb.edu.au
orcid.org/0000-0002-9240-2001

Autophagy is a fundamental and highly conserved eukaryotic process, responsible for maintaining cellular homeostasis and releasing nutrients during times of starvation. An increasingly important function of autophagy is its role in the cell autonomous immune response; a process known as xenophagy. Intracellular pathogens are engulfed by autophagosomes and targeted to lysosomes to eliminate the threat to the host cell. To counteract this, many intracellular bacterial pathogens have developed unique approaches to overcome, evade, or co-opt host autophagy to facilitate a successful infection. The intracellular bacteria *Legionella pneumophila* and *Coxiella burnetii* are able to avoid destruction by the cell, causing Legionnaires' disease and Q fever, respectively. Despite being related and employing homologous Dot/Icm type 4 secretion systems (T4SS) to translocate effector proteins into the host cell, these pathogens have developed their own unique intracellular niches. *L. pneumophila* evades the host endocytic pathway and instead forms an ER-derived vacuole, while *C. burnetii* requires delivery to mature, acidified endosomes which it remodels into a large, replicative vacuole. Throughout infection, *L. pneumophila* effectors act at multiple points to inhibit recognition by xenophagy receptors and disrupt host autophagy, ensuring it avoids fusion with destructive lysosomes. In contrast, *C. burnetii* employs its effector cohort to control autophagy, hypothesized to facilitate the delivery of nutrients and membrane to support the growing vacuole and replicating bacteria. In this review we explore the effector proteins that these two organisms utilize to modulate the host autophagy pathway in order to survive and replicate. By better understanding how these pathogens manipulate this highly conserved pathway, we can not only develop better treatments for these important human diseases, but also better understand and control autophagy in the context of human health and disease.

Keywords: autophagy, type 4 secretion system (T4SS), effector protein, *Coxiella burnetii*, *Legionella pneumophila*

INTRODUCTION

Autophagy is an essential cellular pathway which, at its most basic, acts to degrade unwanted molecules and recover nutrients for the cell. Autophagy is classified into three broad categories: macroautophagy, microautophagy, and chaperone-mediated autophagy. Microautophagy involves the direct engulfment of cytoplasmic cargo by lysosomes and is largely non-specific (Li et al., 2012b), while chaperone-mediated autophagy entails the specific selection of cytosolic proteins by chaperones which directly target them to lysosomes for degradation (Kaushik and Cuervo, 2012). In contrast, macroautophagy (hereafter autophagy) involves the targeted degradation of molecules or organelles in the cell, which are first engulfed by a double-membraned autophagosome that later fuses with a proteolytic lysosome to degrade its contents. Autophagy is a fundamental and highly conserved process in all eukaryotes, with common autophagy proteins (ATG) found in plants, fungi, mammals, and amoeba (King, 2012). It is also involved in a myriad of essential cellular processes, including the elimination of damaged organelles and protein aggregates, nutrient recovery, and disease suppression (Codogno and Meijer, 2005; Nixon, 2013; Anding and Baehrecke, 2015; Ravanan et al., 2017; Yun and Lee, 2018; Wang et al., 2019).

Autophagy is also an important host defence against intracellular pathogens, where it acts to direct invading bacteria to autolysosomes for degradation (a process known as xenophagy) (Rikihisa, 1984; Gutierrez et al., 2004; Nakagawa et al., 2004). Perhaps expectedly, this has led to an evolutionary arms race between host and pathogens, which have developed a range of mechanisms to evade destruction. These host-pathogen interactions have been the subject of much research, providing valuable information not only about key virulence factors and important infections, but also about the underlying host molecular mechanisms that they subvert (Campoy and Colombo, 2009; Kimmey and Stallings, 2016; Casanova, 2017; McEwan, 2017; Sharma et al., 2018; Siqueira et al., 2018; Sudhakar et al., 2019; Wu and Li, 2019; Xiong et al., 2019; Hu et al., 2020).

The autophagy pathway has been thoroughly reviewed previously (Glick et al., 2010; Bento et al., 2016; Yin et al., 2016; Dikic and Elazar, 2018; Yu et al., 2018; Levine and Kroemer, 2019; Wang et al., 2019), however a summary of key events is provided here (**Figure 1**). Autophagy is largely regulated by the mammalian target of rapamycin (TOR) complex 1 (mTORC1), which when active phosphorylates and inhibits autophagy initiation factors, but is inactivated by physiological stresses such as energy depletion, hypoxia, and starvation (Laplante and Sabatini, 2009; Rabanal-Ruiz et al., 2017). Upon the inactivation of mTORC1, Unc-51-like kinase 1 (ULK1) and ATG13 are dephosphorylated, forming a stable complex with focal adhesion kinase family-interacting protein of 200 kDa (FIP200) and ATG101 to form the ULK1 complex (Kim et al., 2011). The active ULK1 complex then activates class III phosphatidylinositol 3-kinase (PI3K) complex I, consisting of the lipid kinase VPS34, Beclin-1, VPS15, and ATG14L, recruiting it and ATG9 to the isolation membrane (IM) - the precursor of the autophagosome membrane (Mack et al., 2012; Russell et al., 2013).

The PI3K complex I produces phosphatidylinositol 3-phosphate (PI(3)P) at the site of autophagosome formation by phosphorylating phosphatidylinositol (PI) (Ktistakis et al., 2012). While the exact source of the IM is unknown, it is likely that multiple sources contribute membrane (Chan and Tang, 2013). The association of the IM with markers of autophagosome initiation at cup-shaped sites at the ER, termed omegasomes, indicates that the ER plays an essential role (Ktistakis et al., 2008; Karanasios et al., 2013). The transient interaction of ATG9-containing vesicles at the IM promotes its expansion to a phagophore (Orsi et al., 2012), potentially *via* the delivery of lipids for membrane formation. Two ubiquitin-like (Ubl) systems then work in concert to attach lipidated LC3 (ATG8) proteins to the phagophore. The first system is the ATG12 Ubl conjugation system, consisting of ATG12, ATG7, ATG10, ATG5, and ATG16L1, results in the formation of an ATG12-ATG5-ATG16L1 complex which is recruited to PI(3)P positive membranes by WIPI2 (Polson et al., 2010; Dooley et al., 2014). The second system is the LC3 Ubl conjugation system, in which ATG4 cleaves the LC3 precursor to produce LC3-I, which is then activated by ATG3 and ATG7 before being conjugated to phosphatidylethanolamine (PE) by the ATG12-ATG5-ATG16L1 complex to produce the membrane associated LC3-PE (LC3-II) (Klionsky and Schulman, 2014).

Cargo destined for autophagic degradation are bound to LC3-II on the inner surface of the developing phagophore by a range of adaptor proteins which bind specific targets (Johansen and Lamark, 2020). In the case of xenophagy, invading pathogens are ubiquitinated by a range of E3 ligases, which acts as binding targets for adaptor proteins (Chen et al., 2019). Once cargo is bound, the crescent shaped phagophore closes to become an autophagosome. At this time ATG4 disassociates external ATG proteins and PI(3)P kinases convert PI(3)P to PI(3,5)P_2, promoting maturation (Yu et al., 2012; Dall'Armi et al., 2013; Reggiori and Ungermann, 2017). Autophagosomes can fuse with endosomes or directly with lysosomes to create autolysosomes (Berg et al., 1998), where the cargo is degraded by the acidic and proteolytic environment provided by the lysosome. Fusion of the different vesicles is controlled by three protein groups: Rab GTPases, membrane-tethering complexes, and soluble N-ethylmaleimide-sensitive factor attachment protein receptors (SNAREs). Rab GTPases recruit tethering complexes to the target membranes and SNAREs are required for the fusion of the lipid membranes (Nakamura and Yoshimori, 2017; Lorincz and Juhasz, 2020).

Coxiella burnetii and *Legionella pneumophila* are related intracellular pathogens which take very different approaches to replicating within the host cell. Both are Gram-negative pathogens that possess a Dot/Icm Type IV secretion system (T4SS) which they use to translocate hundreds of effector proteins into the host and produce a replicative vacuole. *L. pneumophila* uses a subset of its effectors to inhibit autophagy and prevent delivery of the pathogen to lytic autolysosomes and in stark contrast, *C. burnetii* uses its effectors to co-opt host autophagy, using it to develop a vast

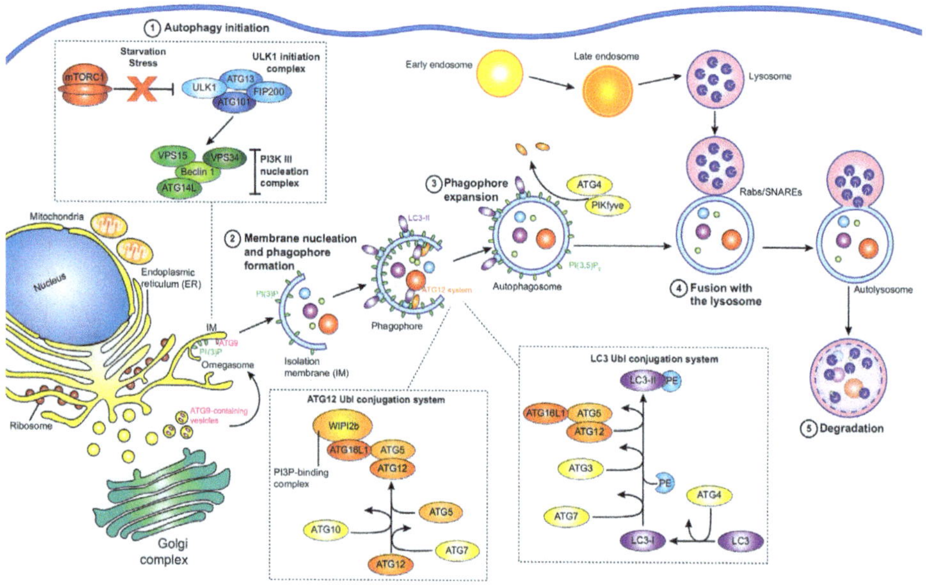

FIGURE 1 | Summary of the autophagy pathway. Under normal conditions, active mTORC1 inhibits autophagy. However, under various stresses such as starvation, mTORC1 is inactivated, and autophagy initiation (1) can proceed. The ULK1 initiation complex consisting of ULK1, ATG13, FIP200, and ATG101 assembles, which then activates the PI3K complex I, containing VPS34, Beclin-1, VPS14, and ATG14L. Along with ATG9 containing vesicles, the PI3K complex I initiates the next step: membrane nucleation and phagophore formation (2). At isolation membranes (IM) on omegasomes at the ER, the PI3K complex I produces PI(3)P, while ATG9 vesicles promote membrane expansion into a phagophore. The ATG12 and LC3 ubiquitin-like (Ubl) conjugation systems act in concert to convert cytosolic LC3 to membrane bound LC3-II, mature LC3 conjugated to PE. WIPI2b recruits the ATG12 Ubl conjugation system to PI(3)P, leading to the accumulation of LC3-II at the growing phagophore, leading to the expansion and maturation of the phagophore (3). Autophagy receptors, such as SQSTM-1 bind targets for degradation and associate with LC3-II at the growing phagophore, which then closes around the cargo. ATG4 then removes ATG proteins from the outside of autophagosome, while PI(3)P kinases convert PI(3)P to PI(3,5)P_2 to favour autophagosome maturation. The remaining LC3-II and PI(3,5)P_2 promote the accumulation of Rab GTPases, membrane tethering complexes, and SNARES, which facilitate the fusion of autophagosomes to late endosomes/lysosomes (4). The autophagosome cargo is then exposed to the proteolytic and acidic contents of the lysosome, resulting in its degradation (5).

replicative vacuole that can occupy the majority of the cellular space.

L. pneumophila is the causative agent of the severe pneumonia-like disease, Legionnaires' Disease. The environmental reservoir of *L. pneumophila* includes natural water sources, such as rivers and lakes, as well as man-made sources, including, but not limited to, cooling towers and air-conditioners, in which the bacteria resides within a wide range of protozoan hosts (Rowbotham, 1980). Opportunistic infection of humans occurs following the inhalation of contaminated aerosols. *L. pneumophila* enters the human host *via* phagocytic uptake by alveolar macrophages (Horwitz, 1983). Immediately, *L. pneumophila* begins to remodel the phagocytic vacuole, to avoid the host endocytic pathway and lysosomal degradation, by creating an ER-like replicative niche for replication, termed the *Legionella*-containing vacuole (LCV, **Figure 2**) (Horwitz, 1983). A key virulence determinant in this process is its T4SS and the repertoire of over 300 bacterial effectors that are translocated into the host cell through this secretion system (Segal et al., 1998; Vogel et al., 1998; Gomez-Valero et al., 2011). Translocation of *L. pneumophila* effector proteins is initiated upon contact with the host cell plasma membrane (Chen et al., 2007; Charpentier et al., 2009), enabling them to immediately manipulate a range of host cell processes in a coordinated manner. Modulated systems include host vesicle trafficking pathways, ubiquitin machinery, autophagy and apoptosis, facilitating bacterial replication (Hubber and Roy, 2010; Isaac and Isberg, 2014; Prashar and Terebiznik, 2014; Ensminger, 2015; Omotade and Roy, 2019). The plasticity of the *Legionella* effector repertoire has been well documented, with comparative genomic studies revealing large variability in effector cohorts between *Legionella* species, with effectors largely unique to each species but presenting significant functional redundancy (Burstein et al., 2016; Gomez-Valero et al., 2019). This is likely a result of its association with a broad range of amoebal hosts, which would drive the evolution of an array of host specific genes that are redundant for infection in human macrophages (Park et al., 2020). Given the conservation of autophagy proteins within protozoan hosts (Calvo-Garrido et al., 2010), there is likely significant evolutionary pressure to evolve and maintain the subset of effectors that regulate host autophagy to facilitate bacterial success.

C. burnetii requires delivery to mature endosomes and fusion with lysosomes for a successful infection (Beron et al., 2002; Gutierrez et al., 2005). *C. burnetii* is the causative agent of Q fever, which although symptomatic in around 40% of cases, can develop into chronic Q fever in 1–5% of cases, with a mortality rate of up to 60% if left untreated (Kampschreur et al., 2015). *C. burnetii* is globally distributed, assisted by its high infectivity and broad host range including humans, fish, birds, arthropods, and livestock (Heinzen et al., 1999). As with *L. pneumophila*, *C. burnetii* infects alveolar macrophages where it enters the host cell

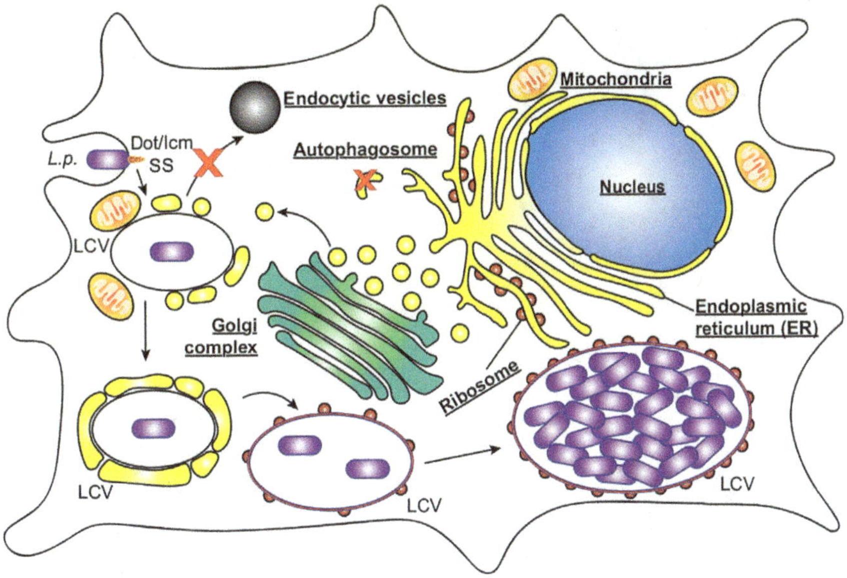

FIGURE 2 | *L. pneumophila* replication within macrophages. *L. pneumophila* (*L.p.*) activates the Dot/Icm secretion system (SS) upon contact with macrophages, and following entry evades endocytic vesicles and autophagosomes to form a replicative niche termed the *Legionella*-containing vacuole (LCV). Initially surrounded by vesicles derived from the endoplasmic reticulum and mitochondria, the LCV eventually resembles rough ER decorated with ribosomes, within which bacteria replicate to high numbers, eventually leading to cell lysis.

by phagocytosis, and its pathogenicity also relies on a functional T4SS. However, *C. burnetii* does not avoid autophagy, instead passing through the host endocytic pathway where the pathogen-containing phagosome fuses with lysosomes. Only upon the subsequent acidification of the phagosome is the *C. burnetii* T4SS activated, translocating approximately 150 effector proteins to modify the host cell and develop the mature phagolysosome into the replication permissive *Coxiella*-containing vacuole (**Figure 3**, CCV) (Howe et al., 2003; Mahapatra et al., 2010; Newton et al., 2013; Burette and Bonazzi, 2020; Newton et al., 2020). The CCV is a spacious vacuole due to its highly fusogenic nature, with multiple CCVs in a cell fusing to form a single large vacuole that promiscuously fuses with host autophagosomes, lysosomes and endocytic vesicles. In this review we will discuss the role that autophagy plays during infections with *L. pneumophila* and *C. burnetii*, focusing on the effector proteins that they use to interfere with the host autophagy system (**Table 1**).

MANIPULATION OF HOST AUTOPHAGY BY *LEGIONELLA* EFFECTORS

Intracellular replication of *L. pneumophila* is dependent on its ability to prevent the rapid fusion of the LCV with lysosomes, a process requiring a functional T4SS (Marra et al., 1992; Berger and Isberg, 1993; Roy et al., 1998; Wiater et al., 1998). *L. pneumophila* was found to associate with the autophagy proteins ATG7 and ATG8 rapidly after infection in a T4SS-dependent manner and to induce autophagosome formation in the host cell (Amer and Swanson, 2005; Amer et al., 2005). However, conflicting data has made it challenging to understand the importance of host autophagy during *L. pneumophila* infection. Early findings demonstrated that inhibition of autophagy, by treatment with the PI3K inhibitor 3-methyladenine (3-MA), increased the degradation of intracellular *L. pneumophila* (Amer and Swanson, 2005), while starvation induced autophagy was found to favour intracellular replication (Swanson and Isberg, 1995). A more recent study found upregulation of autophagy, using the glycolytic inhibitor 2-deoxy-glucose (2DG) or starvation inhibited *L. pneumophila* replication, while siRNA gene silencing of the autophagy protein ATG5 enhanced replication (Matsuda et al., 2009). It was later found that 2DG treatment did not affect bacterial replication in amoebae, and that *L. pneumophila* in macrophages were directly sensitive to 2DG through the hexose-phosphate transporter UhpC (Price et al., 2018). These studies suggest a complex interaction between *L. pneumophila* and the host autophagy pathway, encompassing both beneficial and detrimental aspects, that may be influenced by host specific factors and LCV maturation stage. However, given the observation that *L. pneumophila* actively inhibits host autophagy (Choy et al., 2012), it is likely that the degradative aspect of autophagy is detrimental. In support of this, many *L. pneumophila* effectors have now been identified which inhibit or modulate the host autophagy pathway (**Figure 4**).

RavZ—Targeting LC3-II to Inhibit Autophagy

The first *L. pneumophila* effector recognized as a manipulator of host autophagy was RavZ. RavZ was identified as the Dot/Icm effector responsible for reducing levels of the lipidated

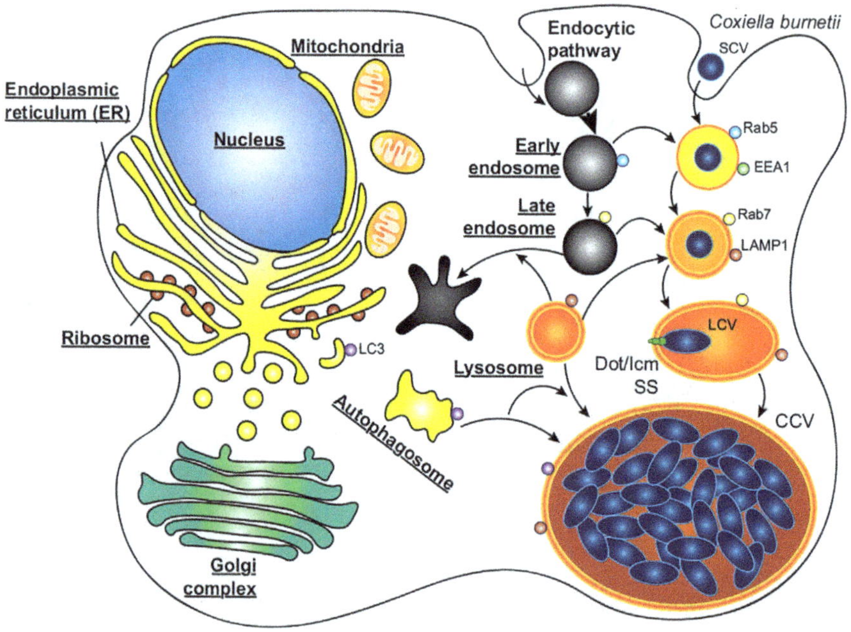

FIGURE 3 | *C. burnetii* replication within macrophages. *C. burnetii* exists in two developmental stages; small cell variants (SCV) which are metabolically inactive but highly stable, and the metabolically active large cell variants (LCV). SCVs enter alveolar macrophages through phagocytosis and proceed through the endocytic pathway. Upon fusion of the pathogen containing phagosome with lysosomes, the SCVs convert to LCVs, and the Dot/Icm secretion system (SS) is activated. Through the activity of effector proteins, the mature phagolysosome is converted into the replication permissive *Coxiella* containing vacuole (CCV). The CCV rapidly expands through heterotypic fusion with autophagosomes, lysosomes, and endocytic vesicles, and homotypic fusion of multiple CCVs. As the CCV expands and fills with bacteria, the LCVs convert back to SCVs, where they will be stable upon release from the cell *via* cell lysis or exocytosis.

TABLE 1 | *L. pneumophila* and *C. burnetii* effectors known to modulate autophagy.

Effector	Size: aa (kDa)	Known mechanism	Notes
***Legionella pneumophila* effectors**			
RavZ (Lpg1683)	502 (56.25)	Cysteine protease, cleaves LC3-II	Possesses PI(3)P binding domain and LIRs Found in 4/41 *Legionella* species[a]
*Lp*SpI (Lpg2176; Lpp2128; LegS2)	608 (67.38)	Degrades sphingolipids	Found in 10/41 *Legionella* species[a]
Lpg1137	322 (35.8)	Degrades Stx17	Found in 4/41 *Legionella* species[a]
Lpg2936	244 (27.29)	Methyltransferase, modifies host DNA	Found in 41/41 *Legionella* species[a]
LegA9 (Lpp2058)	580 (65.17)	Unknown	Found in 18/41 *Legionella* species[a]
SidE family (SidE, SdeA, SdeB, SdeC)	SidE – 1514 (171.69) SdeA – 1499 (169.08) SdeB – 1926 (216.99) SdeC – 1538 (172.93)	Creates non-canonical ubiquitin linkages	Found in 6/41 *Legionella* species[a]
Lgt1, Lgt2, Lgt3	Lgt1 – 525 (59.59) Lgt2 – 636 (71.65) Lgt3 – 873 (99.67)	Glucosylates and inhibits eEF1A	Lgt1 found in 3/41 *Legionella* species[a] Lgt2/3 found in 1/41 *Legionella* species[a]
SetA (Lpg2157)	1506 (169.78)	Glucosylates to activate TFEB	Found in 2/41 *Legionella* species[a]
***Coxiella burnetii* effectors**			
CvpA (CBU0665)	328 (38.01)	Interacts with AP2	Contains endocytic sorting motifs
CvpB (Cig2; CBU0021)	809 (93.1)	Prevents conversion of PI(3)P to PI(3,5)P_2 by PIKfyve	
Cig57 (CBU1751)	420 (48.85)	Interacts with FCHO2	Contains endocytic sorting motifs
CvpF (CBU0626)	695 (79.58)	Interacts with Rab26	Contains endocytic sorting motifs

[a]*(Burstein et al., 2016).*
LIR, LC3 interacting region.

form of LC3 (LC3-II) during infection, thereby inhibiting autophagy (Choy et al., 2012). Infection with RavZ-deficient *L. pneumophila* led to increased levels of LC3-II and LC3-postitive puncta within the host cells however intracellular bacterial replication was not impacted and vacuoles containing Δ*ravZ* mutant were not surrounded by LC3-positive puncta. This provided foundational evidence that other effectors may be secreted by *L. pneumophila* to aid bacterial evasion of the host

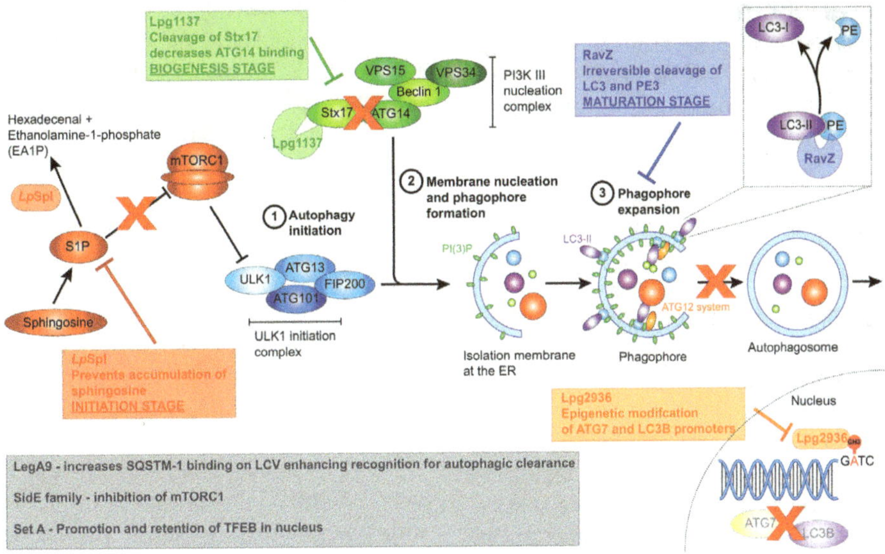

FIGURE 4 | *L. pneumophila* effectors that manipulate host autophagy pathway. *L. pneumophila* effectors interfere with the host autophagy pathway at multiple stages including initiation, biogenesis and maturation stages. The conversion of spingosine-1-phosphate (S1P) to hexadecenal and ethanolamine-1-phosphate by *Lp*Spl prevents sphingosine accumulation resulting in mTORC1 activation, thereby inhibiting the initiation of autophagy. Lpg1137 acts to cleave Stx17, also interfering with the initiation of autophagy. RavZ interrupts the maturation process of autophagy through irreversible cleavage of PE from LC3-II. Finally, Lpg2936 results in epigenetic modification of key autophagy components, including ATG7 and LC3B, resulting in their decreased expression. Grey box indicates effectors that result in the initiation of autophagy, seemingly at odds with *L. pneumophila's* intracellular niche. These including the SideE family of effectors which inhibits mTORC1, SetA which activates TFEB, and LegA9 which targets the LCV for autophagic recognition.

autophagy machinery (Choy et al., 2012). RavZ acts as a cysteine protease that specifically targets lipidated ATG8 family members, including LC3-II. Cleavage of the amide bond between the aromatic tyrosine and the PE-conjugated glycine residue by RavZ results in the irreversible deconjugation of LC3, since the reactive C-terminal glycine normally required for conjugation of PE to LC3 has been removed, ultimately preventing autophagosome formation (Choy et al., 2012) (**Figure 4**). This reaction happens much more rapidly and efficiently compared to its eukaryotic host counterpart, ATG4. ATG4 is also a cysteine protease that deconjugates PE from LC3, although the cleavage occurs after the C-terminal glycine, thereby allowing reconjugation of LC3.

Several research teams have solved the crystal structure of RavZ in order to fully elucidate how RavZ targets membrane-associated ATG8 proteins and catalyses the irreversible deconjugation reaction (Horenkamp et al., 2015; Kwon et al., 2017; Yang et al., 2017). RavZ is composed of two domains, an N-terminal catalytic domain that is responsible for cleavage of LC3-II and a C-terminal domain that binds to PI(3)P on membranes. Recognition of the LC3 substrate occurs through the LC3-interacting region (LIR) motifs located at both the N- and C-terminal regions of RavZ (Horenkamp et al., 2015; Kwon et al., 2017; Yang et al., 2017). Horenkamp et al. (2015) demonstrated that RavZ preferentially binds to high-curvature domains enriched with PI(3)P, commonly associated with phagophore membranes, and that the structural fold of the N-terminal catalytic domain is related to cysteine proteases of the Ubl-specific protease (Ulp) family, with the catalytic triad of Cys258, His176 and Asp197 essential for the function of RavZ. Using numerous biochemical, structural and cell-based analyses, Kwon et al. (2017) observed that both the N- and C-terminal contain LIR motifs that each bind LC3 suggesting that RavZ forms a complex with two molecules of LC3. These LIR motifs possess a higher binding affinity for LC3 than ATG4 and are critical for autophagy inhibition by RavZ. Significantly, Yang et al. (2017) generated semisynthetic LC3 proteins to solve the structure of the RavZ:LC3 interaction and proposed a "lift and cut" mode of action for deconjugation of LC3-II by RavZ. Initially, RavZ targets PI(3)P associated with autophagosome membranes *via* its C-terminal domain and a hydrophobic $\alpha 3$ helix and utilizes the N-terminal LIR motif to bind LC3-II. The $\alpha 3$ helix then aids extraction of the lipid PE moiety and docking of the fatty acid chain within the lipid-binding site of RavZ. Both the LIR interaction and lipid binding are essential for the subsequent proteolytic cleavage of LC3-II (Yang et al., 2017). Further clarification of the specific roles of the binding domains of RavZ was elucidated recently using a series of mutational analyses (Park et al., 2019). Park et al. (2019) observed that targeting autophagosome membranes occurs through two complementary independent pathways: either through PI(3)P-binding (found on early autophagic membranes) or through the LIR motifs binding LC3-II (found on late autophagic membranes), thereby maximising the efficiency of RavZ to inhibit autophagosome formation. Notably, the absence of either domain did not influence the protease function of RavZ.

LpSpl—Influencing Autophagy Through Sphingolipid Metabolism

Interestingly, *L. pneumophila* is also able to inhibit autophagy indirectly by interfering with host sphingolipid metabolism (Rolando et al., 2016). Sphingolipids, and their metabolites sphingosine, sphingosine-1-phosphate (S1P), ceramide and ceramide-1-phosphate, are important components of eukaryotic cell membranes and key signalling molecules involved in a number of host cell processes (Takabe et al., 2008). In particular the leve of S1P, which is tightly regulated by the enzyme sphingosine-1-phosphate lyase 1 (SGPL1), is critical for the balance between sphingolipid induced autophagy and cell death (Takabe et al., 2008).

The *L. pneumophila* effector *Lp*Spl exhibits similarity to eukaryotic SGPL1 in both sequence and activity and has been observed to localize to the mitochondria during infection (Degtyar et al., 2009). Elucidation of the crystal structure of *Lp*Spl confirmed a high level of structural conservation within both the core domain and active site when compared to human SGPL1, in particular, the position of 11 catalytically important residues (Rolando et al., 2016). The lyase activity of *Lp*Spl was confirmed during infection using an *Lp*Spl mutant in mouse embryonic fibroblasts (MEFs) missing endogenous Spl (MEF*spl*$^{-/-}$), in which both wild-type and complemented strains exhibited lyase activity, but not the Δ*spl* *L. pneumophila* strain. Additionally, mutations in key residues of the active site also abolished the enzymatic function of *Lp*Spl (Rolando et al., 2016). In contrast to previous observations that suggested *Lp*Spl localizes to the mitochondria (Degtyar et al., 2009), *Lp*Spl was also observed at the ER (Rolando et al., 2016). The observation of different subcellular localizations of ectopically expressed *Lp*Spl may reflect the different cell lines, expression levels and tags used in these studies. The localization of endogenous, *L. pneumophila* translocated *Lp*Spl remains to be determined.

Using mass spectrometry, Rolando et al. (2016) observed changes in host cell sphingolipid metabolism during infection with *L. pneumophila*. In particular, during wild-type infection, levels of sphingomyelin, ceramide and glycosphingolipids were decreased compared to uninfected cells, although this was not dependent on *Lp*SpI. Cellular levels of sphingosine, however, were significantly increased during infection with the Δ*spl* strain compared to wild-type and complemented strains. Given that translocation of *Lp*Spl by *L. pneumophila* prevents accumulation of sphingosine during infection, and sphingosine stimulates autophagy (Dall'Armi et al., 2013), the impact of *Lp*Spl on the host autophagy machinery was investigated further. Depletion of LC3-II during the ectopic expression of *Lp*Spl under starvation conditions, and no difference in the ratio between autophagosomes and autolysosomes, suggested that *Lp*Spl prevents autophagosome biogenesis, and not autophagosome maturation like RavZ. This action was dependent on the enzymatic function of *Lp*Spl as catalytically inactive mutants did not decrease host autophagy activity (Rolando et al., 2016). Finally, an increase in LC3 puncta was observed during infection of host cells with *L. pneumophila* Δ*spl* compared to wild-type. This increase in LC3 puncta was not as pronounced as the Δ*ravZ* mutant and the Δ*spl/ravZ* double mutant strains, with RavZ contributing more to autophagy inhibition than *Lp*Spl (Rolando et al., 2016). Deletion of *Lp*Spl did not impact intracellular bacterial replication in amoeba or macrophages, similar to RavZ (Choy et al., 2012), but *Lp*Spl is required for efficient replication in a mouse model of infection (Rolando et al., 2016). Ultimately, secretion of *Lp*Spl interferes with S1P, limiting autophagosome biogenesis in the host cell and aiding bacterial survival (**Figure 4**).

Lpg1137—Eliminating the Role of Syntaxin 17

The incidental observation that Syntaxin 17 (Stx17) is degraded during *L. pneumophila* infection in a T4SS-dependent manner led to the discovery of Lpg1137 as the responsible effector (Arasaki et al., 2017). Stx17 is a SNARE protein initially implicated in vesicle trafficking (Steegmaier et al., 2000) but which also has a role in autophagy (Hamasaki et al., 2013; Diao et al., 2015; Kumar et al., 2019; Xian et al., 2019). Stx17, along with ATG14L, SNAP29 and VAMP8, facilitates the fusion of autophagosomes with lysosomes to produce mature autolysosomes (Itakura et al., 2012; Diao et al., 2015; Shen et al., 2020). Additionally, Stx17 has been shown to promote the assembly of the ULK1 initiation complex (Kumar et al., 2019), and has also been found to recruit ATG14L to the mitochondria-associated ER membrane (MAM), which has been implicated in mitophagy (the autophagic degradation of mitochondria) (Xian et al., 2019) and phagophore initiation (Hamasaki et al., 2013). However, the exact role of Stx17 in autophagy initiation is still not entirely known.

Lpg1137 was demonstrated to be a serine protease that localizes to MAM, microsomes, and mitochondria, where it specifically interacts with the cytoplasmic C-terminal of Stx17 (Arasaki et al., 2017). The degradation of Stx17 during *L. pneumophila* infection results in the blockage of autophagosome biogenesis in starvation-induced autophagy (**Figure 4**). The formation of both ATG14-positive and LC3-positive puncta is inhibited in the presence of ectopically expressed Lpg1137 suggesting a failure of PI(3)P formation on omegasomes, similar to silencing Stx17 using siRNA (Arasaki et al., 2017). Despite this, deletion of *lpg1137* did not negatively impact intracellular bacterial replication (Arasaki et al., 2017).

Curiously, bioinformatic analysis and 3D-structure modelling of Lpg1137 suggest this *L. pneumophila* effector is in fact a homologue of mitochondrial SLC25 carrier proteins (Gradowski and Pawlowski, 2017). The authors propose that the cleavage of Stx17 observed by Arasaki *et al.* occurs through either indirect or direct activation of an alternative serine protease in the mitochondrial inner membrane by Lpg1137, or that interaction with Lpg1137 may make Stx17 more prone to cleavage from endogenous proteases or other effector proteases (Gradowski and Pawlowski, 2017). Ultimately, solving the crystal structure of Lpg1137 will be required to fully elucidate its molecular function within host cells during infection.

Lpg2936—Epigenetic Modulator of Autophagy Components

The *L. pneumophila* effector Lpg2936 has recently been implicated as a regulator of autophagosome formation (Abd El

Maksoud et al., 2019). This finding arose from the observation that during *L. pneumophila* infection, the expression of autophagy-related genes ATG7 and LC3B was reduced, alongside a decrease in the expression of unlipidated LC3-I, lipidated LC3-II, and the ATG5-ATG12-ATG16L1 protein complex. Using RNAi against Lpg2936, investigators were able to restore expression of these autophagy-related genes during *L. pneumophila* infection and consequently inhibited bacterial replication (Abd El Maksoud et al., 2019). Bioinformatic and structural analysis previously established that Lpg2936 is a ribosomal RNA protein similar to RsmE-like methyltransferases (Pinotsis and Waksman, 2017). Subsequent examinations demonstrated that Lpg2936 is translocated into the host nucleus where it recognizes the GATC motif in the promoter regions of ATG7 and LC3B and induces irreversible methyladenine changes in this motif from GATC to G(6 mA)TC (Abd El Maksoud et al., 2019). Similar to RNAi against Lpg2936, methylation inhibitors reduced *L. pneumophila* replication and restored expression of autophagy-related genes (Abd El Maksoud et al., 2019). Collectively, this data suggests that Lpg2936 is a transcription factor that translocates into the host cells to regulate autophagosome formation through epigenetic modification of ATG7 and LC3 promoter regions thereby enhancing bacterial replication (**Figure 4**). However, it is worth noting that Lpg2936 may have an autophagy independent role within the bacteria, and therefore silencing Lpg2936 using RNAi or the use of methylation inhibitors has a confounding impact on bacterial viability and replication. Indeed, Pinotsis and Waksman (2017) suggest that the RsmE fold of Lpg2936 is highly specific for bacterial 16S RNAs and would likely target the *Legionella* 16S RNA subunit, rather than eukaryotic ribosomes, enhancing the ability of the bacterium to produce large amounts of effectors during infection. As such, the modification of both ATG7 and LC3 could be inadvertent and unrelated to an effector role for Lpg2936.

LegA9—the Odd One Out

Paradoxically, the *L. pneumophila* genome also encodes for a T4SS effector, ankyrin-containing protein LegA9, that enhances recognition of the LCV for autophagy uptake and clearance (Khweek et al., 2013). The autophagic adapter SQSTM-1 (p62), which binds to both LC3 on autophagosomes and ubiquitinated cargo, is important for targeting intracellular material to the lysosome for clearance (Komatsu et al., 2007; Kirkin et al., 2009) (**Figure 4**). Deletion of *legA9* led not only to a decrease in lysosomal fusion of LCVs but also a reduction in the accumulation of ubiquitin labelling and SQSTM-1 at LCVs compared to wild-type, ultimately promoting bacterial replication within human and mouse-derived macrophages—the latter of which is normally restrictive to *L. pneumophila* replication (Khweek et al., 2013). Although the presence of LegA9 labels the *L. pneumophila* vacuole for autophagy uptake through SQSTM-1 binding, LegA9 does not have a direct role in autophagy activation as no difference in LC3-II levels was observed between wild-type and mutant strains following the stimulation of autophagy by rapamycin treatment (Khweek et al., 2013). Further, the increase in replication of the LegA9 mutant was eliminated by treatment with rapamycin. It is possible that, as with the *L. pneumophila* effector LamA, LegA9 may be adapted to favour infection in alternate hosts such as amoebae, and its induction of autophagy in mammalian cells may be an unintended consequence (Price et al., 2020).

Opposing Roles for the SidE Family

The observation of enhanced SQSTM-1 binding on LCVs has recently been expanded on in a study by Omotade and Roy (2020). Investigations into whether ubiquitin-marked LCVs recruit the necessary adapter receptors for autophagy revealed that despite enrichment of ubiquitin on the LCV, autophagy adapters such as SQSTM-1, NBR1, optineurin and NDP52 are largely absent from the LCV (Omotade and Roy, 2020). The authors proposed that the unique non-canonical ubiquitin linkage created by the SidE family of effectors on proteins on the LCV (Bhogaraju et al., 2016; Qiu et al., 2016), prevents recognition by autophagy adapters. Indeed, a mutant in which all four family members (SidE, SdeA, SdeB, and SdeC) were absent resulted in the increased recruitment of SQSTM-1 to the LCVs. However, no significant increase in LC3B localization to the LCV was detected, even in the absence of RavZ confirming that other mechanisms exist in *L. pneumophila to* block xenophagy (Omotade and Roy, 2020). Additionally, co-infection studies with *Listeria monocytogenes* revealed that the ability of the SidE family to exclude SQSTM-1 occurs specifically at the LCV and does not impact the ability of the cell to target *L. monocytogenes* for autophagic removal through SQSTM-1 binding.

Previously, in addition to the ubiquitylation activities of the SidE family, an effector screen identified the ability of SidE, SdeA, SdeB, and SdeC to promote nuclear translocation of TFEB (transcription factor EB) through inhibition of mTORC1 (De Leon et al., 2017). Active mTORC1 phosphorylates TFEB resulting in retention of TFEB within the cytosol. However, during nutrient limitation, mTORC1 is inactive and the subsequent dephosphorylated TFEB is translocated into the nucleus to induce the transcription of autophagic and lysosomal genes, thereby increasing nutrient availability (Settembre et al., 2012; Roczniak-Ferguson et al., 2012). The activation of TFEB by the SidE family of effectors seems incongruous to the action of other *L. pneumophila* effectors, especially in light of the same effector screen also identifying a role for the Lgt family of effectors (Lgt1, Lgt2, and Lgt3) in preventing TFEB translocation into the nucleus (De Leon et al., 2017). The Lgt family were previously identified as glucosyltransferases that inhibit translation by targeting host elongation factor 1A (eEF1A) (Belyi et al., 2006; Belyi et al., 2008). The authors argue that the two opposing families work together to provide enough nutrients for bacterial replication without promoting autophagy. The SidE family directly ubiquitinylate the Rag small-GTPases that are necessary for mTORC1 to respond to elevated levels of amino acids, effectively blinding mTORC1 to the amino acids liberated by translation inhibitors such as the Lgt effector family (De Leon et al., 2017).

Interestingly, when co-expression of both Lgt and SidE effectors was attempted, expression of SidE was blocked, suggesting temporal regulation of effector translocation is necessary for these effectors to function synergistically during infection (De Leon et al., 2017). Ultimately, inhibition of mTORC1 results in the induction of autophagy; however, secretion of effectors such as RavZ and *Lp*Spl ensure that autophagy is inhibited allowing *L. pneumophila* to acquire nutrients for replication without any detriment to survival. Extensive study of *L. pneumophila* manipulation of Rab1 has demonstrated that a small subset of effectors can control every aspect of Rab1 activity and localization, with effector pairs displaying opposing functions (Murata et al., 2006; Ingmundson et al., 2007; Machner and Isberg, 2007; Müller et al., 2010; Mukherjee et al., 2011; Neunuebel et al., 2011; Tan et al., 2011; Tan and Luo, 2011). Therefore it is possible other effectors can advantageously temporally regulate other key host processes including autophagy.

SetA—Providing Nutrient Control

Recently a similar screen identified a cohort of effectors that promoted translocation of TFEB into the nucleus, including members of the SidE family, SdeA (Lpg2157) and SdeC (Lpg2153) confirming results observed by De Leon et al. (2017), as well as identifying novel effectors: MavH (Lpg2425), VipD (Lpg2831), Lpg2552, Lpg2828, Lpg2888, and SetA (Lpg1978) (Beck et al., 2020). SetA is a mono-*O*-glucosyltransferase containing a DxD catalytic motif that preferentially uses UDP-glucose as a sugar donor targeting a range of host proteins including the small GTPase Rab1a, the chaperonin CCT5 and actin (Jank et al., 2012; Wang et al., 2018; Levanova et al., 2019; Gao et al., 2019). Beck et al. (2020) show that translocation of TFEB into the nucleus is dependent on the glucosyltransferase activity of SetA and mass spectrometry revealed the sites on TFEB that are modified by SetA. In particular, modification of S138 prevented nuclear export (and hence retention within the nucleus) and modification of a cluster of serine and threonine residues near the binding site of 14-3-3 prevents interaction between TFEB and 14-3-3, inhibiting the cytoplasmic retention of TFEB. Despite this, it is currently unknown what role SetA may play during infection, as these experiments were performed exogenously and not in the context of infection.

With the *Legionella* genus predicted to contain >18,000 effectors (Gomez-Valero et al., 2019), it is unsurprising that a significant number have been implicated in influencing autophagy. The high amount of functional redundancy observed in this extensive effector cohort poses a continuing challenge to understanding their collective temporal actions, as well as making it increasingly difficult to identify new effectors modulating this pathway.

MANIPULATION OF HOST AUTOPHAGY BY *COXIELLA* EFFECTORS

Following entry into a host cell, *C. burnetii* traffics through the endocytic pathway to an acidified mature endosome, activating its T4SS to facilitate development of the highly fusogenic CCV (**Figure 3**) (Beron et al., 2002; Romano et al., 2007; Newton et al., 2013; Newton et al., 2020). The CCV also accumulates markers of autophagosomes and lysosomes (Heinzen et al., 1996; Beron et al., 2002; Romano et al., 2007), as well as factors involved in their transport and fusion (Campoy et al., 2011; Campoy et al., 2013; McDonough et al., 2013).

Activation of host autophagy by starvation or rapamycin has been shown to increase the replication and viability of *C. burnetii* following infection (Gutierrez et al., 2005), as well as increasing the size of CCVs (Latomanski and Newton, 2018; Larson et al., 2019). Conversely, inhibition of autophagy by 3-MA impairs the development of the CCV (Beron et al., 2002; Latomanski and Newton, 2018), as does preventing vacuole acidification with bafilomycin A_1 or chloroquine (Heinzen et al., 1996; Newton et al., 2013). Additionally, the essential autophagy genes ATG5, ATG7, ATG12 and STX17 are required for homotypic fusion into one large CCV (McDonough et al., 2013; Newton et al., 2014; Martinez et al., 2016).

While the importance of autophagy to *C. burnetii* is well established, the consequences of host-pathogen interactions on this pathway are less clear. Infection with *C. burnetii* induces an increase in total and lipidated LC3 indicative of an induction of autophagy, although infection also results in an increase in SQSTM-1 (Winchell et al., 2014; Latomanski and Newton, 2018; Larson et al., 2019). The latter finding is curious given that degradation of SQSTM-1 is often monitored as a read out for autophagic flux. However, starvation during infection can still induce autophagic degradation as seen by a decrease in SQSTM-1 (Latomanski and Newton, 2018; Larson et al., 2019). This suggests a more nuanced interference by *C. burnetii* than overt hyperactivation of autophagy or inhibition of autophagic degradation. Interestingly, while it is often hypothesised that *C. burnetii* induces autophagy to delivery nutrients to the expanding CCV, the difficulties in separating out this activity from other actions of autophagy has left this hypothesis without incontrovertible proof. Recent studies have, however, identified several *C. burnetii* effectors proteins involved in the modulation of autophagy (**Figure 5**), although this pathway is likely targeted by more effectors that remain to be characterized.

CvpB—Stabilizing PI(3)P to Facilitate CCV Fusogenicity

One of the most thoroughly characterised *C. burnetii* effector proteins is *Coxiella* vacuolar protein B, named for its association with the CCV membrane (CvpB, Cig2, CBU0021). CvpB was initially identified in a transposon screen as essential for the normal biogenesis of the CCV (Newton et al., 2014). In contrast to the single large CCV of WT *C. burnetii*, CvpB-deficient bacteria instead formed multiple small vacuoles, suggesting a role in the homotypic fusion of CCVs. Interestingly, replication of CvpB transposon mutants was not impaired during infection of HeLa cells, although CvpB mutants were better tolerated by the *Galleria mellonella* infection model (Kohler et al., 2016). Despite replicating to similar numbers as WT, CvpB-deficient

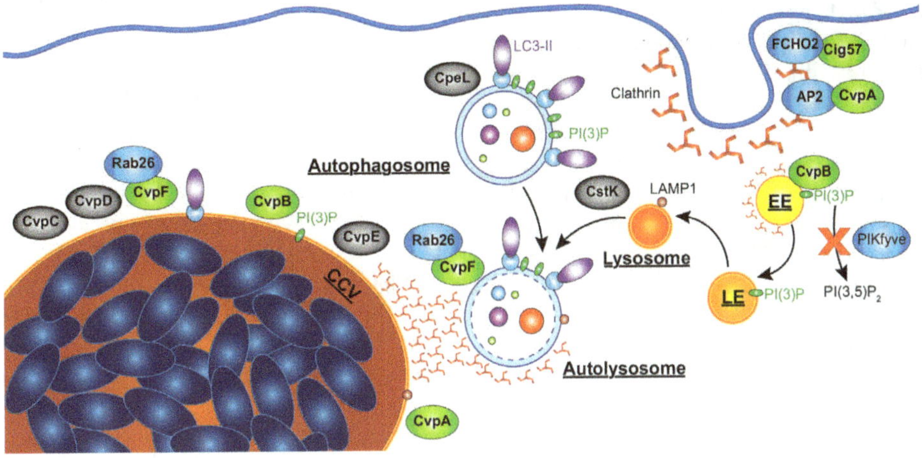

FIGURE 5 | *C. burnetii* effectors that manipulate host autophagy pathway. Effectors of *C. burnetii* co-opt the autophagy pathway to facilitate the development of an extensive CCV. CvpA interacts with AP2, an adaptor involved in clathrin-mediated endocytosis (CME), and is required for CHC accumulation at the CCV. Cig57 associates with FCHO2, another protein involved in CME, and also promotes CHC accumulation at the CCV, as well as being required for *C. burnetii*-induced LC3 lipidation and the delivery of LC3-II positive vesicles to the CCV. CvpB associates with early endosomes and is essential for homotypic fusion of the CCV as well as the accumulation of LC3 at the CCV. CvpB increases cellular PI(3)P by preventing PIKfyve phosphorylating PI(3)P to produce PI(3,5)P_2, facilitating the fusion of autophagosomes with the CCV. CvpF is also required for *C. burnetii*-induced LC3 lipidation, CCV formation, and LC3 delivery to the CCV *via* interactions with Rab26. CvpC, CvpD and CvpE associate with the CCV, and are required for its expansion, CpeL colocalizes with autophagosomes, and CstK was found to localize to vesicles and the CCV. These effectors (grey) have no confirmed activity, but their sub-cellular localizations suggests they may also manipulate autophagy. EE, early endosome; LE, late endosome.

C. burnetii induced slower *Galleria* death. This is a significant observation that may implicate autophagy in disease pathology associated with *C. burnetii* infection.

A multivacuolar phenotype, similar to that of the *cvpB* mutant, was observed with gene silencing of Stx17 (McDonough et al., 2013). Silencing of key autophagy factors ATG5 and ATG12 also produced a multivacuolar phenotype, indicating that a functional host autophagy system is required for homotypic fusion of CCVs (Newton et al., 2014; Kohler et al., 2016). Likewise, *C. burnetii* engineered to translocate the *L. pneumophila* effector RavZ, which inhibits autophagy by cleaving lipidated LC3, also produced a multivacuolar phenotype. LC3 has been observed to be associated with CCVs, indicating the interaction of autophagosomes with the CCV and defining an autolysosomal state for the mature CCV (Beron et al., 2002; Kohler et al., 2016). However, in CvpB transposon mutants LC3 was absent from CCVs, suggesting that CvpB is required for autophagosome fusion with CCVs. CvpB did not interfere with autophagic flux, the delivery of endocytic cargo to the CCV, nor reduce the hydrolytic activity of CCVs (Newton et al., 2014). This indicates that CvpB acts to facilitate the fusion of CCVs and autophagosomes, but does not modulate autophagy itself.

The subcellular localization of ectopically expressed CvpB was examined in high detail using immuno-electron microscopy confirming its localization to the membrane of early endosomes (**Figure 5**). Based on this observation, *in vitro* assays were performed which identified PI(3)P as a binding target for CvpB (Martinez et al., 2016). CvpB lacks any predicted lipid-binding domains but mutational analysis identified the first 500 amino acids as essential for its membrane localization, and that PI(3)P was enriched at the CCV membrane in a CvpB-dependent manner. It was also found that, even in the presence of inhibitors of PI3K, PI(3)P was still detected on vacuolar structures (Martinez et al., 2016). While not a PI3K itself, CvpB instead was shown to inhibit the activity of the PI3 phosphate 5-kinase PIKfyve, which phosphorylates PI(3)P to produce PI(3,5)P_2. Consistent with this, siRNA silencing of PIKfyve was able to correct the multivacuolar phenotype observed in CvpB transposon mutants. Expression of CvpB was found to disassociate PIKfyve from endosomes where it would normally bind and act upon PI(3)P, while CvpB with mutations in the membrane binding domain (MBD) were unable to displace PIKfyve. However, the MBD alone (aa 1-500) was also unable to relocate PIKfyve, suggesting that CvpB does not simply outcompete it for binding to PI(3)P (Martinez et al., 2016). Subsequent studies also found that CvpB is essential for the accumulation of clathrin heavy chain (CHC) at the CCV (Latomanski and Newton, 2018). Interestingly, in the absence of CvpB, clathrin was found associated with LC3B positive autophagosomes near the CCV, which did not fuse with the CCV (Latomanski and Newton, 2018). It is not currently known how CvpB prevents PIKfyve recruitment to PI(3)P positive membranes leading to the accumulation of PI(3)P, nor how this leads to the fusogenic CCV characteristic of *C. burnetii.* However, increased PI(3)P may stabilize LC3 or other pro-fusion molecules (i.e. SNARES) on the CCV membrane, which are normally absent from mature autolysosomes.

Cig57—Enhancing LC3B Lipidation and Clathrin Localization at the CCV

Another effector identified through a transposon mutant screen for CCV defects is Cig57 (CBU1751). Disrupting Cig57 was found to impair the intracellular replication of *C. burnetii* and produce a small CCV phenotype (Newton et al., 2014). Cig57 contains multiple endocytic sorting motifs (two dileucine (DiLeu) and one tyrosine (Tyr)), suggesting it may also interact with clathrin-mediated endocytosis. Indeed, a yeast two-hybrid screen identified FCHO2, a protein that acts to curve the plasma membrane to initiate the formation of clathrin coated vesicles (Henne et al., 2010), as a binding partner of Cig57 (**Figure 5**) (Latomanski et al., 2016).

The identification of FCHO2 suggested an involvement of clathrin, and indeed CHC localization at the CCV was dependent on Cig57 containing functional endosomal sorting motifs (Latomanski et al., 2016). Interestingly, the adapter protein FCHO2 did not alter its sub-cellular localization in the presence or absence of Cig57. FCHO2 knock-out cells presented reduced, but not absent, CHC accumulation at the CCV, suggesting that while FCHO2 enhances the CCV accumulation of CHC, it is not essential. Immunofluorescence imaging found that CHC concentrated in areas where LC3B positive autophagosomes met the CCV, while siRNA silencing of CHC prevented the accumulation of LC3B at CCVs (Latomanski and Newton, 2018). Interestingly, in Cig57 mutants there were no LC3B positive vesicles associated with CHC at the CCV (Latomanski and Newton, 2018). This suggests that clathrin is required for the fusion of autophagosomes to the CCV, while Cig57 is required for the delivery of LC3B positive vesicles to the CCV.

Cig57 was also found to be essential for the *C. burnetii*-dependent LC3B lipidation observed during infection, although Cig57 alone was unable to induce LC3B lipidation (Latomanski and Newton, 2018). Furthermore, *C. burnetii* lacking Cig57 did not accumulate SQSTM-1 during infection as seen in WT strains. While starvation reduced the levels of SQSTM-1 in WT infections to that of non-infected cells, no change was observed in Cig57 transposon mutants. The size of Cig57 transposon mutant CCVs were also not altered by starvation-induced autophagy, while WT CCVs more than doubled (Latomanski and Newton, 2018). CCVs in HeLa cells with siRNA silenced CHC or Stx17 were also unresponsive to induced autophagy. This not only highlights the importance of autophagy to *C. burnetii* infection, but also points to a central role of CHC in mediating autophagy and Cig57 in exploiting this role to support CCV expansion. However, it is currently unclear how Cig57 facilitates LC3 lipidation, SQSTM-1 accumulation, or the fusion of CHC and autophagosomes to the CCV.

CvpF—Inducing LC3 Lipidation *via* Rab26

More recently, an additional *C. burnetii* vacuolar protein, CvpF (CBU0626) was identified as important for intracellular replication of *C. burnetii* and CCV formation while being dispensable for replication in axenic medium (Siadous et al., 2020). During infection of U2OS osteosarcoma cells, both WT and CvpF transposon mutant *C. burnetii* strains developed acidified CCVs decorated with LAMP1, however CCVs produced by CvpF mutants were deficient in LC3B. Further, CvpF transposon mutants did not induce an increase in lipidated LC3B as seen in WT infections (Siadous et al., 2020). However, unlike Cig57, SQSTM-1 was still increased during infection with CvpF transposon mutants and reduced by starvation-induced autophagy. Ectopically expressed CvpF was found to partially co-localize with endosomal sorting complex required for transport (ESCRT), LAMP1, LC3B, and PI(3)P at compartments clustered around the nucleus. Ectopically expressed CvpF was also able to increase LC3B-II and SQSTM-1 levels in bafilomycin A_1 treated cells, indicating an induction of autophagy. Sequence analysis identified three endocytic sorting motifs, of which a single Tyr motif was responsible for CvpF membrane localization as well as inducing LAMP1 and LC3B repositioning to vacuoles. Finally, transfection experiments visualized CvpF localized only to acidified autolysosomes, suggesting a role in autolysosome development (**Figure 5**).

A yeast two-hybrid screen identified the GTPase Rab26 as a binding partner of CvpF, with fluorescence microscopy confirming their colocalization and also observing that CvpF increases the membrane targeting of Rab26 (Siadous et al., 2020). CvpF was found to promote the accumulation of Rab26 at the CCV, although low level accumulation was still observed in CvpF mutants. Rab26 is involved in lysosomal positioning, autophagosome maturation, and the degradation of synaptic vesicles, and has been shown to interact with ATG16L1 (Li et al., 2012a; Jin and Mills, 2014; Binotti et al., 2015). Cells expressing dominant-negative forms of Rab26 possessed reduced LC3 at the CCV, while Rab26 knock-out cell lines supported smaller CCVs and impaired *C. burnetii* intracellular replication (Siadous et al., 2020). Additionally, exogenous expression of CvpF induced the formation of LC3B positive endosomes in transfected cells, which was inhibited by the co-expression of dominant negative Rab26. This data suggests that CvpF interacts with Rab26 to stimulate the accumulation of LC3B at the CCV. However, whether CvpF alters the sub-cellular localization of ATG16L1 or other factors involved in LC3B lipidation or autophagosome formation is currently unknown.

Other Putative Autophagy Modulating Effectors

One of the first CCV-associated effector proteins identified by Larson and colleagues, CvpA, was identified from the *C. burnetii* genome due to the presence of multiple eukaryotic endocytic sorting motifs (Larson et al., 2013). As with other important effector proteins, replication of a CvpA mutant was not compromised in axenic medium, but was significantly reduced during intracellular replication in THP-1 macrophage-like cells (Larson et al., 2013). Additionally, CCVs produced by CvpA mutants were significantly smaller than WT *C. burnetii*. Ectopically expressed CvpA was observed to colocalize at the plasma membrane with CHC and early endosome antigen 1 (EEA1), at pericentrosomal vesicles labelled with LAMP1 (sorting endosomes and recycling endosomes; SE, RE), and in infected cells at the CCV membrane (**Figure 5**). At the plasma membrane, CvpA was found to associate with the transferrin (Tf) receptor (TfR) and inhibited the clathrin mediated uptake of Tf. Following up on this observation, it was demonstrated that CHC accumulated at the CCV in a CvpA-dependent manner.

CvpA contains a leucine-rich repeat associated with protein-protein interactions, and three DiLeu and two Tyr endocytic sorting motifs. Pull-down assays found that the clathrin adapter protein AP2, but not AP1 or AP3, bound to all DiLeu motifs and a peptide containing both Tyr motifs, while CHC was found to associate only to the Tyr-containing peptide, likely mediated by AP2. AP2 is an adaptor found exclusively at the plasma membrane and is involved in clathrin-mediated endocytosis. Cargo recognized by AP2 include the autophagosome initiation factors ATG9 and ATG16L1, and knockdown of clathrin or AP2 inhibits autophagosome formation (Ravikumar et al., 2010; Popovic and Dikic, 2014). siRNA silencing of CHC or AP2 inhibited intracellular replication and led to reduced CCV size in WT infections, suggesting that CvpA co-opts, rather than inhibits clathrin and AP2 activity (Larson et al., 2013).

The exact mechanism by which CvpA facilitates the accumulation of clathrin at the CCV is currently unknown, but it may stabilize the interaction of clathrin/AP2 at endosomes, retaining them to facilitate fusion at the CCV. It is also currently unknown how CvpA may modulate the induction or progression of autophagy, as well as how it may regulate the endocytosis of molecules beyond Tf. Further, given Cig57 and CvpA both interact with host factors early in the endocytic pathway to facilitate the expansion of the CCV and localization of CHC, it is of interest to determine if there is an interaction between them, or if the activity of one effector is reliant on another. It would also be valuable to observe if a more severe defect develops in the absence of both effectors than in the absence of either one alone.

In addition to CvpA and CvpB, Larson and colleagues also identified CvpC-E, which localized to the CCV and whose absence significantly impaired *C. burnetii* intracellular replication and CCV development in THP-1 cells (Larson et al., 2015). The localization of these proteins to the CCV membrane indicates that they may participate in CCV-autophagosome/endosome fusion events, or the manipulation of the autophagy pathway to promote CCV expansion.

CvpC (CBU1556, Cig50) was also identified by Weber et al. (2013) in a screen for effectors required for intracellular replication, where it was found that CvpC interfered with the host secretory pathway in HEK293T cells. However, when HeLa or J774A.1 cells were infected with a CvpC transposon mutant, no defect in replication was observed (Weber et al., 2013). This may suggest that CvpC activity is specific to human macrophage cells, although further research is needed to confirm this.

CvpD (CBU1818) also appeared in a subsequent screen for *C. burnetii* effector proteins (Lifshitz et al., 2014). CvpD was found to inhibit yeast cell growth (Lifshitz et al., 2014), suggesting it interferes with essential eukaryotic pathways. Interestingly, Wachter et al. (2019) identified a *C. burnetii* sRNA CbsR12, which downregulates CvpD transcripts (Wachter et al., 2019). Absence of CbsR12 has a deleterious effect on both axenic and intracellular replication of *C. burnetii*. Loss of CbsR12 also reduced CCV size, although only early in infection; conversely, its overexpression increases CCV size at both early and late time points (Wachter et al., 2019). This may indicate that the activity of effector proteins such as CvpD may be limited to specific infection phases. However, as with CvpC and CvpE, further information about the biochemical activity of CvpD is currently lacking.

A recent screen by Crabill et al. (2018) also used a transposon library to identify effectors required for efficient CCV biogenesis. In addition to confirming previously identified effectors, seven additional effector mutants were shown to produce small CCVs: CBU0414, CBU0513, CBU0987, CBU1387, CBU1524, CBU1752, and CBU2028, with all except CBU2028 also having impaired intracellular replication. Of these, CBU0987, CBU1387, CBU1524, CBU1752, and CBU2028 were shown to accumulate at the CCV membrane. While the activity of these effectors was not explored, it was observed that in CBU0513 mutants LC3 was not present at the CCV as it was in other mutants and WT *C. burnetii* (Crabill et al., 2018), suggesting a role for this effector in autophagosome-CCV fusion.

Beyond those localizing at the CCV, additional proteins within the *C. burnetii* effector repertoire are likely to interact with the autophagy pathway through other mechanisms. For example, when ectopically expressed the effector CpeB (CBUA0013) colocalizes with LC3B, and CpeL (CBUDA0024) partially colocalized with autophagosomes (Voth et al., 2011; Maturana et al., 2013). Additionally, the protein kinase CstK (CBU0175) was found to localize at vesicles and the CCV and interacts with a homologue of mammalian TBC1D5 in an amoeba model (Martinez et al., 2020). TBC1D5 is a GTPase activating protein for Rab7, which is involved in lysosomal biogenesis, positioning and function, as well as fusion with autophagosomes (Guerra and Bucci, 2016). While the sub-cellular localization of these effectors suggests they may be involved in the manipulation of autophagy, further research is needed to confirm whether they contribute to control of this host pathway and what the functional implications of this are.

CONCLUSIONS

Despite their phylogenetic relationship, *L. pneumophila* and *C. burnetii* have adapted two very different approaches to intracellular replication; however, both manipulate common targets in the autophagy pathway in order to remodel the host cell. The convergence of these targets, despite the unique effector cohorts of *L. pneumophila* and *C. burnetii*, may be a result of their shared ancestry, but can also be seen as an indication of key checkpoints along the autophagic pathway and targets amenable to modifications. By better understanding how *L. pneumophila* and *C. burnetii* effectors interact with the host autophagy pathway, we may also identify approaches common to other intracellular pathogens and uncover mechanisms with which to better understand and control autophagy in the context of human health and disease.

L. pneumophila is an intriguing intracellular pathogen whose close association with a broad range of protozoan hosts in the environment has enabled adaptation for survival within human macrophages. Given the known functional redundancy that exists within the large cohort of over 300 effectors translocated by the T4SS, it is not unexpected to discover that multiple effectors are able to influence the host autophagy machinery using a variety of mechanisms (summarised in **Figure 4**). For example, RavZ, which efficiently halts autophagosome maturation through the irreversible cleavage of LC3-II, is not present in all strains and yet the autophagy machinery is still disrupted (Amer and Swanson, 2005). In this instance, it is easy to speculate that the action of effectors such as *Lp*Spl and Lpg1137 that interfere with autophagosome biogenesis, and the transcription factor Lpg2936 would play a greater role in manipulation of host autophagy. It is also equally plausible that these RavZ-deficient strains possess other, strain-specific, autophagy modulating effectors.

Given the importance of evading autophagy to *Legionella*, the existence of a common cohort of autophagy-related effectors may be considered. However, the recent analysis by Gomez-Valero and colleagues which identified a putative 18,000 effector proteins found only eight were present in all *Legionella* genomes assessed (Gomez-Valero et al., 2019). Further, the ability to infect human cells arose independently multiple times throughout the genus, and no specific set of effectors could be attributed to the ability to infect human cells. Finally, 16 Rab-like proteins from eight different *Legionella* species were found to have been acquired by horizontal gene transfer from hosts (Gomez-Valero et al., 2019). This suggests it is unlikely that a core of autophagy-related effectors exists in the *Legionella* genus. Rather, *Legionella* species will have evolved unique cohorts based on genes acquired from their specific host range.

Since deletion of any of these individual autophagy related effectors has no direct impact on intracellular bacterial replication the coordinated interplay between these effectors are important for *L. pneumophila* avoidance of the host autophagic machinery and subsequent success within host cells. The exact role for LegA9 in enhancing recognition of the LCV for autophagy clearance is yet to be fully elucidated, and no doubt controlling the temporal action of this effector is critical for bacterial survival. Given the redundant nature of many *L. pneumophila* effectors, it would be interesting to remove all known effectors that regulate host autophagy. Whether this would affect intracellular survival and replication or not, and if any changes in pathogenicity were host-specific would provide valuable information about the evolutionary pressures that have led to the retention of these effectors. Subsequent restoration of individual autophagy modulating effectors could then also provide a platform to assess the role of these effectors in isolation. This huge assortment of effectors is almost certain to contain more regulators of autophagy, potentially providing a plethora of novel tools, similar to RavZ, that can be used to explore, uncover and manipulate autophagy at a molecular level.

In contrast, *C. burnetii* relies on a functional autophagy pathway in host cells. However, there is also some evidence that *C. burnetii* down-regulates aspects of autophagy, such as increasing the pH of mature endosomes and decreasing cellular lysosome activity and content (Mulye et al., 2017; Samanta et al., 2019). Indeed, overexpression of TFEB was found to decrease CCV size and bacterial replication during infection (Samanta et al., 2019). Conversely, *C. burnetii* has been found to activate TFEB and the related transcription factor TFE3 *via* inhibition of mTORC1 (Larson et al., 2019; Padmanabhan et al., 2020). Consistent with other studies, this did not result in an increase in autophagy or inhibition of autophagic flux, but TFE3/TFEB were required for efficient CCV development. Macrophage cells with TFE3/TFEB knocked out were curiously able to support higher levels of *C. burnetii* replication despite producing smaller CCVs (Larson et al., 2019), while HeLa cells with dual siRNA silencing of TFEB and TFE3 showed no change in replication, despite still exhibiting a significant decrease in CCV size (Padmanabhan et al., 2020). These results highlight the fine balance *C. burnetii* maintains in modulating the host autophagy system to favour infection and indicates the presence of a range of compensatory and complementary modifications to the host cell. They also highlight the uncertainties surrounding the role of autophagy in the *C. burnetii* lifecycle. While it is often hypothesized that autophagy is induced to supply nutrients to the replicating bacteria, the inconsistent results observed regarding regulators of autophagy suggest this may not be a crucial role. Indeed, the capacity of *C. burnetii* to still replicate in the absence of autophagy argues that this process is not required for bacterial nutrient acquisition. It is possible that other functions of autophagy, such as enhanced/altered vesicle trafficking and fusion are important for reasons beyond the delivery of nutrients. However, the interdependent nature of these two roles has made it difficult to disentangle them experimentally. As our understanding of how *C. burnetii* regulates autophagy develops, it will be worth considering how we can adapt these effectors as potential tools in the treatment of diseases associated with the dysregulation of autophagy, including many neurodegenerative disorders.

AUTHOR CONTRIBUTIONS

All authors listed have made a substantial, direct and intellectual contribution to the manuscript and approved it for publication. All authors contributed to the article and approved the submitted version.

REFERENCES

Abd El Maksoud, A.II, Elebeedy, D., Abass, N. H., Awad, A. M., Nasr, G. M., Roshdy, T., et al. (2019). Methylomic Changes of Autophagy-Related Genes by Legionella Effector Lpg2936 in Infected Macrophages. *Front. Cell Dev. Biol.* 7, 390. doi: 10.3389/fcell.2019.00390

Amer, A. O., and Swanson, M. S. (2005). Autophagy is an immediate macrophage response to Legionella pneumophila. *Cell Microbiol.* 7, 765–778. doi: 10.1111/j.1462-5822.2005.00509.x

Amer, A. O., Byrne, B. G., and Swanson, M. S. (2005). Macrophages rapidly transfer pathogens from lipid raft vacuoles to autophagosomes. *Autophagy* 1, 53–58. doi: 10.4161/auto.1.1.1589

Anding, A. L., and Baehrecke, E. H. (2015). Autophagy in Cell Life and Cell Death. *Curr. Top. Dev. Biol.* 114, 67–91. doi: 10.1016/bs.ctdb.2015.07.012

Arasaki, K., Mikami, Y., Shames, S. R., Inoue, H., Wakana, Y., and Tagaya, M. (2017). Legionella effector Lpg1137 shuts down ER-mitochondria communication through cleavage of syntaxin 17. *Nat. Commun.* 8, 15406. doi: 10.1038/ncomms15406

Beck, W. H. J., Kim, D., Das, J., Yu, H., Smolka, M. B., and Mao, Y. (2020). Glucosylation by the Legionella Effector SetA Promotes the Nuclear Localization of the Transcription Factor TFEB. *iScience* 23, 101300. doi: 10.1016/j.isci.2020.101300

Belyi, Y., Niggeweg, R., Opitz, B., Vogelsgesang, M., Hippenstiel, S., Wilm, M., et al. (2006). Legionella pneumophila glucosyltransferase inhibits host elongation factor 1A. *Proc. Natl. Acad. Sci. U.S.A.* 103, 16953–16958. doi: 10.1073/pnas.0601562103

Belyi, Y., Tabakova, I., Stahl, M., and Aktories, K. (2008). Lgt: a family of cytotoxic glucosyltransferases produced by Legionella pneumophila. *J. Bacteriol.* 190, 3026–3035. doi: 10.1128/JB.01798-07

Bento, C. F., Renna, M., Ghislat, G., Puri, C., Ashkenazi, A., Vicinanza, M., et al. (2016). Mammalian Autophagy: How Does It Work? *Annu. Rev. Biochem.* 85, 685–713. doi: 10.1146/annurev-biochem-060815-014556

Berg, T. O., Fengsrud, M., Stromhaug, P. E., Berg, T., and Seglen, P. O. (1998). Isolation and characterization of rat liver amphisomes. Evidence for fusion of autophagosomes with both early and late endosomes. *J. Biol. Chem.* 273, 21883–21892. doi: 10.1074/jbc.273.34.21883

Berger, K. H., and Isberg, R. R. (1993). Two distinct defects in intracellular growth complemented by a single genetic locus in Legionella pneumophila. *Mol. Microbiol.* 7, 7–19. doi: 10.1111/j.1365-2958.1993.tb01092.x

Beron, W., Gutierrez, M. G., Rabinovitch, M., and Colombo, M.II (2002). Coxiella burnetii localizes in a Rab7-labeled compartment with autophagic characteristics. *Infect. Immun.* 70, 5816–5821. doi: 10.1128/IAI.70.10.5816-5821.2002

Bhogaraju, S., Kalayil, S., Liu, Y., Bonn, F., Colby, T., Matic, I., et al. (2016). Phosphoribosylation of Ubiquitin Promotes Serine Ubiquitination and Impairs Conventional Ubiquitination. *Cell* 167, 1636–1649.e13. doi: 10.1016/j.cell.2016.11.019

Binotti, B., Pavlos, N. J., Riedel, D., Wenzel, D., Vorbruggen, G., Schalk, A. M., et al. (2015). The GTPase Rab26 links synaptic vesicles to the autophagy pathway. *Elife* 4, e05597. doi: 10.7554/eLife.05597

Burette, M., and Bonazzi, M. (2020). From neglected to dissected: How technological advances are leading the way to the study of Coxiella burnetii pathogenesis. *Cell Microbiol.* 22, e13180. doi: 10.1111/cmi.13180

Burstein, D., Amaro, F., Zusman, T., Lifshitz, Z., Cohen, O., Gilbert, J. A., et al. (2016). Genomic analysis of 38 Legionella species identifies large and diverse effector repertoires. *Nat. Genet.* 48, 167–175. doi: 10.1038/ng.3481

Calvo-Garrido, J., Carilla-Latorre, S., Kubohara, Y., Santos-Rodrigo, N., Mesquita, A., Soldati, T., et al. (2010). Autophagy in Dictyostelium: genes and pathways, cell death and infection. *Autophagy* 6, 686–701. doi: 10.4161/auto.6.6.12513

Campoy, E., and Colombo, M. I. (2009).. Autophagy in intracellular bacterial infection. *Biochim. Biophys. Acta* 1793, 1465–1477. doi: 10.1016/j.bbamcr.2009.03.003

Campoy, E. M., Zoppino, F. C., and Colombo, M. I. (2011). The early secretory pathway contributes to the growth of the Coxiella-replicative niche. *Infect. Immun.* 79, 402–413. doi: 10.1128/IAI.00688-10

Campoy, E. M., Mansilla, M. E., and Colombo, M. I. (2013). Endocytic SNAREs are involved in optimal Coxiella burnetii vacuole development. *Cell Microbiol.* 15, 922–941. doi: 10.1111/cmi.12087

Casanova, J. E. (2017). Bacterial Autophagy: Offense and Defense at the Host-Pathogen Interface. *Cell Mol. Gastroenterol. Hepatol.* 4, 237–243. doi: 10.1016/j.jcmgh.2017.05.002

Chan, S. N., and Tang, B. L. (2013). Location and membrane sources for autophagosome formation - from ER-mitochondria contact sites to Golgi-endosome-derived carriers. *Mol. Membr. Biol.* 30, 394–402. doi: 10.3109/09687688.2013.850178

Charpentier, X., Gabay, J. E., Reyes, M., Zhu, J. W., Weiss, A., and Shuman, H. A. (2009). Chemical genetics reveals bacterial and host cell functions critical for type IV effector translocation by Legionella pneumophila. *PloS Pathog.* 5, e1000501. doi: 10.1371/journal.ppat.1000501

Chen, J., Reyes, M., Clarke, M., and Shuman, H. A. (2007). Host cell-dependent secretion and translocation of the LepA and LepB effectors of Legionella pneumophila. *Cell Microbiol.* 9, 1660–1671. doi: 10.1111/j.1462-5822.2007.00899.x

Chen, R. H., Chen, Y. H., and Huang, T. Y. (2019). Ubiquitin-mediated regulation of autophagy. *J. BioMed. Sci.* 26, 80. doi: 10.1186/s12929-019-0569-y

Choy, A., Dancourt, J., Mugo, B., O'Connor, T. J., Isberg, R. R., Melia, T. J., et al. (2012). The *Legionella* effector RavZ inhibits host autophagy through irreversible Atg8 deconjugation. *Science* 338, 1072–1076. doi: 10.1126/science.1227026

Codogno, P., and Meijer, A. J. (2005). Autophagy and signaling: their role in cell survival and cell death. *Cell Death Differ.* 12 Suppl 2, 1509–1518. doi: 10.1038/sj.cdd.4401751

Crabill, E., Schofield, W. B., Newton, H. J., Goodman, A. L., and Roy, C. R. (2018). Dot/Icm-Translocated Proteins Important for Biogenesis of the Coxiella burnetii-Containing Vacuole Identified by Screening of an Effector Mutant Sublibrary. *Infect. Immun.* 86, e00758-17. 10.1128/IAI.00758-17

Dall'Armi, C., Devereaux, K. A., and Di Paolo, G. (2013). The role of lipids in the control of autophagy. *Curr. Biol.* 23, R33–R45. doi: 10.1016/j.cub.2012.10.041

De Leon, J. A., Qiu, J., Nicolai, C. J., Counihan, J. L., Barry, K. C., Xu, L., et al. (2017). Positive and Negative Regulation of the Master Metabolic Regulator mTORC1 by Two Families of Legionella pneumophila Effectors. *Cell Rep.* 21, 2031–2038. doi: 10.1016/j.celrep.2017.10.088

Degtyar, E., Zusman, T., Ehrlich, M., and Segal, G. (2009). A Legionella effector acquired from protozoa is involved in sphingolipids metabolism and is targeted to the host cell mitochondria. *Cell Microbiol.* 11, 1219–1235. doi: 10.1111/j.1462-5822.2009.01328.x

Diao, J., Liu, R., Rong, Y., Zhao, M., Zhang, J., Lai, Y., et al. (2015). ATG14 promotes membrane tethering and fusion of autophagosomes to endolysosomes. *Nature* 520, 563–566. doi: 10.1038/nature14147

Dikic, I., and Elazar, Z. (2018). Mechanism and medical implications of mammalian autophagy. *Nat. Rev. Mol. Cell Biol.* 19, 349–364. doi: 10.1038/s41580-018-0003-4

Dooley, H. C., Razi, M., Polson, H. E., Girardin, S. E., Wilson, M. I., and Tooze, S. A. (2014). WIPI2 links LC3 conjugation with PI3P, autophagosome formation, and pathogen clearance by recruiting Atg12-5-16L1. *Mol. Cell* 55, 238–252. doi: 10.1016/j.molcel.2014.05.021

Ensminger, A. W. (2015). *Legionella pneumophila*, armed to the hilt: justifying the largest arsenal of effectors in the bacterial world. *Curr. Opin. Microbiol.* 29, 74–80. doi: 10.1016/j.mib.2015.11.002

Gao, L., Song, Q., Liang, H., Zhu, Y., Wei, T., Dong, N., et al. (2019). Legionella effector SetA as a general O-glucosyltransferase for eukaryotic proteins. *Nat. Chem. Biol.* 15, 213–216. doi: 10.1038/s41589-018-0189-y

Glick, D., Barth, S., and Macleod, K. F. (2010). Autophagy: cellular and molecular mechanisms. *J. Pathol.* 221, 3–12. doi: 10.1002/path.2697

Gomez-Valero, L., Rusniok, C., Cazalet, C., and Buchrieser, C. (2011). Comparative and functional genomics of *legionella* identified eukaryotic like proteins as key players in host-pathogen interactions. *Front. Microbiol.* 2, 208. doi: 10.3389/fmicb.2011.00208

Gomez-Valero, L., Rusniok, C., Carson, D., Mondino, S., Pérez-Cobas, A. E., Rolando, M., et al. (2019). More than 18,000 effectors in the *Legionella* genus genome provide multiple, independent combinations for replication in human cells. *Proc. Natl. Acad. Sci. U.S.A.* 116, 2265–2273. doi: 10.1073/pnas.1808016116

Gradowski, M., and Pawlowski, K. (2017). The Legionella pneumophila effector Lpg1137 is a homologue of mitochondrial SLC25 carrier proteins, not of known serine proteases. *PeerJ* 5, e3849. doi: 10.7717/peerj.3849

Guerra, F., and Bucci, C. (2016). Multiple Roles of the Small GTPase Rab7. *Cells* 5, 34. doi: 10.3390/cells5030034

Gutierrez, M. G., Master, S. S., Singh, S. B., Taylor, G. A., Colombo, M. I., and Deretic, V. (2004). Autophagy is a defense mechanism inhibiting BCG and Mycobacterium tuberculosis survival in infected macrophages. *Cell* 119, 753–766. doi: 10.1016/j.cell.2004.11.038

Gutierrez, M. G., Vazquez, C. L., Munafo, D. B., Zoppino, F. C., Beron, W., Rabinovitch, M., et al. (2005). Autophagy induction favours the generation and maturation of the Coxiella-replicative vacuoles. *Cell Microbiol.* 7, 981–993. doi: 10.1111/j.1462-5822.2005.00527.x

Hamasaki, M., Furuta, N., Matsuda, A., Nezu, A., Yamamoto, A., Fujita, N., et al. (2013). Autophagosomes form at ER-mitochondria contact sites. *Nature* 495, 389–393. doi: 10.1038/nature11910

Heinzen, R. A., Scidmore, M. A., Rockey, D. D., and Hackstadt, T. (1996). Differential interaction with endocytic and exocytic pathways distinguish parasitophorous vacuoles of Coxiella burnetii and Chlamydia trachomatis. *Infect. Immun.* 64, 796–809. doi: 10.1128/IAI.64.3.796-809.1996

Heinzen, R. A., Hackstadt, T., and Samuel, J. E. (1999). Developmental biology of Coxiella burnettii. *Trends Microbiol.* 7, 149–154. doi: 10.1016/S0966-842X(99)01475-4

Henne, W. M., Boucrot, E., Meinecke, M., Evergren, E., Vallis, Y., Mittal, R., et al. (2010). FCHo proteins are nucleators of clathrin-mediated endocytosis. *Science* 328, 1281–1284. doi: 10.1126/science.1188462

Horenkamp, F. A., Kauffman, K. J., Kohler, L. J., Sherwood, R. K., Krueger, K. P., Shteyn, V., et al. (2015). The Legionella Anti-autophagy Effector RavZ Targets the Autophagosome via PI3P- and Curvature-Sensing Motifs. *Dev. Cell* 34, 569–576. doi: 10.1016/j.devcel.2015.08.010

Horwitz, M. A. (1983). Formation of a novel phagosome by the Legionnaires' disease bacterium (Legionella pneumophila) in human monocytes. *J. Exp. Med.* 158, 1319–1331. doi: 10.1084/jem.158.4.1319

Howe, D., Melnicakova, J., Barak, I., and Heinzen, R. A. (2003). Maturation of the Coxiella burnetii parasitophorous vacuole requires bacterial protein synthesis but not replication. *Cell Microbiol.* 5, 469–480. doi: 10.1046/j.1462-5822.2003.00293.x

Hu, W., Chan, H., Lu, L., Wong, K. T., Wong, S. H., Li, M. X., et al. (2020). Autophagy in intracellular bacterial infection. *Semin. Cell Dev. Biol.* 101, 41–50. doi: 10.1016/j.semcdb.2019.07.014

Hubber, A., and Roy, C. R. (2010). Modulation of host cell function by *Legionella pneumophila* type IV effectors. *Annu. Rev. Cell Dev. Biol.* 26, 261–283. doi: 10.1146/annurev-cellbio-100109-104034

Ingmundson, A., Delprato, A., Lambright, D. G., and Roy, C. R. (2007). Legionella pneumophila proteins that regulate Rab1 membrane cycling. *Nature* 450, 365–369. doi: 10.1038/nature06336

Isaac, D. T., and Isberg, R. (2014). Master manipulators: an update on Legionella pneumophila Icm/Dot translocated substrates and their host targets. *Future Microbiol.* 9, 343–359. doi: 10.2217/fmb.13.162

Itakura, E., Kishi-Itakura, C., and Mizushima, N. (2012). The hairpin-type tail-anchored SNARE syntaxin 17 targets to autophagosomes for fusion with endosomes/lysosomes. *Cell* 151, 1256–1269. doi: 10.1016/j.cell.2012.11.001

Jank, T., Böhmer, K. E., Tzivelekidis, T., Schwan, C., Belyi, Y., and Aktories, K. (2012). Domain organization of Legionella effector SetA. *Cell Microbiol.* 14, 852–868. doi: 10.1111/j.1462-5822.2012.01761.x

Jin, R. U., and Mills, J. C. (2014). RAB26 coordinates lysosome traffic and mitochondrial localization. *J. Cell Sci.* 127, 1018–1032. doi: 10.1242/jcs.138776

Johansen, T., and Lamark, T. (2020). Selective Autophagy: ATG8 Family Proteins, LIR Motifs and Cargo Receptors. *J. Mol. Biol.* 432, 80–103. doi: 10.1016/j.jmb.2019.07.016

Kampschreur, L. M., Wegdam-Blans, M. C., Wever, P. C., Renders, N. H., Delsing, C. E., Sprong, T., et al. (2015). Chronic Q fever diagnosis- consensus guideline versus expert opinion. *Emerg. Infect. Dis.* 21, 1183–1188. doi: 10.3201/eid2107.130955

Karanasios, E., Stapleton, E., Manifava, M., Kaizuka, T., Mizushima, N., Walker, S. A., et al. (2013). Dynamic association of the ULK1 complex with omegasomes during autophagy induction. *J. Cell Sci.* 126, 5224–5238. doi: 10.1242/jcs.132415

Kaushik, S., and Cuervo, A. M. (2012). Chaperone-mediated autophagy: a unique way to enter the lysosome world. *Trends Cell Biol.* 22, 407–417. doi: 10.1016/j.tcb.2012.05.006

Khweek, A. A., Caution, K., Akhter, A., Abdulrahman, B. A., Tazi, M., Hassan, H., et al. (2013). A bacterial protein promotes the recognition of the Legionella pneumophila vacuole by autophagy. *Eur. J. Immunol.* 43, 1333–1344. doi: 10.1002/eji.201242835

Kim, J., Kundu, M., Viollet, B., and Guan, K. L. (2011). AMPK and mTOR regulate autophagy through direct phosphorylation of Ulk1. *Nat. Cell Biol.* 13, 132–141. doi: 10.1038/ncb2152

Kimmey, J. M., and Stallings, C. L. (2016). Bacterial Pathogens versus Autophagy: Implications for Therapeutic Interventions. *Trends Mol. Med.* 22, 1060–1076. doi: 10.1016/j.molmed.2016.10.008

King, J. S. (2012). Autophagy across the eukaryotes: is S. cerevisiae the odd one out? *Autophagy* 8, 1159–1162. doi: 10.4161/auto.20527

Kirkin, V., Mcewan, D. G., Novak, I., and Dikic, I. (2009). A role for ubiquitin in selective autophagy. *Mol. Cell* 34, 259–269. doi: 10.1016/j.molcel.2009.04.026

Klionsky, D. J., and Schulman, B. A. (2014). Dynamic regulation of macroautophagy by distinctive ubiquitin-like proteins. *Nat. Struct. Mol. Biol.* 21, 336–345. doi: 10.1038/nsmb.2787

Kohler, L. J., Reed, S. H. C., Sarraf, S. A., Arteaga, D. D., Newton, H. J., and Roy, C. R. (2016). Effector Protein Cig2 Decreases Host Tolerance of Infection by Directing Constitutive Fusion of Autophagosomes with the Coxiella-Containing Vacuole. *mBio* 7, e01127-16. doi: 10.1128/mBio.01127-16

Komatsu, M., Waguri, S., Koike, M., Sou, Y. S., Ueno, T., Hara, T., et al. (2007). Homeostatic levels of p62 control cytoplasmic inclusion body formation in autophagy-deficient mice. *Cell* 131, 1149–1163. doi: 10.1016/j.cell.2007.10.035

Ktistakis, N. T., Griffiths, G., Habermann, A., Roderick, H. L., Chandra, P., Manifava, M., et al. (2008). Autophagosome formation from membrane compartments enriched in phosphatidylinositol 3-phosphate and dynamically connected to the endoplasmic reticulum. *J. Cell Biol.* 182, 685–701. doi: 10.1083/jcb.200803137

Ktistakis, N. T., Manifava, M., Schoenfelder, P., and Rotondo, S. (2012). How phosphoinositide 3-phosphate controls growth downstream of amino acids and autophagy downstream of amino acid withdrawal. *Biochem. Soc. Trans.* 40, 37–43. doi: 10.1042/BST20110684

Kumar, S., Gu, Y., Abudu, Y. P., Bruun, J. A., Jain, A., Farzam, F., et al. (2019). Phosphorylation of Syntaxin 17 by TBK1 Controls Autophagy Initiation. *Dev. Cell* 49, 130–144 e6. doi: 10.1016/j.devcel.2019.01.027

Kwon, D. H., Kim, S., Jung, Y. O., Roh, K. H., Kim, L., Kim, B. W., et al. (2017). The 1:2 complex between RavZ and LC3 reveals a mechanism for deconjugation of LC3 on the phagophore membrane. *Autophagy* 13, 70–81. doi: 10.1080/15548627.2016.1243199

Laplante, M., and Sabatini, D. M. (2009). mTOR signaling at a glance. *J. Cell Sci.* 122, 3589–3594. doi: 10.1242/jcs.051011

Larson, C. L., Beare, P. A., Howe, D., and Heinzen, R. A. (2013). Coxiella burnetii effector protein subverts clathrin-mediated vesicular trafficking for pathogen vacuole biogenesis. *Proc. Natl. Acad. Sci. U.S.A.* 110, E4770–E4779. doi: 10.1073/pnas.1309195110

Larson, C. L., Beare, P. A., Voth, D. E., Howe, D., Cockrell, D. C., Bastidas, R. J., et al. (2015). Coxiella burnetii effector proteins that localize to the parasitophorous vacuole membrane promote intracellular replication. *Infect. Immun.* 83, 661–670. doi: 10.1128/IAI.02763-14

Larson, C. L., Sandoz, K. M., Cockrell, D. C., and Heinzen, R. A. (2019). Noncanonical Inhibition of mTORC1 by Coxiella burnetii Promotes Replication within a Phagolysosome-Like Vacuole. *mBio* 10, e02816–e02818. doi: 10.1128/mBio.02816-18

Latomanski, E. A., and Newton, H. J. (2018). Interaction between autophagic vesicles and the Coxiella-containing vacuole requires CLTC (clathrin heavy chain). *Autophagy* 14, 1710–1725. doi: 10.1080/15548627.2018.1483806

Latomanski, E. A., Newton, P., Khoo, C. A., and Newton, H. J. (2016). The Effector Cig57 Hijacks FCHO-Mediated Vesicular Trafficking to Facilitate Intracellular Replication of Coxiella burnetii. *PloS Pathog.* 12, e1006101. doi: 10.1371/journal.ppat.1006101

Levanova, N., Steinemann, M., Böhmer, K. E., Schneider, S., Belyi, Y., Schlosser, A., et al. (2019). Characterization of the glucosyltransferase activity of Legionella pneumophila effector SetA. *Naunyn Schmiedebergs Arch. Pharmacol.* 392, 69–79. doi: 10.1007/s00210-018-1562-9

Levine, B., and Kroemer, G. (2019). Biological Functions of Autophagy Genes: A Disease Perspective. *Cell* 176, 11–42. doi: 10.1016/j.cell.2018.09.048

Li, C., Fan, Y., Lan, T. H., Lambert, N. A., and Wu, G. (2012a). Rab26 modulates the cell surface transport of alpha2-adrenergic receptors from the Golgi. *J. Biol. Chem.* 287, 42784–42794. doi: 10.1074/jbc.M112.410936

Li, W. W., Li, J., and Bao, J. K. (2012b). Microautophagy: lesser-known self-eating. *Cell Mol. Life Sci.* 69, 1125–1136. doi: 10.1007/s00018-011-0865-5

Lifshitz, Z., Burstein, D., Schwartz, K., Shuman, H. A., Pupko, T., and Segal, G. (2014). Identification of novel Coxiella burnetii Icm/Dot effectors and genetic analysis of their involvement in modulating a mitogen-activated protein kinase pathway. *Infect. Immun.* 82, 3740–3752. doi: 10.1128/IAI.01729-14

Lorincz, P., and Juhasz, G. (2020). Autophagosome-Lysosome Fusion. *J. Mol. Biol.* 432, 2462–2482. doi: 10.1016/j.jmb.2019.10.028

Machner, M. P., and Isberg, R. R. (2007). A bifunctional bacterial protein links GDI displacement to Rab1 activation. *Science* 318, 974–977. doi: 10.1126/science.1149121

Mack, H.II, Zheng, B., Asara, J. M., and Thomas, S. M. (2012). AMPK-dependent phosphorylation of ULK1 regulates ATG9 localization. *Autophagy* 8, 1197–1214. doi: 10.4161/auto.20586

Mahapatra, S., Ayoubi, P., and Shaw, E.II (2010). Coxiella burnetii Nine Mile II proteins modulate gene expression of monocytic host cells during infection. *BMC Microbiol.* 10, 244. doi: 10.1186/1471-2180-10-244

Marra, A., Blander, S. J., Horwitz, M. A., and Shuman, H. A. (1992). Identification of a Legionella pneumophila locus required for intracellular multiplication in human macrophages. *Proc. Natl. Acad. Sci. U.S.A.* 89, 9607–9611. doi: 10.1073/pnas.89.20.9607

Martinez, E., Allombert, J., Cantet, F., Lakhani, A., Yandrapalli, N., Neyret, A., et al. (2016). Coxiella burnetii effector CvpB modulates phosphoinositide metabolism for optimal vacuole development. *Proc. Natl. Acad. Sci. U.S.A.* 113, E3260–E3269. doi: 10.1073/pnas.1522811113

Martinez, E., Huc-Brandt, S., Brelle, S., Allombert, J., Cantet, F., Gannoun-Zaki, L., et al. (2020). The secreted protein kinase CstK from Coxiella burnetii influences vacuole development and interacts with the GTPase-activating host protein TBC1D5. *J. Biol. Chem.* 295, 7391–7403. doi: 10.1074/jbc.RA119.010112

Matsuda, F., Fujii, J., and Yoshida, S. (2009). Autophagy induced by 2-deoxy-D-glucose suppresses intracellular multiplication of Legionella pneumophila in A/J mouse macrophages. *Autophagy* 5, 484–493. doi: 10.4161/auto.5.4.7760

Maturana, P., Graham, J. G., Sharma, U. M., and Voth, D. E. (2013). Refining the plasmid-encoded type IV secretion system substrate repertoire of Coxiella burnetii. *J. Bacteriol.* 195, 3269–3276. doi: 10.1128/JB.00180-13

McDonough, J. A., Newton, H. J., Klum, S., Swiss, R., Agaisse, H., and Roy, C. R. (2013). Host pathways important for Coxiella burnetii infection revealed by genome-wide RNA interference screening. *mBio* 4, e00606–e00612. doi: 10.1128/mBio.00606-12

McEwan, D. G. (2017). Host-pathogen interactions and subversion of autophagy. *Essays Biochem.* 61, 687–697. doi: 10.1042/EBC20170058

Mukherjee, S., Liu, X., Arasaki, K., McDonough, J., Galan, J. E., and Roy, C. R. (2011). Modulation of Rab GTPase function by a protein phosphocholine transferase. *Nature* 477, 103–106. doi: 10.1038/nature10335

Müller, M. P., Peters, H., Blümer, J., Blankenfeldt, W., Goody, R. S., and Itzen, A. (2010). The Legionella effector protein DrrA AMPylates the membrane traffic regulator Rab1b. *Science* 329, 946–949. doi: 10.1126/science.1192276

Mulye, M., Samanta, D., Winfree, S., Heinzen, R. A., and Gilk, S. D. (2017). Elevated Cholesterol in the Coxiella burnetii Intracellular Niche Is Bacteriolytic. *mBio* 8, e02313–16. doi: 10.1128/mBio.02313-16

Murata, T., Delprato, A., Ingmundson, A., Toomre, D. K., Lambright, D. G., and Roy, C. R. (2006). The Legionella pneumophila effector protein DrrA is a Rab1 guanine nucleotide-exchange factor. *Nat. Cell Biol.* 8, 971–977. doi: 10.1038/ncb1463

Nakagawa, I., Amano, A., Mizushima, N., Yamamoto, A., Yamaguchi, H., Kamimoto, T., et al. (2004). Autophagy defends cells against invading group A Streptococcus. *Science* 306, 1037–1040. doi: 10.1126/science.1103966

Nakamura, S., and Yoshimori, T. (2017). New insights into autophagosome-lysosome fusion. *J. Cell Sci.* 130, 1209–1216. doi: 10.1242/jcs.196352

Neunuebel, M. R., Chen, Y., Gaspar, A. H., Backlund, P. S., Yergey, A., and Machner, M. P. (2011). De-AMPylation of the small GTPase Rab1 by the pathogen Legionella pneumophila. *Science* 333, 453–456. doi: 10.1126/science.1207193

Newton, H. J., McDonough, J. A., and Roy, C. R. (2013). Effector Protein Translocation by the Coxiella burnetii Dot/Icm Type IV Secretion System Requires Endocytic Maturation of the Pathogen-Occupied Vacuole. *PloS One* 8, e54566. doi: 10.1371/journal.pone.0054566

Newton, H. J., Kohler, L. J., McDonough, J. A., Temoche-Diaz, M., Crabill, E., Hartland, E. L., et al. (2014). A screen of Coxiella burnetii mutants reveals important roles for Dot/Icm effectors and host autophagy in vacuole biogenesis. *PloS Pathog.* 10, e1004286. doi: 10.1371/journal.ppat.1004286

Newton, P., Thomas, D. R., Reed, S. C. O., Lau, N., Xu, B., Ong, S. Y., et al. (2020). Lysosomal degradation products induce Coxiella burnetii virulence. *Proc. Natl. Acad. Sci. U.S.A.* 117, 6801–6810. doi: 10.1073/pnas.1921344117

Nixon, R. A. (2013). The role of autophagy in neurodegenerative disease. *Nat. Med.* 19, 983–997. doi: 10.1038/nm.3232

Omotade, T. O., and Roy, C. R. (2019). "Manipulation of Host Cell Organelles by Intracellular Pathogens", in *Bacteria and Intracellularity*. Eds. P. Cossart, C. Roy and P. Sansonetti (Washington, DC: ASM Press), 199–214. cellularity. doi: 10.1128/microbiolspec.BAI-0022-2019

Omotade, T. O., and Roy, C. R. (2020). Legionella pneumophila Excludes Autophagy Adaptors from the Ubiquitin-Labeled Vacuole in Which It Resides. *Infect. Immun.* 88, e00793–19. doi: 10.1128/IAI.00793-19

Orsi, A., Razi, M., Dooley, H. C., Robinson, D., Weston, A. E., Collinson, L. M., et al. (2012). Dynamic and transient interactions of Atg9 with autophagosomes, but not membrane integration, are required for autophagy. *Mol. Biol. Cell* 23, 1860–1873. doi: 10.1091/mbc.e11-09-0746

Padmanabhan, B., Fielden, L. F., Hachani, A., Newton, P., Thomas, D. R., Cho, H. J., et al. (2020). Biogenesis of the Spacious Coxiella-Containing Vacuole Depends on Host Transcription Factors TFEB and TFE3. *Infect. Immun.* 88, e00534–19. doi: 10.1128/IAI.00534-19

Park, S. W., Jun, Y. W., Jeon, P., Lee, Y. K., Park, J. H., Lee, S. H., et al. (2019). LIR motifs and the membrane-targeting domain are complementary in the function of RavZ. *BMB Rep.* 52, 700–705. doi: 10.5483/BMBRep.2019.52.12.211

Park, J. M., Ghosh, S., and O'Connor, T. J. (2020). Combinatorial selection in amoebal hosts drives the evolution of the human pathogen Legionella pneumophila. *Nat. Microbiol.* 5, 599–609. doi: 10.1038/s41564-019-0663-7

Pinotsis, N., and Waksman, G. (2017). Crystal structure of the Legionella pneumophila Lpg2936 in complex with the cofactor S-adenosyl-L-methionine reveals novel insights into the mechanism of RsmE family methyltransferases. *Protein Sci.* 26, 2381–2391. doi: 10.1002/pro.3305

Polson, H. E., De Lartigue, J., Rigden, D. J., Reedijk, M., Urbe, S., Clague, M. J., et al. (2010). Mammalian Atg18 (WIPI2) localizes to omegasome-anchored phagophores and positively regulates LC3 lipidation. *Autophagy* 6, 506–522. doi: 10.4161/auto.6.4.11863

Popovic, D., and Dikic, I. (2014). TBC1D5 and the AP2 complex regulate ATG9 trafficking and initiation of autophagy. *EMBO Rep.* 15, 392–401. doi: 10.1002/embr.201337995

Prashar, A., and Terebiznik, M. R. (2014). *Legionella pneumophila*: homeward bound away from the phagosome. *Curr. Opin. Microbiol.* 23C, 86–93. doi: 10.1016/j.mib.2014.11.008

Price, J. V., Jiang, K., Galantowicz, A., Freifeld, A., and Vance, R. E. (2018). Legionella pneumophila Is Directly Sensitive to 2-Deoxyglucose-Phosphate via Its UhpC Transporter but Is Indifferent to Shifts in Host Cell Glycolytic Metabolism. *J. Bacteriol.* 200, e00176–18. doi: 10.1128/JB.00176-18

Price, C., Jones, S., Mihelcic, M., Santic, M., and Abu Kwaik, Y. (2020). Paradoxical Pro-inflammatory Responses by Human Macrophages to an Amoebae Host-Adapted Legionella Effector. *Cell Host Microbe* 27, 571–584 e7. doi: 10.1016/j.chom.2020.03.003

Qiu, J., Sheedlo, M. J., Yu, K., Tan, Y., Nakayasu, E. S., Das, C., et al. (2016). Ubiquitination independent of E1 and E2 enzymes by bacterial effectors. *Nature* 533, 120–124. doi: 10.1038/nature17657

Rabanal-Ruiz, Y., Otten, E. G., and Korolchuk, V.II (2017). mTORC1 as the main gateway to autophagy. *Essays Biochem.* 61, 565–584. doi: 10.1042/EBC20170027

Ravanan, P., Srikumar, I. F., and Talwar, P. (2017). Autophagy: The spotlight for cellular stress responses. *Life Sci.* 188, 53–67. doi: 10.1016/j.lfs.2017.08.029

Ravikumar, B., Moreau, K., Jahreiss, L., Puri, C., and Rubinsztein, D. C. (2010). Plasma membrane contributes to the formation of pre-autophagosomal structures. *Nat. Cell Biol.* 12, 747–757. doi: 10.1038/ncb2078

Reggiori, F., and Ungermann, C. (2017). Autophagosome Maturation and Fusion. *J. Mol. Biol.* 429, 486–496. doi: 10.1016/j.jmb.2017.01.002

Rikihisa, Y. (1984). Glycogen autophagosomes in polymorphonuclear leukocytes induced by rickettsiae. *Anat. Rec.* 208, 319–327. doi: 10.1002/ar.1092080302

Roczniak-Ferguson, A., Petit, C. S., Froehlich, F., Qian, S., Ky, J., Angarola, B., et al. (2012). The transcription factor TFEB links mTORC1 signaling to transcriptional control of lysosome homeostasis. *Sci. Signal* 5, ra42. doi: 10.1126/scisignal.2002790

Rolando, M., Escoll, P., Nora, T., Botti, J., Boitez, V., Bedia, C., et al. (2016). Legionella pneumophila S1P-lyase targets host sphingolipid metabolism and restrains autophagy. *Proc. Natl. Acad. Sci. U.S.A.* 113, 1901–1906. doi: 10.1073/pnas.1522067113

Romano, P. S., Gutierrez, M. G., Beron, W., Rabinovitch, M., and Colombo, M. I. (2007). The autophagic pathway is actively modulated by phase II Coxiella burnetii to efficiently replicate in the host cell. *Cell Microbiol.* 9, 891–909. doi: 10.1111/j.1462-5822.2006.00838.x

Rowbotham, T. J. (1980). Preliminary report on the pathogenicity of Legionella pneumophila for freshwater and soil amoebae. *J. Clin. Pathol.* 33, 1179–1183. doi: 10.1136/jcp.33.12.1179

Roy, C. R., Berger, K. H., and Isberg, R. R. (1998). Legionella pneumophila DotA protein is required for early phagosome trafficking decisions that occur within minutes of bacterial uptake. *Mol. Microbiol.* 28, 663–674. doi: 10.1046/j.1365-2958.1998.00841.x

Russell, R. C., Tian, Y., Yuan, H., Park, H. W., Chang, Y. Y., Kim, J., et al. (2013). ULK1 induces autophagy by phosphorylating Beclin-1 and activating VPS34 lipid kinase. *Nat. Cell Biol.* 15, 741–750. doi: 10.1038/ncb2757

Samanta, D., Clemente, T. M., Schuler, B. E., and Gilk, S. D. (2019). Coxiella burnetii Type 4B Secretion System-dependent manipulation of endolysosomal maturation is required for bacterial growth. *PloS Pathog.* 15, e1007855. doi: 10.1371/journal.ppat.1007855

Segal, G., Purcell, M., and Shuman, H. A. (1998). Host cell killing and bacterial conjugation require overlapping sets of genes within a 22-kb region of the *Legionella pneumophila* genome. *Proc. Natl. Acad. Sci. U.S.A.* 95, 1669–1674. doi: 10.1073/pnas.95.4.1669

Settembre, C., Zoncu, R., Medina, D. L., Vetrini, F., Erdin, S., Huynh, T., et al. (2012). A lysosome-to-nucleus signalling mechanism senses and regulates the lysosome via mTOR and TFEB. *EMBO J.* 31, 1095–1108. doi: 10.1038/emboj.2012.32

Sharma, V., Verma, S., Seranova, E., Sarkar, S., and Kumar, D. (2018). Selective Autophagy and Xenophagy in Infection and Disease. *Front. Cell Dev. Biol.* 6, 147. doi: 10.3389/fcell.2018.00147

Shen, Q., Shi, Y., Liu, J., Su, H., Huang, J., Zhang, Y., et al. (2020). Acetylation of STX17 (syntaxin 17) controls autophagosome maturation. *Autophagy*, 1–13. doi: 10.1080/15548627.2020.1752471

Siadous, F. A., Cantet, F., Van Schaik, E., Burette, M., Allombert, J., Lakhani, A., et al. (2020). Coxiella effector protein CvpF subverts RAB26-dependent autophagy to promote vacuole biogenesis and virulence. *Autophagy*, 1–17. doi: 10.1080/15548627.2020.1728098

Siqueira, M. D. S., Ribeiro, R. M., and Travassos, L. H. (2018). Autophagy and Its Interaction With Intracellular Bacterial Pathogens. *Front. Immunol.* 9, 935. doi: 10.3389/fimmu.2018.00935

Steegmaier, M., Oorschot, V., Klumperman, J., and Scheller, R. H. (2000). Syntaxin 17 is abundant in steroidogenic cells and implicated in smooth endoplasmic reticulum membrane dynamics. *Mol. Biol. Cell* 11, 2719–2731. doi: 10.1091/mbc.11.8.2719

Sudhakar, P., Jacomin, A. C., Hautefort, I., Samavedam, S., Fatemian, K., Ari, E., et al. (2019). Targeted interplay between bacterial pathogens and host autophagy. *Autophagy* 15, 1620–1633. doi: 10.1080/15548627.2019.1590519

Swanson, M. S., and Isberg, R. R. (1995). Association of Legionella pneumophila with the macrophage endoplasmic reticulum. *Infect. Immun.* 63, 3609–3620. doi: 10.1128/IAI.63.9.3609-3620.1995

Takabe, K., Paugh, S. W., Milstien, S., and Spiegel, S. (2008). "Inside-out" signaling of sphingosine-1-phosphate: therapeutic targets. *Pharmacol. Rev.* 60, 181–195. doi: 10.1124/pr.107.07113

Tan, Y., and Luo, Z. Q. (2011). Legionella pneumophila SidD is a deAMPylase that modifies Rab1. *Nature* 475, 506–509. doi: 10.1038/nature10307

Tan, Y., Arnold, R. J., and Luo, Z. Q. (2011). Legionella pneumophila regulates the small GTPase Rab1 activity by reversible phosphorylcholination. *Proc. Natl. Acad. Sci. U.S.A.* 108, 21212–21217. doi: 10.1073/pnas.1114023109

Vogel, J. P., Andrews, H. L., Wong, S. K., and Isberg, R. R. (1998). Conjugative transfer by the virulence system of *Legionella pneumophila*. *Science* 279, 873–876. doi: 10.1126/science.279.5352.873

Voth, D. E., Beare, P. A., Howe, D., Sharma, U. M., Samoilis, G., Cockrell, D. C., et al. (2011). The Coxiella burnetii cryptic plasmid is enriched in genes encoding type IV secretion system substrates. *J. Bacteriol.* 193, 1493–1503. doi: 10.1128/JB.01359-10

Wachter, S., Bonazzi, M., Shifflett, K., Moses, A. S., Raghavan, R., and Minnick, M. F. (2019). A CsrA-Binding, trans-Acting sRNA of Coxiella burnetii Is Necessary for Optimal Intracellular Growth and Vacuole Formation during Early Infection of Host Cells. *J. Bacteriol.* 201, e00524-19. doi: 10.1128/JB.00524-19

Wang, Z., Mccloskey, A., Cheng, S., Wu, M., Xue, C., Yu, Z., et al. (2018). Regulation of the small GTPase Rab1 function by a bacterial glucosyltransferase. *Cell Discovery* 4, 53. doi: 10.1038/s41421-018-0055-9

Wang, L., Ye, X., and Zhao, T. (2019). The physiological roles of autophagy in the mammalian life cycle. *Biol. Rev. Camb. Philos. Soc.* 94, 503–516. doi: 10.1111/brv.12464

Weber, M. M., Chen, C., Rowin, K., Mertens, K., Galvan, G., Zhi, H., et al. (2013). Identification of Coxiella burnetii Type IV Secretion Substrates Required for Intracellular Replication and Coxiella-Containing Vacuole Formation. *J. Bacteriol.* 195, 3914–3924. doi: 10.1128/JB.00071-13

Wiater, L. A., Barbieri, J. T., Dunn, K., Maxfield, F. R., and Shuman, H. A. (1998). Early Events in Phagosome Establishment Are Required for Intracellular Survival of Legionella pneumophila. *Infect. Immun.* 66, 4450–4460. doi: 10.1128/IAI.66.9.4450-4460.1998

Winchell, C. G., Graham, J. G., Kurten, R. C., and Voth, D. E. (2014). Coxiella burnetii type IV secretion-dependent recruitment of macrophage autophagosomes. *Infect. Immun.* 82, 2229–2238. doi: 10.1128/IAI.01236-13

Wu, Y. W., and Li, F. (2019). Bacterial interaction with host autophagy. *Virulence* 10, 352–362. doi: 10.1080/21505594.2019.1602020

Xian, H., Yang, Q., Xiao, L., Shen, H. M., and Liou, Y. C. (2019). STX17 dynamically regulated by Fis1 induces mitophagy via hierarchical macroautophagic mechanism. *Nat. Commun.* 10, 2059. doi: 10.1038/s41467-019-10096-1

Xiong, Q., Yang, M., Li, P., and Wu, C. (2019). Bacteria Exploit Autophagy For Their Own Benefit. *Infect. Drug Resist.* 12, 3205–3215. doi: 10.2147/IDR.S220376

Yang, A., Pantoom, S., and Wu, Y. W. (2017). Elucidation of the anti-autophagy mechanism of the Legionella effector RavZ using semisynthetic LC3 proteins. *Elife* 6, e23905. doi: 10.7554/eLife.23905

Yin, Z., Pascual, C., and Klionsky, D. (2016). Autophagy: machinery and regulation. *Microbial Cell* 3, 588–596. doi: 10.15698/mic2016.12.546

Yu, Z. Q., Ni, T., Hong, B., Wang, H. Y., Jiang, F. J., Zou, S., et al. (2012). Dual roles of Atg8-PE deconjugation by Atg4 in autophagy. *Autophagy* 8, 883–892. doi: 10.4161/auto.19652

Yu, L., Chen, Y., and Tooze, S. A. (2018). Autophagy pathway: Cellular and molecular mechanisms. *Autophagy* 14, 207–215. doi: 10.1080/15548627.2017.1378838

Yun, C. W., and Lee, S. H. (2018). The Roles of Autophagy in Cancer. *Int. J. Mol. Sci.* 19, 3466. doi: 10.3390/ijms19113466

7

Intracellular *Porphyromonas gingivalis* Promotes the Proliferation of Colorectal Cancer Cells *via* the MAPK/ERK Signaling Pathway

Wenxin Mu[1†], Yiqun Jia[2†], Xiaobing Chen[1], Haoyu Li[1], Zhi Wang[1] and Bin Cheng[1*]*

[1] Hospital of Stomatology, Guanghua School of Stomatology, Guangdong Provincial Key Laboratory of Stomatology, Sun Yat-Sen University, Guangzhou, China, [2] Stomatology Center, Shenzhen People's Hospital, The Second Clinical Medical College of Jinan University, The First Affiliated Hospital of Southern University of Science and Technology, Shenzhen, China

***Correspondence:**
Bin Cheng
chengbin@mail.sysu.edu.cn
Zhi Wang
wangzh75@mail.sysu.edu.cn

†These authors have contributed equally to this work

Porphyromonas gingivalis (*P. gingivalis*) is a keystone pathogen in periodontitis. However, several clinical studies have revealed an enrichment of *P. gingivalis* in the stool samples and colorectal mucosa of colorectal cancer patients. Thus, the goal of this study was to determine whether *P. gingivalis* can promote colorectal cancer progression *in vitro*. We established an acute infection model (24 h, multiplicity of infection =100) of *P. gingivalis* invasion of colorectal cancer cells to study the alterations induced by *P. gingivalis* in the proliferation and cell cycle of colorectal cancer cells. We observed that *P. gingivalis* can adhere and invade host cells a few hours after infection. Once invaded, *P. gingivalis* significantly promoted colorectal cancer cell proliferation, and the percentage of S phase cells was increased in the cell cycle assay. However, KDP136, a gingipain-deficient mutant of *P. gingivalis* 33277, showed a decreased ability to promote colorectal cancer cell proliferation, indicating that gingipain is associated with colorectal cancer cell proliferation. Furthermore, we extracted RNA from colorectal cancer cells for high-throughput sequencing analysis and reconfirmed the results by quantitative polymerase chain reaction and western blot analyses. The results suggested that the MAPK/ERK signaling pathway is significantly activated by *P. gingivalis*, while these changes were not observed for KDP136. In conclusion, *P. gingivalis* can invade cells and promote the proliferation of colorectal cancer cells by activating the MAPK/ERK signaling pathway. Gingipain is an essential virulence factor in this interaction.

Keywords: gingipain cysteine endopeptidases, cell cycle, cell proliferation, colorectal neoplasms, *Porphyromonas gingivalis*

INTRODUCTION

The oral microbiota is one of the most complex human microbiomes, second only to that of the gastrointestinal tract, containing 26% of the bacterial species associated with the human body (Group et al., 2009). Furthermore, the results of a recent clinical trial demonstrated that the vast majority of oral microbial species can be transmitted from the oral cavity to the large intestine (Schmidt et al., 2019). Oral bacteria are closely associated with many oral diseases and systemic diseases outside the oral cavity. As the most common opportunistic pathogen in periodontal diseases, Fusobacterium nucleatum (F. nucleatum) is associated with oral squamous cell carcinoma (OSCC) (Al-Hebshi et al., 2017), pregnancy complications (Han et al., 2010), and colorectal cancer (CRC) (Rubinstein et al., 2013). Interestingly, a mixed infection of F. nucleatum and *Porphyromonas gingivalis* (*P. gingivalis*) has been shown to be much more effective than mono-infection in experimental periodontitis (Polak et al., 2009). In addition, *P. gingivalis*, a major pathogen of periodontitis, is also associated with OSCC (Geng et al., 2017; Lafuente Ibanez de Mendoza et al., 2020; Wen et al., 2020), esophageal squamous cell carcinoma (Gao et al., 2016), pancreatic cancer (Michaud et al., 2013), cardiovascular disease (Gibson et al., 2004) and rheumatoid arthritis (Wegner et al., 2010). In the mucosa-adherent and fecal microbiota, *Porphyromonas* has been shown to be enriched in CRC patients (Chen et al., 2012; Ahn et al., 2013; Wu et al., 2013; Zackular et al., 2014).

The virulence factors of *P. gingivalis* include fimbriae, hemagglutinin, capsule, lipopolysaccharide and gingipain. Specially, gingipain plays an essential role in the pathogenicity of the organism in periodontal disease. As a family of unique cysteine endopeptidases, gingipain are abundantly expressed and located on the outer membranes of *P. gingivalis* or secreted into the extracellular milieu (Pike et al., 1994). The gingipain family consists of two types of arginine-specific protease (Rgp; encoded by *rgpA* and *rgpB*) and a lysine-specific protease (Kgp; encoded by *kgp*). Among them, RgpB has been the focus of structural studies aimed at elucidating post-translational processing and maturation of these enzymes because of its simple structure (Eichinger et al., 1999; Nguyen et al., 2007). Gingipain can provide a general proteolytic tool for the degradation of proteinaceous nutrients to *P. gingivalis* for growth. Besides, gingipain had also been proven to be essential in the processing of fimbrial proteins to facilitate bacterial adhesion to the host tissues (Njoroge et al., 1997; Weinberg et al., 1997). Gingipain can also enable bacterial evasion of the host immune response by surface receptor cleavage and cytokine degradation (Brien-Simpson et al., 2003). Previous studies showed that gingipain can activate the ERK1/2-Ets1, p38/HSP27, and PAR2/NFκB pathways to promote cellular invasion and metastasis in OSCC cells (Inaba et al., 2014).

Consequently, we hypothesized that *P. gingivalis* is probably associated with CRC progression and that gingipain is a keystone virulence factor in this process. To test this hypothesis, in this study, we used an acute *in vitro* model of *P. gingivalis* infection of CRC cells.

Abbreviations: F. nucleatum, Fusobacterium nucleatum; OSCC, oral squamous cell carcinoma; CRC, colorectal cancer; *P. gingivalis*, *Porphyromonas gingivalis*; Rgp, arginine-specific protease; Kgp, lysine-specific protease; CFU, colony forming units; MOI, multiplicity of infection; FITC, fluorescein isothiocyanate; TEM, transmission electron microscope; PI, propidium iodide; CCK8, cell counting kit-8; AP1, activator protein-1; U/L, unit/L; KEGG, Kyoto Encyclopedia of Genes and Genomes; DEG, differentially expressed genes; WB, western blot; qPCR, quantitative polymerase chain reaction; proMMP-9, promatrix metalloproteinase-9; PAR2, proteinase-activated receptor 2.

MATERIALS AND METHODS

Bacteria and Cell Culture

The bacterial strains, P. gingivalis ATCC 33277, P. gingivalis W83 and F. nucleatum 25586 were purchased from ATCC. P. gingivalis KDP136 (ΔrgpAΔrgpBΔkgp), a gingipain-deficient mutant of *P. gingivalis* 33277, were kindly provided by Dr. Jinlong Gao from Faculty of Medicine and Health, the University of Sydney. P. gingivalis were grown in BHI broth supplemented with yeast extract (5 mg/ml), cysteine (1 mg/ml), vitamin K1 (0.5 μg/ml) and hemin (5 μg/ml) in the anaerobic chamber (oxygen concentration < 1%). Human CRC cell line S1 (a clone of LS174T cells) and murine colon cancer MC38 cells were purchased from ATCC. The cells were cultured in DMEM medium (Thermo Fisher Scientific Inc., MA, USA) supplemented with 10% fetal bovine serum (FBS) at 37°C in 5%CO_2.

Cell Adhesion Assay

To detect the adhesive ability of P. gingivalis, immunofluorescence microscopy and flow cytometry were used. Cells were infected with P. gingivalis at a MOI of 100 for 6 h incubation. Then infected cells were washed with PBS three times and fixed with 4% paraformaldehyde for 30 min at room temperature. The primary antibody against RgpB (a kind gift from Jinlong Gao, 1:200) were added to cells at 4℃ overnight. Cells were incubated with Alexa Fluor 488 Goat anti-Mouse IgG(H+L) (EMAR, Beijing, China, 1:100) for 1 h at room temperature and photographed by fluorescence microscope (Zeiss Axio observer Z1). Nuclei were stained with DAPI (Solarbio, Beijing, China, 1:100) for 5 min.

P. gingivalis were incubated with FITC (0.1mg/ml) for 30 min at room temperature, followed by washing with PBS three times. Cells were infected with the P. gingivalis labeled with FITC (MOI=100) at 4°C for 30 min, then washed with PBS three times to remove the P. gingivalis in supernatant. The cells were harvested by trypsinization and processed by flow cytometry (Beckman Coulter Cytoflex).

Cell Invasion Assay

S1 and MC38 cells (6×10^5) were seeded in 6-well plates and infected with P. gingivalis (MOI=100) for 24 h. Cells were washed three times with PBS and harvested by trypsinization, then fixed with 2.5% glutaraldehyde in 0.1 M PBS (PH=7.4) at 4 °C overnight. Thin sections were cut and stained with uranyl acetate-lead citrate. Followed by observing in transmission electron microscope (TEM, H7650 Hitachi, Japan). Three fields

(1000 ×, 30–50 CRC cells per fields) were selected randomly and the number of CRC cells with P. gingivalis invasion and without P. gingivalis were counted separately.

The invasion ability of P. gingivalis toward CRC cells were also measured by an antibiotic protection assay (Lamont et al., 1995; Yilmaz et al., 2002). Cells were plated in 6-well plates at a density of 3×10^5 cells/well in complete medium for 12 h, following incubating with P. gingivalis (MOI=100) for 24 h. Remaining external bacteria were killed with gentamicin sulfate (Solarbio, Beijing, China, 1 mg/ml) for 90 min. Cells were washed with PBS three times and lysed with sterile distilled water for 30 min. Internal bacteria were released and plated on blood agar supplemented with hemin and menadione and cultured anaerobically. Colony forming units (CFU) of invasive P. gingivalis were then enumerated and invasion efficiency (Invasion efficiency (%) = CFU of *P. gingivalis* inside CRC cells/CFU of *P. gingivalis* in initial inoculum) were expressed to assess the invasive ability of P. gingivalis (Arjunan et al., 2020).

Cell Counting Kit-8 Assay

Cell proliferation was determined using cell counting kit-8 assay (Telenbiotech, Guangzhou, China). S1 and MC38 cells (3000 cells per well, 100μL culture medium) were seeded in 96-well plates. After 8 h incubation, the cells were infected with P. gingivalis at a MOI of 100 for 24 h. The growth medium in each well was removed, and then filled with cck-8 solutions. After incubated for 1 h at 37℃, the optical density value was detected at 450nm with a versatile microplate absorbance reader (Tecan sunrise, Untersbergstrasse, Austria).

We detect the cell viability of colorectal cancer cells pretreated with *P. gingivalis* 33277 (MOI=100) for 0-96 h. the results showed that *P. gingivalis* can promote the proliferation of CRC cells in a time dependent manner in 0–24 h, and the cell viability of CRC cells reached a plateau in 24–96 h incubation. So the acute infection model in this study was designed as 24 h cocultivation (**Supplementary Figure 1**).

The exogenous gingipains were kindly provided by Prof. Min Liang (Guanghua School of Stomatology, Sun Yat-Sen University), they exogenous it from P. gingivalis W83 strain as described in previous report (Sheets et al., 2005; Zhang et al., 2017; Mo et al., 2020).

Colony Formation Assay

S1 and MC38 were infected with P. gingivalis (MOI=100) for 24 h. Cells were then seeded in 6-well plates (500 cells per well) and cultured for 14 days. Cells were fixed with 4% paraformaldehyde for 30 min at room temperature and stained for 5 min with 0.1%(w/v) crystal violet for 20 min. The number of eukaryotic cell colonies were counted using image processing software (ImageJ 1.48v).

Cell Cycle Analysis

Cells were harvested and fixed with 70% ice-cold ethanol at 4°C for 16–18 h. Cells were washed with cold PBS then re-suspended with mixed medium (PI: 50ug/ml, RNase: 100ug/ml) for incubation at 37°C; for 30 min. The percentage of the cells in each cell cycle phases were detected by a flow cytometer (ModFit LT 4.0).

RNA Extraction and Quantitative Real-time PCR

Cells were infected with *P. gingivalis* at a MOI of 100. Cells were washed by PBS and harvested at 3, 6, 12 and 24 h. Total RNAs were extracted using Trizol reagent (Thermo Fisher Scientific Inc, MA, USA). The concentration and quality were measured using a Nanodrop (Thermo Fisher Scientific Inc, MA, USA). The RNAs were then converted into cDNA using primerscript RT-polymerase (Takara, Shanghai, China). The quantification of selected genes was measured using LightCycler96 (Roche, Shanghai, China) and the data was analyzed using the $2^{-\Delta\Delta Ct}$ method. The primers of KRAS, BRAF, MEK2, ERK2, C-Fos, AP1, and GAPDH (Tsingke, Guangzhou, China) are all listed below in **Supplementary Table 1**. GAPDH was used as an endogenous control and the mRNA levels were normalized to GAPDH.

Western Blot and Antibodies

The total protein was extracted using RIPA Lysis Buffer (Beyotime Biotechnology, Shanghai, China) containing protease and phosphatase inhibitor cocktail (CWBIO, Beijing, China). BCA Protein Assay Kit (BIOTEKE, Beijing, China) was used to quantify the protein. Extracted protein were separated with 8~12% SDS-polyacrylamide gel electrophoresis and transferred to 0.2 μm PVDF Membrane (Roche, Shanghai, China). The membrane then blocked with 5%BSA and incubated with primary antibodies against p-MEK1/2 (CST), t-MEK1/2 (CST), p-ERK1/2 (CST), t-ERK1/2 (CST), p-Akt (308) (CST), p-Akt (473) (CST), Akt (CST), and GAPDH (CST) at 4℃ overnight. Following washing with TBST three times and exposing to secondary antibodies (CST) for 1 h at room temperature. The blots were visualized using an imaging system (BIORAD ChemiDoc Touch) and quantified by an image processing software (ImageJ 1.48v). GAPDH was used as an endogenous control and the protein levels were normalized to GAPDH.

Statistical Analysis

All experiments were repeated at least three times independently and the results were presented as mean ± standard error of mean (SEM). In **Figures 1C, D** and **3B, C, F, H**, comparison between two groups was carried out with student's t-test. In **Figures 2A, C, 3D**, and **5A–D**, **Supplementary Figures 4B, 2A, B, 3B** and **5A, B**, comparison among three or more groups was determined using one-way ANOVA. All data analyses were performed using IBM SPSS Statistics 20 (SPSS Inc. Chicago, IL) and GraphPad Prism 6 (GraphPad Software Inc. la Jolla, CA).

RESULTS

P. gingivalis Can Adhere to and Invade CRC Cells

To assess the adhesive and invasive capabilities of *P. gingivalis*, we established an acute infection model *in vitro*. Two CRC

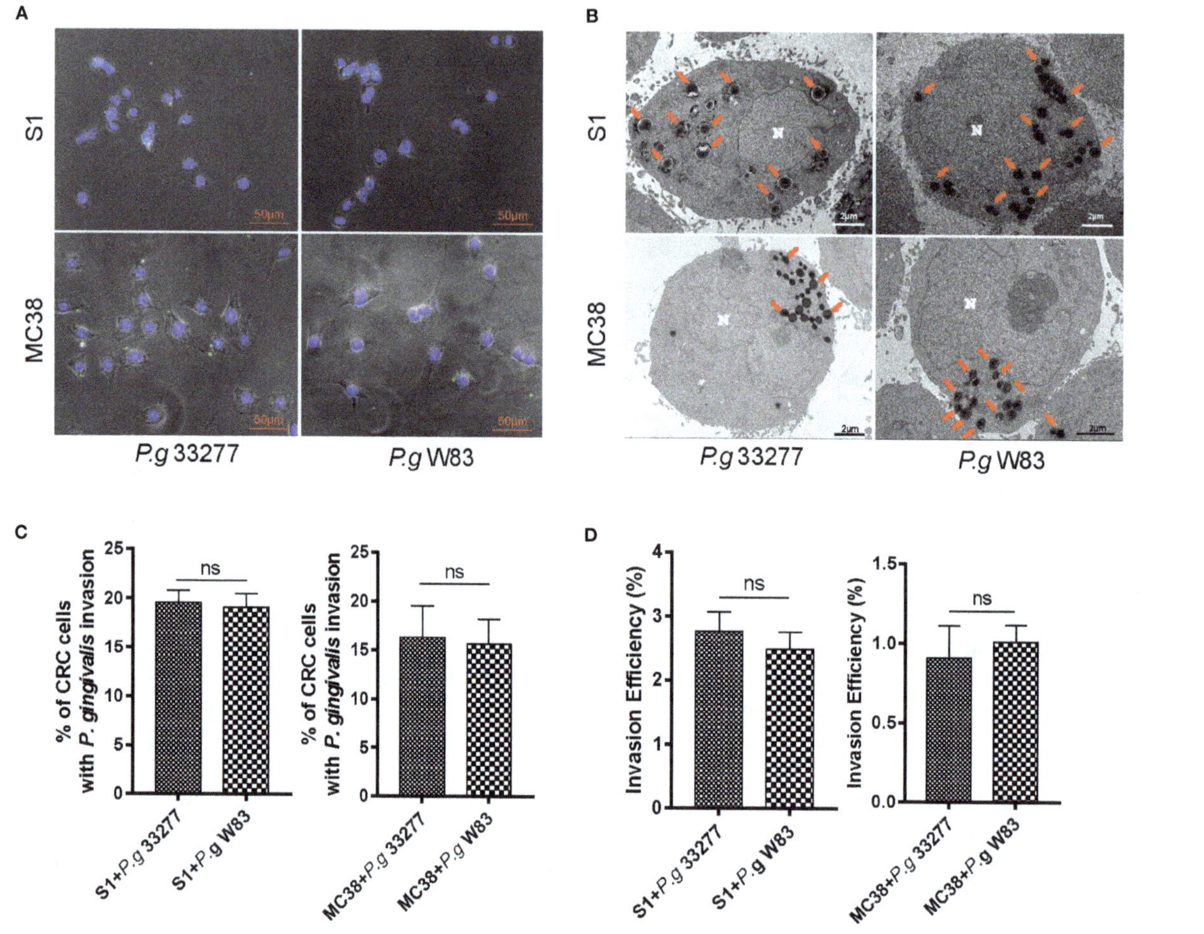

FIGURE 1 | *P. gingivalis* exhibits high adhesion and invasive abilities toward colorectal cancer (CRC) cells. **(A)** Immunofluorescence micrographs show that *P. gingivalis* (green) can adhere to S1 and MC38 cells. **(B)** -Transmission electron microscopy (TEM) results indicate that *P. gingivalis* can intracellularly invade S1 and MC38 cells. **(C)** Percentage of CRC cells with *P. gingivalis* invasion were counted in 3 random fields (1000 ×, 30–50 CRC cells per fields)) by TEM and there is no significant difference between *P. gingivalis* 33277 and *P. gingivalis* W83. **(D)** Antibiotic protection assay shows that *P. gingivalis* can invade S1 and MC38 cells and survive inside the cell at 24 h after infection. The invasion efficiency of *P. gingivalis* 33277 and *P. gingivalis* W83 showed no significant difference. Invasion efficiency (%) = CFU of *P. gingivalis* inside CRC cells/CFU of *P. gingivalis* in initial inoculum. CFU, colony forming units. Red arrows, *P. gingivalis* infection in the cytoplasm. N, nucleus. ns, nonsignificant.

cell lines, MC38 and S1, were infected with *P. gingivalis* at a multiplicity of infection (MOI) of 100 for 6 h. Immunofluorescence tests of MC38 and S1 cells revealed colocalization of *P. gingivalis* marked by Alexa Fluor 488-labeled primary antibodies (**Figure 1A**).

To further verify the subcellular localization of internalized *P. gingivalis* and quantify the frequencies of invasion, we observed the changes in *P. gingivalis*-infected CRC cells by transmission electron microscopy (TEM). *P. gingivalis* appeared as electron-dense objects (0.3 to 0.5-μm in diameter) surrounded by an outer membrane in the cytoplasm of CRC cells (**Figure 1B**). The percentage of CRC cells infected with P. gingivalis 33277 and W83 showed no significant difference in CRC cells (**Figure 1C**). It is shown that around 19.3% S1 cells (S1+P. g 33277: 19.6%; S1 +P. g W83: 19.1%) and 16.0% MC38 cells (MC38+P. g 33277: 16.3%; MC38+P. g W83: 15.6%) were infected with P. gingivalis after 24 h cocultivation. But the percentage will be higher in fact because ultrathin sections (100 nm) cannot present the whole picture of cells.

Antibiotic protection assay was carried to determine the number of *P. gingivalis* survive successfully inside the cell. The results showed that *P. gingivalis* 33277 and W83 can survive inside the cell after 24 h infection and the invasion efficiency of *P. gingivalis* 33277 and W83 showed no significantly difference (**Figure 1D**). There are 2.6% *P. gingivalis* survive successfully in S1 cells after 24 h cocultivation and it is 1.0% in MC38 cells. The invasion efficiency of *P. gingivalis* showed a wide variation in different CRC cell lines.

P. gingivalis Promotes CRC Cell Proliferation and Increases the Percentage of CRC Cells in S Phase *In Vitro*

We performed a cell counting kit-8 (CCK8) assay and found changes in the proliferation of MC38 and S1 cells following

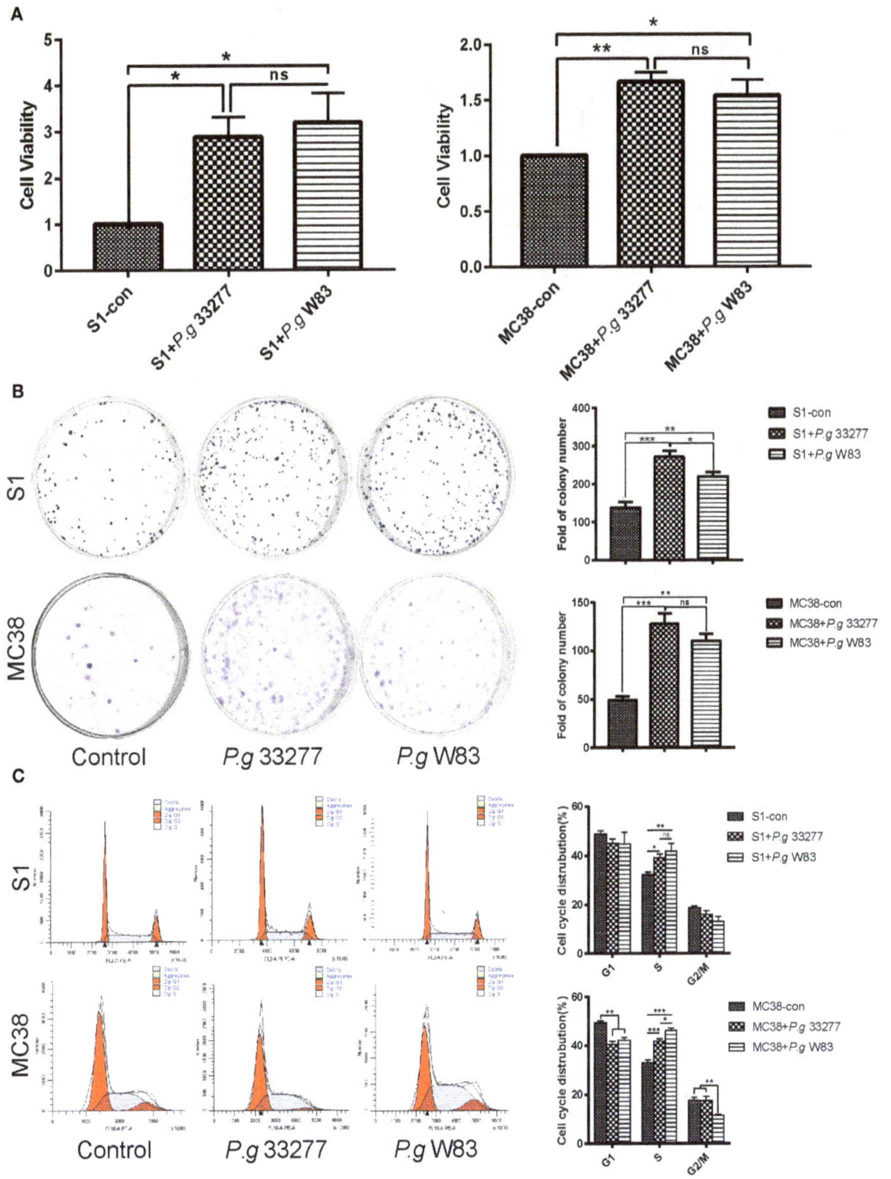

FIGURE 2 | Cell proliferation and cell cycle analyses of CRC cells pretreated with *P. gingivalis*. **(A)** CCK8 assay examination of the proliferation of S1 and MC38 cells pretreated with *P. gingivalis* 33277 and W83 for 24 h. **(B)** Colony formation assay examination of the proliferation of S1 and MC38 cells pretreated with *P. gingivalis* 33277 and W83. **(C)** Cell cycle analysis of S1 and MC38 cells pretreated with *P. gingivalis* 33277 and W83 detected by flow cytometry. The results showed that the fraction of S phase cells was significantly increased. $^{*}P < 0.05$, $^{**}P < 0.01$, $^{***}P < 0.001$. ns, nonsignificant.

infection with *P. gingivalis* 33277 and W83. The results indicated that *P. gingivalis* can significantly promote S1 and MC38 cell proliferation ($P < 0.05$), and no significant difference this activity was observed between the two *P. gingivalis* strains (**Figure 2A**).

Colony formation assays were performed to investigate the proliferative abilities of the CRC cell lines. Interestingly, S1 and MC38 cells co-cultured with *P. gingivalis* formed significantly more total colonies compared with that observed when CRC cells were cultured alone ($P < 0.05$), while no significant difference was observed among the experimental groups in MC38 cells (**Figure 2B**).

The S phase fraction is an important measure of cell proliferative activity. To determine whether *P. gingivalis* specifically associated with CRC cells of a specific cell cycle phase, cells were stained with propidium iodide, and the cell cycle was analyzed by flow cytometry. The ratio of S1 and MC38 cells in S phase was notably higher in the *P. gingivalis* infected groups than that observed in the uninfected control group ($P < 0.05$). No significant difference was observed between the percentage of S phase cells in the *P. gingivalis* 33277 and *P. gingivalis* W83 groups ($P > 0.05$) (**Figure 2C**). Remarkably, *P. gingivalis* 33277 and W83 could promote CRC cells proliferation and increase the percentage of CRC cells in S phase.

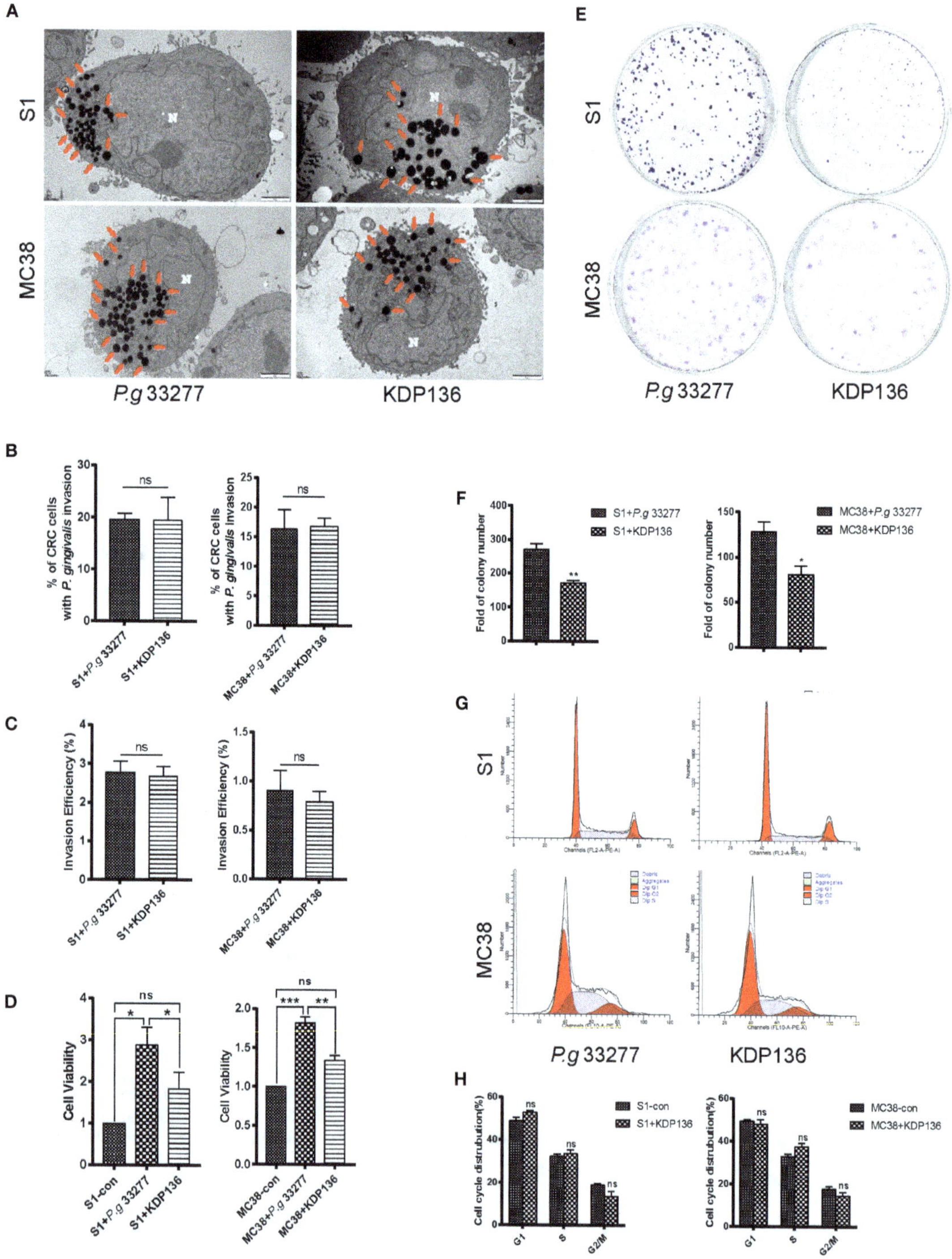

FIGURE 3 | The invasion ability and pro-proliferation ability of KDP136 to S1 and MC38 cells. **(A, B)** TEM results indicate that *P. gingivalis* 33277 and KDP136 can intracellularly invade S1 and MC38 cells and the percentage of CRC cells with *P. gingivalis* invasion showed no significant difference between *P. gingivalis* 33277 and KDP136 groups. **(C)** Antibiotic protection assay shows that P. gingivalis 33277 and KDP136 can survive inside the cell at 24 h after infection and the invasion efficiency of P. gingivalis 33277 and KDP136 showed no significant difference. **(D)** CCK8 assay examination of the proliferation of S1 and MC38 cells pretreated with *P. gingivalis* 33277 and KDP136 for 24 h. **(E, F)** Colony formation assay examination of the proliferation of S1 and MC38 cells pretreated with *P. gingivalis* 33277 and KDP136. **(G, H)** Cell cycle analysis of S1 and MC38 cells pretreated with *P. gingivalis* 33277 and KDP136 detected by flow cytometry. Invasion efficiency (%) = CFU of *P. gingivalis* inside CRC cells/CFU of *P. gingivalis* in initial inoculum. CFU, colony forming units. Red arrows, *P. gingivalis* infection in the cytoplasm. N, nucleus. *$P < 0.05$, **$P < 0.01$, ***$P < 0.001$. ns, nonsignificant. KDP136, a gingipain-deficient mutant of *P. gingivalis* 33277.

P. gingivalis Gingipains Play an Important Role in Promoting the Proliferation of CRC Cells

To assess the effect of gingipains on CRC cell lines, we used the gingipain-deficient mutant KDP136 (*ΔrgpAΔrgpBΔkgp*). The invasion ability of KDP136 was assessed by TEM, and the results indicate that KDP136 can intracellularly invade S1 and MC38 cells and the percentage of CRC cells infected with *P. gingivalis* showed no significant difference between *P. gingivalis* 33277 and KDP136 groups (**Figures 3A, B**). To further quantify the invasion ability of *P. gingivalis*, we performed antibiotic protection assay. The results showed that KDP136 can survive successfully inside the S1 and MC38 cells and the invasive efficiency of KDP136 and *P. gingivalis* 33277 showed no significant difference after infection of 24 h (**Figure 3C**).

For CCK8 assay, the cell viability of the KDP136 group showed no significant differences with that observed in the control group, whereas a significant ($P<0.05$) difference was observed between the KDP136 and *P. gingivalis* 33277 groups (**Figure 3D**). To further verify the role of gingipain, we added exogenous gingipains with different concentrations of 1, 2, 3, 4 and 5 U/L. Exogenous gingipains were added at the beginning of infection. We found that the cell vitality of KDP136 group with exogenous gingipains (5 U/L) is significantly higher than control group (**Supplementary Figure 2**).

For the colony formation assay, KDP136 had a lower ability to promote S1 and MC38 cell proliferation compared with *P. gingivalis* 33277 (**Figures 3E, F**). The percentage of S phase cells in the KDP136 group was significantly lower than that observed in the *P. gingivalis* 33277 group (**Figures 3G, H**). KDP136 can invade CRC cells and survive successfully inside cytoplasm after 24 h cocultivation and the invasive ability of KDP136 are similar to *P. gingivalis* 33277. But as a gingipain-deficient mutant strain of *P. gingivalis* 33277, KDP136 has less of an ability to enhance cell viability and stimulate the cell cycle of CRC cells. Thus, gingipain plays an important role in promoting the proliferation of CRC cells.

P. gingivalis Can Potentially Promote Pathways Associated With Cell Proliferation

RNA was extracted from MC38 cells infected with *P. gingivalis* 33277 (24 h, MOI=100) for RNA-Seq analysis, the results of which were used to perform functional enrichment analysis, including Gene Ontology and Kyoto Encyclopedia of Genes and Genomes (KEGG) analyses. The results showed that there were 3292 differentially expressed genes (DEGs) between the *P. gingivalis* 33277 and control groups, of which 1893 were upregulated and 1399 were downregulated (**Figure 4A**). To determine the functions of the identified DEGs, these genes were analyzed using the Gene Ontology database, resulting in the identification of three ontologies, namely, biological process, cellular component and molecular function (**Figure 4B**). We further analyzed the significantly enriched pathways using the KEGG database. When comparing the *P. gingivalis* 33277 and control groups, 51 pathways were upregulated and 44 pathways were downregulated (**Figure 4C**), and the top 20 pathways are shown in **Figure 4D**. Included among these pathways were proliferation related pathways, including the PI3K-Akt signaling and MAPK signaling pathways (Fang and Richardson, 2005; Schmidt et al., 2020). Moreover, as shown in **Figure 4E**, we compared DEGs associated with the PI3K-Akt and MAPK signaling pathways between *P. gingivalis* 33277 and the control groups (the Log2 Fold Change" and "Adjusted P-Value" information are shown in **Supplementary Table 2**). In addition, compared to *P. gingivalis* 33277, KDP136 lost the ability to activate MAPK pathway-related genes (**Figure 4F**, the Log2 Fold Change" and "Adjusted P-Value" information are shown in **Supplementary Table 3**). Thus, we observed that the PI3K-Akt and MAPK signaling pathways are potentially involved in the interaction between *P. gingivalis* and CRC cells.

P. gingivalis Promotes CRC Cell Proliferation *via* Activation of the MAPK/ERK Pathway

Overexpression and activation of the MAPK/ERK pathway plays an important role in the progression of CRC. Furthermore, activation of the RAS/RAF/MEK/ERK axis is crucial for the ability of the MAPK/ERK pathway to regulate various cellular responses, including the stimulation of C-fos and AP1 (Fang and Richardson, 2005). The RNA-Seq results showed that *Raf1*, *Braf*, and *Fos* levels were upregulated in the *P. gingivalis* 33277 group, whereas changes in the expression of *RAS*, *MEK*, *ERK* and *AP1* were unclear (**Figures 4E, F**). We extracted RNA from MC38 and S1 cells exposed to three *P. gingivalis* stains for quantitative polymerase chain reaction (qPCR) analysis (Gene count of all groups are shown in **Supplementary Material**). In general, our results showed that *KRAS*, *BRAF*, *MEK2*, *ERK2*, *C-fos* and *AP1* levels in the *P. gingivalis* 33277 and *P. gingivalis* W83 groups were higher than those observed in the control group after exposûre to bacteria for different amounts of time. Specifically, the expression levels of *KRAS* and *BRAF*, upstream components of the MAPK/ERK pathway, were upregulated at the early stage (3 and 6 h) of the infection. *MEK2* and *ERK2* levels were significantly higher at 6 and 12 h after infection. As downstream components of the MAPK/ERK pathway, *C-fos* and activator protein-1 (*AP1*) levels were upregulated at 12 and 24 h after infection (**Figures 5A, B**). To assess the protein level and phosphorylation at the early stage of infection, we isolated protein from S1 and MC38 cells after infection with *P. gingivalis* for 3 h. The expression of total MEK1/2 (t- MEK1/2), phospho-MEK1/2 (p-MEK1/2), total ERK1/2 (t- ERK1/2), and phospho-MEK1/2 (p-ERK1/2) was measured by western blot (WB). No significant difference was observed between the experimental and control groups in t- MEK1/2 and t- ERK1/2 expression. However, regarding the phosphorylation levels of these proteins, p-MEK/t-MEK and p-ERK/t-ERK levels were significantly increased (**Figures 5C, D**). Thus, the transcription of genes encoding components in the entire MAPK/ERK pathway was gradually activated by *P. gingivalis* during the first 24 h after infection, while the phosphorylation of components in this pathway was stimulated in the first 3 h after infection.

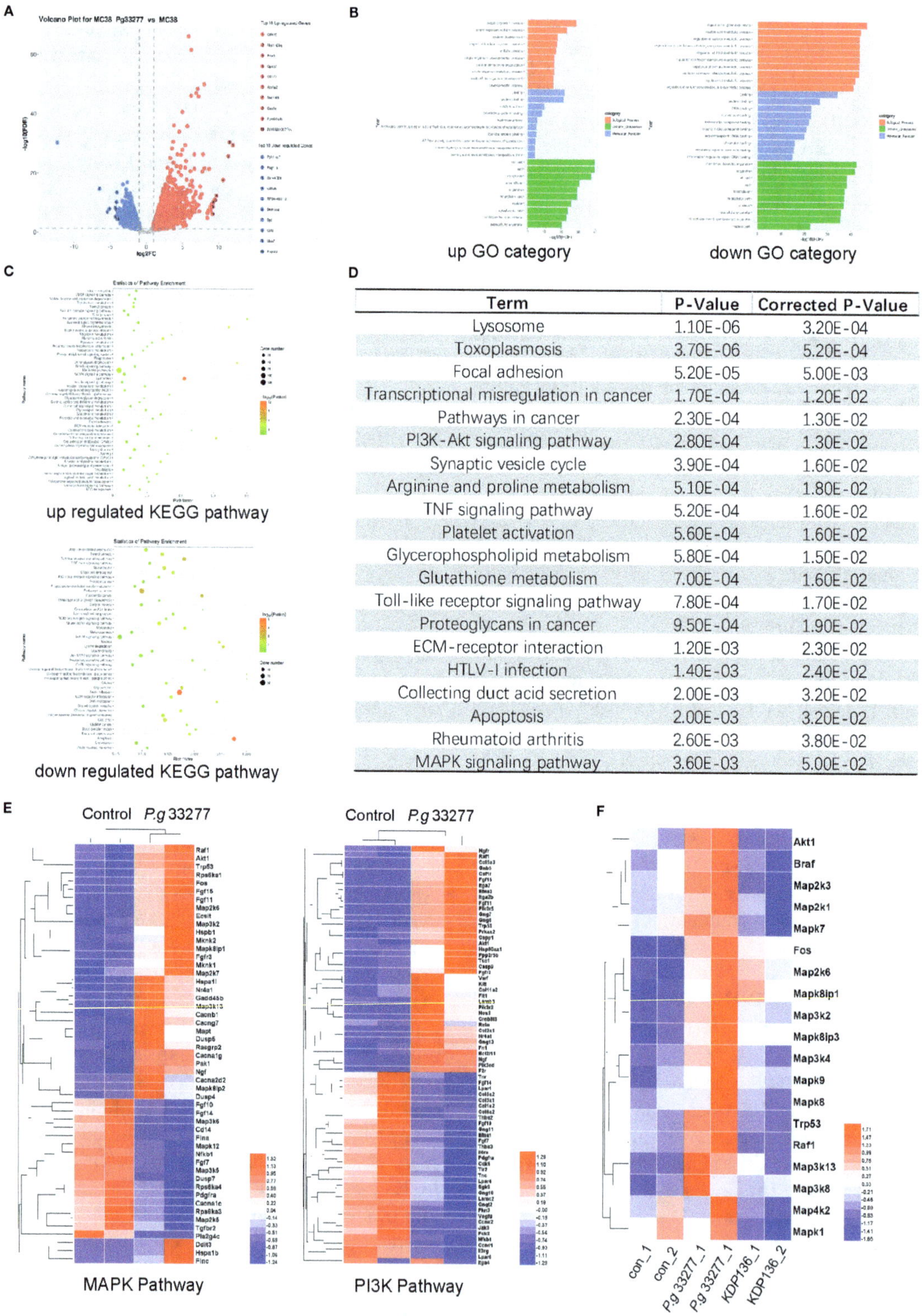

Term	P-Value	Corrected P-Value
Lysosome	1.10E-06	3.20E-04
Toxoplasmosis	3.70E-06	5.20E-04
Focal adhesion	5.20E-05	5.00E-03
Transcriptional misregulation in cancer	1.70E-04	1.20E-02
Pathways in cancer	2.30E-04	1.30E-02
PI3K-Akt signaling pathway	2.80E-04	1.30E-02
Synaptic vesicle cycle	3.90E-04	1.60E-02
Arginine and proline metabolism	5.10E-04	1.80E-02
TNF signaling pathway	5.20E-04	1.60E-02
Platelet activation	5.60E-04	1.60E-02
Glycerophospholipid metabolism	5.80E-04	1.50E-02
Glutathione metabolism	7.00E-04	1.60E-02
Toll-like receptor signaling pathway	7.80E-04	1.70E-02
Proteoglycans in cancer	9.50E-04	1.90E-02
ECM-receptor interaction	1.20E-03	2.30E-02
HTLV-I infection	1.40E-03	2.40E-02
Collecting duct acid secretion	2.00E-03	3.20E-02
Apoptosis	2.00E-03	3.20E-02
Rheumatoid arthritis	2.60E-03	3.80E-02
MAPK signaling pathway	3.60E-03	5.00E-02

FIGURE 4 | RNA-Seq results of MC38 cells pretreated with *P. gingivalis* for 24 h. **(A)** Volcano plot of gene expression differences between the *P. gingivalis* and control groups. **(B)** The distribution of differentially expressed genes (DEGs) within the Gene Ontology (GO) categories. Left, upregulated genes; right, downregulated genes. **(C)** The significantly enriched Kyoto Encyclopedia of Genes and Genomes (KEGG) pathways. Top, upregulated pathways; bottom downregulated pathways. **(D)** Genes with significant differences were analyzed with the KEGG database, and the top 20 pathways that were significantly altered are shown. **(E)** Differentially expressed genes (DEGs) between *P. gingivalis* 33277 group and Control group in the MAPK and PI3K pathways. **(F)** The expression of MAPK pathway-related genes in the MC38, *P. gingivalis* 33277 and KDP136 groups.

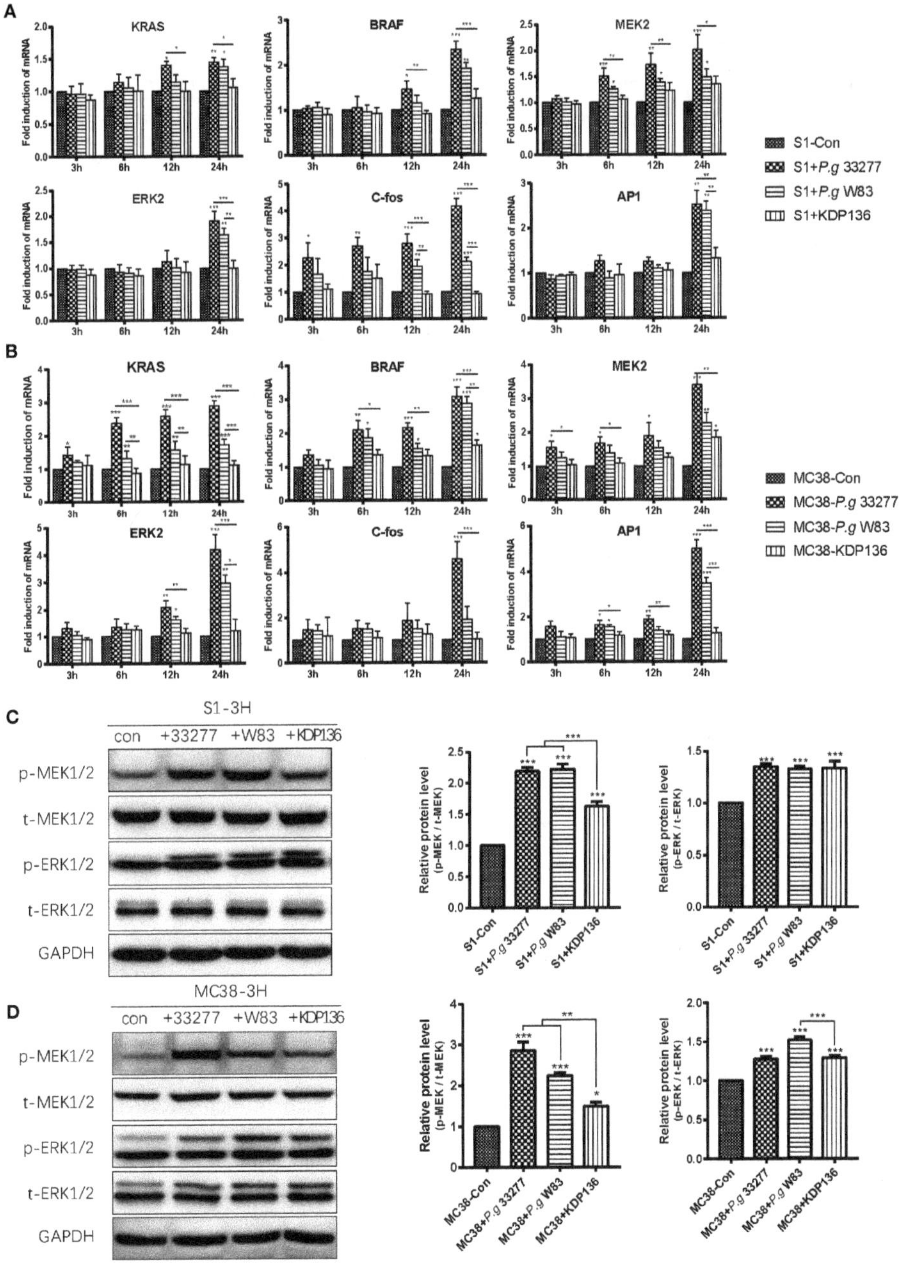

FIGURE 5 | *P. gingivalis* can promote the proliferation of S1 and MC38 cells through the MAPK/ERK pathway. **(A, B)** The mRNA levels of *KRAS, BRAF, MEK2, ERK2, C-Fos* and *AP1* in S1 cells **(A)** and MC38 cells **(B)** were detected by quantitative polymerase chain reaction (qPCR) at 3, 6, 12 and 24 h after infection. **(C, D)** Western blot analysis was performed to assess the levels of the MAPK/ERK pathway-related proteins p-MEK1/2, t-MEK1/2, p-ERK1/2, t-ERK1/2 in S1 cells **(C)** and MC38 cells **(D)** at 3 h after infection. The protein levels were normalized to GAPDH. p-MEK1/2, phospho-MEK1/2. t-MEK1/2, total MEK1/2. p-ERK1/2, phospho-ERK1/2. t-MEK1/2, total ERK1/2. **P* < 0.05, ***P* < 0.01, ****P* < 0.001.

Furthermore, the expression of PI3K-Akt pathway-related proteins in the S1 and MC38 cells was measured by WB, and the results showed that *P. gingivalis* 33277 had no effect on p-Akt (308), p-Akt (473) and Akt (pan) levels after 3, 6 and 12 h of infection (**Supplementary Figure 3**).

Unsurprisingly, KDP136 failed to stimulate the transcription of *KRAS, BRAF, MEK2, ERK2, C-fos, AP1* in S1 and MC38 cells (**Figures 5A, B**). In addition, the levels of p-MEK1/2 in the KDP136 group and p-ERK1/2 in the S1+KDP136 group were significantly lower than those observed in the *P. gingivalis* 33277 and W83 groups (**Figures 5C, D**). Gingipain plays an important role in regulating the MAPK/ERK pathway. Remarkably, KDP136 can significantly increase MEK1/2 and ERK1/2 phosphorylation, prompting the expression of other

P. gingivalis virulence factors that may also influence MEK1/2 and ERK1/2 phosphorylation in addition to gingipain (**Figures 5C, D**). Previous studies showed that *P. gingivalis*-LPS, a bacterial endotoxin located on the lateral lobule of bacterial adventitia, can activate Toll like receptor 4 (TLR4) and Toll like receptor 2 (TLR2) simultaneously after recognized (Darveau et al., 2004; Nichols et al., 2009; Jia et al., 2019). As one of the adaptors of TLR2/4, MYD88 can activate MAPK pathway indirectly (Liu et al., 2014). So the LPS of KDP136 may increase the expression of p-MEK and p-ERK here.

DISCUSSION

P. gingivalis is capable of adhering to and invading host cells, and epithelial cells of the oral mucosa are considered to be the most important intracellular niche for *P. gingivalis* (Colombo et al., 2007; Lee et al., 2018). There are four phases of successful bacterial invasion: (a) entry, (b) survival, (c) replication, and (d) exit from the host cell (Casadevall, 2008). Thus, entering cells is the first step to interact with host cells for intracellular pathogens. Once invaded, this intracellular opportunistic pathogen can manipulate the host machinery to promote its long-term survival. In our results, *P. gingivalis* 33277 could tightly adhere to CRC cells after cocultivation at 4°C for 30 min. Over time, an increasing number of *P. gingivalis* cells entered the host cells, and almost all of them were located in the cytoplasm at 24 h after infection. This finding was consistent with that of a previous study on the invasion ability of *P. gingivalis* toward epithelial cells using a wide-field deconvolution microscopy technique (Belton et al., 1999).

Interestingly, the adhesion ability of *P. gingivalis* W83 and KDP136 which is detected by flow cytometry are obviously lower than *P. gingivalis* 33277 (**Supplementary Figure 4**), but there is no significant difference between *P. gingivalis* 33277 and W83 in invasion detection, promoting the proliferation and activating MAPK/ERK signaling pathway of CRC cells (**Figures 1B, C, Figure 2, Figure 5**). Firstly, the adhesion ability of *P. gingivalis* is related to fimbriae, which is composed of FimA and Mfa1. *P. gingivalis* gingipain were also involved in the biosynthesis process of fimbriae (Weinberg et al., 1997). It is reported that FimA and Mfa1, major subunit proteins of long and short fimbriae, were abundant components of 33277 but not W83 (Mantri et al., 2015). This is why the adhesion ability of *P. gingivalis* W83 (less FimA and Mfa1) is similar to KDP136 (lack of gingipain). Secondly, the cultivation system of *P. gingivalis* and CRC cells *in vitro* is stable while *P. gingivalis* will be washed away by silva and intestinal content in human digestive tract. We supposed that *P. gingivalis* can invade host cells without adhesion *in vitro*, which is different with it *in vivo*. Finally, *P. gingivalis* were co-cultured with CRC cells for 24 h (37°C) in invasion assay while the adhesion assay was carried in 30 min (4°C) after infection. So the variation of adhesion ability after 30 min infection has no influence on our acute infection model (24 h co-culture) *in vitro*. It is reported that around 9–10% of osteoblasts were infected by *P. gingivalis* 33277 after cocultivation (7 d, MOI=200) detected by confocal fluorescence microscopy (Zhang et al., 2010). Another research indicated that 5.4% of gingival fibroblasts were infected by *P. gingivalis* 33277 after cocultivation (4 h, MOI=1000) detected by confocal fluorescence microscopy (Jang et al., 2017). In this study, around 19.3% S1 cells and 16.0% MC38 cells were infected with *P. gingivalis* (S1+P. g 33277: 19.6%, S1+P. g W83: 19.1%, MC38+P. g 33277: 16.3%, MC38+P. g W83: 15.6%) after cocultivation (24 h, MOI=100) which is higher than previous study (**Figures 1C, 3B**). The percentage will be higher in fact because ultrathin sections (100 nm) cannot present the whole picture of cells. We supposed that CRC cells are one of the desirable niches for *P. gingivalis*.

P. gingivalis has been reported to activate the p38MAPK/HSP27, ERK1/2-Ets1, and PAR2/NF-kB pathways to stimulate the expression of promatrix metalloproteinase-9 (proMMP-9), and its ability to invade OSCC cell lines was promoted by gingipain (Inaba et al., 2014). Furthermore, *P. gingivalis* was shown to improve the proliferation ability and upregulated the percentage of S phase cells in human immortalized oral epithelial cells using a long-term infection model (MOI=1, 5–23 weeks) (Geng et al., 2017). Considering that long-term exposure to aerobic environment will dramatically reduce the viability of *P. gingivalis*, we established an acute infection model (MOI=100, <24 h) to ensure that the majority of cells for this obligate anaerobe were alive.

Proteomic analysis showed that *P. gingivalis* elevated the level of Cyclin A to promote the proliferation of gingival epithelial cells (Kuboniwa et al., 2008). Increased expression of cyclin D1 were also detected in OSCC cells infected with *P. gingivalis*. Cyclin D1 contribute to the enhanced proliferation of OSCC cells, which was recognized as an early event of oral carcinogenesis (Ramos-García et al., 2019). Another research showed that *P. gingivalis* promotes the G1/S transition from 6 h to 12 h after infection by up-regulating the expression of cyclin D and cyclin E (Pan et al., 2014). Overall, *P. gingivalis* can regulates cell cycle to enhance the proliferation of OSCC cells and gingival epithelial cells (Kuboniwa et al., 2008). In our model, *P. gingivalis* could also promote the proliferation and improve the percentage of CRC cells in S phase at 24 h after infection. The expression of AP1, a dimer combined with proteins from Fos and Jun sub-families, is up-regulated by *P. gingivalis* in our study (**Figures 5A, B**). As an important transcription factor, AP1 can influence the expression of cyclin D1 (Vartanian et al., 2011) and regulate the cell cycle of CRC cells.

Gingipain is a potent *P. gingivalis* virulence factor that targets several essential components in the human immune system, coagulation cascade, and regulatory pathways (Imamura et al., 2003). *P. gingivalis* can activate the ERK1/2-Ets1, p38/HSP27, and PAR2/NFκB pathways to induce promatrix metalloproteinase-9 expression after invading OSCC cells. Then, proMMP-9 is released into the extracellular environment by gingipain from *P. gingivalis* *via* PAR2 activation to promote cellular invasion and metastasis (Inaba et al., 2014). MMP-9 is known as a type IV collagenase that is associated with various physiological and pathological processes, including reproduction, growth, development, inflammation, and vascular and proliferative diseases (Van den Steen et al., 2002). MMP-9 transcription is positively regulated by E-26 transcription factors, NFκB, polyomavirus enhancer A-binding protein-3, and

AP1 (Crawford and Matrisian, 1996). AP1 was upregulated by gingipain in our acute model in the present study (**Figures 5A, B**). Thus, gingipain may activate proMMP-9 by improving AP1 expression to promote CRC progression.

Increasing evidence has revealed that the MAPK/ERK signaling pathway plays a key role in CRC cell proliferation, migration, invasion, apoptosis and differentiation (Sun et al., 2019; Cheng et al., 2019; Pan et al., 2019). It has been reported that RAF, including its three isoforms (ARAF, BRAF, and CRAF), is activated by RAS and then activates MEK1/2 by increasing its phosphorylation level. Activated MEK1/2 (p-MEK1/2) can increase ERK1/2 phosphorylation. Finally, p-ERK1/2 can increase the expression of AP1, a transcription factor that consists of FOS and JUN, to regulate cell processes (Kim and Choi, 2015; Baxter et al., 2017). According to our RNA-Seq, qPCR and WB results, the RAS/RAF/MEK/ERK signaling pathway was stimulated by *P. gingivalis* through gingipain within 24 h. In addition, the phosphorylation of MEK1/2 and ERK1/2, the core pathway components, was upregulated within 3–6 h. Thus, we can conclude that *P. gingivalis* probably promotes the proliferation of CRC cells by regulating the MAPK/ERK signaling pathway.

It is reported that *F. nucleatum are* always work together with *P. gingivalis* in mouse experimental periodontitis model (Polak et al., 2009). We compare the pro-proliferation ability of *Fusobacterium nucleatum* (*F. nucleatum*) and *P. gingivalis* using CCK8 assay. The results showed there is no significant synergism or summation action between *F. nucleatum* and *P. gingivalis* in our acute infection model *in vitro* (**Supplementary Figure 5**). So we haven't focus on the interaction between *F. nucleatum* and *P. gingivalis* in our study. It is our next step to explore the synergism or summation action between *F. nucleatum* and *P. gingivalis.*

Overall, this study provides a direct evidence for the association of *P. gingivalis* and CRC cells *in vitro*. Animal experiment and retrospective analysis of clinical cases are required to confirm our conclusion. It suggests that *P. gingivalis* can potentially promote CRC progression and CRC patients with *P. gingivalis* infection requires extra attention in clinical work.

AUTHOR CONTRIBUTIONS

ZW and BC contributed to conception and design of the experiment and critically revised the manuscript. WM and YJ performed the experiments and analyzed the data. XC and HL drafted the manuscript. All authors contributed to the article and approved the submitted version.

ACKNOWLEDGMENTS

We would like to thank Dr. Jinlong Gao from Faculty of Medicine and Health for a kind gift of P. gingivalis KDP136 and primary antibody against RgpB.

REFERENCES

Ahn, J., Sinha, R., Pei, Z., Dominianni, C., Wu, J., Shi, J., et al. (2013). Human gut microbiome and risk for colorectal cancer. *J. Natl. Cancer Institute* 105 (24), 1907–1911. doi: 10.1093/jnci/djt300

Al-Hebshi, N. N., Nasher, A. T., Maryoud, M. Y., Homeida, H. E., Chen, T., Idris, A. M., et al. (2017). Inflammatory bacteriome featuring Fusobacterium nucleatum and Pseudomonas aeruginosa identified in association with oral squamous cell carcinoma. *Sci. Rep.* 7 (1), 1834. doi: 10.1038/s41598-017-02079-3

Arjunan, P., Swaminathan, R., Yuan, J., Al-Shabrawey, M., Espinosa-Heidmann, D. G., Nussbaum, J., et al. (2020). Invasion of Human Retinal Pigment Epithelial Cells by Porphyromonas gingivalis leading to Vacuolar/Cytosolic localization and Autophagy dysfunction In-Vitro. *Sci. Rep.* 10 (1), 7468–7468. doi: 10.1038/s41598-020-64449-8

Baxter, D., Perry, S. R., Hill, T. A., Kok, W. M., Zaccai, N. R., Brady, R. L., et al. (2017). Downsizing Proto-oncogene cFos to Short Helix-Constrained Peptides That Bind Jun. *ACS Chem. Biol.* 12 (8), 2051–2061. doi: 10.1021/acschembio.7b00303

Belton, C. M., Izutsu, K. T., Goodwin, P. C., Park, Y., and Lamont, R. J. (1999). Fluorescence image analysis of the association between Porphyromonas gingivalis and gingival epithelial cells. *Cell Microbiol.* 1 (3), 215–223. doi: 10.1046/j.1462-5822.1999.00022.x

Brien-Simpson, N. M. O., Veith, P. D., Dashper, S. G., and Reynolds, E. C. (2003). Porphyromonas gingivalis gingipains: the molecular teeth of a microbial vampire. *Curr. Protein Pept. Sci.* 4 (6), 409–426. doi: 10.2174/1389203033487009

Casadevall, A. (2008). Evolution of intracellular pathogens. *Annu. Rev. Microbiol.* 62, 19–33. doi: 10.1146/annurev.micro.61.080706.093305

Chen, W., Liu, F., Ling, Z., Tong, X., and Xiang, C. (2012). Human intestinal lumen and mucosa-associated microbiota in patients with colorectal cancer. *PloS One* 7 (6), e39743. doi: 10.1371/journal.pone.0039743

Cheng, B., Rong, A., Zhou, Q., and Li, W. (2019). CLDN8 promotes colorectal cancer cell proliferation, migration, and invasion by activating MAPK/ERK signaling. *Cancer Manag. Res.* 11, 3741–3751. doi: 10.2147/CMAR.S189558

Colombo, A. V., da Silva, C. M., Haffajee, A., and Colombo, A. P. (2007). Identification of intracellular oral species within human crevicular epithelial cells from subjects with chronic periodontitis by fluorescence in situ hybridization. *J. Periodontal Res.* 42 (3), 236–243. doi: 10.1111/j.1600-0765.2006.00938.x

Crawford, H. C., and Matrisian, L. M. (1996). Mechanisms controlling the transcription of matrix metalloproteinase genes in normal and neoplastic cells. *Enzyme Protein* 49 (1-3), 20–37. doi: 10.1159/000468614

Darveau, R. P., Pham, T. T., Lemley, K., Reife, R. A., Bainbridge, B. W., Coats, S. R., et al. (2004). Porphyromonas gingivalis lipopolysaccharide contains multiple lipid A species that functionally interact with both toll-like receptors 2 and 4. *Infect. Immun.* 72 (9), 5041–5051. doi: 10.1128/iai.72.9.5041-5051.2004

Eichinger, A., Beisel, H. G., Jacob, U., Huber, R., Medrano, F. J., Banbula, A., et al. (1999). Crystal structure of gingipain R: an Arg-specific bacterial cysteine proteinase with a caspase-like fold. *EMBO J.* 18 (20), 5453–5462. doi: 10.1093/emboj/18.20.5453

Fang, J. Y., and Richardson, B. C. (2005). The MAPK signalling pathways and colorectal cancer. *Lancet Oncol.* 6 (5), 322–327. doi: 10.1016/S1470-2045(05)70168-6

Gao, S., Li, S., Ma, Z., Liang, S., Shan, T., Zhang, M., et al. (2016). Presence of Porphyromonas gingivalis in esophagus and its association with the clinicopathological characteristics and survival in patients with esophageal cancer. *Infect. Agent Cancer* 11, 3. doi: 10.1186/s13027-016-0049-x

Geng, F., Liu, J., Guo, Y., Li, C., Wang, H., Wang, H., et al. (2017). Persistent Exposure to Porphyromonas gingivalis Promotes Proliferative and Invasion Capabilities, and Tumorigenic Properties of Human Immortalized Oral Epithelial Cells. *Front. Cell Infect. Microbiol.* 7, 57. doi: 10.3389/fcimb.2017.00057

Gibson, F.C. ,. 3., Hong, C., Chou, H. H., Yumoto, H., Chen, J., Lien, E., et al. (2004). Innate immune recognition of invasive bacteria accelerates atherosclerosis in apolipoprotein E-deficient mice. *Circulation* 109 (22), 2801–2806. doi: 10.1161/01.CIR.0000129769.17895.F0

Group, Nih Hmp Working, Peterson, J., Garges, S., Giovanni, M., McInnes, P., Wang, L., et al. (2009). The NIH Human Microbiome Project. *Genome Res.* 19 (12), 2317–2323. doi: 10.1101/gr.096651.109

Han, Y. W., Fardini, Y., Chen, C., Iacampo, K. G., Peraino, V. A., Shamonki, J. M., et al. (2010). Term stillbirth caused by oral Fusobacterium nucleatum. *Obstet. Gynecol.* 115 (2 Pt 2), 442–445. doi: 10.1097/AOG.0b013e3181cb9955

Imamura, T., Travis, J., and Potempa, J. (2003). The biphasic virulence activities of gingipains: activation and inactivation of host proteins. *Curr. Protein Pept. Sci.* 4 (6), 443–450. doi: 10.2174/1389203033487027

Inaba, H., Sugita, H., Kuboniwa, M., Iwai, S., Hamada, M., Noda, T., et al. (2014). Porphyromonas gingivalis promotes invasion of oral squamous cell carcinoma through induction of proMMP9 and its activation. *Cell Microbiol.* 16 (1), 131–145. doi: 10.1111/cmi.12211

Jang, J. Y., Baek, K. J., Choi, Y., and Ji, S. (2017). Relatively low invasive capacity of Porphyromonas gingivalis strains into human gingival fibroblasts in vitro. *Arch. Oral. Biol.* 83, 265–271. doi: 10.1016/j.archoralbio.2017.08.007

Jia, L., Han, N., Du, J., Guo, L., Luo, Z., and Liu, Y. (2019). Pathogenesis of Important Virulence Factors of Porphyromonas gingivalis via Toll-Like Receptors. *Front. Cell Infect. Microbiol.* 9, 262. doi: 10.3389/fcimb.2019.00262

Kim, E. K., and Choi, E. J. (2015). Compromised MAPK signaling in human diseases: an update. *Arch. Toxicol.* 89 (6), 867–882. doi: 10.1007/s00204-015-1472-2

Kuboniwa, M., Hasegawa, Y., Mao, S., Shizukuishi, S., Amano, A., Lamont, R. J., et al. (2008). P. gingivalis accelerates gingival epithelial cell progression through the cell cycle. *Microbes Infect.* 10 (2), 122–128. doi: 10.1016/j.micinf.2007.10.011

Lafuente Ibanez de Mendoza, I., Maritxalar Mendia, X., Garcia de la Fuente, A. M., Quindos Andres, G., and Aguirre Urizar, J. M. (2020). Role of Porphyromonas gingivalis in oral squamous cell carcinoma development: A systematic review. *J. Periodontal Res.* 55 (1), 13–22. doi: 10.1111/jre.12691

Lamont, R. J., Chan, A., Belton, C. M., Izutsu, K. T., Vasel, D., and Weinberg, A. (1995). Porphyromonas gingivalis invasion of gingival epithelial cells. *Infect. Immun.* 63 (10), 3878–3885. doi: 10.1128/IAI.63.10.3878-3885

Lee, K., Roberts, J. S., Choi, C. H., Atanasova, K. R., and Yilmaz, O. (2018). Porphyromonas gingivalis traffics into endoplasmic reticulum-rich-autophagosomes for successful survival in human gingival epithelial cells. *Virulence* 9 (1), 845–859. doi: 10.1080/21505594.2018.1454171

Liu, Y., Yin, H., Zhao, M., and Lu, Q. (2014). TLR2 and TLR4 in autoimmune diseases: a comprehensive review. *Clin. Rev. Allergy Immunol.* 47 (2), 136–147. doi: 10.1007/s12016-013-8402-y

Mantri, C. K., Chen, C. H., Dong, X., Goodwin, J. S., Pratap, S., Paromov, V., et al. (2015). Fimbriae-mediated outer membrane vesicle production and invasion of Porphyromonas gingivalis. *Microbiologyopen* 4 (1), 53–65. doi: 10.1002/mbo3.221

Michaud, D. S., Izard, J., Wilhelm-Benartzi, C. S., You, D. H., Grote, V. A., Tjonneland, A., et al. (2013). Plasma antibodies to oral bacteria and risk of pancreatic cancer in a large European prospective cohort study. *Gut* 62 (12), 1764–1770. doi: 10.1136/gutjnl-2012-303006

Mo, W., Luo, H., Wu, J., Xu, N., Zhang, F., Qiu, Q., et al. (2020). Gingipains promote RANKL-induced osteoclastogenesis through the enhancement of integrin beta3 in RAW264.7 cells. *J. Mol. Histol.* 51 (2), 147–159. doi: 10.1007/s10735-020-09865-w

Nguyen, K. A., Travis, J., and Potempa, J. (2007). Does the importance of the C-terminal residues in the maturation of RgpB from Porphyromonas gingivalis reveal a novel mechanism for protein export in a subgroup of Gram-Negative bacteria? *J. Bacteriol.* 189 (3), 833–843. doi: 10.1128/jb.01530-06

Nichols, F. C., Housley, W. J., O'Conor, C. A., Manning, T., Wu, S., and Clark, R. B. (2009). Unique lipids from a common human bacterium represent a new class of Toll-like receptor 2 ligands capable of enhancing autoimmunity. *Am. J. Pathol.* 175 (6), 2430–2438. doi: 10.2353/ajpath.2009.090544

Njoroge, T., Genco, R. J., Sojar, H. T., Hamada, N., and Genco, C. A. (1997). A role for fimbriae in Porphyromonas gingivalis invasion of oral epithelial cells. *Infect. Immun.* 65 (5), 1980–1984. doi: 10.1128/iai.65.5.1980-1984.1997

Pan, C., Xu, X., Tan, L., Lin, L., and Pan, Y. (2014). The effects of Porphyromonas gingivalis on the cell cycle progression of human gingival epithelial cells. *Oral. Dis.* 20 (1), 100–108. doi: 10.1111/odi.12081

Pan, H., Wang, Y., Na, K., Wang, Y., Wang, L., Li, Z., et al. (2019). Autophagic flux disruption contributes to Ganoderma lucidum polysaccharide-induced apoptosis in human colorectal cancer cells via MAPK/ERK activation. *Cell Death Dis.* 10 (6), 456. doi: 10.1038/s41419-019-1653-7

Pike, R., McGraw, W., Potempa, J., and Travis, J. (1994). Lysine- and arginine-specific proteinases from Porphyromonas gingivalis. Isolation, characterization, and evidence for the existence of complexes with hemagglutinins. *J. Biol. Chem.* 269 (1), 406–411.

Polak, D., Wilensky, A., Shapira, L., Halabi, A., Goldstein, D., Weiss, E.II, et al. (2009). Mouse model of experimental periodontitis induced by Porphyromonas gingivalis/Fusobacterium nucleatum infection: bone loss and host response. *J. Clin. Periodontol.* 36 (5), 406–410. doi: 10.1111/j.1600-051X.2009.01393.x

Ramos-García, P., González-Moles, MÁ, Ayén, Á, González-Ruiz, L., Ruiz-Ávila, I., Lenouvel, D., et al. (2019). Asymmetrical proliferative pattern loss linked to cyclin D1 overexpression in adjacent non-tumour epithelium in oral squamous cell carcinoma. *Arch. Oral. Biol.* 97, 12–17. doi: 10.1016/j.archoralbio.2018.10.007

Rubinstein, M. R., Wang, X., Liu, W., Hao, Y., Cai, G., and Han, Y. W. (2013). Fusobacterium nucleatum promotes colorectal carcinogenesis by modulating E-cadherin/beta-catenin signaling via its FadA adhesin. *Cell Host. Microbe* 14 (2), 195–206. doi: 10.1016/j.chom.2013.07.012

Schmidt, T. S., Hayward, M. R., Coelho, L. P., Li, S. S., Costea, P.II, Voigt, A. Y., et al. (2019). Extensive transmission of microbes along the gastrointestinal tract. *Elife* 8, e42693. doi: 10.7554/eLife.42693

Schmidt, S., Denk, S., and Wiegering, A. (2020). Targeting Protein Synthesis in Colorectal Cancer. *Cancers (Basel)* 12 (5), 1298. doi: 10.3390/cancers12051298

Sheets, S. M., Potempa, J., Travis, J., Casiano, C. A., and Fletcher, H. M. (2005). Gingipains from Porphyromonas gingivalis W83 induce cell adhesion molecule cleavage and apoptosis in endothelial cells. *Infect. Immun.* 73 (3), 1543–1552. doi: 10.1128/IAI.73.3.1543-1552.2005

Sun, H., Ou, B., Zhao, S., Liu, X., Song, L., Liu, X., et al. (2019). USP11 promotes growth and metastasis of colorectal cancer via PPP1CA-mediated activation of ERK/MAPK signaling pathway. *EBioMedicine* 48, 236–247. doi: 10.1016/j.ebiom.2019.08.061

Van den Steen, P. E., Dubois, B., Nelissen, I., Rudd, P. M., Dwek, R. A., and Opdenakker, G. (2002). Biochemistry and molecular biology of gelatinase B or matrix metalloproteinase-9 (MMP-9). *Crit. Rev. Biochem. Mol. Biol.* 37 (6), 375–536. doi: 10.1080/10409230290771546

Vartanian, R., Masri, J., Martin, J., Cloninger, C., Holmes, B., Artinian, N., et al. (2011). AP-1 regulates cyclin D1 and c-MYC transcription in an AKT-dependent manner in response to mTOR inhibition: role of AIP4/Itch-mediated JUNB degradation. *Mol. Cancer Res.* 9 (1), 115–130. doi: 10.1158/1541-7786.mcr-10-0105

Wegner, N., Wait, R., Sroka, A., Eick, S., Nguyen, K. A., Lundberg, K., et al. (2010). Peptidylarginine deiminase from Porphyromonas gingivalis citrullinates human fibrinogen and alpha-enolase: implications for autoimmunity in rheumatoid arthritis. *Arthritis Rheum.* 62 (9), 2662–2672. doi: 10.1002/art.27552

Weinberg, A., Belton, C. M., Park, Y., and Lamont, R. J. (1997). Role of fimbriae in Porphyromonas gingivalis invasion of gingival epithelial cells. *Infect. Immun.* 65 (1), 313–316. doi: 10.1128/iai.65.1.313-316.1997

Wen, L., Mu, W., Lu, H., Wang, X., Fang, J., Jia, Y., et al. (2020). Porphyromonas gingivalis Promotes Oral Squamous Cell Carcinoma Progression in an Immune Microenvironment. *J. Dent. Res.* 99 (6), 666–675. doi: 10.1177/0022034520909312

Wu, N., Yang, X., Zhang, R., Li, J., Xiao, X., Hu, Y., et al. (2013). Dysbiosis signature of fecal microbiota in colorectal cancer patients. *Microbial. Ecol.* 66 (2), 462–470. doi: 10.1007/s00248-013-0245-9

Yilmaz, O., Watanabe, K., and Lamont, R. J. (2002). Involvement of integrins in fimbriae-mediated binding and invasion by Porphyromonas gingivalis. *Cell Microbiol.* 4 (5), 305–314. doi: 10.1046/j.1462-5822.2002.00192.x

Zackular, J. P., Rogers, M. A., Ruffin, M. T., and Schloss, P. D. (2014). The human gut microbiome as a screening tool for colorectal cancer. *Cancer Prev. Res. (Philadelphia Pa.)* 7 (11), 1112–1121. doi: 10.1158/1940-6207.capr-14-0129

Zhang, W., Swearingen, E. B., Ju, J., Rigney, T., and Tribble, G. D. (2010). Porphyromonas gingivalis invades osteoblasts and inhibits bone formation. *Microbes Infect.* 12 (11), 838–845. doi: 10.1016/j.micinf.2010.05.011

Zhang, F., Qiu, Q., Song, X., Chen, Y., Wu, J., and Liang, M. (2017). Signal-Regulated Protein Kinases/Protein Kinase B-p53-BH3-Interacting Domain Death Agonist Pathway Regulates Gingipain-Induced Apoptosis in Osteoblasts. *J. Periodontol.* 88 (11), e200–e210. doi: 10.1902/jop.2017.160806

8

The *Salmonella* Effector SseK3 Targets Small Rab GTPases

Jiyao Gan[1,2], Nichollas E. Scott[2], Joshua P. M. Newson[2†], Rachelia R. Wibawa[1,2], Tania Wong Fok Lung[2†], Georgina L. Pollock[1], Garrett Z. Ng[3], Ian van Driel[3], Jaclyn S. Pearson[1,4], Elizabeth L. Hartland[1,4] and Cristina Giogha[1,4*]*

[1] Centre for Innate Immunity and Infectious Diseases, Hudson Institute of Medical Research, Clayton, VIC, Australia, [2] Department of Microbiology and Immunology, University of Melbourne at the Peter Doherty Institute for Infection and Immunity, Melbourne, VIC, Australia, [3] Department of Biochemistry and Molecular Biology, Bio21 Molecular Science and Biotechnology Institute, The University of Melbourne, Parkville, VIC, Australia, [4] Department of Molecular and Translational Science, Monash University, Clayton, VIC, Australia

***Correspondence:**
Elizabeth L. Hartland
elizabeth.hartland@hudson.org.au
Cristina Giogha
cristina.giogha@hudson.org.au

†Present address:
Joshua P. M. Newson,
Department of Biology, Institute of Microbiology, ETH Zurich, Zurich, Switzerland
Tania Wong Fok Lung,
Department of Pediatrics, Columbia University at the Columbia University Medical Center, New York, NY, United States

During infection, *Salmonella* species inject multiple type III secretion system (T3SS) effector proteins into host cells that mediate invasion and subsequent intracellular replication. At early stages of infection, *Salmonella* exploits key regulators of host intracellular vesicle transport, including the small GTPases Rab5 and Rab7, to subvert host endocytic vesicle trafficking and establish the *Salmonella*-containing vacuole (SCV). At later stages of intracellular replication, interactions of the SCV with Rab GTPases are less well defined. Here we report that Rab1, Rab5, and Rab11 are modified at later stages of *Salmonella* infection by SseK3, an arginine *N*-acetylglucosamine (GlcNAc) transferase effector translocated via the *Salmonella* pathogenicity island 2 (SPI-2) type III secretion system. SseK3 modified arginines at positions 74, 82, and 111 within Rab1 and this modification occurred independently of Rab1 nucleotide binding. SseK3 exhibited Golgi localization that was independent of its glycosyltransferase activity but Arg-GlcNAc transferase activity was required for inhibition of alkaline phosphatase secretion in transfected cells. While SseK3 had a modest effect on SEAP secretion during infection of HeLa229 cells, inhibition of IL-1 and GM-CSF cytokine secretion was only observed upon over-expression of SseK3 during infection of RAW264.7 cells. Our results suggest that, in addition to targeting death receptor signaling, SseK3 may contribute to *Salmonella* infection by interfering with the activity of key Rab GTPases.

Keywords: *Salmonella enterica*, Rab, glycosyltransferase, protein secretion, host-pathogen interaction

INTRODUCTION

Salmonella enterica Typhimurium (*S.* Typhimurium) is a common foodborne pathogen that imposes a significant financial burden on healthcare systems in both industrialized and developing countries (Majowicz et al., 2010). Upon ingestion of contaminated food or water, *Salmonellae* that survive the acidic gastric environment colonize the gut by invading both phagocytic and non-phagocytic cells (Clark et al., 1994; Smith, 2003; Geddes et al., 2007; Muller et al., 2012). *Salmonella* pathogenicity island 1 (SPI-1) and *Salmonella* pathogenicity island 2 (SPI-2) both encode type III secretion systems (T3SSs) that deliver virulence effector proteins into host cells that facilitate *Salmonella* pathogenesis. These two T3SSs are regulated in a spatio-temporal dependent manner, and are responsible for delivering two distinct cohorts of effectors into the host cell cytosol to hijack host physiology (LaRock et al., 2015). The SPI-1-encoded T3SS is activated initially, inducing inflammation and host cytoskeletal rearrangements to promote invasion of the bacteria into epithelial cells, while cells such as macrophages internalize *Salmonella* by phagocytosis. Following internalization, the SPI-2-encoded T3SS is activated to establish a membrane-bound replicative niche termed the *Salmonella*-containing vacuole (SCV). The SCV subverts the canonical endolysosomal pathway to facilitate intracellular bacterial survival and replication (Ramsden et al., 2007; Jennings et al., 2017).

SseK1, SseK2, and SseK3 are three SPI-2 translocated effectors which show high amino acid sequence similarity (84, 83, and 80% respectively) to the T3SS effector NleB1 from enteropathogenic *Escherichia coli* (EPEC) (Kujat Choy et al., 2004; Brown et al., 2011; Li et al., 2013; Pearson et al., 2013). NleB1 is a novel glycosyltransferase that blocks apoptotic cell death during EPEC infection by modifying conserved arginine residues with *N*-acetylglucosamine (GlcNAc) within death domain containing proteins, including FADD and TRADD (Li et al., 2013; Pearson et al., 2013; Scott et al., 2017). The post-translational modifications are termed Arg-GlcNAcylation and are only mediated by members of the NleB1 effector protein homolog family including the SseKs (El Qaidi et al., 2017; Gunster et al., 2017; Esposito et al., 2018; Park et al., 2018; Newson et al., 2019). Similar to NleB1, expression of SseK1 or SseK3, but not SseK2, suppresses NF-κB activation during infection, although these effectors exhibit different Arg-GlcNAcylation profiles on infected host cells (Gunster et al., 2017; Newson et al., 2019). Our group recently showed that overexpression of SseK1 and SseK3 resulted in broadened substrate specificity, suggesting that authentic host targets of these effectors need to be identified under native expression conditions during infection (Newson et al., 2019). When expressed at endogenous levels during *Salmonella* infection, SseK1 preferentially modifies the death domain of TRADD, whereas SseK3 targets death domains in the death receptors, TRAILR and TNFR1 (Newson et al., 2019). Interestingly, both SseK2 and SseK3 localize to the Golgi during infection; while SseK1 localizes to the cytosol (Gunster et al., 2017). Such observations suggest SseK2 and SseK3 may have uncharacterized Golgi-localized targets.

In this study, we sought to identify Golgi-associated proteins that are modified by SseK3 during *S.* Typhimurium infection. Using a mass spectrometry-based approach to examine membrane-enriched fractions of infected macrophages, we identified several Rab GTPases as targets of SseK3, including Rab1, Rab5, and Rab11. Notably, we found SseK3 modified arginine residues in the switch II region of Rab1, and that ectopic expression of either SseK2 or SseK3, but not SseK1, blocked host protein secretion. However, only SseK3 appeared to have the capacity to block host protein secretion during *Salmonella* infection. Collectively, these results suggest that SseK3 may contribute to *Salmonella* infection by GlcNAcylating selected Rab GTPases.

TABLE 1 | List of bacterial strains used in this study.

Strain	Characteristics	Source/References
SL1344	*S. enterica* serovar Typhimurium strain SL1344	Hoiseth and Stocker, 1981
Δ*sseK123*	SL1344 Δ*sseK1*Δ*sseK2*Δ*sseK3*	Brown et al., 2011
Δ*sseK23*	SL1344 Δ*sseK2*Δ*sseK3*	Brown et al., 2011
Δ*sseK12*	SL1344 Δ*sseK1*Δ*sseK2*	Kujat Choy et al., 2004
Δ*sseK23* (pSseK2)	SL1344 Δ*sseK2*Δ*sseK3* supplemented with pTrc99A-SseK2 plasmid	This study
Δ*sseK23* (pSseK3)	SL1344 Δ*sseK2*Δ*sseK3* supplemented with pTrc99A-SseK3 plasmid	This study
Δ*ssaR*	SL1344 Δ*ssaR* (SPI-2 mutant)	Kupz et al., 2012

MATERIALS AND METHODS

Bacterial Strains and Growth Conditions

The bacterial strains used in this study are listed in **Table 1**. All bacteria were grown in Luria-Bertani (LB) broth with shaking at 200 rpm in 37°C incubator in the presence of streptomycin (50 μg/ml), kanamycin (100 μg/ml), or ampicillin (100 μg/ml) when necessary.

DNA Cloning and Purification

The plasmids and primers used in this study are listed in **Tables 2, 3** respectively. All plasmid extractions were performed using QIAGEN QIAprep Miniprep Kit (Qiagen, Valencia, CA). DNA restriction digests were applied according to manufacturer's instructions (New England Biolabs, Ipswich, MA). PCR products, restriction digest products and DNA from agarose gels were purified using the Wizard SV Gel and PCR Clean-Up system (Promega, Madison, WI). Digested inserts and plasmids were ligated with a 6:1 molar ratio at 4°C overnight using T4 DNA Ligase system in accordance with manufacturer's instructions (Promega, Madison, WI). The pF_TRE-SEAP ligation product was transformed into DH5α cells and other resultant ligation products were transformed into XL1-Blue cells. The pF_TRE-SEAP plasmid was sequenced with pFTRE-F/R and other plasmids were extracted and sequenced with sequencing primer pair p3xFlag-Myc-CMV-24F/p3xFlag-Myc-CMV-24R.

The p3xFlag-Rab1a vector was constructed by amplifying *RAB1A* from human cDNA using primer pair Rab1a$_F$/Rab1a$_R$, which was then double digested with *EcoRV* and *BamHI* and ligated into p3xFlag-*Myc*-CMV-24 plasmid that had been digested with same restriction enzymes to generate an N-terminal 3xFlag fusion to Rab1a. p3xFlag-Rab5a and p3xFlag-Rab5b were constructed in a similar manner by amplifying *RAB5A* or *RAB5B* from human cDNA using primer pair Rab5a$_F$/Rab5a$_R$ or Rab5b$_F$/Rab5b$_R$ respectively, which were then double digested with *HindIII* and *BamHI* and ligated into p3xFlag-*Myc*-CMV-24. p3xFlag-Rab5c was constructed by amplifying *RAB5C* from human cDNA using primer pair Rab5c$_F$/Rab5c$_R$, which was then double digested with *EcoRV* and *BamHI* and ligated into p3xFlag-*Myc*-CMV-24. p3xFlag-Rab11b was constructed by amplifying *RAB11B* from human cDNA using primer pair Rab11b$_F$/Rab11b$_R$, which was then double digested with *HindIII*

TABLE 2 | List of plasmids used in this study.

Plasmid	Characteristics	Source/References
pEGFP-C2	Mammalian expression vector expressing EGFP fused to the N terminus of the encoding protein, Kan^R	Clontech
pEGFP-C2-SseK1	*sseK1* from *S.* Typhimurium SL1344 in pEGFP-C2, Kan^R	Newson et al., 2019
pEGFP-C2-SseK2	*sseK2* from *S.* Typhimurium SL1344 in pEGFP-C2, Kan^R	Newson et al., 2019
pEGFP-C2-SseK3	*sseK3* from *S.* Typhimurium SL1344 in pEGFP-C2, Kan^R	Newson et al., 2019
pEGFP-C2-$SseK1_{DXD}$	*sseK1* from *S.* Typhimurium SL1344 in pEGFP-C2, with catalytic motif DxD(229-231) mutated to AAA, Kan^R	This study
pEGFP-C2-$SseK2_{DXD}$	*sseK2* from *S.* Typhimurium SL1344 in pEGFP-C2, with catalytic motif DxD(293-241) mutated to AAA, Kan^R	This study
pEGFP-C2-$SseK3_{DXD}$	*sseK3* from *S.* Typhimurium SL1344 in pEGFP-C2, with catalytic motif DxD(226-228) mutated to AAA, Kan^R	This study
p3xFlag-*Myc*-CMV-24	Mammalian expression vector with Met-3xFlag tagged at N-terminal and *c-myc* tagged at C-terminal, Amp^R	Signa-Aldrich
p3xFlag-Rab1a	Human Rab1a in p3xFlag-*Myc*-CMV-24, Amp^R	This study
p3xFlag-$Rab1a_{R74A}$	Human Rab1a with Arg74 mutated to Ala74 in p3xFlag-*Myc*-CMV-24, Amp^R	This study
p3xFlag-$Rab1a_{R82A}$	Human Rab1a with Arg82 mutated to Ala82 in p3xFlag-*Myc*-CMV-24, Amp^R	This study
p3xFlag-$Rab1a_{R111A}$	Human Rab1a with Arg111 mutated to Ala111 in p3xFlag-*Myc*-CMV-24, Amp^R	This study
p3xFlag-$Rab1a_{R74AR82A}$	Human Rab1a with Arg74 and Arg82 mutated to Ala74 and Ala82, respectively, in p3xFlag-*Myc*-CMV-24, Amp^R	This study
p3xFlag-$Rab1a_{R74AR111A}$	Human Rab1a with Arg74 and Arg111 mutated to Ala74 and Ala111, respectively, in p3xFlag-*Myc*-CMV-24, Amp^R	This study
p3xFlag-$Rab1a_{R82AR111A}$	Human Rab1a with Arg82 and Arg111 mutated to Ala82 and Ala111, respectively, in p3xFlag-*Myc*-CMV-24, Amp^R	This study
p3xFlag-$Rab1a_{R74AR82AR111A}$	Human Rab1a with Arg74, Arg82, and Arg111 mutated to Ala74, Ala82 and Ala111in p3xFlag-*Myc*-CMV-24, Amp^R	This study
p3xFlag-Rab5a	Human Rab5a in p3xFlag-*Myc*-CMV-24, Amp^R	This study
p3xFlag-Rab5b	Human Rab5b in p3xFlag-*Myc*-CMV-24, Amp^R	This study
p3xFlag-Rab5c	Human Rab5c in p3xFlag-*Myc*-CMV-24, Amp^R	This study
p3xFlag-Rab11b	Human Rab11b in p3xFlag-*Myc*-CMV-24, Amp^R	This study
pTrc99A-SseK2	*sseK2* from *S.* Typhimurium SL1344 in pTrc99A, Amp^R	Newson et al., 2019
pTrc99A-SseK3	*sseK3* from *S.* Typhimurium SL1344 in pTrc99A, Amp^R	Newson et al., 2019
pSEAP	Secreted embryonic alkaline phosphatase in a mammalian expression vector	Kagan et al., 2004
p3xFlag-AnkX	*ankX* from *L.pneumophila* in p3xFlag-*Myc*-CMV-24, Amp^R	This study
pF_TRE3G_PGK puro	Lentiviral transduction vector, AMP^R	Yamamoto et al., 2012
pF_TRE-SEAP	*seaP* from pSEAP in pF_TRE3G_PGK puro, Amp^R	This study
pCMV-VSV-G	Mammalian expression vector expressing VSV-G glycoprotein, AMP^R	Stewart et al., 2003
pCMVΔR8.2	Mammalian expression vector expressing HIV-1 Gag/Pol, Tat, and Rev, AMP^R	Stewart et al., 2003

and *BamHI* and ligated into p3xFlag-*Myc*-CMV-24. p3xFlag-AnkX was constructed by amplifying *ankX* from *Legionella pneumophila* Philadelphia 1 genomic DNA using primer pair $AnkX_{F/R}$, double digested with *SalI* and *BamHI* and ligated into p3xFlag-*Myc*-CMV-24 plasmid that has been digested with same restriction enzymes to generate an N-terminal 3xFlag fusion to AnkX. The pF_TRE-SEAP was constructed by amplifying *SEAP* from pSEAP vector using primer pair $SEAP_F$/$SEAP_R$, which was subsequently double digested with *BclI* and *NheI* and ligated into pF_TRE3G_PGK puro that has been digested with same restriction enzymes.

Site-Directed Mutagenesis

All site-directed mutagenesis was performed using QuikChange II Site-Directed Mutagenesis Kit according to manufacturer's protocol. p3xFlag-$Rab1a_{R74A}$ was generated with primer pair $Rab1a_{R74A-F}$/$Rab1a_{R74A-R}$ using p3xFlag-Rab1a vector as a template. $Rab1a_{R82A}$ was generated with primer pair $Rab1a_{R82A-F}$/$Rab1a_{R82A-R}$ using p3xFlag-Rab1a vector as a template. $Rab1a_{R111A}$ was generated with primer pair $Rab1aR_{111A-F}$/$Rab1aR_{111A-R}$ using p3xFlag-Rab1a vector as a template. $Rab1a_{R74AR82A}$ and $Rab1a_{R74AR111A}$ were generated with primer pair $Rab1a_{R82A-F}$/$Rab1a_{R82A-R}$ and $Rab1a_{R111A-F}$/$Rab1a_{R111A-R}$ respectively using p3xFlag-$Rab1a_{R74A}$ vector as a template. $Rab1a_{R82AR111A}$ was generated with primer pair $Rab1a_{R111A-F}$/$Rab1a_{R111A-R}$ using p3xFlag-$Rab1a_{R82A}$ vector as a template. $Rab1a_{R74AR82AR111A}$ was generated with primer pair $Rab1a_{R111A-F}$/$Rab1a_{R111A-R}$ using p3xFlag-$Rab1a_{R74AR82A}$ vector as a template. The sequences of resulting plasmids were confirmed by sequencing using primer pair p3xFlag-Myc-CMV-24F/p3xFlag-Myc-CMV-24R. The pEGFP-C2-$SseK1_{DXD}$, pEGFP-C2-$SseK2_{DXD}$ and

TABLE 3 | List of primers used in this study.

Name	Primer sequences 5'-3'
Rab1a$_F$	CGCGATATCGATGTCCAGCATGAATCCCG
Rab1a$_R$	CGCGGATCCTTAGCAGCAACCTCCACCTG
Rab1a$_{R74A-F}$	AGGCCAGGAAAGATTTGCAACAATCACCTCCAGTT
Rab1a$_{R74A-R}$	AACTGGAGGTGATTGTTGCAAATCTTTCCTGGCCT
Rab1a$_{R82A-F}$	ACCTCCAGTTATTACGCAGGAGCCCATGGCATCA
Rab1a$_{R82A-R}$	TGATGCCATGGGCTCCTGCGTAATAACTGGAGGT
Rab1a$_{R111A-F}$	GGCTGCAGGAAATAGATGCATATGCCAGTGAAAATGT
Rab1a$_{R111A-R}$	ACATTTTCACTGGCATATGCATCTATTTCCTGCAGCC
Rab5a$_F$	CCCAAGCTTATGGCTAGTCGAGGCGCAA
Rab5a$_R$	CGCGGATCCTTAGTTACTACAACACTGATTCCTGGTT
Rab5b$_F$	CCCAAGCTTATGACTAGCAGAAGCACAGCTAGG
Rab5b$_R$	CGCGGATCCTCAGTTGCTACAACACTGGCTCTT
Rab5c$_F$	CGCGATATCGATGGCGGGTCGGGGAGG
Rab5c$_R$	CGCGGATCCTCAGTTGCTGCAGCACTGGCT
Rab11b$_F$	CCCAAGCTTATGGGGACCCGGGACGAC
Rab11b$_R$	CGCGGATCCTCACAGGTTCTGGCAGCACTGC
p3xFlag-Myc-CMV-24F	AATGTCGTAATAACCCCGCCCCGTTGACGC
p3xFlag-Myc-CMV-24R	TATTAGGACAAGGCTGGTGGGCAC
AnkX$_F$	AAAGTCGACATGCCAAATCTACCTGG
AnkX$_R$	TTTGGATCCTTACCATTTTAATTTCAAGG
SseK1$_{AAA-F}$	GGTGTATATATCTTGCTGCTGCTATGATTATCACGGAAAA ACTGG
SseK1$_{AAA-R}$	CCAGTTTTTCCGTGATAATCATAGCAGCAGCAAGATATAT ACACC
SseK2$_{AAA-F}$	GTGGGTGCATATATCTTGCTGCAGCTATGTTACTTACTGAT AAAC
SseK2$_{AAA-R}$	GTTTATCAGTAAGTAACATAGCTGCAGCAAGATATATGCA CCCAC
SseK3$_{AAA-F}$	CTGGAGGTGGCTGCATATATCTTGCTGCTGCTATGTTACTT ACAG
SseK3$_{AAA-R}$	CTGTAAGTAACATAGCAGCAGCAAGATATATGCAGCCACC TCCAG
SEAP$_F$	CGCTGATCAATGCTGCTGCTGCTGCTGCTGCTG
SEAP$_R$	CTAGCTAGCTCATGTCTGCTCGAAGCGG
pFTRE-F	GTGTACGGTGGGCGCC
pFTRE-R	GTTGGCGCCTACCGGTG

pEGFP-C2-SseK3$_{DXD}$ vectors were generated using primer pairs SseK1$_{AAA-F/R}$, SseK2$_{AAA-F/R}$ and SseK3$_{AAA-F/R}$ respectively. All resultant plasmids following PCR reactions were digested with *DpnI* at 37°C overnight and then transformed into XL1-Blue competent cells.

Arg-GlcNAcylation Pull Downs on Insoluble Fractions of *Salmonella* Infected RAW264.7 Cells

RAW264.7 cells were seeded to 24 well plates at a concentration of 3 × 10^5 cells per well 1 day before infection. 10 ml LB broths containing appropriate antibiotic were inoculated with *Salmonella* strains and incubated at 37°C overnight with shaking at 180 rpm. On the day of infection, the OD600 readings of the overnight culture were read and used to estimate bacterial counts. Cells were then infected at a multiplicity of infection (MOI) of 10. 24 well plates were centrifuged at 1500 rpm for 5 min at room temperature to promote and synchronize infection. Infected cells were incubated at 37°C, 5% CO_2 for 1 h. Culture media was replaced with media containing 100 μg/ml gentamicin (Pharmacia, Washington, USA), and cells were incubated at 37°C, 5% CO_2 for a further 1 h. Culture media was replaced with media containing 10 μg/ml gentamicin, and cells were incubated at 37°C, 5% CO_2 for a further 18 h.

Infected cells were washed three times in ice-cold PBS, then collected by scraping in PBS containing cOmplete™ EDTA-free protease inhibitor cocktail (Roche), on a bed of ice. Cells were centrifuged at 4000 rpm for 5 min at 4°C, and the supernatant was carefully removed via aspiration using a vacuum pump. Cells were resuspended in 1 ml of ice-cold lysis buffer (20 mM HEPES (pH 7.5), 100 mM KCl, 2.5 mM $MgCl_2$, 100 mM sucrose, 10% PhosSTOP™ (Sigma-Aldrich), and cOmplete™ EDTA-free protease inhibitor cocktail (Roche). Cells were mixed thoroughly by pipette and incubated on ice for 10 min. Cell lysates were centrifuged at 13000 rpm for 5 min at 4°C, and the supernatant was removed by vacuum-aided aspiration as previously. Pelleted material was resuspended in lysis buffer containing 10% digitonin, mixed thoroughly by pipette and incubated on ice for 1 h with intermittent vortexing. Cell lysates were centrifuged at 13000 rpm for 5 min at 4°C, and the supernatant was collected as the digitonin-soluble membrane fraction, while the remaining pellet was collected as the digitonin-insoluble fraction.

Insoluble membrane pellets were resuspended in 8M urea in 100 mM ammonium bicarbonate and protein concentration determined using a BCA assay. 2.5 mg of protein from each sample type, Δ*sseK12* and Δ*sseK123*, were reduced with 10 mM dithiothreitol for 1 h then alkylated with 50 mM chloroacetamide for a further 1 h in the dark. Samples were digested with Lys-C (1/200 w/w) for 3 h at RT, samples diluted to <2M urea and digested with trypsin (1/100 w/w) overnight at RT. Digested samples were acidified to a final concentration of 0.5% formic acid and desalted with 50 mg tC18 SEP-PAK (Waters corporation, Milford, USA) according to the manufacturer's instructions. Briefly, tC18 SEP-PAKs were conditioned with buffer B (80% acetonitrile, 0.1% formic acid), washed with 10 volumes of Buffer A* (0.1% trifluoroacetic acid, 2% acetonitrile), sample loaded, column washed with 10 volumes of Buffer A* and bound peptides eluted with buffer B then dried.

Peptide affinity purification was accomplished according to the protocol of Scott et al. (2017) for peptide based Arg-GlcNAc enrichment. Briefly, aliquots of 100 μl of Protein A/G plus Agarose beads (Santa Cruz, Santa Cruz CA) were washed three times with 1 ml of immunoprecipitation buffer (IAP, 10 mM Na_2HPO_4, 50 mM NaCl, 50 mM MOPS, pH 7.2) and tumbled overnight with 10 μg of anti-Arg-GlcNAc antibody (ab195033, Abcam) at 4°C. Coupled anti-Arg-GlcNAc beads were then washed three times with 1 ml of 100 mM sodium borate (pH 9) to remove non-bound proteins and cross-linked for 30 min by gently rotating with 20 mM Dimethyl Pimelimidate (Thermo Fisher Scientific) in 100 mM HEPES, pH 8.0. Cross-linking was quenched by washing beads with 200 mM ethanolamine, pH 8.0, three times then rotating the beads in an additional 1 ml 200 mM ethanolamine, pH 8.0 for 2 h at 4°C. Beads were washed three times with IAP buffer and used immediately. Purified peptides were resuspended in 1 ml IAP buffer and the pH checked to ensure compatibility with affinity conditions (~pH7.2). Peptide lysates were then added to the prepared cross-linked anti-Arg-GlcNAc antibody beads and rotated for 3 h at 4°C. Upon completion of the incubation, antibody beads were spun down at 3000 G for 2 min at 4°C and the unbound peptide lysates collected. Antibody beads were then washed six times with 1 ml of ice-cold IAP buffer and Arg-GlcNAc peptides eluted using two rounds of acid elution. For each elution round, 100 μl of 0.2% trifluoroacetic acid was added and antibody beads allowed to stand at room temperature with gentle shaking every minute for 10 min. Peptide supernatants were collected and desalted using C18 stage tips (Rappsilber et al., 2007) before analysis by LC-MS.

Identification of Arginine-Glycosylated Affinity Enriched Peptides and Flag-Tagged Proteins Using Reversed Phase LC-MS

Purified peptides prepared were re-suspend in Buffer A* and separated using a two-column chromatography set up composed of a PepMap100 C18 20 mm × 75 μm trap and a PepMap C18 500 mm × 75 μm analytical column (Thermo Fisher Scientific). Samples were concentrated onto the trap column at 5 μl/min for 5 min and infused into an Orbitrap Fusion™ Lumos™ Tribrid™ Mass Spectrometer (Thermo Fisher Scientific) at 300 nl/minute via the analytical column using a Dionex Ultimate 3000 UPLC (Thermo Fisher Scientific). 125 min gradients were run altering the buffer composition from 1% buffer B to 28% B over 90 min, then from 28% B to 40% B over 10 min, then from 40% B to 100% B over 2 min, the composition was held at 100% B for 3 min, and then dropped to 3% B over 5 min and held at 3% B for another 15 min. The Lumos™ Mass Spectrometer was operated in a data-dependent mode automatically switching between the acquisition of a single Orbitrap MS scan (120,000 resolution) every 3 s and Orbitrap HCD for each selected precursor (maximum fill time 100 ms, AGC $2*10^5$ with a resolution of 30,000 for Orbitrap MS-MS scans).

Mass Spectrometry Data Analysis

Identification of proteins and Arg-glycosylated peptides was accomplished using MaxQuant (v1.5.3.30) (Cox and Mann, 2008). Searches were performed against the Mouse (Uniprot proteome id UP000000589—Mus musculus, downloaded 18-05-2016, 50306 entries) and *Salmonella* Typhimurium SL1344 (Uniprot proteome id UP000008962- *Salmonella* Typhimurium SL1344, downloaded 18-05-2016, 4,657 entries) proteomes with carbamidomethylation of cysteine set as a fixed modification. Searches were performed with trypsin cleavage specificity allowing 2 mis-cleavage events and the variable modifications of oxidation of methionine, *N*-Acetylhexosamine addition to arginine (Arg-GlcNAc) and acetylation of protein N-termini. The precursor mass tolerance was set to 20 parts-per-million (ppm) for the first search and 10 ppm for the main search, with a maximum false discovery rate (FDR) of 1.0% set for protein and peptide identifications. To enhance the identification of peptides between samples the Match Between Runs option was enabled with a precursor match window set to 2 min and an alignment window of 10 min. For label-free quantitation, the MaxLFQ option within Maxquant (Cox et al., 2014) was enabled in addition to the re-quantification module. The resulting protein group output was processed within the Perseus (v1.4.0.6) (Tyanova et al., 2016) analysis environment to remove reverse matches and common protein contaminates prior. For LFQ comparisons missing values were imputed using Perseus. Enrichment analysis of Arg-GlcNAcylated targets was undertaken using a Fisher exact test within Perseus. Data was exported into the R framework for visualization using ggplot2. The mass spectrometry proteomics data have been deposited to the ProteomeXchange Consortium via the PRIDE partner repository with the dataset identifier PXD015082.

Construction of Stable Cell Lines Expressing SEAP in HeLa229 Cells

HEK293T cells were seeded in a 10 cm plate and grown to 60% confluency before being transfected with 1.2 μg pF_TRE-SEAP, 2.0 μg pCMVΔR8.2 and 0.8 μg pCMV-VSV-G. Transfected cells were kept in a tissue culture incubator under 5% CO_2 at 37°C for 24 h before the tissue culture media was replaced. The supernatant containing packaged virus was collected 48 h post transfection and then filtered through a 0.45 μm filter. HeLa229 cells were seeded in a 6 well plate and grown to 50% confluency before infection. 5 μg/ml Polybrene (Sigma) was added to virus containing media. DMEM GlutaMAX (Gibco) supplemented with 10% (v/v) Fetal Bovine Serum (FBS, RBS Thermo scientific) was mixed with virus containing media in a 1:1 ratio before being added to HeLa229 cells which were then centrifuged at 1,500 rpm for 30 min at room temperature. Infected cells were incubated for 48 h and then the tissue culture medium was replaced with DMEM GlutaMAX (Gibco) supplemented with 10% (v/v) Fetal Bovine Serum (FBS, RBS Thermo scientific) containing 5 μg/ml puromycin (Sigma). Infected cells were selected with 5 μg/ml puromycin for 3 passages before use.

Mammalian Cell Culture

HEK293T, RAW264.7, HeLa229 and HeLa229 SEAP cells were cultured in T75 cm^2 or T25 cm^2 tissue culture flasks (Corning) in DMEM GlutaMAX (Gibco) supplemented with 10% (v/v) Fetal Bovine Serum (FBS, RBS Thermo scientific) in a tissue culture incubator under 5% CO_2 at 37°C. All cell lines were kept and passaged to maximum of 40 times. HeLa229 cells, HeLa229 SEAP cells and HEK293T cells were split when cells were grown to 80 to 90% confluency. When splitting cells, cells were washed two times with 1 × PBS followed with 0.4 ml for T25 cm^2 flask or 1.0 ml for T75 cm^2 flask of 0.05% Trypsin-EDTA solution (Gibco, Life Technologies) treatment for 1 min at 37°C. 0.05% Trypsin-EDTA solution treated cells were resuspended with 10 ml of

DMEM media supplemented with 10% FBS. RAW264.7 cells were washed two times with 1 x PBS and detached from tissue culture flasks with cell scrapers and diluted in fresh DMEM media supplemented with 10% FBS.

Immunoblotting

Cells were collected and placed on ice, 1xKal-B cell lysis buffer supplemented with protein inhibitors [50 mM Tris-HCl pH 7.4, 1 mM EDTA, 150 mM NaCl, 1% Triton X-100, 10 mM NaF, 1 mM PMSF, 2 mM Na_3VO_4, and 1 x EDTA-free Complete protease inhibitor mixture (Roche)] was used to lyse harvested cells. Collected samples were placed on ice for 30 min, with pipetting up and down every 10 min to complete lysis cells. Cell lysates were pelleted at 13,000 rpm for 10 min. 4 × Bolt® LDS sample buffer (Life Technologies) and DTT (Astral scientific) at a final concentration of 50 mM were added to supernatants of samples and boiled at 70°C for 10 min. Boiled samples were loaded on Bolt® 4–12% Bis-Tris Plus gels (Life Technologies) and electrophoresis was performed according to manufacturer's instructions. Proteins were then transferred to nitrocellulose membranes using iBlot® nitrocellulose transfer stacks (Life Technologies) following manufacturer's protocol. Transferred nitrocellulose membranes were then blocked with 5% (w/v) skim milk in 1 x TBST buffer (20 mM Tris, 50 mM NaCl, 0.1% (v/v) Tween-20, pH 8.0) at room temperature with shaking at 60 rpm for 1 h. Blocked membranes were washed with 1 x TBST buffer 3 times, 5 min for each wash, before being probed with primary antibodies as required at 4°C, shaking at 60 rpm overnight. Primary antibodies are used as follows: mouse monoclonal anti-GFP (1:2000 dilution) (Roche, Basel, Switzerland), mouse monoclonal anti-Flag M2-HRP (1:2000 dilution, Sigma-Aldrich, St Louis, MO), or mouse monoclonal anti-β-actin (1:5000 dilution, Sigma-Adrich, St Louis, MO), or rabbit monoclonal anti-ArgGlcNAc (1:2000 dilution, Abcam, Cambridge, UK). All primary antibodies were diluted in 5% bovine serum albumin (Sigma-Aldrich) in 1 x TBST buffer. Membranes were then washed 3 times, for 5 min each, in 1 × TBST at room temperature with shaking at 60 rpm. Secondary antibodies, if required, were then incubated on membranes at room temperature with shaking at 60 rpm for 1 h. Secondary antibodies used in this study were horseradish peroxidase (HRP) conjugated anti-mouse (PerkinElmer), or HRP conjugated anti-rabbit (Bio-Rad). All secondary antibodies were diluted at 1:3000 with 5% bovine serum albumin (Sigma-Aldrich) in 1 × TBST buffer. Probed membranes were washed 7 times, for 5 min each, with 1 × TBST at room temperature with shaking at 60 rpm before being developed using ECL Prime Western blotting reagent (Amersham Bioscience) according to manufacturer's instructions. Immunoblots were developed using the Amersham Imager 680 blot and gel imager (GE Healthcare).

Transfection of HEK293T Cells

HEK293T cells were seeded in 24 well plates (Greiner Bio-One) at a concentration of 1 × 10^5 cells per well on a coverslip for immunofluorescence experiments or at 2 × 10^5 cells per well for secreted embryonic alkaline phosphatase assays. For immunoprecipitation, HEK293T cells were seeded in 10 cm cell culture dishes (Greiner Bio-One) at 4 × 10^6 cells per dish. Cells were seeded 1 day before transfection. On the day of transfection, FuGENE 6 Transfection Reagent (Roche) was added into Opti-MEM® I (1x) in GlutaMAX(TM)-I (Gibco, Life Technologies) and incubated for 5 min at room temperature before mixed with relevant plasmids according to manufacturer's instructions. Transfection mixtures were incubated for 20 min before added to seeded cells. Transfected cells were placed in 37°C incubator with 5% CO_2 for 18 h incubation before harvesting for other experiments.

Immunoprecipitation of Flag Tagged Fusion Proteins

Harvested HEK293T cells were collected using 1 × KalB buffer supplemented with protein inhibitors as described above. Samples were placed on ice for 30 min, with pipetting up and down every 10 min, for complete cell lysis. Cell debris was pelleted at 13,000 rpm for 10 min at 4°C. Anti-Flag® M2 Magnetic Beads (Sigma-Aldrich) were washed 3 times, for 5 min each with 1 × Kal-B buffer before being loaded with cell lysates. Cell lysates were incubated with the beads with rotation at 4°C overnight. Flag-tagged proteins were eluted with 80 μl of 150 μg/ml Flag peptide (Sigma-Aldrich) with rotation at 4°C for 30 min. Eluted samples were processed for immunoblotting as described above.

Secreted Embryonic Alkaline Phosphatase Assay

For transfection on HEK293T cells, pSEAP vector was used to co-transfect HEK293T cells with vectors expressing AnkX, SseKs or their catalytic mutants following the transfection method described above. 16 h post transfection or infection, 200 μl DMEM GlutaMAX (Gibco) media was added to each well to replace old media and the 24 well plate was put back at 37°C with 5% CO_2 for a further 8 h incubation. 24 h post transfection or infection, supernatants and cells lysates were collected for processing using Phospha-Light^{TM} SEAP Reporter Gene Assay System (ThermoFisher Scientific) following manufacturer's instructions. Processed samples were plated on a 96-well white flat bottom plate and read via CLARIOstar Plus (BMG LABTECH).

Infection of Mammalian Cell Lines

RAW264.7 cells were seeded in 24 well plates at a concentration of 1 × 10^5 cells per well for cytometric bead array assay 1 day before infection. HeLa229 SEAP cells were seeded in 24 well plates at a concentration of 1 x 10^5 cells per well 1 day before experiment. *Salmonella* strains were inoculated in LB broth with relevant antibiotics 1 day before infection and grown at 37°C with shaking at 200 rpm overnight. On the day of infection, *Salmonella* strains were sub-cultured at a ratio of 1:100 in fresh LB broth with relevant antibiotics and grown at 37°C with shaking at 200 rpm for 3.5 h. OD_{600} of different sub-cultures were recorded using a spectrophotometer (SPECTRONIC^{TM} 200, Thermo Scientific^{TM}) to estimate bacterial density. Sub-cultured *Salmonella* was diluted 10-fold in DMEM GlutaMAX (Gibco) media and added into seeded cells at a multiplicity of infection (MOI) of 10. The plates were centrifuged at 1,500 rpm for 5 min to synchronize the infection and incubated at 37°C with 5% CO_2 for 30 min to allow for invasion. The cells were then washed twice with 1 x PBS and incubated with fresh DMEM GlutaMAX (Gibco) media supplemented with 100 μg/ml gentamicin and further incubated at 37°C with 5% CO_2 for 1 h. After 1 h incubation, the cells were washed twice with 1 x PBS and incubated with fresh DMEM GlutaMAX (Gibco) media supplemented with 10 μg/ml gentamicin, and where necessary, a final concentration of 1 mM IPTG was added into the media. Infected cells were incubated at 37°C with 5% CO_2 for various timepoints. For infection on HeLa229 SEAP cells, the media was replaced and 100 ng/ml doxycycline (Sigma) was used to induce SEAP expression where necessary; for cytometric bead array assays, the media was replaced on infected cells at 16 h post infection, and supernatants were collected at 20 or 24 h post infection when required.

Intracellular Replication of *Salmonella* Strains in HeLa229 SEAP and RAW264.7 Cells

Infection of HeLa229-SEAP and RAW264.7 cells with derivatives of *Salmonella* SL1344 was carried out as described above. Infected cells were collected for enumeration of bacteria at 2 and 24 h post infection. Infected cells were washed twice in PBS before being lysed for 5 min in 250 μl 0.1% Triton X-100 before been scraped and collected. Cell lysates were serially diluted in PBS, and then plated on LB agar plates containing 50 μg/ml streptomycin. Plated LB agar plates were incubated at 37°C overnight. *Salmonella* colonies were enumerated, and the colony-forming units (CFU) were calculated for each time point. Fold replication for each *Salmonella* strain was determined by dividing CFU at 24 h post infection by CFU at 2 h post infection.

Immunofluorescence Microscopy

Cells grown on coverslips were fixed with 4% paraformaldehyde in 1 × PBS for 10 min at room temperature at required time points following transfection. Cells were permeabilized using 0.2% Triton X-100 in PBS for 3 min, and then blocked with 3% BSA in 1 × PBS for 1 h. Primary antibodies were diluted at 1:200 in 3% BSA in 1 × PBS and cells were stained for 1 h at room temperature. Primary antibodies used for immunofluorescence were: monoclonal anti-Flag M2 (Sigma-Aldrich), polyclonal anti-Golgin 97 (Abcam, ab84830). The cells were then washed 3 times with 1 × PBS and incubated with fluorophore-conjugated secondary antibodies as required and supplemented with Hoechst stain (Sigma-Aldrich) for 1 h at room temperature in the dark. All secondary antibodies were diluted 1:2000 and Hoechst stain was diluted 1:4,000 in 3% BSA in 1 × PBS. Secondary antibodies used in this study were: Alexa Fluor 568 and Alexa Fluor 633 (Thermo Fisher Scientific). Samples were then washed 3 times with 1 × PBS and mounted using Prolong Gold mounting medium (Life Technologies). Confocal imaging was performed using the Olympus FV1200 Confocal.

Golgi Disruption During *Salmonella* Infection

HeLa229 cells seeded onto coverslips were infected with derivatives of *Salmonella* SL1344. Samples were fixed with 4% paraformaldehyde in 1 × PBS for 10 min at room temperature at 20 h post infection and prepared for immunofluorescence analysis as described above. Primary antibodies used in this experiment were: polyclonal anti-Golgin 97 (Abcam, ab84830), and anti-*Salmonella* CSA-1 (BacTrace, 5310-0322); secondary antibodies used were: Alexa Fluor 488 and Alexa Fluor 568 (Thermo Fisher Scientific).

Cytometric Bead Array Assay

RAW264.7 cells were infected with S. Typhimurium as above, with cell culture media replaced at 16 h post infection to focus on cytokine secretion at late stages of infection. Supernatants from RAW264.7 cells infected with various *Salmonella* strains were collected at 20 or 24 h post infection for analysis. Samples were processed for analysis using Cytometric Bead Array (CBA) Flex Sets (BD Biosciences) according to manufacturer's instructions. Samples were analyzed on the BD LSRFortessa^{TM} cell analyzer.

RESULTS

SseK3 Glycosylates Rab1, Rab5, and Rab11 During *S.* Typhimurium Infection

Using a mass spectrometry-based approach on total cell lysates, we previously identified TNFR1 and TRAILR as targets of SseK3 during *Salmonella* infection of RAW264.7 cells (Newson et al., 2019). However, given that SseK3 exhibits strong Golgi localization during *Salmonella* infection (Gunster et al., 2017), we postulated that SseK3 may also target Golgi-associated membrane proteins. To test this, we enriched membrane fractions from *Salmonella*-infected RAW264.7 cells at 20 h post infection, and digested these to produce peptides that were then immunoprecipitated using an antibody that recognizes Arg-GlcNAc and then analyzed by mass spectrometry. To identify the Arg-GlcNAcylated targets of SseK3 we analyzed the peptides immunoprecipitated from RAW264.7 cells infected with S. Typhimurium Δ*sseK12* and compared these to RAW264.7 cells infected with S. Typhimurium Δ*sseK123*. This approach revealed Arg-GlcNAcylation of previously identified SseK3 targets, TRAILR (Tnfrsf10b) and TNFR1 (Tnfrsf1a) (Newson et al., 2019), as well as novel host and bacterial protein Arg-GlcNAcylation events (**Figure 1A, Supplementary Tables 1, 2**). Consistent with the Golgi localization of SseK3, the Arg-GlcNAcylated targets identified were enriched for GO terms associated with Golgi-related biological processes including "ER to Golgi vesicle- mediated transport", "Golgi vesicle transport" and "Retrograde transport, endosome to Golgi" (**Figure 1B, Supplementary Table 3**). Among the Golgi-related targets identified were several Rab

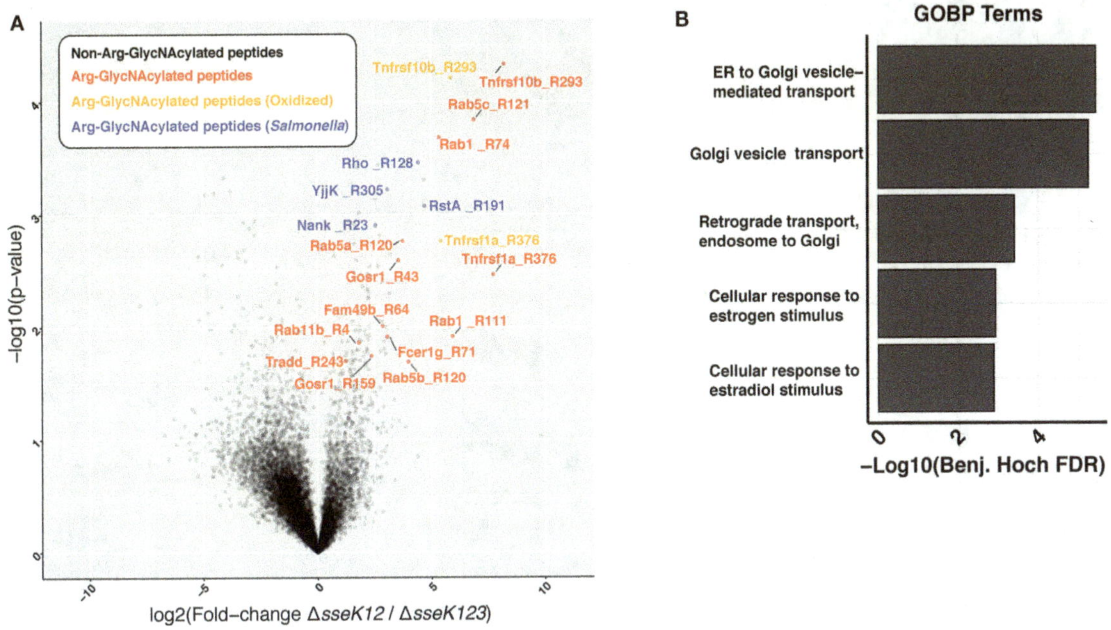

FIGURE 1 | Enrichment of peptides Arg-GlcNAcylated by SseK3 derived from the insoluble fraction of *Salmonella*-infected RAW264.7 cell lysates. Arg-GlcNAcylated peptides were enriched and immunoprecipitated from the insoluble fraction of RAW264.7 cells infected with *S.* Typhimurium SL1344 Δ*sseK12* or *S.* Typhimurium SL1344 Δ*sseK123*. **(A)** The volcano plot depicts the mean ion intensity peptide ratios of Δ*sseK12* vs. Δ*sseK123* plotted against the -logarithmic student *t*-test *p*-values from biological triplicate experiments. Arg-GlcNAcylated peptides with a fold change > 2 and a $p < 0.05$ are highlighted. Peptides are labeled by their gene names followed by the location of the Arg-GlcNacylated arginine. Arg-GlcNAcylated peptides from the host cell are highlighted in red; Methionine oxidized Arg-GlcNAcylated peptides from host cell are highlighted in yellow; Arg-GlcNAcylated peptides from *Salmonella* are highlighted in blue. **(B)** Bar chart of enrichment analysis of GO-terms (Biological processes) associated with Arg-GlcNAcylated peptides compared to all observed peptides from immunoprecipitation experiments. Fisher's exact enrichment analysis demonstrates the over-representation of Golgi-associated processes with Arg-GlcNAcylated proteins in the insoluble membrane immunoprecipitation.

GTPases, including Rab1, Rab5, and Rab11 (**Figure 1A**, **Supplementary Tables 1, 2**).

Rab GTPases are master regulators of intracellular vesicle transport. Rab5 and Rab11 are regulators of early endosomes and recycling endosomes respectively; while Rab1 mediates vesicle transport from the ER to Golgi and can be found on Golgi membranes (Zhen and Stenmark, 2015; Prashar et al., 2017). To confirm that Rabs are modified by SseK3, we co-transfected HEK293T cells with 3xFlag-tagged Rab1a, Rab5a/b/c or Rab11b and GFP tagged SseK3, and then performed anti-Flag immunoprecipitation and subsequent immunoblot analysis using Arg-GlcNAc antibodies. GFP or the catalytically inactive glycosyltransferase motif mutant GFP-SseK3$_{DXD}$ were used as controls. We found that SseK3 does modify human Rab1a (**Figure 2A**), Rab5a/b/c (**Figure 2B**) and Rab11b (**Figure 2C**) with GlcNAc in transfected mammalian cells.

Site-Directed Mutagenesis of Human Rab1 Confirms SseK3 Modifies Arg74, Arg82, and Arg111

Our mass spectrometry analysis revealed three different SseK3-mediated modification sites within Rab1 (corresponding to Arg74, Arg82, and Arg111 in Rab1a), whereas SseK3 modified single arginine residues in Rab5 and Rab11 respectively (**Figure 1A**, **Supplementary Tables 1, 2**). Strikingly, two of the Rab1 modification sites were located in the $_{74}$RTITSSYYR$_{82}$ peptide within the catalytic switch II region (**Figure 3A**). In contrast, the Rab5 and Rab11 modification sites were not located in this region, occurring on Arg120 in the third α-helix and Arg4 at the N-terminus respectively, and the roles of these residues in Rab activity are unknown. The switch II region of Rab1 is a hotspot for post-translational modifications by different bacterial effectors to regulate Rab1 activity (Muller et al., 2010; Mukherjee et al., 2011; Wang et al., 2018). The Rab switch II region, in addition to the switch I region is involved in nucleotide binding, and shift from being unfolded in the GDP-bound state to adopting well-defined conformations in the GTP-bound state to allow for Rab interactions with host effector proteins (Zhen and Stenmark, 2015). Given the importance of the switch II region, we focussed on Rab1a modification by SseK3. To confirm the SseK3 modification sites within Rab1a, we mutated the arginine residues of interest to alanines, which cannot be GlcNAcylated. Individual 3xFlag-tagged Rab1 mutants were co-expressed in HEK293T cells with GFP-tagged SseK3 and then subjected to Flag immunoprecipitation and immunoblot analysis. We found mutation of the arginine residues individually did not significantly impact Arg-GlcNAcylation of Rab1 by SseK3

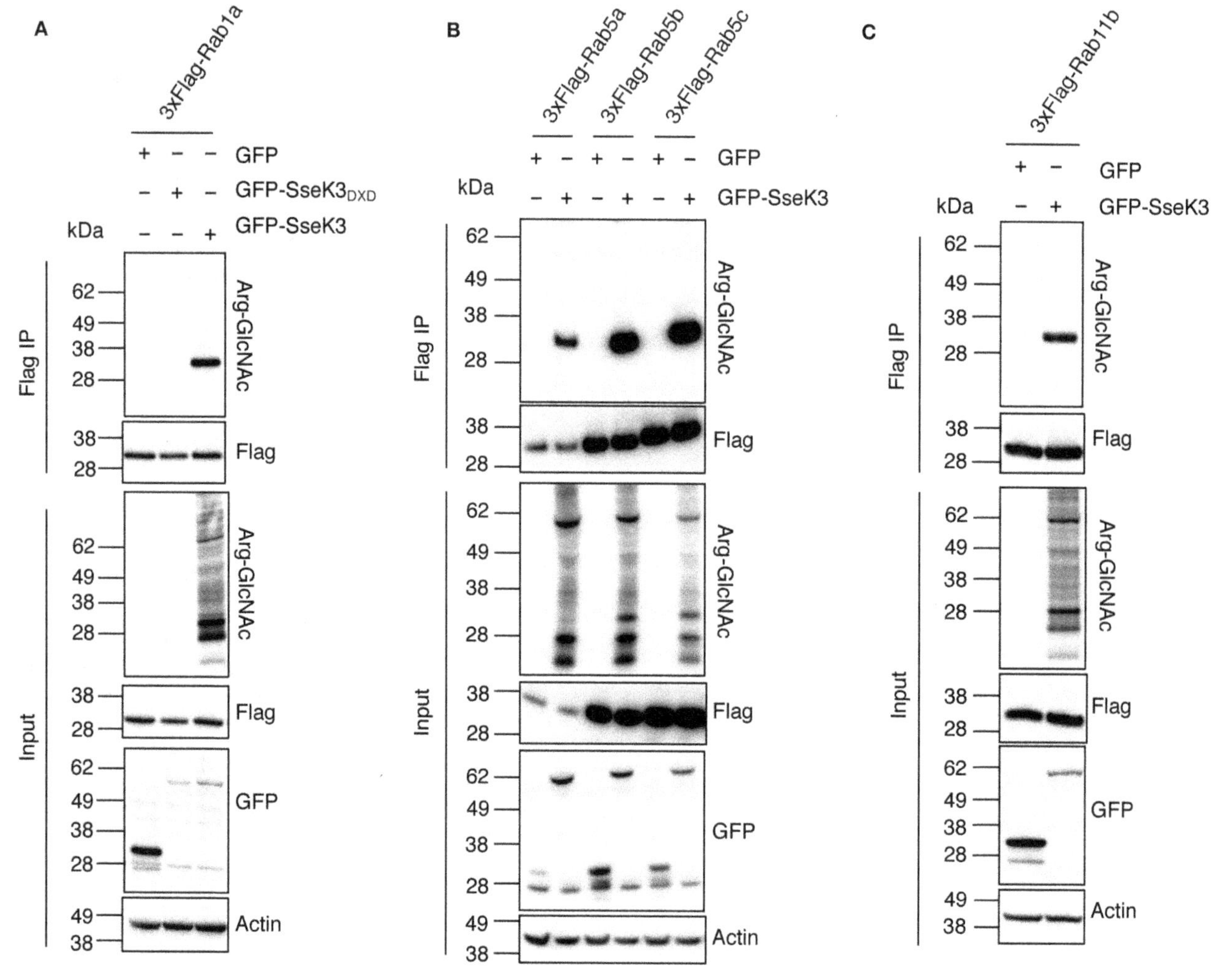

FIGURE 2 | SseK3 modifies Rab1, Rab5, and Rab11 in co-transfected HEK293T cells. pEGFP-C2, pEGFP-C2-SseK3 or pEGFP-C2-SseK3$_{DXD}$ were co-transfected with either p3xFlag-Rab1a **(A)**, p3xFlag-Rab5a/b/c **(B)**, or p3xFlag-Rab11b **(C)** into HEK293T cells before 3xFlag-tagged proteins were immunoprecipitated. Input and immunoprecipitate (IP) were subjected to immunoblot analysis with anti-ArgGlcNAc, anti-Flag M2-HRP, anti-GFP or anti-β-actin antibodies. Representative immunoblots of at least 3 independent experiments.

(**Figure 3B**). Mutating two residues at a time had a modest effect on blocking modification and mutating all of Arg74, Arg82, and Arg111 resulted in complete abrogation of Rab1a Arg-GlcNAcylation, suggesting SseK3 modifies each of these three arginine residues (**Figure 3B**). Rab1a$_{R74AR82A}$ was less efficiently modified by SseK3 compared to the other double site mutants, suggesting SseK3 may preferentially modify these two arginine residues located within the critical switch II region of Rab1a (**Figure 3B**). We then examined the intracellular localization of 3xFlag-tagged Rab1a arginine mutants by confocal microscopy. 3xFlag-tagged Rab1a or arginine mutants were expressed in HEK293T cells before the cells were stained with anti-golgin-97 antibodies for immunofluorescence analysis. 3xFlag-Rab1a showed a staining pattern consistent with Golgi localization (**Supplementary Figure 1**). All other mutants showed similar localization patterns, other than those containing a mutation of Arg82 (**Supplementary Figure 1**). All of 3xFlag-Rab1a$_{R82A}$, 3xFlag-Rab1a$_{R74AR82A}$, 3xFlag-Rab1a$_{R82AR111A}$ and 3xFlag-Rab1a$_{R74AR82AR111A}$ were instead expressed throughout the cell (**Supplementary Figure 1**).

SseK3 Modifies Both GTP-Bound and GDP-Bound Rab1

Rab GTPases cycle between a GTP-bound active state and GDP-bound inactive state to mediate different steps of vesicle trafficking. The Rab switch regions undergo major conformational changes depending on the nucleotide binding state (Pfeffer, 2005). These changes may influence the ability of effectors to bind and modify the Rabs. For example, the *Legionella* glucosyltransferase effector, SetA preferentially modifies GDP-bound Rab1 (Wang et al., 2018), and an endogenous Rab1 regulator, TAK1, also preferentially phosphorylates the GDP-bound form of Rab1 (Levin et al., 2016). Thus, we explored whether SseK3 exhibited a preference for GTP-bound or

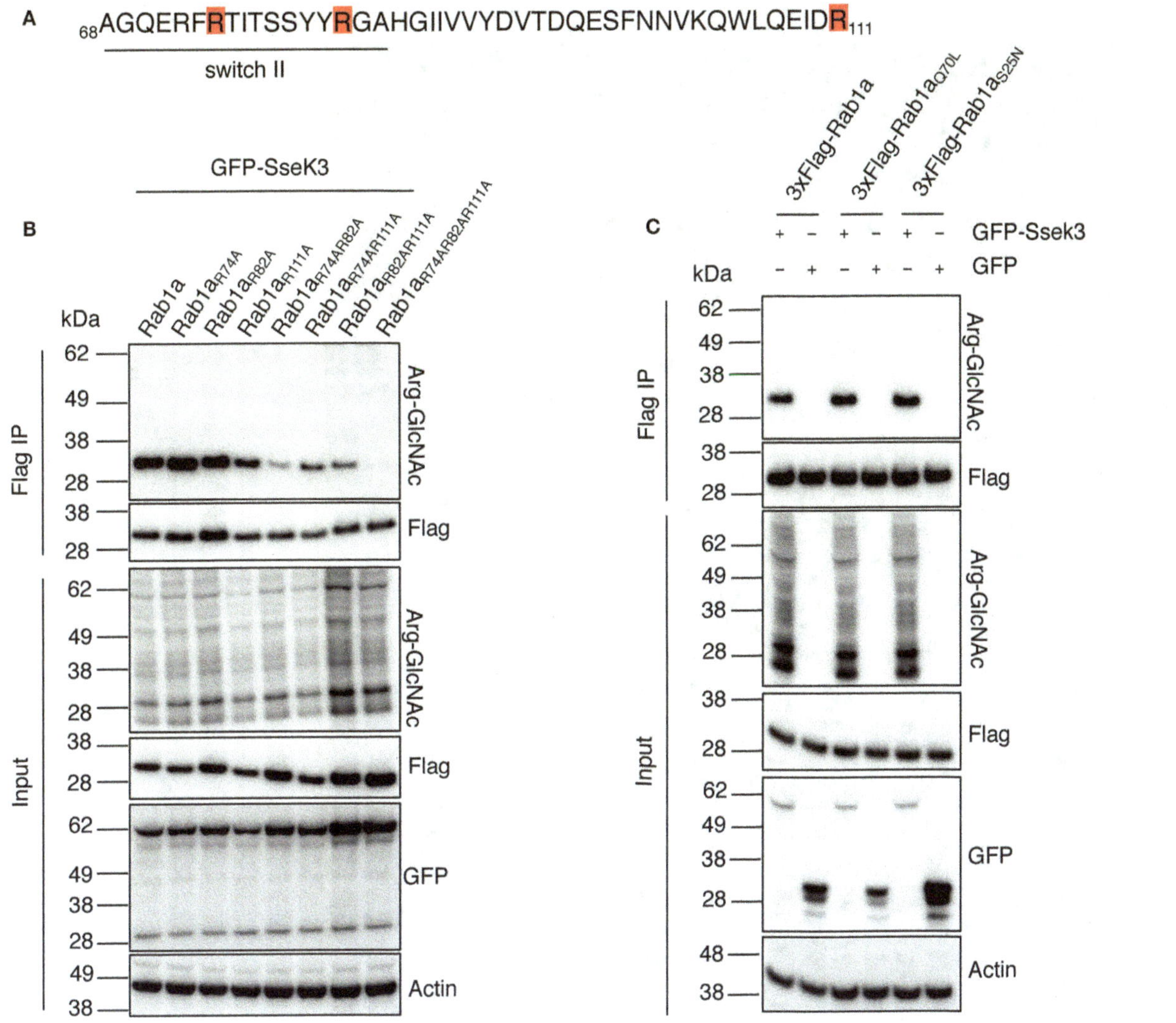

FIGURE 3 | SseK3 modifies Arg74, Arg82, and Arg111 within Rab1a and has no preference for GTP-bound or GDP-bound Rab1a. **(A)** Amino acid sequence of Rab1a containing the arginine residues modified by SseK3 (highlighted in red) and the Rab1 switch II region (underlined). **(B)** GFP-SseK3 was co-expressed with 3xFlag tagged Rab1a arginine site-directed mutants in HEK293T cells. Flag-immunoprecipitation was performed on the cell lysates with subsequent immunoblot analysis using anti-ArgGlcNAc, anti-Flag M2-HRP, anti-GFP or anti-β-actin antibodies. Representative immunoblots of at least 3 independent experiments. **(C)** GFP or GFP-SseK3 together with 3xFlag tagged Rab1a or Rab1a nucleotide binding state mutants (active-state mimetic Rab1a$_{Q70L}$, or constitutively inactive GDP-bound Rab1a$_{S25N}$) were co-expressed in HEK293T cells by transfection. 3xFlag tagged proteins were immunoprecipitated for immunoblot analysis with anti-ArgGlcNAc, anti-Flag M2-HRP, anti-GFP or anti-β-actin antibodies. Representative immunoblots of at least 3 independent experiments.

GDP-bound Rab1a. A constitutively inactive 3xFlag-tagged GDP-bound Rab1a$_{S25N}$ (Nuoffer et al., 1994) or active-state mimetic Rab1a$_{Q70L}$ (Tisdale et al., 1992) were co-transfected with GFP-SseK3 into HEK293T cells before being subjected to Flag-immunoprecipitation and immunoblot analysis. No significant difference in Arg-GlcNAc modification was observed among Rab1a, Rab1a$_{S25N}$ or Rab1a$_{Q70L}$ (**Figure 3C**). Thus, modification of Rab1a by SseK3 was independent of the Rab1a nucleotide binding state. In this way, SseK3 functions differently in comparison to the endogenous Rab1 regulator TAK1, or the glucosyltransferase bacterial effector, SetA.

SseK2 and SseK3 Co-localize With Rab1 at the Golgi

We next investigated the cellular localization of Rab1a in the presence of the SseKs. GFP, GFP-SseK1, GFP-SseK2, GFP-SseK3 or their catalytically inactive mutants were co-expressed with 3xFlag-tagged Rab1 in HEK293T cells before the cells were stained with anti-golgin-97 antibodies for immunofluorescence analysis. 3xFlag-Rab1a showed a Golgi-associated staining pattern in all samples tested, while both GFP and GFP-SseK1 were expressed throughout the cell (**Figure 4A**). Consistent with previous findings, GFP-SseK3 also localized to the Golgi in HEK293T cells (**Figure 4A**) (Gunster et al., 2017). Furthermore,

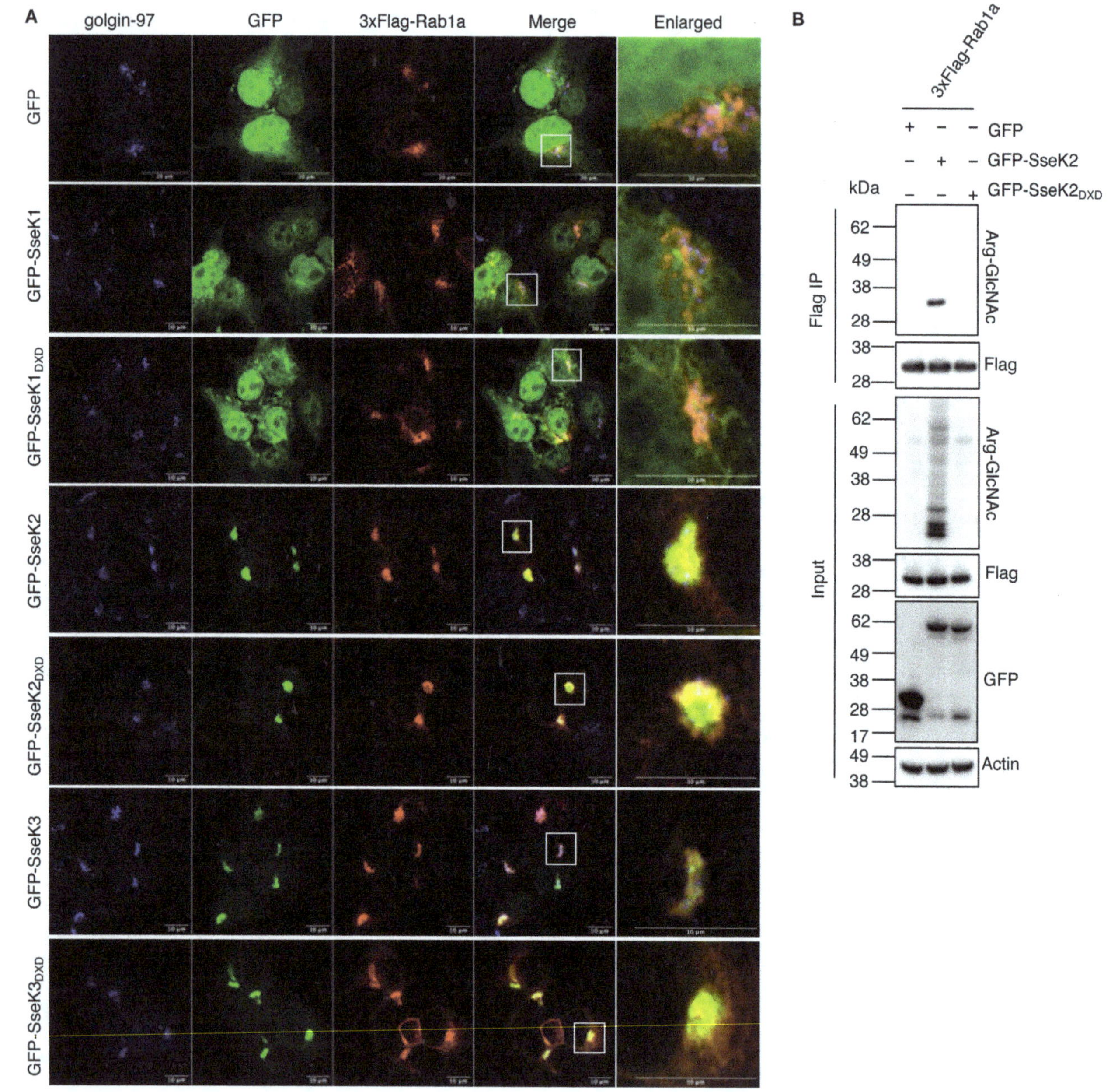

FIGURE 4 | SseK2 modifies Rab1a and both SseK2 and SseK3 co-localize with Rab1a at the Golgi. **(A)** The intracellular localization pattern of 3xFlag-Rab1a was examined in the presence of GFP, GFP-SseK1, GFP-SseK2, GFP-SseK3 and their catalytically inactive mutants by confocal microscopy in transfected HEK293T cells. Anti-golgin-97 and anti-Flag antibodies were used to identify the Golgi and recombinant 3xFlag-Rab1a respectively. White boxes highlight Golgi that are shown enlarged. Representative immunofluorescence fields of at least 3 independent experiments **(B)** SseK2 also modifies Rab1a with Arg-GlcNAc. pEGFP-C2, pEGFP-C2-SseK2 or pEGFP-C2-SseK2$_{DXD}$ were co-transfected with p3xFlag-Rab1a into HEK293T cells before 3xFlag-tagged proteins were immunoprecipitated. Input and immunoprecipitate (IP) were subjected to immunoblot analysis with anti-ArgGlcNAc, anti-Flag M2-HRP, anti-GFP or anti-β-actin antibodies.

we observed that GFP-SseK3$_{DXD}$ also localized to the Golgi with 3xFlag-Rab1a, indicating localization was independent of Arg-GlcNAcylation activity (**Figure 4A**).

SseK2 also localizes to the Golgi during *Salmonella* infection (Gunster et al., 2017), and may therefore share targets with SseK3. We found that SseK2 and Rab1a also co-localized at the Golgi, and this was also independent of SseK2 catalytic activity (**Figure 4A**). This suggested that SseK2 may also target Rab1a for Arg-GlcNAcylation. Flag-immunoprecipitation and immunoblot analysis confirmed Arg-GlcNAcylation of 3xFlag-Rab1a by GFP-SseK2 in co-transfected HEK293T cells (**Figure 4B**).

Inhibition of Rab1 can lead to disruption of the Golgi (Dong et al., 2012). In our co-transfection studies we did not observe

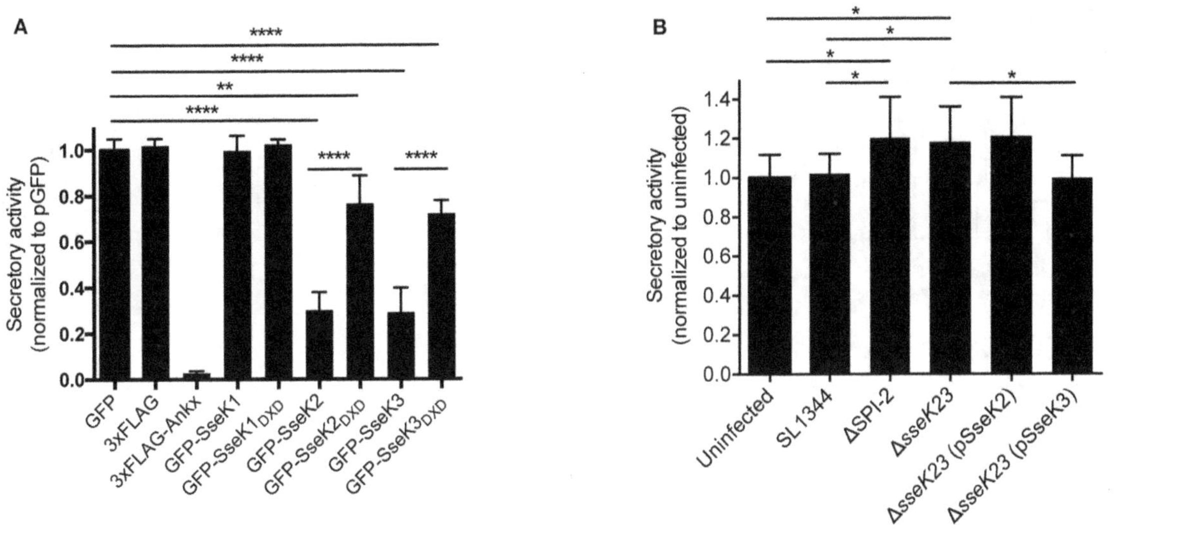

FIGURE 5 | SseK3 inhibits the secretion of alkaline phosphatase in transfected HEK293T cells and during *Salmonella* infection. **(A)** pSEAP was co-transfected with mammalian expression vectors expressing GFP, 3xFlag, 3xFlag-AnkX, GFP-SseK1, GFP-SseK2, GFP-SseK3 or their catalytic mutants into HEK293T cells. Supernatants and cell lysates were then analyzed for alkaline phosphatase activity. Secretory activity was calculated as alkaline phosphatase activity in the supernatants divided by total alkaline phosphatase activity, which includes alkaline phosphatase activity in both supernatants and cell lysates. These were then normalized to the GFP expressing sample. Results are mean + SD of three independent experiments performed in duplicate. ****$P < 0.0001$, **$P < 0.01$; unpaired, two-tailed *t*-test. **(B)** HeLa229 cells stably expressing SEAP were infected with various *Salmonella* SL1344 strains. The cell culture media was replaced at 16 h post infection, and infection allowed to proceed for a further 8 h. Supernatants and cell lysates were analyzed for alkaline phosphatase activity at 24 h of infection. Secretory activity was calculated as alkaline phosphatase activity in the supernatants divided by total alkaline phosphatase activity, normalized to uninfected cells. Results are mean + SD of three independent experiments performed in duplicate. *$P < 0.05$; unpaired, two-tailed *t*-test.

Golgi disruption (**Figure 4A**). However, we also examined the Golgi in cells that were not overexpressing Rab1, and found that after 20 h of infection with wild type S. Typhimurium SL1344, 15% of infected cells showed Golgi disruption, but this was independent of SseK2 or SseK3 (**Supplementary Figure 2**).

SseK3 Inhibits Host Protein Secretion During Transfection and *S.* Typhimurium Infection

As Rab1 mediates vesicle transport from the ER to Golgi early in the secretory pathway, we explored the functional consequences of Arg-GlcNAcylation on Rab1 activity by testing whether the SseK family of proteins inhibited host protein secretion. We employed secreted embryonic alkaline phosphatase (SEAP) as a reporter to examine the activity of the secretory pathway in transfected cells. AnkX, a *Legionella* effector which significantly inhibits SEAP secretion (Mukherjee et al., 2011), was adopted as a positive control for the assay. Vectors expressing 3xFlag, 3xFlag-AnkX, GFP, GFP-SseK1, GFP-SseK2, GFP-SseK3 or their catalytically inactive DXD motif mutants were co-transfected into HEK293T cells with a SEAP expressing vector. As expected, expression of 3xFlag-AnkX significantly inhibited the secretion of SEAP in comparison to cells expressing 3xFlag only (**Figure 5A**). Expression of either GFP-SseK2 or GFP-SseK3 significantly inhibited the secretion of SEAP compared to GFP or GFP-SseK1 expressing cells (**Figure 5A**). Catalytically inactive SseK2 and SseK3 also partially inhibited the secretion of SEAP; however, this was not to the level of inhibition mediated by active SseK2 and SseK3 (**Figure 5A**).

We next explored whether SseK2 and SseK3 inhibited host protein secretion during S. Typhimurium SL1344 infection. A SEAP expressing HeLa229 reporter cell line was constructed and infected with derivatives of S. Typhimurium before a SEAP assay was performed. To analyse protein secretion at a timepoint when SseK2 and SseK3 are likely to be active, the cell media was changed at 16 h post infection, and infection allowed to proceed for a further 8 h before SEAP analysis at 24 h post infection. Wild type S. Typhimurium SL1344-infected cells showed similar levels of SEAP secretion compared to uninfected cells. Complex manipulation of host signaling pathways occurs during infection, thus SL1344 infection may simultaneously activate and interfere with host cell protein secretion. In support of this, modest increases in SEAP secretion were observed from cells infected with either ΔSPI-2 or Δ*sseK23 Salmonella* strains compared to cells infected with S. Typhimurium SL1344 or uninfected cells (**Figure 5B**). However, SEAP secretion returned to the levels observed for SL1344-infected cells only upon complementation of Δ*sseK23* with SseK3, and not SseK2 when over-expressed from a plasmid (**Figure 5B**). Hence, although both SseK2 and SseK3 robustly inhibited the secretion of SEAP when transfected into cells, we were unable to confirm a role for SseK2 in inhibition of the host cell secretory pathway in the context of infection, and SseK3 had only a modest impact on SEAP secretion during *Salmonella* infection. To control for bacterial numbers in these experiments, we examined intracellular replication of

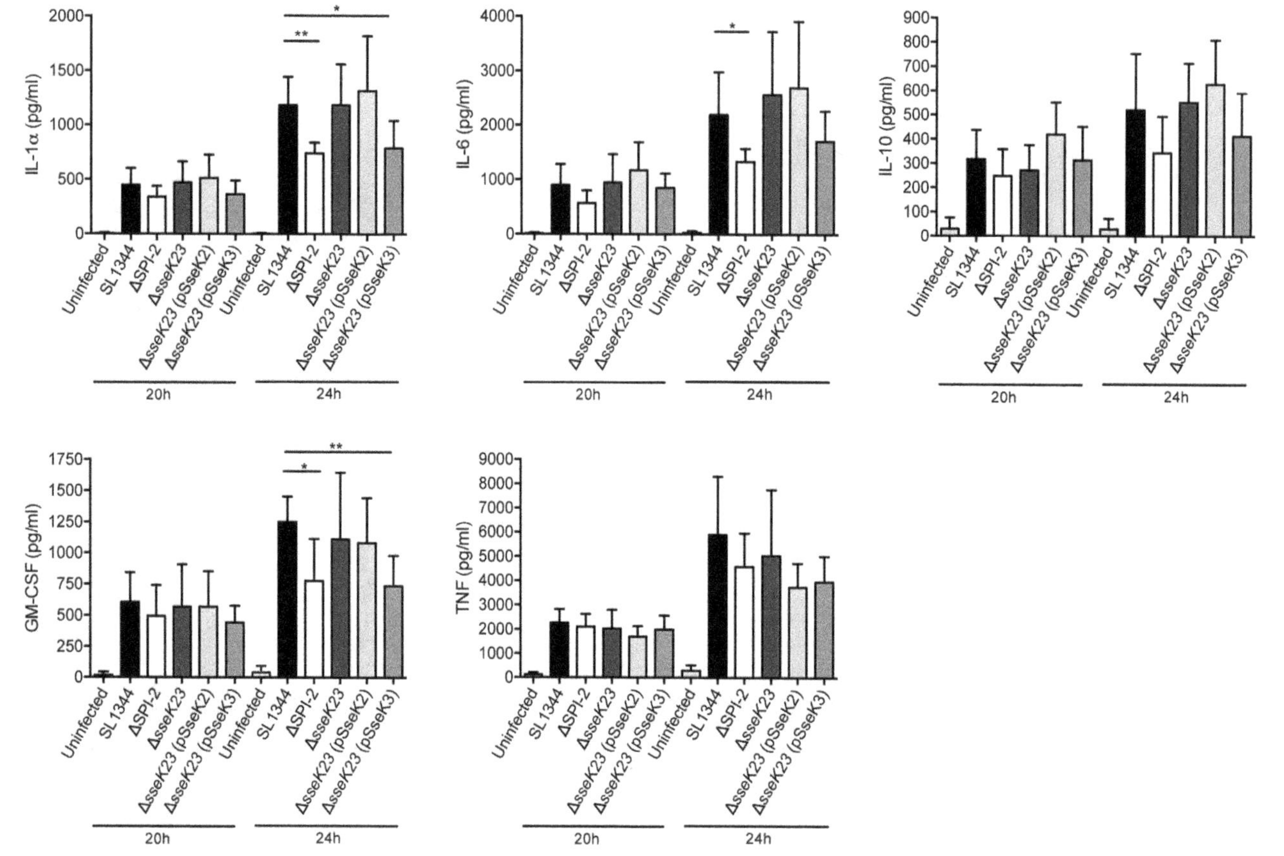

FIGURE 6 | SseK3 does not inhibit the secretion of cytokines during *Salmonella* infection of RAW264.7 cells. Supernatants from RAW264.7 cells infected with wild type, ΔSPI-2, Δ*sseK23*, Δ*sseK23* (pSseK2), or Δ*sseK23* (pSseK3) *S.* Typhimurium SL1344 strains were collected for cytometric bead array analysis. After 16 h of infection, the cell culture media was changed, and supernatants were collected at 20 or 24 h post infection. Concentrations of cytokines in supernatants were determined based on fluorescence intensities from PE conjugated beads by flow cytometry compared to a standard curve and represented as pg/ml. Results are mean + SD of three independent experiments performed in duplicate. $^{*}P < 0.05$, $^{**}P < 0.01$; unpaired, two-tailed *t*-test.

S. Typhimurium SL1344 and its derivatives and found that compared to wild type SL1344, only the SPI-2 mutant showed significantly impaired replication in the HeLa229 SEAP cell line (**Supplementary Figure 3A**).

Effect of SseK3 on Cytokine Secretion During Infection of RAW264.7 Cells

Given their potential effect on host cell protein export, we next explored whether SseK2 or SseK3 inhibited cytokine secretion during *Salmonella* infection. RAW264.7 cells were infected with wild type *S.* Typhimurium SL1344 or mutant derivatives for 16 h before the cell culture media was changed and then supernatants were collected and analyzed for cytokine levels at 20 and 24 h post infection using a cytometric bead array. Replication of the strains at 24 h post infection was also examined, with only the SPI-2 mutant showing impaired replication in RAW264.7 cells (**Supplementary Figure 3B**). Compared to uninfected cells, *S.* Typhimurium SL1344 infection resulted in increased cytokine secretion at both 20 and 24 h post infection (**Figure 6**). In contrast to the SEAP assay, ΔSPI-2 infection of RAW264.7 cells resulted in less IL-1α, IL-6 and GM-CSF cytokine secretion compared to wild type-infected samples at 24 h of infection and no significant differences were observed between wild type SL1344 and Δ*sseK23* infected cells (**Figure 6**). A reduction in IL-1α and GM-CSF secretion was observed when SseK3 was over-expressed in the Δ*sseK23* mutant compared to SL1344-infected cells, but not when compared to Δ*sseK23*-infected cells (**Figure 6**). In summary, during infection SseK3 appeared to have a greater effect on Rab1-dependent host protein secretion than SseK2, but overall SseK3 had only a marginal influence on the secretion levels of some cytokines when overexpressed.

DISCUSSION

Rab GTPases are well known for their role in mediating endocytosis and exocytosis as well as other intracellular membrane trafficking events. Many bacterial pathogens including *Salmonella* hijack Rab-dependent pathways to facilitate infection (Spano and Galan, 2018). For example, maturation of the SCV requires participation of several key intracellular vesicle transport regulators, including Rab5, Rab7, and Rab11 (Knodler and Steele-Mortimer, 2003; Brumell and

Grinstein, 2004; Smith et al., 2005). Upon invasion, *Salmonella* modulates Rab recruitment to the SCV in a SPI-1-dependent manner (Smith et al., 2007). Rab5 is recruited to the SCV by the SPI-1 effector SopB, which is a phosphoinositide phosphatase (Mallo et al., 2008), while another SPI-1 effector, SopE, functions as a guanine exchange factor (GEF) for Rab5 and promotes the formation of GTP-bound active Rab5 on the SCV (Mukherjee et al., 2001). The enhanced retention of active Rab5 on the SCV is hypothesiszed to promote fusion with early endosomes and prevent trafficking to mature lysosomes (Parashuraman and Mukhopadhyay, 2005; Madan et al., 2008). Rab11 is also recruited to early SCVs following *Salmonella* invasion, and is involved in SCV maturation but is not essential for replication (Smith et al., 2005, 2007). Many studies of Rab manipulation by *Salmonella* focus on early time points of infection before the complete repertoire of SPI-2 effectors are expressed and translocated into host cells. Less is known about the impact of *Salmonella* on the host endosomal pathway at later stages of infection, although two SPI-2 effectors, SopD2 and GtgE are known to target Rabs including Rab32. SopD2 functions as a GTPase activating protein (GAP) for Rab32 (Spano et al., 2016) while GtgE is a cysteine protease that cleaves Rab32 (Spano and Galan, 2012, 2018; Wachtel et al., 2018). Together, these effectors inhibit recruitment of Rab32 to the SCV and subsequent Rab32-mediated control of replication (Spano and Galan, 2012; Spano et al., 2016).

In this study, we identified Rab1, Rab5, and Rab11 as host targets of the SPI-2 effector, SseK3 during *Salmonella* infection. These targets were identified by mass spectrometry and confirmed by immunoblot of ectopically expressed Rabs immunoprecipitated from HEK293T cells co-expressing SseK3. Using a SEAP reporter assay we found SseK3 impaired host protein secretion in transfected cells and modestly reduced secretion levels during *Salmonella* infection, suggesting Arg-GlcNAcylation of Rab1 by SseK3 at least partially blocked Rab1 activity. However, SseK3 did not appear to reduce the secretion of selected cytokines during *Salmonella* infection when expressed and translocated at native levels. Interestingly, inactive $SseK3_{DXD}$ and $SseK2_{DXD}$ also partially inhibited SEAP secretion in transfected cells, but the reason for this is unclear.

While this study was under review, another group also reported that Rab1 is Arg-GlcNAcylated by SseK3 (Meng et al., 2020). In contrast to our work, Meng et al. concluded that SseK3 did inhibit cytokine secretion during infection, while the impact of SseK2 on Rab1 and host protein secretion was not examined (Meng et al., 2020). Whereas Meng et al. performed infections with a triple mutant of S. Typhimurium SL1344 lacking all of *sseK1*, *sseK2* and *sseK3*, we used a double *sseK2/sseK3* mutant for our studies. Given that the strains we used retained SseK1, which together with SseK3 can inhibit inflammatory signaling through TNFR1 (Gunster et al., 2017; Newson et al., 2019), differences in the results observed by Meng et al. compared to our study could be due to altered cytokine expression levels due to the presence or absence of SseK1 (Meng et al., 2020). Thus, the SEAP reporter was a more direct measure of the ability of SseK3 to block the host cell secretory pathway as its expression was not influenced by inflammatory signaling. Although a previous study aimed at discovering *Salmonella* effector proteins that interact with host exocytic pathway failed to identify SseK2 and SseK3 as inhibitors of SEAP secretion, (Perrett and Zhou, 2013), this may have been due to insufficient expression of SseK2 and SseK3, which was not determined.

Two Rab1-targeting SPI-2 effectors, SseF and SseG, were recently reported to inhibit host autophagy during infection by abolishing Rab1 activation, indicating Rab1 is targeted by *Salmonella* once the SCV is established (Feng et al., 2018). It is not surprising that *Salmonella* employs multiple effectors to modulate different Rab1-related host cell events, considering intracellular pathogens need to counteract multiple host cell defense pathways. Previous studies on *Legionella* provide a good example of how intracellular pathogens utilize different effectors to modulate Rab1, even with contradictory impacts. During *Legionella* infection, spatio-temporal regulation of Rab1 activity is achieved largely via several post-translational modifications. For example the *Legionella* Dot/Icm effector SidM/DrrA AMPylates Tyr_{80} of Rab1 to retain it in the GTP-bound active state; while this post-translational modification is reversed by another effector, SidD (Muller et al., 2010; Mukherjee et al., 2011; Tan and Luo, 2011). Another *Legionella* effector, AnkX modifies Ser_{79} with a phosphocholine moiety and this modification is eliminated by a dephosphorylcholinase effector, Lem3 (Tan et al., 2011). Furthermore, a *Legionella* glucosyltransferase effector SetA modifies Thr_{75} of Rab1 with a glucose molecule and thus limits GTPase activity (Wang et al., 2018). Interestingly, most sites on Rab1 targeted for post-translational modification by different bacterial effectors, as well as the endogenous Rab1 regulator TAK1 (Levin et al., 2016), are located within the $_{74}$RTITSSYYR$_{82}$ peptide of the switch II region, highlighting the susceptibility of this region to attack and/or regulation. Notably, we found SseK3 modified three different arginine residues in Rab1a, Arg74, Arg82 and Arg111, two of which were located within the Rab1 switch II region. Meng et al. identified the same SseK3-mediated Arg-GlcNAc modification sites within Rab1, and an additional modification at Arg72 when SseK3 was expressed from a multi-copy plasmid during S. Typhimurium infection (Meng et al., 2020).

We did not directly examine the activity of Arg-GlcNAc modified Rab1 in this study, however Meng et al. reported that SseK3-modified Rab1 had reduced GTPase activity, impaired interaction with binding partners and the membrane cycling of Rab1 was also perturbed (Meng et al., 2020). Our observations that SseK3 modified Rab1 regardless of its nucleotide binding state was also supported by Meng et al. (2020) and suggests that Arg-GlcNAc modified Rabs may also be present in the soluble fraction of S. Typhimurium-infected cells, as GDP-bound Rabs are not membrane bound (Zhen and Stenmark, 2015; Prashar et al., 2017). Indeed, we previously identified Arg-GlcNAc modified peptides from Rab1 and Rab5 in total cell lysates from RAW246.7 cells infected with S. Typhimurium Δ*sseK12*, although the modified Rab peptides were either not detected in all three biological replicates or were detected at lower levels compared to TNFR1 and TRAILR2 (Newson et al., 2019).

Similar to Meng et al., we found that SseK3 and SseK2 exhibited Golgi-associated localization with Rab1a, but that this was independent of Arg-GlcNAc activity. SseK3 localized to the *cis*-Golgi independently of its catalytic motif, with localization mediated by a polybasic region within SseK3 that binds phospholipids (Meng et al., 2020). However, whereas Meng *et al.* reported that ectopic expression of SseK3 caused fragmentation of the Golgi (Meng et al., 2020), we did not observe altered Golgi morphology when either SseK2 or SseK3 were translocated at native levels during S. Typhimurium infection. Thus the observations by Meng *et al.* could be artifacts of SseK3 over-expression, making the physiological relevance of SseK3-mediated Golgi disruption during infection unclear (Meng et al., 2020). This may also suggest that Rab1 is not a primary target of SseK3 during infection, or that levels of non-modified Rab1 are still sufficient to maintain Golgi structure. Notably, no host or pathogen factors have been reported to reverse Arg-GlcNAcylation (Scott et al., 2017). As such, the impact of Arg-GlcNAc modification on intracellular vesicle transport could still be significant over time even when only a small portion of target protein is modified. Further work is required to examine the effect of SseK3 on Rab5 or Rab11 function, as they may be preferred targets of SseK3 during S. Typhimurium infection. Interestingly, Meng *et al.* did not report Rab5 or Rab11 as significant targets of SseK3 during infection. Discrepancies between the targets of SseK3 identified may be due to the different approaches and cell lines used. Our Arg-GlcNAc pulldowns were performed on insoluble fractions of infected RAW246.7 cells, while Meng *et al.* identified modified proteins in cell lysates from transfected or infected HEK293T cells, or infected HeLa, iBMDM or MEF cells (Meng et al., 2020).

In addition to Rab1, Rab5, and Rab11, we identified several *Salmonella* proteins that were Arg-GlcNAcylated by SseK3 during infection, including the transcription termination factor Rho and a two-component regulatory system factor, RstA. The consequences of these modifications have not been investigated here, however it is worthwhile noting that NleB of enterohaemorrhagic and enteropathogenic *E. coli* and *C. rodentium* also have intra-bacterial Arg-GlcNAcylation activity, and this impacts bacterial survival in oxidative stress conditions (El Qaidi et al., 2020). NleB1, SseK1, and SseK3 also perform auto-Arg-GlcNAcylation, and this is required for activity against death domain protein targets (Xue et al., 2020).

In summary, we found that the *Salmonella* glycosyltransferase effector SseK3 modified Rab1, Rab5, and Rab11 during *Salmonella* infection. SseK3 targeted critical arginine residues in the switch II region of Rab1, thereby influencing host cell protein secretion during infection. Hence, in addition to its role in blocking death receptor signaling (El Qaidi et al., 2017; Gunster et al., 2017; Newson et al., 2019), SseK3 may modify the host cell secretome during the later stages of *Salmonella* infection.

AUTHOR CONTRIBUTIONS

JG designed and performed experiments and wrote the manuscript. NS and JN designed and performed mass spectrometry experiments. RW, TW, and GP made reagents and GN, ID, and JP provided experimental advice. CG and EH conceptualized and supervised the study and wrote the manuscript. All authors provided critical feedback and edited the manuscript.

ACKNOWLEDGMENTS

We thank Nathaniel Brown (University of British Columbia) and Richard Strugnell (University of Melbourne) for bacterial strains and Craig Roy (Yale University) and John Silke (Walter and Eliza Hall Institute) for plasmids. We thank the Melbourne Mass Spectrometry and Proteomics Facility of The Bio21 Molecular Science and Biotechnology Institute at The University of Melbourne for the support of mass spectrometry analysis. The authors acknowledge Monash Micro Imaging, Monash University, for the provision of instrumentation, training and technical support. The use of a cytometer was supported by the Melbourne Cytometry Platform, The University of Melbourne.

REFERENCES

Brown, N. F., Coombes, B. K., Bishop, J. L., Wickham, M. E., Lowden, M. J., Gal-Mor, O., et al. (2011). Salmonella phage ST64B encodes a member of the SseK/NleB effector family. *PLoS ONE* 6:e17824. doi: 10.1371/journal.pone.0017824

Brumell, J. H., and Grinstein, S. (2004). Salmonella redirects phagosomal maturation. *Curr. Opin. Microbiol.* 7, 78–84. doi: 10.1016/j.mib.2003.12.005

Clark, M. A., Jepson, M. A., Simmons, N. L., and Hirst, B. H. (1994). Preferential interaction of Salmonella typhimurium with mouse Peyer's patch M cells. *Res. Microbiol* 145, 543–552. doi: 10.1016/0923-2508(94)90031-0

Cox, J., Hein, M. Y., Luber, C. A., Paron, I., Nagaraj, N., and Mann, M. (2014). Accurate proteome-wide label-free quantification by delayed normalization and maximal peptide ratio extraction, termed MaxLFQ. *Mol. Cell. Proteomics* 13, 2513–2526. doi: 10.1074/mcp.M113.031591

Cox, J., and Mann, M. (2008). MaxQuant enables high peptide identification rates, individualized p.p.b.-range mass accuracies and proteome-wide protein quantification. *Nat. Biotechnol* 26, 1367–1372. doi: 10.1038/nbt.1511

Dong, N., Zhu, Y., Lu, Q., Hu, L., Zheng, Y., and Shao, F. (2012). Structurally distinct bacterial TBC-like GAPs link Arf GTPase to Rab1 inactivation to counteract host defenses. *Cell* 150, 1029–1041. doi: 10.1016/j.cell.2012.06.050

El Qaidi, S., Chen, K., Halim, A., Siukstaite, L., Rueter, C., Hurtado-Guerrero, R., et al. (2017). NleB/SseK effectors from Citrobacter rodentium, Escherichia coli, and Salmonella enterica display distinct differences in host substrate specificity. *J. Biol. Chem.* 292, 11423–11430. doi: 10.1074/jbc.M117.790675

El Qaidi, S., Scott, N. E., Hays, M. P., Geisbrecht, B. V., Watkins, S., and Hardwidge, P. R. (2020). An intra-bacterial activity for a T3SS effector. *Sci. Rep.* 10:1073. doi: 10.1038/s41598-020-58062-y

Esposito, D., Gunster, R. A., Martino, L., El Omari, K., Wagner, A., Thurston, T. L. M., et al. (2018). Structural basis for the glycosyltransferase activity of the Salmonella effector SseK3. *J. Biol. Chem.* 293, 5064–5078. doi: 10.1074/jbc.RA118.001796

Feng, Z. Z., Jiang, A. J., Mao, A. W., Feng, Y., Wang, W., Li, J., et al. (2018). The Salmonella effectors SseF and SseG inhibit Rab1A-mediated autophagy to facilitate intracellular bacterial survival and replication. *J. Biol. Chem.* 293, 9662–9673. doi: 10.1074/jbc.M117.811737

Geddes, K., Cruz, F., and Heffron, F. (2007). Analysis of cells targeted by Salmonella type III secretion in vivo. *PLoS Pathog.* 3:e196. doi: 10.1371/journal.ppat.0030196

Gunster, R. A., Matthews, S. A., Holden, D. W., and Thurston, T. L. M. (2017). SseK1 and SseK3 type III secretion system effectors inhibit NF-kappaB signaling and necroptotic cell death in salmonella-infected macrophages. *Infect. Immun.* 85:17. doi: 10.1128/IAI.00242-17

Hoiseth, S. K., and Stocker, B. A. (1981). Aromatic-dependent *Salmonella typhimurium* are non-virulent and effective as live vaccines. *Nature* 291, 238–239. doi: 10.1038/291238a0

Jennings, E., Thurston, T. L. M., and Holden, D. W. (2017). Salmonella SPI-2 Type III secretion system effectors: molecular mechanisms and physiological consequences. *Cell Host Microbe* 22, 217–231. doi: 10.1016/j.chom.2017.07.009

Kagan, J. C., Stein, M. P., Pypaert, M., and Roy, C. R. (2004). Legionella subvert the functions of Rab1 and Sec22b to create a replicative organelle. *J. Exp. Med.* 199, 1201–1211. doi: 10.1084/jem.20031706

Knodler, L. A., and Steele-Mortimer, O. (2003). Taking possession: biogenesis of the Salmonella-containing vacuole. *Traffic* 4, 587–599. doi: 10.1034/j.1600-0854.2003.00118.x

Kujat Choy, S. L., Boyle, E. C., Gal-Mor, O., Goode, D. L., Valdez, Y., Vallance, B. A., et al. (2004). SseK1 and SseK2 are novel translocated proteins of Salmonella enterica serovar typhimurium. *Infect. Immun.* 72, 5115–5125. doi: 10.1128/IAI.72.9.5115-5125.2004

Kupz, A., Guarda, G., Gebhardt, T., Sander, L. E., Short, K. R., Diavatopoulos, D. A., et al. (2012). NLRC4 inflammasomes in dendritic cells regulate noncognate effector function by memory CD8(+) T cells. *Nat. Immunol.* 13, 162–169. doi: 10.1038/ni.2195

LaRock, D. L., Chaudhary, A., and Miller, S. I. (2015). Salmonellae interactions with host processes. *Nat. Rev. Microbiol.* 13, 191–205. doi: 10.1038/nrmicro3420

Levin, R. S., Hertz, N. T., Burlingame, A. L., Shokat, K. M., and Mukherjee, S. (2016). Innate immunity kinase TAK1 phosphorylates Rab1 on a hotspot for posttranslational modifications by host and pathogen. *Proc. Natl. Acad. Sci. U.S.A.* 113, E4776–E4783. doi: 10.1073/pnas.1608355113

Li, S., Zhang, L., Yao, Q., Li, L., Dong, N., Rong, J., et al. (2013). Pathogen blocks host death receptor signalling by arginine GlcNAcylation of death domains. *Nature* 501, 242–246. doi: 10.1038/nature12436

Madan, R., Krishnamurthy, G., and Mukhopadhyay, A. (2008). SopE-mediated recruitment of host Rab5 on phagosomes inhibits Salmonella transport to lysosomes. *Methods Mol. Biol.* 445, 417–437. doi: 10.1007/978-1-59745-157-4_27

Majowicz, S. E., Musto, J., Scallan, E., Angulo, F. J., Kirk, M., O'Brien, S. J., et al. (2010). The global burden of nontyphoidal Salmonella gastroenteritis. *Clin. Infect. Dis.* 50, 882–889. doi: 10.1086/650733

Mallo, G. V., Espina, M., Smith, A. C., Terebiznik, M. R., Aleman, A., Finlay, B. B., et al. (2008). SopB promotes phosphatidylinositol 3-phosphate formation on Salmonella vacuoles by recruiting Rab5 and Vps34. *J. Cell Biol.* 182, 741–752. doi: 10.1083/jcb.200804131

Meng, K., Zhuang, X., Peng, T., Hu, S., Yang, J., Wang, Z., et al. (2020). Arginine GlcNAcylation of Rab small GTPases by the pathogen *Salmonella Typhimurium. Commun. Biol.* 3:287. doi: 10.1038/s42003-020-1005-2

Mukherjee, K., Parashuraman, S., Raje, M., and Mukhopadhyay, A. (2001). SopE acts as an Rab5-specific nucleotide exchange factor and recruits non-prenylated Rab5 on Salmonella-containing phagosomes to promote fusion with early endosomes. *J. Biol. Chem.* 276, 23607–23615. doi: 10.1074/jbc.M101034200

Mukherjee, S., Liu, X., Arasaki, K., McDonough, J., Galan, J. E., and Roy, C. R. (2011). Modulation of Rab GTPase function by a protein phosphocholine transferase. *Nature* 477, 103–106. doi: 10.1038/nature10335

Muller, A. J., Kaiser, P., Dittmar, K. E., Weber, T. C., Haueter, S., Endt, K., et al. (2012). Salmonella gut invasion involves TTSS-2-dependent epithelial traversal, basolateral exit, and uptake by epithelium-sampling lamina propria phagocytes. *Cell Host Microbe* 11, 19–32. doi: 10.1016/j.chom.2011.11.013

Muller, M. P., Peters, H., Blumer, J., Blankenfeldt, W., Goody, R. S., and Itzen, A. (2010). The Legionella effector protein DrrA AMPylates the membrane traffic regulator Rab1b. *Science* 329, 946–949. doi: 10.1126/science.1192276

Newson, J. P. M., Scott, N. E., Yeuk Wah Chung, I., Wong Fok Lung, T., Giogha, C., Gan, J., et al. (2019). Salmonella effectors SseK1 and SseK3 target death domain proteins in the TNF and TRAIL signaling pathways. *Mol. Cell. Proteomics* 18, 1138–1156. doi: 10.1074/mcp.RA118.001093

Nuoffer, C., Davidson, H. W., Matteson, J., Meinkoth, J., and Balch, W. E. (1994). A GDP-bound of rab1 inhibits protein export from the endoplasmic reticulum and transport between Golgi compartments. *J. Cell Biol.* 125, 225–237. doi: 10.1083/jcb.125.2.225

Parashuraman, S., and Mukhopadhyay, A. (2005). Assay and functional properties of SopE in the recruitment of Rab5 on Salmonella-containing phagosomes. *Meth. Enzymol.* 403, 295–309. doi: 10.1016/S0076-6879(05)03025-9

Park, J. B., Kim, Y. H., Yoo, Y., Kim, J., Jun, S. H., Cho, J. W., et al. (2018). Structural basis for arginine glycosylation of host substrates by bacterial effector proteins. *Nat. Commun.* 9:4283. doi: 10.1038/s41467-018-06680-6

Pearson, J. S., Giogha, C., Ong, S. Y., Kennedy, C. L., Kelly, M., Robinson, K. S., et al. (2013). A type III effector antagonizes death receptor signalling during bacterial gut infection. *Nature* 501, 247–251. doi: 10.1038/nature 12524

Perrett, C. A., and Zhou, D. (2013). Salmonella type III effector SopB modulates host cell exocytosis. *Emerg. Microbes Infect.* 2:e32. doi: 10.1038/emi.2013.31

Pfeffer, S. R. (2005). Structural clues to Rab GTPase functional diversity. *J. Biol. Chem.* 280, 15485–15488. doi: 10.1074/jbc.R5000 03200

Prashar, A., Schnettger, L., Bernard, E. M., and Gutierrez, M. G. (2017). Rab GTPases in Immunity and Inflammation. *Front. Cell. Infect. Microbiol.* 7:435. doi: 10.3389/fcimb.2017.00435

Ramsden, A. E., Holden, D. W., and Mota, L. J. (2007). Membrane dynamics and spatial distribution of Salmonella-containing vacuoles. *Trends Microbiol.* 15, 516–524. doi: 10.1016/j.tim.2007.10.002

Rappsilber, J., Mann, M., and Ishihama, Y. (2007). Protocol for micro-purification, enrichment, pre-fractionation and storage of peptides for proteomics using StageTips. *Nat. Protoc.* 2, 1896–1906. doi: 10.1038/nprot.2007.261

Scott, N. E., Giogha, C., Pollock, G. L., Kennedy, C. L., Webb, A. I., Williamson, N. A., et al. (2017). The bacterial arginine glycosyltransferase effector NleB preferentially modifies Fas-associated death domain protein (FADD). *J. Biol. Chem.* 292, 17337–17350. doi: 10.1074/jbc.M117.805036

Smith, A. C., Cirulis, J. T., Casanova, J. E., Scidmore, M. A., and Brumell, J. H. (2005). Interaction of the Salmonella-containing vacuole with the endocytic recycling system. *J. Biol. Chem.* 280, 24634–24641. doi: 10.1074/jbc.M500358200

Smith, A. C., Heo, W. D., Braun, V., Jiang, X., Macrae, C., Casanova, J. E., et al. (2007). A network of Rab GTPases controls phagosome maturation and is modulated by Salmonella enterica serovar Typhimurium. *J. Cell Biol.* 176, 263–268. doi: 10.1083/jcb.200611056

Smith, J. L. (2003). The role of gastric acid in preventing foodborne disease and how bacteria overcome acid conditions. *J. Food Prot.* 66, 1292–1303. doi: 10.4315/0362-028X-66.7.1292

Spano, S., and Galan, J. E. (2012). A Rab32-dependent pathway contributes to Salmonella typhi host restriction. *Science* 338, 960–963. doi: 10.1126/science.1229224

Spano, S., and Galan, J. E. (2018). Taking control: Hijacking of Rab GTPases by intracellular bacterial pathogens. *Small GTPases* 9, 182–191. doi: 10.1080/21541248.2017.1336192

Spano, S., Gao, X., Hannemann, S., Lara-Tejero, M., and Galan, J. E. (2016). A Bacterial Pathogen Targets a Host Rab-Family GTPase Defense Pathway with a GAP. *Cell Host Microbe* 19, 216–226. doi: 10.1016/j.chom.2016.01.004

Stewart, S. A., Dykxhoorn, D. M., Palliser, D., Mizuno, H., Yu, E. Y., An, D. S., et al. (2003). Lentivirus-delivered stable gene silencing by RNAi in primary cells. *RNA* 9, 493–501. doi: 10.1261/rna.2192803

Tan, Y., Arnold, R. J., and Luo, Z. Q. (2011). Legionella pneumophila regulates the small GTPase Rab1 activity by reversible phosphorylcholination. *Proc. Natl. Acad. Sci. U.S.A.* 108, 21212–21217. doi: 10.1073/pnas.1114023109

Tan, Y., and Luo, Z. Q. (2011). Legionella pneumophila SidD is a deAMPylase that modifies Rab1. *Nature* 475, 506–509. doi: 10.1038/nature10307

Tisdale, E. J., Bourne, J. R., Khosravi-Far, R., Der, C. J., and Balch, W. E. (1992). GTP-binding mutants of rab1 and rab2 are potent inhibitors of vesicular transport from the endoplasmic reticulum to the Golgi complex. *J. Cell Biol.* 119, 749–761. doi: 10.1083/jcb.119.4.749

Tyanova, S., Temu, T., Sinitcyn, P., Carlson, A., Hein, M. Y., Geiger, T., et al. (2016). The Perseus computational platform for comprehensive analysis of (prote)omics data. *Nat. Methods* 13, 731–740. doi: 10.1038/nmeth.3901

Wachtel, R., Brauning, B., Mader, S. L., Ecker, F., Kaila, V. R. I., Groll, M., et al. (2018). The protease GtgE from Salmonella exclusively targets inactive Rab GTPases. *Nat. Commun* 9, 44. doi: 10.1038/s41467-017-02110-1

Wang, Z., McCloskey, A., Cheng, S., Wu, M., Xue, C., Yu, Z., et al. (2018). Regulation of the small GTPase Rab1 function by a bacterial glucosyltransferase. *Cell Discov* 4:53. doi: 10.1038/s41421-018-0055-9

Xue, J., Pan, X., Peng, T., Duan, M., Du, L., Zhuang, X., et al. (2020). Auto arginine-GlcNAcylation is crucial for bacterial pathogens in regulating host cell death. *Front. Cell. Infect. Microbiol* 10:197. doi: 10.3389/fcimb.2020.00197

Yamamoto, M., Okuyama, M., Ma, J. S., Kimura, T., Kamiyama, N., Saiga, H., et al. (2012). A cluster of interferon-gamma-inducible p65 GTPases plays a critical role in host defense against Toxoplasma gondii. *Immunity* 37, 302–313. doi: 10.1016/j.immuni.2012.06.009

Zhen, Y., and Stenmark, H. (2015). Cellular functions of Rab GTPases at a glance. *J. Cell Sci* 128, 3171–3176. doi: 10.1242/jcs.166074

9

Mechanisms of Effector-Mediated Immunity Revealed by the Accidental Human Pathogen *Legionella pneumophila*

*Tshegofatso Ngwaga, Deepika Chauhan and Stephanie R. Shames**

Division of Biology, Kansas State University, Manhattan, KS, United States

***Correspondence:**
Stephanie R. Shames
sshames@ksu.edu

Many Gram-negative bacterial pathogens employ translocated virulence factors, termed effector proteins, to facilitate their parasitism of host cells and evade host anti-microbial defenses. However, eukaryotes have evolved to detect effector-mediated virulence strategies through a phenomenon termed effector-triggered immunity (ETI). Although ETI was discovered in plants, a growing body of literature demonstrates that metazoans also utilize effector-mediated immunity to detect and clear bacterial pathogens. This mini review is focused on mechanisms of effector-mediated immune responses by the accidental human pathogen *Legionella pneumophila*. We highlight recent advancements in the field and discuss the future prospects of harnessing effectors for the development of novel therapeutics, a critical need due to the prevalence and rapid spread of antibiotic resistance.

Keywords: effector-triggered immunity, effector-mediated immunity, *Legionella pneumophila*, innate immunity, macrophage

EFFECTOR-MEDIATED IMMUNITY ENHANCES HOST DEFENSE AGAINST BACTERIAL PATHOGENS

The evolutionary arms race between host and pathogen has necessitated the use of several complementary innate immune pathways to detect and eradicate pathogens. Initial pathogen recognition occurs through engagement of pathogen associated molecular patterns (PAMPs) by host pattern recognition receptors (PRRs) (Janeway, 1989). PRRs include toll-like-receptors (TLRs), located on either the plasma membrane or endosomal membranes (Medzhitov and Janeway, 2000; Massis and Zamboni, 2011). PRR recognition of PAMPs activates signaling cascades that culminate in production of pro-inflammatory cytokines that contribute to controlling infection (Janeway and Medzhitov, 2002). Inflammasomes are multimeric intracellular protein complexes that activate inflammatory caspases in response to cellular damage or pathogen infection [reviewed in (Martinon et al., 2002)]. Bacterial pathogens have evolved diverse repertoires of virulence factors to promote their survival within hosts by acquisition of host-derived nutrients and avoidance of host defenses. Bacterial effectors are directly injected into host cells through specialized secretion systems and functions within the host cells to facilitate pathogen survival in close association with host cells (Cambronne and Roy, 2006; Galán, 2009). Both intracellular and extracellular pathogens utilize effector proteins, emphasizing their importance in the virulence strategies of diverse bacterial pathogens.

Multiple effector-mediated virulence processes are similar between seemingly diverse bacterial pathogens. Multicellular eukaryotes are able to detect bacterial effectors and/or their virulence processes *via* effector-triggered immunity (ETI) (Stuart et al., 2013; Rajamuthiah and Mylonakis, 2014; Fischer et al., 2019). ETI was first described in immune defense against pathogens in plants as "gene-for-gene resistance" where resistance (R) genes in plants recognize bacterial effectors (Avr) within the plant cell and trigger an immune response (Flor, 1971; Chisholm et al., 2006). However, animals also detect pathogen infection through ETI and effector-mediated responses [reviewed in (Fischer et al., 2019)]. Plant ETI results from either direct recognition of the effector itself or sensing of intracellular effector activity, whereas only the latter has been observed in metazoans (Stuart et al., 2013). In plants, several models have been described of how resistant strains directly or indirectly detect pathogen effectors. The "receptor-ligand model" describes direct recognition of bacterial effectors whereby a host R protein binds and inactivates a bacterial Avr effector (Stuart et al., 2013). The *Pseudomonas syringae* effector AvrPto blocks host pathogen recognition through subversion of receptor-mediated signaling. In resistant plants, AvrPto is directly inactivated by the R protein, Pto (Xiang et al., 2008). The "guard hypothesis," "decoy model," and "bait-and-switch model" are indirect models through which resistant plants detect pathogen effector function (Dangl and Jones, 2001; Stuart et al., 2013). In animal cells, effector function is detected indirectly though cell-autonomous sensing of homeostatic perturbations elicited by the effectors to the benefit of the pathogen (Colaço and Moita, 2016). Examples include cellular detection of effector-mediated translation inhibition, inhibition of Rho GTPases, and pore formation (Fischer et al., 2019). This mini review is focused specifically on mechanisms of effector-mediated and -triggered host defense against the accidental human pathogen *Legionella pneumophila*.

USE OF THE "ACCIDENTAL" PATHOGEN *L. PNEUMOPHILA* AS A MODEL TO UNDERSTAND EFFECTOR-MEDIATED IMMUNE RESPONSES

Several mechanisms of effector-mediated immunity have been uncovered by studying the accidental human pathogen *Legionella pneumophila*. *Legionella* spp. are Gram-negative intracellular bacteria that are ubiquitous in aquatic and soil environments, where they parasitize free-living protozoa (Rowbotham, 1980; Fliermans et al., 1981; Barbaree et al., 1986). Anthropomorphic fresh-water environments such as cooling towers, water fountains and any system that allows for aerosolization of water droplets, have potential to be the source of *Legionella* infection, collectively termed legionellosis (Barbaree et al., 1986). Inhalation or aspiration of *L. pneumophila* can result in an inflammatory pneumonia called Legionnaires' disease, which is fatal in ~10% of cases (Soda et al., 2017). Legionnaires' disease primarily affects elderly and immunocompromised individuals and was named for the initial outbreak, which occurred at the 1976 American Legion Convention in Philadelphia (Fraser et al., 1977; McDade et al., 1980). In immunocompetent individuals, *L. pneumophila* can cause a mild self-limiting flu-like illness called Pontiac Fever (Glick et al., 1978). Legionellosis is a consequence of *L. pneumophila* replication within alveolar macrophages (Nash et al., 1984; Friedman et al., 2002); however, the infection is readily cleared by innate immune responses *in vivo*, owing in part to orchestrated production of pro-inflammatory cytokines (Shin, 2012; Liu et al., 2020). The opportunistic colonization of built freshwater environments, rarity of person-to-person transmission and susceptibility to innate immune responses has led to description of *L. pneumophila* as an "accidental pathogen" (Borges et al., 2016; Boamah et al., 2017).

Virulence strategies evolved by *L. pneumophila* to parasitize free-living protozoa have conferred the ability to replicate within mammalian macrophages (Park et al., 2020). Upon phagocytosis, *L. pneumophila* rapidly remodels its vacuole to prevent lysosomal degradation and establish an intracellular replicative niche called the *Legionella* containing vacuole (LCV) (Horwitz, 1983). For biogenesis of LCV and intracellular replication, *L. pneumophila* employs over three hundred individual effector proteins translocated into host cells by a Dot/Icm type IVB secretion system (T4SS) (Berger and Isberg, 1993; Zhu et al., 2011; Ensminger, 2016). *L. pneumophila* encodes the largest arsenal of translocated effector proteins identified to date, due to its broad and diverse tropism for free-living protozoa (Park et al., 2020). Armed with these effectors, *L. pneumophila* proliferates to high numbers within host phagocytes. Effectors are essential for biogenesis of the LCV and intracellular replication through facilitating nutrient acquisition and prevention of lysosomal degradation. However, several *L. pneumophila* effectors that perform these essential functions paradoxically amplify pro-inflammatory immune responses in macrophages. Thus, *L. pneumophila* has become a useful model pathogen to delineate mechanisms of effector-mediated immune detection and clearance. *Legionella* have also served as a valuable model to study molecular basis of inflammasome activation; however, this aspect of *Legionella* biology has been reviewed previously and will not be discussed here (Mascarenhas and Zamboni, 2017).

Below, we discuss mechanisms of effector-mediated immune defense against *L. pneumophila* and the potential for effector-mediated immunity to be harnessed for development of novel therapeutics to combat infectious diseases.

L. PNEUMOPHILA EFFECTOR-MEDIATED TRANSLATION INHIBITION ENHANCES MACROPHAGE INFLAMMATORY RESPONSES

Effector-mediated host protein translation inhibition, a virulence strategy employed by multiple pathogens, enhances inflammatory signaling in *L. pneumophila* infected macrophages (Fontana et al., 2011; Barry et al., 2013). To replicate intracellularly, *L. pneumophila* is reliant on host-derived amino acids (George et al., 1980; Bruckert et al., 2013; Price et al., 2014; Schunder et al., 2014). Since free amino acid levels are tightly regulated in eukaryotic cells, *L. pneumophila* utilizes several effectors to facilitate acquisition of amino acids from host cells. The effectors Lgt1-3, SidI, SidL, LegK4, and RavX collectively inhibit host protein translation [recently reviewed in (Belyi, 2020)]. The mechanisms by which RavX, SidL, and SidI inhibit translation have not been fully elucidated. However, Lgt1-3 glycosylation of the host translation elongation factor eEF1A on a conserved Ser residue inhibits host polypeptide elongation (Belyi et al., 2006; Belyi et al., 2008) and LegK4 impairs polypeptide refolding through phosphorylation of host Hsp90 (Moss et al., 2019). SidI interacts with eEF1A and eEF1Bγ; however, this interaction is not sufficient for translation inhibition (Shen et al., 2009; Joseph et al., 2020). The collective activity of this redundant family of effectors enhances the inflammatory response to *L. pneumophila* (Shin and Roy, 2008; Barry et al., 2013).

Effector-mediated protein translation inhibition synergizes with PAMP-mediated signaling to enhance inflammation in *L. pneumophila*-infected macrophages. Fontana and colleagues originally discovered that activity of the effectors Lgt1-3, SidI, and SidL induced selective mitogen activated protein kinase (MAPK)-mediated upregulation of interleukin (IL)-1α in *L. pneumophila*-infected macrophages (Fontana et al., 2011; Fontana et al., 2012). MAP kinase signaling cascades culminate in activation of the dimeric activating protein (AP-1) transcription factor—made up of Jun and Fos—which facilitates pro-inflammatory gene expression (Fujioka et al., 2004; Hess et al., 2004; Alonso et al., 2018; Gazon et al., 2018). Interestingly, complementation of a *L. pneumophila* mutant lacking *lgt1-3*, *sidL*, and *sidI* with just *lgt-3* is sufficient to restore MAPK activation during infection (**Figure 1**) (Fontana et al., 2012). Translation inhibition results in selective upregulation of IL-1α, which is critical for host defense against *L. pneumophila* (Barry et al., 2013; Copenhaver et al., 2015; Mascarenhas et al., 2015). The selective upregulation of *Il1a* is a consequence of mRNA superinduction, a phenomenon whereby increased *de novo* transcription of specific genes overcomes bacterial blockade of protein translation and initiates a pro-inflammatory response (Barry et al., 2017). This selective production of IL-1α by infected macrophages results in amplification of pro-inflammatory cytokine production by uninfected translation-competent bystander cells (Copenhaver et al., 2015; Liu et al., 2020) (see below). Translation inhibition also occurs *via* an effector-independent mechanism, which may be a consequence of metabolic reprogramming (Barry et al., 2017; Price et al., 2020) (see below). However, effector-mediated restriction of host protein translation, which liberates amino

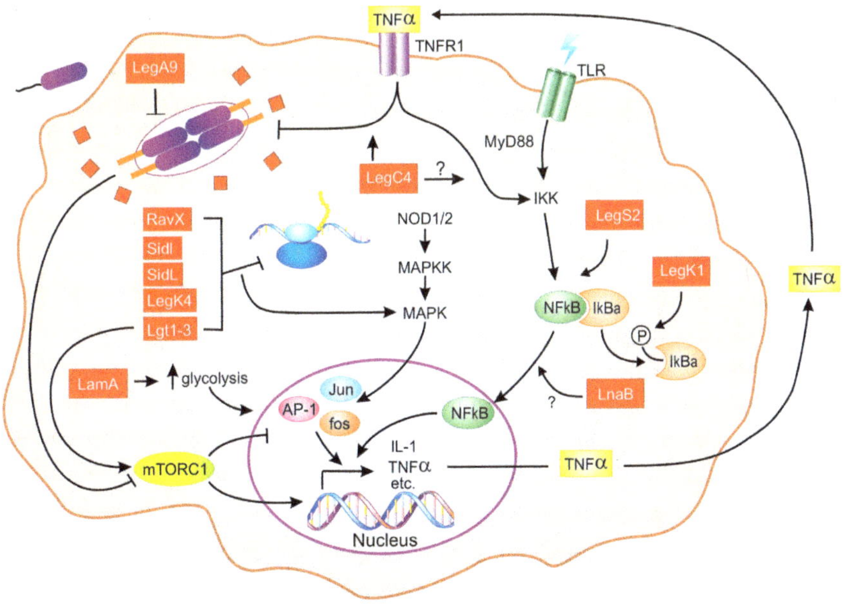

FIGURE 1 | Schematic representation of *L. pneumophila* effector-mediated host defense in macrophages. From the LCV, *L. pneumophila* (purple) translocates hundreds of individual effector proteins (red squares/rectangles) into the host cytosol through the Dot/Icm T4SS (orange). Multiple effectors inhibit host translation elongation (RavX, SidI, SidL, LegK4, and Lgt1-3), which results in activation of MAPK signaling and pro-inflammatory cytokine expression [AP-1 (Jun, Fos)]. The activity of Lgt1-3 also activates the mTORC1 complex, which results in downregulation of pro-inflammatory genes. However, in macrophages, mTOR signaling is attenuated by detection of pathogen-derived molecules. Activation of NF-κB downstream of PRR (TLRs shown) engagement is enhanced by LegS2, LnaB, and LegK1, the latter of which phosphorylates IκBα. *L. pneumophila* replication within macrophages is also impaired by LegA9 and LegC4, the latter of which augments cytokine-mediated restriction. Finally, LamA, a recently characterized *L. pneumophila* effector, degrades cellular glycogen, leading to increased aerobic glycolysis and proinflammatory cytokine production. For clarity, the SidE family of effectors and the role of IL-1 production by infected macrophages are not shown. Question marks indicate unknowns. See text for additional details.

acids for use by *L. pneumophila* (De Leon et al., 2017), contributes to a highly orchestrated pro-inflammatory response in accidental hosts and is an example of canonical ETI.

L. PNEUMOPHILA EFFECTOR-MEDIATED TRANSLATION INHIBITION IMPACTS MECHANISTIC TARGET OF RAPAMYCIN (mTOR) SIGNALING

Modulation of host protein translation also impacts activity of the mechanistic target of rapamycin (mTOR). mTOR is central to many cellular processes and regulates host amino acid metabolism, where availability and dearth of amino acids results in activation or inactivation of mTOR signaling, respectively [**Figure 1**; reviewed in (Condon and Sabatini, 2019)]. Several viral pathogens and protozoan parasites, such as *Leishmania*, have evolved to directly target this pathway and its processes for their own benefit (Buchkovich et al., 2008; Jaramillo et al., 2011; Leroux et al., 2018).

Recent work has revealed the central, albeit complex, role of mTOR in *L. pneumophila* pathogenesis and host defense. Ivanov and Roy initially reported that macrophages detect cytosolic "pathogen signatures," which results in suppression of mTOR and selective production of pro-inflammatory cytokines and independently of translocated effectors (Ivanov and Roy, 2013). Concomitantly, *L. pneumophila* virulence is attenuated in the lungs of mice with mTOR-deficient macrophages. However, subsequent studies uncovered a role for effectors in mTOR regulation during *L. pneumophila* infection (Abshire et al., 2016; De Leon et al., 2017). The mTORC1 complex (a multiprotein complex containing mTOR) is both suppressed and activated by distinct families of *L. pneumophila* effectors (De Leon et al., 2017). Translation inhibition, through the activity of the Lgt effector family (see above), and consequent increases in free amino acids, activates mTORC1 in macrophages. However, mTORC1 is suppressed through ubiquitination and suppression of Rag GTPases by the SidE effector family (SidE/SdeABC), which also inhibit host protein translation (De Leon et al., 2017). Thus, the SidE family of effectors may prevent mTORC1 sensing amino acids that are liberated downstream of Lgt1-3 activity. In macrophages, mTOR activation by the Lgts is downstream of potent translation inhibition. As discussed above, inhibition of protein translation in macrophages results in selective production of a subset of pro-inflammatory mediators, such as IL-1α. Thus, inhibition of mTOR would contribute to selective production of cytokines that orchestrate a robust inflammatory response in the lung through engagement of bystander cells (see below) (Ivanov and Roy, 2013; Copenhaver et al., 2015; Barry et al., 2017; Liu et al., 2020). Together, these studies collectively emphasize the central and complex role of mTOR in *Legionella* pathogenesis and the inflammatory response elicited in accidental hosts.

AN EFFECTOR-MEDIATED STRATEGY FOR REPLICATION IN AMOEBAE LEADS TO PRO-INFLAMMATORY MACROPHAGE RESTRICTION

L. pneumophila is ubiquitous in freshwater environments where it parasitizes and replicates within unicellular eukaryotes, including amoebae (Molmeret et al., 2005; Albert-Weissenberger et al., 2006). When environmental conditions are not optimal for growth and survival, amoeboid trophozoites undergo encystation, a condition where the amoeba turns into a metabolically inactive cyst containing cellulose rich cell wall that is resistant to hostile environmental conditions (Moon and Kong, 2012; Aqeel et al., 2013). Although *L. pneumophila* survives in amoebal cysts, encystation is restrictive to intracellular replication (Kilvington and Price, 1990; Bouyer et al., 2007). Prior to encystation, amoebae accumulate glycogen, which is used for biogenesis of the characteristic cellulose-rich cell wall (Weisman et al., 1970; Fouque et al., 2012; Moon and Kong, 2012; Schaap and Schilde, 2018).

To maintain amoebae as replication-permissive trophozoites, *L. pneumophila* utilizes the effector LamA, an amylase that catalyzes glycogenolysis to limit glycogen accumulation in infection amoebae (Price et al., 2020). LamA alone is not required for *L. pneumophila* replication in the natural hose *Acanthamoeba polyphaga*, likely due to functional redundancy with other effectors (Ghosh and O'Connor, 2017; Park et al., 2020). However, other natural host amoebae were not examined in this study. Thus, it is tempting to speculate that LamA activity is individually important in other species, such as *A. castellanii*, in which encystation is highly restrictive to *L. pneumophila* (Weisman et al., 1970; Bouyer et al., 2007).

LamA-mediated metabolic reprogramming is deleterious to *L. pneumophila* in accidental hosts (Price et al., 2020). Excess cellular glucose results in increased aerobic glycolysis in both *A. polyphaga* and human monocyte-derived macrophages (hMDM) (Price et al., 2020). Aerobic glycolysis in macrophages promotes their activation, M1 polarization and secretion of pro-inflammatory cytokines (Langston et al., 2017). Thus, LamA activity enhances secretion of pro-inflammatory cytokines from hMDMs during infection [**Figure 1** (Price et al., 2020)] and impairs *L. pneumophila* in hMDM through IFN-γ–mediated indolamine-2,3-dioxygenase (IDO) activity (Murray et al., 1989; Price et al., 2020). This result is intriguing since translation inhibition during *L. pneumophila* infection limits production of most cytokines (see above). The authors propose that the amount of IFN-γ produced is sufficient for IDO activation; however, it would be interesting to determine if IFN-γ activation is indeed required for LamA-mediated restriction. In the mouse lung, IL-1α production is severely decreased following infection with a *lamA* mutant compared to wild-type bacteria. However, the *lamA* mutant strain presumably still translocates effector translation inhibitors (see above). Thus, macrophage metabolic reprogramming may contribute to effector-independent translation inhibition and concomitant inflammation (Barry et al., 2017). Moreover, LamA-mediated macrophage activation is a direct consequence of its enzymatic activity, distinguishing this response from canonical ETI.

EFFECTOR-MEDIATED AUGMENTATION OF CYTOKINE-MEDIATED RESTRICTION OF *L. PNEUMOPHILA*

Pro-inflammatory cytokines activate resting macrophages and are critical for restriction of *L. pneumophila* in mammalian hosts (Archer and Roy, 2006). The effector LegC4 was initially identified in a high-throughput forward genetic screen for individual effectors that impact *L. pneumophila* virulence in amoeba and mammalian infection models (Shames et al., 2017; Rolando and Buchrieser, 2018). This screen identified LegC4 as conferring fitness disadvantage on *L. pneumophila* relative to the isogenic parental strain in a mouse model of Legionnaires' disease but not BMMs *ex vivo* (Shames et al., 2017; Ngwaga et al., 2019). Interestingly, LegC4 is individually important for *L. pneumophila* replication in the natural host, *A. castellanii* (Shames et al., 2017). Further investigation revealed that LegC4 is deleterious to *L. pneumophila* specifically within cytokine-activated macrophages (Ngwaga et al., 2019) (**Figure 1**). In cultured mouse BMMs, LegC4-mediated restriction is contingent on autocrine and paracrine TNF receptor 1 (TNFR1)-mediated signaling. However, loss of TNFR1 is insufficient to rescue LegC4-mediated replication defects in the mouse lung, likely due to LegC4-mediated exacerbation of IFN-γ-mediated restriction (Ngwaga et al., 2019). LegC4 additionally enhances secretion of several pro-inflammatory cytokines, including IL-12, IL-6, and TNF-α, from *L. pneumophila*-infected BMMs despite global translation inhibition (Shames et al., 2017; Ngwaga et al., 2019). Whether LegC4-mediated increases in cytokine production from *L. pneumophila*-infected macrophages is due to enhanced transcription or translation is unknown. Revealing the influence of LegC4 on production of IL-1α, TNF-α, and IFN-γ in the lung will provide a foundation for understanding the mechanism of LegC4-mediated restriction.

Interestingly, LegC4 is also augments cytokine-mediated restriction of *L. longbeachae*, and the second leading cause of Legionnaires' disease globally (Gobin et al., 2009). *L. longbeachae* is lethal to mice and is reliant on a Dot/Icm T4SS for intracellular replication; however, *L. longbeachae* does not encode a homolog of *legC4* (Cazalet et al., 2010; Wood et al., 2015; Massis et al., 2016). LegC4 is sufficient to attenuate *L. longbeachae* replication within BMMs activated with either TNF-α or IFN-γ (Ngwaga et al., 2019), demonstrating that LegC4-mediated restriction is not specific to *L. pneumophila*. LegC4—like LamA—enhances *L. pneumophila* virulence in a natural host, but its activity in macrophages is deleterious. The mechanism by which LegC4 impacts *L. pneumophila* fitness in natural and accidental hosts, respectively, and its potential to enhance cytokine-mediated restriction of other intracellular pathogens are currently under investigation in our lab.

EFFECTOR-MEDIATED ACTIVATION OF INFLAMMATORY GENE EXPRESSION AND AUTOPHAGY

Many bacterial pathogens actively attenuate inflammatory signaling by restricting activation of the NF-κB transcription factor (Brodsky and Medzhitov, 2009). However, NF-κB is activated in mammalian cells infected with *L. pneumophila*. Within *L. pneumophila*-infected cells, NF-κB activation occurs in two waves; effector-independent TLR-dependent activation when bacteria first make contact with host cells and effector-mediated activation after several hours of infection (Losick and Isberg, 2006; Asrat et al., 2014). Several *L. pneumophila* effectors contribute to NF-κB activation in mammalian cells (**Figure 1**). LegK1 is a eukaryotic-like serine/threonine kinase that phosphorylates the inhibitor of κBα (IκBα), which results in nuclear localization of NF-κB and consequent upregulation of pro-inflammatory and pro-survival gene (Ge et al., 2009; Rahman and McFadden, 2011). However, LegK1-mediated NF-κB activation occurs only upon ectopic expression of *legK1* in epithelial cells (Ge et al., 2009). LnaB enhances NF-κB-mediated gene expression by an unknown mechanism following ectopic expression and during *L. pneumophila* infection of epithelial cells (Losick et al., 2010). LegK1 does not contribute to NF-κB activity in *L. pneumophila*-infected epithelial cells (Losick et al., 2010), which raises the possibility that this phenotype is a consequence of dose-dependent effect, or mislocalization due to ectopic expression. Neither *lnaB* nor *legK1* are required for *L. pneumophila* replication in mouse macrophages individually or in combination (Losick et al., 2010). Since amoebae lack NF-κB signaling components, direct effector-mediated activation of this pathway is perplexing. It is possible that IκBα phosphorylation by LegK1 is due promiscuous enzymatic activity and/or presence highly conserved target motifs. Identification of LegK1 substrates in amoebae would shed light on this possibility. Effector-mediated NF-κB activation enhances *L. pneumophila* survival in macrophages through prevention of premature apoptosis but also results in expression of pro-inflammatory cytokines, including IL-1α (Losick and Isberg, 2006). NF-κB plays a multifaceted role in *L. pneumophila* infection of accidental hosts, but the evolutionary basis for its activation has yet to be elucidated.

Autophagy is central to cell-autonomous restriction of intracellular bacterial pathogens in phagotrophs. *L. pneumophila* has evolved several effectors capable of regulating host autophagy and two, LegS2 and LegA9, are deleterious to *L. pneumophila* in accidental hosts (Khweek et al., 2016; Rolando et al., 2016; Sherwood and Roy, 2016). LegS2 is a mitochondria-targeted sphingosine-1-phosphate lyase that restricts *L. pneumophila* replication in macrophages, suppresses autophagy, and enhances NF-κB activation (Degtyar et al., 2009; Khweek et al., 2016; Rolando et al., 2016). Suppression of starvation-induced autophagy by LegS2 is facilitated by modulation of host sphingosine metabolism (Rolando et al., 2016), but whether amplification of NF-κB is linked to LegS2-mediated sphingosine metabolism and suppression of autophagy is unknown. LegA9 enhances *L. pneumophila* macrophage clearance by upregulating autophagy in BMMs (Khweek et al., 2013); however, further investigation is required to define the mechanism by which LegA9 augments *L. pneumophila* macrophage clearance.

CONSEQUENCES OF EFFECTOR-MEDIATED IMMUNITY IN A MOUSE MODEL OF LEGIONNAIRES' DISEASE

L. pneumophila replicates robustly in macrophages derived from permissive mice but is efficiently cleared from the lung just days after infection. Restriction of *L. pneumophila* in the mouse lung is due to a rapid and robust pro-inflammatory response orchestrated through engagement of multiple cell types (Blanchard et al., 1987; Blanchard et al., 1989; Brieland et al., 1998; Archer and Roy, 2006). *L. pneumophila*–infected alveolar macrophages are poor producers of TNF-α, IL-6, and IL-12 *in vivo* due to effector-mediated translation inhibition (Copenhaver et al., 2014; Copenhaver et al., 2015). However, selective upregulation of IL-1α by infected translation-impaired cells ultimately results in pro-inflammatory cytokine production by uninfected bystander cells, namely Ly6C^{hi} monocytes and neutrophils (Copenhaver et al., 2015; Barry et al., 2017; Casson et al., 2017). A central role for IL-1α in immune defense against *L. pneumophila* has been well established and is contingent on MyD88-mediated signaling (Barry et al., 2013; Asrat et al., 2014; Copenhaver et al., 2015; Mascarenhas et al., 2015). However, the mechanism by which IL-1α facilitates bacterial clearance *in vivo* was only recently uncovered. IL-1α produced by infected alveolar macrophages engages IL-1R on alveolar epithelial cells, which in turn secrete granulocyte colony stimulating factor (GM-CSF) (Liu et al., 2020). Consequent GM-CSF signaling in inflammatory monocytes upregulates aerobic glycolysis leading to pro-inflammatory cytokine production (Liu et al., 2020). This work exemplifies how effector-driven virulence mechanisms, such as translation inhibition, trigger a highly orchestrated inflammatory response to *L. pneumophila* in the lung.

HARNESSING EFFECTOR-MEDIATED IMMUNITY TO COMBAT BACTERIAL INFECTION

Antimicrobial resistance comprises a major global public health challenge. Thus, to control the emergence and spread of antimicrobial-resistant pathogens, innovative therapeutic strategies are desperately needed. Anti-virulence therapy a promising alternative approach to combat resistant pathogens *via* targeting virulence pathways of the pathogen (Rasko and Sperandio, 2010; Martínez et al., 2019). As pathogen-centric therapeutics are susceptible to evolution of resistance, host-centric therapeutics are an attractive alternative to control bacterial infection. Information gleaned from effector-mediated immune response and the effectors themselves have potential to combat infection by regulated amplification of host immunity.

Bacterial products, including effectors, modulate host immunity. Bacterial PAMPs are promising innate immunologicals, but whether immune-activating effectors can be harnessed directly or indirectly to combat infectious diseases has not been investigated. CpG oligodeoxynucleotides (ODN), a TLR9 agonist, have been used as vaccine adjuvants and can amplify immune responses to multiple bacterial, parasitic and viral pathogens, including *Leishmania major*, *Mycobacterium tuberculosis*, *Francisella tularensis* and, more recently, SARS-CoV-2, the etiological agent of COVID-19 (Zimmermann et al., 1998; Elkins et al., 1999; Juffermans et al., 2002; Scheiermann and Klinman, 2014; Oberemok et al., 2020). Moreover, multiple effectors from well-adapted human pathogens that dampen host immunity have attracted attention as potential drug candidates for the treatment of inflammatory diseases (Rüter and Hardwidge, 2014).

Use of effectors that amplify immunity directly into host cells as possible host-specific therapeutics has not been evaluated. Effector-mediated subversion of host homeostasis triggers ETI but perturbation of host cellular processes poses a major challenge. Global inhibition of translation as a means to enhance anti-microbial immunity would be impractical. However, based on insight from the response to ETI, treatment with IL-1α could initiate an early and robust immune response in the lung against diverse pathogens (see above). The effector LegC4 is an intriguing candidate for host-specific therapeutics based on its ability to amplify cytokine-mediated pathogen restriction. However, further investigation is required to determine the mechanisms of LegC4 function within immune cells and if immune-activating effectors can be harnessed as host-specific therapeutics.

To exert their functions, effectors require access to the cytosol of host cells. Current use of effectors as therapeutics is accomplished by autonomous translocation into host cells *via* fusion to cell-penetrating peptides (Rüter and Schmidt, 2017). However, the N-terminal domain of anthrax lethal factor (LFn), when co-delivered with the anthrax protective antigen, facilitates translocation of cargo protein directly into mammalian cells (Rabideau and Pentelute, 2016). LFn fusion has been successfully utilized to deliver bacterial proteins into host cells and mice (Kofoed and Vance, 2011; Shi et al., 2015). In addition, nanoparticles can be used for specific delivery of nucleic acids for orthologous expression of effectors within target cells or direct delivery of protein cargo (Avila et al., 2016; Rincon-Restrepo et al., 2017).

Thus, potential exists for bacterial effectors to function as therapeutic agents. Effector based therapeutics offer several advantages over conventional biologics such as autonomous translocation, enhanced specificity, efficacy at low concentrations, targeted and topical applications, comparatively fewer side effects, cost-effective and stability at variable pH and temperatures. Moreover, further understanding of novel effectors, mechanisms of effector-mediated immunity, and the development of selective delivery mechanisms offers potential for improved combinatorial therapeutics in the future.

CONCLUSIONS

In addition to sensing pathogens through PRRs, the mammalian immune system has developed additional mechanisms to detect the activity of virulence factors secreted by pathogens. Effector-mediated immunity facilitates detection and/or enhanced clearance of pathogens. This additional mechanism to detect pathogens is important because pathogenic microorganisms have evolved ways to avoid, modulate and hide from the first line of immune defense offered by triggering of PRRs. As an accidental human pathogen, *L. pneumophila* continues to serve as a useful model used to study innate immune mechanisms without the complications of evasion strategies used by mammalian-adapted pathogens. *L. pneumophila* has provided valuable insight into mechanisms of innate immune defense against intracellular bacterial pathogens, including how effector-mediated virulence strategies trigger inflammation. Further investigation of *L. pneumophila* effector function will undoubtedly reveal yet additional mechanisms by which cells of the innate immune system restrict intracellular pathogens. Exploiting effector-mediated immunity to elicit pathogen-centric immunotherapeutics may provide additional treatment or prevention strategies against antimicrobial resistant pathogens.

AUTHOR CONTRIBUTIONS

SS conceived the idea for the manuscript. TN, DC, and SS wrote the paper. All authors contributed to the article and approved the submitted version.

ACKNOWLEDGMENTS

We thank Fern Ness for assistance with figure preparation.

REFERENCES

Abshire, C. F., Dragoi, A.-M., Roy, C. R., and Ivanov, S. S. (2016). MTOR-Driven Metabolic Reprogramming Regulates Legionella pneumophila Intracellular Niche Homeostasis. *PLoS Pathog.* 12, e1006088. doi: 10.1371/journal.ppat.1006088

Albert-Weissenberger, C., Cazalet, C., and Buchrieser, C. (2006). Legionella pneumophila — a human pathogen that co-evolved with fresh water protozoa. *Cell Mol. Life Sci.* 64, 432. doi: 10.1007/s00018-006-6391-1

Alonso, I. G., de los, F., Liang, H.-C., Turner, S. D., Lagger, S., Merkel, O., et al. (2018). The Role of Activator Protein-1 (AP-1) Family Members in CD30-Positive Lymphomas. *Cancers* 10, 93. doi: 10.3390/cancers10040093

Aqeel, Y., Siddiqui, R., Iftikhar, H., and Khan, N. A. (2013). The effect of different environmental conditions on the encystation of Acanthamoeba castellanii belonging to the T4 genotype. *Exp. Parasitol.* 135, 30–35. doi: 10.1016/j.exppara.2013.05.017

Archer, K. A., and Roy, C. R. (2006). MyD88-Dependent Responses Involving Toll-Like Receptor 2 Are Important for Protection and Clearance of Legionella pneumophila in a Mouse Model of Legionnaires' Disease. *Infect. Immun.* 74, 3325–3333. doi: 10.1128/iai.02049-05

Asrat, S., Dugan, A. S., and Isberg, R. R. (2014). The frustrated host response to Legionella pneumophila is bypassed by MyD88-dependent translation of pro-inflammatory cytokines. *PLoS Pathog.* 10, e1004229–e1004229. doi: 10.1371/journal.ppat.1004229

Avila, L. A., Aps, L. R. M. M., Ploscariu, N., Sukthankar, P., Guo, R., Wilkinson, K. E., et al. (2016). Gene delivery and immunomodulatory effects of plasmid DNA associated with Branched Amphiphilic Peptide Capsules. *J. Control Rel.* 241, 15–24. doi: 10.1016/j.jconrel.2016.08.042

Barbaree, J. M., Fields, B. S., Feeley, J. C., Gorman, G. W., and Martin, W. T. (1986). Isolation of protozoa from water associated with a legionellosis outbreak and demonstration of intracellular multiplication of Legionella pneumophila. *Appl. Environ. Microb.* 51, 422–424. doi: 10.1128/AEM.51.2.422-424.1986

Barry, K. C., Fontana, M. F., Portman, J. L., Dugan, A. S., and Vance, R. E. (2013). IL-1α signaling initiates the inflammatory response to virulent Legionella pneumophila in vivo. *J. Immunol.* 190, 6329–6339. doi: 10.4049/jimmunol.1300100

Barry, K. C., Ingolia, N. T., and Vance, R. E. (2017). Global analysis of gene expression reveals mRNA superinduction is required for the inducible immune response to a bacterial pathogen. *eLife* 6, e22707. doi: 10.7554/eLife.22707

Belyi, Y., Niggeweg, R., Opitz, B., Vogelsgesang, M., Hippenstiel, S., Wilm, M., et al. (2006). Legionella pneumophila glucosyltransferase inhibits host elongation factor 1A. *Proc. Natl. Acad. Sci.* 103, 16953–16958. doi: 10.1073/pnas.0601562103

Belyi, Y., Tabakova, I., Stahl, M., and Aktories, K. (2008). Lgt: a family of cytotoxic glucosyltransferases produced by Legionella pneumophila. *J. Bacteriol.* 190, 3026–3035. doi: 10.1128/jb.01798-07

Belyi, Y. (2020). Targeting Eukaryotic mRNA Translation by Legionella pneumophila. *Front. Mol. Biosci.* 7, 80. doi: 10.3389/fmolb.2020.00080

Berger, K. H., and Isberg, R. R. (1993). Two distinct defects in intracellular growth complemented by a single genetic locus in Legionella pneumophila. *Mol. Microbiol.* 7, 7–19. doi: 10.1111/j.1365-2958.1993.tb01092.x

Blanchard, D. K., Djeu, J. Y., Klein, T. W., Friedman, H., and Stewart, W. E. (1987). Induction of tumor necrosis factor by Legionella pneumophila. *Infect. Immun.* 55, 433–437. doi: 10.1128/IAI.55.2.433-437.1987

Blanchard, D. K., Friedman, H., Klein, T. W., and Djeu, J. Y. (1989). Induction of interferon-gamma and tumor necrosis factor by Legionella pneumophila: augmentation of human neutrophil bactericidal activity. *J. Leukocyte Biol.* 45, 538–545. doi: 10.1002/jlb.45.6.538

Boamah, D. K., Zhou, G., Ensminger, A. W., and O'Connor, T. J. (2017). From Many Hosts, One Accidental Pathogen: The Diverse Protozoan Hosts of Legionella. *Front. Cell Infect. Mi.* 7, 477. doi: 10.3389/fcimb.2017.00477

Borges, V., Nunes, A., Sampaio, D. A., Vieira, L., Machado, J., Simões, M. J., et al. (2016). Legionella pneumophila strain associated with the first evidence of person-to-person transmission of Legionnaires' disease: a unique mosaic genetic backbone. *Sci. Rep-uk* 6, 26261. doi: 10.1038/srep26261

Bouyer, S., Imbert, C., Rodier, M., and Héchard, Y. (2007). Long-term survival of Legionella pneumophila associated with Acanthamoeba castellanii vesicles. *Environ. Microbiol.* 9, 1341–1344. doi: 10.1111/j.1462-2920.2006.01229.x

Brieland, J. K., Remick, D. G., LeGendre, M. L., Engleberg, N. C., and Fantone, J. C. (1998). In vivo regulation of replicative Legionella pneumophila lung infection by endogenous interleukin-12. *Infect. Immun.* 66, 65–69. doi: 10.1128/IAI.66.1.65-69.1998

Brodsky, I. E., and Medzhitov, R. (2009). Targeting of immune signalling networks by bacterial pathogens. *Nat. Cell Biol.* 11, 521–526. doi: 10.1038/ncb0509-521

Bruckert, W. M., Price, C. T., and Kwaik, Y. A. (2013). Rapid nutritional remodeling of the host cell upon attachment of Legionella pneumophila. *Infect. Immun.* 82, 72–82. doi: 10.1128/iai.01079-13

Buchkovich, N. J., Yu, Y., Zampieri, C. A., and Alwine, J. C. (2008). The TORrid affairs of viruses: effects of mammalian DNA viruses on the PI3K-Akt-mTOR signalling pathway. *Nat. Rev. Microbiol.* 6, 266–275. doi: 10.1038/nrmicro1855

Cambronne, E. D., and Roy, C. R. (2006). Recognition and Delivery of Effector Proteins into Eukaryotic Cells by Bacterial Secretion Systems. *Traffic* 7, 929–939. doi: 10.1111/j.1600-0854.2006.00446.x

Casson, C. N., Doerner, J. L., Copenhaver, A. M., Ramirez, J., Holmgren, A. M., Boyer, M. A., et al. (2017). Neutrophils and Ly6Chi monocytes collaborate in generating an optimal cytokine response that protects against pulmonary Legionella pneumophila infection. *PLoS Pathog.* 13, e1006309. doi: 10.1371/journal.ppat.1006309

Cazalet, C., Gomez-Valero, L., Rusniok, C., Lomma, M., Dervins-Ravault, D., Newton, H. J., et al. (2010). Analysis of the Legionella longbeachae genome and transcriptome uncovers unique strategies to cause Legionnaires' disease. *PLoS Genet.* 6, e1000851. doi: 10.1371/journal.pgen.1000851

Chisholm, S. T., Coaker, G., Day, B., and Staskawicz, B. J. (2006). Host-Microbe Interactions: Shaping the Evolution of the Plant Immune Response. *Cell* 124, 803–814. doi: 10.1016/j.cell.2006.02.008

Colaço, H. G., and Moita, L. F. (2016). Initiation of innate immune responses by surveillance of homeostasis perturbations. *FEBS J.* 283, 2448–2457. doi: 10.1111/febs.13730

Condon, K. J., and Sabatini, D. M. (2019). Nutrient regulation of mTORC1 at a glance. *J. Cell Sci.* 132, jcs222570. doi: 10.1242/jcs.222570

Copenhaver, A. M., Casson, C. N., Nguyen, H. T., Fung, T. C., Duda, M. M., Roy, C. R., et al. (2014). Alveolar macrophages and neutrophils are the primary reservoirs for Legionella pneumophila and mediate cytosolic surveillance of type IV secretion. *Infect. Immun.* 82, 4325–4336. doi: 10.1128/iai.01891-14

Copenhaver, A. M., Casson, C. N., Nguyen, H. T., Duda, M. M., and Shin, S. (2015). IL-1R signaling enables bystander cells to overcome bacterial blockade of host protein synthesis. *Proc. Natl. Acad. Sci. USA* 112, 7557–7562. doi: 10.1073/pnas.1501289112

Dangl, J. L., and Jones, J. D. G. (2001). Plant pathogens and integrated defence responses to infection. *Nature* 411, 826–833. doi: 10.1038/35081161

De Leon, J. A., Qiu, J., Nicolai, C. J., Counihan, J. L., Barry, K. C., Xu, L., et al. (2017). Positive and Negative Regulation of the Master Metabolic Regulator mTORC1 by Two Families of Legionella pneumophila Effectors. *Cell Rep.* 21, 2031–2038. doi: 10.1016/j.celrep.2017.10.088

Degtyar, E., Zusman, T., Ehrlich, M., and Segal, G. (2009). A Legionella effector acquired from protozoa is involved in sphingolipids metabolism and is targeted to the host cell mitochondria. *Cell Microbiol.* 11, 1219–1235. doi: 10.1111/j.1462-5822.2009.01328.x

Elkins, K. L., Rhinehart-Jones, T. R., Stibitz, S., Conover, J. S., and Klinman, D. M. (1999). Bacterial DNA containing CpG motifs stimulates lymphocyte-dependent protection of mice against lethal infection with intracellular bacteria. *J. Immunol. Baltim. Md. 1950* 162, 2291–2298.

Ensminger, A. W. (2016). Legionella pneumophila, armed to the hilt: justifying the largest arsenal of effectors in the bacterial world. *Curr. Opin. Microbiol.* 29, 74–80. doi: 10.1016/j.mib.2015.11.002

Fischer, N. L., Naseer, N., Shin, S., and Brodsky, I. E. (2019). Effector-triggered immunity and pathogen sensing in metazoans. *Nat. Microbiol.* 5, 14–26. doi: 10.1038/s41564-019-0623-2

Fliermans, C. B., Cherry, W. B., Orrison, L. H., Smith, S. J., Tison, D. L., and Pope, D. H. (1981). Ecological distribution of Legionella pneumophila. *Appl. Environ. Microb.* 41, 9–16. doi: 10.1128/AEM.41.1.9-16.1981

Flor, H. H. (1971). Current Status of the Gene-For-Gene Concept. *Annu. Rev. Phytopathol.* 9, 275–296. doi: 10.1146/annurev.py.09.090171.001423

Fontana, M. F., Banga, S., Barry, K. C., Shen, X., Tan, Y., Luo, Z.-Q., et al. (2011). Secreted bacterial effectors that inhibit host protein synthesis are critical for induction of the innate immune response to virulent Legionella pneumophila. *PLoS Pathog.* 7, e1001289–e1001289. doi: 10.1371/journal.ppat.1001289

Fontana, M. F., Shin, S., and Vance, R. E. (2012). Activation of host mitogen-activated protein kinases by secreted Legionella pneumophila effectors that inhibit host protein translation. *Infect. Immun.* 80, 3570–3575. doi: 10.1128/iai.00557-12

Fouque, E., Trouilhé, M.-C., Thomas, V., Hartemann, P., Rodier, M.-H., and Héchard, Y. (2012). Cellular, Biochemical, and Molecular Changes during Encystment of Free-Living Amoebae. *Eukaryot. Cell* 11, 382–387. doi: 10.1128/ec.05301-11

Fraser, D. W., Tsai, T. R., Orenstein, W., Parkin, W. E., Beecham, H. J., Sharrar, R. G., et al. (1977). Legionnaires' disease: description of an epidemic of pneumonia. *New Engl. J. Med.* 297, 1189–1197. doi: 10.1056/nejm197712012972201

Friedman, H., Yamamoto, Y., and Klein, T. W. (2002). Legionella pneumophila pathogenesis and immunity. *Semin. Pediatr. Infect. Dis.* 13, 273–279. doi: 10.1053/spid.2002.127206

Fujioka, S., Niu, J., Schmidt, C., Sclabas, G. M., Peng, B., Uwagawa, T., et al. (2004). NF-κB and AP-1 Connection: Mechanism of NF-κB-Dependent Regulation of AP-1 Activity. *Mol. Cell Biol.* 24, 7806. doi: 10.1128/mcb.24.17.7806-7819.2004

Galán, J. E. (2009). Common Themes in the Design and Function of Bacterial Effectors. *Cell Host Microbe* 5, 571–579. doi: 10.1016/j.chom.2009.04.008

Gazon, H., Barbeau, B., Mesnard, J.-M., and Peloponese, J.-M. (2018). Hijacking of the AP-1 Signaling Pathway during Development of ATL. *Front. Microbiol.* 08, 2686. doi: 10.3389/fmicb.2017.02686

Ge, J., Xu, H., Li, T., Zhou, Y., Zhang, Z., Li, S., et al. (2009). A Legionella type IV effector activates the NF-kappaB pathway by phosphorylating the IkappaB family of inhibitors. *P Natl. Acad. Sci. U.S.A.* 106, 13725–13730. doi: 10.1073/pnas.0907200106

George, J. R., Pine, L., Reeves, M. W., and Harrell, W. K. (1980). Amino acid requirements of Legionella pneumophila. *J. Clin. Microbiol.* 11, 286–291. doi: 10.1128/JCM.11.3.286-291.1980

Ghosh, S., and O'Connor, T. J. (2017). Beyond Paralogs: The Multiple Layers of Redundancy in Bacterial Pathogenesis. *Front. Cell Infect. Mi.* 7:467:467. doi: 10.3389/fcimb.2017.00467

Glick, T. H., Gregg, M. B., Berman, B., Mallison, G., Rhodes, W. W., and Kassanoff, I. (1978). Pontiac fever. An epidemic of unknown etiology in a health department: I. Clinical and epidemiologic aspects. *Am. J. Epidemiol.* 107, 149–160. doi: 10.1093/oxfordjournals.aje.a112517

Gobin, I., Newton, P. R., Hartland, E. L., and Newton, H. J. (2009). Infections caused by nonpneumophila species of Legionella. *Rev. Med. Microbiol.* 20, 1–11. doi: 10.1097/mrm.0b013e32832e82da

Hess, J., Angel, P., and Schorpp-Kistner, M. (2004). AP-1 subunits: quarrel and harmony among siblings. *J. Cell Sci.* 117, 5965–5973. doi: 10.1242/jcs.01589

Horwitz, M. A. (1983). Formation of a novel phagosome by the Legionnaires' disease bacterium (Legionella pneumophila) in human monocytes. *J. Exp. Med.* 158, 1319–1331. doi: 10.1084/jem.158.4.1319

Ivanov, S. S., and Roy, C. R. (2013). Pathogen signatures activate a ubiquitination pathway that modulates the function of the metabolic checkpoint kinase mTOR. *Nat. Immunol.* 14, 1219–1228. doi: 10.1038/ni.2740

Janeway, C. A., and Medzhitov, R. (2002). Innate immune recognition. *Immunology* 20, 197–216. doi: 10.1146/annurev.immunol.20.083001.084359

Janeway, C. (1989). Approaching the asymptote? Evolution and revolution in immunology. *Cold Spring Harb. Sym.* 54 (Pt 1), 1–13. doi: 10.1101/sqb.1989.054.01.003

Jaramillo, M., Gomez, M. A., Larsson, O., Shio, M. T., Topisirovic, I., Contreras, I., et al. (2011). Leishmania Repression of Host Translation through mTOR Cleavage Is Required for Parasite Survival and Infection. *Cell Host Microbe* 9, 331–341. doi: 10.1016/j.chom.2011.03.008

Joseph, A. M., Pohl, A. E., Ball, T. J., Abram, T. G., Johnson, D. K., Geisbrecht, B. V., et al. (2020). The Legionella pneumophila metaeffector Lpg2505 (MesI) regulates SidI-mediated translation inhibition and novel glycosyl hydrolase activity. *Infect. Immun.* 88, e00853–19. doi: 10.1128/iai.00853-19

Juffermans, N. P., Leemans, J. C., Florquin, S., Verbon, A., Kolk, A. H., Speelman, P., et al. (2002). CpG Oligodeoxynucleotides Enhance Host Defense during Murine Tuberculosis. *Infect. Immun.* 70, 147–152. doi: 10.1128/iai.70.1.147-152.2002

Khweek, A. A., Caution, K., Akhter, A., Abdulrahman, B. A., Tazi, M., Hassan, H., et al. (2013). A bacterial protein promotes the recognition of the Legionella pneumophila vacuole by autophagy. *Eur. J. Immunol.* 43, 1333–1344. doi: 10.1002/eji.201242835

Khweek, A. A., Kanneganti, A., D, D. C. G., and Amer, A. O. (2016). The Sphingosine-1-Phosphate Lyase (LegS2) Contributes to the Restriction of Legionella pneumophila in Murine Macrophages. *PLoS One* 11, e0146410. doi: 10.1371/journal.pone.0146410

Kilvington, S., and Price, J. (1990). Survival of Legionella pneumophila within cysts of Acanthamoeba polyphaga following chlorine exposure. *J. Appl. Microbiol.* 68, 519–525. doi: 10.1111/j.1365-2672.1990.tb02904.x

Kofoed, E. M., and Vance, R. E. (2011). Innate immune recognition of bacterial ligands by NAIPs determines inflammasome specificity. *Nature* 477, 592–595. doi: 10.1038/nature10394

Langston, P. K., Shibata, M., and Horng, T. (2017). Metabolism Supports Macrophage Activation. *Front. Immunol.* 8, 61. doi: 10.3389/fimmu.2017.00061

Leroux, L.-P., Lorent, J., Graber, T. E., Chaparro, V., Masvidal, L., Aguirre, M., et al. (2018). The Protozoan Parasite Toxoplasma gondii Selectively Reprograms the Host Cell Translatome. *Infect. Immun.* 86, e00244–e00218. doi: 10.1128/iai.00244-18

Liu, X., Boyer, M. A., Holmgren, A. M., and Shin, S. (2020). Legionella-Infected Macrophages Engage the Alveolar Epithelium to Metabolically Reprogram Myeloid Cells and Promote Antibacterial Inflammation. *Cell Host Microbe.* 28, 683–698. doi: 10.1016/j.chom.2020.07.019

Losick, V. P., and Isberg, R. R. (2006). NF-κB translocation prevents host cell death after low-dose challenge by Legionella pneumophila. *J. Exp. Med.* 203, 2177–2189. doi: 10.1084/jem.20060766

Losick, V. P., Haenssler, E., Moy, M., and Isberg, R. R. (2010). LnaB: a Legionella pneumophila activator of NF-κB. *Cell Microbiol.* 12, 1083–1097. doi: 10.1111/j.1462-5822.2010.01452.x

Martínez, O. F., Cardoso, M. H., Ribeiro, S. M., and Franco, O. L. (2019). Recent Advances in Anti-virulence Therapeutic Strategies With a Focus on Dismantling Bacterial Membrane Microdomains, Toxin Neutralization, Quorum-Sensing Interference and Biofilm Inhibition. *Front. Cell Infect. Mi.* 9, 74. doi: 10.3389/fcimb.2019.00074

Martinon, F., Burns, K., and Tschopp, J. (2002). The inflammasome: a molecular platform triggering activation of inflammatory caspases and processing of proIL-beta. *Mol. Cell* 10, 417–426. doi: 10.1016/s1097-2765(02)00599-3

Mascarenhas, D. P. A., and Zamboni, D. S. (2017). Inflammasome biology taught by Legionella pneumophila. *J. Leukocyte Biol.* 101, 841–849. doi: 10.1189/jlb.3mr0916-380r

Mascarenhas, D. P. A., Pereira, M. S. F., Manin, G. Z., Hori, J.II, and Zamboni, D. S. (2015). Interleukin 1 Receptor-Driven Neutrophil Recruitment Accounts to MyD88-Dependent Pulmonary Clearance of Legionella pneumophila Infection In Vivo. *J. Infect. Dis.* 211, 322–330. doi: 10.1093/infdis/jiu430

Massis, L. M., and Zamboni, D. S. (2011). Innate Immunity to Legionella Pneumophila. *Front. Microbiol.* 2, 109. doi: 10.3389/fmicb.2011.00109

Massis, L. M., Assis-Marques, M. A., Castanheira, F. V. S., Capobianco, Y. J., Balestra, A. C., Escoll, P., et al. (2016). Legionella longbeachae is immunologically silent and highly virulent in vivo. *J. Infect. Dis.* 215, 440–451. doi: 10.1093/infdis/jiw560

McDade, J. E., Shepard, C. C., Fraser, D. W., Tsai, T. R., Redus, M. A., and Dowdle, W. R. (1980). Legionnaires' disease: isolation of a bacterium and demonstration of its role in other respiratory disease. *New Engl. J. Med.* 297, 1197–1203. doi: 10.1056/nejm197712012972202

Medzhitov, R., and Janeway, C. (2000). Innate immune recognition: mechanisms and pathways. *Immunol. Rev.* 173, 89–97. doi: 10.1034/j.1600-065X.2000.917309.x

Molmeret, M., Horn, M., Wagner, M., Santic, M., and Kwaik, Y. A. (2005). Amoebae as Training Grounds for Intracellular Bacterial Pathogens. *Appl. Environ. Microb.* 71, 20–28. doi: 10.1128/aem.71.1.20-28.2005

Moon, E.-K., and Kong, H.-H. (2012). Short-Cut Pathway to Synthesize Cellulose of Encysting Acanthamoeba. *Korean J. Parasitol.* 50, 361–364. doi: 10.3347/kjp.2012.50.4.361

Moss, S. M., Taylor, I. R., Ruggero, D., Gestwicki, J. E., Shokat, K. M., and Mukherjee, S. (2019). A Legionella pneumophila Kinase Phosphorylates the Hsp70 Chaperone Family to Inhibit Eukaryotic Protein Synthesis. *Cell Host Microbe* 25, 454–462.e6. doi: 10.1016/j.chom.2019.01.006

Murray, H. W., Szuro-Sudol, A., Wellner, D., Oca, M. J., Granger, A. M., Libby, D. M., et al. (1989). Role of tryptophan degradation in respiratory burst-independent antimicrobial activity of gamma interferon-stimulated human macrophages. *Infect. Immun.* 57, 845–849. doi: 10.1128/iai.57.3.845-849.1989

Nash, T. W., Libby, D. M., and Horwitz, M. A. (1984). Interaction between the legionnaires' disease bacterium (Legionella pneumophila) and human alveolar macrophages. Influence of antibody, lymphokines, and hydrocortisone. *J. Clin. Invest.* 74, 771–782. doi: 10.1172/jci111493

Ngwaga, T., Hydock, A. J., Ganesan, S., and Shames, S. R. (2019). Potentiation of Cytokine-Mediated Restriction of Legionella Intracellular Replication by a Dot/Icm-Translocated Effector. *J. Bacteriol.* 201, 13. doi: 10.1128/jb.00755-18

Oberemok, V. V., Laikova, K. V., Yurchenko, K. A., Marochkin, N. A., Fomochkina, I.II, and Kubyshkin, A. V. (2020). SARS-CoV-2 will constantly sweep its tracks: a vaccine containing CpG motifs in 'lasso' for the multi-faced virus. *Inflammation Res.* 69, 801–812. doi: 10.1007/s00011-020-01377-3

Park, J. M., Ghosh, S., and O'Connor, T. J. (2020). Combinatorial selection in amoebal hosts drives the evolution of the human pathogen Legionella pneumophila. *Nat. Microbiol.* 5, 599–609. doi: 10.1038/s41564-019-0663-7

Price, C. T. D., Richards, A. M., Dwingelo, J. E. V., Samara, H. A., and Kwaik, Y. A. (2014). Amoeba host-Legionella synchronization of amino acid auxotrophy and its role in bacterial adaptation and pathogenic evolution. *Environ. Microbiol.* 16, 350–358. doi: 10.1111/1462-2920.12290

Price, C., Jones, S., Mihelcic, M., Santic, M., and Kwaik, Y. A. (2020). Paradoxical Pro-inflammatory Responses by Human Macrophages to an Amoebae Host-Adapted Legionella Effector. *Cell Host Microbe* 27, 571–584.e7. doi: 10.1016/j.chom.2020.03.003

Rabideau, A. E., and Pentelute, B. L. (2016). Delivery of Non-Native Cargo into Mammalian Cells Using Anthrax Lethal Toxin. *ACS Chem. Biol.* 11, 1490–1501. doi: 10.1021/acschembio.6b00169

Rahman, M. M., and McFadden, G. (2011). Modulation of NF-κB signalling by microbial pathogens. *Nat. Rev. Microbiol.* 9, 291–306. doi: 10.1038/nrmicro2539

Rajamuthiah, R., and Mylonakis, E. (2014). Effector triggered immunity. *Virulence* 5, 697–702. doi: 10.4161/viru.29091

Rasko, D. A., and Sperandio, V. (2010). Anti-virulence strategies to combat bacteria-mediated disease. *Nat. Rev. Drug Discovery* 9, 117–128. doi: 10.1038/nrd3013

Rincon-Restrepo, M., Mayer, A., Hauert, S., Bonner, D. K., Phelps, E. A., Hubbell, J. A., et al. (2017). Vaccine nanocarriers: Coupling intracellular pathways and cellular biodistribution to control CD4 vs CD8 T cell responses. *Biomaterials* 132, 48–58. doi: 10.1016/j.biomaterials.2017.03.047

Rolando, M., and Buchrieser, C. (2018). Legionella Effectors Explored with INSeq: New Functional Insights. *Trends Microbiol.* 26, 169–170. doi: 10.1016/j.tim.2018.01.003

Rolando, M., Escoll, P., Nora, T., Botti, J., Boitez, V., Bedia, C., et al. (2016). Legionella pneumophila S1P-lyase targets host sphingolipid metabolism and restrains autophagy. *Proc. Natl. Acad. Sci.* 113, 1901–1906. doi: 10.1073/pnas.1522067113

Rowbotham, T. J. (1980). Preliminary report on the pathogenicity of Legionella pneumophila for freshwater and soil amoebae. *J. Clin. Pathol.* 33:1179. doi: 10.1136/jcp.33.12.1179

Rüter, C., and Hardwidge, P. R. (2014). "Drugs from bugs": bacterial effector proteins as promising biological (immune-) therapeutics. *FEMS Microbiol. Lett.* 351, 126–132. doi: 10.1111/1574-6968.12333

Rüter, C., and Schmidt, M. A. (2017). Cell-Penetrating Bacterial Effector Proteins: Better Tools than Targets. *Trends Biotechnol.* 35, 109–120. doi: 10.1016/j.tibtech.2016.08.002

Schaap, P., and Schilde, C. (2018). Encystation: the most prevalent and underinvestigated differentiation pathway of eukaryotes. *Microbiology+* 164, 727–739. doi: 10.1099/mic.0.000653

Scheiermann, J., and Klinman, D. M. (2014). Clinical evaluation of CpG oligonucleotides as adjuvants for vaccines targeting infectious diseases and cancer. *Vaccine* 32, 6377–6389. doi: 10.1016/j.vaccine.2014.06.065

Schunder, E., Gillmaier, N., Kutzner, E., Herrmann, V., Lautner, M., Heuner, K., et al. (2014). Amino Acid Uptake and Metabolism of Legionella pneumophila

Hosted by Acanthamoeba castellanii. *J. Biol. Chem.* 289, 21040–21054. doi: 10.1074/jbc.m114.570085

Shames, S. R., Liu, L., Havey, J. C., Schofield, W. B., Goodman, A. L., and Roy, C. R. (2017). Multiple Legionella pneumophila effector virulence phenotypes revealed through high-throughput analysis of targeted mutant libraries. *Proc. Natl. Acad. Sci.* 63, 201708553. doi: 10.1073/pnas.1708553114

Shen, X., Banga, S., Liu, Y., Xu, L., Gao, P., Shamovsky, I., et al. (2009). Targeting eEF1A by a Legionella pneumophila effector leads to inhibition of protein synthesis and induction of host stress response. *Cell Microbiol.* 11, 911–926. doi: 10.1111/j.1462-5822.2009.01301.x

Sherwood, R. K., and Roy, C. R. (2016). Autophagy Evasion and Endoplasmic Reticulum Subversion: The Yin and Yang of Legionella Intracellular Infection. *Annu. Rev. Microbiol.* 70, 413–433. doi: 10.1146/annurev-micro-102215-095557

Shi, J., Zhao, Y., Wang, K., Shi, X., Wang, Y., Huang, H., et al. (2015). Cleavage of GSDMD by inflammatory caspases determines pyroptotic cell death. *Nature* 526, 660–665. doi: 10.1038/nature15514

Shin, S., and Roy, C. R. (2008). Host cell processes that influence the intracellular survival of Legionella pneumophila. *Cell Microbiol.* 10, 1209–1220. doi: 10.1111/j.1462-5822.2008.01145.x

Shin, S. (2012). Innate Immunity to Intracellular Pathogens: Lessons Learned from Legionella pneumophila. *Adv. Appl. Microbiol.* 79, 43–71. doi: 10.1016/b978-0-12-394318-7.00003-6

Soda, E. A., Barskey, A. E., Shah, P. P., Schrag, S., Whitney, C. G., Arduino, M. J., et al. (2017). Vital Signs: Health Care–Associated Legionnaires' Disease Surveillance Data From 20 States and a Large Metropolitan Area—United States 2015. *Am. J. Transplant.* 17, 2215–2220. doi: 10.1111/ajt.14407

Stuart, L. M., Paquette, N., and Boyer, L. (2013). Effector-triggered versus pattern-triggered immunity: how animals sense pathogens. *Nat. Rev. Immunol.* 13, 199–206. doi: 10.1038/nri3398

Weisman, R. A., Spiegel, R. S., and McCauley, J. G. (1970). Differentiation in acanthamoeba: Glycogen levels and glycogen synthetase activity during encystment. *Biochim. Et Biophys. Acta Bba Gen. Subj.* 201, 45–53. doi: 10.1016/0304-4165(70)90008-5

Wood, R. E., Newton, P., Latomanski, E. A., and Newton, H. J. (2015). Dot/Icm Effector Translocation by Legionella longbeachae Creates a Replicative Vacuole Similar to That of Legionella pneumophila despite Translocation of Distinct Effector Repertoires. *Infect. Immun.* 83, 4081–4092. doi: 10.1128/iai.00461-15

Xiang, T., Zong, N., Zou, Y., Wu, Y., Zhang, J., Xing, W., et al. (2008). Pseudomonas syringae Effector AvrPto Blocks Innate Immunity by Targeting Receptor Kinases. *Curr. Biol.* 18, 74–80. doi: 10.1016/j.cub.2007.12.020

Zhu, W., Banga, S., Tan, Y., Zheng, C., Stephenson, R., Gately, J., et al. (2011). Comprehensive identification of protein substrates of the Dot/Icm type IV transporter of Legionella pneumophila. *PLoS One* 6, e17638. doi: 10.1371/journal.pone.0017638

Zimmermann, S., Egeter, O., Hausmann, S., Lipford, G. B., Röcken, M., Wagner, H., et al. (1998). CpG oligodeoxynucleotides trigger protective and curative Th1 responses in lethal murine leishmaniasis. *J. Immunol. Baltim. Md. 1950* 160, 3627–3630.

10

Advances and Challenges in Studying Type III Secretion Effectors of Attaching and Effacing Pathogens

*Sabrina L. Slater and Gad Frankel**

MRC Centre for Molecular Bacteriology and Infection, Department of Life Sciences, Imperial College London, London, United Kingdom

Keywords: enteropathogenic *E. coli*, EPEC, enterohaemorrhagic *E. coli*, EHEC, T3SS, effector, *C. rodentium*, machine learning

Correspondence:
Gad Frankel
g.frankel@imperial.ac.uk

INTRODUCTION

Outbreaks of the diarrhoeal disease caused by enteropathogenic *Escherichia coli* (EPEC) and enterohaemorrhagic *E. coli* (EHEC) present a significant burden to public health in countries with low and high human development indices (HDIs) alike. Over the course of evolution, horizontal gene transfer events have expanded the 4.1 Mb core genome of the human colonic commensal *E. coli* by 1 Mb, resulting in the emergence of pathogens (Croxen and Finlay, 2010; Clements et al., 2012). Each pathogenic strain is characterized by their unique virulence factor repertoire and clinical epidemiology (Nataro and Kaper, 1998; Gomes et al., 2016). EPEC causes infantile diarrhea in countries with a low HDI. Indeed, from 2007 to 2015 the World Health Organization (WHO) estimate 230,000 cases of death from diarrhoeal disease, of which EPEC was responsible for 16%, disproportionately affecting children under five (WHO, 2015). EHEC, on the other hand, is defined by its ability to produce Shiga toxins (Stx) (Melton-Celsa, 2014; Krause et al., 2018). Distinguishing between EPEC and EHEC is clinically important, as treatment of EHEC with antibiotics can incite Stx expression (Zhang et al., 2000) and consequently acute kidney failure, a sequela of haemolytic uremic syndrome (HUS) (Pacheco and Sperandio, 2012).

EPEC and EHEC are united in their ability to intimately adhere to human enterocytes, causing elongation and loss of microvilli, and the formation of actin-rich pedestals at the site of bacterial attachment (Finlay et al., 1992; Frankel and Phillips, 2008) As such, EPEC and EHEC are members of the attaching and effacing (A/E) family of pathogens, which also include *E. albertii* (Bhatt et al., 2019), the murine-restricted *Citrobacter rodentium* (Mullineaux-Sanders et al., 2019) and rabbit enteropathogenic *E. coli* (REPEC) (Milon et al., 1999). The formation of A/E lesions is facilitated by proteins encoded on the locus of enterocyte effacement (LEE), a largely conserved 35.6 kb pathogenicity island that encodes components of the Type 3 Secretion System (T3SS). The T3SS of A/E pathogens is ~3.5 MDa and includes several elements. Its cytoplasmic complex is equipped with an ATPase (EscN), secretion regulators (SepL and SepD) and chaperones (such as CesT). A basal body spans the inner and outer membranes, tethering the sheathed extracellular EscF needle to EspA filaments (Knutton et al., 1998) and culminates in the translocation pore (comprising EspB and EspD) in the host membrane (reviewed by Slater et al., 2018). The integration of signals from the gut environment, the microbiome and chaperones facilitate T3SS assembly, and translocation of effector proteins from the bacterium directly into the host cytosol (McDaniel et al., 1995; Connolly et al., 2015; Furniss and Clements, 2017; Katsowich et al., 2017; Serapio-Palacios and Finlay, 2020).

The effectors of A/E pathogens are encoded on either the LEE, prophages or insertion elements. While all the effectors rely on an N-terminal translocation sequence and specific chaperone-binding motifs to guide secretion (Deng et al., 2017; Slater et al., 2018; Wagner et al., 2018), their sequences are otherwise highly adapted to intercept specific host processes. Additionally, effector

TABLE 1 | Reported repertoire of A/E effectors, their method of discovery and inferred role during infection.

Effector	Discovery method	Inferred function during infection
Cif	Transposon mutagenesis in rabbit ileal loop model	Promotes bacterial survival/ cell cycle arrest
Tir/EspE	Anti-phosphotyrosine blotting	Host adherence, actin polymerisation and pyroptosis
EspF	LEE inspection	Induces apoptosis and disrupts tissue architecture
EspG	LEE inspection/ effector homology	Manipulates small GTPases
EspH	LEE inspection	Cytoskeletal remodeling, inhibition of phagocytosis
EspI/NleA	Coomassie of secreted proteins	Inhibits inflammasome activation and protein secretion
EspJ	Transcriptome analysis of adhered bacteria	Inhibits phagocytosis
EspK	Transposon mutagenesis in calf model	Unknown
EspL	Mass spectrometry (MS) of secreted proteins	Inhibits necroptosis
EspM1/2	MS of secreted proteins	Cytoskeletal remodeling
EspN	MS of secreted proteins	Unknown
EspO	MS of secreted proteins	Promotes tissue integrity by promoting IL-22 secretion
EspR	MS of secreted proteins	Unknown
EspS/Ibe/OspB	Effector homology	Suppresses colonic crypt hyperplasia
EspT	Effector homology	Cytoskeletal remodeling, NF-κB modulation
EspW	MS of secreted proteins	Cytoskeletal remodeling
EspX/NleL	MS of secreted proteins	Ligates ubiquitin
EspY	MS of secreted proteins	Unknown
EspZ/SepZ	Transposon mutagenesis in cell culture	Limits effector translocation
Map	LEE inspection	Cytoskeletal remodeling, mitochondrial disruption and colonic oxygenisation
NleB	MS of secreted proteins	Inhibits pro-inflammatory signaling and necroptosis
NleC	MS of secreted proteins	NF-κB inhibition
NleD	Transposon mutagenesis in bovine gastrointestinal tract model	MAPK signaling inhibition
NleE	MS of secreted proteins	NF-κB inhibition
NleF	MS of secreted proteins	Inhibits caspase-4/8/9 activity to limit cell death
NleG	MS of secreted proteins	Ligates ubiquitin
NleH/OspG	Proximity to known effector	Inhibits cell death and NF-κB signaling
NleJ	MS of secreted proteins	Unknown
NleK	MS of secreted proteins	Unknown
TccP/EspF$_U$	Transcriptome analysis of adhered bacteria	Actin polymerisation

functions can be antagonistic or cooperative (Shenoy et al., 2018), which taken together with their low abundance and continuous acquisition, underpins the challenge of identifying and studying effectors in a meaningful context.

Current research into the effector biology of A/E pathogens can be considered in three phases: discovery, *in vitro* functional characterization, and defining the contribution of each effector to the pathogen's infection strategy *in vivo*. To date, 30 families of effectors have been identified in A/E pathogens (**Table 1**). However, as with many pathogens, the rate of effector discovery has surpassed their biological characterization, and the contribution of many effectors to pathogenesis remains unknown. Here we highlight recent advances in technical and conceptual approaches to characterize effectors in A/E pathogens in the context of human disease.

PREDICTING AND VERIFYING TRANSLOCATION SUBSTRATES

Several resources have been developed to identify new effectors. Effector-encoding genes can be predicted *in silico* to varying degrees of accuracy (McDermott et al., 2011; Hobbs et al., 2016; Xue et al., 2019). These algorithms harness experimental knowledge of typical type III effector features, such as N-terminal enrichment of small polar amino acids (e.g., serine and threonine; Arnold et al., 2009), conservation of regulatory motifs upstream of the gene, a differing GC content to the rest of the genome, lack of gene homology to non-T3SS-encoding strains, and gene proximity to known effectors (Teper et al., 2016). Indeed, many novel type III effectors have been identified and validated using algorithm-based approaches, including *Pseudomonas syringae* and *P. fluorescens* (Vinatzer et al., 2005; Samudrala et al., 2009), *Ralstonia spp.* (Sabbagh et al., 2019), *S.* Typhimurium (Samudrala et al., 2009), *Chlamydia trachomatis* and *C. psittaci* (Hovis et al., 2013), *Xanthomonas euvesicatoria* (Teper et al., 2016) and *Pantoea agglomerans* (Nissan et al., 2018).

Despite these successes, algorithm-based approaches have yet to be applied to A/E genomes. Instead, effectors in A/E pathogens were historically discovered through manually curating pathogenicity island genes and mutagenesis screening (Dziva et al., 2004; Mundy et al., 2004; Kanack et al., 2005), homology searches to other T3SS effectors in different species

(Bulgin et al., 2009; Petty et al., 2010), mRNA profiling during infection (Dahan et al., 2005), and peptide discovery mass spectrometry (MS) of secreted proteins, notably in combination with 2D gel electrophoresis and effector hypersecretion mutants (Kresse et al., 2000; Creasey et al., 2003; Deng et al., 2004, 2010, 2012; Gruenheid et al., 2004; O'Connell et al., 2004; Tobe et al., 2006; Orton et al., 2013). Moving forward, the employment of techniques that do not rely on homology offer less bias and are therefore preferable. Additionally, A/E pathogens that encode a second functional T3SS (named ETT2) may also secrete its own cognate effectors (Fox et al., 2020), and secretion substrates could be shared between the two T3SSs, as there is evidence for regulatory crossover (Zhang et al., 2004; Luzader et al., 2016).

Once identified, the T3SS-dependent translocation of candidate effectors must be experimentally confirmed. A common approach, developed in 2004 by Charpentier and Oswald, is to C-terminally tag the effector with the TEM-1 β-lactamase and infect CCF2-loaded cells (Charpentier and Oswald, 2004); alternative and refined protocols have since been developed that decrease the tag size, minimize cell toxicity and offer single cell resolution. Collectively, these approaches benefit from their capacity to support different modes of analysis depending on the infection setup, such as enzymatic assays, optical readouts in a 96-well plate, flow cytometry and immunofluorescence microscopy (Mills et al., 2008; Miyake et al., 2008; Gawthorne et al., 2016; O'Boyle et al., 2018).

IN VITRO CHARACTERIZATION OF THE EFFECTORS

Approaches to characterize the role of each new effector during infection are ever-developing. At its most fundamental, effector functionality can be investigated under overexpression conditions *in vitro*, where amenable cells are transfected for ectopic effector expression or microinjected with purified protein. Overexpression protocols can provide readouts for drastic visual phenotypes, such as the radical cytoskeletal rearrangements resulting from the transfection of EspV (Arbeloa et al., 2011). Non-mammalian eukaryotic systems such as *Saccharomyces cerevisiae* have also been instrumental in the definition of eukaryotic interaction partners for A/E effectors (Hardwidge et al., 2004; Popa et al., 2016), as well as delineating interfaces for substrate interaction and catalytic residues (Blasche et al., 2013, 2014; Sandu et al., 2017).

These approaches, however, share a common weakness: effectors localize differently when not natively translocated through the injectisome. As such, the infection of appropriate mammalian cells (i.e., colonic epithelial cells) with bacteria translocating a tagged effector protein can provide a more physiologically relevant readout for effector localization. Indeed, native expression of a tagged effector is readily achievable by introducing a C-terminal tag onto the chromosome for effector visualization by immunofluorescence, or for use in tandem with co-immunoprecipitation and/or MS to probe for host protein interactors upon infection. Chromosomal manipulation by triparental conjugation works efficiently in A/E pathogens and other enteric pathogens (Mullineaux-Sanders et al., 2017; Watson et al., 2019; Wong et al., 2019). This conjugation protocol can similarly be used to generate scarless isogenic effector mutants in lieu of traditional gene disruption with antibiotic resistance cassettes or transposon elements (Cepeda-Molero et al., 2017).

DEFINING ROLES FOR THE EFFECTORS DURING INFECTION

Despite the ease of culturing, transfection and microscopy offered by non-polarized and non-colonic cells, the integrated use of more relevant models circumvents cell-line-specific phenotypes. A shift in practice toward more native models for infection is therefore evolving, using differentiated, polarized colonic cell lines, explants and organoids, primary tissues and laboratory animals (Carvalho et al., 2005; Law et al., 2013; Lewis et al., 2016; Cepeda-Molero et al., 2017). Increasingly, infections with A/E pathogens are also modeled in immune-associated cells, whose distinctive protein expression profiles allows researchers to probe the impact of effector delivery on immune-specific pathways (Pearson et al., 2017; Goddard et al., 2019), providing an alternative insight into the impact of effectors in human infection.

One particularly useful model for probing effector function is the infection of mice with *C. rodentium. C. rodentium* is a natural murine pathogen which causes transmissible colonic crypt hyperplasia (CCH) and A/E lesions that are indistinguishable from those cause by EPEC and EHEC in humans (Barthold et al., 1978). Critically, *C. rodentium* shares 67% homology with EPEC and EHEC genomes, most notably in the LEE (Petty et al., 2010, 2011), making it an invaluable tool for the study of the role of the T3SS and its cognate effectors *in vivo* (Mundy et al., 2005; Borenshtein et al., 2008; Collins et al., 2014; Mullineaux-Sanders et al., 2019).

The contribution of single or multiple effectors to pathogenesis can be assayed *via* infection with *C. rodentium* deletion or point mutants (Crepin et al., 2016). Key to revealing novel and physiological phenotypes is the selection of an appropriate mouse strain, or knock-out mice (Simmons et al., 2002; Zheng et al., 2008; Carson et al., 2019). To highlight some examples, the mouse model has delineated Tir, NleA and NleB as essential effectors for efficient colonization (Deng et al., 2003; Mundy et al., 2004; Kelly et al., 2006), demonstrated the impact of individual effectors deletions on host physiology, such as EspO and EspS impacting CCH (Berger et al., 2018; Connolly et al., 2018), and substantiated *in vitro* data indicating Map impacts colonic oxygen availability through mitochondrial disruption (Berger et al., 2017). Recently, mouse-specific differences in infection signatures have been identified through RNAseq and proteomics (Kang et al., 2018; Carson et al., 2019); it remains to be seen whether these differences are fine-tuned by the synergistic action of the effectors.

There are important genetic differences between *C. rodentium* and human A/E pathogens. While *C. rodentium* encodes a type IV pilus named colonization factor *Citrobacter* (CFC), which is related to the EPEC bundle forming pilus (BFP) (Mundy et al., 2003), it does not encode some strain-specific effectors, nor does it express the Shiga toxin or a flagellum, and it likely benefits from mouse-specific host adaptations. Nevertheless, modifications to this model can be implemented to investigate specific aspects of infection by human A/E pathogens, such as HUS and diarrhea (Vallance et al., 2003; Mallick et al., 2012), and drawing parallels from *C. rodentium* studies offers invaluable insight into human infections.

At the other end of the spectrum, the use of minimal effector models should prove instrumental for delineating the complex interplay between effectors, where the creation and complementation of isogenic strains cannot. As has been employed for *Yersinia pestis* (Palace et al., 2018), all known effectors were recently removed from the EPEC E2348/69 genome to investigate the contribution of select effectors and intact pathogenicity islands to A/E lesion formation on the human intestinal mucosa *ex vivo*, confirming that while Tir is essential, it is not sufficient and other elusive effectors are required (Cepeda-Molero et al., 2017). Undoubtedly the same approach would be of great use for other T3SS-encoding pathogens. Building on the decades of fundamental biochemical research into individual effector proteins, the mutation of clusters of effectors with similar functions *in vitro* should also be considered.

Despite the wealth of technical and biological knowledge unearthed over the last two decades, the synergies and redundancies in the function of the effectors hamper the comparison of *in vitro* research to *in vivo* scenarios. However, large-scale sequencing efforts of human pathogenic *E. coli* isolates have revealed the correlation between the presence of effectors, and other virulence factors, and the severity of human pathology through comparative genomics (Donnenberg et al., 2015; Hazen et al., 2016). This can be extended to assaying the prevalence of effectors in strains from different environments (Xu et al., 2017). Although challenging to integrate, these datasets offer unparalleled insight into the relative importance of effector proteins during human infection. Finally, following translocation, effectors form tight interaction networks. As clinical EPEC and EHEC isolates encode strain-specific effector gene combinations, it would be interesting to test the robustness of these networks *in vivo*. This will offer insight into the virulence potential of alternative effector combinations present in clinical isolates. *C. rodentium* provides an ideal model to address this.

CONCLUSIONS AND FUTURE PERSPECTIVES

With technological advances, our conceptual approaches must keep up. For example, moving beyond the idea that effectors act exclusively once translocated has illuminated greater functionality for *C. rodentium* NleB, which was recently demonstrated to GlcNAcylate an intrabacterial glutathione synthase GshB to promote bacterial survival (El Qaidi et al., 2020); other effectors with enzymatic functions may also have intrabacterial targets. In the same vein, addressing effector functionality in a top-down fashion (from mammalian phenotype to bacterial genotype) has recently unveiled both a unique mode of EPEC-induced inflammatory cell death and a further role for Tir during infection (Goddard et al., 2019).

Effector proteins exhibit fascinating diversity and specificity. Consequently, their study is warranted not only by their importance in human disease. Effectors can demonstrate first-in-class functions, the characterization of which broadens our understanding of protein biochemistry. Additionally, as effectors have witnessed the complexity of eukaryotic signaling cascades for considerably longer than modern researchers, they offer a unique insight into the breadth of function our own cells.

AUTHOR CONTRIBUTIONS

SS and GF wrote the manuscript. All authors listed have made a substantial, direct and intellectual contribution to the work, and approved it for publication.

ACKNOWLEDGMENTS

We thank Caroline Mullineaux-Sanders for critical reading of the manuscript.

REFERENCES

Arbeloa, A., Oates, C. V., Marchès, O., Hartland, E. L., and Frankel, G. (2011). Enteropathogenic and enterohemorrhagic *Escherichia coli* type III secretion effector EspV induces radical morphological changes in eukaryotic cells. *Infect. Immun.* 79, 1067–1076. doi: 10.1128/IAI.01003-10

Arnold, R., Brandmaier, S., Kleine, F., Tischler, P., Heinz, E., Behrens, S., et al. (2009). Sequence-based prediction of type III secreted proteins. *PLoS Pathog.* 5:e1000376. doi: 10.1371/journal.ppat.1000376

Barthold, S. W., Coleman, G. L., Jacoby, R. O., Livstone, E. M., and Jonas, A. M. (1978). Transmissible murine colonic hyperplasia. *Vet. Pathol.* 15, 223–236. doi: 10.1177/030098587801500209

Berger, C. N., Crepin, V. F., Roumeliotis, T. I., Wright, J. C., Carson, D., Pevsner-Fischer, M., et al. (2017). Citrobacter rodentium Subverts ATP flux and cholesterol homeostasis in intestinal epithelial cells *in vivo*. *Cell Metab.* 26, 738–752. doi: 10.1016/j.cmet.2017.09.003

Berger, C. N., Crepin, V. F., Roumeliotis, T. I., Wright, J. C., Serafini, N., Pevsner-Fischer, M., et al. (2018). The Citrobacter rodentium type III secretion system effector EspO affects mucosal damage repair and antimicrobial responses. *PLoS Pathog.* 14, 1–23. doi: 10.1371/journal.ppat.1007406

Bhatt, S., Egan, M., Critelli, B., Kouse, A., Kalman, D., and Upreti, C. (2019). The evasive enemy: Insights into the virulence and epidemiology of the emerging attaching and effacing pathogen escherichia albertii. *Infect. Immun.* 87, 1–15. doi: 10.1128/IAI.00254-18

Blasche, S., Arens, S., Ceol, A., Siszler, G., Schmid, M. A., Häuser, R., et al. (2014). The EHEC-host interactome reveals novel targets for the translocated intimin receptor. *Sci. Rep.* 4, 22–26. doi: 10.1038/srep07531

Blasche, S., Mörtl, M., Steuber, H., Siszler, G., Nisa, S., Schwarz, F., et al. (2013). The E. coli effector protein NleF is a caspase inhibitor. *PLoS ONE* 8:e58937. doi: 10.1371/journal.pone.0058937

Borenshtein, D., McBee, M. E., and Schauer, D. B. (2008). Utility of the Citrobacter rodentium infection model in laboratory mice. *Curr. Opin. Gastroenterol.* 24, 32–37. doi: 10.1097/MOG.0b013e3282f2b0fb

Bulgin, R. R., Arbeloa, A., Chung, J. C. S., and Frankel, G. (2009). EspT triggers formation of lamellipodia and membrane ruffles through activation of Rac-1 and Cdc42. *Cell. Microbiol.* 11, 217–229. doi: 10.1111/j.1462-5822.2008.01248.x

Carson, D., Barry, R., Hopkins, E. G. D., Roumeliotis, T. I., García-Weber, D., Mullineaux-Sanders, C., et al. (2019). Citrobacter rodentium induces rapid and unique metabolic and inflammatory responses in mice suffering from severe disease. *Cell. Microbiol.* 22, 1–17. doi: 10.1111/cmi.13126

Carvalho, H. M., Teel, L. D., Goping, G., and O'Brien, A. D. (2005). A three-dimensional tissue culture model for the study of attach and efface lesion formation by enteropathogenic and enterohaemorrhagic *Escherichia coli*. *Cell. Microbiol.* 7, 1771–1781. doi: 10.1111/j.1462-5822.2004.00594.x

Cepeda-Molero, M., Berger, C. N., Walsham, A. D. S., Ellis, S. J., Wemyss-Holden, S., Schüller, S., et al. (2017). Attaching and effacing (A/E) lesion formation by enteropathogenic E. coli on human intestinal mucosa is dependent on non-LEE effectors. *PLOS Pathog.* 13:e1006706. doi: 10.1371/journal.ppat.1006706

Charpentier, X., and Oswald, E. (2004). Identification of the secretion and translocation domain of the enteropathogenic and enterohemorrhagic. *J. Bacteriol.* 186, 5486–5495. doi: 10.1128/JB.186.16.5486-5495.2004

Clements, A., Young, J. C., Constantinou, N., and Frankel, G. (2012). Infection strategies of enteric pathogenic *Escherichia coli*. *Gut Microbes* 3, 71–87. doi: 10.4161/gmic.19182

Collins, J. W., Keeney, K. M., Crepin, V. F., Rathinam, V. A. K., Fitzgerald, K. A., Finlay, B. B., et al. (2014). Citrobacter rodentium: infection, inflammation and the microbiota. *Nat. Rev. Microbiol.* 12, 612–623. doi: 10.1038/nrmicro3315

Connolly, J. P. R., Brett Finlay, B., and Roe, A. J. (2015). From ingestion to colonization: The influence of the host environment on regulation of the LEE encoded type III secretion system in enterohaemorrhagic *Escherichia coli*. *Front. Microbiol.* 6:568. doi: 10.3389/fmicb.2015.00568

Connolly, J. P. R., Slater, S. L., O'Boyle, N., Goldstone, R. J., Crepin, V. F., Ruano-Gallego, D., et al. (2018). Host-associated niche metabolism controls enteric infection through fine-tuning the regulation of type 3 secretion. *Nat. Commun.* 9:4187. doi: 10.1038/s41467-018-06701-4

Creasey, E. A., Delahay, R. M., Daniell, S. J., and Frankel, G. (2003). Yeast two-hybrid system survey of interactions between LEE-encoded proteins of enteropathogenic *Escherichia coli*. *Microbiology* 149, 2093–2106. doi: 10.1099/mic.0.26355-0

Crepin, V. F., Collins, J. W., Habibzay, M., and Frankel, G. (2016). Citrobacter rodentium mouse model of bacterial infection. *Nat. Protoc.* 11, 1851–1876. doi: 10.1038/nprot.2016.100

Croxen, M. A., and Finlay, B. B. (2010). Molecular mechanisms of *Escherichia coli* pathogenicity. *Nat. Rev. Microbiol.* 8, 26–38. doi: 10.1038/nrmicro2265

Dahan, S., Wiles, S., La Ragione, R. M., Best, A., Woodward, M. J., Stevens, M. P., et al. (2005). EspJ is a prophage-carried type III effector protein of attaching and effacing pathogens that modulates infection dynamics. *Infect. Immun.* 73, 679–686. doi: 10.1128/IAI.73.2.679-686.2005

Deng, W., De Hoog, C. L., Yu, H. B., Li, Y., Croxen, M. A., Thomas, N. A., et al. (2010). A comprehensive proteomic analysis of the type III secretome of Citrobacter rodentium. *J. Biol. Chem.* 285, 6790–6800. doi: 10.1074/jbc.M109.086603

Deng, W., Marshall, N. C., Rowland, J. L., McCoy, J. M., Worrall, L. J., Santos, A. S., et al. (2017). Assembly, structure, function and regulation of type III secretion systems. *Nat. Rev. Microbiol.* 15, 323–337. doi: 10.1038/nrmicro.2017.20

Deng, W., Puente, J. L., Gruenheid, S., Li, Y., Vallance, B. A., Vázquez, A., et al. (2004). Dissecting virulence: Systematic and functional analyses of a pathogenicity island. *Proc. Natl. Acad. Sci. U.S.A.* 101, 3597–3602. doi: 10.1073/pnas.0400326101

Deng, W., Vallance, B. A., Li, Y., Puente, J. L., and Finlay, B. B. (2003). Citrobacter rodentium translocated intimin receptor (Tir) is an essential virulence factor needed for actin condensation, intestinal colonization and colonic hyperplasia in mice. *Mol. Microbiol.* 48, 95–115. doi: 10.1046/j.1365-2958.2003.03429.x

Deng, W., Yu, H. B., de Hoog, C. L., Stoynov, N., Li, Y., Foster, L. J., et al. (2012). Quantitative proteomic analysis of type iii secretome of enteropathogenic *Escherichia coli* reveals an expanded effector repertoire for attaching/effacing bacterial pathogens. *Mol. Cell. Proteomics* 11, 692–709. doi: 10.1074/mcp.M111.013672

Donnenberg, M. S., Hazen, T. H., Farag, T. H., Panchalingam, S., Antonio, M., Hossain, A., et al. (2015). Bacterial factors associated with lethal outcome of enteropathogenic *Escherichia coli* infection: genomic case-control studies. *PLoS Negl. Trop. Dis.* 9:e0003791. doi: 10.1371/journal.pntd.0003791

Dziva, F., van Diemen, P. M., Stevens, M. P., Smith, A. J., and Wallis, T. S. (2004). Identification of Escherichia coli O157:H7 genes influencing colonization of the bovine gastrointestinal tract using signature-tagged mutagenesis. *Microbiology* 150, 3631–3645. doi: 10.1099/mic.0.27448-0

El Qaidi, S., Scott, N. E., Hays, M. P., Geisbrecht, B. V., Watkins, S., and Hardwidge, P. R. (2020). An intra-bacterial activity for a T3SS effector. *Sci. Rep.* 10:1073. doi: 10.1038/s41598-020-58062-y

Finlay, B. B., Rosenshine, I., Donnenberg, M. S., and Kaper, J. B. (1992). Cytoskeletal composition of attaching and effacing lesions associated with enteropathogenic *Escherichia coli* adherence to HeLa cells. *Infect. Immun.* 60, 2541–2543. doi: 10.1128/IAI.60.6.2541-2543.1992

Fox, S., Goswami, C., Holden, M., Connolly, J. P. R., Mordue, J., O'Boyle, N., et al. (2020). A highly conserved complete accessory *Escherichia coli* type III secretion system 2 is widespread in bloodstream isolates of the ST69 lineage. *Sci. Rep.* 10, 1–11. doi: 10.1038/s41598-020-61026-x

Frankel, G., and Phillips, A. D. (2008). Attaching effacing *Escherichia coli* and paradigms of Tir-triggered actin polymerization: getting off the pedestal. *Cell. Microbiol.* 10, 549–556. doi: 10.1111/j.1462-5822.2007.01103.x

Furniss, R. C. D., and Clements, A. (2017). Regulation of the locus of enterocyte effacement in attaching and effacing pathogens. *J. Bacteriol.* 200, e00336-17. doi: 10.1128/JB.00336-17

Gawthorne, J. A., Audry, L., McQuitty, C., Dean, P., Christie, J. M., Enninga, J., et al. (2016). Visualizing the translocation and localization of bacterial type III effector proteins by using a genetically encoded reporter system. *Appl. Environ. Microbiol.* 82, 2700–2708. doi: 10.1128/AEM.03418-15

Goddard, P. J., Sanchez-Garrido, J., Slater, S. L., Kalyan, M., Ruano-Gallego, D., Marchès, O., et al. (2019). Enteropathogenic *Escherichia coli* stimulates effector-driven rapid caspase-4 activation in human macrophages. *Cell Rep.* 27, 1008–1017. doi: 10.1016/j.celrep.2019.03.100

Gomes, T. A. T., Elias, W. P., Scaletsky, I. C. A., Guth, B. E. C., Rodrigues, J. F., Piazza, R. M. F., et al. (2016). Diarrheagenic *Escherichia coli*. *Brazilian J. Microbiol.* 47, 3–30. doi: 10.1016/j.bjm.2016.10.015

Gruenheid, S., Sekirov, I., Thomas, N. A., Deng, W., O'Donnell, P., Goode, D., et al. (2004). Identification and characterization of NleA, a non-LEE-encoded type III translocated virulence factor of enterohaemorrhagic Escherichia coli O157:H7. *Mol. Microbiol.* 51, 1233–1249. doi: 10.1046/j.1365-2958.2003.03911.x

Hardwidge, P. R., Rodriguez-Escudero, I., Goode, D., Donohoe, S., Eng, J., Goodlett, D. R., et al. (2004). Proteomic analysis of the intestinal epithelial cell response to enteropathogenic *Escherichia coli*. *J. Biol. Chem.* 279, 20127–20136. doi: 10.1074/jbc.M401228200

Hazen, T. H., Donnenberg, M. S., Panchalingam, S., Antonio, M., Hossain, A., Mandomando, I., et al. (2016). Genomic diversity of EPEC associated with clinical presentations of differing severity. *Nat. Microbiol.* 1:15014. doi: 10.1038/nmicrobiol.2015.14

Hobbs, C. K., Porter, V. L., Stow, M. L. S., Siame, B. A., Tsang, H. H., and Leung, K. Y. (2016). Computational approach to predict species-specific type III secretion system (T3SS) effectors using single and multiple genomes. *BMC Genomics* 17:1048. doi: 10.1186/s12864-016-3363-1

Hovis, K. M., Mojica, S., McDermott, J. E., Pedersen, L., Simhi, C., Rank, R. G., et al. (2013). Genus-optimized strategy for the identification of chlamydial type III secretion substrates. *Pathog. Dis.* 69, 213–222. doi: 10.1111/2049-632X.12070

Kanack, K. J., Crawford, J. A., Tatsuno, I., Karmali, M. A., and Kaper, J. B. (2005). SepZ/EspZ is secreted and translocated into HeLa cells by the enteropathogenic *Escherichia coli* type III secretion system. *Infect. Immun.* 73, 4327–4337. doi: 10.1128/IAI.73.7.4327-4337.2005

Kang, E., Zhou, G., Yousefi, M., Cayrol, R., Xia, J., and Gruenheid, S. (2018). Loss of disease tolerance during Citrobacter rodentium infection is associated with impaired epithelial differentiation and hyperactivation of T cell responses. *Sci. Rep.* 8, 1–14. doi: 10.1038/s41598-017-17386-y

Katsowich, N., Elbaz, N., Pal, R. R., Mills, E., Kobi, S., Kahan, T., et al. (2017). Host cell attachment elicits posttranscriptional regulation in infecting enteropathogenic bacteria. *Science* 355, 735–739. doi: 10.1126/science.aah4886

Kelly, M., Hart, E., Mundy, R., Marchès, O., Wiles, S., Badea, L., et al. (2006). Essential role of the type III secretion system effector NleB in colonization of mice by Citrobacter rodentium. *Infect. Immun.* 74, 2328–2337. doi: 10.1128/IAI.74.4.2328-2337.2006

Knutton, S., Rosenshine, I., Pallen, M. J., Nisan, I., Neves, B. C., Bain, C., et al. (1998). A novel EspA-associated surface organelle of enteropathogenic Escherichia coli involved in protein translocation into epithelial cells. *EMBO J.* 17, 2166–2176. doi: 10.1093/emboj/17.8.2166

Krause, M., Barth, H., and Schmidt, H. (2018). Toxins of locus of enterocyte effacement-negative shiga toxin-producing *Escherichia coli. Toxins (Basel).* 10:241. doi: 10.3390/toxins10060241

Kresse, A. U., Beltrametti, F., Muller, A., Ebel, F., and Guzman, C. A. (2000). Characterization of SepL of enterohemorrhagic *Escherichia coli. J. Bacteriol.* 182, 6490–6498. doi: 10.1128/JB.182.22.6490-6498.2000

Law, R. J., Gur-Arie, L., Rosenshine, I., and Brett Finlay, B. (2013). *In vitro* and *in vivo* model systems for studying enteropathogenic *Escherichia coli* infections. *Cold Spring Harb. Perspect. Med.* 3:a009977. doi: 10.1101/cshperspect.a009977

Lewis, S. B., Prior, A., Ellis, S. J., Cook, V., Chan, S. S. M., Gelson, W., et al. (2016). Flagellin induces β-defensin 2 in human colonic *Ex vivo* infection with enterohemorrhagic *Escherichia coli. Front. Cell. Infect. Microbiol.* 6:68. doi: 10.3389/fcimb.2016.00068

Luzader, D. H., Willsey, G. G., Wargo, M. J., and Kendall, M. M. (2016). The ETT2-encoded regulator EtrB modulates enterohemorrhagic Escherichia coli virulence gene expression. *Infect. Immun.* 84, 2555–2565. doi: 10.1128/IAI.00407-16

Mallick, E. M., McBee, M. E., Vanguri, V. K., Melton-Celsa, A. R., Schlieper, K., Karalius, B. J., et al. (2012). A novel murine infection model for Shiga toxin-producing *Escherichia coli. J. Clin. Invest.* 122, 4012–4024. doi: 10.1172/JCI62746

McDaniel, T. K., Jarvis, K. G., Donnenberg, M. S., and Kaper, J. B. (1995). A genetic locus of enterocyte effacement conserved among diverse enterobacterial pathogens. *Proc. Natl. Acad. Sci. U.S.A.* 92, 1664–1668. doi: 10.1073/pnas.92.5.1664

McDermott, J. E., Corrigan, A., Peterson, E., Oehmen, C., Niemann, G., Cambronne, E. D., et al. (2011). Minireview: computational prediction of type III and IV secreted effectors in gram-negative bacteria. *Infect. Immun.* 79, 23–32. doi: 10.1128/IAI.00537-10

Melton-Celsa, A. R. (2014). Shiga toxin (Stx) classification, structure, and function. *Microbiol. Spectr.* 2, 1–13. doi: 10.1128/microbiolspec.EHEC-0024-2013

Mills, E., Baruch, K., Charpentier, X., Kobi, S., and Rosenshine, I. (2008). Real-time analysis of effector translocation by the type III secretion system of enteropathogenic *Escherichia coli. Cell Host Microbe* 3, 104–113. doi: 10.1016/j.chom.2007.11.007

Milon, A., Oswald, E., and De Rycke, J. (1999). Rabbit EPEC: a model for the study of enteropathogenic *Escherichia coli. Vet. Res.* 30, 203–219.

Miyake, M., Sakane, S., Kobayashi, C., Hanajima-Ozawa, M., Fukui, A., Kamitani, S., et al. (2008). A colorimetric assay for studying effector secretion through the bacterial type III secretion system. *FEMS Microbiol. Lett.* 278, 36–42. doi: 10.1111/j.1574-6968.2007.00943.x

Mullineaux-Sanders, C., Collins, J. W., Ruano-Gallego, D., Levy, M., Pevsner-Fischer, M., Glegola-Madejska, I. T., et al. (2017). Citrobacter rodentium relies on commensals for colonization of the colonic mucosa. *Cell Rep.* 21, 3381–3389. doi: 10.1016/j.celrep.2017.11.086

Mullineaux-Sanders, C., Sanchez-Garrido, J., Hopkins, E. G. D., Shenoy, A. R., Barry, R., and Frankel, G. (2019). Citrobacter rodentium–host–microbiota interactions: immunity, bioenergetics and metabolism. *Nat. Rev. Microbiol.* 17, 701–715. doi: 10.1038/s41579-019-0252-z

Mundy, R., MacDonald, T. T., Dougan, G., Frankel, G., and Wiles, S. (2005). Citrobacter rodentium of mice and man. *Cell. Microbiol.* 7, 1697–1706. doi: 10.1111/j.1462-5822.2005.00625.x

Mundy, R., Petrovska, L., Smollett, K., Simpson, N., Wilson, R. K., Yu, J., et al. (2004). Identification of a novel citrobacter rodentium type III secreted protein, espi, and roles of this and other secreted proteins in infection. *Infect. Immun.* 72, 2288–2302. doi: 10.1128/IAI.72.4.2288-2302.2004

Mundy, R., Pickard, D., Wilson, R. K., Simmons, C. P., Dougan, G., and Frankel, G. (2003). Identification of a novel type IV pilus gene cluster required for gastrointestinal colonization of Citrobacter rodentium. *Mol. Microbiol.* 48, 795–809. doi: 10.1046/j.1365-2958.2003.03470.x

Nataro, J. P., and Kaper, J. B. (1998). Diarrheagenic *Escherichia coli. Clin. Microbiol. Rev.* 11, 142–201. doi: 10.1128/CMR.11.1.142

Nissan, G., Gershovits, M., Morozov, M., Chalupowicz, L., Sessa, G., Manulis-Sasson, S., et al. (2018). Revealing the inventory of type III effectors in Pantoea agglomerans gall-forming pathovars using draft genome sequences and a machine-learning approach. *Mol. Plant Pathol.* 19, 381–392. doi: 10.1111/mpp.12528

O'Boyle, N., Connolly, J. P. R., and Roe, A. J. (2018). Tracking elusive cargo: illuminating spatio-temporal Type 3 effector protein dynamics using reporters. *Cell. Microbiol.* 20, 1–9. doi: 10.1111/cmi.12797

O'Connell, C. B., Creasey, E. A., Knutton, S., Elliott, S., Crowther, L. J., Luo, W., et al. (2004). SepL, a protein required for enteropathogenic *Escherichia coli* type III translocation, interacts with secretion component SepD. *Mol. Microbiol.* 52, 1613–1625. doi: 10.1111/j.1365-2958.2004.04101.x

Orton, D. J., Arsenault, D. J., Thomas, N. A., and Doucette, A. A. (2013). GELFrEE fractionation combined with mass spectrometry for proteome analysis of secreted toxins from Enteropathogenic Escherichia coli (EPEC). *Mol. Cell. Probes* 27, 200–207. doi: 10.1016/j.mcp.2013.06.004

Pacheco, A. R., and Sperandio, V. (2012). Shiga toxin in enterohemorrhagic E.coli: regulation and novel anti-virulence strategies. *Front. Cell. Infect. Microbiol.* 2:81. doi: 10.3389/fcimb.2012.00081

Palace, S. G., Proulx, M. K., Szabady, R. L., and Goguen, J. D. (2018). Gain-of-function analysis reveals important virulence roles for the yersinia pestis type iii secretion system effectors YopJ, YopT, and YpkA. *Infect. Immun.* 86, 1–11. doi: 10.1128/IAI.00318-18

Pearson, J. S., Giogha, C., Mühlen, S., Nachbur, U., Pham, C. L. L., Zhang, Y., et al. (2017). EspL is a bacterial cysteine protease effector that cleaves RHIM proteins to block necroptosis and inflammation. *Nat. Microbiol.* 2:258. doi: 10.1038/nmicrobiol.2016.258

Petty, N. K., Bulgin, R., Crepin, V. F., Cerdeño-Tárraga, A. M., Schroeder, G. N., Quail, M. A., et al. (2010). The citrobacter rodentium genome sequence reveals convergent evolution with human pathogenic *Escherichia coli. J. Bacteriol.* 192, 525–538. doi: 10.1128/JB.01144-09

Petty, N. K., Feltwell, T., Pickard, D., Clare, S., Toribio, A. L., Fookes, M., et al. (2011). Citrobacter rodentium is an unstable pathogen showing evidence of significant genomic flux. *PLoS Pathog.* 7:2018. doi: 10.1371/journal.ppat.1002018

Popa, C., Coll, N. S., Valls, M., and Sessa, G. (2016). Yeast as a heterologous model system to uncover type iii effector function. *PLoS Pathog.* 12:5360. doi: 10.1371/journal.ppat.1005360

Sabbagh, C. R. R., Carrere, S., Lonjon, F., Vailleau, F., Macho, A. P., Genin, S., et al. (2019). Pangenomic type III effector database of the plant pathogenic *Ralstonia* spp. *PeerJ.* 2019:e7346. doi: 10.7717/peerj.7346

Samudrala, R., Heffron, F., and McDermott, J. E. (2009). Accurate prediction of secreted substrates and identification of a conserved putative secretion signal for type III secretion systems. *PLoS Pathog.* 5:e1000375. doi: 10.1371/journal.ppat.1000375

Sandu, P., Crepin, V. F., Drechsler, H., McAinsh, A. D., Frankel, G., and Berger, C. N. (2017). The enterohemorrhagic *Escherichia coli* effector EspW triggers actin remodeling in a Rac1-dependent manner. *Infect. Immun.* 85, e00244–e00217. doi: 10.1128/IAI.00244-17

Serapio-Palacios, A., and Finlay, B. B. (2020). Dynamics of expression, secretion and translocation of type III effectors during enteropathogenic Escherichia coli infection. *Curr. Opin. Microbiol.* 54, 67–76. doi: 10.1016/j.mib.2019.12.001

Shenoy, A. R., Furniss, R. C. D., Goddard, P. J., and Clements, A. (2018). Modulation of host cell processes by T3SS effectors. *Curr. Top. Microbiol. Immunol.* 416, 73–115. doi: 10.1007/82_2018_106

Simmons, C. P., Goncalves, N. S., Ghaem-Maghami, M., Bajaj-Elliott, M., Clare, S., Neves, B., et al. (2002). Impaired resistance and enhanced pathology during infection with a noninvasive, attaching-effacing enteric bacterial pathogen, citrobacter rodentium, in mice lacking IL-12 or IFN-γ. *J. Immunol.* 168, 1804–1812. doi: 10.4049/jimmunol.168.4.1804

Slater, S. L., Sågfors, A. M., Pollard, D. J., Ruano-Gallego, D., and Frankel, G. (2018). The type III secretion system of pathogenic *Escherichia coli. Curr. Top. Microbiol. Immunol.* 416, 51–72. doi: 10.1007/82_2018_116

Teper, D., Burstein, D., Salomon, D., Gershovitz, M., Pupko, T., and Sessa, G. (2016). Identification of novel Xanthomonas euvesicatoria type III effector proteins by a machine-learning approach. *Mol. Plant Pathol.* 17, 398–411. doi: 10.1111/mpp.12288

Tobe, T., Beatson, S. A., Taniguchi, H., Abe, H., Bailey, C. M., Fivian, A., et al. (2006). An extensive repertoire of type III secretion effectors in Escherichia coli O157 and the role of lambdoid phages in their dissemination. *Proc. Natl. Acad. Sci. U.S.A.* 103, 14941–14946. doi: 10.1073/pnas.0604891103

Vallance, B. A., Deng, W., Jacobson, K., and Finlay, B. B. (2003). Host susceptibility to the attaching and effacing bacterial pathogen Citrobacter rodentium. *Infect. Immun.* 71, 3443–3453. doi: 10.1128/IAI.71.6.3443-3453.2003

Vinatzer, B. A., Jelenska, J., and Greenberg, J. T. (2005). Bioinformatics correctly identifies many type III secretion substrates in the plant pathogen Pseudomonas syringae and the biocontrol isolate P. fluorescens SBW25. *Mol. Plant-Microbe Interact.* 18, 877–888. doi: 10.1094/MPMI-18-0877

Wagner, S., Grin, I., Malmsheimer, S., Singh, N., Torres-Vargas, C. E., and Westerhausen, S. (2018). Bacterial type III secretion systems: A complex device for the delivery of bacterial effector proteins into eukaryotic host cells. *FEMS Microbiol. Lett.* 365, 1–13. doi: 10.1093/femsle/fny201

Watson, J. L., Sanchez-Garrido, J., Goddard, P. J., Torraca, V., Mostowy, S., Shenoy, A. R., et al. (2019). Shigella sonnei O-antigen inhibits internalization, vacuole escape, and inflammasome activation. *MBio* 10, 1–14. doi: 10.1128/mBio.02654-19

WHO (2015). *WHO Estimates of the Global Burden of Foodborne Diseases.* Geneva: WHO Press.

Wong, J. L. C., Romano, M., Kerry, L. E., Kwong, H.-S., Low, W.-W., Brett, S. J., et al. (2019). OmpK36-mediated Carbapenem resistance attenuates ST258 Klebsiella pneumoniae *in vivo. Nat. Commun.* 10:3957. doi: 10.1038/s41467-019-11756-y

Xu, Y., Bai, X., Jin, Y., Hu, B., Wang, H., Sun, H., et al. (2017). High prevalence of virulence genes in specific genotypes of atypical enteropathogenic *Escherichia coli. Front. Cell. Infect. Microbiol.* 7:109. doi: 10.3389/fcimb.2017.00109

Xue, L., Tang, B., Chen, W., and Luo, J. (2019). DeepT3: Deep convolutional neural networks accurately identify Gram-negative bacterial type III secreted effectors using the N-Terminal sequence. *Bioinformatics* 35, 2051–2057. doi: 10.1093/bioinformatics/bty931

Zhang, L., Chaudhuri, R. R., Constantinidou, C., Hobman, J. L., Patel, M. D., Jones, A. C., et al. (2004). Regulators encoded in the *Escherichia coli* type III secretion system 2 gene cluster influence expression of genes within the locus for enterocyte effacement in enterohemorrhagic *E. coli* O157:H7. *Infect. Immun.* 72, 7282–7293. doi: 10.1128/IAI.72.12.7282-7293.2004

Zhang, X., McDaniel, A. D., Wolf, L. E., Keusch, G. T., Waldor, M. K., and Acheson, D. W. K. (2000). Quinolone antibiotics induce shiga toxin-encoding bacteriophages, toxin production, and death in mice. *J. Infect. Dis.* 181, 664–670. doi: 10.1086/315239

Zheng, Y., Valdez, P. A., Danilenko, D. M., Hu, Y., Sa, S. M., Gong, Q., et al. (2008). Interleukin-22 mediates early host defense against attaching and effacing bacterial pathogens. *Nat. Med.* 14, 282–289. doi: 10.1038/nm1720

11

Bordetella Type III Secretion Injectosome and Effector Proteins

*Jana Kamanova**

Institute of Microbiology of the Czech Academy of Sciences, Prague, Czechia

***Correspondence:**
Jana Kamanova
kamanova@biomed.cas.cz

Pertussis, also known as whooping cough, is a resurging acute respiratory disease of humans primarily caused by the Gram-negative coccobacilli *Bordetella pertussis,* and less commonly by the human-adapted lineage of *B. parapertussis*$_{HU}$. The ovine-adapted lineage of *B. parapertussis*$_{OV}$ infects only sheep, while *B. bronchiseptica* causes chronic and often asymptomatic respiratory infections in a broad range of mammals but rarely in humans. A largely overlapping set of virulence factors inflicts the pathogenicity of these bordetellae. Their genomes also harbor a pathogenicity island, named *bsc* locus, that encodes components of the type III secretion injectosome, and adjacent *btr* locus with the type III regulatory proteins. The Bsc injectosome of bordetellae translocates the cytotoxic BteA effector protein, also referred to as BopC, into the cells of the mammalian hosts. While the role of type III secretion activity in the persistent colonization of the lower respiratory tract by *B. bronchiseptica* is well recognized, the functionality of the type III secretion injectosome in *B. pertussis* was overlooked for many years due to the adaptation of laboratory-passaged *B. pertussis* strains. This review highlights the current knowledge of the type III secretion system in the so-called classical *Bordetella* species, comprising *B. pertussis*, *B. parapertussis,* and *B. bronchiseptica,* and discusses its functional divergence. Comparison with other well-studied bacterial injectosomes, regulation of the type III secretion on the transcriptional and post-transcriptional level, and activities of BteA effector protein and BopN protein, homologous to the type III secretion gatekeepers, are addressed.

Keywords: pertussis, *Bordetella*, type III secretion system, effector protein, BteA/BopC, BopN

INTRODUCTION

The genus *Bordetella* (*Alcaligenaceae*, phylum *Betaproteobacteria*) currently embraces 16 species of the Gram-negative coccobacilli. Its members include important respiratory pathogens of mammals and humans with limited genetic diversity, the so-called classical *Bordetella* species, namely *B. pertussis, B. parapertussis,* and *B. bronchiseptica.* The genus further comprises less extensively studied and phylogenetically distinct *Bordetella* species collectively referred to as non-classical bordetellae, comprising pathogens, opportunistic pathogens, and environmental isolates (**Table 1**). The phylogenetic analysis of the genus suggests that the animal-associated species likely evolved from their ancestors living in soil and/or water (Hamidou Soumana et al., 2017). The diversification and speciation in the genus were accompanied by the gain and loss of multiple genes, including genes for bacterial protein toxins, protein secretion systems, and other virulence factors (Linz et al., 2016, 2019). The presence of the genes encoding *Bordetella* protein toxins, consisting of adenylate cyclase toxin, pertussis toxin, and dermonecrotic toxin distinguishes classical *Bordetella* species

TABLE 1 | Species and lineages of the *Bordetella* genus and their characteristics.

Species/lineages	Host range/Source	Disease	References
B. pertussis	Humans	Pertussis	Mattoo and Cherry, 2005
B. parapertussis$_{HU}$	Humans	Pertussis-like disease	Bergfors et al., 1999; Cherry and Seaton, 2012
B. parapertussis$_{OV}$	Sheep	Asymptomatic chronic infections, pneumonia	Porter et al., 1994
B. bronchiseptica	Various mammals	Asymptomatic chronic infections, respiratory disease	Goodnow, 1980; Gueirard et al., 1995; Mattoo and Cherry, 2005
B. holmesii	Humans	Pertussis-like disease, septicemia	Weyant et al., 1995; Yih et al., 1999
B. avium	Birds	Respiratory disease—bordetellosis	Kersters et al., 1984
B. hinzii	Poultry, rabbits, rodents, humans	Opportunistic infections in humans: respiratory disease, septicemia	Vandamme et al., 1995; Register et al., 2015
B. pseudohinzii	Rodents	Respiratory tract infections	Ivanov et al., 2016
B. trematum	Humans	Opportunistic infections: wound infections, otitis	Vandamme et al., 1996
B. ansorpii	Humans	Opportunistic infections: epidermal cyst, blood sample	Ko et al., 2005; Fry et al., 2007
B. petrii	Environment, humans	Opportunistic infections: bone infections, respiratory tract infections	Von Wintzingerode et al., 2001; Fry et al., 2005; Le Coustumier et al., 2011
B. bronchialis	Humans	Opportunistic infections: respiratory specimen	Vandamme et al., 2015
B. flabilis	Humans	Opportunistic infections: respiratory specimen	Vandamme et al., 2015
B. sputigena	Humans	Opportunistic infections: respiratory specimen	Vandamme et al., 2015
B. muralis	Environment		Tazato et al., 2015
B. tumulicola	Environment		Tazato et al., 2015
B. tumbae	Environment		Tazato et al., 2015

from the non-classical bordetellae (Linz et al., 2016). However, pertussis toxin is solely produced by *B. pertussis* due to mutations in the *ptx* promoter (Arico and Rappuoli, 1987; Parkhill et al., 2003), and dermonecrotic toxin appears to be also imported into *B. avium* that causes respiratory disease of birds called bordetellosis (Linz et al., 2016). For detailed characterization of acquisition and loss of virulence-associated factors during the evolution of the genus *Bordetella* the reader is referred to a recent work of Linz B. and colleagues (Linz et al., 2016). This review aims to explore and discuss the type III secretion system (T3SS) in classical *Bordetella* species, its regulation and mechanism of action, and its role in *Bordetella* infections that is yet to be explored for *B. pertussis*. Remarkably, *bsc-btr* loci encoding *Bordetella* T3SS and its regulatory proteins are also present in *B. ansorpii*, but absent from genomes of all other non-classical bordetellae. A comparison of the genetic organization of *bsc-btr* loci of *B. ansorpii* and classical *Bordetella* species will also be provided.

The human-adapted *B. pertussis* is the primary causative agent of pertussis, also known as whooping cough, a contagious, prolonged respiratory illness that used to be the major cause of infant mortality in the pre-vaccine era (Mattoo and Cherry, 2005). Pertussis remains one of the least controlled vaccine-preventable infectious diseases. In the recent years, an increase in pertussis incidence and/or pertussis outbreaks have been experienced in a number of most developed countries with high vaccine coverage, including the Czech Republic, U.S., U.K., Netherlands, and Australia (Fabianova et al., 2010; Spokes et al., 2010; Burns et al., 2014; Sealey et al., 2016). Unrecognized or mildly symptomatic *B. pertussis* infections in adolescents and adults are common and represent a threat to unvaccinated infants to whom the disease can be fatal (Cherry, 2019). The key contributing factors of increased pertussis incidence are still under debate. Besides greater awareness, improved diagnostics and genetic changes in circulating *B. pertussis* strains, the major cause of pertussis resurgence appears to be the switch from whole-cell pertussis (wP) to the less reactogenic but less effective acellular pertussis (aP) vaccines. These confer significantly shorter-lasting protection and are not efficient in preventing the colonization of the vaccinated individuals by *B. pertussis* and pathogen spread in the population (reviewed in Mooi et al., 2014; Cherry, 2019; Kapil and Merkel, 2019).

Other *Bordetella* species, the human-adapted lineage of *B. parapertussis$_{HU}$* and the non-classical species *B. holmesii* can also cause pertussis-like disease in humans, although generally accompanied by milder symptoms and shorter illness duration (Bergfors et al., 1999; Yih et al., 1999; Cherry and Seaton, 2012; Rodgers et al., 2013). The ovine-adapted lineage of *B. parapertussis*$_{OV}$ colonizes only sheep with no or little transmission to humans (Van Der Zee et al., 1996). In contrast, *B. bronchiseptica* infects a variety of mammals and causes diverse pathologies that range from typical chronic and often asymptomatic respiratory infections up to more acute diseases, such as the kennel cough in dogs, bronchitis in cats, bronchopneumonia and atrophic rhinitis in piglets, and snuffles in rabbits (Goodnow, 1980; Mattoo and Cherry, 2005). *B. bronchiseptica* infections in humans are rare and occur mostly in immunocompromised patients, children, and in elderly that are in contact with animals (Goodnow, 1980; Gueirard et al., 1995; Mattoo and Cherry, 2005). Nevertheless, clustering of *B. bronchiseptica* strains into two distinct *B. bronchiseptica* subpopulations, complex I, primarily of animal origin (68%), and complex IV that is primarily isolated from humans (80%) was reported based on multilocus sequence typing (MLST), distribution of insertion sequence elements (ISEs) and whole-genome sequence comparisons (Diavatopoulos et al., 2005; Park et al., 2012).

The classical *Bordetella* species are phylogenetically closely related, despite a different range of their mammalian hosts and diverse pathologies they cause. Hence, it was proposed to classify them as subspecies, rather than species (Musser et al., 1986). The *B. pertussis*, *B. parapertussis*$_{HU}$, and *B. parapertussis*$_{OV}$ likely evolved very recently and independently from different lineages of *B. bronchiseptica*-like ancestors (Van Der Zee et al., 1996, 1997; Parkhill et al., 2003). Specifically, *B. parapertussis*$_{HU}$ and *B. parapertussis*$_{OV}$ appear to have evolved from a *B. bronchiseptica* complex I-like ancestor, whereas *B. pertussis* may have shared a common ancestor with complex IV strains of human-associated lineages of *B. bronchiseptica* (Diavatopoulos et al., 2005; Park et al., 2012). The speciation of *B. pertussis* and *B. parapertussis* has been accompanied by a large-scale gene loss and inactivation. This resulted in a respective reduction of their genome size by 22 and 9%, compared to *B. bronchiseptica* species (cf. *B. pertussis* Tohama I $\sim$ 4.1 Mbp, *B. parapertussis*$_{HU}$ 12822 $\sim$ 4.8 Mbp, *B. parapertussis*$_{OV}$ Bpp5 $\sim$ 4.9 Mbp, *B. bronchiseptica* RB50 $\sim$ 5.3 Mbp). These genomic rearrangements were apparently mediated by the acquisition and expansion of insertion sequence elements (ISEs) and recombination between their copies (Parkhill et al., 2003; Preston et al., 2004; Weigand et al., 2019). Interestingly, host adaptation and speciation of classical *Bordetella* species likely was a consequence of loss of function, rather than a gain of function. As a result, the differences in the virulence of the classical *Bordetella* species are related to the loss of regulatory or control functions and/or sequence polymorphism (Parkhill et al., 2003; Cummings et al., 2004).

A BRIEF OVERVIEW OF THE VIRULENCE FACTORS OF CLASSICAL BORDETELLAE

The three classical *Bordetella* species produce a largely overlapping array of virulence factors that are involved in the colonization of the host respiratory tract, immune evasion, and transmission to new hosts. These comprise (i) adhesins, such as filamentous hemagglutinin and fimbriae, (ii) a large number of autotransporters involved in the adhesion and/or resistance to complement, including the adhesion molecule pertactin and complement evasion factor Vag8, and (iii) protein toxins consisting of adenylate cyclase toxin and dermonecrotic toxin (Mattoo et al., 2001; Parkhill et al., 2003; Hovingh et al., 2017). The adenylate cyclase toxin is a potent immunomodulatory toxin that subverts host innate and adaptive immune defenses by its adenyl cyclase activity (reviewed in Fedele et al., 2017) while the dermonecrotic toxin is associated with induction of turbinate atrophy in pigs and appears to have neurotoxic activity (Brockmeier et al., 2002; Teruya et al., 2020). By contrast, pertussis toxin (PTX) that catalyzes the ADP-ribosylation of the alpha subunit of heterotrimeric G proteins of the $G_{i/o}$ class is produced exclusively by *B. pertussis* species. The PTX production is responsible for systemic symptoms of pertussis disease, such as leukocytosis that is associated with the mortality in infants (Pierce et al., 2000; Carbonetti, 2010). The classical *Bordetella* species also differ in the levels of expression of the type III secretion system (T3SS) *in vitro*. Unlike *B. bronchiseptica* and ovine *B. parapertussis*$_{OV}$, the human-adapted *B. pertussis* and *B. parapertussis*$_{HU}$ species have the expression of their T3SS blocked at a post-transcriptional level when grown in Stainer-Scholte medium (Mattoo et al., 2004). Therefore, the function of the T3SS in human-adapted *Bordetella* species was overlooked for many years. However, isolates of *B. pertussis* express the T3SS effector protein BteA/BopC (Hegerle et al., 2013). In addition, they produce a functional T3SS upon passage on an infected animal or eukaryotic cells (Fennelly et al., 2008; Gaillard et al., 2011; Bibova et al., 2015), or when grown in media with limiting glutamate and/or iron concentrations (Brickman et al., 2011; Hanawa et al., 2016).

BORDETELLA INJECTOSOME

The T3SS injectosome is a sophisticated protein-export apparatus that enables the delivery of bacterial effector proteins directly from bacterial cytosol into the cytosol of the host cells through a conduit spanning the two bacterial membranes and the plasma membrane of the target cell. It consists of an extracellular needle-like appendage with a central channel of $\sim$2 nm in diameter, which protrudes from the bacterial surface and is linked to a cell wall-embedded secretion system machinery steered by the associated cytoplasmic components, as depicted in **Figure 1** (reviewed in Galan et al., 2014; Notti and Stebbins, 2016).

The genes for the components of the T3SS injectosome are generally clustered on a mobile genetic element, a pathogenicity island or a plasmid, and appear to have been acquired as an intact genetic block by horizontal gene transfer (Hueck, 1998; Hacker and Kaper, 2000). T3SS injectosomes are widespread among gram-negative bacteria. They are present in not only animal and plant pathogens but also in insect and amoeba pathogens, and are also essential to some symbionts (Troisfontaines and Cornelis, 2005). First genes in *Bordetella* spp. with a high degree of similarity to the genes coding for components of *Yersinia* injectosome were reported by Yuk et al. (1998). In the

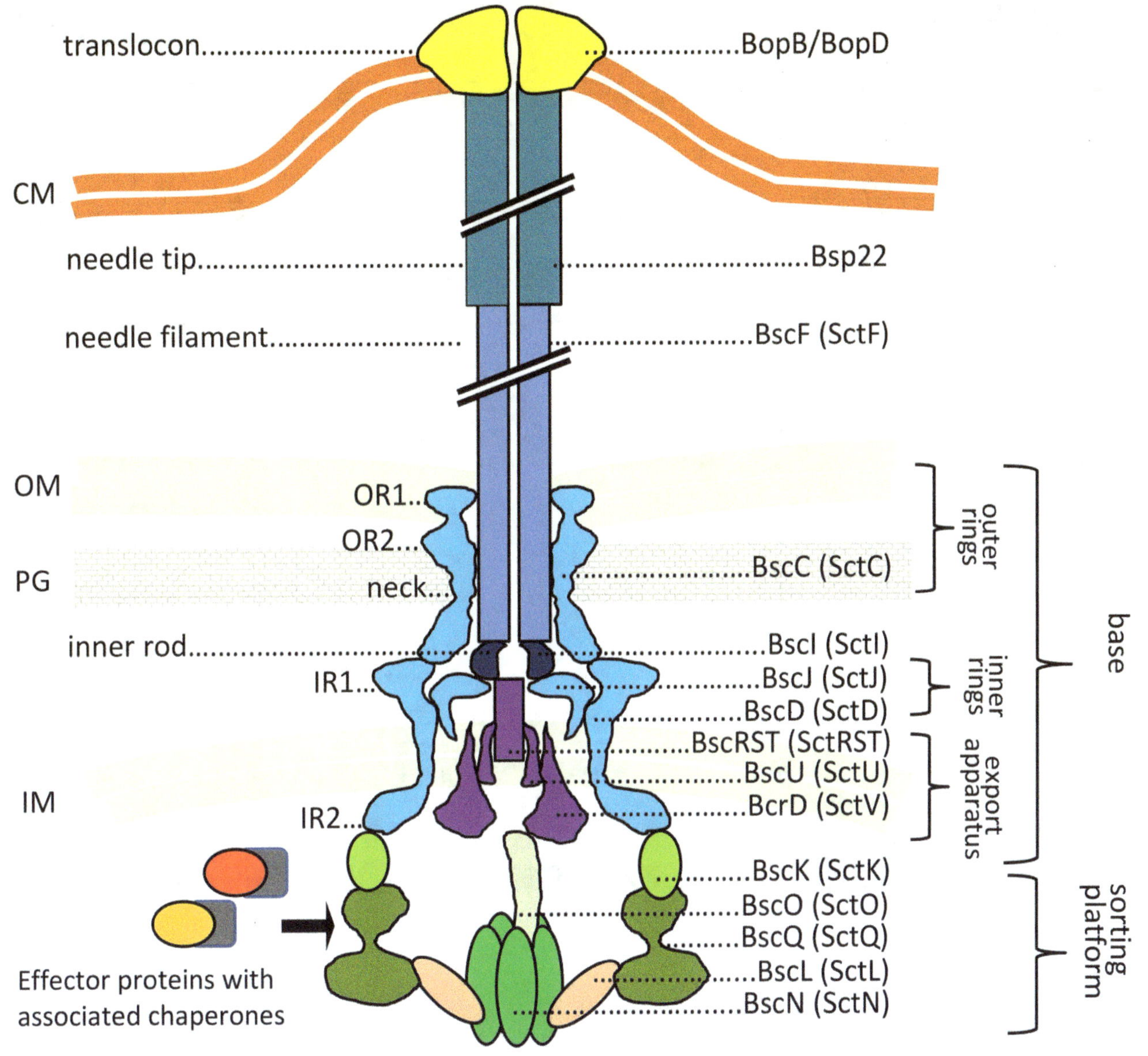

FIGURE 1 | The predicted structure of *Bordetella* injectosome. The diagram is based on Galan et al. (2014), Hu et al. (2017), Park et al. (2018). Known or predicted functions and locations of *Bordetella* proteins are indicated. In *Bordetella* spp., the injectosome needle tip is formed by a helical assembly of the Bsp22 protein (Medhekar et al., 2009). See the text for further details. CM, cytoplasmic membrane; OM, outer membrane; PG, peptidoglycan; IM, inner membrane. The proteins are color-coded based on coding genes shown in **Figure 3A**.

following years, the whole 22.5 kbp *bsc* (*Bordetella* secretion) locus encoding the *Bordetella* injectosome, and the adjacent 11 kbp *btr* (*Bordetella* type III regulation) locus, encoding injectosome-regulatory proteins, have been described (Kerr et al., 1999; Fauconnier et al., 2001; Mattoo et al., 2004; Ahuja et al., 2016). Genomic analysis of classical bordetellae showed that *Bordetella* injectosome is their most conserved secretion system (Park et al., 2012). The gene positions within the 33.5 kbp *btr-bcs* loci *(btr* BP2226-2234, *bsc* BP2235-BP2265) of *B. pertussis* Tohama I match the organization of the genes in *bsc-btr* loci (*bsc* BB1608-BB1637, *btr* BB1638-BB1646) of *B. bronchiseptica* RB50. In addition, the vast majority of nucleotide substitutions are silent or result in conservative amino acid substitutions, which implies the evolutionary pressure for the preservation of the T3SS in *B. pertussis* (Mattoo et al., 2004). However, compared to *B. bronchiseptica* RB50 genome, the *B. pertussis* Tohama I genomic region harboring the *btr-bsc* underwent inversion, likely due to an ISE-mediated rearrangement (see **Figure 2**). Interestingly as also depicted in **Figure 2**, although the *bsc-btr* loci (*bsc* BPP5_1370-BPP5_1399, *btr* BPP5_1400-BPP5_1408) of *B. parapertussis*$_{OV}$ Bpp5 are homologous and intact, the *bsc-btr*

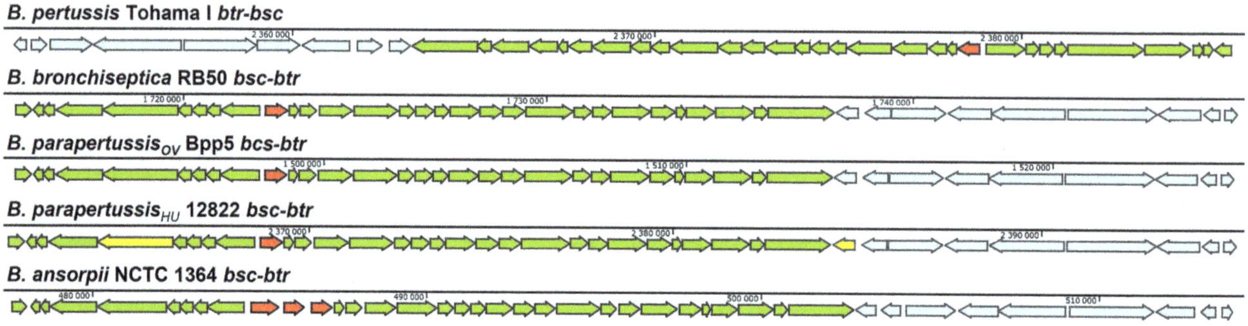

FIGURE 2 | Genetic organization of the type III regulation *btr* and secretion *bsc* loci in different *Bordetella* species. The genetic organization of *btr-bsc* loci of *B. pertussis* Tohama I (NC_002929.2, *btr* BP2226–BP2234, *bsc* BP2235–BP2265) and *bsc-btr* loci of *B. bronchiseptica* RB50 (NC_002927.3, *bsc* BB1608-BB1637, *btr* BB1638-BB1646), *B. parapertussis*$_{OV}$ Bpp5 (NC_018828.1, *bsc* BPP5_1370-BPP5_1399, *btr* BPP5_1400-BPP5_1408), *B. parapertussis*$_{HU}$ 12822 (NC_002928.3, *bsc* BPP2211-BPP2240, *btr* BPP2241-BPP2249) and *B. ansorpii* NCTC13364 (FKBS01000014.1, contig: ERS088257SCcontig000014, *bsc* SAMEA1982600_01974-SAMEA1982600_02005, *btr* SAMEA1982600_02006-SAMEA1982600_02014) is shown. The genes of *btr* and *bsc* loci are depicted in blue and green, respectively. The putative pseudogenes BPP2215 and BPP2241 in *bsc-btr* loci of *B. parapertussis*$_{HU}$ 12822 are highlighted in yellow (Parkhill et al., 2003) whereas the Bsp22 tip filament protein gene in *bsc* loci of classical bordetellae and 3 orfs that replaced this gene in *B. ansorpii* are highlighted in red. Arrows indicate the direction of transcription.

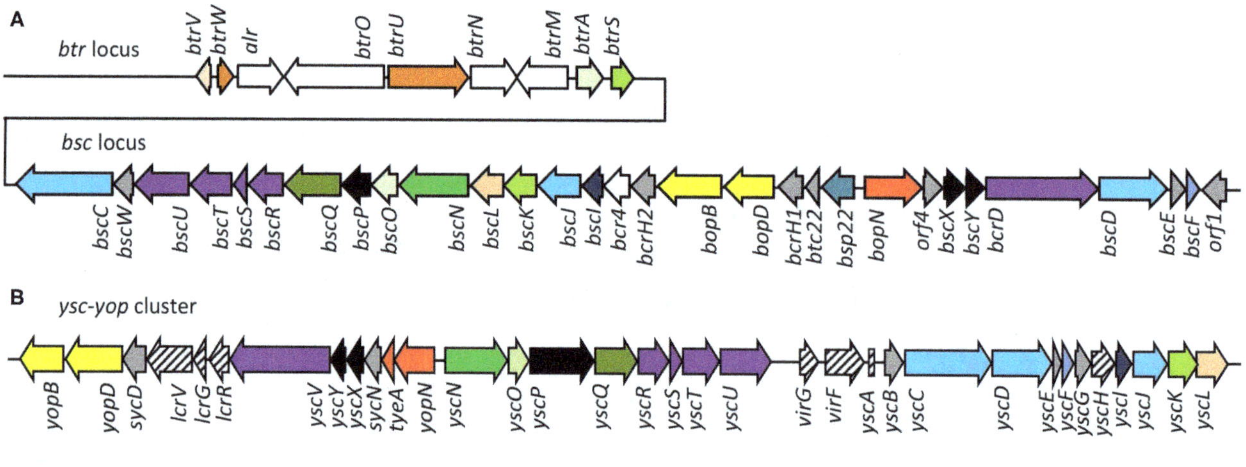

FIGURE 3 | Comparison of genetic organization of *B. pertussis btr-bsc* loci and *Yersinia* Yop-Ysc cluster. **(A)** The type III regulation *btr* and secretion *bsc* loci of the *B. pertussis* Tohama I genome, accession number NC_002929.2, *btr* genes BP2226 – BP2234, *bsc* genes BP2235–BP2265. **(B)** The Yop-Ysc cluster of *Yersinia enterocolitica* plasmid pYVe227, accession number AF102990.1, between *yopB* and *ycL*. Arrows indicate the direction of transcription. Homologies of the encoded injectosome proteins are indicated by the same color code, chaperones are colored in gray.

loci (*bsc* BPP2211-BPP2240, *btr* BPP2241-BPP2249) of the human-adapted *B. parapertussis*$_{HU}$ 12882 was suggested to contain pseudogenes for regulatory (BPP2241) and structural (BPP2215) proteins (Parkhill et al., 2003; Linz et al., 2016). The functionality of T3SS in human-adapted *B. parapertussis*$_{HU}$ thus remains to be clarified. Remarkably, the *bsc-btr* loci are also present in one non-classical *Bordetella* species, *B. ansorpii* that has so far been isolated only twice (Ko et al., 2005; Fry et al., 2007). The analysis of these loci in *B. ansorpii* NCTC1364 (*bsc* SAMEA1982600_01974-SAMEA1982600_02005, *btr* SAMEA1982600_02006-SAMEA1982600_02014) reveals a homologous organization, suggesting that *B. ansorpii* and classical bordetellae had a common ancestor. However, three different genes (tagged as SAMEA1982600_01983, SAMEA1982600_01984, SAMEA1982600_01985) are present at the position of the gene encoding the Bsp22 tip filament protein of classical bordetellae as highlighted in **Figure 2**.

The injectosome of *Bordetella* spp. encoded in the aforementioned *bsc* locus does not fit comfortably into any of the seven phylogenetic families of the non-flagellar T3SS being classified by loci organization and amino acid sequences of the encoded proteins, and consisting of families Ysc, Inv-Mxi-Spa, Ssa-Esc, Hrc-Hrp 1, Hrc-Hrp 2, the Rhizobiales and the Chlamydiales (Troisfontaines and Cornelis, 2005). The *Bsc* system of *Bordetella* spp. could nevertheless form a subgroup within the Ysc family of injectosomes, which comprises the Ysc system of *Yersinia* spp., Asc system of *Aeromonas* spp., Lsc system of *Photorhabdus luminescens*, Psc system of *Pseudomonas aeruginosa*, and the Vsc system of *Vibrio parahaemolyticus* (Pallen et al., 2005; Troisfontaines and Cornelis, 2005). As

TABLE 2 | *Bordetella bsc*-encoded genes, corresponding proteins, and predicted functions.

Locus tag *Bp* Tohama I	Protein name *Bordetella* spp.	Protein name *Yersinia* spp.[a]	Sct common nomenclature[b]	Predicted function/functional name
BP2235	BscC	YscC	SctC	Needle complex outer rings
BP2236	BscW	–	–	T3SS chaperone
BP2237	BscU	YscU	SctU	Export apparatus switch protein
BP2238	BscT	YscT	SctT	Minor export apparatus protein
BP2239	BscS	YscS	SctS	Minor export apparatus protein
BP2240	BscR	YscR	SctR	Minor export apparatus protein
BP2241	BscQ	YscQ	SctQ	C -ring protein
BP2243	BscP	YscP	SctP	Needle length regulator
BP2244	BscO	YscO	SctO	Stalk
BP2245	BscN	YscN	SctN	ATPase
BP2246	BscL	YscL	SctL	Stator
BP2247	BscK	YscK	SctK	Accessory sorting platform protein
BP2248	BscJ	YscJ	SctJ	Needle complex inner rings
BP2249	BscI	YscI	SctI	Inner rod component
BP2250	Bcr4	–	–	?
BP2251	BcrH2	–	–	Class II translocator chaperone
BP2252	BopB	YopB	–	Effector translocator, pore protein
BP2253	BopD	YopD	–	Effector translocator, pore protein
BP2254	BcrH1	–	–	Class II translocator chaperone
BP2255	Btc22 (Orf6)[c]	–	–	Bsp22 chaperone
BP2256	Bsp22	–	–	Tip filament protein
BP2257	BopN	YopN / TyeA	SctW	Gatekeeper
BP2258	Orf4	–	–	T3SS chaperone
BP2259	BscX (Orf3)[d]	YscX	–	T3SS protein X
BP2260	BscY (Orf2)[d]	YscY	–	T3SS protein Y
BP2261	BcrD	YscV (LcrD)	SctV	Major export apparatus protein
BP2262	BscD	YscD	SctD	Needle complex inner rings
BP2263	BscE	–	–	T3SS chaperone
BP2264	BscF	YscF	SctF	Needle filament protein
BP2265	Orf1	–	–	T3SS chaperone

[a]*Homologues from Yersinia spp. T3SS are given for reference;* [b]*Where applicable, the universal nomenclature for T3SS structural proteins is listed to allow for comparison; -, not present;* [c]*(Kurushima et al., 2012a);* [d]*(Gurung et al., 2018).*

depicted in **Figures 3A,B**, several homologous genes in the *Bordetella bsc* locus (e.g., *bscI* to *bscL*, as well as the *bscN* to *bscU* genes of *Bordetella* spp.) exhibit the same relative positions as the genes of the plasmid-encoded *ysc-yop* cluster of *Y. enterocolitica* W22703. However, the relative positions of other homologous genes differ, with *bscN* immediately following *bscL* in *Bordetella*, while *yscN* is not adjacent to *yscL* in *Y. enterocolitica* (Fauconnier et al., 2001). However, the *bsc*-encoded proteins still show strong sequence similarity to the well-described homologous proteins found in *Yersinia* spp. as listed in **Table 2** (Yuk et al., 1998; Kerr et al., 1999; Fauconnier et al., 2001). The precise composition and function of the *Bordetella* injectosome still need to be verified experimentally. Nevertheless, based on studies of the homologous injectosomes of *Yersinia* and *Salmonella* (reviewed in Dewoody et al., 2013; Galan et al., 2014; Galan and Waksman, 2018; Wagner et al., 2018), its structural organization can be predicted with a high degree of confidence to resemble the arrangement depicted in **Figure 1**.

By analogy, the base or the so-called basal body of *Bordetella* injectosome, embedded in the bacterial envelope, would consist of two membrane-spanning ring structures. The outer membrane ring would be built by oligomerization of BscC and the two concentric inner membrane rings would be formed by oligomerization of BscD on the outside and BscJ on the inside. The basal body would connect with the inner membrane export apparatus, formed by the BscRSTU and BcrD (SctV homolog) components. The latter would interact with the cytosolic sorting platform and additional regulatory proteins to allow for hierarchy in protein secretion, substrate unfolding, and export. By analogy to *Salmonella*, the sorting platform would be composed of three scaffolding proteins BscK, BscQ, BscL, and an ATPase BscN, linked to the export apparatus through another component, BscO (Hu et al., 2017). The needle filament composed of polymerized BscF would then attach to the basal body through the inner rod made from the BscI protein, and extend into the extracellular milieu forming a rigid hollow conduit for the secretion of proteins (**Figure 1**). The *Bordetella* needle filament on its distal end appears to be capped by another hollow helical assembly composed of the filament tip protein Bsp22 (Medhekar et al., 2009). The Bsp22 protein undergoes spontaneous polymerization and requires its chaperone Btc22 (formerly Orf6) for stabilization in the bacterial cytoplasm (Kurushima et al., 2012a; Villarino Romero et al., 2013). Bsp22 also binds directly to BopD, a component of the *Bordetella* translocon pore and it is essential for effector protein delivery into the target cells (Medhekar et al., 2009). It is thus assumed that the Bsp22 polymer forms a long flexible connecting channel that links the needle filament to the translocon pore inserted within the target cell membrane, as depicted in **Figure 1**. Indeed, similar needle extensions were described to be formed by the filament tip protein EspA of enteropathogenic *Escherichia coli* (EPEC) (Daniell et al., 2001; Sekiya et al., 2001; Wang et al., 2006). Intriguingly, the EspA filament appears to be eliminated upon attachment of EPEC to target cells (Knutton et al., 1998) and the function of the Bsp22 and EspA-formed needle extensions is currently unknown. It has been speculated that EspA filaments may cross the barrier of the mucous layer and help in adhesion to epithelial cells and/or in biofilm formation (Daniell et al., 2001; Cleary et al., 2004; Moreira et al., 2006). The *Bordetella* proteins BopD and BopB hetero-oligomerize with unknown stochiometry within the host plasma membrane and form the translocon pore that functions as the conduit for translocation of T3SS effector proteins into host cell cytosol (Kuwae et al., 2003; Nogawa et al., 2004), as depicted in **Figure 1**. Both BopD and BopB proteins are required for the pore-forming hemolytic activity of the injectosome on red blood cells but are not needed for *in vitro* secretion of other T3SS substrates (Kuwae et al., 2003; Nogawa et al., 2004).

Additional regulatory and structural components, presumably involved in the timing of protein secretion by the injectosome, are encoded within the *Bordetella bsc* locus. These include proteins BscX (formerly Orf3, homologous to T3SS protein X) and BscY (formerly Orf2, homologous to T3SS protein

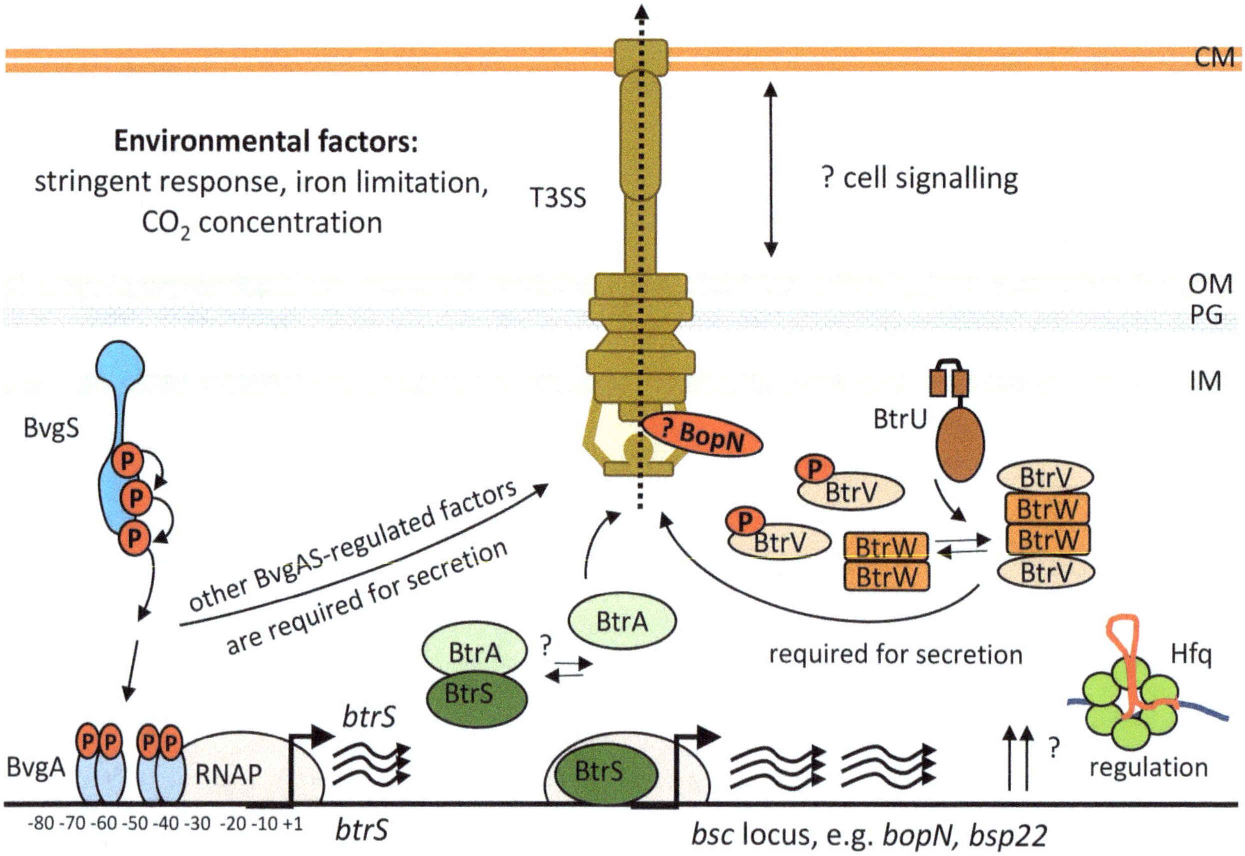

FIGURE 4 | Scheme of transcriptional and post-transcriptional regulation of T3SS in *Bordetella* species. The two-component system, BvgAS, composed of the sensor kinase BvgS and the response regulator BvgA activates expression of the extracytoplasmic function (ECF) sigma factor BtrS and regulates additional factors necessary for T3SS function (Mattoo et al., 2004; Moon et al., 2017). The activity of BtrS leads to transcription of several *bsc*-encoded genes while being counteracted by the T3SS-exported anti-sigma factor, BtrA (Mattoo et al., 2004; Kurushima et al., 2012c; Ahuja et al., 2016). See text for further details. Besides, the partner-switcher proteins BtrU, BtrW, and BtrV control secretion through the injectosome via a series of serine phosphorylation and dephosphorylation events by a yet unknown mechanism (Mattoo et al., 2004; Kozak et al., 2005). BopN protein homologous to the gatekeeper StcW might also regulate the secretion post-transcriptionally. Also RNA chaperone Hfq is required for the expression of T3SS genes in *B. pertussis* (Bibova et al., 2015; Dienstbier et al., 2019). CM, cytoplasmic membrane; OM, outer membrane; PG, peptidoglycan; IM, inner membrane.

Y), BscP, and BopN (**Table 2**). The BscX and BscY proteins would by analogy with *Yersinia* YscX and YscY orchestrate the secretion of early substrates through their interaction with BcrD (Diepold et al., 2012). The YscX- and YscY-like proteins are unique to the Ysc family of injectosomes and are not encoded within other injectosome families (Gurung et al., 2018). BscP would by analogy with SctP protein control the length of BscF needle filament by a poorly understood mechanism (reviewed in Diepold and Wagner, 2014). Finally, the BopN protein would activate effector protein secretion upon contact with the host cell, in the so-called "second substrate switching event," as deduced from its homology to the gatekeeper protein, SctW (reviewed in Portaliou et al., 2016). The BopN function in *Bordetella* injectosome, however, remains unclear and will be discussed in the section on effector proteins. The *bsc* locus further encodes the chaperones for the respective components of the injectosome (**Table 2**), e.g., Bsp22 chaperone Btc22 (formerly Orf6), the putative chaperone BcrH2 that co-immunoprecipitates with the BopB-BopD complex from bacterial cytosol, and one additional protein, called Bcr4, with unclear activity (Nogawa et al., 2004; Kurushima et al., 2012a; Nishimura et al., 2018).

TYPE III SECRETION REGULATION IN *BORDETELLA*

The *bsc*-encoded genes of the *Bordetella* injectosome are induced during infection, and are responsive to blood or serum exposure, and increased CO_2 concentrations (Gaillard et al., 2011; Hester et al., 2012; Bibova et al., 2015; Gestal et al., 2018; Van Beek et al., 2018; Wong et al., 2019). Other stimuli that can activate T3SS expression and secretion in bordetellae are the stringent response induced by iron limitation and/or starvation for carbon source, as depicted in **Figure 4** (Brickman et al., 2011; Kurushima et al., 2012b; Hanawa et al., 2016). However, upon the internalization of bordetellae into macrophages the transcription of injectosome genes is down-regulated (Rivera et al., 2019; Petrackova et al., 2020). The *bsc*-encoded genes are under the control of *Bordetella* master virulence regulatory system BvgAS. This two-component system is composed of the membrane-bound sensor kinase BvgS and of its phosphorylation substrate, the DNA-binding response regulator protein BvgA, which coordinates expression of hundreds of genes (Hot et al., 2003; Cummings et al., 2006; Nicholson, 2007; Moon et al., 2017). The system appears to function as a "rheostat" controlling a spectrum of phenotypic modes in response to environmental clues. Nevertheless, the exact nature of the signals and the mechanism by which these signals are perceived and integrated into the BvgAS regulon remains unknown (reviewed in Mattoo et al., 2001; Chen and Stibitz, 2019).

The BvgA-mediated activation of injectosome genes occurs at least for part of them indirectly through an extracytoplasmic function (ECF) sigma factor BtrS, annotated as BrpL, which is encoded in the *btr* locus (**Figure 3A**, **Table 3**) (Mattoo et al., 2004; Moon et al., 2017). As shown in **Figure 4**, in the BvgS-active mode, the membrane-bound phosphorelay sensor kinase BvgS phosphorylates the transcriptional activator BvgA, which in turn directly activates *btrS* transcription by binding as head-to-head BvgA dimers at positions centered at −41.5 and −63.5 upstream of the transcriptional start site (Moon et al., 2017). The BtrS can then activate transcription of *bsp22* and *bopN* genes encoded in the *bsc* locus (Mattoo et al., 2004). However, the ectopic expression of *btrS* in the Bvg$^-$ locked mutant of *B. bronchiseptica* RB50 strain (Δ*bvgS*) does not allow for T3SS secretion. Additional unknown Bvg$^+$ factors are thus required for the secretion process (Mattoo et al., 2004). Interestingly, two differently regulated gene clusters in the *bsc* locus of *B. bronchiseptica* RB50 strain were identified. The first cluster required the presence of BtrS for transcription (*bscN* to *bsp22*, and *bopN* to *orf2*) whereas genes in the other cluster (*bscC* to *bscO*, and *bcrD* to *bscF*) showed very little dependence, if any, on the presence of BtrS (Ahuja et al., 2016).

TABLE 3 | *Bordetella btr*-encoded genes, corresponding proteins, and functions.

Locus tag *Bp* Tohama I	Protein name	Functional name	Function	References
BP2226	BtrV	STAS domain-containing protein (Anti-sigma factor antagonist)	T3SS regulator	Mattoo et al., 2004
BP2227	BtrW	Serine kinase	T3SS regulator	Mattoo et al., 2004
BP2228	alr	Alanine racemase	unknown	–
BP2229	BtrO	MFS transporter	unknown	–
BP2230	BtrU	Serine phosphatase, SpoIIE family	T3SS regulator	Mattoo et al., 2004
BP2231	BtrN	ABC transporter	unknown	–
BP2232	BtrM	Gamma-glutamylcysteine synthetase	unknown	–
BP2233	BtrA (BspR)	Secreted anti-ECF sigma factor	Binds BtrS, T3SS inhibitor	Kurushima et al., 2012c; Ahuja et al., 2016
BP2234	BtrS (BrpL)	ECF sigma factor	T3SS activator	Mattoo et al., 2004; Ahuja et al., 2016

Besides of the ECF sigma factor BtrS, which controls also other regulatory networks (Gestal et al., 2019), additional genes encoded in the *btr* locus are linked to the injectosome function. These comprise the anti-sigma factor BtrA and the partner-switcher proteins BtrU, BtrW, and BtrV, respectively (**Figure 3A**, **Table 3**). As depicted in **Figure 4**, the secreted anti-sigma factor BtrA, also called BspR (*Bordetella* secreted protein regulator) was suggested to function as a secreted BtrS antagonist and establish a positive feedback loop that couples the injectosome secretion with the expression of T3SS genes in both *B. pertussis* and *B. bronchiseptica* species (Ahuja et al., 2016). The deletion of BtrA, indeed, enhances T3SS-dependent secretion and tissue culture phenotypes of both bacterial species (Kurushima et al., 2012c; Ahuja et al., 2016), and differential control over BtrA intracellular levels was suggested to contribute to the distinct T3SS activities of *B. pertussis* and *B. bronchiseptica* (Ahuja et al., 2016). However, the exact mechanism of BtrA-mediated inhibition of T3SS expression in *Bordetella* species remains to fully clarified. The work of Ahuja and colleagues used yeast two-hybrid system to show that BtrA interacts with BtrS, however, Kurushima and colleagues failed to detect this interaction using GST-pull down assay (Kurushima et al., 2012c; Ahuja et al., 2016).

Although studied so far only in *B. bronchiseptica*, the cascade that regulates T3SS secretion in bordetellae likely involves the partner-switcher regulatory proteins BtrU, BtrW, and BtrV (**Figure 4**, **Table 3**). These proteins exhibit homologies to partner-switching complexes of other bacteria that consist of a phosphatase, homologous to BtrU, a protein kinase/anti-sigma factor, homologous to BtrW, and an antagonist protein/anti-anti-sigma factor, homologous to BtrV. Nevertheless, the precise mechanism of regulation in *Bordetella* seems to differ from other bacteria, as all of the BtrU, BtrW, and BtrV proteins are required for *Bordetella* T3SS secretion and none of them acts as a negative regulator. BtrV seems to exert post-transcriptional control required for translation and/or protein stability, whereas BtrU and BtrW are assumed to specifically govern the secretion process (Mattoo et al., 2004; Kozak et al., 2005). Another level of complexity of regulation of T3SS production comes from the post-transcriptional regulator Hfq, a small hexameric RNA-binding protein (reviewed in Chao and Vogel, 2010). Indeed, the Hfq chaperone that anneals small RNA (sRNA) molecules to mRNA targets was found to be required for the expression of some of the T3SS genes (**Figure 4**) (Bibova et al., 2015; Dienstbier et al., 2019).

EFFECTOR PROTEINS OF *BORDETELLA* INJECTOSOME

The function of the T3SS injectosome consists in the transport of bacterial effector proteins into the host cell, where these modulate host cell functions by diverse molecular activities for the benefit

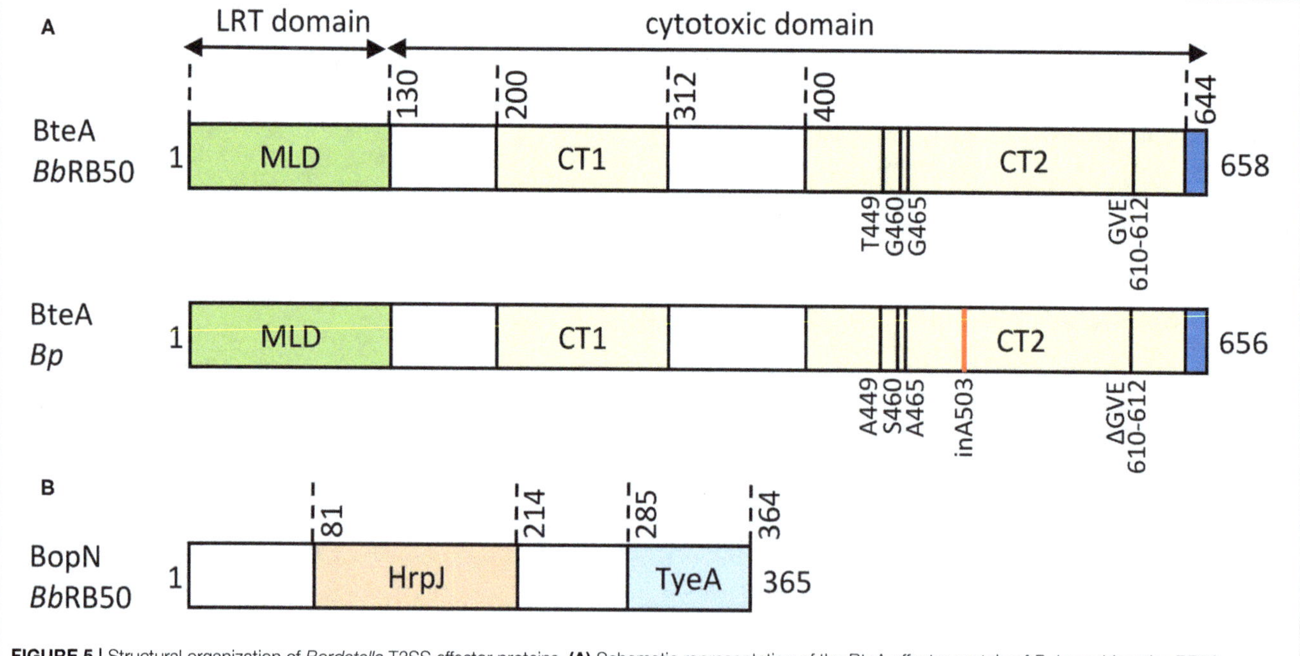

FIGURE 5 | Structural organization of *Bordetella* T3SS effector proteins. **(A)** Schematic representation of the BteA effector protein of *B. bronchiseptica* RB50 (*Bb*RB50 BteA, 658 aa) and *B. pertussis* Tohama I (*Bp* BteA, 656 aa). BteA is composed of two primary domains: a lipid-raft targeting domain (LRT) and a cytotoxic domain. The N-terminal LRT domain of ~ 130 residues is a multifunctional four helical bundle membrane localization domain (MLB) that harbors the secretion signal, comprises BtcA chaperone binding sites, and binds phosphatidylinositol 4,5-bisphosphate (PIP_2) (French et al., 2009; Guttman et al., 2013; Yahalom et al., 2019). The C-terminal cytotoxic domain of ~ 528 aa contains 2 segments associated with cytotoxicity (*Bb*RB50 BteA, CT1, aa 200-312, CT2, aa 400-658; *Bp* BteA, CT1, aa 200-312, CT2, aa 400-656) (Kuwae et al., 2016). In addition, the last 14 aa were shown to be critical for BteA cytotoxicity (French et al., 2009). Differences in the primary structure of *Bb*RB50 BteA and *Bp* BteA are shown, insertion of A503 within *Bp* BteA is highlighted in red. **(B)** Schematic representation of the 365 residue-long BopN protein of *B. bronchiseptica* RB50 (*Bb*RB50 BopN). The HrpJ (aa 81-214) and TyeA (aa 285-364) domains are indicated.

of the bacteria. Only two effector proteins were so far reported to be present in classical bordetellae, namely the effector protein BteA, also called BopC, and the BopN protein, a homolog of a T3SS regulator (Panina et al., 2005; Kuwae et al., 2006; Nagamatsu et al., 2009).

The BteA effector protein was originally identified by Panina EM and colleagues in 2005 using a computational screen for chaperone-effector loci in *B. bronchiseptica* (Panina et al., 2005). Although the chaperone-effector pair, designated *btcA-bteA*, is located 2.5 Mbp away from the *bsc* locus, *bteA* expression is coordinated with the expression of injectosome genes and is activated by the BvgAS system and ECF sigma factor BtrS (Panina et al., 2005; Ahuja et al., 2016). As expected, BteA secretion depends on the T3SS ATPase BscN, and also on the BtrU, BtrW, BtrV partner-switcher proteins (Panina et al., 2005). Upon translocation into the host cells, BteA of *B. bronchiseptica* (*Bb* BteA) induces potent cytotoxicity. Compared to the wild type strains, inducing caspase 1-independent necrotic cell death, *bteA*-deficient strains of *B. bronchiseptica* exhibit negligible cytotoxicity levels similar to the type III secretion-deficient Δ*bscN* strains (Stockbauer et al., 2003; Panina et al., 2005; Kuwae et al., 2006; Ahuja et al., 2012). The *Bb* BteA effector protein alone is capable of inducing potent cytotoxicity in tissue culture and also yeast cells since even trace amounts of *Bb* BteA (undetectable by fluorescence microscopy or Western blot) are cytotoxic (Panina, 2007, dissertation thesis; French et al., 2009).

The 69 kDa BteA effector protein exhibits a modular architecture and is composed of an N-terminal multifunctional lipid raft targeting domain (LRT) of ~ 130 amino acid residues, and a cytotoxic C-terminal domain of ~528 amino acid residues, as depicted in **Figure 5A**. The LRT domain is rich in highly hydrophobic (~20% are Ile, Leu, Val) and positively charged (>10% are Arg, Lys) amino acid residues, resembling other known membrane localization domains (MLD) but targeting a specific portion of the plasma membrane. Therefore, the membrane localization domain of BteA was called the lipid raft targeting (LRT) domain (reviewed in Geissler, 2012). Within bacteria, the LRT domain binds the cognate chaperone BtcA that guides BteA for injectosome secretion upon recognition of the N-terminal secretion signal of the LRT (Panina et al., 2005; Kuwae et al., 2006). It is assumed that BtcA-BteA complex has a stoichiometry of 2:1 (chaperon:effector) like the chaperon-effector pair, InvB-SipA (Lilic et al., 2006; Guttman et al., 2013). Upon T3SS-mediated translocation into target cell cytosol, the LRT appears to mediate BteA localization into the cytosolic leaflet of lipid rafts of cell plasma membrane via phosphatidylinositol 4,5-bisphosphate (PIP2) binding (French et al., 2009; Yahalom et al., 2019). The crystal structure shows that LRT is an elongated four-helix bundle packed against two shorter perpendicular helices, the second of which caps the domain in a tip motif. The continuous positively charged surface of the second bundle helix was proposed to mediate a direct electrostatic interaction with the negatively charged PIP2 head while being supported by the structural tip helix (Yahalom et al., 2019). Interestingly, homologous domains responsible for lipid raft targeting, but surrounded by other domains, are also present in several known and predicted T3SS effectors and MARTX (multifunctional autoprocessing repeats-in-toxin) toxins, including Plu4750 and Plu3217 from *Photorhabdus luminescens*. The membrane-localization LRT domain thus seems to have been reshuffled during evolution (Panina et al., 2005; French et al., 2009). The C-terminal domain of BteA of ~ 528 amino acid residues (**Figure 5A**) is solely responsible for the BteA-mediated cytotoxicity (French et al., 2009; Kuwae et al., 2016). Whereas no reliable predictions of the mechanism of BteA-mediated cytotoxicity can be obtained by sequence homology searches, the deletion mutagenesis suggests that the cytotoxic domain of BteA harbors two separate cytotoxic activity-related regions that span over the amino acid residues 200–312 and 400–658, respectively (Kuwae et al., 2016). The last 14 amino acid residues of BteA are also critical for full cytotoxic activity of BteA (French et al., 2009). However, the mechanism underlying the cytotoxic action of BteA as well as its cellular targets remain unknown. The targets of BteA action were proposed to be associated with the cholesterol-rich domains of the host cell membrane since depletion of membrane cholesterol protected cells from the T3SS-dependent cytotoxic action of BteA (French et al., 2009). Nevertheless, the observed protection could have also been due to diminished translocation of BteA into cholesterol-depleted cells (Hayward et al., 2005). Besides, the previously reported T3SS-mediated dephosphorylation of tyrosine residues of proteins in infected mammalian cells appears to be a rather indirect consequence of the cytotoxic action of BteA (Yuk et al., 1998; Kuwae et al., 2006).

The BteA effector proteins of classical *Bordetella* species were claimed to be functionally interchangeable (French et al., 2009). However, our recent study by Bayram and colleagues demonstrated that compared to its BteA homolog from the *B. bronchiseptica*, the BteA effector of *B. pertussis* (*Bp* BteA) exhibits a significantly reduced specific cytotoxic activity toward cultured cells (Bayram et al., 2020). This activity difference could be unambiguously attributed to the insertion of a single alanine residue at position 503 of the *Bp* BteA protein (**Figure 5A**). Indeed, the specific cytotoxic activity of the *Bp* BteA protein was strongly increased upon deletion of the A503 residue, and the activity of the *Bb* BteA protein was strongly reduced by insertion of an A503 residue, respectively (Bayram et al., 2020). This explains why low cytotoxicity was observed in cells infected by *B. pertussis* translocating BteA through a functional T3SS injectosome (Han et al., 2011; Ahuja et al., 2016). Remarkably, the analysis of amino acid sequences of BteA of the classical bordetellae revealed that the A503 residue is conserved across all *B. pertussis* lineages but is absent in the BteA of all distinct subpopulations and lineages of *B. bronchiseptica* and *B. parapertussis* species. This suggests that the acquisition of the A503 residue in the *Bp* BteA protein occurred early in the *B. pertussis* speciation to human hosts (Bayram et al., 2020). Interestingly, *B. ansorpii* does not appear to encode BteA effector protein homolog nor *Bordetella* master virulence regulatory system BvgAS.

The BopN protein was originally identified as a *Bordetella* T3SS-secreted protein that exhibits homology to the gatekeeper proteins, StcW (**Table 2**) (Yuk et al., 2000). Remarkably, in the Ysc family of injectosomes, the StcW protein is encoded as two polypeptides, e.g., YopN and its C-terminus-binding chaperone TyeA. There is a single protein corresponding to a chimeric product of YopN-TyeA in the other T3SS systems, including classical *Bordetella* species. The 39 kDa BopN protein (**Figure 5B**) thus highlights the divergence of *Bordetella* Bsc system away from the Ysc family of injectosomes (Pallen et al., 2005). The function of the StcW is to regulate translocator secretion and/or prevent a premature secretion of the effector proteins presumably through binding to the export apparatus in a protein complex that is released upon activating signal (sensing of the contact to a host cell) (Schubot et al., 2005; Portaliou et al., 2017; Yu et al., 2018). The SctW protein is then either secreted as is the case of *Yersinia* YopN and *Shigella* MxiC, or is degraded like the *Salmonella* SPI-II SsaL protein (Cheng et al., 2001; Botteaux et al., 2009; Yu et al., 2010). The fate of SctW protein if injected into the target host cell is not clear. It is assumed that the SctW does not exert any effector function, except for the *Chlamydia* CopN protein that was shown to induce G2/M cell cycle arrest by inhibiting tubulin polymerization (Huang et al., 2008; Archuleta et al., 2011; Nawrotek et al., 2014).

The function of the *Bordetella* BopN protein remains controversial (**Figure 4**). The protein was reported to be required for manifestation of the full BteA-mediated cytotoxicity in *B. bronchiseptica* infections of rat L2 pulmonary epithelial cells but was not required for cytotoxicity in mouse DC2.4 dendritic cells (Nagamatsu et al., 2009; Abe et al., 2017). Besides, the deletion of BopN did not affect the secretion of BteA into the culture supernatants of *B. bronchiseptica* grown *in vitro* (Abe et al., 2017). This would indicate that effector secretion in *B. bronchiseptica* does not require host cell contact and/or gatekeeper function, or that the used culture medium artificially mimics the contact to a cell, as previously observed for other T3SS-expressing bacterial species, e.g., enteropathogenic *Escherichia coli* (EPEC), *Vibrio parahaemolyticus* or *Shigella flexnerii* (Deng et al., 2005; Botteaux et al., 2009; Tandhavanant et al., 2018). Remarkably, the BopN protein was reported to be translocated into cells where it was suggested to localize to the cell nucleus (Nagamatsu et al., 2009; Abe et al., 2017). Its activity was also proposed to down-regulate MAPK signaling, block nuclear translocation of the NF-kBp65 subunit while promoting translocation of the NF-kBp50 subunit, and enhancing IL-10 production (Nagamatsu et al., 2009). However, the molecular basis of these processes remain unknown.

TYPE III SECRETION IN INFECTIONS WITH BORDETELLAE

The role of T3SS-mediated delivery of effector proteins is thoroughly characterized in animal infections with *B. bronchiseptica*. The elimination of T3SS function due to the deletion of the T3SS ATPse BscN results in defects of *B. bronchiseptica* persistence in the lower respiratory tract of rats, mice, and pigs (**Figure 6**) (Yuk et al., 1998, 2000; Nicholson

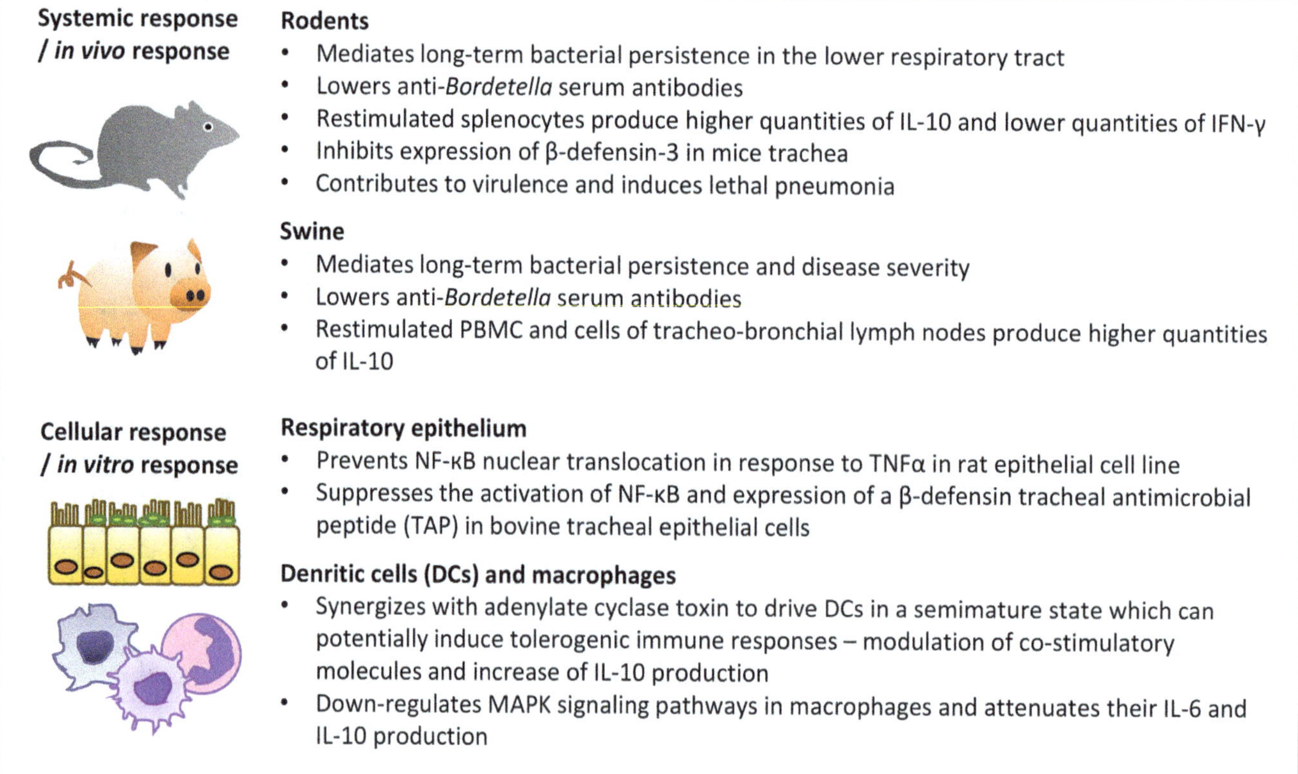

FIGURE 6 | Summary of the most important activities of *B. bronchiseptica* T3SS in animal models *in vivo* and on cells *in vitro*. See the text for further details and references.

et al., 2014). The functional T3SS of *B. bronchiseptica* was also reported to lower titres of anti-*Bordetella* serum antibodies and inhibit the generation of *IFN-γ*-producing splenocytes while enhancing the production of immunosuppressive IL-10 (**Figure 6**) (Yuk et al., 2000; Skinner et al., 2005; Pilione and Harvill, 2006; Nicholson et al., 2014). Since natural clearance of *B. bronchiseptica* depends on antibodies and production of IFN-γ while IL-10 promotes bacterial colonization, these results suggest that T3SS activity could favor bacterial persistence by altering the balance between IL-10 and IFN-γ, and hindering the antibody production (Kirimanjeswara et al., 2003; Skinner et al., 2005; Pilione and Harvill, 2006). Nevertheless, it appears that T3SS primarily targets innate immunity functions since Δ*bscN* strain was found to be hypervirulent in SCID-beige mice that are devoid of functional B cells and T cells (SCID) and NK cells (beige mutation) (Yuk et al., 2000). The T3SS may synergize with adenylate cyclase toxin to modulate macrophage and dendritic cell phenotypes and thereby subvert adaptive immune responses (**Figure 6**) (Skinner et al., 2004, 2005; Reissinger et al., 2005; Siciliano et al., 2006). Alternatively, or in combination, T3SS action may consist in inhibition of NF-kB activation in epithelial cells of the respiratory epithelia (**Figure 6**) (Yuk et al., 2000; Legarda et al., 2005; Ryan et al., 2018). Indeed, infection with wild type *B. bronchiseptica,* but not with the Δ*bscN* mutant, suppressed the activation of NF-κB and induction of β-defensin in primary bovine tracheal epithelial cells and/or during mice infection (Legarda et al., 2005; Ryan et al., 2018). Interestingly, though, T3SS-dependent phenotypes of *B. bronchiseptica* may vary between different phylogenetic lineages and/or isolates (**Figure 6**) (Buboltz et al., 2009; Ahuja et al., 2012). For example, enhanced expression of T3SS genes was reported to be partially responsible for the increased virulence of complex I *B. bronchiseptica* 1289 strain isolated from diseased host, as compared to the virulence of the RB50 strain isolated from an asymptomatic host (Cotter and Miller, 1994; Buboltz et al., 2009). Along the same line, *in vitro* T3SS-mediated hyper-cytotoxicity of a subset of complex IV *B. bronchiseptica* strains correlated with the increased ability of these strains to cause lethal pulmonary infections in mice (Ahuja et al., 2012).

The potential role of T3SS in *B. pertussis* infections was overlooked for many years due to adaptional block of injectosome expression in *B. pertussis* until the seminal studies of Fennelly et al. (2008), Gaillard et al. (2011). The activity of *B. pertussis* T3SS was then suggested to dampen the inflammatory responses in infected mice by inhibiting proinflammatory cytokine production and enhancing anti-inflammatory IL-10 production in the lungs early after infection, which correlated with lower antigen-specific IFN-γ, IL-17 and IgG responses later in the infection (Fennelly et al., 2008). These data, however, warrant further investigation. Surprisingly, compared with wild type bacteria, the T3SS-deficient *B. pertussis* mutant was significantly impaired in the capacity to colonize the lungs already 3 h after respiratory challenge (Fennelly et al., 2008).

The functionality of *B. pertussis* T3SS injectosome is intriguing, given the acute nature of pertussis disease, which is relatively short-lived. This contrasts with the rather chronic infections caused by *B. bronchiseptica*, which have been related to the action of the T3SS (Yuk et al., 1998, 2000; Nicholson et al., 2014) and/or BteA effector protein (Panina, 2007, dissertation thesis). The adaptation of T3SS of *B. pertussis* may have occurred in signaling cascades activating T3SS and/or BteA effector protein expression during *in vivo* infection. Alternatively, or in combination, a functional divergence could originate in the level of effector specific activities as recently described by Bayram et al. (2020). The *Bp* BteA has only residual cytotoxic activity as compared to *Bb* BteA, and the deletion of a differing A503 residue from the wild type *Bp* BteA (**Figure 5A**) increased T3SS-mediated *B. pertussis* cytotoxicity. Besides, the mutant *B. pertussis bteA*ΔA503 exhibited a higher virulence in the mouse model of intranasal infection, while at a sublethal challenge dose it accounted for a reduced pathology in *B. pertussis*-infected mouse lungs. These data show that a more active BteAΔA503 was able to importantly intervene in the interactions of *B. pertussis* with the host defense and its action shaped the course and outcome of the infection (Bayram et al., 2020).

CONCLUDING REMARKS AND OUTSTANDING QUESTIONS

Since 1906, when the founding member of the *Bordetella* genus, *B. pertussis,* was isolated by Bordet and Gengou from Bordet's son suffering from pertussis, our comprehension of the bordetellae species has come a long way. Nevertheless, many unknowns remain and dozens of outstanding questions can be formulated in the very specific area of research on the type III secretion in bordetellae. Perhaps the most important question is the role of T3SS activity in the pathophysiology of human pertussis. Hopefully, the availability of a non-human primate in which the human pathology of pertussis and transmission of the pathogen can be reproduced in olive baboons will yield an answer (Pinto and Merkel, 2017; Zimmerman et al., 2018). It remains to be determined if the variation in the T3SS contributed to *B. pertussis* evolution and pertussis pathogenesis. A better insight into the regulation of the T3SS expression and injectosome functionality is also needed. It remains unclear what are the host signals that *Bordetella* spp. respond to and whether these signals and/or signaling cascades are different for *B. pertussis* and *B. bronchiseptica.* We still do not understand by which mechanism do *B. pertussis* bacteria turn off the T3SS activity upon passage on laboratory media (Gaillard et al., 2011) and why despite the potent cytotoxicity in cultured cells the *in vivo B. bronchiseptica*-colonized respiratory epithelia shows no damage (Cotter and Miller, 1994; Panina et al., 2005). Future research should also focus on the understanding of the *Bordetella* injectosome that differs by the presence of the Bsp22 helical assembly and YscX-like and YscY-like proteins from the well-studied injectosomes of *Salmonella enterica* and *Shigella* spp. Last not least, we need to get a better comprehension of the molecular details behind the actions of *Bordetella* effector protein/s, their function and interplay with other virulence factors of classical *Bordetella* species.

AUTHOR CONTRIBUTIONS

The author confirms being the sole contributor of this work and has approved it for publication.

ACKNOWLEDGMENTS

The author wants to thank Peter Šebo (Insitute of Microbiology, Prague, Czech Republic) for critical reading of the manuscript and helpful advice.

REFERENCES

Abe, A., Nishimura, R., and Kuwae, A. (2017). Bordetella effector BopN is translocated into host cells via its N-terminal residues. *Microbiol. Immunol.* 61, 206–214. doi: 10.1111/1348-0421.12489

Ahuja, U., Liu, M., Tomida, S., Park, J., Souda, P., Whitelegge, J., et al. (2012). Phenotypic and genomic analysis of hypervirulent human-associated Bordetella bronchiseptica. *BMC Microbiol.* 12:167. doi: 10.1186/1471-2180-12-167

Ahuja, U., Shokeen, B., Cheng, N., Cho, Y., Blum, C., Coppola, G., et al. (2016). Differential regulation of type III secretion and virulence genes in Bordetella pertussis and Bordetella bronchiseptica by a secreted anti-sigma factor. *Proc. Natl. Acad. Sci. U.S.A.* 113, 2341–2348. doi: 10.1073/pnas.1600320113

Archuleta, T. L., Du, Y., English, C. A., Lory, S., Lesser, C., Ohi, M. D., et al. (2011). The Chlamydia effector chlamydial outer protein N (CopN) sequesters tubulin and prevents microtubule assembly. *J. Biol. Chem.* 286, 33992–33998. doi: 10.1074/jbc.M111.258426

Arico, B., and Rappuoli, R. (1987). Bordetella parapertussis and Bordetella bronchiseptica contain transcriptionally silent pertussis toxin genes. *J. Bacteriol.* 169, 2847–2853. doi: 10.1128/JB.169.6.2847-2853.1987

Bayram, J., Malcova, I., Sinkovec, L., Holubova, J., Streparola, G., Jurnecka, D., et al. (2020). Cytotoxicity of the effector protein BteA was attenuated in Bordetella pertussis by insertion of an alanine residue. *PLoS Pathog.* doi: 10.1371/journal.ppat.1008512

Bergfors, E., Trollfors, B., Taranger, J., Lagergard, T., Sundh, V., and Zackrisson, G. (1999). Parapertussis and pertussis: differences and similarities in incidence, clinical course, and antibody responses. *Int. J. Infect. Dis.* 3, 140–146. doi: 10.1016/S1201-9712(99)90035-8

Bibova, I., Hot, D., Keidel, K., Amman, F., Slupek, S., Cerny, O., et al. (2015). Transcriptional profiling of Bordetella pertussis reveals requirement of RNA chaperone Hfq for Type III secretion system functionality. *RNA Biol.* 12, 175–185. doi: 10.1080/15476286.2015.1017237

Botteaux, A., Sory, M. P., Biskri, L., Parsot, C., and Allaoui, A. (2009). MxiC is secreted by and controls the substrate specificity of the Shigella flexneri type III secretion apparatus. *Mol. Microbiol.* 71, 449–460. doi: 10.1111/j.1365-2958.2008.06537.x

Brickman, T. J., Cummings, C. A., Liew, S. Y., Relman, D. A., and Armstrong, S. K. (2011). Transcriptional profiling of the iron starvation response in Bordetella pertussis provides new insights into siderophore utilization and virulence gene expression. *J. Bacteriol.* 193, 4798–4812. doi: 10.1128/JB.05136-11

Brockmeier, S. L., Register, K. B., Magyar, T., Lax, A. J., Pullinger, G. D., and Kunkle, R. A. (2002). Role of the dermonecrotic toxin of Bordetella bronchiseptica in the pathogenesis of respiratory disease in swine. *Infect. Immun.* 70, 481–490. doi: 10.1128/IAI.70.2.481-490.2002

Buboltz, A. M., Nicholson, T. L., Weyrich, L. S., and Harvill, E. T. (2009). Role of the type III secretion system in a hypervirulent lineage of Bordetella bronchiseptica. *Infect. Immun.* 77, 3969–3977. doi: 10.1128/IAI.01362-08

Burns, D. L., Meade, B. D., and Messionnier, N. E. (2014). Pertussis resurgence: perspectives from the Working Group Meeting on pertussis on the causes, possible paths forward, and gaps in our knowledge. *J. Infect. Dis.* 209 (Suppl 1), S32–S35. doi: 10.1093/infdis/jit491

Carbonetti, N. H. (2010). Pertussis toxin and adenylate cyclase toxin: key virulence factors of Bordetella pertussis and cell biology tools. *Future Microbiol.* 5, 455–469. doi: 10.2217/fmb.09.133

Chao, Y., and Vogel, J. (2010). The role of Hfq in bacterial pathogens. *Curr. Opin. Microbiol.* 13, 24–33. doi: 10.1016/j.mib.2010.01.001

Chen, Q., and Stibitz, S. (2019). The BvgASR virulence regulon of Bordetella pertussis. *Curr. Opin. Microbiol.* 47, 74–81. doi: 10.1016/j.mib.2019.01.002

Cheng, L. W., Kay, O., and Schneewind, O. (2001). Regulated secretion of YopN by the type III machinery of Yersinia enterocolitica. *J. Bacteriol.* 183, 5293–5301. doi: 10.1128/JB.183.18.5293-5301.2001

Cherry, J. D. (2019). The 112-year odyssey of pertussis and pertussis vaccines-mistakes made and implications for the future. *J. Pediatric Infect. Dis. Soc.* 8, 334–341. doi: 10.1093/jpids/piz005

Cherry, J. D., and Seaton, B. L. (2012). Patterns of Bordetella parapertussis respiratory illnesses: 2008-2010. *Clin. Infect. Dis* 54, 534–537. doi: 10.1093/cid/cir860

Cleary, J., Lai, L. C., Shaw, R. K., Straatman-Iwanowska, A., Donnenberg, M. S., Frankel, G., et al. (2004). Enteropathogenic Escherichia coli (EPEC) adhesion to intestinal epithelial cells: role of bundle-forming pili (BFP), EspA filaments and intimin. *Microbiology (Reading,. Engl)* 150, 527–538. doi: 10.1099/mic.0.26740-0

Cotter, P. A., and Miller, J. F. (1994). BvgAS-mediated signal transduction: analysis of phase-locked regulatory mutants of Bordetella bronchiseptica in a rabbit model. *Infect. Immun.* 62, 3381–3390. doi: 10.1128/IAI.62.8.3381-3390.1994

Cummings, C. A., Bootsma, H. J., Relman, D. A., and Miller, J. F. (2006). Species- and strain-specific control of a complex, flexible regulon by Bordetella BvgAS. *J. Bacteriol.* 188, 1775–1785. doi: 10.1128/JB.188.5.1775-1785.2006

Cummings, C. A., Brinig, M. M., Lepp, P. W., Van De Pas, S., and Relman, D. A. (2004). Bordetella species are distinguished by patterns of substantial gene loss and host adaptation. *J. Bacteriol.* 186, 1484–1492. doi: 10.1128/JB.186.5.1484-1492.2004

Daniell, S. J., Takahashi, N., Wilson, R., Friedberg, D., Rosenshine, I., Booy, F. P., et al. (2001). The filamentous type III secretion translocon of enteropathogenic *Escherichia coli*. *Cell. Microbiol.* 3, 865–871. doi: 10.1046/j.1462-5822.2001.00168.x

Deng, W., Li, Y., Hardwidge, P. R., Frey, E. A., Pfuetzner, R. A., Lee, S., et al. (2005). Regulation of type III secretion hierarchy of translocators and effectors in attaching and effacing bacterial pathogens. *Infect. Immun.* 73, 2135–2146. doi: 10.1128/IAI.73.4.2135-2146.2005

Dewoody, R. S., Merritt, P. M., and Marketon, M. M. (2013). Regulation of the Yersinia type III secretion system: traffic control. *Front. Cell. Infect. Microbiol.* 3:4. doi: 10.3389/fcimb.2013.00004

Diavatopoulos, D. A., Cummings, C. A., Schouls, L. M., Brinig, M. M., Relman, D. A., and Mooi, F. R. (2005). Bordetella pertussis, the causative agent of whooping cough, evolved from a distinct, human-associated lineage of B. bronchiseptica. *PLoS Pathog.* 1:e45. doi: 10.1371/journal.ppat.0010045

Dienstbier, A., Amman, F., Stipl, D., Petrackova, D., and Vecerek, B. (2019). Comparative integrated omics analysis of the Hfq regulon in *Bordetella pertussis*. *Int. J. Mol. Sci.* 20:3073. doi: 10.3390/ijms20123073

Diepold, A., and Wagner, S. (2014). Assembly of the bacterial type III secretion machinery. *FEMS Microbiol. Rev.* 38, 802–822. doi: 10.1111/1574-6976.12061

Diepold, A., Wiesand, U., Amstutz, M., and Cornelis, G. R. (2012). Assembly of the Yersinia injectisome: the missing pieces. *Mol. Microbiol.* 85, 878–892. doi: 10.1111/j.1365-2958.2012.08146.x

Fabianova, K., Benes, C., and Kriz, B. (2010). A steady rise in incidence of pertussis since nineties in the Czech Republic. *Epidemiol. Mikrobiol. Imunol.* 59, 25–33.

Fauconnier, A., Veithen, A., Gueirard, P., Antoine, R., Wacheul, L., Locht, C., et al. (2001). Characterization of the type III secretion locus of Bordetella pertussis. *Int. J. Med. Microbiol.* 290, 693–705. doi: 10.1016/S1438-4221(01)80009-6

Fedele, G., Schiavoni, I., Adkins, I., Klimova, N., and Sebo, P. (2017). Invasion of dendritic cells, macrophages and neutrophils by the bordetella adenylate cyclase toxin: a subversive move to fool host immunity. *Toxins (Basel)* 9. doi: 10.3390/toxins9100293

Fennelly, N. K., Sisti, F., Higgins, S. C., Ross, P. J., Van Der Heide, H., Mooi, F. R., et al. (2008). Bordetella pertussis expresses a functional type III secretion system that subverts protective innate and adaptive immune responses. *Infect. Immun.* 76, 1257–1266. doi: 10.1128/IAI.00836-07

French, C. T., Panina, E. M., Yeh, S. H., Griffith, N., Arambula, D. G., and Miller, J. F. (2009). The Bordetella type III secretion system effector BteA contains a conserved N-terminal motif that guides bacterial virulence factors to lipid rafts. *Cell. Microbiol.* 11, 1735–1749. doi: 10.1111/j.1462-5822.2009.01361.x

Fry, N. K., Duncan, J., Malnick, H., and Cockcroft, P. M. (2007). The first UK isolate of 'Bordetella ansorpii' from an immunocompromised patient. *J. Med. Microbiol.* 56, 993–995. doi: 10.1099/jmm.0.47078-0

Fry, N. K., Duncan, J., Malnick, H., Warner, M., Smith, A. J., Jackson, M. S., et al. (2005). Bordetella petrii clinical isolate. *Emerging Infect. Dis.* 11, 1131–1133. doi: 10.3201/eid1107.050046

Gaillard, M. E., Bottero, D., Castuma, C. E., Basile, L. A., and Hozbor, D. (2011). Laboratory adaptation of Bordetella pertussis is associated with the loss of type three secretion system functionality. *Infect. Immun.* 79, 3677–3682. doi: 10.1128/IAI.00136-11

Galan, J. E., Lara-Tejero, M., Marlovits, T. C., and Wagner, S. (2014). Bacterial type III secretion systems: specialized nanomachines for protein delivery into target cells. *Annu. Rev. Microbiol.* 68, 415–438. doi: 10.1146/annurev-micro-092412-155725

Galan, J. E., and Waksman, G. (2018). Protein-injection machines in bacteria. *Cell* 172, 1306–1318. doi: 10.1016/j.cell.2018.01.034

Geissler, B. (2012). Bacterial toxin effector-membrane targeting: outside in, then back again. *Front. Cell. Infect. Microbiol.* 2:75. doi: 10.3389/fcimb.2012.00075

Gestal, M. C., Howard, L. K., Dewan, K., Johnson, H. M., Barbier, M., Bryant, C., et al. (2019). Enhancement of immune response against *Bordetella* spp. by disrupting immunomodulation. *Sci. Rep.* 9:20261. doi: 10.1038/s41598-019-56652-z

Gestal, M. C., Rivera, I., Howard, L. K., Dewan, K. K., Soumana, I. H., Dedloff, M., et al. (2018). Blood or serum exposure induce global transcriptional changes, altered antigenic profile, and increased cytotoxicity by classical bordetellae. *Front. Microbiol* 9:1969. doi: 10.3389/fmicb.2018.01969

Goodnow, R. A. (1980). Biology of Bordetella bronchiseptica. *Microbiol. Rev.* 44, 722–738. doi: 10.1128/MMBR.44.4.722-738.1980

Gueirard, P., Weber, C., Le Coustumier, A., and Guiso, N. (1995). Human Bordetella bronchiseptica infection related to contact with infected animals: persistence of bacteria in host. *J. Clin. Microbiol.* 33, 2002–2006. doi: 10.1128/JCM.33.8.2002-2006.1995

Gurung, J. M., Amer, A. A., Francis, M. K., Costa, T. R. D., Chen, S., Zavialov, A. V., et al. (2018). Heterologous complementation studies with the YscX and YscY protein families reveals a specificity for Yersinia pseudotuberculosis type III secretion. *Front. Cell. Infect. Microbiol.* 8:80. doi: 10.3389/fcimb.2018.00080

Guttman, C., Davidov, G., Yahalom, A., Shaked, H., Kolusheva, S., Bitton, R., et al. (2013). BtcA, A class IA type III chaperone, interacts with the BteA N-terminal domain through a globular/non-globular mechanism. *PLoS ONE* 8:e81557. doi: 10.1371/journal.pone.0081557

Hacker, J., and Kaper, J. B. (2000). Pathogenicity islands and the evolution of microbes. *Annu. Rev. Microbiol.* 54, 641–679. doi: 10.1146/annurev.micro.54.1.641

Hamidou Soumana, I., Linz, B., and Harvill, E. T. (2017). Environmental origin of the genus bordetella. *Front. Microbiol* 8:28. doi: 10.3389/fmicb.2017.00028

Han, H. J., Kuwae, A., Abe, A., Arakawa, Y., and Kamachi, K. (2011). Differential expression of type III effector BteA protein due to IS481 insertion in Bordetella pertussis. *PLoS ONE* 6:e17797. doi: 10.1371/journal.pone.0017797

Hanawa, T., Kamachi, K., Yonezawa, H., Fukutomi, T., Kawakami, H., and Kamiya, S. (2016). Glutamate limitation, BvgAS activation, and (p)ppGpp regulate the expression of the bordetella pertussis Type 3 secretion system. *J. Bacteriol.* 198, 343–351. doi: 10.1128/JB.00596-15

Hayward, R. D., Cain, R. J., Mcghie, E. J., Phillips, N., Garner, M. J., and Koronakis, V. (2005). Cholesterol binding by the bacterial type III translocon is essential for virulence effector delivery into mammalian cells. *Mol. Microbiol.* 56, 590–603. doi: 10.1111/j.1365-2958.2005.04568.x

Hegerle, N., Rayat, L., Dore, G., Zidane, N., Bedouelle, H., and Guiso, N. (2013). In-vitro and in-vivo analysis of the production of the Bordetella type three secretion system effector A in Bordetella pertussis, Bordetella parapertussis and Bordetella bronchiseptica. *Microbes Infect.* 15, 399–408. doi: 10.1016/j.micinf.2013.02.006

Hester, S. E., Lui, M., Nicholson, T., Nowacki, D., and Harvill, E. T. (2012). Identification of a CO_2 responsive regulon in Bordetella. *PLoS ONE* 7:e47635. doi: 10.1371/journal.pone.0047635

Hot, D., Antoine, R., Renauld-Mongenie, G., Caro, V., Hennuy, B., Levillain, E., et al. (2003). Differential modulation of Bordetella pertussis virulence genes as evidenced by DNA microarray analysis. *Mol. Genet. Genomics* 269, 475–486. doi: 10.1007/s00438-003-0851-1

Hovingh, E. S., Van Den Broek, B., Kuipers, B., Pinelli, E., Rooijakkers, S. H. M., and Jongerius, I. (2017). Acquisition of C1 inhibitor by Bordetella pertussis virulence associated gene 8 results in C2 and C4 consumption away from the bacterial surface. *PLoS Pathog.* 13:e1006531. doi: 10.1371/journal.ppat.1006531

Hu, B., Lara-Tejero, M., Kong, Q., Galan, J. E., and Liu, J. (2017). *In situ* molecular architecture of the salmonella Type III secretion machine. *Cell* 168, 1065–1074. doi: 10.1016/j.cell.2017.02.022

Huang, J., Lesser, C. F., and Lory, S. (2008). The essential role of the CopN protein in Chlamydia pneumoniae intracellular growth. *Nature* 456, 112–115. doi: 10.1038/nature07355

Hueck, C. J. (1998). Type III protein secretion systems in bacterial pathogens of animals and plants. *Microbiol. Mol. Biol. Rev.* 62, 379–433. doi: 10.1128/MMBR.62.2.379-433.1998

Ivanov, Y. V., Linz, B., Register, K. B., Newman, J. D., Taylor, D. L., Boschert, K. R., et al. (2016). Identification and taxonomic characterization of *Bordetella pseudohinzii* sp. nov. isolated from laboratory-raised mice. *Int. J. Syst. Evol. Microbiol.* 66, 5452–5459. doi: 10.1099/ijsem.0.001540

Kapil, P., and Merkel, T. J. (2019). Pertussis vaccines and protective immunity. *Curr. Opin. Immunol.* 59, 72–78. doi: 10.1016/j.coi.2019.03.006

Kerr, J. R., Rigg, G. P., Matthews, R. C., and Burnie, J. P. (1999). The Bpel locus encodes type III secretion machinery in Bordetella pertussis. *Microb. Pathog* 27, 349–367. doi: 10.1006/mpat.1999.0307

Kersters, K., Hinz, K.-H., Hertle, A., Segers, P., Lievens, A., Siegmann, O., et al. (1984). *Bordetella avium* sp. nov., isolated from the respiratory tracts of turkeys and other birds. Intern. *J. System. Evol. Microbiol.* 34, 56–70. doi: 10.1099/00207713-34-1-56

Kirimanjeswara, G. S., Mann, P. B., and Harvill, E. T. (2003). Role of antibodies in immunity to Bordetella infections. *Infect. Immun.* 71, 1719–1724. doi: 10.1128/IAI.71.4.1719-1724.2003

Knutton, S., Rosenshine, I., Pallen, M. J., Nisan, I., Neves, B. C., Bain, C., et al. (1998). A novel EspA-associated surface organelle of enteropathogenic *Escherichia coli* involved in protein translocation into epithelial cells. *EMBO J.* 17, 2166–2176. doi: 10.1093/emboj/17.8.2166

Ko, K. S., Peck, K. R., Oh, W. S., Lee, N. Y., Lee, J. H., and Song, J. H. (2005). New species of Bordetella, *Bordetella ansorpii* sp. nov., isolated from the purulent exudate of an epidermal cyst. *J. Clin. Microbiol.* 43, 2516–2519. doi: 10.1128/JCM.43.5.2516-2519.2005

Kozak, N. A., Mattoo, S., Foreman-Wykert, A. K., Whitelegge, J. P., and Miller, J. F. (2005). Interactions between partner switcher orthologs BtrW and BtrV regulate type III secretion in Bordetella. *J. Bacteriol.* 187, 5665–5676. doi: 10.1128/JB.187.16.5665-5676.2005

Kurushima, J., Kuwae, A., and Abe, A. (2012a). Btc22 chaperone is required for secretion and stability of the type III secreted protein Bsp22 in Bordetella bronchiseptica. *FEMS Microbiol. Lett.* 331, 144–151. doi: 10.1111/j.1574-6968.2012.02561.x

Kurushima, J., Kuwae, A., and Abe, A. (2012b). Iron starvation regulates the type III secretion system in Bordetella bronchiseptica. *Microbiol. Immunol.* 56, 356–362. doi: 10.1111/j.1348-0421.2012.00442.x

Kurushima, J., Kuwae, A., and Abe, A. (2012c). The type III secreted protein BspR regulates the virulence genes in Bordetella bronchiseptica. *PLoS ONE* 7:e38925. doi: 10.1371/journal.pone.0038925

Kuwae, A., Matsuzawa, T., Ishikawa, N., Abe, H., Nonaka, T., Fukuda, H., et al. (2006). BopC is a novel type III effector secreted by Bordetella bronchiseptica and has a critical role in type III-dependent necrotic cell death. *J. Biol. Chem.* 281, 6589–6600. doi: 10.1074/jbc.M512711200

Kuwae, A., Momose, F., Nagamatsu, K., Suyama, Y., and Abe, A. (2016). BteA secreted from the *Bordetella bronchiseptica* type III secetion system induces necrosis through an actin cytoskeleton signaling pathway and inhibits phagocytosis by macrophages. *PLoS ONE* 11:e0148387. doi: 10.1371/journal.pone.0148387

Kuwae, A., Ohishi, M., Watanabe, M., Nagai, M., and Abe, A. (2003). BopB is a type III secreted protein in Bordetella bronchiseptica and is required for cytotoxicity against cultured mammalian cells. *Cell. Microbiol.* 5, 973–983. doi: 10.1046/j.1462-5822.2003.00341.x

Le Coustumier, A., Njamkepo, E., Cattoir, V., Guillot, S., and Guiso, N. (2011). Bordetella petrii infection with long-lasting persistence in human. *Emerging Infect. Dis.* 17, 612–618. doi: 10.3201/eid1704.101480

Legarda, D., Klein-Patel, M. E., Yim, S., Yuk, M. H., and Diamond, G. (2005). Suppression of NF-kappaB-mediated beta-defensin gene expression in the mammalian airway by the Bordetella type III secretion system. *Cell. Microbiol.* 7, 489–497. doi: 10.1111/j.1462-5822.2004.00473.x

Lilic, M., Vujanac, M., and Stebbins, C. E. (2006). A common structural motif in the binding of virulence factors to bacterial secretion chaperones. *Mol. Cell* 21, 653–664. doi: 10.1016/j.molcel.2006.01.026

Linz, B., Ivanov, Y. V., Preston, A., Brinkac, L., Parkhill, J., Kim, M., et al. (2016). Acquisition and loss of virulence-associated factors during genome evolution and speciation in three clades of Bordetella species. *BMC Genomics* 17:767. doi: 10.1186/s12864-016-3112-5

Linz, B., Ma, L., Rivera, I., and Harvill, E. T. (2019). Genotypic and phenotypic adaptation of pathogens: lesson from the genus Bordetella. *Curr. Opin. Infect. Dis.* 32, 223–230. doi: 10.1097/QCO.0000000000000549

Mattoo, S., and Cherry, J. D. (2005). Molecular pathogenesis, epidemiology, and clinical manifestations of respiratory infections due to *Bordetella pertussis* and other *Bordetella subspecies*. *Clin. Microbiol. Rev.* 18, 326–382. doi: 10.1128/CMR.18.2.326-382.2005

Mattoo, S., Foreman-Wykert, A. K., Cotter, P. A., and Miller, J. F. (2001). Mechanisms of *Bordetella pathogenesis*. *Front. Biosci.* 6:741. doi: 10.2741/Mattoo

Mattoo, S., Yuk, M. H., Huang, L. L., and Miller, J. F. (2004). Regulation of type III secretion in Bordetella. *Mol. Microbiol.* 52, 1201–1214. doi: 10.1111/j.1365-2958.2004.04053.x

Medhekar, B., Shrivastava, R., Mattoo, S., Gingery, M., and Miller, J. F. (2009). Bordetella Bsp22 forms a filamentous type III secretion system tip complex and is immunoprotective *in vitro* and *in vivo*. *Mol. Microbiol.* 71, 492–504. doi: 10.1111/j.1365-2958.2008.06543.x

Mooi, F. R., Van Der Maas, N. A., and De Melker, H. E. (2014). Pertussis resurgence: waning immunity and pathogen adaptation - two sides of the same coin. *Epidemiol. Infect.* 142, 685–694. doi: 10.1017/S0950268813000071

Moon, K., Bonocora, R. P., Kim, D. D., Chen, Q., Wade, J. T., Stibitz, S., et al. (2017). The BvgAS regulon of *Bordetella pertussis*. *MBio* 8, 1–15. doi: 10.1128/mBio.01526-17

Moreira, C. G., Palmer, K., Whiteley, M., Sircili, M. P., Trabulsi, L. R., Castro, A. F., et al. (2006). Bundle-forming pili and EspA are involved in biofilm formation by enteropathogenic *Escherichia coli*. *J. Bacteriol.* 188, 3952–3961. doi: 10.1128/JB.00177-06

Musser, J. M., Hewlett, E. L., Peppler, M. S., and Selander, R. K. (1986). Genetic diversity and relationships in populations of Bordetella spp. *J. Bacteriol.* 166, 230–237. doi: 10.1128/JB.166.1.230-237.1986

Nagamatsu, K., Kuwae, A., Konaka, T., Nagai, S., Yoshida, S., Eguchi, M., et al. (2009). Bordetella evades the host immune system by inducing IL-10 through a type III effector, BopN. *J. Exp. Med.* 206, 3073–3088. doi: 10.1084/jem.20090494

Nawrotek, A., Guimaraes, B. G., Velours, C., Subtil, A., Knossow, M., and Gigant, B. (2014). Biochemical and structural insights into microtubule perturbation by CopN from Chlamydia pneumoniae. *J. Biol. Chem.* 289, 25199–25210. doi: 10.1074/jbc.M114.568436

Nicholson, T. L. (2007). Construction and validation of a first-generation Bordetella bronchiseptica long-oligonucleotide microarray by transcriptional profiling the Bvg regulon. *BMC Genomics* 8:220. doi: 10.1186/1471-2164-8-220

Nicholson, T. L., Brockmeier, S. L., Loving, C. L., Register, K. B., Kehrli, M. E., Jr., et al. (2014). The Bordetella bronchiseptica type III secretion system is required for persistence and disease severity but not transmission in swine. *Infect. Immun.* 82, 1092–1103. doi: 10.1128/IAI.01115-13

Nishimura, R., Abe, A., Sakuma, Y., and Kuwae, A. (2018). Bordetella bronchiseptica Bcr4 antagonizes the negative regulatory function of BspR via its role in type III secretion. *Microbiol. Immunol.* 62, 743–754. doi: 10.1111/1348-0421.12659

Nogawa, H., Kuwae, A., Matsuzawa, T., and Abe, A. (2004). The type III secreted protein BopD in Bordetella bronchiseptica is complexed with BopB for pore formation on the host plasma membrane. *J. Bacteriol.* 186, 3806–3813. doi: 10.1128/JB.186.12.3806-3813.2004

Notti, R. Q., and Stebbins, C. E. (2016). The structure and function of type III secretion systems. *Microbiol. Spectr* 4, 1–18. doi: 10.1128/9781555819286.ch9

Pallen, M. J., Beatson, S. A., and Bailey, C. M. (2005). Bioinformatics, genomics and evolution of non-flagellar type-III secretion systems: a Darwinian perspective. *FEMS Microbiol. Rev* 29, 201–229. doi: 10.1016/j.femsre.2005.01.001

Panina, E. M. (2007). *Identification and Characterization of Type III Secretion Effector Proteins in Gram-negative Bacteria.* [dissertation thesis]. Los Angeles, CA: University of California Los Angeles.

Panina, E. M., Mattoo, S., Griffith, N., Kozak, N. A., Yuk, M. H., and Miller, J. F. (2005). A genome-wide screen identifies a Bordetella type III secretion effector and candidate effectors in other species. *Mol. Microbiol.* 58, 267–279. doi: 10.1111/j.1365-2958.2005.04823.x

Park, J., Zhang, Y., Buboltz, A. M., Zhang, X., Schuster, S. C., Ahuja, U., et al. (2012). Comparative genomics of the classical Bordetella subspecies: the evolution and exchange of virulence-associated diversity amongst closely related pathogens. *BMC Genomics* 13:545. doi: 10.1186/1471-2164-13-545

Park, D., Lara-Tejero, M., Waxham, M. N., Li, W., Hu, B., Galan, J. E., et al. (2018). Visualization of the type III secretion mediated Salmonella-host cell interface using cryo-electron tomography. *Elife* 7, 1–15. doi: 10.7554/eLife.39514.018

Parkhill, J., Sebaihia, M., Preston, A., Murphy, L. D., Thomson, N., Harris, D. E., et al. (2003). Comparative analysis of the genome sequences of Bordetella pertussis, Bordetella parapertussis and Bordetella bronchiseptica. *Nat. Genet.* 35, 32–40. doi: 10.1038/ng1227

Petrackova, D., Farman, M. R., Amman, F., Linhartova, I., Dienstbier, A., Kumar, D., et al. (2020). Transcriptional profiling of human macrophages during infection with *Bordetella pertussis*. *RNA Biol.* 17, 731–742. doi: 10.1080/15476286.2020.1727694

Pierce, C., Klein, N., and Peters, M. (2000). Is leukocytosis a predictor of mortality in severe pertussis infection? *Intensive Care Med.* 26, 1512–1514. doi: 10.1007/s001340000587

Pilione, M. R., and Harvill, E. T. (2006). The Bordetella bronchiseptica type III secretion system inhibits gamma interferon production that is required for efficient antibody-mediated bacterial clearance. *Infect. Immun.* 74, 1043–1049. doi: 10.1128/IAI.74.2.1043-1049.2006

Pinto, M. V., and Merkel, T. J. (2017). Pertussis disease and transmission and host responses: insights from the baboon model of pertussis. *J. Infect.* (74 Suppl 1), S114–S119. doi: 10.1016/S0163-4453(17)30201-3

Portaliou, A. G., Tsolis, K. C., Loos, M. S., Balabanidou, V., Rayo, J., Tsirigotaki, A., et al. (2017). Hierarchical protein targeting and secretion is controlled by an affinity switch in the type III secretion system of enteropathogenic *Escherichia coli*. *EMBO J.* 36, 3517–3531. doi: 10.15252/embj.201797515

Portaliou, A. G., Tsolis, K. C., Loos, M. S., Zorzini, V., and Economou, A. (2016). Type III secretion: building and operating a remarkable nanomachine. *Trends Biochem. Sci.* 41, 175–189. doi: 10.1016/j.tibs.2015.09.005

Porter, J. F., Connor, K., and Donachie, W. (1994). Isolation and characterization of Bordetella parapertussis-like bacteria from ovine lungs. *Microbiology* 140 (Pt 2), 255–261. doi: 10.1099/13500872-140-2-255

Preston, A., Parkhill, J., and Maskell, D. J. (2004). The bordetellae: lessons from genomics. *Nat. Rev. Microbiol* 2, 379–390. doi: 10.1038/nrmicro886

Register, K. B., Ivanov, Y. V., Harvill, E. T., Brinkac, L., Kim, M., and Losada, L. (2015). Draft genome sequences of six bordetella hinzii isolates acquired from avian and mammalian hosts. *Genome Announc.* 3, 1–2. doi: 10.1128/genomeA.00081-15

Reissinger, A., Skinner, J. A., and Yuk, M. H. (2005). Downregulation of mitogen-activated protein kinases by the Bordetella bronchiseptica Type III secretion system leads to attenuated nonclassical macrophage activation. *Infect. Immun.* 73, 308–316. doi: 10.1128/IAI.73.1.308-316.2005

Rivera, I., Linz, B., Dewan, K. K., Ma, L., Rice, C. A., Kyle, D. E., et al. (2019). Conservation of ancient genetic pathways for intracellular persistence among animal pathogenic bordetellae. *Front. Microbiol.* 10:2839. doi: 10.3389/fmicb.2019.02839

Rodgers, L., Martin, S. W., Cohn, A., Budd, J., Marcon, M., Terranella, A., et al. (2013). Epidemiologic and laboratory features of a large outbreak of pertussis-like illnesses associated with cocirculating Bordetella holmesii and Bordetella pertussis–Ohio, 2010-2011. *Clin. Infect. Dis.* 56, 322–331. doi: 10.1093/cid/cis888

Ryan, L. K., Wu, J., Schwartz, K., Yim, S., and Diamond, G. (2018). beta-defensins coordinate *in vivo* to inhibit bacterial infections of the trachea. *Vaccines (Basel)* 6:57. doi: 10.3390/vaccines6030057

Schubot, F. D., Jackson, M. W., Penrose, K. J., Cherry, S., Tropea, J. E., Plano, G. V., et al. (2005). Three-dimensional structure of a macromolecular assembly that regulates type III secretion in Yersinia pestis. *J. Mol. Biol.* 346, 1147–1161. doi: 10.1016/j.jmb.2004.12.036

Sealey, K. L., Belcher, T., and Preston, A. (2016). Bordetella pertussis epidemiology and evolution in the light of pertussis resurgence. *Infect. Genet. Evol.* 40, 136–143. doi: 10.1016/j.meegid.2016.02.032

Sekiya, K., Ohishi, M., Ogino, T., Tamano, K., Sasakawa, C., and Abe, A. (2001). Supermolecular structure of the enteropathogenic Escherichia coli type III secretion system and its direct interaction with the EspA-sheath-like structure. *Proc. Natl. Acad. Sci. U.S.A.* 98, 11638–11643. doi: 10.1073/pnas.191378598

Siciliano, N. A., Skinner, J. A., and Yuk, M. H. (2006). Bordetella bronchiseptica modulates macrophage phenotype leading to the inhibition of CD4+ T cell proliferation and the initiation of a Th17 immune response. *J. Immunol.* 177, 7131–7138. doi: 10.4049/jimmunol.177.10.7131

Skinner, J. A., Pilione, M. R., Shen, H., Harvill, E. T., and Yuk, M. H. (2005). Bordetella type III secretion modulates dendritic cell migration resulting in immunosuppression and bacterial persistence. *J. Immunol.* 175, 4647–4652. doi: 10.4049/jimmunol.175.7.4647

Skinner, J. A., Reissinger, A., Shen, H., and Yuk, M. H. (2004). Bordetella type III secretion and adenylate cyclase toxin synergize to drive dendritic cells into a semimature state. *J. Immunol.* 173, 1934–1940. doi: 10.4049/jimmunol.173.3.1934

Spokes, P. J., Quinn, H. E., and Mcanulty, J. M. (2010). Review of the 2008-2009 pertussis epidemic in NSW: notifications and hospitalisations. *N. S. W. Public Health Bull.* 21, 167–173. doi: 10.1071/NB10031

Stockbauer, K. E., Foreman-Wykert, A. K., and Miller, J. F. (2003). Bordetella type III secretion induces caspase 1-independent necrosis. *Cell. Microbiol.* 5, 123–132. doi: 10.1046/j.1462-5822.2003.00260.x

Tandhavanant, S., Matsuda, S., Hiyoshi, H., Iida, T., and Kodama, T. (2018). Vibrio parahaemolyticus Senses Intracellular K(+) To Translocate Type III secretion system 2 effectors Effectively. *MBio* 9, 1–14. doi: 10.1128/mBio.01366-18

Tazato, N., Handa, Y., Nishijima, M., Kigawa, R., Sano, C., and Sugiyama, J. (2015). Novel environmental species isolated from the plaster wall surface of mural paintings in the Takamatsuzuka tumulus: Bordetella muralis sp. *nov., Bordetella tumulicola* sp. nov. and *Bordetella tumbae* sp. nov. *Int. J. Syst. Evol. Microbiol.* 65, 4830–4838. doi: 10.1099/ijsem.0.000655

Teruya, S., Hiramatsu, Y., Nakamura, K., Fukui-Miyazaki, A., Tsukamoto, K., Shinoda, N., et al. (2020). Bordetella dermonecrotic toxin is a neurotropic virulence factor that uses CaV3.1 as the cell surface receptor. *MBio* 11, 1–15. doi: 10.1128/mBio.03146-19

Troisfontaines, P., and Cornelis, G. R. (2005). Type III secretion: more systems than you think. *Physiology (Bethesda)* 20, 326–339. doi: 10.1152/physiol.00011.2005

Van Beek, L. F., De Gouw, D., Eleveld, M. J., Bootsma, H. J., De Jonge, M. I., Mooi, F. R., et al. (2018). Adaptation of Bordetella pertussis to the respiratory tract. *J. Infect. Dis.* 217, 1987–1996. doi: 10.1093/infdis/jiy125

Van Der Zee, A., Groenendijk, H., Peeters, M., and Mooi, F. R. (1996). The differentiation of Bordetella parapertussis and Bordetella bronchiseptica from humans and animals as determined by DNA polymorphism mediated by two different insertion sequence elements suggests their phylogenetic relationship. *Int. J. Syst. Bacteriol.* 46, 640–647. doi: 10.1099/00207713-46-3-640

Van Der Zee, A., Mooi, F., Van Embden, J., and Musser, J. (1997). Molecular evolution and host adaptation of *Bordetella* spp.: phylogenetic analysis using multilocus enzyme electrophoresis and typing with three insertion sequences. *J. Bacteriol.* 179, 6609–6617. doi: 10.1128/JB.179.21.6609-6617.1997

Vandamme, P., Heyndrickx, M., Vancanneyt, M., Hoste, B., De Vos, P., Falsen, E., et al. (1996). Bordetella trematum sp. nov., isolated from wounds and ear infections in humans, and reassessment of Alcaligenes denitrificans Ruger and Tan 1983. *Int. J. Syst. Bacteriol.* 46, 849–858. doi: 10.1099/00207713-46-4-849

Vandamme, P., Hommez, J., Vancanneyt, M., Monsieurs, M., Hoste, B., Cookson, B., et al. (1995). Bordetella hinzii sp. nov., isolated from poultry and humans. *Int. J. Syst. Bacteriol.* 45, 37–45. doi: 10.1099/00207713-45-1-37

Vandamme, P. A., Peeters, C., Cnockaert, M., Inganas, E., Falsen, E., Moore, E. R. B., et al. (2015). Bordetella bronchialis sp. nov., Bordetella flabilis sp. nov. and Bordetella sputigena sp. nov., isolated from human respiratory specimens, and reclassification of Achromobacter sediminum Zhang et al. 2014 as *Verticia sediminum* gen. nov., comb. nov. *Int. J. Syst. Evol. Microbiol.* 65, 3674–3682. doi: 10.1099/ijsem.0.000473

Villarino Romero, R., Bibova, I., Cerny, O., Vecerek, B., Wald, T., Benada, O., et al. (2013). The Bordetella pertussis type III secretion system tip complex protein Bsp22 is not a protective antigen and fails to elicit serum antibody responses during infection of humans and mice. *Infect. Immun.* 81, 2761–2767. doi: 10.1128/IAI.00353-13

Von Wintzingerode, F., Schattke, A., Siddiqui, R. A., Rosick, U., Gobel, U. B., and Gross, R. (2001). Bordetella petrii sp. nov., isolated from an anaerobic bioreactor, and emended description of the genus Bordetella. *Int. J. Syst. Evol. Microbiol.* 51, 1257–1265. doi: 10.1099/00207713-51-4-1257

Wagner, S., Grin, I., Malmsheimer, S., Singh, N., Torres-Vargas, C. E., and Westerhausen, S. (2018). Bacterial type III secretion systems: a complex device for the delivery of bacterial effector proteins into eukaryotic host cells. *FEMS Microbiol. Lett.* 365, 1–13. doi: 10.1093/femsle/fny201

Wang, Y. A., Yu, X., Yip, C., Strynadka, N. C., and Egelman, E. H. (2006). Structural polymorphism in bacterial EspA filaments revealed by cryo-EM and an improved approach to helical reconstruction. *Structure* 14, 1189–1196. doi: 10.1016/j.str.2006.05.018

Weigand, M. R., Peng, Y., Batra, D., Burroughs, M., Davis, J. K., Knipe, K., et al. (2019). Conserved patterns of symmetric inversion in the genome evolution of bordetella respiratory pathogens. *mSystems* 4, 1–15. doi: 10.1128/mSystems.00702-19

Weyant, R. S., Hollis, D. G., Weaver, R. E., Amin, M. F., Steigerwalt, A. G., O'connor, S. P., et al. (1995). *Bordetella holmesii* sp. nov., a new gram-negative species associated with septicemia. *J. Clin. Microbiol.* 33, 1–7. doi: 10.1128/JCM.33.1.1-7.1995

Wong, T. Y., Hall, J. M., Nowak, E. S., Boehm, D. T., Gonyar, L. A., Hewlett, E. L., et al. (2019). Analysis of the *in vivo* transcriptome of bordetella pertussis during infection of mice. *mSphere* 4, 1–14. doi: 10.1128/mSphereDirect.00154-19

Yahalom, A., Davidov, G., Kolusheva, S., Shaked, H., Barber-Zucker, S., Zarivach, R., et al. (2019). Structure and membrane-targeting of a Bordetella pertussis effector N-terminal domain. *Biochim Biophys Acta Biomembr* 1861:183054. doi: 10.1016/j.bbamem.2019.183054

Yih, W. K., Silva, E. A., Ida, J., Harrington, N., Lett, S. M., and George, H. (1999). Bordetella holmesii-like organisms isolated from Massachusetts patients with pertussis-like symptoms. *Emerging Infect. Dis.* 5, 441–443. doi: 10.3201/eid0503.990317

Yu, X. J., Mcgourty, K., Liu, M., Unsworth, K. E., and Holden, D. W. (2010). pH sensing by intracellular Salmonella induces effector translocation. *Science* 328, 1040–1043. doi: 10.1126/science.1189000

Yu, X. J., Grabe, G. J., Liu, M., Mota, L. J., and Holden, D. W. (2018). SsaV interacts with SsaL to control the translocon-to-effector switch in the Salmonella SPI-2 type three secretion system. *MBio* 9, 1–16. doi: 10.1128/mBio.01149-18

Yuk, M. H., Harvill, E. T., Cotter, P. A., and Miller, J. F. (2000). Modulation of host immune responses, induction of apoptosis and inhibition of NF-kappaB activation by the Bordetella type III secretion system. *Mol. Microbiol.* 35, 991–1004. doi: 10.1046/j.1365-2958.2000.01785.x

Yuk, M. H., Harvill, E. T., and Miller, J. F. (1998). The BvgAS virulence control system regulates type III secretion in Bordetella bronchiseptica. *Mol. Microbiol.* 28, 945–959. doi: 10.1046/j.1365-2958.1998.00850.x

12

An Overview of Anti-Eukaryotic T6SS Effectors

*Julia Monjarás Feria and Miguel A. Valvano**

Wellcome-Wolfson Institute for Experimental Medicine, Queen's University Belfast, Belfast, United Kingdom

***Correspondence:**
Miguel A. Valvano
m.valvano@qub.ac.uk

The type VI secretion system (T6SS) is a transmembrane multiprotein nanomachine employed by many Gram-negative bacterial species to translocate, in a contact-dependent manner, effector proteins into adjacent prokaryotic or eukaryotic cells. Typically, the T6SS gene cluster encodes at least 13 conserved core components for the apparatus assembly and other less conserved accessory proteins and effectors. It functions as a contractile tail machine comprising a TssB/C sheath and an expelled puncturing device consisting of an Hcp tube topped by a spike complex of VgrG and PAAR proteins. Contraction of the sheath propels the tube out of the bacterial cell into a target cell and leads to the injection of toxic proteins. Different bacteria use the T6SS for specific roles according to the niche and versatility of the organism. Effectors are present both as cargo (by non-covalent interactions with one of the core components) or specialized domains (fused to structural components). Although several anti-prokaryotic effectors T6SSs have been studied, recent studies have led to a substantial increase in the number of characterized anti-eukaryotic effectors. Against eukaryotic cells, the T6SS is involved in modifying and manipulating diverse cellular processes that allows bacteria to colonize, survive and disseminate, including adhesion modification, stimulating internalization, cytoskeletal rearrangements and evasion of host innate immune responses.

Keywords: bacterial pathogenesis, secretion system, T6SS, host manipulation, effector

INTRODUCTION

Gram-negative bacteria depend on specific secretion systems, numbered Type I through Type VII, to transport proteins outside the cell for survival and fitness. It has been estimated that >25% of pathogenic and non-pathogenic proteobacteria encode between one and six Type VI secretion systems (T6SS) (Bingle et al., 2008; Boyer et al., 2009). The T6SS is a dynamic contractile protein nanomachine, evolutionarily related to bacteriophage tails, which delivers protein effectors in a contact-dependent manner into diverse cellular types, including other bacteria, fungi, and host eukaryotic cells. As reviewed elsewhere (Records, 2011; Basler, 2015; Cianfanelli et al., 2016b; Nguyen et al., 2018; Cherrak et al., 2019; Navarro-Garcia et al., 2019; Hernandez et al., 2020), the T6SS gene cluster encodes 13 core components for apparatus assembly. The system can be divided in three substructures, (i) a membrane complex (TssJLM) anchored to the inner membrane and associated to the outer membrane (Aschtgen et al., 2008; Ma et al., 2009b; Durand et al., 2015; Logger et al., 2016; Rapisarda et al., 2019; Yin et al., 2019), (ii) a baseplate complex assembled by a wedge (TssEFGK) (Brunet et al., 2015; Cherrak et al., 2018; Nazarov et al., 2018) and a spike (VgrG and, in some cases PAAR proteins) (Shneider et al., 2013; Brunet et al., 2015; Renault et al., 2018) and (iii) the dynamic tail complex that comprises the inner tube (Hcp)

(Ballister et al., 2008; Brunet et al., 2014; Douzi et al., 2014) and the contractile sheath (TssBC) that wraps around the Hcp tube and propels the spike (Bonemann et al., 2009; Basler et al., 2012; Broms et al., 2013; Zhang et al., 2013; Kube et al., 2014).

The T6SS can translocate effector proteins in two modular ways: binding of an additional protein domain to structural components of the needle, Hcp, PAAR, or VgrG (specialized or evolved effectors) or by non-covalent direct or indirect interactions, via adaptor proteins, with any of the components of the needle (cargo effectors) (Shneider et al., 2013; Durand et al., 2014; Whitney et al., 2014; Alcoforado Diniz et al., 2015; Ma et al., 2017; Pissaridou et al., 2018). There are T6SS effector chaperone (TEC), or adaptor (Tap-1), proteins that are essential for toxin loading and delivery through binding to VgrG and effector proteins (Liang et al., 2015; Bondage et al., 2016; Flaugnatti et al., 2016; Jana and Salomon, 2019). TEC and Tap-1 proteins share a highly conserved domain of unknown function (DUF4123) and are not secreted; they exhibit a low pI values and are often genetically encoded upstream of their cognate effector genes or downstream of *vgrG* genes (Liang et al., 2015; Unterweger et al., 2015). Proteins containing DUF2169 domains are commonly found downstream of *vgrG* and upstream of DUF4150-containing effector genes and also serve as adaptor or chaperone in binding the N-terminal PAAR or PAAR-like domains of its cognate effector to the tip for translocation (Bondage et al., 2016; Santos et al., 2019). The DUF1795 containing proteins, namely Eag proteins, bind and stabilize the N-terminal PAAR-containing domains of their cognate effectors. Eag chaperone family members are frequently encoded adjacent to putative effectors with predicted transmembrane domains (Cianfanelli et al., 2016a; Quentin et al., 2018).

It has been reported that the T6SS mainly functions as a device for inter-bacterial competition to inject toxic antibacterial proteins into rival bacterial cells, thus modulating polymicrobial communities. More recently, the range of known functions of the T6SS has extended, including action against microbial fungi, biofilm formation and transport of ions. The T6SS also functions as a classical virulence factor by delivering toxins that allow bacteria to manipulate and subvert eukaryotic cells.

The T6SS toxins targeting eukaryotic cells are varied in biological and biochemical functions (Hachani et al., 2016). In general, different bacterial species use and adapt their T6SS for specific roles according to the host, niche or survival strategy of the organism and there is also considerable diversity in effector portfolio. In this review, we discuss and summarize the activity, target and mode of delivery of eukaryotic cell-targeting T6SS toxins important in pathogenicity, which interact and manipulate different components of the host cell. The effectors below revised are categorized accordingly to the bacterial species that encodes them.

T6SS EUKARYOTIC EFFECTORS

Table 1 and **Figure 1** give a general overview of functionalities of the eukaryotic T6SS effectors described in the text.

Vibrio

Vibrio cholerae is a natural free-living bacterium widely distributed in aquatic environments and also the environment within human hosts. *V. cholerae* is a non-invasive intestinal pathogen; O1 and O139 serogroup cause the diarrheal disease cholera. The first reported T6SS toxin targeting eukaryotic cells was VgrG-1. Mutants unable to produce this protein lack the ability to secrete Hcp or to infect amoebae and mammalian macrophages, suggesting that rather in addition to be an essential component of the T6SS apparatus, VgrG-1 is a genuine effector (Pukatzki et al., 2007; Zheng et al., 2011). VgrG-1 carries a large (395 amino acids) C-terminal extension with homology to the actin cross-linking domain (ACD) of the RtxA toxin, a member of the MARTX family (Durand et al., 2012). VgrG-1 catalyzes *in vitro* the covalent cross-linking of two G-actin monomers in a Mg^{2+}/Mn^{2+}-ATP dependent manner and *in vivo* induced massive cross-linking of cytosolic actin in macrophages and from harvested intestines in an infant mice model of infection (Pukatzki et al., 2007; Ma and Mekalanos, 2010; Durand et al., 2012). Actin oligomers disrupt the normal inter-subunit interface in the actin filament and prevent polymerization (Satchell, 2009; Heisler et al., 2015). Bacterial internalization by endocytosis is needed for VgrG-1 ACD domain translocation into phagocytic cells to impair their function and cause cell death, preventing bacterial clearance from the gut (Ma et al., 2009a). VgrG-1 also possesses an actin binding motif (ABM) on the surface of the ACD similar to WH2 domain. Actin nucleation is inhibited by this ACD-ABM because the motif can bind and sequester actin monomers; this binding domain is also indispensable for ACD mediated actin cross-linking (Dutta et al., 2019). VgrG-1 forms homotrimeric and heterotrimeric complexes by interacting with VgrG-2 and VgrG-3 (Pukatzki et al., 2007). The crystal structure of VgrG-1-ACD (PDB 4DTD) reveals a V-shaped structure formed of 12 β-strands and 9 α-helices and an active site composed of 5 residues; one of these, Glu-16, is the critical residue for the cross-linking activity (Durand et al., 2012).

Another noteworthy T6SS *V. cholerae* protein is the virulence-associated secretion protein X (VasX or VCA0020), encoded in the T6SS gene cluster downstream of *hcp* and *vgrG-2*. The 121-kDa protein VasX requires the T6SS transcriptional activator VasH for expression, and a functional T6SS apparatus for secretion with the VgrG spike as carrier for its delivery. VasX plays a role in T6SS mediated virulence, killing amoebae by a mechanism that depends on actin cross-linking (Zheng et al., 2011; Dong et al., 2013). VasX carries an N-terminal Pleckstrin-homology (PH) domain that binds to membrane lipids including phosphatidic acid (PA) and each of the phosphatidylinositol phosphates (PIP). Since inositol phosphates are rarely found in bacteria the PH domain of VasX may have a role in binding to host membrane lipids and is thought to form pores in lipid bilayers (Miyata et al., 2011). VasX also carries a motif named MIX (marker for type six effectors) and a C-terminal colicin domain important for its secretion and T6SS assembly (Salomon et al., 2014; Liang et al., 2019).

Vibrio proteolyticus (Vpr) is a marine bacterium that has been previously isolated from corals with yellow band disease.

TABLE 1 | List of anti-eukaryotic T6SS effectors and their functions.

Organism	Effector	Function (biochemical activity)	References
Vibrio			
V. cholerae	**VgrG1**	Contains an actin cross-linking domain (ACD) that binds and covalently cross-links actin, leading to an accumulation of toxic actin oligomers and altering host cell morphology, preventing host cell cytoskeletal rearrangements and disabling phagocytosis.	Pukatzki et al., 2007; Ma et al., 2009a; Ma and Mekalanos, 2010; Durand et al., 2012; Heisler et al., 2015; Dutta et al., 2019
	VasX	Required for virulence toward *Dictyostelium discoideum*. The PH domain binds host membrane lipids.	Miyata et al., 2011; Zheng et al., 2011; Dong et al., 2013
V. proteolyticus	**Vpr01580**	Predicted MIX-effector with a C-terminal domain homologous to cytotoxic proteins and other T6SS effectors that contain Rhs repeats.	Ray et al., 2017
	Vpr01570	MIX V effector containing a CNF1 domain that targets Rho GTPases resulting in actin cytoskeleton rearrangements in macrophages and toxicity to yeast.	Ray et al., 2017
	Vpr00400	Predicted effector homologous to the C-terminal domain of the insecticidal toxin Txp40 of *Xenorhabdus* and *Photorhabdus*.	Ray et al., 2017
Escherichia coli			
Enterohemorrhagic *E. coli*	**KatN**	Mn^{2+}-containing catalase secreted into the host cell's cytosol after phagocytosis. It decreases the level of intracellular reactive oxygen species, enabling bacterial survival in macrophages.	Wan et al., 2017
Extra-intestinal pathogenic *E. coli*	**VgrG1**	Involved in bacterial adherence, multiplication, and evasion of innate immune responses.	Zong et al., 2019
Pseudomonas			
P. aeruginosa	**PldA**	Phospholipase D effector; it induces PI3K activation by interacting with Akt1 and Akt2 and promotes bacterial internalization into non-phagocytic cells.	Wilderman et al., 2001; Russell et al., 2013; Bleves et al., 2014; Jiang et al., 2014; Wettstadt et al., 2019
	PldB	Phospholipase D effector; it promotes bacterial internalization into epithelial cells *via* the induction of the PI3K/Akt pathway.	Bleves et al., 2014; Jiang et al., 2014
	VgrG2b	Enables entry into non-phagocytic cells by interacting with members of the microtubule γ-TuRC complex.	Sana et al., 2015; Wood et al., 2019
	TplE	Contains a eukaryotic PGAP1-like domain, which targets the host cell's ER leading to an unfolded protein response through the IRE1α-XBP1 pathway, which in turns induces stress and autophagy.	Jiang et al., 2016
Klebsiella			
K. pneumoniae	**Pld1**	Essential phospholipase for bacterial virulence in mice that plays a role in pathogenesis. It is encoded within a T6SS core gene cluster.	Lery et al., 2014
	VgrG4	Plays a role in T6SS-mediated intoxication of fungal cells.	Storey et al., 2020
Francisella			
F. tularensis	**PdpC**	Plays a role in phagosomal escape, trafficking to lysosomes, intramacrophage replication and is important for virulence *in vivo*. It is required for replication of bacteria in the liver and spleen of mice and for AIM2 inflammasome activation.	Lindgren et al., 2013a,b; Long et al., 2013; Uda et al., 2014; Eshraghi et al., 2016; Ozanic et al., 2016; Brodmann et al., 2017
	PdpD	Contributes to intramacrophage growth and phagosomal rupture. It is required to activate the AIM2 inflammasome.	Ludu et al., 2008; Eshraghi et al., 2016; Brodmann et al., 2017
	OpiA	Contributes to intramacrophage bacterial growth by promoting bacterial endosomal escape into the cytoplasm. It belongs to a family of bacterial PI3K enzymes and also plays a role in evasion of innate immunity in host cells by reducing the levels of TNF-α.	Eshraghi et al., 2016; Ledvina et al., 2018; Cantlay et al., 2020
	OpiB	Contributes to intracellular growth in phagocytic cells. The C-terminus is homologous to the ankyrin repeat domains and the N-terminus corresponds to an evolutionarily conserved cysteine protease.	Eshraghi et al., 2016
	IglE	It is translocated into macrophages and associates to microtubule organizing centers modulating membrane trafficking for bacterial intracellular growth.	Broms et al., 2012; Shimizu et al., 2019
Edwardsiella			
E. tarda	**EvpP**	The C-terminal domain interacts with EvpC and suppresses activation of the NLRP3 inflammasome by inhibiting the Ca^{2+}-dependent MAPK-Jnk pathway. NLRP3 inhibition promotes bacterial colonization.	Zheng and Leung, 2007; Wang et al., 2009; Hu et al., 2014; Chen et al., 2017
E. ictaluri	**EvpP**	Plays a role in host cell colonization, apoptosis and necrosis in macrophages. Promotes adhesion and internalization.	Kalindamar et al., 2020

(Continued)

TABLE 1 | Continued

Organism	Effector	Function (biochemical activity)	References
E. piscicida	**EvpP**	EvpP-inhibits the Jnk-MAPK pathway and Jnk-caspy inflammasome signaling pathways suppressing recruitment of neutrophils to infection sites and promoting bacterial colonization. Interacts with ribosomal protein S5 (RPS5) to regulate apoptosis.	Tan et al., 2019; Qin et al., 2020
Burkholderia			
B. cenocepacia	**TecA**	Disrupts macrophage actin cytoskeleton by deamidating Rho GTPases, which results in the activation of the Pyrin inflammasome.	Aubert et al., 2016
B. pseudomallei and *B. thailandensis*	**VgrG5**	The C-terminal domain is involved in mediating multinucleated giant cell formation, membrane fusion and virulence in mice.	Schwarz et al., 2014; Toesca et al., 2014
Serratia			
S. marcescens	**Tfe1**	Acts against fungal cells causing plasma membrane depolarization leading to cell death.	Trunk et al., 2018
	Tfe2	Acts against target fungal cells, leading to fungal cell death. Disrupts nutrient uptake and amino acid metabolism leading to the induction of autophagy.	Trunk et al., 2018
Aeromonas			
A. hydrophila	**VgrG1**	Targets the actin cytoskeleton. Has a vegetative insecticidal protein-2 domain with actin ADP-ribosyl transferase activity.	Suarez et al., 2010
Yersinia			
Y. pseudotuberculosis	**YezP**	Zn^{2+}-binding effector that protects the pathogen from ROS and plays a role in virulence.	Wang et al., 2015

Three T6SS effectors with putative anti-eukaryotic activities were identified by analyzing the Vpr secretome. Vpr01570 contains an N-terminal MIX V domain and a C-terminal CNF1 (cytotoxic necrotizing factor 1) deamidase domain that targets and activates Rho GTPases. Vpr01570 exogenously expressed in macrophages induces actin cytoskeleton rearrangements, including assembly of contractile actin stress fibers and ruffles at the top of the cells in a T6SS-dependent manner. Vpr01570 induces toxicity when expressed in yeast and these effects depend on the CNF1 domain (Ray et al., 2017). Vpr01580 is encoded next to the Vpr01570 encoding gene and also contains a MIX V domain; its homologous proteins are cytotoxic and contain Rhs repeats. Vpr00400 is homologous to the C-terminal domain of the toxic protein Txp40 which has insecticidal activity. Additional studies are required to elucidate the role of Vpr01580 and Vpr00400 (Ray et al., 2017).

Escherichia coli

Enterohemorrhagic Escherichia coli (EHEC) is a human intestinal pathogen responsible for outbreaks of bloody diarrhea and hemolytic uremic syndrome worldwide. KatN is 84% identical to the Mn^{2+}-containing catalase KatN of *Salmonella enterica* and the specific activity of KatN is 268.3 U/mg protein (Wan et al., 2017). KatN contributes to the EHEC response to oxidative stress *in vitro*; OxyR and RpoS are involved in *katN* transcription activation and H-NS, a global regulator, in its repression. After phagocytosis, EHEC induces the expression of T6SS, and translocated KatN contributes to the survival of intracellular bacteria in macrophages by hydrolyzing and decreasing the levels of reactive oxygen species (ROS) providing an ideal niche for bacterial growth and further infection (Wan et al., 2017).

Extra-intestinal pathogenic *Escherichia coli* (ExPEC) strains can cause urinary tract, bloodstream, prostate, and other infections at non-intestinal sites, leading to disease in humans and other animals. They are a serious threat to human public health and high risk for food safety. Porcine ExPEC causes meningitis, pneumonia, arthritis, and septicemia and is multidrug-resistant. The VgrG protein, a core component and a T6SS effector, performs diverse functions as an effector in addition to its structural component role. ExPEC VgrG1 plays a role in bacterial adherence, multiplication, and also a main role in evasion of innate immune response. In the absence of VgrG1, the serum level of IL-1β in mice is significantly reduced (Zong et al., 2019).

Pseudomonas

One of the most virulent opportunistic pathogens is *Pseudomonas aeruginosa*, commonly found in soil and water as well as in plants and humans. *P. aeruginosa* is metabolically versatile and can cause a wide range of severe opportunistic infections in patients with cancer, cystic fibrosis and burns. The *P. aeruginosa* genome encodes three evolutionary distinct T6SS clusters, the H1–3-T6SSs, which are expressed simultaneously, each secreting a variable set of toxins. The H1-T6SS targets bacteria, while H2-3-T6SS targets bacteria and are also involved in internalization into eukaryotic cells (Mougous et al., 2006; Sana et al., 2012, 2016).

Phospholipases D (PLDs) are found in only a very limited number of prokaryotic organisms but, when present, they often play a role in bacterial pathogenesis. The 122-kDa protein PldA (Tle5a) from *P. aeruginosa* has high homology with eukaryotic PLDs; the protein is secreted via H2-T6SS and delivered as a cargo effector via their cognate VgrG4b. PldA possesses two

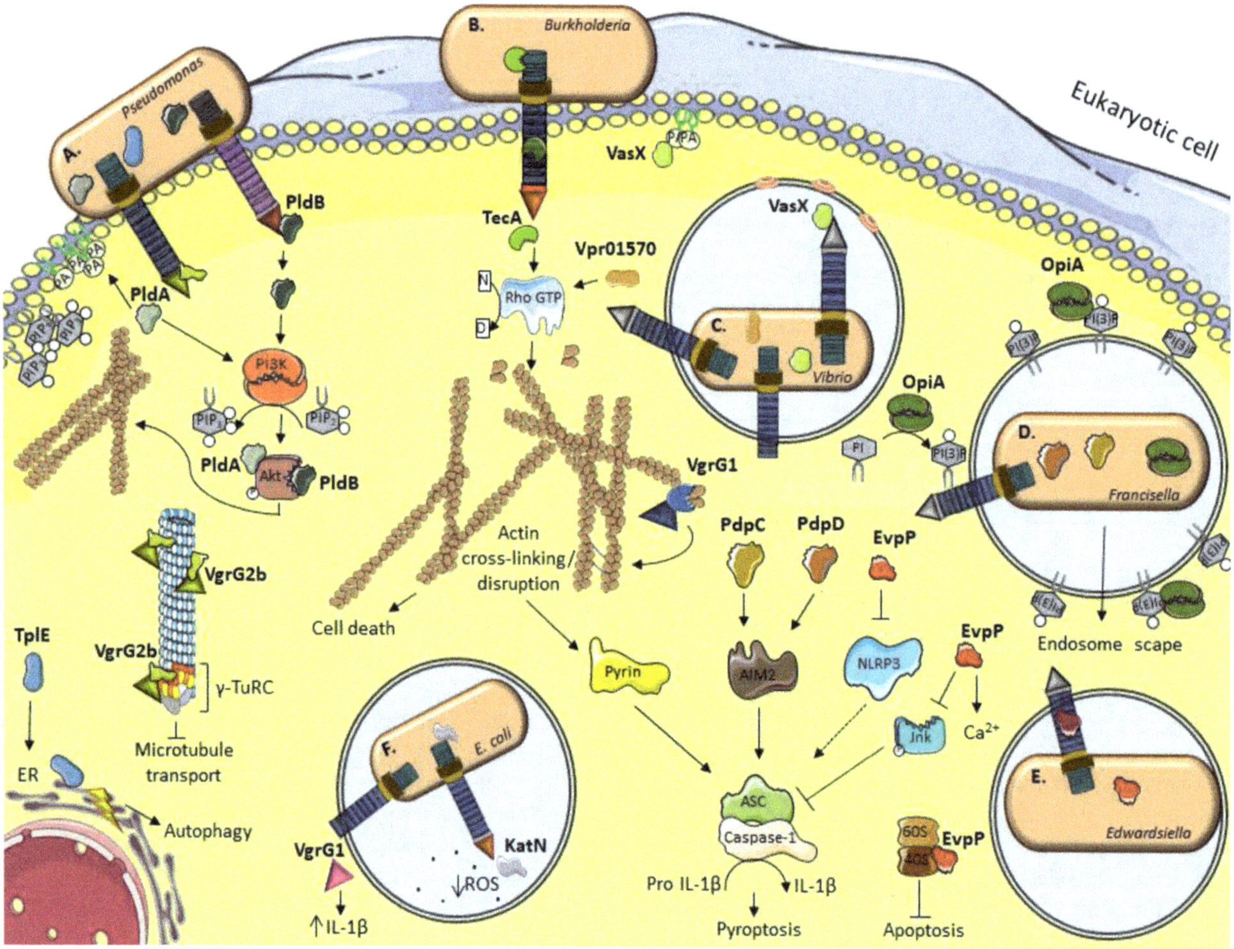

FIGURE 1 | Schematic illustration of current models for the role of some anti-eukaryotic effectors. **(A)** *P. aeruginosa* delivers PldA and PldB which bind Akt to allow bacterial internalization via the induction of the PI3K pathway. VgrG2b enables bacterial internalization by interacting with members of the microtubule γ-TuRC complex. TplE targets the endoplasmic reticulum (ER) and induces stress and autophagy. **(B)** *B. cenocepacia* TecA is a deamidase that disrupts actin cytoskeleton by deamidating Rho GTPases and activates the Pyrin inflammasome. **(C)** The *V. cholerae* evolved VgrG1 interacts with and cross-link actin, leading to an accumulation of toxic actin oligomers and altering host cell morphology. VasX binds the lipid membrane phosphatidic acid (PA) and is thought to form pores in lipid bilayers. *V. proteolyticus* Vpr01570 contains a deamidase domain that activates Rho GTPases. **(D)** *F. tularensis* OpiA is a kinase able to phosphorylate phosphatidylinositol (PI) and binding to phosphatidylinositol trisphosphate [PI(3)P] for its recruitment to endosomal membranes. PdpC and PdpD activate AIM2 inflammasome. **(E)** *E. tarda* translocates EvpP which modifies calcium flux and has an inhibitory role in NLRP3 inflammasome by reducing Jnk phosphorylation and ASC oligomerization. *E. ictalurid* EvpP interacts with ribosomal protein S5 to negatively regulate apoptosis. **(F)** The EHEC effector KatN is a catalase that contributes to the survival in macrophages by hydrolyzing and decreasing the levels of reactive oxygen species (ROS). VgrG1 is secreted by ExPEC that alters the IL-1β levels. This figure was prepared using free templates on the Servier medical art website (https://smart.servier.com/).

HXKXXXXD catalytic motifs and it has phospholipase calcium-regulated activity *in vitro*. PldA enzymatic activity resulting in phosphatidylcholine hydrolysis depends on a catalytic histidine residue (H855) (Wilderman et al., 2001; Russell et al., 2013; Wettstadt et al., 2019). PldA can induce cell death through PA accumulation via PLD activity, primarily aimed against phosphatidylethanolamine (Russell et al., 2013; Jiang et al., 2014).

The 83-kDa protein PldB (Tle5b) is a *P. aeruginosa* H2- and H3-T6SS-dependent PLD effector delivered via their cognate VgrG5 and is able to translocate into human epithelial cells. PldB possesses two HXKXXXXD catalytic motifs that play a crucial role in toxicity. PldA and PldB do not share homology, suggesting that they have developed similar functions by convergent evolution (Jiang et al., 2014; Wettstadt et al., 2019). A study deciphering the prevalence of genes encoding T6SS effectors in clinical isolates found that the prevalence of *pldA* was increased in isolates responsible for severe acute pulmonary infection and septicemia. In contrast, *pldB* prevalence was high in all isolates (Boulant et al., 2018). PldA and PldB are not involved in bacterial adhesion but promote intracellular invasion of host eukaryotic cells by activation of the phosphatidylinositol 3-kinase (PI3K)/Akt signaling pathway that is crucial for cell growth, proliferation, and programmed cell death. After injection into epithelial cells, PldA and PldB directly interact with Akt1 and/or Akt2 kinase, resulting in activation of the PI3K-Akt pathway. Indeed, Akt phosphorylation at serine 473 promotes remodeling of the apical membrane in which protrusions enriched in phosphatidylinositol-3,4,5-triphosphate (PIP_3) and actin enables bacterial entry (Bleves et al., 2014; Jiang et al., 2014).

VgrG2b is conserved in all *P. aeruginosa* strains present in the *Pseudomonas* genome database. VgrG2b is a 113-kDa protein that contains the conserved VgrG domain homologous to gp27 and gp5 phage-tail proteins followed by a domain of unknown function, DUF2345, and a C-terminal extension with a Zn^{2+}-dependent metallopeptidase domain (LFIHEMTHVW). It is an evolved VgrG with double function as a structural component of the secretion machinery and a true effector translocated via the H2-T6SS required for invasion of host cells. VgrG2b injection precedes internalization; its C-terminal domain interacts with α- and β-tubulin complexes and with the γ-tubulin complexes, such as the γ-tubulin small complex (γTuSC) and the γ-tubulin ring complex (γ-TuRC) involved in microtubule nucleation. This interaction allows bacterial uptake into epithelial cells to be mediated by actin cytoskeletal rearrangement (Sana et al., 2015; Wood et al., 2019). The crystal structure of Vgr2b C-terminal encompassing residues 833-1019 (PDB 6H56) presents a metallopeptidase fold (Wood et al., 2019).

Sana et al. (2016) proposed a working model for the interplay of T6SS effectors PldA, PldB and VgrG2b in *P. aeruginosa* internalization. First, VgrG2b is translocated via H2-T6SS, causing the polarization of epithelial cells by targeting the microtubule network, promoting microtubule nucleation at the membrane by interacting with γ-TuRC. These novel sites of non-radial microtubule nucleation interfere with the transport of microtubule-dependent cargoes in the cell, like PI3K. Simultaneously, PldA and PldB are translocated by the different H2 and H3-T6SSs, activating Akt which allows actin-dependent membrane protrusion that enables bacterial internalization into the epithelial cells.

Another effector, TplE, contains a eukaryotic PGAP1 (post-glycosylphosphatidylinositol attachment to proteins 1)-like domain. TplE is translocated into epithelial cells in an H2-T6SS-dependent manner and localizes to host endoplasmic reticulum (ER), causing a contraction of the ER surrounding the nuclear periphery. TplE phospholipase activity is not involved in localization but is required for disruption of ER structure. TplE induces the upregulation of Bip and CHOP chaperones that are biomarkers for ER stress and induces the splicing of XBP1 mRNA, suggesting that the TplE-induced unfolded protein response is dependent on the IRE1α-XBP1 signaling pathway. It was also reported that autophagic flux is induced by TplE delivery into human epithelial cells (Jiang et al., 2016).

Klebsiella

Klebsiella pneumoniae is a ubiquitous species in nature, a gut commensal, and an opportunistic pathogen in humans. As a prominent nosocomial pathogen, it can cause a wide range of infections, including urinary tract, respiratory tract or blood infections, bacteremia and liver abscesses. Due to the regular occurrence of multiple antibiotic-resistant isolates, *K. pneumoniae* is considered a global public health concern. In *K. pneumoniae* three different T6SS loci were defined, and a gene encoding a PLD family protein Pld1 is located within a type VI secretion system locus (Sarris et al., 2011; Lery et al., 2014). Pld1 is a Tle5 homolog, has two conserved HXKXXXXD motif and is expressed during *K. pneumoniae* virulence in a mouse model of pneumonia. The *pld1* phospholipase mutant was strongly attenuated *in vivo*, suggesting an effect on lipid metabolism in *K. pneumoniae* pathogenesis (Lery et al., 2014).

VgrG4 encodes a C-terminal domain of unknown function DUF2345. VgrG4 is needed for bacteria-induced killing of the fungal pathogen *Candida albicans* and *Saccharomyces cerevisiae*, implicating the T6SS in intoxication of fungal cells. The DUF2345 domain is sufficient for the anti-eukaryotic activity (Storey et al., 2020).

Francisella

Francisella tularensis is one of the most infectious intracellular pathogens known. After entering the body via the skin, mucous membranes, or respiratory or gastrointestinal tracts, it causes tularemia, a necrotizing bronchopneumonia that leads to sepsis and death. The T6SS encoded by the *Francisella* pathogenicity island (FPI) is critical for the virulence of this bacterium. In contrast *F. tularensis* subsp. *novicida* (*F. novicida*) has low virulence in humans, but is highly virulent in mice and thus often used as a laboratory model for tularemia (Eshraghi et al., 2016). PdpC (pathogenicity determinant protein C) is a 156-kDa protein encoded within the FPI that contributes to phagosomal escape, trafficking to lysosomes and intramacrophage replication. PdpC plays a role in virulence in the mouse model, as demonstrated by the Δ*pdpC* mutant causing significantly lower mortality in mice with a corresponding reduction in bacterial burden in organs. PdpC is required to activate the AIM2 inflammasome and Δ*pdpC* induces lower levels of type I interferon production (Lindgren et al., 2013a,b; Long et al., 2013; Uda et al., 2014; Eshraghi et al., 2016; Ozanic et al., 2016; Brodmann et al., 2017).

PdpD is a protein encoded within the FPI; its export requires VgrG and PdpA. This effector contributes to intramacrophage growth and phagosome rupture. PdpD is also required to activate the AIM2 inflammasome (Eshraghi et al., 2016; Brodmann et al., 2017).

OpiA and OpiB are encoded by open reading frames located outside of the FPI and recently identified as T6SS substrates. They contribute to intracellular growth. There are no homologs of OpiA found outside of *Francisella*, and *in silico* analyses were unable to identify characterized domains or motifs within the protein. The OpiB C-terminus is homologous to the ankyrin repeat domains mediating protein-protein interactions that are normally found in eukaryotic proteins. The OpiB N-terminus constitutes an evolutionarily plastic cysteine protease (Eshraghi et al., 2016). OpiA belongs to a family of wortmannin-resistant bacterial PI3K enzymes with members found in a wide range of intracellular pathogens. OpiA can phosphorylate PI but not PIP_2. OpiA binds phosphatidylinositol 3-phosphate [PI(3)P] in a selective and high-affinity manner serving as a mechanism for the specific recruitment of OpiA to endosomal membranes.

OpiA acts on the *Francisella*-containing phagosome, leading to efficient bacterial escape from late endosomes into the cytoplasm of infected cells (Ledvina et al., 2018). The protein is translocated into phagocytic cells and reduces the levels of TNF-α, a pro-inflammatory cytokine from monocytes required to block intracellular replication. OpiA contributes to the pathogenesis of *F. tularensis*, as demonstrated using a chicken embryo infection model (Cantlay et al., 2020).

The protein IglE (intracellular growth locus E) is translocated into macrophages (Broms et al., 2012). The Δ*iglE* mutant has a slower intracellular growth rate in human macrophages, suggesting a role for this protein in intracellular replication. IglE interacts with β-tubulin, pericentrin and with microtubule organizing centers. It inhibits the dynein- based intracellular trafficking in host cells, allowing *F. novicida* to escape from fusion with lysosomes (Shimizu et al., 2019).

Edwardsiella

Edwardsiella tarda infects a wide range of hosts including fish, birds, reptiles and humans. In humans, it causes both intestinal and extra-intestinal infections, mainly in individuals with impaired immune systems. Edwardsiellosis in fish is a devastating disease predominant in worldwide aquaculture industries, making it of particular importance to the fishing industry (Zheng and Leung, 2007). EvpP (*E. tarda* virulence protein P) transcription is iron-dependent. EvpP is a 20-kDa protein that is not conserved in other bacteria and contains no conserved domains or motifs. It is secreted via T6SS and the EvpP C-terminus interacts with EvpV (Hcp homolog) (Zheng and Leung, 2007; Hu et al., 2014). In an *in vivo* fish model, EvpP plays a role in proliferation and infection. This toxin also mediates hemolytic activity in sheep erythrocytes and contributes to mucus adhesion and serum resistance of Japanese flounder. EvpP is important for internalization into epithelial papilloma of carp cells (Wang et al., 2009). The protein localizes in the membrane after injection and has an inhibitory role in NLRP3 inflammasome activation by reducing Jnk phosphorylation and ASC oligomerization. It was reported that Δ*evpP* induced higher intracellular calcium flux than wildtype *E. tarda* indicating that EvpP-mediated manipulation of the Jnk-ASC could be traced upstream to intracellular Ca^{2+} signaling (Chen et al., 2017).

Edwardsiella ictaluri causes enteric septicemia of catfish and is the most important endemic infectious disease in catfish aquaculture industry. EvpP toxin is involved in adhesion and internalization of *E. ictaluri* in catfish ovary cells. EvpP plays a role in growth regulation in the phagolysosome where oxidative stress and limited nutrients are present, and also favors survival and increases apoptosis and necrosis in catfish anterior kidney macrophages (Kalindamar et al., 2020).

Edwardsiella piscicida is abundant in water and causes food and waterborne infections in fish, animals and humans (Leung et al., 2019). Using an *in vivo* zebrafish larvae infection model EvpP inhibits immune cells recruitment via Jnk-MAPK signaling cascades. EvpP reduces the expression of *cxcl8a* (chemokine ligand 8) and *mmp13* (matrix metallopeptidase 13) transcripts, indicating that EvpP plays a role in inhibiting the recruitment of neutrophils. Meanwhile, EvpP also inhibits the Jnk-caspy inflammasome and IL-1β expression to suppress neutrophil recruitment, thereby promoting bacterial colonization (Tan et al., 2019). EvpP is also able to reduce Annexin V binding and activation of cleaved caspase-3 involved in apoptosis. This effector interacts with ribosomal protein S5 (RPS5), most likely resulting in downregulation of apoptosis-associated pathways in macrophages (Qin et al., 2020).

Burkholderia

Burkholderia cenocepacia is widespread in the environment, particularly within the rhizosphere. *B. cenocepacia* is also an opportunistic pathogen causing chronic lung infections in patients with cystic fibrosis as well as in other immunocompromised patients (Loutet and Valvano, 2010). The 17-kDa protein TecA is a non-VgrG T6SS effector responsible for actin disruption *in vivo*. TecA and other bacterial homologs bear a cysteine protease-like catalytic triad, which inactivates Rho GTPases by deamidating a conserved asparagine in the GTPase switch-I region. RhoA deamidation induces Pyrin inflammasome activation (Aubert et al., 2016).

Burkholderia thailandensis is a soil saprophyte of low virulence. *Burkholderia pseudomallei* is the causative agent of melioidosis, a serious and often fatal human infection. These species, referred as the Bptm group, encode several T6SSs but the type VI secretion system 5 (T6SS-5) is the one required for virulence in mammalian infection models. VgrG-5 is a substrate of T6SS-5 and is translocated into macrophages. VgrG-5 C-terminal domain is involved in mediating multinucleated giant cell formation, membrane fusion and virulence in mice (Schwarz et al., 2014; Toesca et al., 2014).

Serratia

Serratia marcescens occurs naturally in soil and water. It is associated with urinary and respiratory infections, endocarditis, osteomyelitis, septicemia, wound infections, eye infections, and meningitis. Tfe1 (T6SS antifungal effector 1) is an antifungal small T6SS toxin (20 kDa), deletion of Tfe1 encoding gene resulted in a four-fold increase in recovery of viable *Candida albicans* target cells compared with the wild type bacteria. Tfe1 causes cell distortion and lysis in both the budding and filamentous forms of *C. albicans*. Tfe1 inhibits growth of *S. cerevisiae* and induces abnormally large vacuoles and cell lysis, confirming the fungicidal role of this effector. Tfe1 intoxication results in membrane depolarization by loss of membrane potential, which is not due to pore formation but can lead to a loss of membrane integrity and cell death (Trunk et al., 2018, 2019). Removal of Tfe2 (T6SS antifungal effector 2) encoding gene, resulted in almost complete loss of activity against *S. cerevisiae* or *Candida glabrata* and reduced activity against *C. albicans*. Tfe2 is a small protein (26 kDa) which, when expressed in *S. cerevisiae*, is able to inhibit its growth. Tfe2 intoxication disrupts nutrient uptake and amino acid metabolism and causes autophagy. Tfe1 and Tfe2 act on different cellular targets in fungal cells (Trunk et al., 2018, 2019).

Aeromonas

Aeromonas hydrophila is common in freshwater environments and causes disease in fish, reptiles, amphibians, and humans. It causes a broad spectrum of infections (including septicemia, meningitis, endocarditis) in humans and severe motile septicemia in warmwater fishes. The 103-kDa protein VgrG1 is translocated by the T6SS. VgrG1 contains a vegetative insecticidal protein domain at its C-terminus with actin ADP-ribosyltransferase activity. This effector alters the actin cytoskeleton and induces apoptosis in epithelial cells (Suarez et al., 2010).

Yersinia

Yersinia pseudotuberculosis is an enteric pathogen, which usually grows in the environment and can be transmitted to mammalian hosts through ingestion of contaminated food or water. It typically causes a broad range of gastrointestinal diseases, from enteritis to mesenteric lymphadenitis (Yang et al., 2018). *Y. pseudotuberculosis* contains four T6SS clusters. The T6SS-4 secreted substrate YezP (*Yersinia* extracellular zinc-binding protein) is a Zn^{2+}-binding protein that has the ability to rescue the sensitivity to oxidative stress exhibited by T6SS mutants when added to extracellular milieu. YezP plays a role in virulence for mice but its contribution to the infection process requires additional investigation (Wang et al., 2015).

CONCLUDING REMARKS

Bacterial pathogens employ many strategies to invade mammalian hosts, damage tissues, organelles and prevent the immune system from responding. One strategy is the secretion of proteins (effectors) across membranes. As we described in this review, these toxins are secreted and injected into host cells via the T6SS and exist both as evolved VgrGs and cargo effectors. Translocated effectors can play many roles in eukaryotic cells, which promote bacterial virulence ranging from attachment to directly intoxicating target cells and disrupting their functions to finally establishing a replicative niche and successful colonization.

The clearance of pathogens depends on the host innate immune responses that take place at early stages of infection and in which macrophages and neutrophils are the essential players. Once inside the macrophage, intracellular bacteria can reside in vacuoles or in the cytosol, depending on their effector repertoire which help them to evade host defense and continue the infection cycle and replicate. Here, we described 27 T6SS effectors employed by several bacterial species to promote virulence in eukaryotic cells. These effectors can display similar or complementary functions into host cells and modulate the same central pathway of the host cell (e.g., inflammasome) or having different roles. Moreover, a pathogen may secrete several proteins to produce the same outcome (e.g., PldA and PldB).

Vibrio, *Pseudomonas*, *Burkholderia* and *Aeromonas* species translocate toxins in a T6SS-dependent manner leading to resistance to phagocytosis, inflammasome activation, as well as bacterial internalization by manipulating the actin cytoskeleton. *Vibrio*, *Pseudomonas*, and *Burkholderia* in particular, disrupt the host cell cytoskeleton, targeting actin, although *Pseudomonas* and *Francisella* effectors target the microtubules. The interference with immunity pathways is a hallmark function achieved by T6SS-dependent effectors. *Burkholderia*, *Vibrio*, *Francisella* and *Edwardsiella* inject toxins involved in the activation of the different inflammasomes that lead to the secretion of proinflammatory cytokines. In this context, activation of the inflammasome can be important for the clearance of the pathogen, suggesting the possibility that the T6SS effectors may also have a role as anti-virulence factors. This notion is supported from results using TecA deficient mutants in experimental mice infection whereby the mutant bacteria were able to kill infected mice while the parental strain was cleared. This clearance effect was abolished in infections using Pyrin inflammasome-defective mice (Aubert et al., 2016).

Another important mechanism of host defense is the generation of reactive oxygen species to eradicate intracellular bacteria. *E. coli* and *Yersinia* T6SSs deliver effectors with the ability to modulate the oxidative stress and protect the pathogen from ROS and allowing growth. Finally, *Serratia* delivers effectors into fungal cells, causing depolarization of the plasma membrane and metabolism disrupted, leading to cell death.

In recent years, remarkable progress has been made toward elucidating the function of eukaryotic effectors of the T6SS, which has contributed to better understand several aspects of bacterial pathogenesis. However, our understanding of the molecular mechanism of many T6SS-secreted toxins awaits detailed functional analysis, including biochemical, biophysical, immunological and structural studies. The kinetics of effector delivery is also an open question since very little is known on whether their translocation is regulated in a temporal and spatial manner and the signals that triggers their secretion.

AUTHOR CONTRIBUTIONS

JM and MV: wrote the manuscript.

ACKNOWLEDGMENTS

The authors would like to thank Dr. Keren Turton for critical review of this manuscript.

REFERENCES

Alcoforado Diniz, J., Liu, Y. C., and Coulthurst, S. J. (2015). Molecular weaponry: diverse effectors delivered by the Type VI secretion system. *Cell. Microbiol.* 17, 1742–1751. doi: 10.1111/cmi.12532

Aschtgen, M. S., Bernard, C. S., De Bentzmann, S., Lloubes, R., and Cascales, E. (2008). SciN is an outer membrane lipoprotein required for type VI secretion in enteroaggregative *Escherichia coli*. *J. Bacteriol.* 190, 7523–7531. doi: 10.1128/JB.00945-08

Aubert, D. F., Xu, H., Yang, J., Shi, X., Gao, W., Li, L., et al. (2016). A burkholderia type VI effector deamidates Rho GTPases to activate the pyrin inflammasome and trigger inflammation. *Cell Host Microbe* 19, 664–674. doi: 10.1016/j.chom.2016.04.004

Ballister, E. R., Lai, A. H., Zuckermann, R. N., Cheng, Y., and Mougous, J. D. (2008). *In vitro* self-assembly of tailorable nanotubes from a simple protein building block. *Proc. Natl. Acad. Sci. U.S.A.* 105, 3733–3738. doi: 10.1073/pnas.0712247105

Basler, M. (2015). Type VI secretion system: secretion by a contractile nanomachine. *Philos. Trans. R. Soc. Lond. B. Biol. Sci.* 370. doi: 10.1098/rstb.2015.0021

Basler, M., Pilhofer, M., Henderson, G. P., Jensen, G. J., and Mekalanos, J. J. (2012). Type VI secretion requires a dynamic contractile phage tail-like structure. *Nature* 483, 182–186. doi: 10.1038/nature10846

Bingle, L. E., Bailey, C. M., and Pallen, M. J. (2008). Type VI secretion: a beginner's guide. *Curr. Opin. Microbiol.* 11, 3–8. doi: 10.1016/j.mib.2008.01.006

Bleves, S., Sana, T. G., and Voulhoux, R. (2014). The target cell genus does not matter. *Trends Microbiol.* 22, 304–306. doi: 10.1016/j.tim.2014.04.011

Bondage, D. D., Lin, J. S., Ma, L. S., Kuo, C. H., and Lai, E. M. (2016). VgrG C terminus confers the type VI effector transport specificity and is required for binding with PAAR and adaptor-effector complex. *Proc. Natl. Acad. Sci. U.S.A.* 113, E3931–E3940. doi: 10.1073/pnas.1600428113

Bonemann, G., Pietrosiuk, A., Diemand, A., Zentgraf, H., and Mogk, A. (2009). Remodelling of VipA/VipB tubules by ClpV-mediated threading is crucial for type VI protein secretion. *EMBO J.* 28, 315–325. doi: 10.1038/emboj.2008.269

Boulant, T., Boudehen, Y. M., Filloux, A., Plesiat, P., Naas, T., and Dortet, L. (2018). Higher prevalence of PldA, a *Pseudomonas aeruginosa* trans-kingdom H2-type VI secretion system effector, in clinical isolates responsible for acute infections and in multidrug resistant strains. *Front. Microbiol* 9:2578. doi: 10.3389/fmicb.2018.02578

Boyer, F., Fichant, G., Berthod, J., Vandenbrouck, Y., and Attree, I. (2009). Dissecting the bacterial type VI secretion system by a genome wide in silico analysis: what can be learned from available microbial genomic resources? *BMC Genomics* 10:104. doi: 10.1186/1471-2164-10-104

Brodmann, M., Dreier, R. F., Broz, P., and Basler, M. (2017). Francisella requires dynamic type VI secretion system and ClpB to deliver effectors for phagosomal escape. *Nat. Commun* 8:15853. doi: 10.1038/ncomms15853

Broms, J. E., Ishikawa, T., Wai, S. N., and Sjostedt, A. (2013). A functional VipA-VipB interaction is required for the type VI secretion system activity of Vibrio cholerae O1 strain A1552. *BMC Microbiol.* 13:96. doi: 10.1186/1471-2180-13-96

Broms, J. E., Meyer, L., Sun, K., Lavander, M., and Sjostedt, A. (2012). Unique substrates secreted by the type VI secretion system of Francisella tularensis during intramacrophage infection. *PLoS ONE* 7:e50473. doi: 10.1371/journal.pone.0050473

Brunet, Y. R., Henin, J., Celia, H., and Cascales, E. (2014). Type VI secretion and bacteriophage tail tubes share a common assembly pathway. *EMBO Rep.* 15, 315–321. doi: 10.1002/embr.201337936

Brunet, Y. R., Zoued, A., Boyer, F., Douzi, B., and Cascales, E. (2015). The type VI secretion TssEFGK-VgrG phage-like baseplate is recruited to the TssJLM membrane complex via multiple contacts and serves as assembly platform for tail tube/sheath polymerization. *PLoS Genet.* 11:e1005545. doi: 10.1371/journal.pgen.1005545

Cantlay, S., Haggerty, K., and Horzempa, J. (2020). OpiA, a type six secretion system substrate, localizes to the cell pole and plays a role in bacterial growth and viability in francisella tularensis LVS. *J. Bacteriol.* 202:e00048-20. doi: 10.1128/JB.00048-20

Chen, H., Yang, D., Han, F., Tan, J., Zhang, L., Xiao, J., et al. (2017). The bacterial T6SS effector EvpP prevents NLRP3 inflammasome activation by inhibiting the Ca(2+)-dependent MAPK-Jnk pathway. *Cell Host Microbe* 21, 47–58. doi: 10.1016/j.chom.2016.12.004

Cherrak, Y., Flaugnatti, N., Durand, E., Journet, L., and Cascales, E. (2019). "Structure and activity of the type VI secretion system," in *Protein Secretion in Bacteria*, eds M. Sandkvist, E. Cascales, and P. J. Christie. 7. doi: 10.1128/9781683670285.ch26

Cherrak, Y., Rapisarda, C., Pellarin, R., Bouvier, G., Bardiaux, B., Allain, F., et al. (2018). Biogenesis and structure of a type VI secretion baseplate. *Nat. Microbiol.* 3, 1404–1416. doi: 10.1038/s41564-018-0260-1

Cianfanelli, F. R., Alcoforado Diniz, J., Guo, M., de Cesare, V., Trost, M., and Coulthurst, S. J. (2016a). VgrG and PAAR proteins define distinct versions of a functional type VI secretion system. *PLoS Pathog.* 12:e1005735. doi: 10.1371/journal.ppat.1005735

Cianfanelli, F. R., Monlezun, L., and Coulthurst, S. J. (2016b). Aim, load, fire: the type VI secretion system, a bacterial nanoweapon. *Trends Microbiol.* 24, 51–62. doi: 10.1016/j.tim.2015.10.005

Dong, T. G., Ho, B. T., Yoder-Himes, D. R., and Mekalanos, J. J. (2013). Identification of T6SS-dependent effector and immunity proteins by Tn-seq in Vibrio cholerae. *Proc. Natl. Acad. Sci. U.S.A.* 110, 2623–2628. doi: 10.1073/pnas.1222783110

Douzi, B., Spinelli, S., Blangy, S., Roussel, A., Durand, E., Brunet, Y. R., et al. (2014). Crystal structure and self-interaction of the type VI secretion tail-tube protein from enteroaggregative *Escherichia coli*. *PLoS ONE* 9:e86918. doi: 10.1371/journal.pone.0086918

Durand, E., Cambillau, C., Cascales, E., and Journet, L. (2014). VgrG, Tae, Tle, and beyond: the versatile arsenal of Type VI secretion effectors. *Trends Microbiol.* 22, 498–507. doi: 10.1016/j.tim.2014.06.004

Durand, E., Derrez, E., Audoly, G., Spinelli, S., Ortiz-Lombardia, M., Raoult, D., et al. (2012). Crystal structure of the VgrG1 actin cross-linking domain of the *Vibrio cholerae* type VI secretion system. *J. Biol. Chem.* 287, 38190–38199. doi: 10.1074/jbc.M112.390153

Durand, E., Nguyen, V. S., Zoued, A., Logger, L., Pehau-Arnaudet, G., Aschtgen, M. S., et al. (2015). Biogenesis and structure of a type VI secretion membrane core complex. *Nature* 523, 555–560. doi: 10.1038/nature14667

Dutta, P., Jijumon, A. S., Mazumder, M., Dileep, D., Mukhopadhyay, A. K., Gourinath, S., et al. (2019). Presence of actin binding motif in VgrG-1 toxin of *Vibrio cholerae* reveals the molecular mechanism of actin cross-linking. *Int. J. Biol. Macromol.* 133, 775–785. doi: 10.1016/j.ijbiomac.2019.04.026

Eshraghi, A., Kim, J., Walls, A. C., Ledvina, H. E., Miller, C. N., Ramsey, K. M., et al. (2016). Secreted effectors encoded within and outside of the *Francisella* pathogenicity island promote intramacrophage growth. *Cell Host Microbe* 20, 573–583. doi: 10.1016/j.chom.2016.10.008

Flaugnatti, N., Le, T. T., Canaan, S., Aschtgen, M. S., Nguyen, V. S., Blangy, S., et al. (2016). A phospholipase A1 antibacterial Type VI secretion effector interacts directly with the C-terminal domain of the VgrG spike protein for delivery. *Mol. Microbiol.* 99, 1099–1118. doi: 10.1111/mmi.13292

Hachani, A., Wood, T. E., and Filloux, A. (2016). Type VI secretion and anti-host effectors. *Curr. Opin. Microbiol.* 29, 81–93. doi: 10.1016/j.mib.2015.11.006

Heisler, D. B., Kudryashova, E., Grinevich, D. O., Suarez, C., Winkelman, J. D., Birukov, K. G., et al. (2015). Actin-directed toxin. ACD toxin-produced actin oligomers poison formin-controlled actin polymerization. *Science* 349, 535–539. doi: 10.1126/science.aab4090

Hernandez, R. E., Gallegos-Monterrosa, R., and Coulthurst, S. J. (2020). Type VI secretion system effector proteins: effective weapons for bacterial competitiveness. *Cell. Microbiol.* 22, e13241. doi: 10.1111/cmi.13241

Hu, W., Anand, G., Sivaraman, J., Leung, K. Y., and Mok, Y. K. (2014). A disordered region in the EvpP protein from the type VI secretion system of *Edwardsiella tarda* is essential for EvpC binding. *PLoS ONE* 9:e110810. doi: 10.1371/journal.pone.0110810

Jana, B., and Salomon, D. (2019). Type VI secretion system: a modular toolkit for bacterial dominance. *Future Microbiol.* 14, 1451–1463. doi: 10.2217/fmb-2019-0194

Jiang, F., Wang, X., Wang, B., Chen, L., Zhao, Z., Waterfield, N. R., et al. (2016). The *Pseudomonas aeruginosa* type VI secretion PGAP1-like effector induces host autophagy by activating endoplasmic reticulum stress. *Cell Rep.* 16, 1502–1509. doi: 10.1016/j.celrep.2016.07.012

Jiang, F., Waterfield, N. R., Yang, J., Yang, G., and Jin, Q. (2014). A *Pseudomonas aeruginosa* type VI secretion phospholipase D effector targets both prokaryotic and eukaryotic cells. *Cell Host Microbe* 15, 600–610. doi: 10.1016/j.chom.2014.04.010

Kalindamar, S., Kordon, A. O., Abdelhamed, H., Tan, W., Pinchuk, L. M., and Karsi, A. (2020). Edwardsiella ictaluri evpP is required for colonisation of channel catfish ovary cells and necrosis in anterior kidney macrophages. *Cell. Microbiol.* 22:e13135. doi: 10.1111/cmi.13135

Kube, S., Kapitein, N., Zimniak, T., Herzog, F., Mogk, A., and Wendler, P. (2014). Structure of the VipA/B type VI secretion complex suggests a contraction-state-specific recycling mechanism. *Cell Rep.* 8, 20–30. doi: 10.1016/j.celrep.2014.05.034

Ledvina, H. E., Kelly, K. A., Eshraghi, A., Plemel, R. L., Peterson, S. B., Lee, B., et al. (2018). A phosphatidylinositol 3-kinase effector alters phagosomal maturation to promote intracellular growth of francisella. *Cell Host Microbe 24*, 285–295.e288. doi: 10.1016/j.chom.2018.07.003

Lery, L. M., Frangeul, L., Tomas, A., Passet, V., Almeida, A. S., Bialek-Davenet, S., et al. (2014). Comparative analysis of *Klebsiella pneumoniae* genomes identifies a phospholipase D family protein as a novel virulence factor. *BMC Biol* 12:41. doi: 10.1186/1741-7007-12-41

Leung, K. Y., Wang, Q., Yang, Z., and Siame, B. A. (2019). Edwardsiella piscicida: a versatile emerging pathogen of fish. *Virulence* 10, 555–567. doi: 10.1080/21505594.2019.1621648

Liang, X., Kamal, F., Pei, T. T., Xu, P., Mekalanos, J. J., and Dong, T. G. (2019). An onboard checking mechanism ensures effector delivery of the type VI secretion system in Vibrio cholerae. *Proc. Natl. Acad. Sci. U. S. A.* 116, 23292–23298. doi: 10.1073/pnas.1914202116

Liang, X., Moore, R., Wilton, M., Wong, M. J., Lam, L., and Dong, T. G. (2015). Identification of divergent type VI secretion effectors using a conserved chaperone domain. *Proc. Natl. Acad. Sci. U.S.A.* 112, 9106–9111. doi: 10.1073/pnas.1505317112

Lindgren, M., Broms, J. E., Meyer, L., Golovliov, I., and Sjostedt, A. (2013a). The *Francisella tularensis* LVS DeltapdpC mutant exhibits a unique phenotype during intracellular infection. *BMC Microbiol* 13:20. doi: 10.1186/1471-2180-13-20

Lindgren, M., Eneslatt, K., Broms, J. E., and Sjostedt, A. (2013b). Importance of PdpC, IglC, IglI, and IglG for modulation of a host cell death pathway induced by *Francisella tularensis. Infect. Immun.* 81, 2076–2084. doi: 10.1128/IAI.00275-13

Logger, L., Aschtgen, M. S., Guerin, M., Cascales, E., and Durand, E. (2016). Molecular dissection of the interface between the type VI secretion TssM cytoplasmic domain and the TssG baseplate component. *J. Mol. Biol.* 428, 4424–4437. doi: 10.1016/j.jmb.2016.08.032

Long, M. E., Lindemann, S. R., Rasmussen, J. A., Jones, B. D., and Allen, L. A. (2013). Disruption of *Francisella tularensis* Schu S4 iglI, iglJ, and pdpC genes results in attenuation for growth in human macrophages and *in vivo* virulence in mice and reveals a unique phenotype for pdpC. *Infect. Immun.* 81, 850–861. doi: 10.1128/IAI.00822-12

Loutet, S. A., and Valvano, M. A. (2010). A decade of *Burkholderia cenocepacia* virulence determinant research. *Infect. Immun.* 78, 4088–4100. doi: 10.1128/IAI.00212-10

Ludu, J. S., de Bruin, O. M., Duplantis, B. N., Schmerk, C. L., Chou, A. Y., Elkins, K. L., et al. (2008). The *Francisella* pathogenicity island protein PdpD is required for full virulence and associates with homologues of the type VI secretion system. *J. Bacteriol.* 190, 4584–4595. doi: 10.1128/JB.00198-08

Ma, A. T., McAuley, S., Pukatzki, S., and Mekalanos, J. J. (2009a). Translocation of a *Vibrio cholerae* type VI secretion effector requires bacterial endocytosis by host cells. *Cell Host Microbe* 5, 234–243. doi: 10.1016/j.chom.2009.02.005

Ma, A. T., and Mekalanos, J. J. (2010). *In vivo* actin cross-linking induced by *Vibrio cholerae* type VI secretion system is associated with intestinal inflammation. *Proc. Natl. Acad. Sci. U.S.A.* 107, 4365–4370. doi: 10.1073/pnas.0915156107

Ma, J., Pan, Z., Huang, J., Sun, M., Lu, C., and Yao, H. (2017). The Hcp proteins fused with diverse extended-toxin domains represent a novel pattern of antibacterial effectors in type VI secretion systems. *Virulence* 8, 1189–1202. doi: 10.1080/21505594.2017.1279374

Ma, L. S., Lin, J. S., and Lai, E. M. (2009b). An IcmF family protein, ImpLM, is an integral inner membrane protein interacting with ImpKL, and its walker a motif is required for type VI secretion system-mediated Hcp secretion in *Agrobacterium tumefaciens. J. Bacteriol.* 191, 4316–4329. doi: 10.1128/JB.00029-09

Miyata, S. T., Kitaoka, M., Brooks, T. M., McAuley, S. B., and Pukatzki, S. (2011). *Vibrio cholerae* requires the type VI secretion system virulence factor VasX to kill dictyostelium discoideum. *Infect. Immun.* 79, 2941–2949. doi: 10.1128/IAI.01266-10

Mougous, J. D., Cuff, M. E., Raunser, S., Shen, A., Zhou, M., Gifford, C. A., et al. (2006). A virulence locus of *Pseudomonas aeruginosa* encodes a protein secretion apparatus. *Science* 312, 1526–1530. doi: 10.1126/science.1128393

Navarro-Garcia, F., Ruiz-Perez, F., Cataldi, A., and Larzabal, M. (2019). Type VI secretion system in pathogenic *Escherichia coli*: structure, role in virulence, and acquisition. *Front. Microbiol.* 10:1965. doi: 10.3389/fmicb.2019.01965

Nazarov, S., Schneider, J. P., Brackmann, M., Goldie, K. N., Stahlberg, H., and Basler, M. (2018). Cryo-EM reconstruction of type VI secretion system baseplate and sheath distal end. *EMBO J.* 37:e97103. doi: 10.15252/embj.201797103

Nguyen, V. S., Douzi, B., Durand, E., Roussel, A., Cascales, E., and Cambillau, C. (2018). Towards a complete structural deciphering of Type VI secretion system. *Curr. Opin. Struct. Biol.* 49, 77–84. doi: 10.1016/j.sbi.2018.01.007

Ozanic, M., Marecic, V., Lindgren, M., Sjostedt, A., and Santic, M. (2016). Phenotypic characterization of the *Francisella tularensis* DeltapdpC and DeltaiglG mutants. *Microbes Infect.* 18, 768–776. doi: 10.1016/j.micinf.2016.07.006

Pissaridou, P., Allsopp, L. P., Wettstadt, S., Howard, S. A., Mavridou, D. A. I., and Filloux, A. (2018). The *Pseudomonas aeruginosa* T6SS-VgrG1b spike is topped by a PAAR protein eliciting DNA damage to bacterial competitors. *Proc. Natl. Acad. Sci. U.S.A.* 115, 12519–12524. doi: 10.1073/pnas.1814181115

Pukatzki, S., Ma, A. T., Revel, A. T., Sturtevant, D., and Mekalanos, J. J. (2007). Type VI secretion system translocates a phage tail spike-like protein into target cells where it cross-links actin. *Proc. Natl. Acad. Sci. U.S.A.* 104, 15508–15513. doi: 10.1073/pnas.0706532104

Qin, L., Wang, X., Gao, Y., Bi, K., and Wang, W. (2020). Roles of EvpP in *Edwardsiella* piscicida-macrophage interactions. *Front. Cell. Infect. Microbiol.* 10:53. doi: 10.3389/fcimb.2020.00053

Quentin, D., Ahmad, S., Shanthamoorthy, P., Mougous, J. D., Whitney, J. C., and Raunser, S. (2018). Mechanism of loading and translocation of type VI secretion system effector Tse6. *Nat. Microbiol.* 3, 1142–1152. doi: 10.1038/s41564-018-0238-z

Rapisarda, C., Cherrak, Y., Kooger, R., Schmidt, V., Pellarin, R., Logger, L., et al. (2019). *In situ* and high-resolution cryo-EM structure of a bacterial type VI secretion system membrane complex. *EMBO J.* 38:e100886. doi: 10.15252/embj.2018100886

Ray, A., Schwartz, N., de Souza Santos, M., Zhang, J., Orth, K., and Salomon, D. (2017). Type VI secretion system MIX-effectors carry both antibacterial and anti-eukaryotic activities. *EMBO Rep.* 18, 1978–1990. doi: 10.15252/embr.201744226

Records, A. R. (2011). The type VI secretion system: a multipurpose delivery system with a phage-like machinery. *Mol. Plant Microbe Interact.* 24, 751–757. doi: 10.1094/MPMI-11-10-0262

Renault, M. G., Zamarreno Beas, J., Douzi, B., Chabalier, M., Zoued, A., Brunet, Y. R., et al. (2018). The gp27-like hub of VgrG serves as adaptor to promote hcp tube assembly. *J. Mol. Biol.* 430, 3143–3156. doi: 10.1016/j.jmb.2018.07.018

Russell, A. B., LeRoux, M., Hathazi, K., Agnello, D. M., Ishikawa, T., Wiggins, P. A., et al. (2013). Diverse type VI secretion phospholipases are functionally plastic antibacterial effectors. *Nature* 496, 508–512. doi: 10.1038/nature 12074

Salomon, D., Kinch, L. N., Trudgian, D. C., Guo, X., Klimko, J. A., Grishin, N. V., et al. (2014). Marker for type VI secretion system effectors. *Proc. Natl. Acad. Sci. U.S.A.* 111, 9271–9276. doi: 10.1073/pnas.1406110111

Sana, T. G., Baumann, C., Merdes, A., Soscia, C., Rattei, T., Hachani, A., et al. (2015). Internalization of *Pseudomonas aeruginosa* strain PAO1 into epithelial cells is promoted by interaction of a T6SS effector with the microtubule network. *MBio* 6:e00712-15. doi: 10.1128/mBio.00712-15

Sana, T. G., Berni, B., and Bleves, S. (2016). The T6SSs of *Pseudomonas aeruginosa* strain PAO1 and their effectors: beyond bacterial-cell targeting. *Front. Cell. Infect. Microbiol.* 6:61. doi: 10.3389/fcimb.2016.00061

Sana, T. G., Hachani, A., Bucior, I., Soscia, C., Garvis, S., Termine, E., et al. (2012). The second type VI secretion system of *Pseudomonas aeruginosa* strain PAO1 is regulated by quorum sensing and Fur and modulates internalization in epithelial cells. *J. Biol. Chem.* 287, 27095–27105. doi: 10.1074/jbc.M112.376368

Santos, M. N. M., Cho, S. T., Wu, C. F., Chang, C. J., Kuo, C. H., and Lai, E. M. (2019). Redundancy and specificity of type VI secretion vgrG Loci in antibacterial activity of *Agrobacterium tumefaciens* 1D1609 strain. *Front. Microbiol.* 10:3004. doi: 10.1101/740209

Sarris, P. F., Zoumadakis, C., Panopoulos, N. J., and Scoulica, E. V. (2011). Distribution of the putative type VI secretion system core genes in *Klebsiella* spp. *Infect. Genet. Evol.* 11, 157–166. doi: 10.1016/j.meegid.2010.09.006

Satchell, K. J. (2009). Actin crosslinking toxins of gram-negative bacteria. *Toxins* 1, 123–133. doi: 10.3390/toxins1020123

Schwarz, S., Singh, P., Robertson, J. D., LeRoux, M., Skerrett, S. J., Goodlett, D. R., et al. (2014). VgrG-5 is a Burkholderia type VI secretion system-exported protein required for multinucleated giant cell formation and virulence. *Infect. Immun.* 82, 1445–1452. doi: 10.1128/IAI.01368-13

Shimizu, T., Otonari, S., Suzuki, J., Uda, A., Watanabe, K., and Watarai, M. (2019). Expression of *Francisella* pathogenicity island protein intracellular growth locus E (IglE) in mammalian cells is involved in intracellular trafficking, possibly through microtubule organizing center. *Microbiologyopen* 8:e00684. doi: 10.1002/mbo3.684

Shneider, M. M., Buth, S. A., Ho, B. T., Basler, M., Mekalanos, J. J., and Leiman, P. G. (2013). PAAR-repeat proteins sharpen and diversify the type VI secretion system spike. *Nature* 500, 350–353. doi: 10.1038/nature12453

Storey, D., McNally, A., Astrand, M., Sa-Pessoa Graca Santos, J., Rodriguez-Escudero, I., Elmore, B., et al. (2020). *Klebsiella pneumoniae* type VI secretion system-mediated microbial competition is PhoPQ controlled and reactive oxygen species dependent. *PLoS Pathog.* 16:e1007969. doi: 10.1371/journal.ppat.1007969

Suarez, G., Sierra, J. C., Erova, T. E., Sha, J., Horneman, A. J., and Chopra, A. K. (2010). A type VI secretion system effector protein, VgrG1, from Aeromonas hydrophila that induces host cell toxicity by ADP ribosylation of actin. *J. Bacteriol.* 192, 155–168. doi: 10.1128/JB.01260-09

Tan, J., Yang, D., Wang, Z., Zheng, X., Zhang, Y., and Liu, Q. (2019). EvpP inhibits neutrophils recruitment via Jnk-caspy inflammasome signaling *in vivo*. *Fish Shellfish Immunol.* 92, 851–860. doi: 10.1016/j.fsi.2019.05.051

Toesca, I. J., French, C. T., and Miller, J. F. (2014). The Type VI secretion system spike protein VgrG5 mediates membrane fusion during intercellular spread by pseudomallei group *Burkholderia* species. *Infect. Immun.* 82, 1436–1444. doi: 10.1128/IAI.01367-13

Trunk, K., Coulthurst, S. J., and Quinn, J. (2019). A new front in microbial warfare-delivery of antifungal effectors by the type VI secretion system. *J. Fungi.* 5:50. doi: 10.3390/jof5020050

Trunk, K., Peltier, J., Liu, Y. C., Dill, B. D., Walker, L., Gow, N. A. R., et al. (2018). The type VI secretion system deploys antifungal effectors against microbial competitors. *Nat. Microbiol.* 3, 920–931. doi: 10.1038/s41564-018-0191-x

Uda, A., Sekizuka, T., Tanabayashi, K., Fujita, O., Kuroda, M., Hotta, A., et al. (2014). Role of pathogenicity determinant protein C (PdpC) in determining the virulence of the *Francisella tularensis* subspecies tularensis SCHU. *PLoS ONE* 9:e89075. doi: 10.1371/journal.pone.0089075

Unterweger, D., Kostiuk, B., Otjengerdes, R., Wilton, A., Diaz-Satizabal, L., and Pukatzki, S. (2015). Chimeric adaptor proteins translocate diverse type VI secretion system effectors in *Vibrio cholerae*. *EMBO J.* 34, 2198–2210. doi: 10.15252/embj.201591163

Wan, B., Zhang, Q., Ni, J., Li, S., Wen, D., Li, J., et al. (2017). Type VI secretion system contributes to Enterohemorrhagic *Escherichia coli* virulence by secreting catalase against host reactive oxygen species (ROS). *PLoS Pathog.* 13:e1006246. doi: 10.1371/journal.ppat.1006246

Wang, T., Si, M., Song, Y., Zhu, W., Gao, F., Wang, Y., et al. (2015). Type VI secretion system transports Zn2+ to combat multiple stresses and host immunity. *PLoS Pathog.* 11:e1005020. doi: 10.1371/journal.ppat.1005020

Wang, X., Wang, Q., Xiao, J., Liu, Q., Wu, H., Xu, L., et al. (2009). *Edwardsiella* tarda T6SS component evpP is regulated by esrB and iron, and plays essential roles in the invasion of fish. *Fish Shellfish Immunol.* 27, 469–477. doi: 10.1016/j.fsi.2009.06.013

Wettstadt, S., Wood, T. E., Fecht, S., and Filloux, A. (2019). Delivery of the *Pseudomonas aeruginosa* phospholipase effectors PldA and PldB in a VgrG- and H2-T6SS-dependent manner. *Front. Microbiol.* 10:1718. doi: 10.3389/fmicb.2019.01718

Whitney, J. C., Beck, C. M., Goo, Y. A., Russell, A. B., Harding, B. N., De Leon, J. A., et al. (2014). Genetically distinct pathways guide effector export through the type VI secretion system. *Mol. Microbiol.* 92, 529–542. doi: 10.1111/mmi.12571

Wilderman, P. J., Vasil, A. I., Johnson, Z., and Vasil, M. L. (2001). Genetic and biochemical analyses of a eukaryotic-like phospholipase D of *Pseudomonas aeruginosa* suggest horizontal acquisition and a role for persistence in a chronic pulmonary infection model. *Mol. Microbiol.* 39, 291–303. doi: 10.1046/j.1365-2958.2001.02282.x

Wood, T. E., Howard, S. A., Forster, A., Nolan, L. M., Manoli, E., Bullen, N. P., et al. (2019). The *Pseudomonas aeruginosa* T6SS delivers a periplasmic toxin that disrupts bacterial cell morphology. *Cell Rep. 29*, 187–201.e187. doi: 10.1016/j.celrep.2019.08.094

Yang, X., Pan, J., Wang, Y., and Shen, X. (2018). Type VI secretion systems present new insights on pathogenic yersinia. *Front. Cell. Infect. Microbiol.* 8:260. doi: 10.3389/fcimb.2018.00260

Yin, M., Yan, Z., and Li, X. (2019). Architecture of type VI secretion system membrane core complex. *Cell Res.* 29, 251–253. doi: 10.1038/s41422-018-0130-7

Zhang, X. Y., Brunet, Y. R., Logger, L., Douzi, B., Cambillau, C., Journet, L., et al. (2013). Dissection of the TssB-TssC interface during type VI secretion sheath complex formation. *PLoS ONE* 8:e81074. doi: 10.1371/journal.pone.0081074

Zheng, J., Ho, B., and Mekalanos, J. J. (2011). Genetic analysis of anti-amoebae and anti-bacterial activities of the type VI secretion system in *Vibrio cholerae*. *PLoS ONE* 6:e23876. doi: 10.1371/journal.pone.0023876

Zheng, J., and Leung, K. Y. (2007). Dissection of a type VI secretion system in *Edwardsiella* tarda. *Mol. Microbiol.* 66, 1192–1206. doi: 10.1111/j.1365-2958.2007.05993.x

Zong, B., Zhang, Y., Wang, X., Liu, M., Zhang, T., Zhu, Y., et al. (2019). Characterization of multiple type-VI secretion system (T6SS) VgrG proteins in the pathogenicity and antibacterial activity of porcine extra-intestinal pathogenic *Escherichia coli*. *Virulence* 10, 118–132. doi: 10.1080/21505594.2019.1573491

13

Divergence of Legionella Effectors Reversing Conventional and Unconventional Ubiquitination

Tomoe Kitao [1], *Hiroki Nagai* [1,2] *and Tomoko Kubori* [1,2*]

[1] *Department of Microbiology, Graduate School of Medicine, Gifu University, Gifu, Japan,* [2] *G-CHAIN, Gifu University, Gifu, Japan*

Correspondence:
Tomoko Kubori
tkubori@gifu-u.ac.jp

The intracellular bacterial pathogen *Legionella pneumophila* employs bacteria-derived effector proteins in a variety of functions to exploit host cellular systems. The ubiquitination machinery constitutes a crucial eukaryotic system for the regulation of numerous cellular processes, and is a representative target for effector-mediated bacterial manipulation. *L. pneumophila* transports over 300 effector proteins into host cells through its Dot/Icm type IV secretion system. Among these, several effector proteins have been found to function as ubiquitin ligases, including unprecedented enzymes that catalyze ubiquitination through unconventional mechanisms. Recent studies have identified many *L. pneumophila* effector proteins that can interfere with ubiquitination. These effectors include proteins that are distantly related to the ovarian tumor protein superfamily described as deubiquitinases (DUBs), which regulate important signaling cascades in human cells. Intriguingly, *L. pneumophila* DUBs are not limited to enzymes that exhibit canonical DUB activity. Some *L. pneumophila* DUBs can catalyze the cleavage of the unconventional linkage between ubiquitin and substrates. Furthermore, novel mechanisms have been found that adversely affect the function of specific ubiquitin ligases; for instance, effector-mediated posttranslational modifications of ubiquitin ligases result in the inhibition of their activity. In the context of *L. pneumophila* infection, the existence of enzymes that reverse ubiquitination primarily relates to a fine tuning of biogenesis and remodeling of the *Legionella*-containing vacuole as a replicative niche. The complexity of the effector arrays reflects sophisticated strategies that bacteria have adopted to adapt their host environment and enable their survival in host cells. This review summarizes the current state of knowledge on the divergent mechanisms of the *L. pneumophila* effectors that can reverse ubiquitination, which is mediated by other effectors as well as the host ubiquitin machinery.

Keywords: *Legionella*, deubiquitinases, ubiquitin, effectors, metaeffectors, type IV secretion, *Legionella*-containing vacuoles, membrane traffic

INTRODUCTION

Ubiquitination is one of the most versatile posttranslational modifications found in eukaryotic cells, and it controls a wide array of cellular processes (Hershko and Ciechanover, 1998; Hochstrasser, 2009). *Legionella pneumophila*, the causative agent for Legionnaires' disease, which is a severe form of pneumonia, employs diverse strategies to utilize or modulate its host's ubiquitin system for its own benefit (Hubber et al., 2013; Lin and Machner, 2017; Qiu and Luo, 2017a,b; Kubori et al., 2019). This bacterium transports a vast array of effector proteins into the host cell cytosol via the type IV secretion system (T4SS), which is encoded by *dot/icm* genes in the bacterial genome (Nagai and Kubori, 2011; Kubori and Nagai, 2016; Galán and Waksman, 2018). Using the effector proteins, *L. pneumophila* establishes a replicative organelle called the *Legionella*-containing vacuole (LCV), counteracts host immune defense, and survives and replicates in host cells (Isberg et al., 2009; Hubber and Roy, 2010; Asrat et al., 2014). The LCV originates from the plasma membrane-derived phagosome, and is remodeled into an endoplasmic reticulum (ER)-associated compartment. In this way, the LCV co-opts several host factors to facilitate the recruitment and fusion of ER-derived vesicles by intercepting membrane traffic between the ER and the Golgi apparatus (Asrat et al., 2014; Hilbi et al., 2017). It has been shown that LCVs are highly decorated with ubiquitin through the Dot/Icm T4SS-dependent process, and a significant amount of ubiquitinated proteins are degraded by the host proteasome (Dorer et al., 2006). One possible explanation for ubiquitin accumulation on the bacterial vacuoles is that degraded ubiquitinated proteins serve as a source of nutrition for bacterial proliferation (Price et al., 2011). On the other hand, it has been well established that ubiquitinated proteins in cytosolic bacteria or bacterial vacuoles are readily recognized by host immune systems, including pathogen-associated selective autophagy, the commonly named xenophagy (Perrin et al., 2004).

Interestingly, recent studies have discovered that *L. pneumophila* deploys a novel form of ubiquitination to establish LCV biogenesis (**Figure 1**). SidE, SdeA, SdeB, and SdeC are paralogues, known as the SidE family of proteins (SidEs). SdeA, SdeB, and SdeC are encoded in the same locus on the *L. pneumophila* chromosome, while SidE is encoded in a separate locus (**Figure 1A**). SidEs catalyze the ubiquitin ligation of host proteins with no requirement of E1 and E2 enzymes, which are essential for canonical ubiquitination (Qiu et al., 2016; Kotewicz et al., 2017). The reaction proceeds in two steps: first, a mono ADP-ribosyl transferase (mART) domain of SidE covalently attaches the ADP-ribose moiety of the cofactor nicotinamide adenine dinucleotide to the Arg42 residue of ubiquitin, resulting in the formation of an ADP-ribosylated ubiquitin (ADPR-Ub) intermediate. Then ADPR-Ub is catalyzed by a phosphodiesterase (PDE) domain of SidE to release AMP, resulting in the conversion to phospho-ribosylated ubiquitin (PR-Ub), which can attach to the serine residue of a substrate (**Figure 1B**). The substrates for PR-ubiquitination include reticulon 4 (RTN4) (Kotewicz et al., 2017) and Rab33b (Qiu et al., 2016). RTN4 is an ER-associated protein that regulates the dynamics of tubular ER (English et al., 2009). This protein is recruited to the LCV depending on the Dot/Icm T4SS (Haenssler et al., 2015; Kotewicz et al., 2017; Steiner et al., 2017). The enzymatic activity of SidEs is responsible for RTN4 recruitment to the LCV (Kalayil et al., 2018), and for ER remodeling as mediated by RTN4, which facilitates the initial step of LCV biogenesis (**Figure 2**) (Kotewicz et al., 2017). Rab33b is a small GTPase that is involved in retrograde transport from the Golgi apparatus to the ER (Starr et al., 2010). These mechanisms suggest that PR-ubiquitination mediated by SidEs is a key event in LCV formation.

Analyses of several *L. pneumophila* effector proteins have revealed a multi-tiered functional interaction among them. A class of proteins called metaeffectors has emerged; these are effectors that regulate other effectors (Kubori et al., 2010; Havey and Roy, 2015; Jeong et al., 2015; Urbanus et al., 2016; Shames et al., 2017; Valleau et al., 2018). In addition to the effector/metaeffector pairs, *L. pneumophila* possesses effector pairs with opposing activities against the same targets (Neunuebel et al., 2011; Tan and Luo, 2011; Tan et al., 2011). The existence of the effector/metaeffector and agonist/antagonist pairs provides another layer for the bacterial regulation of cellular processes. By utilizing metaeffectors, bacteria can finely regulate the activity of effectors in a temporal and spatial manner inside their host cells, striking a balance between perturbing cellular functions and maintaining a replicative niche (Kubori and Nagai, 2011; Hilbi et al., 2017). The first-identified metaeffector, LubX, is an E3 ubiquitin ligase that polyubiquitinates another *L. pneumophila* effector, SidH, and subjects it to proteasomal degradation (Kubori et al., 2008, 2010). Current knowledge indicates that, *L. pneumophila* possesses at least 10 proteins that have ubiquitin ligase activity (Qiu and Luo, 2017a), and a significant number of proteins have been inferred to function as deubiquitinases (DUBs), which have been rarely identified in other bacteria (Kubori et al., 2019). In this review, we highlight the fascinating role that *L. pneumophila* effectors play in reversing both conventional and unconventional ubiquitination in various ways to accomplish intracellular survival.

SidE FAMILY PROTEINS (SidE, SdeA, SdeB, AND SdeC)

Sheedlo *et al.* utilized the suicide inhibitor of DUBs, the hemagglutinin-tagged ubiquitin vinyl methyl ester (HA-Ub-VME), which can form a covalent linkage with an active site of DUBs, to isolate *L. pneumophila* proteins that have potential DUB activity from bacterial lysates, which led to the first discovery of the *L. pneumophila* DUB, SdeA (Sheedlo et al., 2015). The crystal structures of the SdeA DUB module ($SdeA_{DUB}$), both alone and in complex with Ub-VME, indicate that SdeA possesses a Cys–His–Asp catalytic triad. This triad is commonly found in the bacterial cysteine proteases belonging to the CE clan, which mainly consist of ubiquitin-like (Ubl) proteases (ULPs) and DUBs (Pruneda et al., 2016). Structural data indicate that $SdeA_{DUB}$ most prominently engages Gln40 of ubiquitin, instead

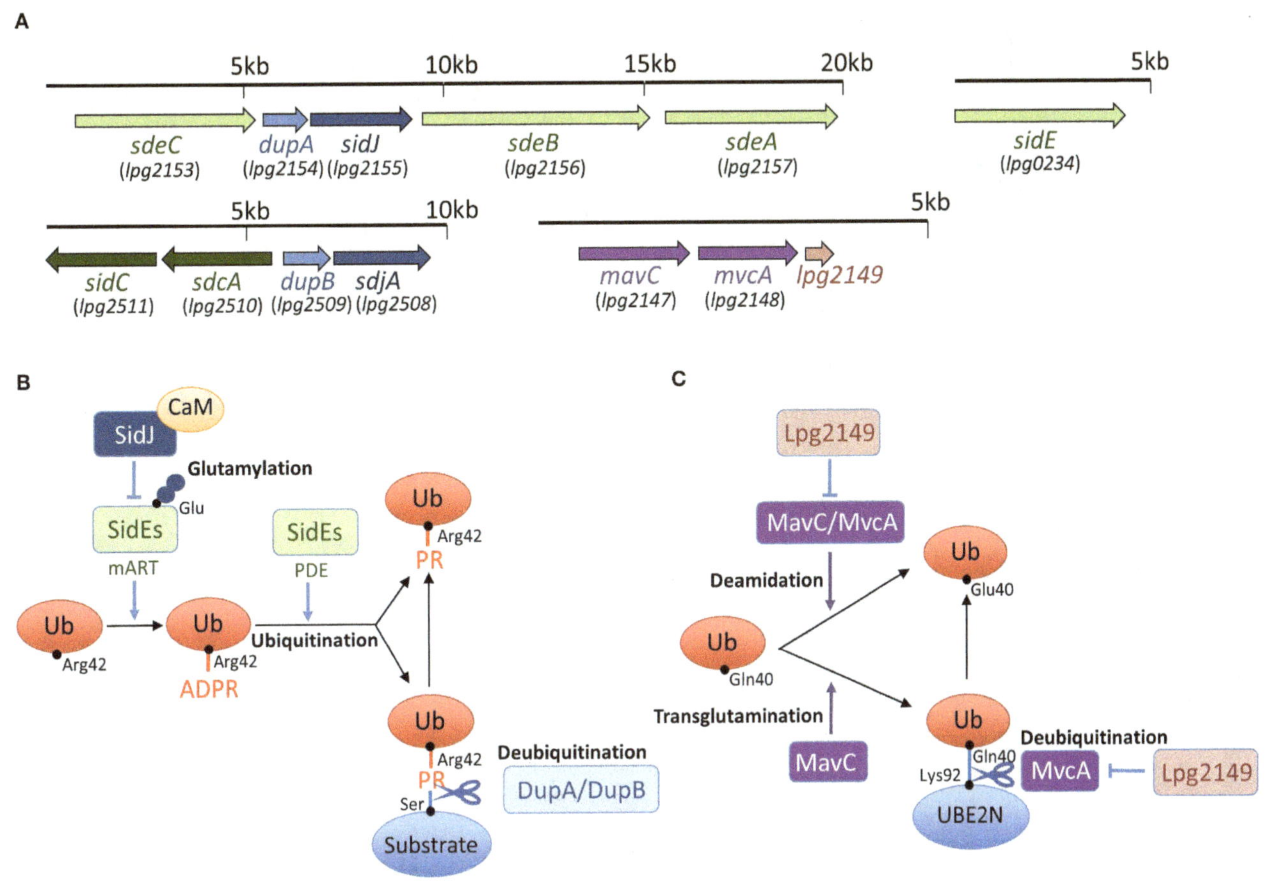

FIGURE 1 | Schemes of effector interplay in the regulation of unconventional ubiquitination and gene organization of *sidEs*, *dupA/dupB*, *sidJ/sdjA*, and *mavC/mvcA* loci in *L. pneumophila*. **(A)** SidEs are encoded in two loci in the *L. pneumophila* genome. *sdeA*, *sdeB*, and *sdeC* genes are in a cluster, and the *sidE* gene is in a separate locus. DupA is encoded in the same locus as *sdeA*, *sdeB*, and *sdeC*, while DupB is encoded in a different locus. SidJ is encoded in a locus with *sdeA*, *sdeB*, *sdeC*, and *dupA*. *sdjA* is thought to be a paralogue of *sidJ*, representing in the same locus with *dupB*. *mavC* and *mvcA* are tandem-arrayed paralogous genes adjacent to *lpg2149*. **(B)** SidEs have an equivalent function in PR-ubiquitination mediated by mART and PDE activities. Products of the paralogous genes *dupA* and *dupB* can reverse PR-ubiquitination with their unique deubiquitination activity. Together with the cofactor CaM, SidJ executes an enzymatic effect to glutamylate the active site of the mART domain of SidEs, thereby inhibiting SidEs-mediated ubiquitination (see text for details). **(C)** MavC and MvcA display deamidation activity against ubiquitin. MavC, but not MvcA, possesses a transglutamination activity and forms a covalent bond between ubiquitin and the E2 enzyme UBE2N. The unconventional ubiquitination of UBE2N can be reversed by the unique DUB activity of MvcA. Lpg2149 can prevent the deamidase activity of both MavC and MvcA. It can also prevent the DUB activity of MvcA (see text for details).

of the Ile44 patch that is widely used by eukaryotic DUBs for interaction, suggesting that some bacterial DUBs have evolved in different ways from eukaryotic DUBs, in terms of substrate preference. SdeA was shown to possess the ability to cleave Lys11/48/63-linked ubiquitin chains, with a distinct preference for Lys63-linkage. The architecture of the active-site cleft presents an open arrangement that has an advantage in accommodating varying structures of the three most abundant polyubiquitin chains. SdeA$_{\text{DUB}}$ is also able to recognize and cleave the chains of neural precursor cell-expressed developmental downregulated 8 (NEDD8) but not of the Ubl protein small ubiquitin-related modifier 1 (SUMO1). Interestingly, structural similarity analyses of SdeA$_{\text{DUB}}$ have identified the mammalian ULP family of cysteine protease Den1, which exhibits an exclusive specificity toward NEDD8. As an indicator of the functional significance of SdeA, polyubiquitin enrichment on *L. pneumophila* phagosomes is regulated by SdeA in infected mouse bone marrow-derived macrophages (**Figure 2**). The Cys–His–Asp catalytic triad is conserved among the DUB modules of SidE, SdeA, SdeB, and SdeC, and the DUB activity has been confirmed by *in vitro* reactions, not only for SdeA but also for SdeB and SdeC (Sheedlo et al., 2015).

In addition to the DUB activity of SdeA, it has been demonstrated that SdeA-mediated phosphoribosylation of ubiquitin impairs the conventional ubiquitination cascade (Bhogaraju et al., 2016). In an *in vitro* experimental system, the activities of the model E3 ubiquitin ligases, Parkin, tripartite motif-containing 56, and *Salmonella* effector protein SopA, were abrogated in the presence of SdeA, even without the DUB domain. Bhogaraju *et al.* reasoned that the apparent lack of

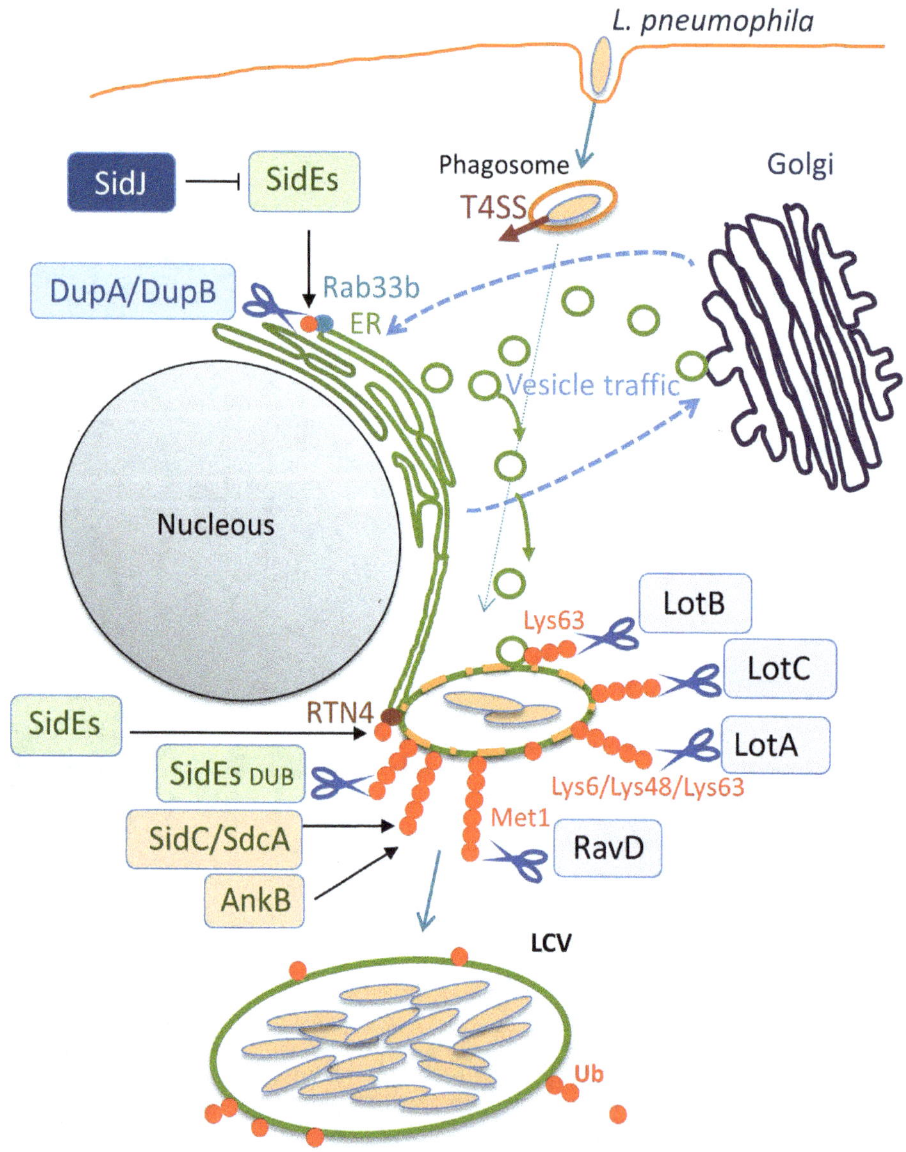

FIGURE 2 | The ubiquitin system-modulating *L. pneumophila* effectors that are associated with LCV biogenesis. Once inside the host cell, *L. pneumophila* intercepts membrane traffic between the ER and the Golgi apparatus to acquire secretory vesicles to form the LCV. Among the *L. pneumophila* effector proteins delivered via the Dot/Icm T4SS, many enzymes associate with the LCV and catalyze ubiquitination, deubiquitination, or reactions that modulate the function of other effectors. In addition to the conventional ubiquitin ligases SidC, SdcA (Luo and Isberg, 2004; Ragaz et al., 2008; Hsu et al., 2014; Wasilko et al., 2018), and AnkB (Price et al., 2009, 2011), the unconventional ubiquitin ligases SidEs have been found to associate with the LCV. SidEs conjugate PR-Ub to many host proteins that function in the secretory pathway, such as Rab33b and RTN4, to modulate their functions. DupA and DupB remove the SidEs-conjugated PR-Ub from substrates with their DUB activity. SidJ negatively regulates the function of SidEs, using its enzymatic activity to modify the active site of SidEs. The amount of ubiquitin on LCVs can be reduced by the function of the bacterial DUBs LotA, LotB, LotC, and RavD. These can localize to the LCV and execute their function on the LCV, depending on their catalytic activities to cleave distinctive linkage-type ubiquitin chains. SidEs also act as canonical DUBs to remove ubiquitin from the LCV.

activity among the E3 ligases was due to defective E1 or E2 reactions, because SdeA-modified ubiquitin cannot be effectively loaded onto E1 or E2 enzymes (Bhogaraju et al., 2016). They also demonstrated that cellular ubiquitin systems easily shut down when a pool of ubiquitin is utilized by SdeA. Therefore, it has been suggested that *L. pneumophila* can effectively control the overall ubiquitin systems in infected cells.

DupA AND DupB

Recently, two paralogues of *L. pneumophila* effectors, DupA and DupB, that possess the PDE domain, were identified as enzymes that can specifically reverse unconventional ubiquitination mediated by SidEs (Wan et al., 2019a; Shin et al., 2020b) (**Figure 1B**). The PDE domain is conserved in nine effectors of the Philadelphia strain of *L. pneumophila*; SidE, SdeA, SdeB, SdeC, DupA, DupB (SdeD), Lpg1496, Lpg2239, and Lpg2523 (Wan et al., 2019a). DupA and DupB are relatively small proteins, consisting of only the PDE domain. DupA is encoded by a gene located in the cluster *sdeA*, *sdeB*, and *sdeC* (**Figure 1A**). DupB was originally named SdeD, as a product of a *sidE* paralog, and it is encoded by a gene upstream of the paralogs *sdcA* and *sidC* genes, which encode the E3 ubiquitin ligases (**Figure 1A**).

The crystal structure of DupA has a resemblance to the PDE domain of SdeA ($SdeA_{PDE}$) (Shin et al., 2020b). A comparison of the structure of $SdeA_{PDE}$ to the crystal structures of DupB (SdeD) in complex with Ub and ADPR-Ub revealed that SdeA and DupB can interact with ubiquitin similarly using conserved residues (Akturk et al., 2018). Consistent with these facts, DupA and DupB have the ability to process ADPR-Ub to PR-Ub, as SidEs do. However, unlike SidEs, they do not transfer PR-Ub to the model substrate, Rab33b (Wan et al., 2019a; Shin et al., 2020b). Biochemical analyses revealed that these PDE domains exhibit hydrolase activity that cleaves the phosphoester bond between PR-Ub and the serine residue in the substrate (**Figure 1B**). In this sense, DupA and DupB are DUBs that remove PR-Ub from substrates specifically targeted by unconventional *L. pneumophila* ubiquitin ligases.

The enzymatic activities of DupA and DupB have been shown to suppress the Golgi fragmentation (Wan et al., 2019a) that has been reported to be caused by SdeA (Jeong et al., 2015). Furthermore, the proteomic identification of host proteins subjected to PR-ubiquitination was conducted utilizing a *dupA* and *dupB* double knock-out *L. pneumophila* strain (Wan et al., 2019a; Shin et al., 2020b). The identified proteins include Golgi reassembly-stacking protein 2, torsin-1A-interacting protein 2 (Wan et al., 2019a) and FAM134C, RTN1, RTN3, lunapark 1, and testis expressed 264 (Shin et al., 2020b), in addition to the previously reported RTN4 (Kotewicz et al., 2017). Validation of the PR-ubiquitination of the selected proteins has shown that multiple ER proteins are PR-ubiquitinated during *L. pneumophila* infection, which can contribute to LCV remodeling (**Figure 2**).

MavC AND MvcA

In addition to SidEs, MavC is another example of *L. pneumophila* effectors that catalyze non-canonical ubiquitination (Valleau et al., 2018; Gan et al., 2019a). MavC, a structural homolog of cycle inhibiting factor (Cif), a bacterial ubiquitin deamidase, reveals deamidase activity against ubiquitin but not against the Ubl protein NEDD8 (Valleau et al., 2018). This protein also catalyzes the covalent linkage of the Gln40 residue of ubiquitin to Lys92 or Lys94 residues of the host E2 enzyme UBE2N through its transglutaminase activity (**Figure 1C**). Thus MavC inhibits the enzymatic activity of UBE2N to form Lys63-linked ubiquitin chains that activate the nuclear factor κB (NF-κB) pathway (Gan et al., 2019a) (**Figure 3**). In this sense, MavC can be considered as a non-canonical ubiquitin ligase, and an enzyme that has inhibitory activity against UBE2N-mediated ubiquitination.

As can be seen in **Figure 1A**, *mvcA* is a *mavC* paralogue (50% amino-acid identity) (**Figure 1A**). Recently, Gan *et al.* identified the unique enzymatic activity of MvcA that hydrolyzes the Gln40-Lys92 isopeptide bond between ubiquitin and UBE2N that is formed by the transglutaminase activity of MavC (Gan et al., 2020) (**Figure 1C**). The prolonged inhibition of UBE2N activity by MavC-mediated transglutamination dampens the NF-κB pathway and restricts intracellular *L. pneumophila* replication (Gan et al., 2020). Gan *et al.* showed that MvcA functions as a specific DUB to restore UBE2N activity, thereby allowing NF-κB activation (**Figure 3**). Like MavC, MvcA possesses the catalytic cysteine residue required for ubiquitin deamidase activity (Valleau et al., 2018) (**Figure 1C**). The same residue Cys83 is required for the cleavage between ubiquitin and UBE2N. To gain structural insights into the unprecedented deubiquitination, the complex of $MvcA_{C83A}$ and UBE2N-Ub (the wild-type MavC-catalyzed non-canonical ubiquitination product) was crystallized. Structural analyses revealed that the same catalytic center executes both deamidation and unique deubiquitination and that, as with MavC, the "insertion" domain of MvcA plays a vital role in the recognition of UBE2N-Ub (Gan et al., 2020). The insertion domain is not present in canonical ubiquitin deamidases such as Cif and a Cif homolog in *Burkholderia pseudomallei* (CHBP) (Yao et al., 2009; Valleau et al., 2018). It has been proposed that the insertion domain was acquired during evolution from the canonical ubiquitin deamidases to enable the ability to recognize the specific substrate (Gan et al., 2020).

Interestingly, the effector Lpg2149, which is encoded in a gene adjacent to *mavC* and *mvcA* (**Figure 1A**), is a metaeffector that negatively regulates the deamidase activity of both MavC and MvcA (Valleau et al., 2018) (**Figure 1C**). In combination with yeast two-hybrid analyses, the Lpg2149 structure, solved as a homodimer crystal, revealed a potential interface with MavC and MvcA, possibly masking catalytic residues (Valleau et al., 2018). It was further shown that Lpg2149 could inhibit the unique DUB activity of MvcA (Gan et al., 2020) (**Figure 1C**).

More recently, the MavC structure was extensively analyzed in three studies (Guan et al., 2020; Mu et al., 2020; Puvar et al., 2020). The importance of the insertion domain in MavC is highlighted in these studies in terms of the recognition of UBE2N. Guan *et al.* prepared the complex composed of the catalytic mutant of MavC ($MavC_{C74A}$) and UBE2N-Ub. Solving the crystal structure of the MavC/UBE2N-Ub binary complex showed that the insertion and "tail" domains in MavC undergo rotational movement during the formation of the ternary complex, conferring the necessary flexibility to access UBE2N (Guan et al., 2020). They confirmed that the insertion domain is not essential for the ubiquitin deamidation activity of MavC but is essential for UBE2N ubiquitination. By comparison with the reported structures of MvcA (Valleau et al., 2018; Gan et al., 2019a), the structure of MavC/UBE2N-Ub revealed the distinctive features of MavC and MvcA in reactions, which explains why the highly similar proteins execute opposing functions. The residues in proximity to the active sites (Trp255 in MavC and Phe268 in MvcA)

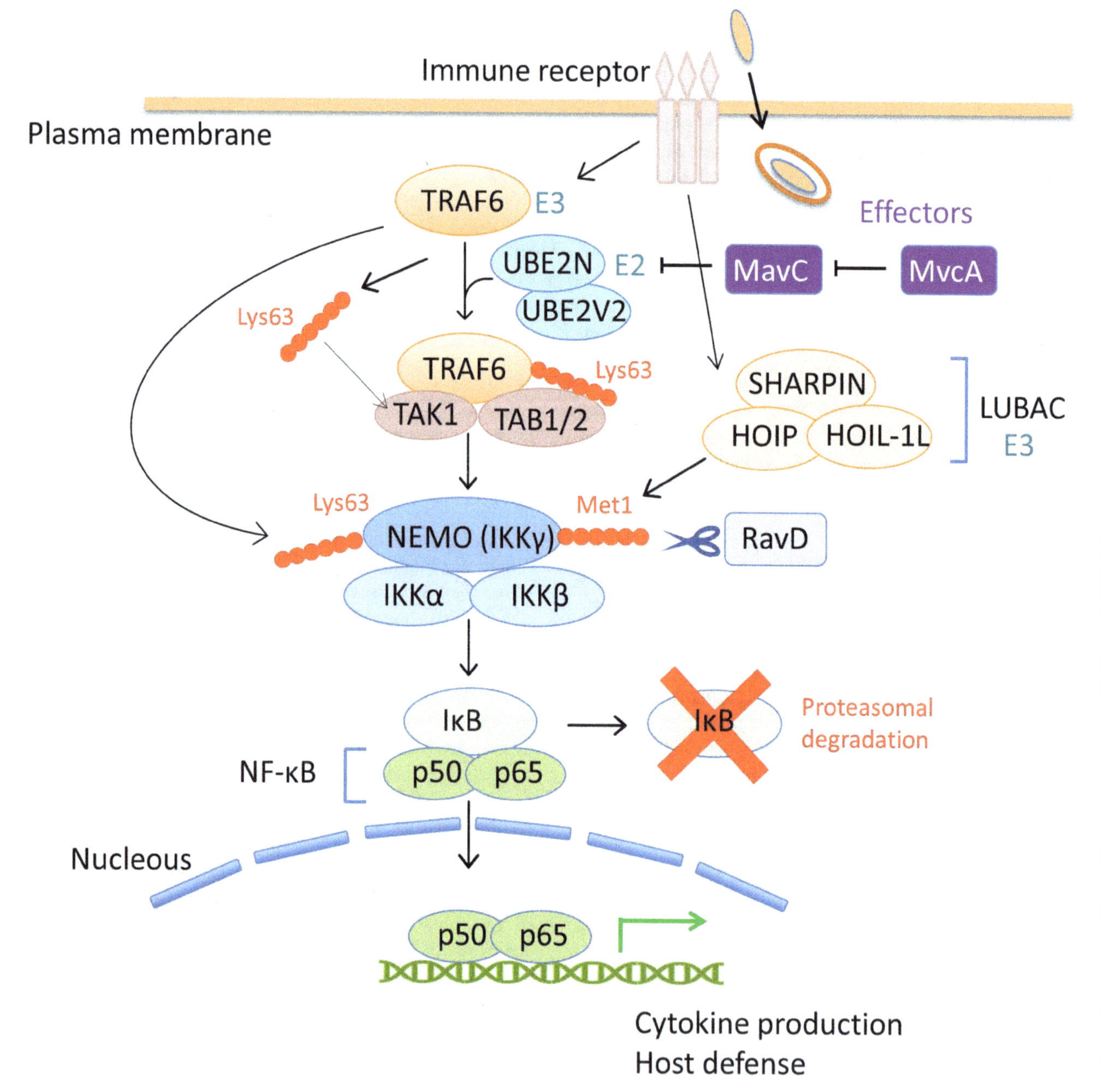

FIGURE 3 | The coordinated regulation of the host NF-κB pathway by *L. pneumophila* effectors via ubiquitin modulation. Upon bacterial infection, the NF-κB family of transcription factors play a central role in controlling the host immune response, such as the production of proinflammatory cytokines, including tumor necrosis factor α (TNFα) and interleukin-1 (IL-1) and the induction of host defense signaling (Li and Verma, 2002). The inhibitor of NF-κB (IκB) suppresses the activity of NF-κB by binding and preventing nuclear translocation. The NF-κB pathway is mainly mediated by the engagement of host immune receptors that can recognize pathogen-associated molecular pattern molecules of extracellular or intracellular bacteria. One of the downstream pathways in which TNFα-receptor-associated factor 6 (TRAF6) is involved, is modulated by a *L. pneumophila* effector, MavC. TRAF6 is an E3 ubiquitin ligase that functions with the E2 enzyme hetero-dimer complex UBE2N/UBE2V2 (also known as Ubc13/Uev1A) to induce its auto-polyubiquitination and polyubiquitination of other neighboring proteins via Lys63-linked ubiquitin chains (Wang et al., 2012). The Lys63-linked ubiquitin chains of TRAF6 serve as docking sites for TGF-β-activated kinase 1 (TAK1)-binding protein 2 (TAB2). TAB1 and TAB2 promote the recruitment of the serine-threonine kinase TAK1 into the complex to activate the pathway. Activated TAK1 phosphorylates IκB kinase β (IKKβ) to promote the proteasomal degradation of IκB, thereby allowing NF-κB nuclear translocation. TRAF6-mediated Lys63-linked ubiquitin chains are associated with the NF-κB essential modifier (NEMO, also known as IKKγ) in the IKKα/IKKβ/NEMO complex and play a crucial role in IKK signaling. TRAF6 also produces free Lys63-linked polyubiquitin chains, which directly activate TAK1 kinase by binding to TAB2. In this scheme, the inhibitory activity of MavC against UBE2N dampens the NF-κB signaling pathway (including many other signaling branches not shown in this figure) by preventing overall Lys63-linked ubiquitination. MvcA, the metaeffector of MavC, can counteract the activity of MavC in the later stage of infection, presumably to restore cellular conditions that optimize bacterial replication. RavD, the other *L. pneumophila* effector, functions as a DUB, and contributes to negative regulation of NF-κB signaling. The host multi-subunit E3 ligase LUBAC, composed of

(Continued)

FIGURE 3 | HOIL-1L, HOIP, and SHARPIN, catalyzes the formation of Met1-linked linear ubiquitin chains. NEMO harbors a Met1 ubiquitin-specific binding domain that is important for NF-κB signaling (Komander et al., 2009; Rahighi et al., 2009). RavD, which is similar to OTULIN (Keusekotten et al., 2013), cleaves LUBAC-assembled linear ubiquitin chains, thereby inhibiting the host NF-κB response.

possess distinct properties that result in the differing affinity between the enzyme and the substrate. In MvcA, the lower binding affinity promotes the dissociation of the substrate, leading to deubiquitination. They reasoned, therefore, that transglutamination and deubiquitination mediated by MavC and MvcA, respectively, are forward and backward versions of the same catalytic reaction (Guan et al., 2020). Mu *et al.* also crystalized $MavC_{C74A}$ with UBE2N-Ub (MavC/UBE2N-Ub binary complex) as well as the MavC/UBE2N/Ub ternary complex (Mu et al., 2020). Compared to the structure of the ternary complex, the binary complex structure revealed that the loop containing the UBE2N modification site, Lys92, undergoes a marked conformational change during ubiquitination, extending toward the deamidated Gln40 residue of ubiquitin. It has been suggested that this conformational change allows the reaction to be completed by MavC. They further analyzed the structure of MavC in complex with Lpg2149. This showed that Lpg2149 has an overlapping binding site with ubiquitin in MavC, suggesting that Lpg2149 can inhibit the ubiquitination of UBE2N by competing with ubiquitin to bind MavC.

Puvar et al. took a unique approach utilizing ubiquitin tethered to UBE2N via a disulfide bond (UBE2N-SS-Ub) that mimics the charged E2~Ub conjugate, the form of ubiquitin intrinsically present in the ubiquitin-reaction cascade (Puvar et al., 2020). Their biochemical and NMR-based interaction analyses showed that free ubiquitin binds only weakly to MavC, while UBE2N-SS-Ub has significantly higher affinity to MavC. This argues that MavC targets the UBE2N~Ub conjugate to catalyze an intramolecular transglutaminase reaction to form an isopeptide bond between UBE2N and ubiquitin. The crystal structure of the MavC/UBE2N-SS-Ub complex revealed that the conformational dynamics of the MavC insertion domain is crucial for the remodeling of the UBE2N active site. These results demonstrate that the mechanism of the MavC-mediated catalytic reaction underlies the protein dynamics that favor transamidation over deamidation. Considering that deamidated ubiquitin was undetected in cells infected with wild-type *L. pneumophila* (Valleau et al., 2018; Gan et al., 2019a), the ubiquitin deamidase activity associated with MavC and MvcA is not likely to be biologically significant. The reported analyses of MavC and MvcA give insights into the structural and functional divergence of the bacterial Cif family proteins.

SidJ AND SdjA

SidJ is one of a few *L. pneumophila* effectors that can result in substantial growth defects when deleted from the genome (Liu and Luo, 2007; Jeong et al., 2015). This protein is encoded in the same locus as *sde* genes (**Figure 1A**). The initial observation that SidJ can counteract the function of SidEs was reported by Jeong *et al.* The intracellular growth of *L. pneumophila* in *Acanthamoeba castellanii* was inhibited upon the over-expression of SdeA; however, this was only the case when SidJ was absent (Jeong et al., 2015). They found that the ectopic expression of *L. pneumophila* SidEs in both yeast and mammalian cells caused cellular growth defects, and toxicity was suppressed by co-expression of SidJ. From these results, they proposed that SidJ could function as a metaeffector that controls the activity of SidEs. Their study showed that SidJ could alter the cellular localization of SidE, resulting in the disappearance of SidE from the LCV.

Having observed that SidJ can suppress the yeast toxicity of SdeA, Qiu *et al.* conducted mutagenesis on SidJ to identify the residues responsible for this activity (Qiu et al., 2017). In relation to the critical residues Asp542 and Asp545, SidJ showed presumable enzymatic activity to interfere with the SdeA-mediated ubiquitination of Rab33b in eukaryotic cells. The DXXD motif has previously been shown to be required to counteract the cellular growth defect caused by SdeA expression and the displacement of SidE from the LCV (Jeong et al., 2015). It has also been demonstrated that $SdeA_{DUB}$ does not remove ubiquitin from Rab33b. *In vitro* reactions using purified SidJ with di-ubiquitins (di-Ubs) or cellular ubiquitinated proteins suggested that SidJ has a canonical DUB activity. However, covalent binding between SidJ and Ub-VME was not detected, and potential catalytic cysteine residues were found not to be essential for the cleavage of ubiquitin chains (Qiu et al., 2017). These results suggested that SidJ does not possess DUB activity, or has a catalytic activity very different from that of classic DUBs. Further biochemical analyses by Qiu *et al.* revealed that SidJ, which was purified from *L. pneumophila*, has the ability to cleave the phosphodiester bond between PR-Ub and the serine residue of Rab33b, which is formed by SdeA (Qiu et al., 2017). Controversially, it was indicated that SidJ, purified from *Escherichia coli*, cannot hydrolyze PR-Ub from Rab33b (Bhogaraju et al., 2019; Black et al., 2019; Wan et al., 2019a). These results raised the possibility that SidJ isolated from *L. pneumophila* might associate with other proteins with DUB activity against PR-Ub.

Four independent studies opened the door to identifying the unexpected enzymatic activity of SidJ (Bhogaraju et al., 2019; Black et al., 2019; Gan et al., 2019b; Sulpizio et al., 2019). It has been shown that SidJ adopts a protein kinase fold. The putative kinase domain in SidJ is crucial for the inhibition of SdeA-catalyzed ubiquitination (Black et al., 2019). The key residues conserved in protein kinases, namely, a metal-binding residue Asp542 in the DXXD motif and an active-site ion-pair Lys367 residue, are both essential for the enzymatic activity of SidJ (Black et al., 2019; Sulpizio et al., 2019). However, an *in vitro* reaction using purified proteins and liquid chromatography-tandem mass spectrometry (LC-MS) has shown that the SidJ-dependent enzymatic modification of SdeA was not phosphorylation but glutamylation. LC-MS and mutagenetic analyses revealed that Glu860 is the major target residue in SdeA, which is the catalytic residue of mART activity (Bhogaraju et al., 2019; Black et al., 2019; Gan et al., 2019b; Sulpizio et al., 2019). Each of these studies has demonstrated that SidJ could catalyze the formation of isopeptide bonds between the amino group of glutamate and the γ-carboxyl group of the catalytic Glu860 residue in the mART domain of SdeA, thereby inhibiting SdeA-mediated PR-ubiquitination (**Figure 1B**).

The glutamylation activity of SidJ requires ATP/Mg^{2+} (Black et al., 2019). It was reported that an evolutionarily conserved pseudokinase is not "inactive," but has the ability to transfer AMP from ATP to a protein substrate (AMPylation) (Sreelatha et al., 2018). Considering this remarkable finding, it is presumable that SidJ activates the glutamylation reaction, and is mediated by its AMPylation activity of the pseudokinase domain. Indeed, it was shown that SidJ undergoes self-AMPylation in the absence of glutamate or SdeA (Gan et al., 2019b). The crystal structure of SidJ captured the enzyme in complex with the AMP moiety (Gan et al., 2019b). The importance of the AMPylation reaction was demonstrated by structure-guided mutagenesis; mutations in residues involved in the binding of AMP abolished the activity of SidJ to inhibit the SdeA-mediated PR-ubiquitination (Gan et al., 2019b). The results of these studies suggest that the AMPylation activity of SidJ is primarily used for the activation of the reaction to glutamylate SdeA by adding an AMP moiety to the target Glu860 residue in SdeA, as proposed by Black et al. (2019), or by self-AMPylation of SidJ.

It was shown that Glu860 was polyglutamylated; Glu860 in SdeA was modified with a various number of glutamates, with two in the majority (Black et al., 2019). In addition, not only SdeA but also SdeB, SdeC and SidE were glutamylated by SidJ (Black et al., 2019). Interestingly, the glutamylation activity of SidJ requires an interaction with the host protein calmodulin (CaM) (**Figure 1B**). The structures of SidJ in complex with CaM determined by X-ray crystallography (Black et al., 2019; Gan et al., 2019b; Sulpizio et al., 2019) and cryo-electron microscopy (Bhogaraju et al., 2019) indicate that the interaction with CaM relies on the IQ motif located in the C-terminal region of SidJ. It is plausible that the requirement for the eukaryotic cofactor CaM is to ensure that inactivation of the SidEs would only occur in infected cells, but not in the bacteria where the proteins are untranslocated, as in the case of the adenylate cyclase toxin from *Bordetella pertussis* (Wolff et al., 1980). The CaM requirement also suggests that SidJ activity could be regulated by the dynamics of the intracellular Ca^{2+} concentration upon infection (Bhogaraju et al., 2019), but the significance of this remains to be fully elucidated.

In contrast with SidJ, its paralogue SdjA does not show any activity to rescue yeast toxicity or to affect the ubiquitination of Rab33b (Qiu et al., 2017), suggesting that SdjA has a different function from SidJ. The gene that encodes SdjA is located just downstream of *dupB*, which encodes the PDE protein that functions in the PR-Ub-specific DUB (**Figure 1A**). This fact suggests a possible linkage between SdjA and DupB in regulating PR-ubiquitination, although the role of SdjA has remained elusive.

Lot-DUBs

Humans encode approximately 100 DUBs, 16 of which belong to the ovarian tumor (OTU) family, which displays great variation in structure and function (Du et al., 2019). The subfamily of OTU DUBs mediates important signaling cascades in eukaryotic cells, such as NF-κB signaling, interferon signaling, DNA damage repair, and immunity (Mevissen et al., 2013; Swatek and Komander, 2016). Unlike other DUBs, most OTU DUBs exhibit an intrinsic specificity against ubiquitin linkage types (Mevissen et al., 2013; Mevissen and Komander, 2017). For instance, otubain 1 preferentially cleaves Lys48-linked ubiquitin chains (Edelmann et al., 2009; Wang et al., 2009), while ovarian tumor deubiquitinase with linear linkage specificity (OTULIN, also known as FAM105B) exclusively cleaves Met1-linked linear ubiquitin chains (Keusekotten et al., 2013). Recent findings on OTU-family DUBs in *L. pneumophila*, Lot (*Legionella* OTU-like)-DUBs, have revealed the existence of a novel group of DUBs that are related to but functionally different from eukaryotic OTU proteins, as described in the following subsections.

LotA

LotA (Lpg2248, also known as Lem21) was the first Lot-DUB identified among *L. pneumophila* Dot/Icm T4SS effectors (Kubori et al., 2018). The notable feature of LotA is the dual catalytic activity that it possesses; this protein has two distinctive catalytic cysteine residues, Cys13 and Cys303. Biochemical experiments demonstrated that Cys13 preferentially cleaves Lys6-linked ubiquitin chains, while Cys303 revealed a preference for polyubiquitin than di-Ubs. Infection analyses showed that the level of Lys48/Lys63-linked polyubiquitin chains decorated on LCVs was reduced predominantly by the enzymatic activity depending on C303 (**Figure 2**). Furthermore, LotA possesses the ability to bind to the early endosome-marker phosphatidylinositol 3-phosphate (PI(3)P) and the late endosome marker PI(3,5)P2 via the C-terminal region, and is located on the LCV throughout the early and late stages of infection. Mutagenetic analyses have found that the lipid-binding ability of LotA is crucial for both ubiquitin removal from- and localization to the LCV. Interestingly, a quintuple mutant strain lacking *lotA* and the four *sidEs* genes showed a significant growth defect relative to the quadruple mutant strain lacking only the four *sidEs* genes, while the *lotA* single-deletion mutant strain did not show a growth defect compared to the wild-type strain. Importantly, a LotA deficiency in PI(3)P-binding did not complement the growth defect of the quintuple mutant strain, but full-length LotA did. In addition, the catalytically inactive LotA C13S mutant, but not the C303S mutant, restored the growth defect. These data demonstrate that LotA plays an important role in *L. pneumophila* infection through lipid binding and deubiquitination of a protein decorated with polyubiquitin chains on the LCV. Further analyses exploring the target proteins for LotA should clarify its role in infection, as well as the biological significance of LotA in catalyzing the cleavage of Lys6-linked ubiquitin chains as a DUB. Structural analyses conducted with these points in mind would aid our understanding of the mechanisms by which LotA can differentially use two cysteine residues to execute its functions.

LotB

LotB (Lpg1621, also known as Ceg23) is a recently characterized Lot-DUB that exhibits high specificity toward Lys63-linked ubiquitin chains (Ma et al., 2020; Schubert et al., 2020; Shin et al., 2020a). This protein was originally identified as a Dot/Icm T4SS effector regulated by the transcriptional activator PmrA (Zusman et al., 2007). However, neither its enzymatic activity nor physiological role had been elucidated. Recently, three independent research groups have analyzed its structure, and the catalytic motif in the N-terminal region is now seen as distantly similar to those associated with the members of the OTU protein subfamily (Ma et al., 2020; Schubert et al., 2020; Shin et al., 2020a). The crystal structure showed that the DUB domain of LotB has a papain-like OTU core fold with a typical catalytic triad, consisting of Asp–Cys–His (Ma et al., 2020). Intriguingly, structural analyses that compare it to eukaryotic OTU DUBs have shown that there is an inserted helical region that extends from the central catalytic Cys-loop (Ma et al., 2020; Schubert et al., 2020; Shin et al., 2020a). Interestingly, the recent report in a preprint suggests that the ubiquitin binding S1 site is located at the helical arm in this inserted region, and that LotB has an additional ubiquitin binding site (S1') that enables it to specifically cleave the Lys63-linked polyubiquitin chains (Shin et al., 2020a). It has also been shown that LotB has a potential dual catalytic activity against ubiquitin and SUMO1 (Schubert et al., 2020) and can bind itself to NEDD8 (Shin et al., 2020a).

The report additionally showed the co-localization of LotB with the resident-ER protein Calnexin, but not with the Golgi apparatus or with mitochondrial markers in the host cell, (Shin et al., 2020a). Importantly, LotB localizes to the LCV, and might regulate the proteins associated with Lys63-linked polyubiquitin chains on the LCV during infection (Ma et al., 2020) (**Figure 2**). Comprehensive MS-based interactome analyses have shown that LotB interacts with OTU deubiquitinase (OTUD) 4, which deubiquitinates K63-linked ubiquitin chains of myeloid differentiation primary response 88 (MYD88) and downregulates NF-κB-dependent inflammation (Shin et al., 2020a). Ma *et al.* explored whether LotB might influence the NF-κB signaling pathway and found that it does not associate with the pathway (Ma et al., 2020). Future identification of the host protein(s) targeted by the DUB activity of LotB could promote an understanding of the physiological role of LotB in *L. pneumophila* infection.

LotC

Concomitantly with the analyses of LotB, another Lot-DUB LotC (Lpg2529, also known as Lem27) was also characterized in the studies recently reported in preprints (Liu et al., 2020; Shin et al., 2020a). The crystal structures of LotC (Shin et al., 2020a) and the LotC-ubiquitin-propargylamide (Ub-PA) complex (Liu et al., 2020) were solved, and revealed that LotC features the Asp–Cys–His catalytic triad, that is also present in LotB. The crystal structures revealed that, similar to LotB, LotC has an extended helical region between the Cys-loop and the variable loop, which is commonly shorter in eukaryotic OTU DUBs (Shin et al., 2020a), and that extensive hydrogen bonding can be used for the recognition of ubiquitin (Liu et al., 2020). However, LotC lacks the S1' site, which defines specificity in the Lys63-linked polyubiquitin chains in LotB (Shin et al., 2020a).

Liu et al. demonstrated that LotC (Lem27) localizes to the LCVs depending on the functional Dot/Icm T4SS and regulates the association of ubiquitin on the LCV (Liu et al., 2020) (**Figure 2**). Recently the host's small GTPase Rab10, which plays a vital role in *L. pneumophila* intracellular replication, was shown to be ubiquitinated and recruited to the LCV by the function of the E3 ligases, SidC and SdcA (Jeng et al., 2019). LotC (Lem27) was found to reverse the SidC-induced Rab10 ubiquitination and decrease its association with the LCV, showing the interplay between *L. pneumophila* ubiquitin ligases and a DUB (Liu et al., 2020). Intriguingly, Shin *et al.* determined that host proteins interact with catalytically inactive LotC in the presence or absence of the *L. pneumophila* ubiquitin ligases SidC and SdcA (Shin et al., 2020a). These results indicate that a significant number of ribosomal proteins interact with LotC in the presence of SdcA but not in the presence of SidC, suggesting distinct physiological roles for SidC and SdcA. Exhibiting 71% sequence similarity, SidC and SdcA had been thought to have an equivalent function, based on the observation that recruitment of Calnexin, which was impaired in the vacuoles containing SidC-deficient *L. longbeachae*, was restored either by *L. pneumophila* SidC or SdcA (Dolinsky et al., 2014).

Common Features of Lot-DUBs

Among the three Lot-DUBs (LotA, LotB, and LotC), LotA is the only one without a solved three-dimensional structure. Notably, sequence analyses have shown that LotA also contains an extended region that LotB and LotC commonly possess (Shin et al., 2020a), which means that the three Lot-DUBs share a structural feature that distinguishes them from eukaryotic OTU DUBs. This raises the possibility that a novel class evolved from the eukaryotic OTU family. Like LotA (Kubori et al., 2018), LotC (Liu et al., 2020; Shin et al., 2020a) exhibits broad activity toward several polyubiquitin chain-types including Lys6-linkage. LotB (Ceg23) also possesses the Lys6-linkage preference with a weaker tendency (Schubert et al., 2020). In addition, another potential Lot-DUB, Ceg7 (Lpg0227), as well as OTU proteins encoded by other bacteria (*E. coli, Burkholdaria ambifaria, Rickettsia massiliae,* and *Wolbachia pipientis*) have a preference for the broad chain types, including Lys48-, Lys63-, K11- and Lys-6 linkages (Schubert et al., 2020). This implies that the evolution of the enzymes from eukaryotic OTU DUBs adapted to various types of ubiquitin chains and is not limited to *L. pneumophila* but widely occurs in pathogenic bacteria. An enterohemorrhagic *E. coli* (EHEC)-encoded HECT (homologous to E6AP carboxyl terminus)-like E3 ligase, NleL, is the only currently known bacterial enzyme that assembles Lys6-linked ubiquitin chains (Lin et al., 2011). NleL activity restricts the formation of actin pedestals under the adherent bacterium on the host cell surface (Lin et al., 2011) and enhances EHEC infection by ubiquitinating and inactivating JNKs (human c-Jun NH2-terminal kinases)

(Sheng et al., 2017). Interestingly, NleL can assemble not only Lys6-linked but also Lys48-linked ubiquitin chains (Lin et al., 2011), and has the propensity to assemble branched ubiquitin chains (Hospenthal et al., 2013). Taken together, pathogenic bacteria are thought to have acquired the ability to finely regulate various types of ubiquitin conjugates. Atypical ubiquitin chains, including Lys6-linked polyubiquitin, may be detrimental to the intracellular survival of *L. pneumophila*; hence the evolution of Lot-DUBs, which have a multiplicity of enzymatic functions for adapting to specific circumstances, may have been induced by *L. pneumophila* infection.

RavD

A recent exciting update regarding bacterial DUB is the discovery of RavD (Lpg0160), which specifically cleaves Met1-linked linear polyubiquitin chains (Wan et al., 2019b). Utilizing a unique assay, in which linear ubiquitin chains were reacted with lysates prepared from bacterial species of 19 different genera *in vitro*, Wan *et al.* found that only the lysate of the *L. pneumophila* strain Philadelphia 1 has the activity of cleaving linear ubiquitin chains (Wan et al., 2019b). Further *in vitro* DUB assays using purified proteins and the bacterial lysate of deletion mutant strains demonstrated that no other *L. pneumophila* effectors cleave linear ubiquitin chains. Sequence alignment analyses revealed that RavD orthologs are present in two different *Legionella* species, *L. clemsonensis* and *L. bozemanni*, in addition to nine *L. pneumophila* serogroups. The crystal structure of the N-terminal region of RavD from the *L. pneumophila* Corby strain, indicates that RavD adopts a papain-like fold with an unconventional Cys–His–Ser catalytic triad (Wan et al., 2019b). This triad is present in ubiquitin-specific proteases (USPs), including USP30, which shows Lys6-linkage specificity (Gersch et al., 2017). The conformation of linear di-Ubs in complex with RavD, as well as the RavD-interacting residues, are largely shared with those in complex with OTULIN, a linear-ubiquitin-specific host DUB. OTULIN negatively regulates the NF-κB signaling pathway, which is partly mediated by the host multi-subunit E3 ligase linear ubiquitin chain assembly complex (LUBAC), which is composed of heme-oxidized iron-responsive element-binding protein 2 ubiquitin ligase-1L (HOIL-1L), HOIL-1L-interacting protein (HOIP), and SHANK-associated RH-domain-interacting protein (SHARPIN) (Kirisako et al., 2006; Keusekotten et al., 2013) (**Figure 3**). However, the overall structure of RavD is distinct from that of OTULIN. Nevertheless, RavD has a function that closely resembles that of OTULIN, inhibiting host NF-κB signaling during infection (**Figure 3**). In addition, RavD localizes to the LCV and prevents the accumulation of linear ubiquitin chains on the LCV (Wan et al., 2019b) (**Figure 2**), which is implicated in the avoidance of xenophagy (Damgaard and Pruneda, 2019). Although the PI(3)P binding region at the C-terminus was predicted to be required for the localization of RavD to the LCV, this localization was independent from PI(3)P binding (Pike et al., 2019). This raises the question of whether RavD interacts with another membrane component to remain attached to the LCV.

LupA

LupA (Lpg1148) is suggested to be a metaeffector that has the function of inactivating another *L. pneumophila* effector, LegC3. This protein silences the cytotoxicity caused by LegC3 in yeast (Urbanus et al., 2016). Urbanus *et al.* determined the crystal structure of LupA and identified the canonical Cys–His–Asp catalytic triad, suggesting that it is a plausible DUB (Urbanus et al., 2016). Using immunoprecipitation assays with the lysate of human cells co-expressing LupA and LegC3, ubiquitinated species of LegC3 were detected in the presence of catalytically inactive LupA variants, but not in the presence of wild-type LupA. This result shows that the plausible metaeffector activity of LupA, where it counteracts the function of LegC3, is mediated by its deubiquitination ability (Urbanus et al., 2016). However, unlike other reported effector-metaeffector pairs, LupA and LegC3 are not encoded in the same locus of the *L. pneumophila* chromosome. LegC3 is a glutamine (Q)-SNARE (soluble NSF attachment protein receptor)-like protein that interacts with host arginine (R)-SNARE vesicle-associated membrane protein 4, together with two other *L. pneumophila* Q-SNAREs, the effectors LegC7/YlfA and LegC2/YlfB, to hijack host vesicular trafficking for LCV remodeling during infection (De Felipe et al., 2008; Bennett et al., 2013; Yao et al., 2014; Shi et al., 2016). Investigating a mechanism by which the DUB activity of LupA against LegC3 contributes to the manipulation of host organelle trafficking could lead to interesting results.

DISCUSSION

L. pneumophila is a unique bacterial pathogen that possesses an extraordinary number of effector proteins with complicated functional networks in the infected host cells. Recent analyses of the *Legionella* effector proteins, focusing on the regulation of the ubiquitin systems, have found not only a mimicry of eukaryotic enzymes, but also many unprecedented chemical reactions. Here, we focused on *L. pneumophila* effector proteins, which are negative regulators of the ubiquitin system. Some of the identified *L. pneumophila* DUBs have the canonical catalytic triads that characterize the OTU superfamily proteases. However, structural analyses have found distinctive features of bacterial OTUs compared to the eukaryotic DUBs. One notable highlight in the field of *Legionella*-effector research is the discovery of non-canonical chemistries of ubiquitination, namely, SidEs mART/PDE-mediated PR-ubiquitination and MavC transglutaminase-mediated ubiquitination. Accordingly, reversing enzymes have been identified among the effectors, namely DupA/B, MvcA, and SidJ. These findings provide knowledge of unexpected enzymatic activities of bacterial effector proteins. In this series of analyses, structural and MS approaches have been adopted, which are powerful and indispensable tools for the identification of novel chemical reactions.

The studies examined in this review represent paradigms of the effector/metaeffector or opposing functional relationship. It is conceivable that, by utilizing these effector sets, *Legionella* can achieve fine-tuning of the cellular environment to maximize its own growth. The exact roles of the effectors discussed here remain to be fully explored. It can be expected that the functions of *Legionella* effectors characterized thus far are just the tip of the iceberg. More effectors with functional relevance for the host ubiquitination machinery are expected to be identified in the near future, including other genes that encode possible OTU domains and PDE domains. Further analyses will elucidate the comprehensive effector networks that regulate the ubiquitin system following the stages of *Legionella* infection.

AUTHOR CONTRIBUTIONS

TKu and TKi prepared the first draft of the manuscript. TKu prepared the figures and figure legends. HN and TKi discussed the contents and revised the manuscript. TKu revised and finalized the manuscript. All authors contributed to the article and approved the submitted version.

ACKNOWLEDGMENTS

We would like to thank all members of the laboratory who took part in discussion.

REFERENCES

Akturk, A., Wasilko, D. J., Wu, X., Liu, Y., Zhang, Y., Qiu, J., et al. (2018). Mechanism of phosphoribosyl-ubiquitination mediated by a single Legionella effector. *Nature* 557, 729–733. doi: 10.1038/s41586-018-0147-6

Asrat, S., De Jesús, D. A., Hempstead, A. D., Ramabhadran, V., and Isberg, R. R. (2014). Bacterial pathogen manipulation of host membrane trafficking. *Annu. Rev. Cell Dev. Biol.* 30, 79–109. doi: 10.1146/annurev-cellbio-100913-013439

Bennett, T. L., Kraft, S. M., Reaves, B. J., Mima, J., O'brien, K. M., and Starai, V. J. (2013). LegC3, an effector protein from legionella pneumophila, inhibits homotypic yeast vacuole fusion *in vivo* and *in vitro*. *PLoS ONE* 8:e56798. doi: 10.1371/journal.pone.0056798

Bhogaraju, S., Bonn, F., Mukherjee, R., Adams, M., Pfleiderer, M. M., Galej, W. P., et al. (2019). Inhibition of bacterial ubiquitin ligases by SidJ–calmodulin catalysed glutamylation. *Nature* 572, 382–386. doi: 10.1038/s41586-019-1440-8

Bhogaraju, S., Kalayil, S., Liu, Y., Bonn, F., Colby, T., Matic, I., et al. (2016). Phosphoribosylation of ubiquitin promotes serine ubiquitination and impairs conventional ubiquitination. *Cell* 167, 1636–1649. doi: 10.1016/j.cell.2016.11.019

Black, M. H., Osinski, A., Gradowski, M., Servage, K. A., Pawlowski, K., Tomchick, D. R., et al. (2019). Bacterial pseudokinase catalyzes protein polyglutamylation to inhibit the SidE-family ubiquitin ligases. *Science* 364, 787–792. doi: 10.1126/science.aaw7446

Damgaard, R. B., and Pruneda, J. N. (2019). Legionella dismantles linear ubiquitin. *Nat. Microbiol.* 4, 1244–1245. doi: 10.1038/s41564-019-0515-5

De Felipe, K. S., Glover, R. T., Charpentier, X., Anderson, O. R., Reyes, M., Pericone, C. D., et al. (2008). Legionella eukaryotic-like type IV substrates interfere with organelle trafficking. *PLoS Pathog.* 4:e1000117. doi: 10.1371/journal.ppat.1000117

Dolinsky, S., Haneburger, I., Cichy, A., Hannemann, M., Itzen, A., and Hilbi, H. (2014). The Legionella longbeachae Icm/Dot substrate SidC selectively binds phosphatidylinositol 4-phosphate with nanomolar affinity and promotes pathogen vacuole-endoplasmic reticulum interactions. *Infect. Immun.* 82, 4021–4033. doi: 10.1128/IAI.01685-14

Dorer, M. S., Kirton, D., Bader, J. S., and Isberg, R. R. (2006). RNA interference analysis of Legionella in Drosophila cells: exploitation of early secretory apparatus dynamics. *PLoS Pathog.* 2:e34. doi: 10.1371/journal.ppat.0020034

Du, J., Fu, L., Sui, Y., and Zhang, L. (2019). The function and regulation of OTU deubiquitinases. *Front. Med.* doi: 10.1007/s11684-019-0734-4. [Epub ahead of print].

Edelmann, M. J., Iphofer, A., Akutsu, M., Altun, M., Di Gleria, K., Kramer, H. B., et al. (2009). Structural basis and specificity of human otubain 1-mediated deubiquitination. *Biochem. J.* 418, 379–390. doi: 10.1042/BJ20081318

English, A. R., Zurek, N., and Voeltz, G. K. (2009). Peripheral ER structure and function. *Curr. Opin. Cell Biol.* 21, 596–602. doi: 10.1016/j.ceb.2009.04.004

Galán, J. E., and Waksman, G. (2018). Protein-injection machines in bacteria. *Cell* 172, 1306–1318. doi: 10.1016/j.cell.2018.01.034

Gan, N., Guan, H., Huang, Y., Yu, T., Fu, J., Nakayasu, E. S., et al. (2020). Legionella pneumophila regulates the activity of UBE2N by deamidase-mediated deubiquitination. *EMBO J.* 39:e102806. doi: 10.15252/embj.2019102806

Gan, N., Nakayasu, E. S., Hollenbeck, P. J., and Luo, Z.-Q. (2019a). Legionella pneumophila inhibits immune signalling via MavC-mediated transglutaminase-induced ubiquitination of UBE2N. *Nat. Microbiol.* 4, 134–143. doi: 10.1038/s41564-018-0282-8

Gan, N., Zhen, X., Liu, Y., Xu, X., He, C., Qiu, J., et al. (2019b). Regulation of phosphoribosyl ubiquitination by a calmodulin-dependent glutamylase. *Nature* 572, 387–391. doi: 10.1038/s41586-019-1439-1

Gersch, M., Gladkova, C., Schubert, A. F., Michel, M. A., Maslen, S., and Komander, D. (2017). Mechanism and regulation of the Lys6-selective deubiquitinase USP30. *Nat. Struct. Mol. Biol.* 24, 920–930. doi: 10.1038/nsmb.3475

Guan, H., Fu, J., Yu, T., Wang, Z. X., Gan, N., Huang, Y., et al. (2020). Molecular basis of ubiquitination catalyzed by the bacterial transglutaminase MavC. *Adv. Sci.* 7:2000871. doi: 10.1002/advs.202000871

Haenssler, E., Ramabhadran, V., Murphy, C. S., Heidtman, M. I., and Isberg, R. R. (2015). Endoplasmic reticulum tubule protein reticulon 4 associates with the legionella pneumophila vacuole and with translocated substrate Ceg9. *Infect. Immun.* 83, 3479–3489. doi: 10.1128/IAI.00507-15

Havey, J. C., and Roy, C. R. (2015). Toxicity and SidJ-mediated suppression of toxicity require distinct regions in the SidE family of legionella pneumophila effectors. *Infect. Immun.* 83, 3506–3514. doi: 10.1128/IAI.00497-15

Hershko, A., and Ciechanover, A. (1998). The ubiquitin system. *Annu. Rev. Biochem.* 67, 425–479. doi: 10.1146/annurev.biochem.67.1.425

Hilbi, H., Nagai, H., Kubori, T., and Roy, C. R. (2017). Subversion of host membrane dynamics by the legionella Dot/Icm Type IV secretion system. *Curr. Top. Microbiol. Immunol.* 413, 221–242. doi: 10.1007/978-3-319-75241-9_9

Hochstrasser, M. (2009). Origin and function of ubiquitin-like proteins. *Nature* 458, 422–429. doi: 10.1038/nature07958

Hospenthal, M. K., Freund, S. M. V., and Komander, D. (2013). Assembly, analysis and architecture of atypical ubiquitin chains. 20, 555–565. doi: 10.1038/nsmb.2547

Hsu, F., Luo, X., Qiu, J., Teng, Y. B., Jin, J., Smolka, M. B., et al. (2014). The Legionella effector SidC defines a unique family of ubiquitin ligases important for bacterial phagosomal remodeling. *Proc. Natl. Acad. Sci. U.S.A.* 111, 10538–10543. doi: 10.1073/pnas.1402605111

Hubber, A., Kubori, T., and Nagai, H. (2013). Modulation of the ubiquitination machinery by Legionella. *Curr. Top. Microbiol. Immunol.* 376, 227–247. doi: 10.1007/82_2013_343

Hubber, A., and Roy, C. R. (2010). Modulation of host cell function by Legionella pneumophila type IV effectors. *Annu. Rev. Cell Dev. Biol.* 26, 261–283. doi: 10.1146/annurev-cellbio-100109-104034

Isberg, R. R., O'connor, T. J., and Heidtman, M. (2009). The Legionella pneumophila replication vacuole: making a cosy niche inside host cells. *Nat. Rev. Microbiol.* 7, 13–24. doi: 10.1038/nrmicro1967

Jeng, E. E., Bhadkamkar, V., Ibe, N. U., Gause, H., Jiang, L., Chan, J., et al. (2019). Systematic identification of host cell regulators of legionella pneumophila pathogenesis using a genome-wide CRISPR screen. *Cell Host Microbe* 26, 551–563. doi: 10.1016/j.chom.2019.08.017

Jeong, K. C., Sexton, J. A., and Vogel, J. P. (2015). Spatiotemporal regulation of a legionella pneumophila T4SS substrate by the metaeffector SidJ. *PLoS Pathog.* 11:e1004695. doi: 10.1371/journal.ppat.1004695

Kalayil, S., Bhogaraju, S., Bonn, F., Shin, D., Liu, Y., Gan, N., et al. (2018). Insights into catalysis and function of phosphoribosyl-linked serine ubiquitination. *Nature* 557, 734–738. doi: 10.1038/s41586-018-0145-8

Keusekotten, K., Paul, G.L., Berthe, R., and Kulathu, Y. (2013). OTULIN antagonizes LUBAC signaling by specifically hydrolyzing Met1-linked polyubiquitin. *Cell* 153, 1312–1326. doi: 10.1016/j.cell.2013.05.014

Kirisako, T., Kamei, K., Murata, S., Kato, M., Fukumoto, H., Kanie, M., et al. (2006). A ubiquitin ligase complex assembles linear polyubiquitin chains. *EMBO J.* 25, 4877–4887. doi: 10.1038/sj.emboj.7601360

Komander, D., Reyes-Turcu, F., Licchesi, J. D. F., Odenwaelder, P., Wilkinson, K. D., and Barford, D. (2009). Molecular discrimination of structurally equivalent Lys 63-linked and linear polyubiquitin chains. *EMBO Rep.* 10, 466–473. doi: 10.1038/embor.2009.55

Kotewicz, K. M., Ramabhadran, V., Sjoblom, N., Vogel, J. P., Haenssler, E., Zhang, M., et al. (2017). A single legionella effector catalyzes a multistep ubiquitination pathway to rearrange tubular endoplasmic reticulum for replication. *Cell Host Microbe* 21, 169–181. doi: 10.1016/j.chom.2016.12.007

Kubori, T., Hyakutake, A., and Nagai, H. (2008). Legionella translocates an E3 ubiquitin ligase that has multiple U-boxes with distinct functions. *Mol. Microbiol.* 67, 1307–1319. doi: 10.1111/j.1365-2958.2008.06124.x

Kubori, T., Kitao, T., Ando, H., and Nagai, H. (2018). LotA, a Legionella deubiquitinase, has dual catalytic activity and contributes to intracellular growth. *Cell Microbiol* 20:e12840. doi: 10.1111/cmi.12840

Kubori, T., Kitao, T., and Nagai, H. (2019). Emerging insights into bacterial deubiquitinases. *Curr. Opin. Microbiol.* 47, 14–19. doi: 10.1016/j.mib.2018.10.001

Kubori, T., and Nagai, H. (2011). Bacterial effector-involved temporal and spatial regulation by hijack of the host ubiquitin pathway. *Front. Microbiol.* 2:145. doi: 10.3389/fmicb.2011.00145

Kubori, T., and Nagai, H. (2016). The Type IVB secretion system: an enigmatic chimera. *Curr. Opin. Microbiol.* 29, 22–29. doi: 10.1016/j.mib.2015.10.001

Kubori, T., Shinzawa, N., Kanuka, H., and Nagai, H. (2010). Legionella metaeffector exploits host proteasome to temporally regulate cognate effector. *PLoS Pathogens* 6:e1001216. doi: 10.1371/journal.ppat.1001216

Li, Q., and Verma, I. M. (2002). NF-κB regulation in the immune system. *Nat. Rev. Immunol.* 2, 725–734. doi: 10.1038/nri910

Lin, D. Y. W., Diao, J., Zhou, D., and Chen, J. (2011). Biochemical and structural studies of a HECT-like ubiquitin ligase from *Escherichia coli* O157:H7. 286, 441–449. doi: 10.1074/jbc.M110.167643

Lin, Y. H., and Machner, M. P. (2017). Exploitation of the host cell ubiquitin machinery by microbial effector proteins. *J. Cell Sci.* 130, 1985–1996. doi: 10.1242/jcs.188482

Liu, S., Luo, J., Zhen, X., Qiu, J., Ouyang, S., and Luo, Z.-Q. (2020). Interplay between bacterial deubiquitinase and ubiquitin E3 ligase regulates ubiquitin dynamics on Legionella phagosomes. *bioRxiv* 2020:104554. doi: 10.1101/2020.05.19.104554

Liu, Y., and Luo, Z.-Q. (2007). The legionella pneumophila effector SidJ Is required for efficient recruitment of endoplasmic reticulum proteins to the bacterial phagosome. *Infect. Immun.* 75, 592–603. doi: 10.1128/IAI.01278-06

Luo, Z. Q., and Isberg, R. R. (2004). Multiple substrates of the Legionella pneumophila Dot/Icm system identified by interbacterial protein transfer. *Proc. Natl. Acad. Sci. U.S.A.* 101, 841–846. doi: 10.1073/pnas.0304916101

Ma, K., Zhen, X., Zhou, B., Gan, N., Cao, Y., Fan, C., et al. (2020). The bacterial deubiquitinase Ceg23 regulates the association of Lys-63-linked polyubiquitin molecules on the Legionella phagosome. *J. Biol. Chem.* 295, 1646–1657. doi: 10.1074/jbc.RA119.011758

Mevissen, T. E., Hospenthal, M. K., Geurink, P. P., Elliott, P. R., Akutsu, M., Arnaudo, N., et al. (2013). OTU deubiquitinases reveal mechanisms of linkage specificity and enable ubiquitin chain restriction analysis. *Cell* 154, 169–184. doi: 10.1016/j.cell.2013.05.046

Mevissen, T. E. T., and Komander, D. (2017). Mechanisms of deubiquitinase specificity and regulation. *Annu. Rev. Biochem.* 86, 159–192. doi: 10.1146/annurev-biochem-061516-044916

Mu, Y., Wang, Y., Huang, Y., Li, D., Han, Y., Chang, M., et al. (2020). Structural insights into the mechanism and inhibition of transglutaminase-induced ubiquitination by the Legionella effector MavC. *Nat. Commun.* 11, 1774. doi: 10.1038/s41467-020-15645-7

Nagai, H., and Kubori, T. (2011). Type IVB secretion systems of legionella and other gram-negative bacteria. *Front Microbiol* 2:136. doi: 10.3389/fmicb.2011.00136

Neunuebel, M. R., Chen, Y., Gaspar, A. H., Backlund, P. S., Yergey, A., and Machner, M. P. (2011). De-AMPylation of the small GTPase Rab1 by the pathogen legionella pneumophila. *Science* 333, 453–456. doi: 10.1126/science.1207193

Perrin, A. J., Jiang, X., Birmingham, C. L., So, N. S. Y., and Brumell, J. H. (2004). Recognition of bacteria in the cytosol of mammalian cells by the ubiquitin system. *Curr. Biol.* 14, 806–811. doi: 10.1016/j.cub.2004.04.033

Pike, C. M., Boyer-Andersen, R., Kinch, L. N., Caplan, J. L., and Neunuebel, M. R. (2019). Legionella effector RavD binds phosphatidylinositol-3-phosphate and helps suppress endolysosomal maturation of the Legionella-containing vacuole. *J. Biol. Chem.* 294, 6405–6415. doi: 10.1074/jbc.RA118.007086

Price, C. T., Al-Khodor, S., Al-Quadan, T., Santic, M., Habyarimana, F., Kalia, A., et al. (2009). Molecular mimicry by an F-Box effector of legionella pneumophila hijacks a conserved polyubiquitination machinery within macrophages and protozoa. *PLoS Pathog.* 5:e1000704. doi: 10.1371/journal.ppat.1000704

Price, C. T., Al-Quadan, T., Santic, M., Rosenshine, I., and Abu Kwaik, Y. (2011). Host proteasomal degradation generates amino acids essential for intracellular bacterial growth. *Science* 334, 1553–1557. doi: 10.1126/science.1212868

Pruneda, J. N., Durkin, C. H., Geurink, P. P., Ovaa, H., Santhanam, B., Holden, D. W., et al. (2016). The molecular basis for ubiquitin and ubiquitin-like specificities in bacterial effector proteases. *Mol Cell* 63, 261–276. doi: 10.1016/j.molcel.2016.06.015

Puvar, K., Iyer, S., Fu, J., Kenny, S., Negrón Terón, K. I., Luo, Z.-Q., et al. (2020). Legionella effector MavC targets the Ube2N~Ub conjugate for noncanonical ubiquitination. *Nat. Commun.* 11. doi: 10.1038/s41467-020-16211-x

Qiu, J., and Luo, Z.-Q. (2017a). Hijacking of the host ubiquitin network by legionella pneumophila. *Front. Cell. Infect. Microbiol.* 7:487. doi: 10.3389/fcimb.2017.00487

Qiu, J., and Luo, Z. Q. (2017b). Legionella and Coxiella effectors: strength in diversity and activity. *Nat. Rev. Microbiol.* 15, 591–605. doi: 10.1038/nrmicro.2017.67

Qiu, J., Sheedlo, M. J., Yu, K., Tan, Y., Nakayasu, E. S., Das, C., et al. (2016). Ubiquitination independent of E1 and E2 enzymes by bacterial effectors. *Nature* 533, 120–124. doi: 10.1038/nature17657

Qiu, J., Yu, K., Fei, X., Liu, Y., Nakayasu, E. S., Piehowski, P. D., et al. (2017). A unique deubiquitinase that deconjugates phosphoribosyl-linked protein ubiquitination. *Cell Res.* 27, 865–881. doi: 10.1038/cr.2017.66

Ragaz, C., Pietsch, H., Urwyler, S., Tiaden, A., Weber, S. S., and Hilbi, H. (2008). The Legionella pneumophila phosphatidylinositol-4 phosphate-binding type IV substrate SidC recruits endoplasmic reticulum vesicles to a replication-permissive vacuole. *Cell Microbiol* 10, 2416–2433. doi: 10.1111/j.1462-5822.2008.01219.x

Rahighi, S., Ikeda, F., Kawasaki, M., Akutsu, M., Suzuki, N., Kato, R., et al. (2009). Specific recognition of linear ubiquitin chains by NEMO is important for NF-κB activation. *Cell* 136, 1098–1109. doi: 10.1016/j.cell.2009.03.007

Schubert, A. F., Nguyen, J. V., Franklin, T. G., Geurink, P. P., Roberts, C. G., Sanderson, D. J., et al. (2020). Identification and characterization of diverse OTU deubiquitinases in bacteria. *EMBO J.* 39:e105127. doi: 10.15252/embj.2020105127

Shames, S. R., Liu, L., Havey, J. C., Schofield, W. B., Goodman, A. L., and Roy, C. R. (2017). Multiple Legionella pneumophila effector virulence phenotypes revealed through high-throughput analysis of targeted mutant libraries. *Proc. Natl. Acad. Sci. U.S.A.* 114, 10446–10454. doi: 10.1073/pnas.1708553114

Sheedlo, M. J., Qiu, J., Tan, Y., Paul, L. N., Luo, Z.-Q., and Das, C. (2015). Structural basis of substrate recognition by a bacterial deubiquitinase important for dynamics of phagosome ubiquitination. *Proc. Natl. Acad. Sci. U.S.A.* 112, 15090–15095. doi: 10.1073/pnas.1514568112

Sheng, X., You, Q., Zhu, H., Chang, Z., Li, Q., Wang, H., et al. (2017). Bacterial effector NleL promotes enterohemorrhagic E. coli-induced attaching and effacing lesions by ubiquitylating and inactivating JNK. *PLOS Pathogens* 13:e1006534. doi: 10.1371/journal.ppat.1006534

Shi, X., Halder, P., Yavuz, H., Jahn, R., and Shuman, H. A. (2016). Direct targeting of membrane fusion by SNARE mimicry: convergent evolution of Legionella effectors. *Proc. Natl. Acad. Sci. U.S.A.* 113, 8807–8812. doi: 10.1073/pnas.1608755113

Shin, D., Mukherjee, R., Liu, Y., Gonzalez, A., Bonn, F., Liu, Y., et al. (2020b). Regulation of phosphoribosyl-linked serine ubiquitination by deubiquitinases DupA and DupB. *Mol. Cell* 77, 164–179. doi: 10.1016/j.molcel.2019.10.019

Shin, D., Bhattacharya, A., Cheng, Y.-L., Alonso, M. C., Mehdipour, A. R., Noort, G. J. V. D. H. V., et al. (2020a). Novel class of OTU deubiquitinases regulate substrate ubiquitination upon Legionella infection. *bioRxiv.* doi: 10.1101/2020.04.25.060954

Sreelatha, A., Yee, S. S., Lopez, V. A., Park, B. C., Kinch, L. N., Pilch, S., et al. (2018). Protein AMPylation by an evolutionarily conserved pseudokinase. *Cell* 175, 809–821. doi: 10.1016/j.cell.2018.08.046

Starr, T., Sun, Y., Wilkins, N., and Storrie, B. (2010). Rab33b and Rab6 are functionally overlapping regulators of golgi homeostasis and trafficking. *Traffic* 11, 626–636. doi: 10.1111/j.1600-0854.2010.01051.x

Steiner, B., Swart, A. L., Welin, A., Weber, S., Personnic, N., Kaech, A., et al. (2017). ER remodeling by the large GTPase atlastin promotes vacuolar growth of Legionella pneumophila. *EMBO Rep.* 18, 1817–1836. doi: 10.15252/embr.201743903

Sulpizio, A., Minelli, M. E., Wan, M., Burrowes, P. D., Wu, X., Sanford, E. J., et al. (2019). Protein polyglutamylation catalyzed by the bacterial calmodulin-dependent pseudokinase SidJ. *eLife* 8:e51162. doi: 10.7554/eLife.51162

Swatek, K. N., and Komander, D. (2016). Ubiquitin modifications. *Cell Res.* 26, 399–422. doi: 10.1038/cr.2016.39

Tan, Y., Arnold, R. J., and Luo, Z. Q. (2011). Legionella pneumophila regulates the small GTPase Rab1 activity by reversible phosphorylcholination. *Proc. Natl. Acad. Sci. U.S.A.* 108, 21212–21217. doi: 10.1073/pnas.1114023109

Tan, Y., and Luo, Z.-Q. (2011). Legionella pneumophila SidD is a deAMPylase that modifies Rab1. *Nature* 475, 506–509. doi: 10.1038/nature10307

Urbanus, M. L., Quaile, A. T., Stogios, P. J., Morar, M., Rao, C., Di Leo, R., et al. (2016). Diverse mechanisms of metaeffector activity in an intracellular bacterial pathogen, Legionella pneumophila. *Mol. Syst. Biol.* 12:893. doi: 10.15252/msb.20167381

Valleau, D., Quaile, A. T., Cui, H., Xu, X., Evdokimova, E., Chang, C., et al. (2018). Discovery of ubiquitin deamidases in the pathogenic arsenal of legionella pneumophila. *Cell Rep.* 23, 568–583. doi: 10.1016/j.celrep.2018.03.060

Wan, M., Sulpizio, A. G., Akturk, A., Beck, W. H. J., Lanz, M., Faca, V. M., et al. (2019a). Deubiquitination of phosphoribosyl-ubiquitin conjugates by phosphodiesterase-domain-containing Legionella effectors. *Proc. Natl. Acad. Sci. U.S.A.* 116, 23518–23526. doi: 10.1073/pnas.1916287116

Wan, M., Wang, X., Huang, C., Xu, D., Wang, Z., Zhou, Y., et al. (2019b). A bacterial effector deubiquitinase specifically hydrolyses linear ubiquitin chains to inhibit host inflammatory signalling. *Nat. Microbiol.* 4, 1282–1293. doi: 10.1038/s41564-019-0454-1

Wang, G., Gao, Y., Li, L., Jin, G., Cai, Z., Chao, J.-I., et al. (2012). K63-Linked ubiquitination in kinase activation and cancer. *Front Oncol.* 2:5. doi: 10.3389/fonc.2012.00005

Wang, T., Yin, L., Cooper, E. M., Lai, M.-Y., Dickey, S., Pickart, C. M., et al. (2009). Evidence for bidentate substrate binding as the Basis for the K48 linkage specificity of otubain 1. *J. Mol. Biol.* 386, 1011–1023. doi: 10.1016/j.jmb.2008.12.085

Wasilko, D. J., Huang, Q., and Mao, Y. (2018). Insights into the ubiquitin transfer cascade catalyzed by the Legionella effector SidC. *eLife* 7:e36154. doi: 10.7554/eLife.36154

Wolff, J., Cook, G. H., Goldhammer, A. R., and Berkowitz, S. A. (1980). Calmodulin activates prokaryotic adenylate cyclase. *Proc. Natl. Acad. Sci. U.S.A.* 77, 3841–3844. doi: 10.1073/pnas.77.7.3841

Yao, D., Cherney, M., and Cygler, M. (2014). Structure of the N-terminal domain of the effector protein LegC3 from Legionella pneumophila. *Acta Crystallogr. D Biol. Crystallogr.* 70, 436–441. doi: 10.1107/S13990047130 2991X

Yao, Q., Cui, J., Zhu, Y., Wang, G., Hu, L., Long, C., et al. (2009). A bacterial type III effector family uses the papain-like hydrolytic activity to arrest the host cell cycle. *Proc. Natl. Acad. Sci. U.S.A.* 106, 3716–3721. doi: 10.1073/pnas.0900212106

Zusman, T., Aloni, G., Halperin, E., Kotzer, H., Degtyar, E., Feldman, M., et al. (2007). The response regulator PmrA is a major regulator of the icm/dot type IV secretion system in Legionella pneumophila and Coxiella burnetii. *Mol. Microbiol.* 63, 1508–1523. doi: 10.1111/j.1365-2958.2007.05604.x

14

Interesting Biochemistries in the Structure and Function of Bacterial Effectors

Hazel Mak and Teresa L. M. Thurston *

MRC Centre for Molecular Bacteriology and Infection, Imperial College London, London, United Kingdom

Correspondence:
Teresa L. M. Thurston
t.thurston@imperial.ac.uk

Bacterial effector proteins, delivered into host cells by specialized multiprotein secretion systems, are a key mediator of bacterial pathogenesis. Following delivery, they modulate a range of host cellular processes and functions. Strong selective pressures have resulted in bacterial effectors evolving unique structures that can mimic host protein biochemical activity or enable novel and distinct biochemistries. Despite the protein structure-function paradigm, effectors from different bacterial species that share biochemical activities, such as the conjugation of ubiquitin to a substrate, do not necessarily share structural or sequence homology to each other or the eukaryotic proteins that carry out the same function. Furthermore, some bacterial effectors have evolved structural variations to known protein folds which enable different or additional biochemical and physiological functions. Despite the overall low occurrence of intrinsically disordered proteins or regions in prokaryotic proteomes compared to eukaryotes proteomes, bacterial effectors appear to have adopted intrinsically disordered regions that mimic the disordered regions of eukaryotic signaling proteins. In this review, we explore examples of the diverse biochemical properties found in bacterial effectors that enable effector-mediated interference of eukaryotic signaling pathways and ultimately support pathogenesis. Despite challenges in the structural and functional characterisation of effectors, recent progress has been made in understanding the often unusual and fascinating ways in which these virulence factors promote pathogenesis. Nevertheless, continued work is essential to reveal the array of remarkable activities displayed by effectors.

Keywords: bacterial effectors, secretion systems, structure-function, pathogenesis, protein organization, host-pathogen

INTRODUCTION

The pathogenesis of many Gram-negative bacteria is highly dependent on specialized multiprotein machines that deliver a repertoire of bacterial effectors in a spatiotemporally coordinated manner into the host cell, where they modulate a range of eukaryotic cellular processes. These specialized multiprotein machines are known as secretion systems. Of the seven known secretion systems found in Gram-negative bacteria, the direct delivery of effectors across a host cell membrane can only be achieved by the type III, type IV, and type VI secretion systems (T3SS, T4SS, and T6SS respectively) (Galán and Waksman, 2018). The type VII secretion system present in Gram-positive bacteria and mycobacteria will not be discussed here. Through horizontal gene transfer (HGT) and selective pressure from the host(s), pathogenic bacteria have acquired bacterial effectors, mainly encoded on pathogenicity islands and virulence plasmids. The function of bacterial effectors is varied but broadly they promote bacterial invasion and colonisation of host cells, as well as bacterial survival, growth, and replication. Other key effector functions include modulating host immune signaling and establishing a bacteria-beneficial niche within the host. This alters the relationship between pathogen and host, with bacterial effectors reprogramming complex eukaryotic processes to promote a parasitic relationship, where the pathogen is supported by its host. Parasitic relationships result in injury, disease, and potentially death of the host. However, host cells detect pathogen-associated molecular patterns (PAMPs) from invading organisms, and this activates a cascade of pro-inflammatory signaling and defence mechanisms that can protect the host from the invading pathogen (Janeway, 1989). In addition, effector-triggered immunity (ETI), which is well characterized in plant cells, is now reported in metazoans and refers to the initiation of a protective immune response upon detection of bacterial toxins, secreted proteins or the detection of their activities (Lopes Fischer et al., 2020). In this way, ETI provides another layer of immune defence that detects pathogen manipulation of key cellular processes, including the subversion of host immune responses. This refocuses the relationship from being parasitic to one where the host gains immunity, in order to resist and clear the pathogen. As a result of adaption and the evolution of this bacteria-host relationship, different bacterial species have acquired specific repertoires of effectors. Nevertheless, common themes, related to effector structure and function, exist among effectors from diverse pathogens.

In general, bacterial proteins that are not secreted, have an individual biochemical activity associated to a physiologically relevant function and structure within the bacteria. This follows the structure-function paradigm, where the function of a protein is directly related to its three-dimensional structure. However, effectors are distinct from other bacterial proteins, as they primarily function and exert their biochemical activity within the target cell, rather than within the bacteria. Of the different secretion systems that deliver bacterial proteins across a mammalian cell membrane, the T6SS is the most recently identified and still remains poorly described in terms of effector delivery and the action of T6SS effectors within eukaryotic host cells. Therefore, we will not discuss T6SS effectors further here. Prior to effector delivery through the multiprotein T3SS, many but not all T3SS effectors are chaperoned to the base of the secretion system. Here, they are secreted in an ATP-dependent, unfolded or partially folded and inactive state, which allows passage through the narrow secretion system tunnel (Radics et al., 2014; Dohlich et al., 2014). Once delivered into a eukaryotic host cell, the folding of effectors into their active conformation may or may not require host proteins and additional, host-mediated, post-translational modifications. This supports a hypothesis whereby T3SS effectors primarily function within host cells. However, there is at least one example describing effector catalytic activity within the pathogen (Qaidi et al., 2020). The T3SS effector and *N*-acetylglucosamine transferase, NleB, from pathogenic *Escherichia coli*, modifies bacterial glutathione synthetase (GshB) to promote GshB activity and bacterial survival to oxidative stress (Qaidi et al., 2020). Given this unexpected observation, further experiments should investigate how widespread this phenomenon is.

T4SSs are versatile systems capable of secreting protein and DNA into target cells that include other bacteria and eukaryotic cells. Relevant to this review is the delivery of proteins through the T4SS into host eukaryotic cells from pathogens such as *Legionella pneumophila* and *Helicobacter pylori*. The translocation of T4SS effectors is similar to T3SS effectors; most effectors are unfolded or partially folded and in complex with a chaperone for secretion (Costa et al., 2015; Sgro et al., 2019). There are also examples of folded proteins that need to be unfolded for T4SS translocation (Trokter and Waksman, 2018).

Secretion systems have evolved to deliver bacterial effectors in a spatiotemporally regulated and coordinated manner (Selkrig et al., 2020), which enables effectors to work in concert with each other. For example, the *Salmonella* T3SS effectors SseF and SseG, function together to anchor *Salmonella*-containing vacuoles (SCV) to the Golgi Network (Yu et al., 2016) and the global mapping of *Salmonella*-host protein-protein interactions revealed that SseJ and SseL collaborate in order to redirect cholesterol to the SCV (Walch et al., 2020). Effectors working in opposition to each other have also been described, for example, *Legionella* LubX targets the bacterial effector SidH for degradation *via* the host proteasome in the later stages of infection and this finding gave rise to the term "meta-effectors", or "effectors of effectors" (Kubori et al., 2010; Urbanus et al., 2016). Another key difference between bacterial proteins and secretion system effectors is that many effectors have more than one host cellular target and hence may have multiple biochemical activities and biological functions (Galán, 2009; Walch et al., 2020).

The relatively low concentration of many effectors within host cells is likely to drive the evolution of enzymatic activities, yet some effectors appear to function as adaptors. Many effectors that lack their own enzymatic activity function by recruiting and redirecting host enzymes to indirectly modify target protein(s) and modulate host cell signaling (Ohlson et al., 2008; Bayer-Santos et al., 2016; Panagi et al., 2020). Like eukaryotes, prokaryotes exploit the use of post translational modifications (PTMs) to increase the functional diversity of their proteome in a dynamic way. However, the repertoire of PTMs is divergent to that found in eukaryotes. For example, prokaryotes lack the full array of enzymes required for the conjugation of ubiquitin to target proteins. Nevertheless, bacteria have evolved to exploit host machinery to carry out ubiquitination *via* bacterial E3 ligase effectors as well as various other PTMs that are not required for the regulation of bacterial physiology.

In this review, we will examine common themes of effectors from diverse bacterial species in terms of structure and biochemical activity. We will take examples from several pathogens that utilize type III and type IV secretion systems yet acknowledge that we are unable to review the vast array of effector-mediated functions and biochemical activities. The first group of effectors we will consider are those that have adopted structural similarities to eukaryotic proteins that enable them to mimic the biochemical activities of host proteins. For example, effectors that act as proteases, phosphatases, kinases, glycosylases, and more, have been described. Other bacterial effectors have unique structures that lack homology to our current knowledge of eukaryotic proteins and have interesting biochemical activities that perform unique and alternative PTMs and functions in comparison to those exhibited by normal eukaryotic processes. The final structural property of bacterial effectors we will consider is the occurrence of intrinsically disordered regions (IDRs). In general, prokaryotic proteomes show a low degree of IDRs when compared to eukaryotes (Dunker et al., 2000). However, bacterial effectors seem comparatively enriched with IDRs that mimic those found in mammalian proteins (Iakoucheva et al., 2002; Marín et al., 2012). As described below, these IDRs are likely to mediate specific host-pathogen protein-protein interactions. Together, these structural aspects enhance the potency of the effector. Enzymatic mimicry and novel biochemistry are unlikely to be directly inhibited or reversed by the host and precise protein-protein interactions ensure a high degree of specificity for effector activity.

We will end our review by highlighting some of the current challenges in characterising the structure and function of effectors in different bacterial species as well as the advances in experimental techniques that may be used to improve our knowledge and characterization of bacterial effectors. Understanding the structures and functions of diverse effectors improves our understanding of the mechanisms that drive bacterial pathogenesis. Furthermore, uncovering unique biochemical mechanisms, which appear to be absent from normal host cell biology, provides potentially new targets for the development of antimicrobials that will not interfere with host biochemistry.

STRUCTURED "ORDERED" BACTERIAL EFFECTORS MIMICKING HOST PROTEIN FUNCTION

A large majority of characterized bacterial effectors are well-defined, structured, and ordered proteins with a stable, fixed three-dimensional structure that influences the effector function. Within this group of structured effectors, there are numerous examples where effectors have evolved to mimic the biochemical activity or structural properties of host cell proteins without significant sequence or structural homology to any particular host protein. The use of eukaryotic-like domains to mimic endogenous cellular proteins could represent a selective advantage for bacteria as the activity of these effectors might not be directly inhibited by host proteins and the diversification in protein structure might also result in additional physiological functions that enhance the virulence potential of the effector. In addition, bacterial effectors that fine-tune host cell processes using eukaryotic-like biochemistry might promote the silent manipulation of host cell signaling without triggering ETI. Finally, it is interesting to consider that some effectors have also evolved that mimic eukaryotic-like protein biochemistry but act towards other bacterial proteins as well as host proteins. The *Legionella* meta-effector and E3 ubiquitin ligase, LubX, acts to spatiotemporally regulate the activity of the effector SidH (Kubori et al., 2010). LubX contains two U-box domains, one that serves as an E2-binding site and a second U-box that functions as a substrate binding site. In this way, LubX has evolved to exploit the host ubiquitination machinery and proteasome in order to regulate one of its own effectors within host cells (Kubori et al., 2010).

The *Salmonella* T3SS effector SopA and the Enterohemorrhagic *E. coli* (EHEC) effector NleL represent homologous proteins that both exhibit ubiquitin ligase activity. SopA and NleL, which do not share any sequence homology, contain some structural similarities to each other and the host cell protein domain, eukaryotic E3 ubiquitin ligase homologous to E6-AP carboxy terminus (HECT), which mediates the addition of ubiquitin on to target proteins (Zhang et al., 2006; Lin et al., 2011). Crystal structures of both SopA and NleL show the distinguishing bi-lobal structure of HECT domains (Lin et al., 2012) (**Figure 1A**). Both SopA and NleL also contain a conserved C-terminal region cysteine residue that is required to form the thioester-linked intermediate prior to ubiquitin transfer to the substrate. The N-lobe contains the E2-binding site and is attached to a structurally flexible C-lobe in SopA and NleL (**Figures 1B, C**). Similar to the HECT domain, the structural flexibility between N- and C-lobes most likely enables SopA and NleL to interact with E2 ubiquitin-conjugating enzymes and to ubiquitinate target host proteins. Through molecular mimicry, both SopA and NleL bind the canonical surface of the E2 UbcH7, hijacking the host ubiquitination machinery despite showing little similarity to the E2-interacting surface of eukaryotic HECT E3 ligases (Lin et al., 2012). In comparison to the HECT domain, there is an additional N-terminal β-helix domain in SopA and NleL (**Figure 1C**). While functionally uncharacterized, this β-helix domain may act as a substrate binding site (Fiskin et al., 2017). The lack of sequence similarity between SopA and NleL results in differing molecular surfaces which might explain the functional differences observed (Diao et al., 2008; Lin et al., 2012). SopA modulates *Salmonella*-induced intestinal inflammation and stimulation of transepithelial migration of polymorphonuclear leucocytes (Lin et al., 2011; Kamanova et al., 2016). SopA appears to function through interaction with and ubiquitination of TRIM56 and TRIM65, however the precise molecular mechanism remains controversial (Kamanova et al., 2016; Fiskin et al., 2017). In contrast, NleL inhibits formation of actin membrane protrusions, called pedestals, on the surface of host cells by the EHEC effector Tir by an unknown mechanism (Piscatelli et al., 2011). Alternatively, NleL might promote EHEC adherence by targeting host c-Jun NH2-terminal kinases (JNKs) (Sheng et al., 2017) and overexpressed NleL inhibits NF-κB signaling (Sheng et al., 2020). This highlights that although effectors might share structural and biochemical similarities, unique physiological functions are likely to have evolved.

Another example of effectors exploiting molecular mimicry is the family of effector zinc metalloproteases. This family consists

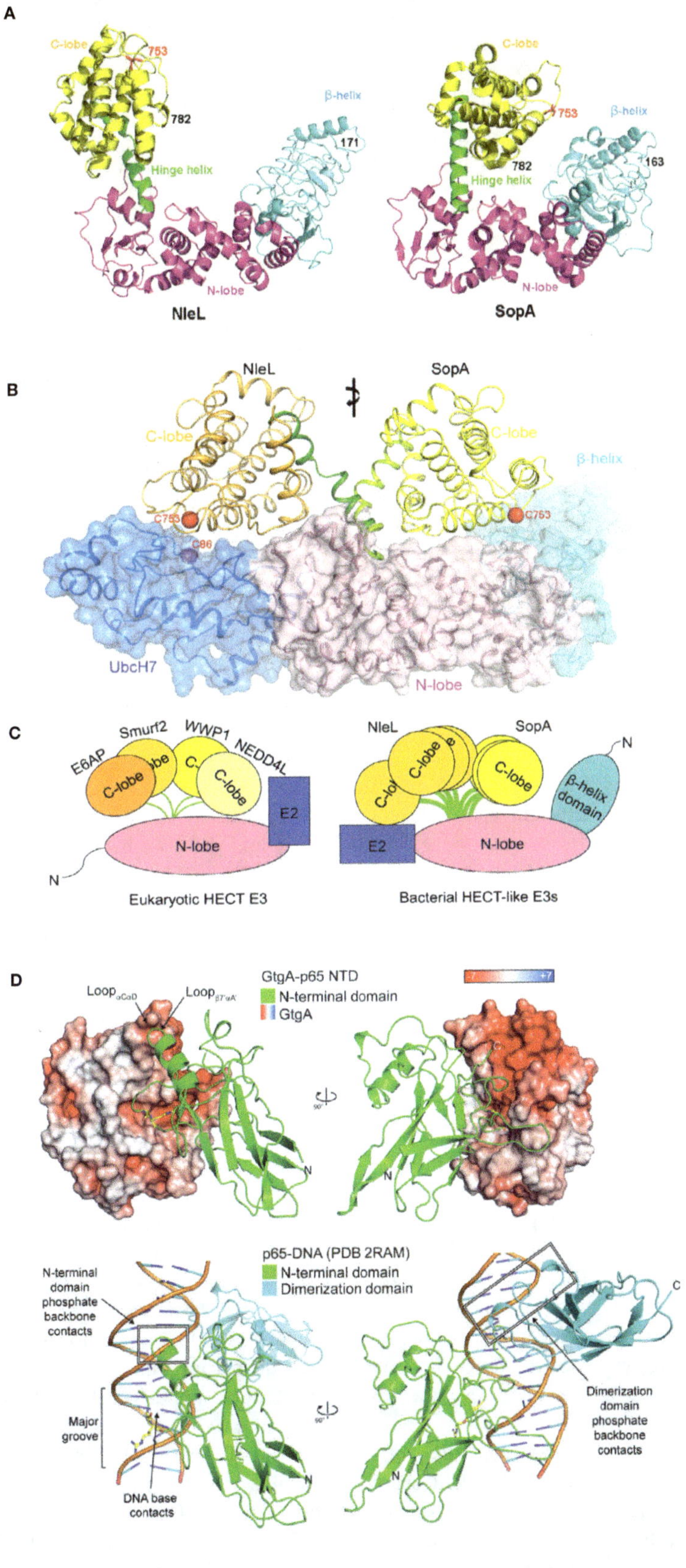

FIGURE 1 | Continued

FIGURE 1 | Structured bacterial effectors mimicking host cell proteins. **(A)** Structural comparison between Enterohemorrhagic *E. coli* (EHEC) effector NleL (residue 170-782) and *Salmonella* effector SopA (residue 163-782) in ribbon cartoon representation. Structures consist of the N-lobe (magenta), the C-lobe (yellow) and the β-helix domain (cyan). The catalytic cysteine (Cys) residue 753 are labelled in red. The hinge helix is labelled and shown in green (taken from Lin et al., 2011). **(B)** Structural superposition of two bacterial HECT-like E3 ligases, SopA from *Salmonella* and NleL from EHEC, bound to human E2 protein UbcH7 (shown in dark blue). The E3 ligase N-lobes and β-helix domains are shown in pink and light blue respectively, as ribbon representation and transparent surface. The C-lobe is structurally flexible and shown in orange for NleL and in yellow for SopA. The catalytic cysteine (Cys) residues are shown in red (taken from Lin et al., 2012). **(C)** Schematic of structural comparison between eukaryotic HECT E3 ligases and bacterial HECT-like E3 ligases. Eukaryotic HECT E3 ligases include E6AP, Smurf2, WWP1, and NEDD4L. Bacterial HECT-like E3 ligases include SopA from *Salmonella* and NleL from EHEC. Structural flexibility is shown in the C-lobe of NleL and SopA (taken from Lin et al., 2012). **(D)** Structural mimicry of DNA by *Salmonella* effector GtgA. Top structure shows GtgA in complex with the N-terminal domain (NTD) of p65. Bottom structure shows DNA in complex with the NTD and the dimerization domain of p65. GtgA is shown in surface representation and coloured according to its electrostatic surface potential (red is negative; white is neutral; blue is positive). The NTD of p65 is shown in green and the dimerization domain of p65 is shown in cyan. The cleavage site residues in p65 (Gly-40/Arg-41) are shown with yellow sticks (taken from Jennings et al., 2018).

of GtgA, GogA, and PipA from *Salmonella enterica*, NleC, and NleD from enteropathogenic *E. coli* and EHEC and RipAX2 from *Ralstonia solanacearum*, which all contain a conserved short metal binding HEXXH motif essential for catalytic activity. Although these proteins do not share significant sequence homology to known host zinc metalloproteases, they structurally retain the catalytic core of the Zincin superfamily (Jennings et al., 2018). Due to the diverse sequence homology found within the family, its members target different host proteins. NleD directly cleaves mitogen-activated protein kinases (MAPKs), JNK, and p38, in the flexible activation loop, thereby inhibiting activator protein-1 (AP-1)-dependent gene transcription and the JNK-dependent apoptosis (Baruch et al., 2011; Gur-Arie et al., 2020). In contrast, GtgA, GogA, PipA, and NleC specifically and directly cleave subunits of NF-κB to suppresses host pro-inflammatory immune responses (Baruch et al., 2011; Sun et al., 2016; Jennings et al., 2018) and NleC also cleaves and degrades the host acetyltransferase and transcriptional coactivator, p300 (Shames et al., 2011). Despite superficial similarity in targeting NF-κB subunits, GtgA, GogA, and PipA only cleave p65, RelB and cRel, whereas NleC can also hydrolyse p105/p50 and p100/p52 (Jennings et al., 2018). This molecular specificity arises from different cleavage sites, with GtgA, GogA, and PipA cleaving p65 between Gly40 and Arg41 (Sun et al., 2016). Arg41, in the P1' position, is conserved in p65, RelB and cRel and is accommodated by a negatively charged pocket within GtgA, but p105/p50 and p100/p52 encode a proline at the corresponding residue which prevents cleavage (Jennings et al., 2018). In contrast, NleC cleaves p65 between residue Cys-38 and Glu-39 and the P1' residue is conserved in all five NF-κB subunits. Despite these differences, both NleC and GtgA target NF-κB subunits through mimicry of the major groove of DNA (**Figure 1D**), which represents the normal binding target for nuclear NF-κB (Turco and Sousa, 2014; Jennings et al., 2018). Therefore, GtgA, GogA, PipA, and NleC show two forms of structural mimicry; first, functionally they act as zinc metalloproteases without sharing significant sequence homology to other known zinc metalloproteases and second, they mimic DNA in order to mediate substrate binding.

Although many bacterial effectors do not resemble eukaryotic host cell proteins in their overall structure, they might share short sequence motifs that are found in both eukaryotic proteins and effectors from different bacterial species. For example, a group of T3SS effectors that arose from convergent evolution share a conserved tryptophan (W)-xxx-glutamine (E) motif, which is found among effectors from diverse species and among TIR (Toll/Interleukin-1 receptor)-domain containing eukaryotic proteins (Felix et al., 2014). The WxxxE family of T3SS effectors include *Shigella* effectors IpgB1 and IpgB2, *Salmonella* effectors SifA and SifB (Alto et al., 2006), and EPEC and EHEC effectors Map (Kenny et al., 2002), EspM (Arbeloa et al., 2008), and EspT (Bulgin et al., 2009). In addition, despite not containing a WxxxE motif, *Salmonella* effectors SopE and SopE2, and BopE from *Burkholderia* share similar 3D structures to the WxxxE effectors and are therefore grouped into a larger family of effectors, known as the WxxxE effector and SopE-like family (Bulgin et al., 2010). Within this family of effectors, the WxxxE motif appears to have a structural role in maintaining the conformation of the putative catalytic loop, which mediates intrinsic guanine nucleotide exchange factor (GEF) activity towards Rho GTPases (Felix et al., 2014). Mechanistically, GDP to GTP exchange appears to mimic the "push and pull" mechanism exhibited by certain eukaryotic Rho GTPase GEFs. That is, interactions between the effector catalytic motif with the switch I and switch II regions on the target Rho GTPase lead to a conformational change that ejects GDP. Functionally, this effector-mediated manipulation of Rho GTPases controls host actin dynamics, with each effector showing specificity for different GTPases that mediate differential function (Bulgin et al., 2010). Interestingly, although SifA contains the conserved WxxxE motif in its C-terminal domain, SifA lacks the catalytic residues in the putative catalytic loop required for GEF activity, and does not show GEF activity *in vivo* (Ohlson et al., 2008). Instead, the N-terminal domain and C-terminal domain of SifA interact with protein partners independently, suggesting that SifA may have evolved from a GEF to an adaptor protein related to GTPase activity (Ohlson et al., 2008; Jennings et al., 2017). As in the above examples, the study of the WxxxE/SopE-like effectors illustrates how functional mimicry, in this case Rho GTPase GEF activity, is achieved without structural homology to host enzymes.

Of note, some bacterial effectors share modular sequence homology with diverse effectors and have more than one biochemical activity. For example, the *Salmonella* effector SptP contains two biochemical activities; the N-terminal domain contains a GTPase-activating protein (GAP) domain that is similar to YopE of *Yersinia* and ExoS of *Pseudomonas aeruginosa*, whereas the C-terminal domain shows sequence similarity to the protein tyrosine phosphatase YopH of *Yersinia* (Zhou and Galán, 2001). Both the GAP and tyrosine phosphatase activity contribute to SptP inhibiting Raf activation and the subsequent ERK MAPK signaling pathway (Lin et al., 2003). This dual activity is key in promoting proinflammatory cytokine release and dampens innate immune signaling and pathogen clearance in the host. Such dual activity is rare among non-secreted and non-virulent bacterial proteins.

In this section, we have highlighted examples whereby bacterial effectors with ordered structures mimic host cell protein biochemical activity and function but in the absence of significant sequence similarity. Next, we will consider effectors that mediate eukaryotic-like covalent modification through entirely novel protein folds as well as previously unseen post-translational modifications that have not been described in the study of eukaryotic biochemistry.

STRUCTURED "ORDERED" BACTERIAL EFFECTORS WITH NEW PROTEIN FOLDS AND BIOCHEMISTRY

In recent years, an array of distinct and novel biochemical mechanisms that are catalyzed by bacterial effectors, but seemingly not eukaryotic proteins, have been identified. This may be advantageous to the pathogen as the effectors and/or their modified host targets are less likely to be controlled by host feedback loops and/or regulatory proteins and evolution of host resistance mechanisms is likely to require significant time. These PTMs manipulate host cell signaling and cause detrimental downstream effects to host responses.

Effector-Mediated Ubiquitination and Phosphorylation

Protein ubiquitination is key in regulating many eukaryotic (but not prokaryotic) cellular processes. Interestingly however, bacteria have evolved different types of effector biochemistry that uniquely mediate, target and modify eukaryotic protein ubiquitination. The first example we will consider is the evolution of a family of "novel E3 ligases" (NELs) found among effectors that are structurally unique from mammalian E3 ligases (**Figure 2A**). This diversifies the repertoire of ubiquitin ligases that target host proteins, in addition to the bacterial E3 ligases mimicking eukaryotic E3 ligase domains (Ashida and Sasakawa, 2017). The NEL family contains members from at least six pathogenic bacteria, including *Salmonella* effectors SlrP (Bernal-Bayard and Ramos-Morales, 2009), SspH1 and SspH2 (Haraga and Miller, 2006; Quezada et al., 2009; Levin et al., 2010; Keszei et al., 2014), YopM of *Yersinia* (Soundararajan et al., 2011), and the IpaH effector family from *Shigella* (Singer et al., 2008). Although structurally unlike known eukaryotic E3 ligases, NELs show three key similarities that support ubiquitination. Similar to eukaryotic HECT E3 ligases, NELs contain a conserved catalytic cysteine residue that forms the ubiquitin thioester intermediate prior to ubiquitin transfer onto the substrate. NEL domains also contain a potential E2-interacting surface (Quezada et al., 2009), which enables these NEL effectors to bind to host E2 enzymes charged with ubiquitin, and compete with host E3 ligase proteins. Finally, representing a defining point of NEL family effectors is the presence of a canonical leucine-rich repeat (LRR) domain (**Figure 2B**) that interacts with the NEL domain to form an autoinhibitory fold. This prevents premature activation of the ligase, providing exquisite control of effector activity and might also prevent cellular toxicity induced by the NEL domain, which when expressed alone is toxic (Quezada et al., 2009; Chou et al., 2012; Keszei et al., 2014). This highly specific target bound activity might enable NEL effectors to limit the degree of ETI in mammalian and plant-adapted pathogens. Mechanistically, structural studies reveal that target binding causes the NEL effector to undergo a substantial conformational change, exposing the catalytic site. In addition, the variable length of the LRR domain enables the recognition of a range of different host targets, supporting target diversification (Quezada et al., 2009). Whether NEL effector activity is detected and/or regulated by host proteins remains to be determined, but their unique structural properties represent an opportunity for the development of inhibitors that specifically target the bacterial virulence factors without affecting host E3 ligases.

In contrast to NEL ligases mediating ubiquitination through hijacking of host machinery, *Legionella* effector MavC uses a remarkable E1-independent ubiquitin ligation method to block eukaryotic ubiquitination mediated *via* a specific E2 protein, Ube2N. Although MavC has some structural and functional similarities to the bacterial deamidases Cif from EPEC and CHBP from *Burkholderia pseudomallei*, MavC also contains a unique "insertion" domain which recognizes and interacts with the Ube2N-ubiquitin conjugate (**Figure 2C**) (Yao et al., 2009; Cui et al., 2010; Valleau et al., 2018; Puvar et al., 2020). Rather than exhibiting deamidase activity, MavC catalyzes an intramolecular covalent transglutamination reaction between ubiquitin (Ub) and the host E2 Ube2N, resulting in a γ-glutamyl-ϵ-Lys (Gln40^{Ub}-Lys92^{Ube2N}) isopeptide crosslink. Functionally, this inactivates the E2-ubiquitin conjugate in the Uev1a:Ube2N-Ub complex, where Uev1a is a non-catalytic partner protein, preventing Lys63-linked poly-ubiquitin chains, and ultimately NF-κB (nuclear factor κ-light-chain-enhancer of activated B cells) activation (Valleau et al., 2018; Puvar et al., 2020). Notably, structural analysis and biochemical functional assays were key in revealing how the deamidase core appended to an insertion domain enabled ubiquitination in a MavC-dependent manner without nucleotide-dependent activation of ubiquitin.

The *Legionella* effector SdeA, part of the SidE effector family, adopts another novel variation of ubiquitination known as phosphoribosyl-linked (PR) ubiquitination. SdeA catalyzes the conjugation of ubiquitin to a target protein on serine residues in an E1 and E2 independent manner (Bhogaraju et al., 2016; Qiu et al., 2016; Akturk et al., 2018; Kim et al., 2018). Structural analysis showed that the mono-ADP-ribosyltransferase (mART) and phosphodiesterase (PDE) domains are the key catalytic domains in SdeA. The mART domain binds to ubiquitin with

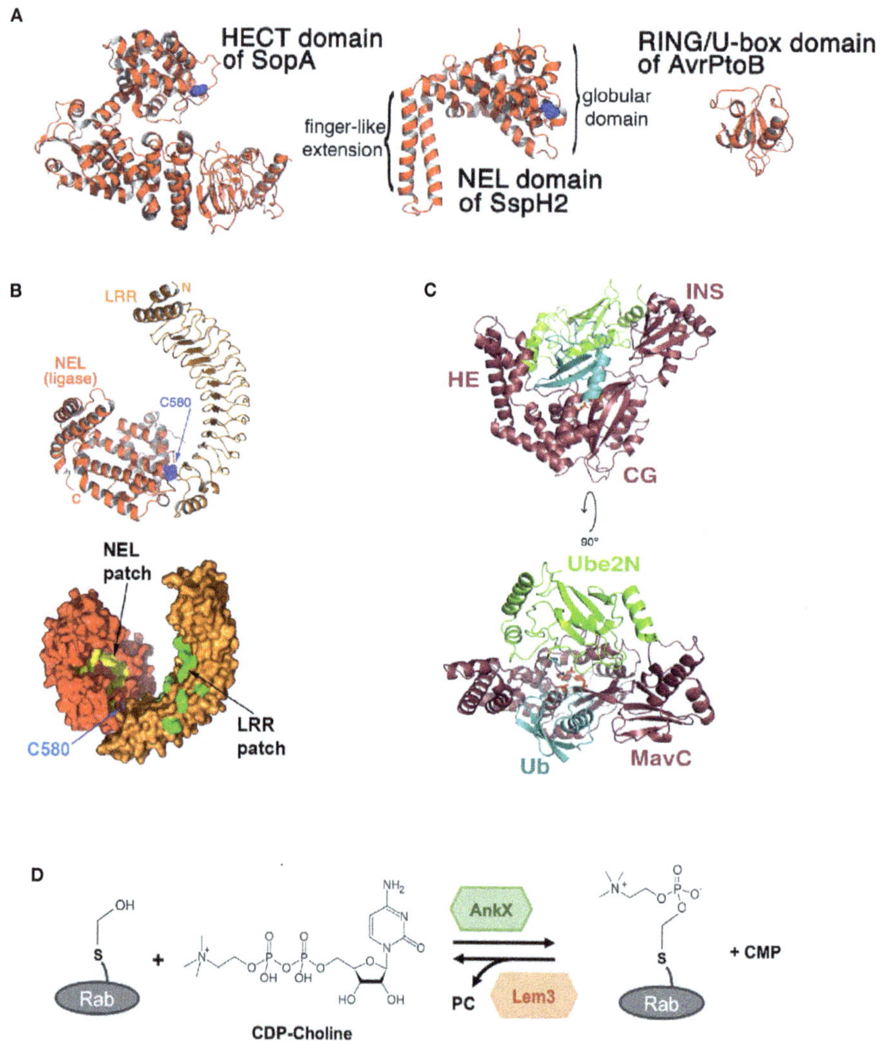

FIGURE 2 | Novel structures and biochemical activities mediated by bacterial effectors. **(A)** Structural comparison between the Novel E3 Ligase (NEL) domain in *Salmonella* effector SspH2 with bacterial E3 ligases that mimic eukaryotic E3 ligase domains: the homologous to E6-AP carboxy terminus (HECT) domain in SopA from *Salmonella* and the Really Interesting New Gene (RING/U-box) domain in AvrPtoB from *Pseudomonas Syringae*. Catalytic cysteine residues are shown in blue (taken from Quezada et al., 2009). **(B)** Crystal structure of *Salmonella* effector SspH2 shown in ribbon representation (top) and molecular surface representation (bottom). The Novel E3 Ligase (NEL) domain is shown in red and the leucine-rich repeat (LRR) domain in orange. The catalytic cysteine residue in SspH2 (C580) is shown in blue. Hydrophobic patches are labelled and shown in yellow for the NEL domain and in green for the LRR domain (taken and adapted from Quezada et al., 2009). **(C)** Crystal structure of the *Legionella* effector MavC bound to E2 Ube2N-ubiquitin conjugate. MavC is shown in dark pink, Ube2N in green and ubiquitin (Ub) in blue. The domains of MavC are labelled: helical extension (HE), core globular domain (CG), and insertion domain (INS). The active site residues in MavC (C74, H231, and Q252) are shown as red sticks (taken from Puvar et al., 2020). **(D)** Phosphocholination and dephosphocholination of Rab GTPase protein by the *Legionella* effector AnkX and Lem3. AnkX catalyzes phosphocholination, the transfer of the phosphocholine moiety from cytidine diphosphate (CDP)-choline onto the hydroxyl group of a serine residue in certain Rab GTPase proteins. Lem3 catalyzes the dephosphocholination by removing the phosphocholine (PC) (adapted from Heller et al., 2015).

a novel binding model that is distinct from known eukaryotic ubiquitin-protein interactions and undergoes significant conformational changes in order to ADP-ribosylate arginine-42 of ubiquitin (Ub^{R42}) with cofactor NAD^+. Subsequently, the PDE domain cleaves the phosphodiester bond of ADP-ribosylated ubiquitin, resulting in phosphoribosyl ubiquitin (PR-Ub), which can be linked to the hydroxyl group of serine residues in target substrate proteins (Dong et al., 2018). Interestingly, this non-canonical PR serine ubiquitination in the host cell is tightly regulated and can be reversed by *Legionella* effectors encoding deubiquitinases for PR-linked ubiquitination (DUPs; DupA and DupB) (Shin et al., 2020). Meta-effectors also regulate SidE family member activity. SidJ, a pseudokinase that is activated upon calmodulin binding in the host cell, polyglutamylates SidE ubiquitin ligases, regulating the function of these effectors within the host. Structural analysis convey that the protein kinase fold in SidJ catalyzes the ATP-dependent isopeptide bond formation between the free glutamate amino group and the SidE active site glutamate γ-carboxyl group (Black et al., 2019).

Other effectors also show a high degree of specificity in terms of activity, only becoming activated upon binding to host proteins. *Shigella* T3SS effector OspG, and its homologous effectors NleH1 and NleH2 from EPEC, are atypical serine/threonine kinases that share sequence homology to eukaryotic kinase subdomains I-VII (Zhou et al., 2013). However, they appear to lack other kinase components, including the kinase core and the activation loop. This results in a low or undetectable kinase activity that makes the protein inactive in the bacteria and may prevent non-specific activity in host cells that could initiate ETI. Only upon binding to host ubiquitin, including poly-ubiquitin chains and ubiquitin-conjugated proteins, through hydrophobic interactions mediated by the C-terminal region, is the autophosphorylation and intrinsic ATP hydrolysis activity of OspG stimulated (Zhou et al., 2013). This stimulation enables OspG to phosphorylate host ubiquitin-conjugating enzyme (e.g., UbcH5), which subsequently prevents canonical degradation of phosphorylated inhibitor of NF-κB type α (phospho-IκBα) and TNF-α stimulated NF-κB activation (Kim et al., 2005; Zhou et al., 2013). In contrast, NleH1 and NleH2, which also contain an atypical kinase in the C-terminal domain, autophosphorylate serine and threonine residues in their N-terminal domain independent of ubiquitin binding (Zhou et al., 2013). Autophosphorylation promotes interaction with and phosphorylation of target proteins. The substrate specificities and functional differences may be a result of sequence variation in the N-terminal domains NleH1 and NleH2. NleH1 phosphorylates and inhibits host MAPK (mitogen-activated protein kinase) proteins, ERK1/2 (extracellular signal-regulated kinase 1/2) and p38, to suppress NF-κB activation and apoptosis, whereas NleH2 only inhibits p38 and apoptosis (Zhou et al., 2013; Kralicek et al., 2018). As well as atypical kinases, canonical serine/threonine effector kinases, such as *Yersinia* spp. effector YpkA are also regulated by autophosphorylation. Stimulated YpkA binds and phosphorylates the heterotrimeric G protein complex (Gαq), inhibiting G protein-coupled receptor signaling in the host cell (Navarro et al., 2007; Pha et al., 2014). Overall, it is clear that regulation of effector activity *via* interaction with host proteins and PTM represents a key mechanism by which effector function is tightly regulated within host cells.

Novel Effector-Mediated Post-Translational Modifications

NEL family E3 ligases and kinase effectors carry out biochemical processes (ubiquitination and phosphorylation respectively) that are found in eukaryotic cells and can therefore be reversed by host cell enzymes. In contrast, other effectors have evolved biochemical activities that appear to mediate irreversible PTMs. For example, the *Shigella* T3SS effector OspF was identified as a phosphothreonine lyase through mass spectrometry of host targets (Li et al., 2007). OspF shares 63% sequence identity with the *Salmonella* T3SS effector SpvC and both proteins catalyze an irreversible phosphate elimination reaction (Mazurkiewicz et al., 2008). Phosphate elimination converts phosphothreonine or phosphoserine to dehydrobutyrine or dehydroalanine respectively and prevents re-phosphorylation of the residue, unlike dephosphorylation. Residues in various MAPKs, including ERK, p38 and JNK, with the dual-phosphorylated pT-X-pY motif, are targeted (Zhu et al., 2007; Mazurkiewicz et al., 2008). This leads to the impairment of MAPK signaling and blocks the activation of pro-inflammatory NF-κB regulated genes and the expression of pro-inflammatory cytokines (Li et al., 2007; Mazurkiewicz et al., 2008). This is likely to be a highly potent method for interference of host cell signaling as the protein cannot be reactivated and instead requires *de novo* protein synthesis. Of the phospholyases described to date, there is a high degree of homology, with the residues required for catalysis fully conserved. However, there is substrate specificity among this effector family which presumably reflects the differing niches of the pathogens that encode them.

Another family of bacterial effectors mediating a variation to known eukaryotic biochemistry is the family of NleB glycosyltransferases. EPEC T3SS effector NleB and the orthologs, SseK1, SseK2 and Ssek3 from *Salmonella*, catalyze the transfer of *N*-acetylglucosamine (GlcNAc) (Gao et al., 2013; Pearson et al., 2013; Li et al., 2013). In the case of NleB, modification of various host death domain-containing proteins, such as FADD, TRADD, and RIPK1, disrupts NF-κB signaling and apoptosis, presumably by preventing host protein dimerization (Pearson et al., 2013; Li et al., 2013). The conserved DXD motif is key in coordinating a metal divalent cation required for the transfer of GlcNAc onto the guanidino group of a target arginine residue (Esposito et al., 2018; Park J. B. et al., 2018). Mechanistically, this is divergent to other known enzymes, which mediate *N*-linked or *O*-linked GlcNAcylation in eukaryotes, with GlcNAc attached to the amide nitrogen in asparagine residues or the hydroxyl oxygen of serine or threonine residues respectively. Structurally, SseK3 displays a classical retaining glycosyltransferases-A (GT-A) Rossman like fold, where substrates are retained until the transfer reaction is complete (Esposito et al., 2018; Newson et al., 2019). Functionally, some substrates of NleB and SseK family members overlap, for example SseK1 modifies FADD and TRADD, whereas SseK3 modifies only TRADD. Together, SseK1, and SseK3 prevent necroptosis of infected macrophages, suggesting some redundancy in effector function (Günster et al., 2017). In addition, SseK3 appears to have evolved to modify small Rab GTPases such as Rab1 (Meng et al., 2020). Intriguingly, during macrophage infection, SseK2 does not appear to show significant arginine-GlcNAcylation at all (Günster et al., 2017). Differences in the surface electrostatic charge distribution between the SseK and NleB family members likely mediates the observed variation in host targets (Günster et al., 2017; Newson et al., 2019). Therefore, as seen with the NEL family of effectors, shared structural and biochemical activities do not always result in functional homology from effectors of different species. This underpins the importance of characterising each bacterial effector in the physiologically relevant context and not relying on the structure-function model to predict the function of any individual effector. In summary, effector-mediated arginine-GlcNAcylation represents a highly potent PTM, irreversible by host cell enzymes. Therefore, as with NEL effectors, inhibitors can be developed to target this unusual and distinct modification, providing a potential alternative to antibiotic therapy that targets virulence factors of the pathogen.

FIC (family of filamentation induced by cyclic adenosine monophosphate) domains, which typically bind ATP and transfer adenosine monophosphate (AMPylation) onto target proteins (Yarbrough et al., 2009), are found in bacterial effectors from diverse species including *Vibrio*, *Legionella*, and *Bartonella* and are also conserved from bacteria to humans (Worby et al., 2009). However, as seen above, evolutionary pressures exerted by the host, drives functional, and biochemical diversification of effectors. In the case of *Legionella* type IV effector AnkX, the FIC domain of AnkX appears unique. Instead of mediating nucleotidyl transferase activity, the FIC domain of AnkX mediates the covalent attachment of a phosphocholine moiety onto a serine residue of host Rab GTPases, including Rab1 and Rab35, modifying Rab function in the host cell (Mukherjee et al., 2011). Similar to AMPylation, the donor molecule for phosphocholination is a nucleotide-based substrate. However, rather than interacting with ATP to transfer the nucleotide moiety, AnkX interacts with cytidine diphosphate (CDP)-choline and transfers the phosphocholine moiety onto hydroxyl-containing residues of target proteins (**Figure 2D**). Structural analysis of AnkX reveals that the orientation of CDP-choline provides an explanation for FIC-motif-mediated transfer of phosphocholine (Campanacci et al., 2013; Ernst et al., 2020). Unexpectedly, the Ankyrin repeats, which normally mediate protein-protein interactions, mediate intramolecular interactions within AnkX. Of note, a second effector from *Legionella*, Lpg0696 (Lem3) has the ability to remove the phosphocholine group from Rab1, restoring the GTPase to its unmodified state (Tan et al., 2011; Goody et al., 2012; Heller et al., 2015). Presumably, this allows for the exquisite control of Rab activity during *Legionella* infection of cells, exploiting an unconventional posttranslational modification that has otherwise only been described for secreted placental polypeptides (Lovell et al., 2007).

This section has described examples of bacterial effectors that mediate covalent modifications with interesting biochemistry. We described effectors that carry out PTMs commonly found in both eukaryotes and bacteria, such as phosphorylation as well as PTMs that, despite only occurring in eukaryotes, have been adopted by bacterial effectors in order to manipulate eukaryotic intracellular signaling, such as ubiquitination. Some of these effectors mediate eukaryotic PTMs *via* non-canonical mechanisms that have not previously been described, such as PR-ubiquitination and transglutamination. Whereas other effectors mediate variations of known chemistry, for example the GlcNAcylation of non-canonical residues and the NEL effectors functioning like eukaryotic E3 ligases. There are also effectors that mediate unconventional PTMs that have not been seen before or are rare in eukaryotes, such as phosphocholination and irreversible phosphate elimination. Together, these studies provide intriguing perspectives and comparisons to traditional eukaryotic-like PTMs. The analysis of novel effector biochemistry also provides fresh and exciting research potential for the development of anti-microbial therapies that target bacteria-specific mechanisms within the host cell.

INTRINSIC DISORDER IN BACTERIAL EFFECTORS

Although the function of a protein is generally related to its three-dimensional structure, the classical structure-function paradigm is not applicable to all proteins. A lack of globular structure is found in many proteins, including bacterial effectors. Proteins that are unstructured and flexible with little or no secondary and/or tertiary structure under physiological conditions are referred to as intrinsically disordered proteins (IDPs). IDPs range from fully unstructured to partially unstructured proteins, which contain intrinsically disordered regions (IDRs) (Uversky, 2011). In contrast to ordered protein sequences, IDPs and IDRs have low sequence complexity and contain residues with low mean hydrophobicity and high net charge at neutral pH (Romero et al., 2001; Dyson and Wright, 2005; Vacic et al., 2007b), which leads to intrinsic disorder.

In a study consisting of a large group of structures (16,370 structures) from 5,434 different proteins and from 910 different organisms, only ~7% contained no disorder and only ~25% of structures had >95% of their sequence resolved (Gall et al., 2007). A lack in structural resolution can arise for several reasons, including the presence of transmembrane domains and other factors that impact crystal packing, yet this finding suggests that a large proportion of PDB (protein data bank) structures contain disordered and flexible regions that are not observed in electron density maps. The prevalence of IDRs (in >50 residues) in eukaryotic proteomes is relatively high, with an average of 20%, whereas the IDR abundance is lower in bacterial proteomes (8% on average) and other prokaryotic proteomes (Dunker et al., 2000). In the human genome, cell signaling proteins are particularly enriched with IDRs; with approximately 70% of them predicted to contain long IDRs (Iakoucheva et al., 2002; Marín et al., 2012; Marín et al., 2013). This likely reflects the biological importance of intrinsic disorder in regulating protein-protein interactions for signaling proteins.

Despite the overall low abundance of IDRs in bacterial proteomes, a large number of secreted bacterial effectors are enriched with IDRs. Long disordered regions in the middle and/or in the C-terminal regions are found in roughly 60% of *Salmonella enterica* effectors, 52% of *Pseudomonas syringae* effectors and 71% in *Xanthomonas* subspecies (Marín et al., 2013). This enrichment of IDRs in effectors when compared to the rest of the bacterial proteome suggests the existence of strong selective pressures. For example, structural flexibility is likely to be important for secretion, with disordered regions reducing the need for active unfolding prior to delivery through the narrow T3SS. The poorly described secretion signal for T3SS effectors often represents a disordered region (Samudrala et al., 2009). In many cases, intrinsically disordered N-terminal regions of T3SS effectors have been described to undergo partial folding when bound to a chaperone, as seen for *Yersinia* effector YopE and its chaperone SycE (**Figure 3A**) (Rodgers et al., 2008; Aepfelbacher et al., 2011). This disorder-to-order transition forms a three-dimensional targeting signal that promotes the translocation of YopE through the T3SS (Rodgers et al., 2008).

In addition to effector secretion, the structural flexibility in IDPs/IDRs provides additional advantages. Increased structural flexibility enables IDPs and IDRs to interact with multiple proteins once inside the host cell, adopting various conformations that depend on the structurally divergent interaction partners (Dyson and Wright, 2002; Dyson and Wright, 2005). This is due to the lack of compactness enabling IDRs to expose more surface area per residue, resulting in the exposure of more potential binding sites, at a lower energetic cost to the cell compared to ordered regions of the same residue length (Cortese et al., 2008; Nishikawa and Hatakeyama, 2017). These regions may have evolved to functionally mimic the disordered regions of eukaryotic proteins and support effector-mediated interference with host cell signaling. Indeed, analysis of eukaryotic linear motifs and bacterial motifs show that motif mimicry of eukaryotic motifs is commonly found in bacterial effector proteins (Kumar et al., 2020). Moreover, the flexible nature allows the IDR of proteins to easily access and fit into the catalytic cleft of protein-modifying enzymes for post-translational modification, which enables further regulation of protein-protein interactions and therefore function. For example, the *Helicobacter pylori* type IV secreted effector CagA, contains an intrinsically disordered C-terminal region that acts as a scaffold for multiple interactions with host proteins. Such short linear motifs in the intrinsically disordered C-terminal region, including the Glu-Pro-Ile-Tyr-Ala (EPIYA) motif and the CagA multimerization (CM) motif, are essential for the biological activity of CagA as an oncogenic virulence factor that promotes the transformation of gastric epithelial cells into gastric cancer cells (Nishikawa et al., 2016). Interestingly, the CM motif mimics host substrates in terms of sequence and structure. However, rather than being a substrate, CagA binding actually inhibits the activity of the PAR1 kinase (polarity-regulating serine/threonine kinase in partitioning-defective 1, also known as MARK). This leads to defective cell junctions and polarity of epithelial cells (Saadat et al., 2007; Nešić et al., 2010; Nishikawa et al., 2016). Furthermore, the EPIYA motif undergoes tyrosine phosphorylation by host kinases and this promotes interaction with the Src homology 2 (SH2) domain-containing proteins, such as the pro-oncogenic tyrosine phosphatase SHP2. This induces a conformational change in

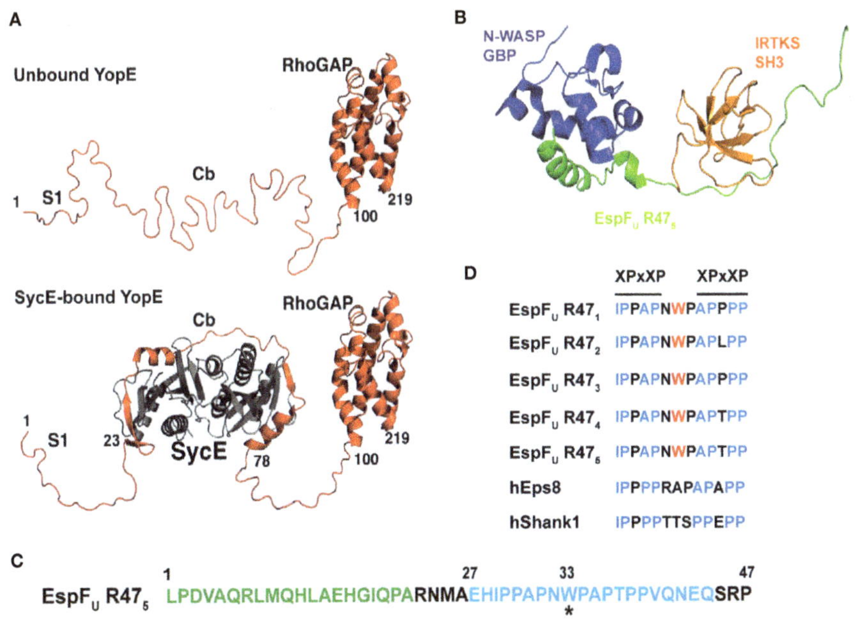

FIGURE 3 | Intrinsic disorder in bacterial effector proteins. **(A)** Crystal structure of *Yersinia* YopE (red) as unbound 'free' and bound to the chaperone SycE (grey) in ribbon representation. Functional regions of YopE are labelled: N-terminal secretion signal 1 (S1), chaperone-binding (Cb) and Rho-GAP domain. Intrinsic disorder occurs in the first 100 residues of YopE, which includes the S1 and Cb regions. The chaperone, SycE (gray), binds to the Cb region and cause disorder-to-order conformational change in the Cb region of YopE (taken and adapted from Rodgers et al., 2008). **(B)** Structure of the tri-molecular complex consisting of the GTPase binding domain (GBD) domain of N-WASP, the fifth consecutive 47-residue repeat of EspF$_U$ (EspF$_U$ R47$_5$) and the SH3 domain of IRTKS at the lowest energy conformation in ribbon representation. Structure obtained from NMR spectroscopy. N-WASP GBD, EspF$_U$ R47$_5$, and IRTKS SH3 are shown and labelled in dark blue, green and orange, respectively (PDB accession number 2LNH from Aitio et al., 2012). **(C)** Amino acid sequence of the fifth repeat of EspF$_U$ (EspF$_U$ R47$_5$). This repeat is one of the highly conserved consecutive 47-residue repeats in EspF$_U$. The GBD domain of N-WASP interacts and binds to the N-terminal helix binding region shown in green, and the SH3 domain of IRTKS binds to the C-terminal Proline-rich region shown in blue. The asterisk (*) indicates the tryptophan switch in the linker region (adapted from Aitio et al., 2012). **(D)** Sequence alignment of the linker between the two XPxXP motif in the EspF$_U$ repeats and in other IRTKS binding interaction partners, including human Eps8 (hEps8) and human Shank1, showing the tryptophan switch in the linker region. In the XPxXP motif, "X" is a hydrophobic residue and "x" is any residue and P is proline (adapted from Aitio et al., 2012).

SHP2 that unlocks the autoinhibitory conformation resulting in deregulated and aberrant SHP2 phosphatase activity, promoting pro-oncogenic mitogenic Ras-ERK signaling and abnormal cell morphology and motility (Higashi et al., 2002; Nishikawa et al., 2016; Hatakeyama, 2017).

In fact, the EPIYA motif is found in IDRs of several bacterial effectors (Hayashi et al., 2013). Each functioning in a tyrosine phosphorylation-dependent manner where the phosphorylated EPIYA or EPIYA-related motifs interact with SH2 domain-containing proteins, causing abnormal and aberrant host cell signaling. Most of these effectors are structurally poorly characterized. However, CD spectroscopy and analytical ultracentrifugation of EPEC effector Tir revealed a monomeric, highly elongated conformation at physiological conditions with a lack of secondary structures. This suggests that Tir is natively unfolded and disordered in solution. Upon phosphorylation of Ser434 and Ser463 by cAMP kinase, Tir undergoes conformational changes that may promote membrane insertion and possibly intermolecular interactions that are required for biological function (Race et al., 2007). Upon tyrosine phosphorylation at the EPIYA-related motif by host kinases, Tir interacts and forms a complex with the SH2 domain containing adaptor protein Nck, promoting actin polymerization (Hayashi et al., 2013).

Furthermore, structural flexibility in IDPs/IDRs may also play a part in protein evolution, potentially providing a selective advantage in comparison to ordered folded protein regions in bacterial effectors. The disordered motifs of CagA are more exposed to the host cell cytoplasm and more prone to sequence polymorphism. These polymorphisms influence the binding affinity to host target proteins and determines the pro-oncogenic degree exhibited by each CagA variant (Nishikawa and Hatakeyama, 2017; Hatakeyama, 2017). Therefore, the disordered nature of IDPs means they are more susceptible to mutational changes as natural selection drives the reduction of molecular disorder, or entropy in thermodynamic terms. This means intrinsically disordered effectors will evolve faster than ordered proteins, with subsequent mutations possibly altering effector interaction partners or enabling the acquisition of new effector functions (Brown et al., 2011; Nishikawa and Hatakeyama, 2017).

Aromatic residues are often involved in protein-protein interactions, but their overall representation is rare in IDPs. However, aromatic residues are found to be enriched in linear motifs or short molecular recognition elements of IDPs. This enables a high degree of specificity to be achieved in a stimulus-dependent manner, and is often observed in proteins involved in signal transduction (Vacic et al., 2007a; Uversky, 2011). Although IDPs typically form weak molecular interactions, the intrinsically disordered EHEC T3SS effector $EspF_u$ (also known as TccP) forms a high affinity tri-molecular complex with host proteins N-WASP and insulin receptor tyrosine kinase substrate (IRTKS) (**Figure 3B**). As revealed by NMR spectroscopy, $EspF_u$ is a 337-residue IDP that consist of a N-terminal secretion signal followed by highly conserved consecutive 47-residue repeats. Each highly conserved repeat contains the GTPase binding domain (GBD) that interacts with N-WASP and the XPxXP motifs that promote interaction with the Src homology 3 (SH3) domain of IRTKS (**Figure 3C**). Interestingly, there is a tryptophan residue in the linker between the two XPxXP motifs in the $EspF_U$ sequence that is absent in known host SH3 interaction partners of IRTKS (**Figure 3D**). This tryptophan appears to have evolved to enable superior binding affinity to outcompete host cellular targets. In this manner, a high affinity tri-molecular complex forms that stimulates actin polymerization for intestinal colonization of EHEC (Aitio et al., 2012). As IDPs can adopt multiple conformations and the conformation constantly changes depending on its interactions and biochemical environment (Marín et al., 2013), it is interesting to postulate that the lack of structure within IDPs enables these bacterial effectors to avoid direct recognition by host inhibitory proteins and the initiation of ETI.

Overall, structural analysis of IDPs or IDRs in proteins is difficult due to the flexible and disordered nature, tendency for degradation, and presence of multiple conformations. Hence, the majority of available structural data on bacterial effectors is limited to structured, ordered, and folded protein regions. Despite this, more work is required as undoubtedly IDRs are important with respect to effector function and therefore their analysis will provide a better understanding of pathogenesis.

CONCLUDING REMARKS AND CHALLENGES

Although significant progress has been made over the past two decades, many effectors across diverse pathogens await structural and functional characterization. In different bacterial species, the repertoire of effectors varies in terms of both the number of effectors and the degree to which they have been characterized. For example, 28 out of the 44 identified *Salmonella* T3SS effectors are characterized to a large extent in terms of structure, physiological function and biochemical activity, with only a handful of effectors being completely elusive (Ramos-Morales, 2012; Jennings et al., 2017). A similar degree of characterization has been achieved for EPEC, EHEC and *Shigella*. In contrast, there are many T2SS, T3SS, and T6SS effectors in *Burkholderia pseudomallei* that await validation and characterization (Broek and Stevens, 2017). Work with *B. pseudomallei* requires access to a category III containment laboratory and this may in part explain the poor degree of effector characterization for this species. Perhaps the most striking case is the effector repertoire of *Legionella* species, where more than 18,000 effectors have been identified across the entire genus through genomic analysis. Within this repertoire, 137 different eukaryotic domains were identified with more than 200 effectors containing these eukaryotic-like protein features (Gomez-Valero et al., 2019). This suggests there will be an enormous degree of novel protein domains present among *Legionella* effectors that await characterization.

So, why do so many effectors remain functionally uncharacterized? One significant reason is that primary amino acid sequence is a poor indicator of secondary and tertiary structure and hence biochemical function. For this reason, X-ray crystallography represents an important method for determining the overall tertiary structure and hence the putative biochemical activity of an effector. Yet, even this represents just the start of the road. As described above, a previously unstudied effector may structurally resemble a given eukaryotic enzyme, but actually carry out a modified or an entirely new biochemical function. This makes it difficult to functionally characterize bacterial effectors, for both those with no similarity and even for those where similar effectors have already been studied. Uncovering novel effector functions therefore often requires a combination of structural, proteomic and biochemical studies along with infection work and an open mind. Additional complexities then exist as structural studies of effectors are often hampered due to protein insolubility, cytotoxicity, the presence of intrinsically disordered regions, and the fact that numerous effectors are membrane proteins.

Another important consideration is whether a host protein is required in order for the effector to exist in its active conformation; in this case, it may first be necessary to identify physiologically relevant binding partners prior to acquisition of protein complexes. On the other hand, as bacterial effectors tend to show limited structural homology to known proteins, crystallography of effectors in complex with a host interaction partner of known structure might help resolve the "phase problem" and therefore determination of the structure from the diffraction data. For these reasons, alternative structural and biophysical techniques, such as NMR spectroscopy and CD spectroscopy, could be explored in tandem. Alternatively, an emerging structural technique, cryogenic electron microscopy (cryo-EM), can be used to determine biomolecular structures at near-atomic resolution. Cryo-EM is mainly limited to larger biomolecules and complexes and has been instrumental in solving the structures of many bacterial secretion systems (Kooger et al., 2018; Lunelli et al., 2020; Park D. et al., 2018). However, recent advances show that small proteins of less than 50 kDa can be assembled into large symmetric cage complexes or attached to rigid symmetrical scaffolds for cryo-EM imaging (Liu et al., 2018). However, while cryo-EM represents a promising alternative to X-ray crystallography, there are still potential problems; prior structural knowledge on the protein of interest is required and scaffolds may distort the structure of the protein of interest, particularly disordered regions, resulting in physiologically irrelevant structures.

As the overall number of characterized effectors remains relatively low, it is likely that new effector-mediated biochemistries await discovery. Therefore, these challenges should not deter from continued attempts to structurally and functionally determine effectors from diverse pathogens. This is essential for the continued understanding of how bacterial virulence factors manipulate the host system to promote pathogenesis.

AUTHOR CONTRIBUTIONS

Conceptualization: HM and TT. Investigation: HM. Writing—original draft: HM. Reviewing and editing: HM and TT. Supervision: TT. All authors contributed to the article and approved the submitted version.

ACKNOWLEDGMENTS

We thank members from the Thurston laboratory for scientific discussion. We also thank Ioanna Panagi and Rachel Ibbott for reading the manuscript and providing valuable comments.

REFERENCES

Aepfelbacher, M., Roppenser, B., Hentschke, M., and Ruckdeschel, K. (2011). Activity Modulation of the Bacterial Rho GAP YopE: An Inspiration for the Investigation of Mammalian Rho GAPs. *Eur. J. Cell Biol.* 90 (11), 951–954. doi: 10.1016/j.ejcb.2010.12.004

Aitio, O., Hellman, M., Skehan, B., Kesti, T., Leong, J. M., Saksela, K., et al. (2012). Enterohaemorrhagic Escherichia Coli Exploits a Tryptophan Switch to Hijack Host F-Actin Assembly. *Structure* 20 (10), 1692–1703. doi: 10.1016/j.str.2012.07.015

Akturk, A., Wasilko, D. J., Wu, X., Liu, Y., Zhang, Y., Qiu, J., et al. (2018). Mechanism of Phosphoribosyl-Ubiquitination Mediated by a Single Legionella Effector. *Nature* 55 (7707), 729–745. doi: 10.1038/s41586-018-0147-6

Alto, N. M., Shao, F., Lazar, C. S., Brost, R. L., Chua, G., Mattoo, S., et al. (2006). Identification of a Bacterial Type III Effector Family with G Protein Mimicry Functions. *Cell* 124 (1), 133–145. doi: 10.1016/j.cell.2005.10.031

Arbeloa, A., Bulgin, R. R., Mackenzie, G., Shaw, R. K., Pallen, M. J., Crepin, V. F., et al. (2008). Subversion of Actin Dynamics by EspM Effectors of Attaching and Effacing Bacterial Pathogens. *Cell. Microbiol.* 10 (7), 1429–1441. doi: 10.1111/j.1462-5822.2008.01136.x

Ashida, H., and Sasakawa, C. (2017). Bacterial E3 Ligase Effectors Exploit Host Ubiquitin Systems. *Curr. Opin. Microbiol.* 35 (1), 16–22 .doi: 10.1016/j.mib.2016.11.001

Baruch, K., Gur-Arie, L., Nadler, C., Koby, S., Yerushalmi, G., Ben-Neriah, Y., et al. (2011). Metalloprotease Type III Effectors That Specifically Cleave JNK and NF-KB. *EMBO J.* 30 (1), 221–231. doi: 10.1038/emboj.2010.297

Bayer-Santos, E., Durkin, C. H., Rigano, L. A., Kupz, A., Alix, E., Cerny, O., et al. (2016). The Salmonella Effector SteD Mediates MARCH8-Dependent Ubiquitination of MHC II Molecules and Inhibits T Cell Activation. *Cell Host Microbe* 20 (5), 584–595. doi: 10.1016/j.chom.2016.10.007

Bernal-Bayard, J., and Ramos-Morales, F. (2009). Salmonella Type III Secretion Effector SlrP Is an E3 Ubiquitin Ligase for Mammalian Thioredoxin. *J. Biol. Chem.* 284 (40), 27587–27595. doi: 10.1074/jbc.M109.010363

Bhogaraju, S., Kalayil, S., Liu, Y., Bonn, F., Colby, T., Matic, I., et al. (2016). Phosphoribosylation of Ubiquitin Promotes Serine Ubiquitination and Impairs Conventional Ubiquitination. *Cell* 167 (6), 1636–1649.e13. doi: 10.1016/j.cell.2016.11.019

Black, M. H., Osinski, A., Gradowski, M., Servage, K. A., Pawłowski, K., Tomchick, D. R., et al. (2019). Bacterial Pseudokinase Catalyzes Protein Polyglutamylation to Inhibit the SidE-Family Ubiquitin Ligases. *Science (New York N.Y.)* 364 (6442), 787–792. doi: 10.1126/science.aaw7446

Broek, C.W.V., and Stevens, J. M. (2017). Type III Secretion in the Melioidosis Pathogen Burkholderia Pseudomallei. *Front. Cell. Infect. Microbiol.* 7, 255. doi: 10.3389/fcimb.2017.00255

Brown, C. J., Johnson, A. K., Dunker, A. K., and Daughdrill, G. W. (2011). Evolution and Disorder. *Curr. Opin. Struct. Biol.* 21 (3), 441–46. doi: 10.1016/j.sbi.2011.02.005

Bulgin, R. R., Arbeloa, A., Chung, J. C. S., and Frankel, G. (2009). EspT Triggers Formation of Lamellipodia and Membrane Ruffles through Activation of Rac-1 and Cdc42. *Cell. Microbiol.* 11 (2), 217–229. doi: 10.1111/j.1462-5822.2008.01248.x

Bulgin, R., Raymond, B., Garnett, J. A., Frankel, G., Crepin, V. F., Berger, C. N., et al. (2010). Bacterial Guanine Nucleotide Exchange Factors SopE-Like and WxxxE Effectors. *Infect. Immun.* 78 (4), 1417–1425. doi: 10.1128/IAI.01250-09

Campanacci, V., Mukherjee, S., Roy, C. R., and Cherfils, J. (2013). Structure of the Legionella Effector AnkX Reveals the Mechanism of Phosphocholine Transfer by the FIC Domain. *EMBO J.* 32 (10), 1469–1477. doi: 10.1038/emboj.2013.82

Chou, Y. C., Keszei, A. F. A., Rohde, J. R., Tyers, M., and Sicheri, F. (2012). Conserved Structural Mechanisms for Autoinhibition in IpaH Ubiquitin Ligases. *J. Biol. Chem.* 287 (1), 268–275. doi: 10.1074/jbc.M111.316265

Cortese, M. S., Uversky, V. N., and Keith Dunker, A. (2008). Intrinsic Disorder in Scaffold Proteins: Getting More from Less. *Prog. Biophys. Mol. Biol.* 98 (1), 85–106 .doi: 10.1016/j.pbiomolbio.2008.05.007

Costa, T. R. D., Felisberto-Rodrigues, C., Meir, A., Prevost, M. S., Redzej, A., Trokter, M., et al. (2015). Secretion Systems in Gram-Negative Bacteria: Structural and Mechanistic Insights. *Nat. Rev. Microbiol.* 13, 343. doi: 10.1038/nrmicro3456

Cui, J., Yao, Q., Li, S., Ding, X., Lu, Q., Mao, H., et al. (2010). Glutamine Deamidation and Dysfunction of Ubiquitin/NEDD8 Induced by a Bacterial Effector Family. *Science* 329 (5996), 1215–1218. doi: 10.1126/science.1193844

Diao, J., Zhang, Y., Huibregtse, J. M., Zhou, D., and Chen, J. (2008). Crystal Structure of SopA, a Salmonella Effector Protein Mimicking a Eukaryotic Ubiquitin Ligase. *Nat. Struct. Mol. Biol.* 15 (1), 65–70. doi: 10.1038/nsmb1346

Dohlich, K., Zumsteg, A. B., Goosmann, C., and Kolbe, M. (2014). A Substrate-Fusion Protein Is Trapped inside the Type III Secretion System Channel in Shigella Flexneri. *PLoS Pathog.* 10 (1), 1003881. doi: 10.1371/journal.ppat.1003881

Dong, Y., Mu, Y., Xie, Y., Zhang, Y., Han, Y., Zhou, Y., et al. (2018). Structural Basis of Ubiquitin Modification by the Legionella Effector SdeA. *Nature* 557 (7707), 674–678. doi: 10.1038/s41586-018-0146-7

Dunker, A. K., Obradovic, Z., Romero, P., Garner, E. C., and Brown, C. J. (2000). Intrinsic Protein Disorder in Complete Genomes. *Genome Inform. Workshop Genome Inf.* 11, 161–171. doi: 10.11234/gi1990.11.161

Dyson, H. J., and Wright, P. E. (2002). Coupling of Folding and Binding for Unstructured Proteins. *Curr. Opin. Struct. Biol.* 12 (1), 54–60. doi: 10.1016/s0959-440x(02)00289-0

Dyson, H. J., and Wright, P. E. (2005). Intrinsically Unstructured Proteins and Their Functions. *Nat. Rev. Mol. Cell Biol.* 6 (3), 197–208. doi: 10.1038/nrm1589

Ernst, S., Ecker, F., Kaspers, M. S., Ochtrop, P., Hedberg, C., Groll, M., et al. (2020). Legionella Effector AnkX Displaces the Switch II Region for Rab1b Phosphocholination. *Sci. Adv.* 6 (20), eaaz8041. doi: 10.1126/sciadv.aaz8041

Esposito, D., Günster, R. A., Martino, L., Omari, K., Wagner, A., Thurston, T. L. M., et al. (2018). Structural Basis for the Glycosyltransferase Activity of the Salmonella Effector SseK3. *J. Biol. Chem.* 293 (14), 5064–5078. doi: 10.1074/jbc.RA118.001796

Felix, C., Türköz, B. K., Ranaldi, S., Koelblen, T., Terradot, L., O'Callaghan, D., et al. (2014). The Brucella TIR Domain Containing Proteins BtpA and BtpB Have a Structural WxxxE Motif Important for Protection against Microtubule Depolymerisation. *Cell Commun. Signaling* 12 (1), 53. doi: 10.1186/s12964-014-0053-y

Fiskin, E., Bhogaraju, S., Herhaus, L., Kalayil, S., Hahn, M., and Dikic, I. (2017). Structural Basis for the Recognition and Degradation of Host TRIM Proteins by Salmonella Effector SopA. *Nat. Commun.* 8, 14004. doi: 10.1038/ncomms14004

Galán, J. E. (2009). Common Themes in the Design and Function of Bacterial Effectors. *Cell Host Microbe.* 5 (6), 571–579. doi: 10.1016/j.chom.2009.04.008

Galán, J. E., and Waksman, G. (2018). Protein-Injection Machines in Bacteria. *Cell.* 172 (6), 1306–1318. doi: 10.1016/j.cell.2018.01.034

Gall, T., Romero, P. R., Cortese, M. S., Uversky, V. N., and Dunker, A. K. (2007). Intrinsic Disorder in the Protein Data Bank. *J. Biomol. Struct. Dyn.* 24 (4), 325–341. doi: 10.1080/07391102.2007.10507123

Gao, X., Wang, X., Pham, T. H., Feuerbacher, L. A., Lubos, M. L., Huang, M., et al. (2013). NleB, a Bacterial Effector with Glycosyltransferase Activity, Targets GADPH Function to Inhibit NF-KB Activation. *Cell Host Microbe* 13 (1), 87–99. doi: 10.1016/j.chom.2012.11.010

Gomez-Valero, L., Rusniok, C., Carson, D., Mondino, S., Pérez-Cobas, A. E., Rolando, M., et al. (2019). More than 18,000 Effectors in the Legionella Genus Genome Provide Multiple, Independent Combinations for Replication in Human Cells. *Proc. Natl. Acad. Sci. U. S. A.* 116 (6), 2265–2273. doi: 10.1073/pnas.1808016116

Goody, P. R., Heller, K., Oesterlin, L. K., Müller, M. P., Itzen, A., and Goody, R. S. (2012). Reversible Phosphocholination of Rab Proteins by *Legionella Pneumophila* Effector Proteins. *EMBO J.* 31 (7), 1774–1784. doi: 10.1038/emboj.2012.16

Günster, R. A., Matthews, S. A., Holden, D. W., and Thurston, T. L. M. (2017). SseK1 and SseK3 Type III Secretion System Effectors Inhibit NF-KB Signaling and Necroptotic Cell Death in Salmonella-Infected Macrophages. *Infect. Immun.* 85 (3), e00010-17. doi: 10.1128/IAI.00010-17

Gur-Arie, L., Eitan-Wexler, M., Weinberger, N., Rosenshine, I., and Livnah, O. (2020). The Bacterial Metalloprotease NleD Selectively Cleaves Mitogen-Activated Protein Kinases That Have High Flexibility in Their Activation Loop. *J. Biol. Chem.* 295 (28), 9409–20. doi: 10.1074/jbc.ra120.013590

Haraga, A., and Miller, S. I. (2006). A Salmonella Type III Secretion Effector Interacts with the Mammalian Serine/Threonine Protein Kinase PKN1. *Cell. Microbiol.* 8 (5), 837–846. doi: 10.1111/j.1462-5822.2005.00670.x

Hatakeyama, M. (2017). Structure and Function of Helicobacter Pylori Caga, the First-Identified Bacterial Protein Involved in Human Cancer. *Proc. Jpn. Acad. Ser. B, Phys. Biol. Sci.* 93 (4), 196–219. doi: 10.2183/pjab.93.013

Hayashi, T., Morohashi, H., and Hatakeyama, M. (2013). Bacterial EPIYA Effectors - Where Do They Come from? What Are They? Where Are They Going? *Cell. Microbiol.* 15 (3), 377–385. doi: 10.1111/cmi.12040

Heller, K., Ochtrop, P., Albers, M. F., Zauner, F. B., Itzen, A., and Hedberg, C. (2015). Covalent Protein Labeling by Enzymatic Phosphocholination. *Angew. Chem. Int. Edit.* 54 (35), 10327–10330. doi: 10.1002/anie.201502618

Higashi, H., Tsutsumi, R., Muto, S., Sugiyama, T., Azuma, T., Asaka, M., et al. (2002). SHP-2 Tyrosine Phosphatase as an Intracellular Target of Helicobacter Pylori CagA Protein. *Science* 295 (5555), 683–686. doi: 10.1126/science.1067147

Iakoucheva, L. M., Brown, C. J., Lawson, J. D., Obradović, Z., and Dunker, A. K. (2002). Intrinsic Disorder in Cell-Signaling and Cancer-Associated Proteins. *J. Mol. Biol.* 323 (3), 573–584. doi: 10.1016/S0022-2836(02)00969-5

Janeway, C. A. (1989). Approaching the Asymptote? Evolution and Revolution in Immunology. *Cold Spring Harb. Symp. Quant. Biol.* 54 (1), 1–13. doi: 10.1101/sqb.1989.054.01.003

Jennings, E., Thurston, T. L. M., and Holden, D. W. (2017). Salmonella SPI-2 Type III Secretion System Effectors: Molecular Mechanisms And Physiological Consequences. *Cell Host Microbe* 22 (2), 217–231. doi: 10.1016/J.CHOM.2017.07.009

Jennings, E., Esposito, D., Rittinger, K., and Thurston, T. L. M. (2018). Structure-Function Analyses of the Bacterial Zinc Metalloprotease Effector Protein GtgA Uncover Key Residues Required for Deactivating NF-KB. *J. Biol. Chem.* 293 (39), 15316–15329. doi: 10.1074/jbc.RA118.004255

Kamanova, J., Sun, H., Lara-Tejero, M., and Galán, J. E. (2016). The Salmonella Effector Protein SopA Modulates Innate Immune Responses by Targeting TRIM E3 Ligase Family Members. *PLoS Pathog.* 12 (4), e1005552. doi: 10.1371/journal.ppat.1005552

Kenny, B., Ellis, S., Leard, A. D., Warawa, J., Mellor, H., and Jepson, M. A. (2002). Co-Ordinate Regulation of Distinct Host Cell Signalling Pathways by Multifunctional Enteropathogenic Escherichia Coli Effector Molecules. *Mol. Microbiol.* 44 (4), 1095–1107. doi: 10.1046/j.1365-2958.2002.02952.x

Keszei, A. F. A., Tang, X., McCormick, C., Zeqiraj, E., Rohde, J. R., Tyers, M., et al. (2014). Structure of an SspH1-PKN1 Complex Reveals the Basis for Host Substrate Recognition and Mechanism of Activation for a Bacterial E3 Ubiquitin Ligase. *Mol. Cell. Biol.* 34 (3), 362–373. doi: 10.1128/mcb.01360-13

Kim, D. W., Lenzen, G., Page, A. L., Legrain, P., Sansonetti, P. J., and Parsot, C. (2005). The Shigella Flexneri Effector OspG Interferes with Innate Immune Responses by Targeting Ubiquitin-Conjugating Enzymes. *Proc. Natl. Acad. Sci. U. S. A.* 102 (39), 14046–14051. doi: 10.1073/pnas.0504466102

Kim, L., Kwon, D. H., Kim, B. H., Kim, J., Park, M. R., Park, Z. Y., et al. (2018). Structural and Biochemical Study of the Mono-ADP-Ribosyltransferase Domain of SdeA, a Ubiquitylating/Deubiquitylating Enzyme from Legionella Pneumophila. *J. Mol. Biol.* 430 (17), 2843–2856. doi: 10.1016/j.jmb.2018.05.043

Kooger, R., Szwedziak, P., Böck, D., and Pilhofer, M. (2018). CryoEM of Bacterial Secretion Systems. *Curr. Opin. Struct. Biol.* 52, 64–70. doi: 10.1016/j.sbi.2018.08.007

Kralicek, S. E., Nguyen, M., Rhee, K. J., Tapia, R., and Hecht, G. (2018). EPEC NleH1 Is Significantly More Effective in Reversing Colitis and Reducing Mortality than NleH2 via Differential Effects on Host Signaling Pathways. *Lab. Invest.* 98 (4), 477–488. doi: 10.1038/s41374-017-0016-1

Kubori, T., Shinzawa, N., Kanuka, H., and Nagai, H. (2010). Legionella Metaeffector Exploits Host Proteasome to Temporally Regulate Cognate Effector. *PLoS Pathog.* 6 (12), e1001216. doi: 10.1371/journal.ppat.1001216

Kumar, M., Gouw, M., Michael, S., Sámano-Sánchez, H., Pancsa, R., Glavina, J., et al. (2020). ELM-the Eukaryotic Linear Motif Resource in 2020. *Nucleic Acids Res.* 48 (D1), D296–D306. doi: 10.1093/nar/gkz1030

Levin, I., Eakin, C., Blanc, M. P., Klevit, R. E., Miller, S. I., and Brzovic, P. S. (2010). Identification of an Unconventional E3 Binding Surface on the UbcH5 ~ Ub Conjugate Recognized by a Pathogenic Bacterial E3 Ligase. *Proc. Natl. Acad. Sci. U. S. A.* 107 (7), 2848–2853. doi: 10.1073/pnas.0914821107

Li, H., Xu, H., Zhou, Y., Zhang, J., Long, C., Li, S., et al. (2007). The Phosphothreonine Lyase Activity of a Bacterial Type III Effector Family. *Science* 315 (5814), 1000–1003. doi: 10.1126/science.1138960

Li, S., Zhang, L., Yao, Q., Li, L., Dong, N., Rong, J., et al. (2013). Pathogen Blocks Host Death Receptor Signalling by Arginine GlcNAcylation of Death Domains. *Nature* 501 (7466), 242–246. doi: 10.1038/nature12436

Lin, S. L., Le, T. X., and Cowen, D. S. (2003). SptP, a Salmonella Typhimurium Type III-Secreted Protein, Inhibits the Mitogen-Activated Protein Kinase Pathway by Inhibiting Raf Activation. *Cell. Microbiol.* 5 (4), 267–275. doi: 10.1046/j.1462-5822.2003.t01-1-00274.x

Lin, D. Y. W., Diao, J., Zhou, D., and Chen, J. (2011). Biochemical and Structural Studies of a HECT-like Ubiquitin Ligase from Escherichia Coli O157:H7. *J. Biol. Chem.* 286 (1), 441–449. doi: 10.1074/jbc.M110.167643

Lin, D. Y. W., Diao, J., and Chen, J. (2012). Crystal Structures of Two Bacterial HECT-like E3 Ligases in Complex with a Human E2 Reveal Atomic Details of Pathogen-Host Interactions. *Proc. Natl. Acad. Sci. U. S. A.* 109 (6), 1925–1930. doi: 10.1073/pnas.1115025109

Liu, Y., Gonen, S., Gonen, T., and Yeates, T. O. (2018). Near-Atomic Cryo-EM Imaging of a Small Protein Displayed on a Designed Scaffolding System. *Proc. Natl. Acad. Sci. U. S. A.* 115 (13), 3362–3367. doi: 10.1073/pnas.1718825115

Lopes Fischer, N., Naseer, N., Shin, S., and Brodsky, I. E. (2020). Effector-Triggered Immunity and Pathogen Sensing in Metazoans. *Nat. Microbiol.* 5 (1), 14–26. doi: 10.1038/s41564-019-0623-2

Lovell, T. M., Woods, R. J., Butlin, D. J., Brayley, K. J., Manyonda, I. T., Jarvis, J., et al. (2007). Identification of a Novel Mammalian Post-Translational Modification, Phosphocholine, on Placental Secretory Polypeptides. *J. Mol. Endocrinol.* 39 (3–4), 189–198. doi: 10.1677/JME-07-0007

Lunelli, M., Kamprad, A., Bürger, J., Mielke, T., Spahn, C. M. T., and Kolbe, M. (2020). Cryo-EM Structure of the Shigella Type III Needle Complex. *PLoS Pathog.* 16 (2), e1008263. doi: 10.1371/journal.ppat.1008263

Marín, M., Thallmair, V., and Ott, T. (2012). The Intrinsically Disordered N-Terminal Region of AtREM1.3 Remorin Protein Mediates Protein-Protein Interactions. *J. Biol. Chem.* 287 (47), 39982–39991. doi: 10.1074/jbc.M112.414292

Marín, M., Uversky, V. N., and Ott, T. (2013). Intrinsic Disorder in Pathogen Effectors: Protein Flexibility as an Evolutionary Hallmark in a Molecular Arms Race. *Plant Cell.* 25 (9), 3153–3157. doi: 10.1105/tpc.113.116319

Mazurkiewicz, P., Thomas, J., Thompson, J. A., Liu, M., Arbibe, L., Sansonetti, P., et al. (2008). SpvC Is a Salmonella Effector with Phosphothreonine Lyase Activity on Host Mitogen-Activated Protein Kinases. *Mol. Microbiol.* 67 (6), 1371–1383. doi: 10.1111/j.1365-2958.2008.06134.x

Meng, K., Zhuang, X., Peng, T., Hu, S., Yang, J., Wang, Z., et al. (2020). Arginine GlcNAcylation of Rab Small GTPases by the Pathogen Salmonella Typhimurium. *Commun. Biol.* 3 (1), 1–13. doi: 10.1038/s42003-020-1005-2

Mukherjee, S., Liu, X., Arasaki, K., McDonough, J., Galán, J. E., and Roy, C. R. (2011). Modulation of Rab GTPase Function by a Protein Phosphocholine Transferase. *Nature* 477 (7362), 103–108. doi: 10.1038/nature10335

Navarro, L., Koller, A., Nordfelth, R., Wolf-Watz, H., Taylor, S., and Dixon, J. E. (2007). Identification of a Molecular Target for the Yersinia Protein Kinase A. *Mol. Cell* 26 (4), 465–477. doi: 10.1016/j.molcel.2007.04.025

Nešić, D., Miller, M. C., Quinkert, Z. T., Stein, M., Chait, B. T., and Stebbins, C. E. (2010). Helicobacter Pylori CagA Inhibits PAR1-MARK Family Kinases by Mimicking Host Substrates. *Nat. Struct. Mol. Biol.* 17 (1), 130–132. doi: 10.1038/nsmb.1705

Newson, J. P. M., Scott, N. E., Chung, I. Y. W., Lung, T. W. F., Giogha, C., Gan, J., et al. (2019). Salmonella Effectors SseK1 and SseK3 Target Death Domain Proteins in the TNF and TRAIL Signaling Pathways. *Mol. Cell. Proteomics* 18 (6), 1138–1156. doi: 10.1074/mcp.RA118.001093

Nishikawa, H., Hayashi, T., Arisaka, F., Senda, T., and Hatakeyama, M. (2016). Impact of Structural Polymorphism for the Helicobacter Pylori CagA Oncoprotein on Binding to Polarity-Regulating Kinase PAR1b. *Sci. Rep.* 6 (1), 1–13. doi: 10.1038/srep30031

Nishikawa, H., and Hatakeyama, M. (2017). Sequence Polymorphism and Intrinsic Structural Disorder as Related to Pathobiological Performance of the Helicobacter Pylori CagA Oncoprotein. *Toxins.* 9 (4), 136. doi: 10.3390/toxins9040136

Ohlson, M. B., Huang, Z., Alto, N. M., Blanc, M. P., Dixon, J. E., Chai, J., et al. (2008). Structure and Function of Salmonella SifA Indicate That Its Interactions with SKIP, SseJ, and RhoA Family GTPases Induce Endosomal Tubulation. *Cell Host Microbe* 4 (5), 434–446. doi: 10.1016/j.chom.2008.08.012

Panagi, I., Jennings, E., Zeng, J., Günster, R. A., Stones, C. D., Mak, H., et al. (2020). Salmonella Effector SteE Converts the Mammalian Serine/Threonine Kinase GSK3 into a Tyrosine Kinase to Direct Macrophage Polarization. *Cell Host Microbe* 27 (1), 41–53.e6. doi: 10.1016/j.chom.2019.11.002

Park, D., Lara-Tejero, M., Waxham, M. N., Li, W., Hu, B., Galán, J. E., et al. (2018). Visualization of the Type III Secretion Mediated Salmonella–Host Cell Interface Using Cryo-Electron Tomography. *ELife* 7, e39514. doi: 10.7554/eLife.39514

Park, J. B., Kim, Y. H., Yoo, Y., Kim, J., Jun, S. H., Cho, J. W., et al. (2018). Structural Basis for Arginine Glycosylation of Host Substrates by Bacterial Effector Proteins. *Nat. Commun.* 9 (1), 1–15. doi: 10.1038/s41467-018-06680-6

Pearson, J. S., Giogha, C., Ong, S. Y., Kennedy, C. L., Kelly, M., Robinson, K. S., et al. (2013). A Type III Effector Antagonizes Death Receptor Signalling during Bacterial Gut Infection. *Nature* 501 (7466), 247–251. doi: 10.1038/nature12524

Pha, K., Wright, M. E., Barr, T. M., Eigenheer, R. A., and Navarro, L. (2014). Regulation of Yersinia Protein Kinase A (YpkA) Kinase Activity by Multisite Autophosphorylation and Identification of an n-Terminal Substrate-Binding Domain in YpkA. *J. Biol. Chem.* 289 (38), 26167–26177. doi: 10.1074/jbc.M114.601153

Piscatelli, H., Kotkar, S. A., McBee, M. E., Muthupalani, S., Schauer, D. B., Mandrell, R. E., et al. (2011). The Ehec Type Iii Effector Nlel Is an E3 Ubiquitin Ligase That Modulates Pedestal Formation. *PLoS One* 6 (4), e19331. doi: 10.1371/journal.pone.0019331

Puvar, K., Iyer, S., Fu, J., Kenny, S., Negrón Terón, K. I., Luo, Z. Q., et al. (2020). Legionella Effector MavC Targets the Ube2N~Ub Conjugate for Noncanonical Ubiquitination. *Nat. Commun.* 11 (1), 1–14. doi: 10.1038/s41467-020-16211-x

Qaidi, S., Scott, N. E., Hays, M. P., Geisbrecht, B. V., Watkins, S., and Hardwidge, P. R. (2020). An Intra-Bacterial Activity for a T3SS Effector. *Sci. Rep.* 10 (1), 1073. doi: 10.1038/s41598-020-58062-y

Qiu, J., Sheedlo, M. J., Yu, K., Tan, Y., Nakayasu, E. S., Das, C., et al. (2016). Ubiquitination Independent of E1 and E2 Enzymes by Bacterial Effectors. *Nature* 533 (7601), 120–124. doi: 10.1038/nature17657

Quezada, C. M., Hicks, S. W., Galán, J. E., and Erec Stebbins, C. (2009). A Family of Salmonella Virulence Factors Functions as a Distinct Class of Autoregulated E3 Ubiquitin Ligases. *Proc. Natl. Acad. Sci. U. S. A.* 106 (12), 4864–4869. doi: 10.1073/pnas.0811058106

Race, P. R., Solovyova, A. S., and Banfield, M. J. (2007). Conformation of the EPEC Tir Protein in Solution: Investigating the Impact of Serine Phosphorylation at Positions 434/463. *Biophys. J.* 93 (2), 586–596. doi: 10.1529/biophysj.106.101766

Radics, J., Königsmaier, L., and Marlovits, T. C. (2014). Structure of a Pathogenic Type 3 Secretion System in Action. *Nat. Struct. Mol. Biol.* 21 (1), 82–87. doi: 10.1038/nsmb.2722

Ramos-Morales, F. (2012). Impact of Salmonella Enterica Type III Secretion System Effectors on the Eukaryotic Host Cell. *Int. Scholarly Res. Netw. ISRN Cell Biol.* 2012, 36. doi: 10.5402/2012/787934

Rodgers, L., Gamez, A., Riek, R., and Ghosh, P. (2008). The Type III Secretion Chaperone SycE Promotes a Localized Disorder-to-Order Transition in the Natively Unfolded Effector YopE. *J. Biol. Chem.* 283 (30), 20857–20863. doi: 10.1074/jbc.M802339200

Romero, P., Obradovic, Z., Li, X., Garner, E. C., Brown, C. J., and Dunker, A. K. (2001). Sequence Complexity of Disordered Protein. *Proteins* 42 (1), 38–48. doi: 10.1002/1097-0134(20010101)42:1<38::AID-PROT50>3.0.CO;2-3

Saadat, I., Higashi, H., Obuse, C., Umeda, M., Murata-Kamiya, N., Saito, Y., et al. (2007). Helicobacter Pylori CagA Targets PAR1/MARK Kinase to Disrupt Epithelial Cell Polarity. *Nature* 447 (7142), 330–333. doi: 10.1038/nature05765

Samudrala, R., Heffron, F., and McDermott, J. E. (2009). Accurate Prediction of Secreted Substrates and Identification of a Conserved Putative Secretion Signal for Type III Secretion Systems. *PLoS Pathog.* 5 (4), e1000375. doi: 10.1371/journal.ppat.1000375

Selkrig, J., Li, N., Hausmann, A., Mangan, M. S. J., Zietek, M., Mateus, A., et al. (2020). Spatiotemporal Proteomics Uncovers Cathepsin-Dependent Macrophage Cell Death during Salmonella Infection. *Nat. Microbiol.* 5 (1), 119–33. doi: 10.1038/s41564-020-0736-7

Sgro, G. G., Oka, G. U., Souza, D. P., Cenens, W., Bayer-Santos, E., Matsuyama, B. Y., et al. (2019). Bacteria-Killing Type IV Secretion Systems. *Front. Microbiol.* 10 (1), 1078. doi: 10.3389/fmicb.2019.01078

Shames, S. R., Bhavsar, A. P., Croxen, M. A., Law, R. J., Mak, S. H. C., Deng, W., et al. (2011). The Pathogenic Escherichia Coli Type III Secreted Protease NleC Degrades the Host Acetyltransferase P300. *Cell. Microbiol.* 13 (10), 1542–1557. doi: 10.1111/j.1462-5822.2011.01640.x

Sheng, X., You, Q., Zhu, H., Chang, Z., Li, Q., Wang, H., et al. (2017). Bacterial Effector NleL Promotes Enterohemorrhagic E. Coli-Induced Attaching and Effacing Lesions by Ubiquitylating and Inactivating JNK. *PLoS Pathog.* 13 (7), e1006534. doi: 10.1371/journal.ppat.1006534

Sheng, X., You, Q., Zhu, H., Li, Q., Gao, H., Wang, H., et al. (2020). Coli Effector NleL Disrupts Host NF-KB Signaling by Targeting Multiple Host Proteins. *J. Mol. Cell Biol.* 12 (4), 318–321. doi: 10.1093/jmcb/mjaa003

Shin, D., Mukherjee, R., Liu, Y., Gonzalez, A., Bonn, F., Liu, Y., et al. (2020). Regulation of Phosphoribosyl-Linked Serine Ubiquitination by Deubiquitinases DupA and DupB. *Mol. Cell* 77 (1), 164–179.e6. doi: 10.1016/j.molcel.2019.10.019

Singer, A. U., Rohde, J. R., Lam, R., Skarina, T., Kagan, O., DiLeo, R., et al. (2008). Structure of the Shigella T3SS Effector IpaH Defines a New Class of E3 Ubiquitin Ligases. *Nat. Struct. Mol. Biol.* 15 (12), 1293–1301. doi: 10.1038/nsmb.1511

Soundararajan, V., Patel, N., Subramanian, V., Sasisekharan, V., and Sasisekharan, R. (2011). The Many Faces of the YopM Effector from Plague Causative Bacterium Yersinia Pestis and Its Implications for Host Immune Modulation. *Innate Immun.* 17 (6), 548–557. doi: 10.1177/1753425910377099

Sun, H., Kamanova, J., Lara-Tejero, M., and Galán, J. E. (2016). A Family of Salmonella Type III Secretion Effector Proteins Selectively Targets the NF-KB Signaling Pathway to Preserve Host Homeostasis. *PLoS Pathog.* 12 (3), e1005484. doi: 10.1371/journal.ppat.1005484

Tan, Y., Arnold, R. J., and Luo, Z. Q. (2011). Legionella Pneumophila Regulates the Small GTPase Rab1 Activity by Reversible Phosphorylcholination. *Proc. Natl. Acad. Sci. U. S. A.* 108 (52), 21212–21217. doi: 10.1073/pnas.1114023109

Trokter, M., and Waksman, G. (2018). Translocation through the Conjugative Type IV Secretion System Requires Unfolding of Its Protein Substrate. *J. Bacteriol.* 200 (6), e00615–17. doi: 10.1128/JB.00615-17

Turco, M. M., and Sousa, M. C. (2014). The Structure and Specificity of the Type III Secretion System Effector NleC Suggest a DNA Mimicry Mechanism of Substrate Recognition. *Biochemistry* 53 (31), 5131–5139. doi: 10.1021/bi500593e

Urbanus, M. L., Quaile, A. T., Stogios, P. J., Morar, M., Rao, C., Leo, R., et al. (2016). Diverse Mechanisms of Metaeffector Activity in an Intracellular Bacterial Pathogen, Legionella Pneumophila. *Mol. Syst. Biol.* 12 (12), 893. doi: 10.15252/msb.20167381

Uversky, V. N. (2011). Intrinsically Disordered Proteins from A to Z. *Int. J. Biochem. Cell Biol.* 43 (8), 1090–1093. doi: 10.1016/j.biocel.2011.04.001

Vacic, V., Oldfield, C. J., Mohan, A., Radivojac, P., Cortese, M. S., Uversky, V. N., et al. (2007a). Characterization of Molecular Recognition Features, MoRFs, and Their Binding Partners. *J. Proteome Res.* 6 (6), 2351–2366. doi: 10.1021/pr0701411

Vacic, V., Uversky, V. N., Dunker, A. K., and Lonardi, S. (2007b). Composition Profiler: A Tool for Discovery and Visualization of Amino Acid Composition Differences. *BMC Bioinformatics* 8 (1), 211. doi: 10.1186/1471-2105-8-211

Valleau, D., Quaile, A. T., Cui, H., Xu, X., Evdokimova, E., Chang, C., et al. (2018). Discovery of Ubiquitin Deamidases in the Pathogenic Arsenal of Legionella Pneumophila. *Cell Rep.* 23 (2), 568–583. doi: 10.1016/j.celrep.2018.03.060

Walch, P., Selkrig, J., Knodler, L., Rettel, M., Stein, F., Fernandez, K., et al. (2020). Global Mapping of Salmonella Enterica -Host Protein-Protein Interactions during Infection. *BioRxiv* 2020.05.04.075937. doi: 10.1101/2020.05.04.075937

Worby, C. A., Mattoo, S., Kruger, R. P., Corbeil, L. B., Koller, A., Mendez, J. C., et al. (2009). The Fic Domain: Regulation of Cell Signaling by Adenylylation. *Mol. Cell* 34 (1), 93–103. doi: 10.1016/j.molcel.2009.03.008

Yao, Q., Cui, J., Zhu, Y., Wang, G., Hu, L., Long, C., et al. (2009). A Bacterial Type III Effector Family Uses the Papain-like Hydrolytic Activity to Arrest the Host Cell Cycle. *Proc. Natl. Acad. Sci. U. S. A.* 106 (10), 3716–3721. doi: 10.1073/pnas.0900212106

Yarbrough, M. L., Li, Y., Kinch, L. N., Grishin, N. V., Ball, H. L., and Orth, K. (2009). AMPylation of Rho GTPases by Vibrio VopS Disrupts Effector Binding and Downstream Signaling. *Science* 323 (5911), 269–272. doi: 10.1126/science.1166382

Yu, X.-J., Liu, M., Holden, D. W., and Hultgren, S. J. (2016). Salmonella Effectors SseF and SseG Interact with Mammalian Protein ACBD3 (GCP60) To Anchor Salmonella-Containing Vacuoles at the Golgi Network. *mBio* 7(4):e00474–16. doi: 10.1128/mBio.00474-16

Zhang, Y., Higashide, W. M., McCormick, B. A., Chen, J., and Zhou, D. (2006). The Inflammation-Associated Salmonella SopA Is a HECT-like E3 Ubiquitin Ligase. *Mol. Microbiol.* 62 (3), 786–793. doi: 10.1111/j.1365-2958.2006.05407.x

Zhou, Y., Dong, N., Hu, L., and Shao, F. (2013). The Shigella Type Three Secretion System Effector OspG Directly and Specifically Binds to Host Ubiquitin for Activation. *PLoS One* 8 (2), e57558. doi: 10.1371/journal.pone.0057558

Zhou, D., and Galán, J. (2001). Salmonella Entry into Host Cells: The Work in Concert of Type III Secreted Effector Proteins. *Microbes Infect.* 3 (14–15), 1293–1298. doi: 10.1016/s1286-4579(01)01489-7

Zhu, Y., Li, H., Long, C., Hu, L., Xu, H., Liu, L., et al. (2007). Structural Insights into the Enzymatic Mechanism of the Pathogenic MAPK Phosphothreonine Lyase. *Mol. Cell* 28 (5), 899–913. doi: 10.1016/j.molcel.2007.11.011

15

Interferons: Tug of War Between Bacteria and Their Host

Noémie Alphonse[1,2†], Ruth E. Dickenson[1†] and Charlotte Odendall[1]*

[1] *Department of Infectious Diseases, School of Immunology and Microbial Sciences, King's College London, London, United Kingdom,* [2] *Immunoregulation Laboratory, Francis Crick Institute, London, United Kingdom*

***Correspondence:**
Charlotte Odendall
Charlotte.odendall@kcl.ac.uk

†These authors have contributed equally to this work

Type I and III interferons (IFNs) are archetypally antiviral cytokines that are induced in response to recognition of foreign material by pattern recognition receptors (PRRs). Though their roles in anti-viral immunity are well established, recent evidence suggests that they are also crucial mediators of inflammatory processes during bacterial infections. Type I and III IFNs restrict bacterial infection *in vitro* and in some *in vivo* contexts. IFNs mainly function through the induction of hundreds of IFN-stimulated genes (ISGs). These include PRRs and regulators of antimicrobial signaling pathways. Other ISGs directly restrict bacterial invasion or multiplication within host cells. As they regulate a diverse range of anti-bacterial host responses, IFNs are an attractive virulence target for bacterial pathogens. This review will discuss the current understanding of the bacterial effectors that manipulate the different stages of the host IFN response: IFN induction, downstream signaling pathways, and target ISGs.

Keywords: bacterial effectors, interferons, interferon-stimulated genes, janus kinase signal transducer and activator of transcription signaling, immunity, microbial pathogenesis, host-pathogen interactions

INTRODUCTION

The first interferons (IFNs) were discovered in 1957 by Alick Isaacs and Jean Lindenmann when they noticed that tissues inoculated with inactivated virus produced a soluble substance that "interfered" with subsequent viral infection (Isaacs and Lindenmann, 1957). Over 20 years before Charles Janeway's (Janeway, 1989) predictions on pattern recognition, Isaacs and Lindenmann had recognized the fundamental properties of IFN: a cytokine produced in response to detection of pathogen associated molecular patterns (PAMPs) by pattern recognition receptors (PRRs), that can protect tissues from subsequent microbial infections.

There are three families of IFN: type I, II, and III, which bind to the IFNα receptor (IFNAR), IFNγ receptor (IFNGR) and IFNλ receptor (IFNLR), respectively. Type I IFNs include IFNα, IFNβ, and the lesser-studied IFNϵ, κ, τ, δ, ζ, and ω and are expressed by almost all cells (Pestka et al., 2004; Hertzog and Williams, 2013). IFNγ is the sole type II IFN family member and is not induced in response to pattern recognition, but rather by immune cells in response to other cytokines (Pien et al., 2000; Salazar-Mather et al., 2000; Schindler et al., 2001). Type III IFNs were discovered most recently (Sheppard et al., 2002; Kotenko et al., 2003) and include four members in humans: IFNλ1-4. IFNAR and IFNGR are expressed on all nucleated cells, however the expression of IFNLR is restricted to epithelial cells (ECs) and some immune cells, including neutrophils (Blazek et al., 2015; Broggi et al., 2017). As such, type I and II IFNs have systemic functions (De Weerd and Nguyen, 2012; Chen et al., 2017; Lazear et al., 2019) while type III IFNs are crucial for host defense at barrier sites including the gut and lung (Durbin et al., 2013; Broggi et al., 2017; Lazear et al., 2019; Broggi et al., 2020). IFNs signal in an autocrine and paracrine manner through their respective receptors to activate janus kinase/signal transducer and activator of transcription (JAK/STAT) signaling cascades, resulting in the expression of hundreds of genes, known collectively as interferon-stimulated genes (ISGs). Although the roles of IFNs and ISGs have been extensively studied in the context of viral infection, there is a dearth of knowledge of their role during bacterial infections. Type I and III IFNs are strongly induced upon recognition of bacterial ligands, and play diverse and context-dependent roles during infection (Kagan et al., 2008; Pandey et al., 2009).

To promote their survival within the host, bacteria have evolved virulence factors such as secretion systems; apparatuses that translocate effector proteins across host membranes (Green and Mecsas, 2016). These effectors enable the pathogen to evade and perturb the host response or even use it to their advantage. This review will examine the interplay between bacterial effectors and the IFN response. As modulation of IFNγ has been previously discussed (Kak et al., 2018) we have focussed our discussion on the lesser understood type I and III IFNs. We will first summarize the current understanding of the complex functions of type I and III IFNs in the defense against pathogenic bacteria.

THE CAPRICIOUS ROLES OF TYPES I AND III INTERFERONS DURING BACTERIAL INFECTIONS

Type I and III IFNs have anti-bacterial properties in most *in vitro* tissue culture models. For example, treatment of polarized ECs by type I and/or III IFNs protects epithelial barriers from damage caused by enteropathogenic *Escherichia coli* (EPEC) (Long et al., 2014), *Salmonella enterica* serovar Typhimurium, and *Shigella flexneri* (Odendall et al., 2017). In addition, type I IFN restricts the replication of several intracellular bacteria, including *S.* Typhimurium and *S. flexneri* (Bukholm et al., 1984; Niesel et al., 1986; Duménil et al., 1998; Helbig et al., 2019), *Chlamydia trachomatis* (Snyder et al., 2017), *Mycobacterium tuberculosis* (Ranjbar et al., 2015), *Legionella pneumophila* (Lippmann et al., 2011), *Listeria monocytogenes* (Zwaferink et al., 2008; Radoshevich et al., 2015), *Francisella novicida* (Henry et al., 2007; Henry et al., 2010), and *Rickettsia parkeri* (Burke et al., 2020). The inhibition of bacterial replication is often the result of the action of ISGs. For example, viperin is a highly evolutionarily conserved ISG that restricts *S. flexneri* infection in HeLa cells (Helbig et al., 2019).

The roles of type I and III IFNs are more complex to delineate *in vivo* and can vary depending on the pathogen and biological context. Type I IFNs can restrict infections by pathogens such as *Bacillus anthracis* (Walberg et al., 2008), *L. pneumophila* (Lippmann et al., 2011; Naujoks et al., 2016), *Helicobacter pylori* (Watanabe et al., 2010), *Streptococcus pyogenes* (Castiglia et al., 2016), group B *Streptococcus*, *E. coli* (Mancuso et al., 2007), and *Streptococcus pneumoniae* (LeMessurier et al., 2013). In particular, IFNs prevented bacterial migration across endothelial and epithelial barriers (LeMessurier et al., 2013). Whether this protection extends to type III IFNs is unclear, but type III IFNs were demonstrated to be protective in murine models of colitis (Rauch et al., 2015; Broggi et al., 2017). In contrast, in other infection models, type I IFNs can be detrimental to hosts infected with bacterial pathogens. When compared to wild-type (WT) mice, *Ifnar*$^{-/-}$ animals were more resistant to systemic infection with *L. monocytogenes* (Auerbuch et al., 2004; Brzoza-Lewis et al., 2012), and *S.* Typhimurium (Robinson et al., 2012; Perkins et al., 2015; Wilson et al., 2019; Zhang et al., 2020). Similarly, IFNβ treatment exacerbated infection with *M. tuberculosis* (Manca et al., 2001; Manca et al., 2005) and *B. anthracis* (Gold et al., 2007). The detrimental nature of type I IFN responses during *M. tuberculosis* infection is reflected in human disease. The blood transcriptome of patients with active disease has demonstrated a correlation between the abundance of Type I IFN-inducible transcripts and disease pathogenesis (Berry et al., 2010; Maertzdorf et al., 2011; Ottenhoff et al., 2012; Cliff et al., 2013). In addition, individuals with an inherited defect in ISG15 have an increased susceptibility to mycobacterial, but not viral, disease (Bogunovic et al., 2012).

Finally, type I and III IFNs induced by viral infection were shown to exacerbate subsequent respiratory superinfections with *S. pneumoniae* or *Staphylococcus aureus*. Indeed, *Ifnar*$^{-/-}$ and *Ifnlr*$^{-/-}$ mice showed improved bacterial control in virus-bacteria superinfection models (Shahangian et al., 2009; Broggi et al., 2020; Major et al., 2020). Similarly, administration of recombinant type I or III IFN resulted in increased bacterial burdens following viral infection or activation of antiviral pathways with viral ligands. Recent work has shown that uncontrolled type III IFN responses, such as those observed during SARS-CoV-2 infection, led to damage of epithelial barriers and increased susceptibility to bacterial superinfection, a known complication of COVID19 and Influenza (Broggi et al., 2020; Langford et al., 2020; Major et al., 2020).

TARGETING OF TYPES I AND III INTERFERON EXPRESSION BY BACTERIAL EFFECTORS

In order to control their host, pathogenic bacteria secrete effectors that manipulate different stages of the IFN response, from its production to signaling and even ISG functions.

Host recognition of bacterial PAMPs occurs *via* PRR ligation, leading to the assembly of signaling complexes and the activation of intracellular adaptor proteins [reviewed in (Odendall and Kagan, 2017)]. Cytosolic RNA sensors, retinoic acid-inducible gene-I (RIG-I)-like receptors (RLRs), signal *via* the adaptor mitochondria-antiviral signaling protein (MAVS) (Kawai et al., 2005; Meylan et al., 2005; Seth et al., 2005; Xu et al., 2005; Dixit et al., 2010), while the DNA sensor cyclic GMP–AMP synthase (cGAS) signals *via* stimulator of interferon genes (STING), localized on the endoplasmic reticulum (ER) (Gao et al., 2012; Wu et al., 2017). Toll-like receptors (TLRs) present on the cell surface and endosomes, signal *via* the adaptors MyD88 and/or TRIF (Kawasaki and Kawai, 2014). Although only endosomal TLRs efficiently induce type I IFNs, all TLRs were shown to strongly induce type III IFNs in response to bacterial ligands (Odendall et al., 2017). These signaling pathways culminate in the activation and nuclear translocation of transcription factors, including interferon regulatory factors (IRFs) that control IFN expression (Takeuchi and Akira, 2010; Odendall and Kagan, 2017; Odendall and Kagan, 2019). IFN expression also requires the NFκB and MAP-kinase activated transcription factor AP-1 (Odendall and Kagan, 2017).

As IFNs can have both beneficial and detrimental effects on bacterial pathogens, some species have evolved effectors that promote or inhibit their production (summarized in **Figure 1** and **Table 1**). For example, *L. monocytogenes* secretes *Zea*, a ribonucleoprotein, that binds RIG-I and potentiates type I IFN responses in ECs (Pagliuso et al., 2019). In contrast, many bacterial effectors block the induction of IFN. The effects of some of these effectors on IFN induction may be considered as indirect as they either interfere with PAMP detection or block general innate sensing pathways. *L. monocytogenes* secretes an effector, PgdA, that deacetylates peptidoglycan, conferring a resistance to host lysozyme. This prevents the release of PAMPs and dampens the induction of cytokines and type I IFN (Boneca et al., 2007). Similarly, the *L. pneumophila* effector SdhA inhibits RLR activation in mouse Bone marrow-derived macrophage (BMDM) by maintaining the integrity of the *Legionella*-containing vacuole (Monroe et al., 2009). The cGAS/STING pathway has since been shown to mediate IFN expression following *Legionella* infection (Ruiz-Moreno et al., 2018).

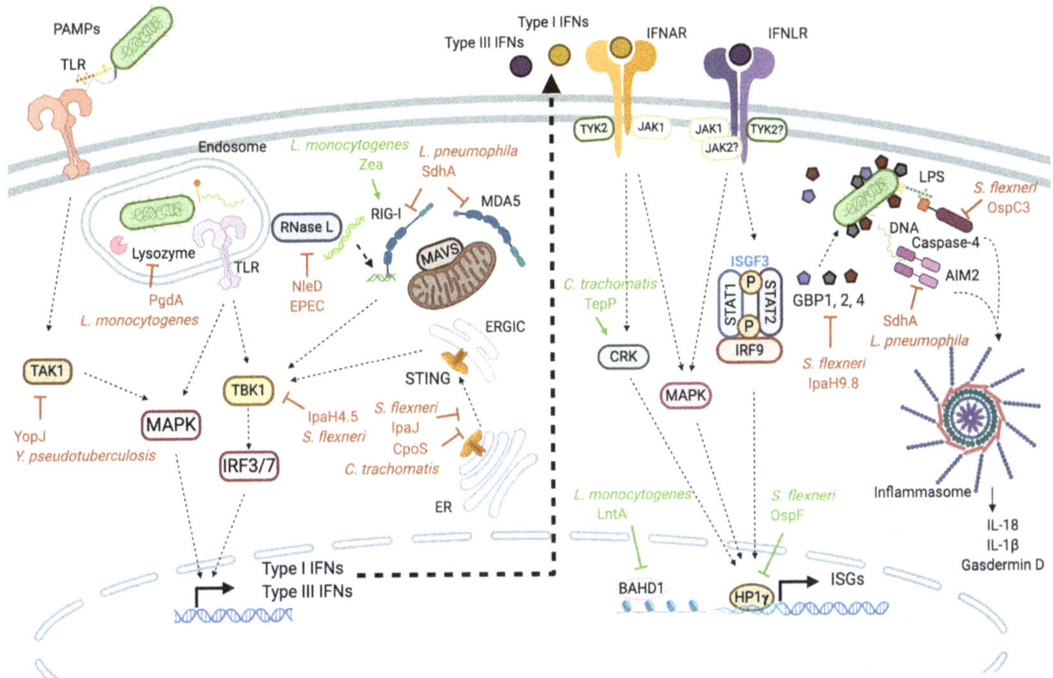

FIGURE 1 | Modulation of types I and III interferon responses by bacterial effectors. In many different cell types, the recognition of bacterial PAMPs by PRRs leads to the production of type I and III IFNs (left side). Type I and III IFNs bind their respective receptors initiating signaling pathways which trigger the expression of ISGs (right side). Bacteria have evolved effectors that either inhibit (depicted in red) or stimulate (depicted in green) type I and III IFN responses. *Yersinia pseudotuberculosis* YopJ effector interferes with the MAP kinase activator TAK1. When internalized in endosomes, *Listeria monocytogenes* secretes PgdA which modifies its peptidoglycan, hiding it from the host lysozyme. *Listeria monocytogenes* also secretes Zea which binds RIG-I and potentiates the production of IFNs. The *Legionella pneumophila* protein SdhA blocks RIG-I and MDA5 cytosolic PRRs preventing them to interact with MAVS. The *Shigella flexneri* protease IpaJ interferes with STING translocation to the ERGIC while IpaH4.5 inhibits TBK1 kinase. Similarly, *Chlamydia trachomatis* CpoS effector affects STING migration. EPEC, through the secretion of NleD inhibits RNase L which counteracts the protective effects of type I IFN on epithelial barriers. Additionally, some bacteria perturb type I and III IFN signaling. *L. monocytogenes* LntA and *S. flexneri* OspF effectors enter the cell nucleus and potentiate ISGs transcription. LntA sequesters BAHD1, a negative regulator of ISGs transcription therefore stimulating ISGs expression. OspF interacts with HP1γ preventing its phosphorylation, which promotes ISGs expression. *C. trachomatis* TepP effector inhibits CRK, a component and regulators of type I and III IFN signaling, stimulating ISGs transcription. Finally, *S. flexneri* IpaH9.8 and OspC3 as well as *L. pneumophila* SdhA effectors inhibit inflammasome components; GBP1, caspase-4, and AIM2 respectively. PRRs, pattern recognition receptors; PAMPs, pathogen-associated molecular patterns; TLR, Toll like receptor; STING, stimulator of interferon genes; ER, endoplasmic reticulum; ERGIC, ER-Golgi intermediate compartments; IRF, interferon regulatory factor; RIG-I, retinoic acid-inducible gene-I; MDA5, melanoma differentiation-associated protein 5; MAVS, mitochondria-antiviral signaling protein; TAK1, TGFbeta activated kinase 1; TYK2, tyrosine kinase 2; JAK1, Janus kinase 1; STAT, signal transducer and activator of transcription; ISGF3, interferon-stimulated gene factor 3; MAPK, mitogen-activated protein kinase; GBP, guanylate-binding protein; AIM2, absent in melanoma-2; LPS, lipopolysaccharide; ISG, interferon stimulated gene; IFNAR, IFNα receptor; IFNLR,IFNλ receptor. **Figure 1** was created with Biorender.com.

TABLE 1 | Summary of effectors that target IFN pathways.

Bacteria	Effector	Secretion system	Biochemical activity	Function	References
Bacillus anthracis	UN	T4SS-like	UN	Inhibits the assembly of ISGF3	(Gold et al., 2004)
Chlamydia pneumoniae	UN	T3SS	Protease	Degrades TRAF3	(Wolf and Fields, 2013)
Chlamydia trachomatis	CpoS	T3SS	UN	Counteracts STING-mediated IFN responses and perturbs ISRE and ISG expression	(Sixt et al., 2017)
Chlamydia trachomatis	TepP	T3SS	Scaffold	Binds Crk-I and Crk-II, initiating and amplifying signaling cascades (*IFIT1*, *IFIT2*)	(Chen et al., 2014)
Enteropathogenic Escherichia coli	NleD	T3SS	Zinc metallo-protease	Targets the endoribonuclease RNase L, a key mediator of IFNβ induction and action, compromising the barrier integrity	(Long et al., 2014)
Legionella pneumophila	SdhA	T4SS	UN	Helps maintain vacuolar integrity	(Monroe et al., 2009) (Creasey and Isberg, 2012)
		T4SS	UN	Negatively regulates AIM2 inflammasome activation	(Ge et al., 2012)
Listeria monocytogenes	LntA	Sec SS	Nucleo-modulin	Interacts with BAHD1, inducing ISG expression	(Lebreton et al., 2011)
	PgdA	?	Peptido-glycan N-deacetylase	Deacetylates bacterial peptidoglycan, conferring a resistance to host lysozyme which ultimately prevents the release of PAMPs	(Boneca et al., 2007)
	Zea	Sec SS	Ribonucleo-protein	Triggers a RIG-I dependent type I IFN response	(Pagliuso et al., 2019)
***Salmonella* Typhimurium**	UN	T3SS	UN	Represses the TRIF-dependent type I IFN response	(Owen et al., 2016)
Shigella flexneri	IpaJ	T3SS	Cysteine protease	Blocks STING translocation from the ER to the ERGIC compartment	(Dobbs et al., 2015)
	IpaH4.5	T3SS	E3 ubiquitin ligase	Promotes proteasome-dependent degradation of TBK1	(Zheng et al., 2016)
	OspC3	T3SS	UN	Binds and inhibits Caspase 4	(Kobayashi et al., 2013)
	OspF	T3SS	Phospho-threonine lyase	Inhibits HP1γ; phosphorylation, repressing the expression of some IFN-regulated genes: *Ifi44*, *Ifit3*, and *Oas1b*	(Lavigne et al., 2009; Harouz et al., 2014)
	IpaH9.8	T3SS	E3 ubiquitin ligase	Targets GBP1, GBP2 and GBP4 for proteasomal degradation	(Li et al., 2017; Wandel et al., 2017)
Yersinia pseudo-tuberculosis	YopJ	T3SS	De- ubiquitinase/ Acetyl transferase	Targets TAK1 to inhibit IFN production	(Rosadini et al., 2015) (Paquette et al., 2012)

UN, unknown.

Although this has not been formally tested, it is likely that SdhA also protects from detection by cGAS. *Yersinia pseudotuberculosis* YopJ targets MAP-kinase signaling in dendritic cells and macrophages, disrupting MyD88- and TRIF-dependent signaling downstream of TLR4, which prevents type I IFN and pro-inflammatory cytokine expression (Paquette et al., 2012; Rosadini et al., 2015). EPEC is able to counteract the protective effects of type I IFN on epithelial barriers; its effector NleD inhibits RNase L, an endoribonuclease that enhances RLR-mediated production of IFNβ (Long et al., 2014).

Other effectors affect the localization of adaptor proteins that are crucial to PRR signaling cascades. For example, STING translocates from the ER to the ER-Golgi intermediate compartments (ERGIC) to mediate cGAS signaling (Ishikawa et al., 2009; Burdette and Vance, 2013). IpaJ is a *S. flexneri* protease that targets the small guanosine triphosphatases (GTPases) required for this migration event. This blocks STING translocation, abolishing type I IFN and ISG expression (Dobbs et al., 2015). Similarly CpoS from *C. trachomatis* affects STING migration and blocks type I and III IFN and ISG induction (Sixt et al., 2017).

The kinase TBK1 lies downstream of many PRRs and phosphorylates IRF3 and IRF7, leading to the expression of type I and III IFNs (Clark et al., 2011; Tanaka and Chen, 2012; Schneider et al., 2014; Liu et al., 2015; Bakshi et al., 2017). It can also activate IRF1, which may specifically drive type III IFN induction (Odendall et al., 2014). The *S. flexneri* effector IpaH4.5 is an E3 ubiquitin ligase that targets TBK1 for proteasomal degradation, leading to the suppression of IFN production (Zheng et al., 2016). Interestingly, mice infected with Δ*ipaH4.5* mutants had lower bacterial burdens than those infected with WT *S. flexneri*.

Finally, some bacteria modulate IFN production *via* yet-unknown mechanisms. *Salmonella* represses TRIF-dependent type I IFN induction in macrophages (Owen et al., 2016). Likewise, *Chlamydia pneumoniae* infection of ECs does not lead to IFNβ production and very efficiently blocks IRF3 phosphorylation. Although the effector responsible was not identified, *Chlamydia* was shown to induce the degradation of TRAF3, a crucial TBK1 activator (Wolf and Fields, 2013). Conversely, some pathogens such as *Legionella* (Ruiz-Moreno et al., 2018) or *Burkholderia pseudomallei* (Ku et al., 2020), were shown to induce IFNβ by activating the cGAS-STING pathway in a secretion system-dependent manner.

Although the pathways that drive type I and III IFN expression overlap significantly, the majority of this research has focused on type I IFNs. It will be interesting to investigate whether bacterial effectors can specifically target the production of one family or the other.

EFFECTOR-DRIVEN MANIPULATION OF SIGNALING DOWNSTREAM OF INTERFERON RECEPTORS

JAK/STAT signaling activated by type I or III IFN receptor binding leads to the formation of a STAT1/STAT2/IRF9 complex called the interferon-stimulated gene factor 3 (ISGF3). This complex translocates to the nucleus and binds interferon stimulated response elements (ISREs) to transcribe ISGs. Instead of targeting IFN production, some bacterial effectors target signaling cascades downstream of IFN receptors to alter the expression of ISGs. For example, *B. anthracis* inhibits IFNβ-mediated STAT1 phosphorylation and the formation of ISGF3, but the bacterial effectors remain to be identified (Gold et al., 2004).

Bacteria can also affect transcription of ISGs by altering epigenetic regulators of transcription. Heterochromatin protein 1 (HP1) family members are epigenetic regulators that bind methylated histone H3 to influence the expression of a wide variety of genes (Saint-André et al., 2011; Ameyar-Zazoua et al., 2012; Smallwood et al., 2012). In a guinea pig model of *Shigella* enterocolitis, *S. flexneri* inhibited HP1γ phosphorylation *via* OspF. Next-generation sequencing of mouse embryonic fibroblast-derived cell lines revealed that HP1γ represses the expression of some IFN-regulated genes, namely *Ifi44*, *Ifit3*, and *Oas1b* (Harouz et al., 2014), confirming previous observations (Lavigne et al., 2009). This suggests that *S. flexneri* effectors are capable of modifying epigenetic regulators of subsets of ISGs to alter the outcome of IFN signaling. Likewise, the *C. trachomatis* effector TepP, was shown to bind the adaptor protein CRK and affect expression of ISGs in human cell lines (Chen et al., 2014).

Some bacterial effectors enter the nucleus to alter chromatin remodeling in their mammalian hosts (Bhavsar et al., 2007; Lebreton et al., 2011). Infection of cells with an *L. monocytogenes* strain constitutively expressing the effector LntA resulted in enrichment of expression of a number of genes, 83% of which were ISGs. Interestingly, when *lntA* was deleted, there was a reduction in bacterial burden in the spleens and livers of mice. LntA was shown to interact with BAHD1; a heterochromatin protein that acts as a transcriptional repressor (Bierne et al., 2009). Chromatin immunoprecipitation analysis showed that association of BAHD1 with key ISGs was reduced in cells infected with *L. monocytogenes* constitutively expressing LntA. Therefore, *Listeria* induces the expression of ISGs by inhibiting the negative regulator BAHD1 (Lebreton et al., 2011).

BACTERIAL EFFECTORS AND TYPE I AND III INTERFERON-STIMULATED GENES

All three families of IFNs induce the expression of ISGs. As discussed above, the set of 300–600 genes transcribed by type I and III IFNs contain ISRE sequences. In contrast, IFNγ induces ISGs containing a gamma interferon activation site (GAS) element. ISGs can contain both ISRE and GAS sequences and therefore be induced by all three families. As they are induced in concert, the individual roles of many ISGs are not well understood. Although more is known about the ISGs targeted by viral virulence factors, there are few examples of bacterial factors that can alter their integrity or function.

Among ISGs are several components of inflammasomes. These are large multi-protein complexes that are assembled upon recognition of PAMPs or changes in host homeostasis. Inflammasome activation leads to pyroptosis, a highly inflammatory form of cell death. Inflammasomes are key to host defense against intracellular bacteria. As such, some bacteria have evolved strategies to evade inflammasome activation. For example, absent in melanoma-2 (AIM2) is an IFN-inducible inflammasome receptor that detects the presence of bacterial DNA in the cytosol (Man et al., 2015; Meunier et al., 2015). The *L. pneumophila* effector SdhA, in addition to suppressing type I IFN responses (Monroe et al., 2009; Creasey and Isberg, 2012), negatively regulates AIM2 inflammasome activation in human macrophages (Ge et al., 2012).

Another inflammasome 'receptor' induced by IFNs is murine caspase 11, which detects intracellular LPS and induces the non-canonical inflammasome (Ding and Shao, 2017). The human orthologs of caspase 11 are caspases 4 and 5, but although caspase 4 was shown to be transcriptionally induced by type I IFN, whether this occurs at the protein level is still poorly understood (Knodler et al., 2014; Casson et al., 2015; Schmid-Burgk et al., 2015). However, it is clear that the caspase 4 inflammasome is "primed" by IFN *via* the action of guanylate-binding proteins (GBPs) (Pilla et al., 2014; Feng and Man, 2020; Santos et al., 2020). GBPs are among the best-characterized antibacterial ISGs. This family of GTPases has an ever-growing list of functions, many of which contribute to defenses against intracellular pathogens. There are seven *GBP* genes in humans, whose expression is most robustly induced by IFNγ, but can also be induced by treatment with type I IFNs (Cheng et al., 1985; Kim et al., 2011; Tretina et al., 2019). Some GBPs can also be produced in response to treatment with type III IFNs (Alase et al., 2015; Tretina et al., 2019). GBPs are recruited to intracellular gram-negative bacteria (*S.* Typhimurium, *S. flexneri*, *Burkholderia thailandensis*, *Brucella abortus*, *F. novicida*, *C. trachomatis*, and *L. pneumophila*) and liberate LPS into the cytosol (Kim et al., 2011; Kim et al., 2012; Shenoy et al., 2012; Haldar et al., 2014; Meunier et al., 2014; Man et al., 2015; Meunier and Broz, 2015; Pilla et al., 2014; Shi et al., 2014; Man et al., 2016; Feeley et al., 2017; Finethy et al., 2017; Lindenberg et al., 2017; Cerqueira et al., 2018; Fisch et al., 2019; Bass and Shin, 2020; Santos et al., 2020). Recent work on *S. flexneri* and *S.* Typhimurium has shown that GBP1 directly binds bacterial LPS through electrostatic interactions and assembles a signaling platform containing other GBPs. This platform recruits and activates human caspase 4, leading to assembly of the non-canonical inflammasome (Kutsch et al., 2020; Santos et al., 2020; Wandel et al., 2020). In addition, GBPs inhibit intracellular motility and cell-to-cell spread of *B. thailandensis* and *S. flexneri* (Ostler et al., 2014; Piro et al., 2017; Wandel et al., 2017).

S. flexneri has evolved at least two effector proteins to bypass these important host defense mechanisms; firstly, OspC3, which binds and inhibits caspase 4. *ospC3* mutants have a growth defect in IFNγ-treated cells and induce rapid cell death.

These mutants were also attenuated in WT mice, but not in mice lacking GBP1, GBP2, or caspase 11 (Kobayashi et al., 2013; Wandel et al., 2020). Whether they are more sensitive to type I or III IFNs is unclear. Secondly, the E3 ubiquitin ligase IpaH9.8 targets hGBP1, hGBP2, and hGBP4 for proteasomal degradation (Li et al., 2017; Wandel et al., 2017). *Shigella* strains lacking *ipaH9.8* displayed an increased recruitment of GBP1, 2, 3, and 4, as well as caspase 4, compared to WT (Wandel et al., 2017; Wandel et al., 2020). In addition, cell-to-cell spread of *ipaH9.8* mutant bacteria was lowered (Wandel et al., 2017). These data highlight the importance of GBPs to entrap intracellular bacteria and act as platforms for the activation of innate signaling. It is therefore highly probable that other intracellular bacteria have evolved mechanisms to counteract GBPs.

CONCLUDING REMARKS

IFNs are crucial mediators of inflammation, playing complex, yet key roles in both systemic and localized bacterial infections. In order to survive and cause disease, bacteria secrete effectors that interfere with type I and III IFN production and signaling, as well as ISG expression and function. As there is still much to learn about the role of IFN and ISGs, perhaps understanding the mechanisms that bacterial pathogens have evolved to evade or enhance these responses will lead to insights into their function in host defense. These discussions certainly mandate further studies into the interactions between IFNs and bacteria.

AUTHOR CONTRIBUTIONS

NA and RD contributed equally. All authors contributed to the article and approved the submitted version.

ACKNOWLEDGMENTS

We thank our colleagues Joseph Wanford, Andrew Voak and Abderrahman Hachani for ideas and critical reading of this manuscript.

REFERENCES

Alase, A. A., El-Sherbiny, Y. M., Vital, E. M., Tobin, D. J., Turner, N. A., and Wittmann, M. (2015). IFNλ stimulates MxA production in human dermal fibroblasts via a MAPK-dependent STAT1-independent mechanism. *J. Invest. Dermatol.* 135, 2935–2943. doi: 10.1038/jid.2015.317

Ameyar-Zazoua, M., Rachez, C., Souidi, M., Robin, P., Fritsch, L., Young, R., et al. (2012). Argonaute proteins couple chromatin silencing to alternative splicing. *Nat. Struct. Mol. Biol.* 19, 998–1005. doi: 10.1038/nsmb.2373

Auerbuch, V., Brockstedt, D. G., Meyer-Morse, N., O'Riordan, M., and Portnoy, D. A. (2004). Mice lacking the type I interferon receptor are resistant to Listeria monocytogenes. *J. Exp. Med.* 200, 527–533. doi: 10.1084/jem.20040976

Bakshi, S., Taylor, J., Strickson, S., McCartney, T., and Cohen, P. (2017). Identification of TBK1 complexes required for the phosphorylation of IRF3 and the production of interferon β. *Biochem. J.* 474, 1163–1174. doi: 10.1042/BCJ20160992

Bass, A. R., and Shin, S. (2020). Human GBP1 promotes pathogen vacuole rupture and inflammasome activation during. doi: 10.1101/2020.05.27.120477

Berry, M. P. R., Graham, C. M., McNab, F. W., Xu, Z., Bloch, S. A. A., Oni, T., et al. (2010). An interferon-inducible neutrophil-driven blood transcriptional signature in human tuberculosis. *Nature* 466, 973–977. doi: 10.1038/nature09247

Bhavsar, A. P., Guttman, J. A., and Finlay, B. B. (2007). Manipulation of host-cell pathways by bacterial pathogens. *Nature* 449, 827–834. doi: 10.1038/nature06247

Bierne, H., Tham, T. N., Batsche, E., Dumay, A., Leguillou, M., Kernéis-Golsteyn, S., et al. (2009). Human BAHD1 promotes heterochromatic gene silencing. *Proc. Natl. Acad. Sci. U. S. A.* 106, 13826–13831. doi: 10.1073/pnas.0901259106

Blazek, K., Eames, H. L., Weiss, M., Byrne, A. J., Perocheau, D., Pease, J. E., et al. (2015). IFN-λ resolves inflammation via suppression of neutrophil infiltration and IL-1β production. *J. Exp. Med.* 212, 845–853. doi: 10.1084/jem.20140995

Bogunovic, D., Byun, M., Durfee, L. A., Abhyankar, A., Sanal, O., Mansouri, D., et al. (2012). Mycobacterial disease and impaired IFN-γ immunity in humans with inherited ISG15 deficiency. *Science* 337, 1684–1688. doi: 10.1126/science.1224026

Boneca, I. G., Dussurget, O., Cabanes, D., Nahori, M.-A., Sousa, S., Lecuit, M., et al. (2007). A critical role for peptidoglycan N-deacetylation in Listeria evasion from the host innate immune system. *Proc. Natl. Acad. Sci. U. S. A.* 104, 997–1002. doi: 10.1073/pnas.0609672104

Broggi, A., Tan, Y., Granucci, F., and Zanoni, I. (2017). IFN-λ suppresses intestinal inflammation by non-translational regulation of neutrophil function. *Nat. Immunol.* 18, 1084–1093. doi: 10.1038/ni.3821

Broggi, A., Ghosh, S., Sposito, B., Spreafico, R., Balzarini, F., Lo Cascio, A., et al. (2020). Type III interferons disrupt the lung epithelial barrier upon viral recognition. *Science* 369, 706–712. doi: 10.1126/science.abc3545

Brzoza-Lewis, K. L., Jason Hoth, J., and Hiltbold, E. M. (2012). Type I interferon signaling regulates the composition of inflammatory infiltrates upon infection with Listeria monocytogenes. *Cell. Immunol.* 273, 41–51. doi: 10.1016/j.cellimm.2011.11.008

Bukholm, G., Berdal, B. P., Haug, C., and Degre, M. (1984). Mouse fibroblast interferon modifies Salmonella typhimurium infection in infant mice. *Infect. Immun.* 45, 62–66. doi: 10.1128/iai.45.1.62-66.1984

Burdette, D. L., and Vance, R. E. (2013). STING and the innate immune response to nucleic acids in the cytosol. *Nat. Immunol.* 14, 19–26. doi: 10.1038/ni.2491

Burke, T. P., Engström, P., Chavez, R. A., Fonbuena, J. A., Vance, R. E., and Welch, M. D. (2020). Inflammasome-mediated antagonism of type I interferon enhances Rickettsia pathogenesis. *Nat. Microbiol.* 5, 688–696. doi: 10.1038/s41564-020-0673-5

Casson, C. N., Yu, J., Reyes, V. M., Taschuk, F. O., Yadav, A., Copenhaver, A. M., et al. (2015). Human caspase-4 mediates noncanonical inflammasome activation against gram-negative bacterial pathogens. *Proc. Natl. Acad. Sci. U. S. A.* 112, 6688–6693. doi: 10.1073/pnas.1421699112

Castiglia, V., Piersigilli, A., Ebner, F., Janos, M., Goldmann, O., Damböck, U., et al. (2016). Type i Interferon Signaling Prevents IL-1β-Driven Lethal Systemic Hyperinflammation during Invasive Bacterial Infection of Soft Tissue. *Cell Host Microbe* 19, 375–387. doi: 10.1016/j.chom.2016.02.003

Cerqueira, D. M., Gomes, M. T. R., Silva, A. L. N., Rungue, M., Assis, N. R. G., Guimarães, E. S., et al. (2018). Guanylate-binding protein 5 licenses caspase-11 for Gasdermin-D mediated host resistance to Brucella abortus infection. *PloS Pathog.* 14, e1007519. doi: 10.1371/journal.ppat.1007519

Chen, Y.-S., Bastidas, R. J., Saka, H. A., Carpenter, V. K., Richards, K. L., Plano, G. V., et al. (2014). The Chlamydia trachomatis Type III Secretion Chaperone Slc1 Engages Multiple Early Effectors, Including TepP, a Tyrosine-phosphorylated Protein Required for the Recruitment of CrkI-II to Nascent Inclusions and Innate Immune Signaling. *PloS Pathog.* 10, e1003954. doi: 10.1371/journal.ppat.1003954

Chen, C., Chen, L., Chen, C., Chen, Q., Zhao, Q., and Dong, Y. (2017). The Distribution Frequency of Interferon-Gamma Receptor 1 Gene Polymorphisms in Interferon- γ Release Assay-Positive Patients. *Dis. Markers* 2017, 5. doi: 10.1155/2017/4031671

Cheng, Y. S. E., Becker-Manley, M. F., Chow, T. P., and Horan, D. C. (1985). Affinity purification of an interferon-induced human guanylate-binding protein and its characterization. *J. Biol. Chem.* 260, 15834–15839. doi: 10.1016/S0021-9258(17)36334-2

Clark, K., Takeuchi, O., Akira, S., and Cohen, P. (2011). The TRAF-associated protein TANK facilitates cross-talk within the IκB kinase family during Toll-like receptor signaling. *Proc. Natl. Acad. Sci. U. S. A.* 108, 17093–17098. doi: 10.1073/pnas.1114194108

Cliff, J. M., Lee, J. S., Constantinou, N., Cho, J. E., Clark, T. G., Ronacher, K., et al. (2013). Distinct phases of blood gene expression pattern through tuberculosis treatment reflect modulation of the humoral immune response. *J. Infect. Dis.* 207, 18–29. doi: 10.1093/infdis/jis499

Creasey, E. A., and Isberg, R. R. (2012). The protein SdhA maintains the integrity of the Legionella-containing vacuole. *Proc. Natl. Acad. Sci. U. S. A.* 109, 3481–3486. doi: 10.1073/pnas.1121286109

De Weerd, N. A., and Nguyen, T. (2012). The interferons and their receptors-distribution and regulation. *Immunol. Cell Biol.* 90, 483–491. doi: 10.1038/icb.2012.9

Ding, J., and Shao, F. (2017). SnapShot: The Noncanonical Inflammasome. *Cell* 168, 544–544.e1. doi: 10.1016/j.cell.2017.01.008

Dixit, E., Boulant, S., Zhang, Y., Lee, A. S. Y., Odendall, C., Shum, B., et al. (2010). Peroxisomes Are Signaling Platforms for Antiviral Innate Immunity. *Cell* 141, 668–681. doi: 10.1016/j.cell.2010.04.018

Dobbs, N., Burnaevskiy, N., Chen, D., Gonugunta, V. K., Alto, N. M., and Yan, N. (2015). STING activation by translocation from the ER is associated with infection and autoinflammatory disease. *Cell Host Microbe* 176, 139–148. doi: 10.1016/j.chom.2015.07.001

Duménil, G., Olivo, J. C., Pellegrini, S., Fellous, M., Sansonetti, P. J., and Van Tran Nhieu, G. (1998). Interferon α inhibits a Src-mediated pathway necessary for Shigella- induced cytoskeletal rearrangements in epithelial cells. *J. Cell Biol.* 143, 1003–1012. doi: 10.1083/jcb.143.4.1003

Durbin, R. K., Kotenko, S. V., and Durbin, J. E. (2013). Interferon induction and function at the mucosal surface. *Immunol. Rev.* 255, 25–39. doi: 10.1111/imr.12101

Feeley, E. M., Pilla-Moffett, D. M., Zwack, E. E., Piro, A. S., Finethy, R., Kolb, J. P., et al. (2017). Galectin-3 directs antimicrobial guanylate binding proteins to vacuoles furnished with bacterial secretion systems. *Proc. Natl. Acad. Sci. U. S. A.* 114, E1698–E1706. doi: 10.1073/pnas.1615771114

Feng, S., and Man, S. M. (2020). Captain GBP1: inflammasomes assemble, pyroptotic endgame. *Nat. Immunol.* 21, 823–834. doi: 10.1038/s41590-020-0727-0

Finethy, R., Luoma, S., Orench-Rivera, N., Feeley, E. M., Haldar, A. K., Yamamoto, M., et al. (2017). Inflammasome activation by bacterial outer membrane vesicles requires guanylate binding proteins. *MBio* 8, 1–11. doi: 10.1128/mBio.01188-17

Fisch, D., Bando, H., Clough, B., Hornung, V., Yamamoto, M., Shenoy, A. R., et al. (2019). Human GBP 1 is a microbe-specific gatekeeper of macrophage apoptosis and pyroptosis. *EMBO J.* 38, 1–19. doi: 10.15252/embj.2018100926

Gao, D., Wu, J., Wu, Y., Du, F., Aroh, C., Yan, N., et al. (2012). Cyclic GMP-AMP Synthase Is a Cytosolic DNA Sensor That Activates the Type I Interferon Pathway. *Science* 339, 786–791. doi: 10.1126/science.1232458

Ge, J., Gong, Y. N., Xu, Y., and Shao, F. (2012). Preventing bacterial DNA release and absent in melanoma 2 inflammasome activation by a Legionella effector functioning in membrane trafficking. *Proc. Natl. Acad. Sci. U. S. A.* 109, 6193–6198. doi: 10.1073/pnas.1117490109

Gold, J. A., Hoshino, Y., Hoshino, S., Jones, M. B., Nolan, A., and Weiden, M. D. (2004). Exogenous Gamma and Alpha/Beta Interferon Rescues Human Macrophages from Cell Death Induced by Bacillus anthracis. *Infect. Immun.* 72, 1291–1297. doi: 10.1128/IAI.72.3.1291-1297.2004

Gold, J. A., Hoshino, Y., Jones, M. B., Hoshino, S., Nolan, A., and Weiden, M. D. (2007). Exogenous interferon-α and interferon-γ increase lethality of murine inhalational anthrax. *PloS One* 72, 1291–1297. doi: 10.1371/journal.pone.0000736

Green, E. R., and Mecsas, J. (2016). Bacterial Secretion Systems: An Overview. *Virulence Mech. Bact. Pathog.* 4, 213–239. doi: 10.1128/9781555819286.ch8

Haldar, A. K., Piro, A. S., Pilla, D. M., Yamamoto, M., and Coers, J. (2014). The E2-like conjugation enzyme Atg3 promotes binding of IRG and Gbp proteins to Chlamydia- and Toxoplasma- containing vacuoles and host resistance. *PloS One* 9, e86684. doi: 10.1371/journal.pone.0086684

Harouz, H., Rachez, C., Meijer, B. M., Marteyn, B., Donnadieu, F., Cammas, F., et al. (2014). Shigella flexneri targets the HP 1γ subcode through the phosphothreonine lyase OspF. *EMBO J.* 33, 2606–2622. doi: 10.15252/embj.201489244

Helbig, K. J., Teh, M. Y., Crosse, K. M., Monson, E. A., Smith, M., Tran, E. N., et al. (2019). The interferon stimulated gene viperin, restricts Shigella. flexneri in vitro. *Sci. Rep.* 9, 1–12. doi: 10.1038/s41598-019-52130-8

Henry, T., Brotcke, A., Weiss, D. S., Thompson, L. J., and Monack, D. M. (2007). Type I interferon signaling is required for activation of the inflammasome during Francisella infection. *J. Exp. Med.* 204, 987–994. doi: 10.1084/jem.20062665

Henry, T., Kirimanjeswara, G. S., Ruby, T., Jones, J. W., Perret, M., Ho, L., et al. (2010). Type I Interferon signaling constrains IL-17A/F secretion by γδ T cells during bacterial infections. *J. Immunol.* 184, 3755–3767. doi: 10.4049/jimmunol.0902065

Hertzog, P. J., and Williams, B. R. G. (2013). Fine tuning type I interferon responses. *Cytokine Growth Factor Rev.* 24, 217–225. doi: 10.1016/j.cytogfr.2013.04.002

Isaacs, A., and Lindenmann, J. (1957). Virus interference. I. The interferon. *Proc. R. Soc London Ser. B - Biol. Sci.* 147, 258–267. doi: 10.1098/rspb.1957.0048

Ishikawa, H., Ma, Z., and Barber, G. N. (2009). STING regulates intracellular DNA-mediated, type I interferon-dependent innate immunity. *Nature* 461, 788–792. doi: 10.1038/nature08476

Janeway, C. A. (1989). Approaching the asymptote? Evolution and revolution in immunology. *Cold Spring Harb. Symp. Quant. Biol.* 54, 1–13. doi: 10.1101/sqb.1989.054.01.003

Kagan, J. C., Su, T., Horng, T., Chow, A., Akira, S., and Medzhitov, R. (2008). TRAM couples endocytosis of Toll-like receptor 4 to the induction of interferon-β. *Nat. Immunol.* 9, 361–368. doi: 10.1038/ni1569

Kak, G., Raza, M., and Tiwari, B. K. (2018). Interferon-gamma (IFN-γ): Exploring its implications in infectious diseases. *Biomol. Concepts* 9, 64–79. doi: 10.1515/bmc-2018-0007

Kawai, T., Takahashi, K., Sato, S., Coban, C., Kumar, H., Kato, H., et al. (2005). IPS-1, an adaptor triggering RIG-I- and Mda5-mediated type I interferon induction. *Nat. Immunol.* 6, 981–988. doi: 10.1038/ni1243

Kawasaki, T., and Kawai, T. (2014). Toll-like receptor signaling pathways. *Front. Immunol.* 5, 461. doi: 10.3389/fimmu.2014.00461

Kim, B. H., Shenoy, A. R., Kumar, P., Das, R., Tiwari, S., and MacMicking, J. D. (2011). A family of IFN-γ-inducible 65-kD GTPases protects against bacterial infection. doi: 10.1126/science.1201711

Kim, B. H., Shenoy, A. R., Kumar, P., Bradfield, C. J., and MacMicking, J. D. (2012). IFN-inducible GTPases in host cell defense. *Cell Host Microbe* 12, 432–444. doi: 10.1016/j.chom.2012.09.007

Knodler, L. A., Crowley, S. M., Sham, H. P., Yang, H., Wrande, M., Ma, C., et al. (2014). Non-canonical inflammasome activation of caspase-4/ caspase-11 mediates epithelial defenses against enteric bacterial pathogens. *Cell Host Microbe* 16, 249–256. doi: 10.1016/j.chom.2014.07.002

Kobayashi, T., Ogawa, M., Sanada, T., Mimuro, H., Kim, M., Ashida, H., et al. (2013). The Shigella OspC3 effector inhibits caspase-4, antagonizes inflammatory cell death, and promotes epithelial infection. *Cell Host Microbe* 13, 570–583. doi: 10.1016/j.chom.2013.04.012

Kotenko, S. V., Gallagher, G., Baurin, V. V., Lewis-Antes, A., Shen, M., Shah, N. K., et al. (2003). IFN-λs mediate antiviral protection through a distinct class II cytokine receptor complex. *Nat. Immunol.* 4, 69–77. doi: 10.1038/ni875

Ku, J. W. K., Chen, Y., Lim, B. J. W., Gasser, S., Crasta, K. C., and Gan, Y. H. (2020). Bacterial-induced cell fusion is a danger signal triggering cGAS-STING pathway via micronuclei formation. *Proc. Natl. Acad. Sci. U. S. A.* 117, 15923–15934. doi: 10.1073/pnas.2006908117

Kutsch, M., Coers, J., Sistemich, L., Lesser, C. F., Goldberg, M. B., and Herrmann, C. (2020). Direct binding of polymeric GBP1 to LPS disrupts bacterial cell envelope functions. *Embo*, 39, 1–22. doi: 10.15252/embj.2020104926

Langford, B. J., So, M., Raybardhan, S., Leung, V., Westwood, D., MacFadden, D. R., et al. (2020). Bacterial co-infection and secondary infection in patients with COVID-19: a living rapid review and meta-analysis. *Clin. Microbiol. Infect.* 26, 1622–1629. doi: 10.1016/j.cmi.2020.07.016

Lavigne, M., Eskeland, R., Azebi, S., Saint-André, V., Jang, S. M., Batsché, E., et al. (2009). Interaction of HP1 and Brg1/Brm with the globular domain of histone H3 is required for HP1-mediated repression. *PloS Genet.* 5, e1000769. doi: 10.1371/journal.pgen.1000769

Lazear, H. M., Schoggins, J. W., and Diamond, M. S. (2019). Shared and Distinct Functions of Type I and Type III Interferons. *Immunity* 50, 907–923. doi: 10.1016/j.immuni.2019.03.025

Lebreton, A., Lakisic, G., Job, V., Fritsch, L., Tham, T. N., Camejo, A., et al. (2011). A bacterial protein targets the BAHD1 chromatin complex to stimulate type III interferon response. *Science* 331, 1319–1321. doi: 10.1126/science.1200120

LeMessurier, K. S., Häcker, H., Chi, L., Tuomanen, E., and Redecke, V. (2013). Type I Interferon Protects against Pneumococcal Invasive Disease by Inhibiting Bacterial Transmigration across the Lung. *PloS Pathog.* 9, e1003727. doi: 10.1371/journal.ppat.1003727

Li, P., Jiang, W., Yu, Q., Liu, W., Zhou, P., Li, J., et al. (2017). Ubiquitination and degradation of GBPs by a Shigella effector to suppress host defence. *Nature* 551, 378–383. doi: 10.1038/nature24467

Lindenberg, V., Mölleken, K., Kravets, E., Stallmann, S., Hegemann, J. H., Degrandi, D., et al. (2017). Broad recruitment of mGBP family members to Chlamydia trachomatis inclusions. *PloS One* 12, 1–14. doi: 10.1371/journal.pone.0185273

Lippmann, J., Müller, H. C., Naujoks, J., Tabeling, C., Shin, S., Witzenrath, M., et al. (2011). Dissection of a type I interferon pathway in controlling bacterial intracellular infection in mice. *Cell. Microbiol.* 13, 1668–1682. doi: 10.1111/j.1462-5822.2011.01646.x

Liu, S., Cai, X., Wu, J., Cong, Q., Chen, X., Li, T., et al. (2015). Phosphorylation of innate immune adaptor proteins MAVS, STING, and TRIF induces IRF3 activation. *Science* 347, aaa2630. doi: 10.1126/science.aaa2630

Long, T. M., Nisa, S., Donnenberg, M. S., and Hassel, B. A. (2014). Enteropathogenic Escherichia coli inhibits type I interferon- and RNase L-mediated host defense to disrupt intestinal epithelial cell Barrier function. *Infect. Immun.* 82, 2802–2814. doi: 10.1128/IAI.00105-14

Maertzdorf, J., Repsilber, D., Parida, S. K., Stanley, K., Roberts, T., Black, G., et al. (2011). Human gene expression profiles of susceptibility and resistance in tuberculosis. *Genes Immun.* 12, 15–22. doi: 10.1038/gene.2010.51

Major, J., Crotta, S., Llorian, M., McCabe, T. M., Gad, H. H., Hartmann, R., et al. (2020). Type I and III interferons disrupt lung epithelial repair during recovery from viral infection. *Science* 369, 712–717. doi: 10.1126/science.abc2061

Man, S. M., Karki, R., Malireddi, R. K. S., Neale, G., Vogel, P., Yamamoto, M., et al. (2015). The transcription factor IRF1 and guanylate-binding proteins target activation of the AIM2 inflammasome by Francisella infection. *Nat. Immunol.* 16, 467–475. doi: 10.1038/ni.3118

Man, S. M., Karki, R., Sasai, M., Place, D. E., Kesavardhana, S., Temirov, J., et al. (2016). IRGB10 Liberates Bacterial Ligands for Sensing by the AIM2 and Caspase-11-NLRP3 Inflammasomes. *Cell* 167, 382–396.e17. doi: 10.1016/j.cell.2016.09.012

Manca, C., Tsenova, L., Bergtold, A., Freeman, S., Tovey, M., Musser, J. M., et al. (2001). Virulence of a Mycobacterium tuberculosis clinical isolate in mice is determined by failure to induce Th1 type immunity and is associated with induction of IFN-α/β. *Proc. Natl. Acad. Sci. U. S. A.* 98, 5752–5757. doi: 10.1073/pnas.091096998

Manca, C., Tsenova, L., Freeman, S., Barczak, A. K., Tovey, M., Murray, P. J., et al. (2005). Hypervirulent M. tuberculosis W/Beijing strains upregulate Type I IFNs and increase expression of negative regulators of the jak-stat pathway. *J. Interf. Cytokine Res.* 25, 694–701. doi: 10.1089/jir.2005.25.694

Mancuso, G., Midiri, A., Biondo, C., Beninati, C., Zummo, S., Galbo, R., et al. (2007). Type I IFN Signaling Is Crucial for Host Resistance against Different Species of Pathogenic Bacteria. *J. Immunol.* 178, 3126–3133. doi: 10.4049/jimmunol.178.5.3126

Meunier, E., and Broz, P. (2015). Interferon-inducible GTPases in cell autonomous and innate immunity. *Cell. Microbiol.* 18, 168–180. doi: 10.1111/cmi.12546

Meunier, E., Dick, M. S., Dreier, R. F., Schürmann, N., Broz, D. K., Warming, S., et al. (2014). Caspase-11 activation requires lysis of pathogen-containing vacuoles by IFN-induced GTPases. *Nature* 509, 366–370. doi: 10.1038/nature13157

Meunier, E., Wallet, P., Dreier, R. F., Costanzo, S., Anton, L., Rühl, S., et al. (2015). Guanylate-binding proteins promote AIM2 inflammasome activation during Francisella novicida infection by inducing cytosolic bacteriolysis and DNA release. *Nat. Immunol.* 16, 476–484. doi: 10.1038/ni.3119

Meylan, E., Curran, J., Hofmann, K., Moradpour, D., Binder, M., Bartenschlager, R., et al. (2005). Cardif is an adaptor protein in the RIG-I antiviral pathway and is targeted by hepatitis C virus. *Nature* 437, 1167–1172. doi: 10.1038/nature04193

Monroe, K. M., McWhirter, S. M., and Vance, R. E. (2009). Identification of host cytosolic sensors and bacterial factors regulating the type I interferon response to Legionella pneumophila. *PloS Pathog.* 5, e1000665. doi: 10.1371/journal.ppat.1000665

Naujoks, J., Tabeling, C., Dill, B. D., Hoffmann, C., Brown, A. S., Kunze, M., et al. (2016). IFNs Modify the Proteome of Legionella-Containing Vacuoles and Restrict Infection Via IRG1-Derived Itaconic Acid. *PloS Pathog.* 12, e1005408. doi: 10.1371/journal.ppat.1005408

Niesel, D. W., Hess, C. B., Cho, Y. J., Klimpel, K. D., and Klimpel, G. R. (1986). Natural and recombinant interferons inhibit epithelial cell invasion by Shigella spp. *Infect. Immun.* 52, 828–833. doi: 10.1128/IAI.52.3.828-833.1986

Odendall, C., and Kagan, J. C. (2017). Activation and pathogenic manipulation of the sensors of the innate immune system. *Microbes Infect.* 176, 139–148. doi: 10.1016/j.micinf.2017.01.003

Odendall, C., and Kagan, J. C. (2019). Host-encoded sensors of bacteria—our windows into the microbial world. *Microbiol. Spectr.* 176, 139–148. doi: 10.1128/microbiolspec.BAI-0011-2019

Odendall, C., Dixit, E., Stavru, F., Bierne, H., Franz, K. M., Durbin, A. F., et al. (2014). Diverse intracellular pathogens activate type III interferon expression from peroxisomes. *Nat. Immunol.* 15, 717–726. doi: 10.1038/ni.2915

Odendall, C., Voak, A. A., and Kagan, J. C. (2017). Type III interferons are commonly induced by bacteria-sensing TLRs, and reinforce epithelial barriers during infection. *J. Immunol.* 176, 139–148. doi: 10.4049/jimmunol.1700250

Ostler, N., Britzen-Laurent, N., Liebl, A., Naschberger, E., Lochnit, G., Ostler, M., et al. (2014). Gamma Interferon-Induced Guanylate Binding Protein 1 Is a Novel Actin Cytoskeleton Remodeling Factor. *Mol. Cell. Biol.* 34, 196–209. doi: 10.1128/mcb.00664-13

Ottenhoff, T. H. M., Dass, R. H., Yang, N., Zhang, M. M., Wong, H. E. E., Sahiratmadja, E., et al. (2012). Genome-Wide Expression Profiling Identifies Type 1 Interferon Response Pathways in Active Tuberculosis. *PloS One* 7, e45839. doi: 10.1371/journal.pone.0045839

Owen, K. A., Anderson, C. J., and Casanova, J. E. (2016). Salmonella suppresses the TRIF-dependent type I interferon response in macrophages. *MBio* 7, 1–15. doi: 10.1128/mBio.02051-15

Pagliuso, A., Tham, T. N., Allemand, E., Robertin, S., Dupuy, B., Bertrand, Q., et al. (2019). An RNA-Binding Protein Secreted by a Bacterial Pathogen Modulates RIG-I Signaling. *Cell Host Microbe* 26, 823–835.e11. doi: 10.1016/j.chom.2019.10.004

Pandey, A. K., Yang, Y., Jiang, Z., Fortune, S. M., Coulombe, F., Behr, M. A., et al. (2009). Nod2, Rip2 and Irf5 play a critical role in the type I interferon response to Mycobacterium tuberculosis. *PloS Pathog.* 5, e1000500. doi: 10.1371/journal.ppat.1000500

Paquette, N., Conlon, J., Sweet, C., Rus, F., Wilson, L., Pereira, A., et al. (2012). Serine/threonine acetylation of TGFβ-activated kinase (TAK1) by Yersinia pestis YopJ inhibits innate immune signaling. *Proc. Natl. Acad. Sci. U. S. A.* 109, 12710–12715. doi: 10.1073/pnas.1008203109

Perkins, D. J., Rajaiah, R., Tennant, S. M., Ramachandran, G., Higginson, E. E., Dyson, T. N., et al. (2015). Salmonella Typhimurium Co-opts the Host Type I Interferon System to Restrict Macrophage Innate Immune Transcriptional Responses Selectively. *J. Immunol.* 176, 139–148. doi: 10.4049/jimmunol.1500105

Pestka, S., Krause, C. D., and Walter, M. R. (2004). Interferons, interferon-like cytokines, and their receptors. *Immunol. Rev.* 202, 8–32. doi: 10.1111/j.0105-2896.2004.00204.x

Pien, G. C., Satoskar, A. R., Takeda, K., Akira, S., and Biron, C. A. (2000). Cutting Edge: Selective IL-18 Requirements for Induction of Compartmental IFN-γ

Responses During Viral Infection. *J. Immunol.* 165, 4787–4791. doi: 10.4049/jimmunol.165.9.4787

Pilla, D. M., Hagar, J. A., Haldar, A. K., Mason, A. K., Degrandi, D., Pfeffer, K., et al. (2014). Guanylate binding proteins promote caspase-11-dependent pyroptosis in response to cytoplasmic LPS. *Proc. Natl. Acad. Sci.* 111, 6046–6051. doi: 10.1073/pnas.1321700111

Piro, A. S., Hernandez, D., Luoma, S., Feeley, E. M., Finethy, R., Yirga, A., et al. (2017). Detection of cytosolic shigella flexneri via a C-terminal triple-arginine motif of GBP1 inhibits actin-based motility. *MBio* 8, e01979–17. doi: 10.1128/mBio.01979-17

Radoshevich, L., Impens, F., Ribet, D., Quereda, J. J., Tham, T. N., Nahori, M. A., et al. (2015). ISG15 counteracts Listeria monocytogenes infection. *Elife* 4, 1–23. doi: 10.7554/eLife.06848

Ranjbar, S., Haridas, V., Jasenosky, L. D., Falvo, J. V., and Goldfeld, A. E. (2015). A Role for IFITM Proteins in Restriction of Mycobacterium tuberculosis Infection. *Cell Rep.* 13, 874–883. doi: 10.1016/j.celrep.2015.09.048

Rauch, I., Rosebrock, F., Hainzl, E., Heider, S., Majoros, A., Wienerroither, S., et al. (2015). Noncanonical Effects of IRF9 in Intestinal Inflammation: More than Type I and Type III Interferons. *Mol. Cell. Biol.* 35, 2332–2343. doi: 10.1128/mcb.01498-14

Robinson, N., McComb, S., Mulligan, R., Dudani, R., Krishnan, L., and Sad, S. (2012). Type i interferon induces necroptosis in macrophages during infection with Salmonella enterica serovar Typhimurium. *Nat. Immunol.* 13, 954–962. doi: 10.1038/ni.2397

Rosadini, C. V., Zanoni, I., Odendall, C., Green, E. R., Paczosa, M. K., Philip, N. H., et al. (2015). A Single Bacterial Immune Evasion Strategy Dismantles Both MyD88 and TRIF Signaling Pathways Downstream of TLR4. *Cell Host Microbe* 18, 682–693. doi: 10.1016/j.chom.2015.11.006

Ruiz-Moreno, J. S., Hamann, L., Shah, J. A., Verbon, A., Mockenhaupt, F. P., Puzianowska-Kuznicka, M., et al. (2018). The common HAQ STING variant impairs cGAS-dependent antibacterial responses and is associated with susceptibility to Legionnaires' disease in humans. *PloS Pathog.* 14, 1–22. doi: 10.1371/journal.ppat.1006829

Saint-André, V., Batsché, E., Rachez, C., and Muchardt, C. (2011). Histone H3 lysine 9 trimethylation and HP1γ favor inclusion of alternative exons. *Nat. Struct. Mol. Biol.* 18, 337–344. doi: 10.1038/nsmb.1995

Salazar-Mather, T. P., Hamilton, T. A., and Biron, C. A. (2000). A chemokine-to-cytokine-to-chemokine cascade critical in antiviral defense. *J. Clin. Invest.* 105, 985–993. doi: 10.1172/JCI9232

Santos, J. C., Boucher, D., Schneider, L. K., Demarco, B., Dilucca, M., Shkarina, K., et al. (2020). Human GBP1 binds LPS to initiate assembly of a caspase-4 activating platform on cytosolic bacteria. *Nat. Commun.* 11, 3276. doi: 10.1038/s41467-020-16889-z

Schindler, H., Lutz, M. B., Röllinghoff, M., and Bogdan, C. (2001). The Production of IFN-γ by IL-12/IL-18-Activated Macrophages Requires STAT4 Signaling and Is Inhibited by IL-4. *J. Immunol.* 166, 3075–3082. doi: 10.4049/jimmunol.166.5.3075

Schmid-Burgk, J. L., Gaidt, M. M., Schmidt, T., Ebert, T. S., Bartok, E., and Hornung, V. (2015). Caspase-4 mediates non-canonical activation of the NLRP3 inflammasome in human myeloid cells. *Eur. J. Immunol.* 45, 2911–2917. doi: 10.1002/eji.201545523

Schneider, W. M., Chevillotte, M. D., and Rice, C. M. (2014). Interferon-Stimulated Genes: A Complex Web of Host Defenses. *Annu. Rev. Immunol.* 32, 513–545. doi: 10.1146/annurev-immunol-032713-120231

Seth, R. B., Sun, L., Ea, C.-K., and Chen, Z. J. (2005). Identification and characterization of MAVS, a mitochondrial antiviral signaling protein that activates NF-κB and IRF3. *Cell* 122, 669–682. doi: 10.1016/j.cell.2005.08.012

Shahangian, A., Chow, E. K., Tian, X., Kang, J. R., Ghaffari, A., Liu, S. Y., et al. (2009). I IFNs mediate development of postinfluenza bacterial pneumonia in mice. *J. Clin. Invest.* 119, 1910–1920. doi: 10.1172/JCI35412.1910

Shenoy, A. R., Wellington, D. A., Kumar, P., Kassa, H., Booth, C. J., Cresswell, P., et al. (2012). GBP5 Promotes NLRP3 inflammasome assembly and immunity in mammals. *Science* 336, 481–485. doi: 10.1126/science.1217141

Sheppard, P., Kindsvogel, W., Xu, W., Henderson, K., Schlutsmeyer, S., Whitmore, T. E., et al. (2002). IL-28, IL-29 and their class II cytokine receptor IL-28R. *Nat. Immunol.* 4, 63–68. doi: 10.1038/ni873

Shi, J., Zhao, Y., Wang, Y., Gao, W., Ding, J., Li, P., et al. (2014). Inflammatory caspases are innate immune receptors for intracellular LPS. *Nature* 514, 187–192. doi: 10.1038/nature13683

Sixt, B. S., Bastidas, R. J., Finethy, R., Baxter, R. M., Carpenter, V. K., Kroemer, G., et al. (2017). The Chlamydia trachomatis Inclusion Membrane Protein CpoS Counteracts STING-Mediated Cellular Surveillance and Suicide Programs. *Cell Host Microbe* 21, 113–121. doi: 10.1016/j.chom.2016.12.002

Smallwood, A., Hon, G. C., Jin, F., Henry, R. E., Espinosa, J. M., and Ren, B. (2012). CBX3 regulates efficient RNA processing genome-wide. *Genome Res.* 22, 1426–1436. doi: 10.1101/gr.124818.111

Snyder, D. T., Hedges, J. F., and Jutila, M. A. (2017). Getting "inside" Type I IFNs: Type I IFNs in Intracellular Bacterial Infections. *J. Immunol. Res.* 2017, 17. doi: 10.1155/2017/9361802

Takeuchi, O., and Akira, S. (2010). Pattern Recognition Receptors and Inflammation. *Cell* 140, 805–820. doi: 10.1016/j.cell.2010.01.022

Tanaka, Y., and Chen, Z. J. (2012). STING specifies IRF3 phosphorylation by TBK1 in the cytosolic DNA signaling pathway. *Sci. Signal.* 5, ra20. doi: 10.1126/scisignal.2002521

Tretina, K., Park, E. S., Maminska, A., and MacMicking, J. D. (2019). Interferon-induced guanylate-binding proteins: Guardians of host defense in health and disease. *J. Exp. Med.* 216, 482–500. doi: 10.1084/jem.20182031

Walberg, K., Baron, S., Poast, J., Schwartz, B., Izotova, L., Pestka, S., et al. (2008). Interferon protects mice against inhalation anthrax. *J. Interf. Cytokine Res.* 28, 597–601. doi: 10.1089/jir.2007.0143

Wandel, M. P., Pathe, C., Werner, E. I., Ellison, C. J., Boyle, K. B., von der Malsburg, A., et al. (2017). GBPs Inhibit Motility of Shigella flexneri but Are Targeted for Degradation by the Bacterial Ubiquitin Ligase IpaH9.8. *Cell Host Microbe* 22, 507–518.e5. doi: 10.1016/j.chom.2017.09.007

Wandel, M. P., Kim, B., Park, E., Boyle, K. B., Nayak, K., Lagrange, B., et al. (2020). Guanylate-binding proteins convert cytosolic bacteria into caspase-4 signaling platforms. *Nat. Immunol.* 21, 880–891. doi: 10.1038/s41590-020-0697-2

Watanabe, T., Asano, N., Fichtner-feigl, S., Gorelick, P. L., Tsuji, Y., Matsumoto, Y., et al. (2010). NOD1 contributes to mouse host defense against Helicobacter pylori via induction of type I IFN and activation of the ISGF3 signaling pathway. *J. Clin. Invest.* 120, 1645–1662. doi: 10.1172/JCI39481DS1

Wilson, R. P., Tursi, S. A., Rapsinski, G. J., Medeiros, N. J., Le, L. S., Kotredes, K. P., et al. (2019). STAT2 dependent type i interferon response promotes dysbiosis and luminal expansion of the enteric pathogen Salmonella Typhimurium. *PloS Pathog.* 15, 1–22. doi: 10.1371/journal.ppat.1007745

Wolf, K., and Fields, K. A. (2013). Chlamydia pneumoniae Impairs the Innate Immune Response in Infected Epithelial Cells by Targeting TRAF3. *J. Immunol.* 190, 1695–1701. doi: 10.4049/jimmunol.1202443

Wu, J., Sun, L., Chen, X., Du, F., Shi, H., Chen, C., et al. (2017). Cyclic GMP-AMP as an endogenous second messenger in innate immune signaling by cytosolic DNA. *Annu. Rev. Biochem.* 86, 541–566. doi: 10.1146/annurev-biochem-061516-044813

Xu, L. G., Wang, Y. Y., Han, K. J., Li, L. Y., Zhai, Z., and Shu, H.-B. (2005). VISA is an adapter protein required for virus-triggered IFN-β signaling. *Mol. Cell* 19, 727–740. doi: 10.1016/j.molcel.2005.08.014

Zhang, H., Zoued, A., Liu, X., Sit, B., and Waldor, M. K. (2020). Functional remodeling of lysosomes by type I interferon modifies host defense. *bioRxiv Cell Biol.* doi: 10.1101/2020.02.25.965061

Zheng, Z., Wei, C., Guan, K., Yuan, Y., Zhang, Y., Ma, S., et al. (2016). Bacterial E3 Ubiquitin Ligase IpaH4.5 of Shigella flexneri Targets TBK1 To Dampen the Host Antibacterial Response. *J. Immunol.* 196, 1199–1208. doi: 10.4049/jimmunol.1501045

Zwaferink, H., Stockinger, S., Reipert, S., and Decker, T. (2008). Stimulation of inducible nitric oxide synthase expression by beta interferon increases necrotic death of macrophages upon Listeria monocytogenes infection. *Infect. Immun.* 76, 1649–1656. doi: 10.1128/IAI.01251-07

16

Persistence of Intracellular Bacterial Pathogens—With a Focus on the Metabolic Perspective

Wolfgang Eisenreich[1*], *Thomas Rudel*[2], *Jürgen Heesemann*[3] *and Werner Goebel*[3]

[1] *Department of Chemistry, Chair of Biochemistry, Technische Universität München, Garching, Germany,* [2] *Chair of Microbiology, Biocenter, University of Würzburg, Würzburg, Germany,* [3] *Max von Pettenkofer-Institute, Ludwig Maximilian University of Munich, München, Germany*

Correspondence:
Wolfgang Eisenreich
wolfgang.eisenreich@ch.tum.de

Persistence has evolved as a potent survival strategy to overcome adverse environmental conditions. This capability is common to almost all bacteria, including all human bacterial pathogens and likely connected to chronic infections caused by some of these pathogens. Although the majority of a bacterial cell population will be killed by the particular stressors, like antibiotics, oxygen and nitrogen radicals, nutrient starvation and others, a varying subpopulation (termed persisters) will withstand the stress situation and will be able to revive once the stress is removed. Several factors and pathways have been identified in the past that apparently favor the formation of persistence, such as various toxin/antitoxin modules or stringent response together with the alarmone (p)ppGpp. However, persistence can occur stochastically in few cells even of stress-free bacterial populations. Growth of these cells could then be induced by the stress conditions. In this review, we focus on the persister formation of human intracellular bacterial pathogens, some of which belong to the most successful persister producers but lack some or even all of the assumed persistence-triggering factors and pathways. We propose a mechanism for the persister formation of these bacterial pathogens which is based on their specific intracellular bipartite metabolism. We postulate that this mode of metabolism ultimately leads, under certain starvation conditions, to the stalling of DNA replication initiation which may be causative for the persister state.

Keywords: persistence, mechanisms of persister formation, intracellular bacterial pathogens, stress conditions, ATP-DnaA complex, DNA replication initiation

INTRODUCTION

Bacterial Survival Strategies

Bacteria have evolved several strategies by which subpopulations are able to survive life-threatening conditions that are lethal for most members of bacterial populations. The best characterized strategies are: (a) formation of endospores (Higgins and Dworkin, 2012; Hutchison et al., 2014; Moir and Cooper, 2015), observed among Gram-positive bacteria, especially in the Bacillales and Clostridiales orders (all belonging to the phylum Firmicutes), (b) formation of exospores (Ohnishi et al., 2002; Sigle et al., 2015; Jones and Elliot, 2018) mainly found among members of the

Actinomycetales, (c) formation of persister cells (Lewis, 2007; Balaban et al., 2013; Harms et al., 2016; Fisher et al., 2017; Radzikowski et al., 2017; Van Den Bergh et al., 2017; Kim J. S. et al., 2018; Wilmaerts et al., 2019) occurring in most bacteria, and (d) formation of the (apparently) related viable but not-culturable cells (VBNCs) (Oliver, 2010; Li et al., 2014; Ramamurthy et al., 2014). In all four survival states, metabolism and cell division either stop completely (as in case of endospores) or are at least highly reduced (as in case of persistent bacterial cells and VBNC). Common to all of these survival states is the ability of the "dormant" cells to rewake under favorable conditions and the unaltered genotype compared to the original cell population. It has been argued that the entry of a subset of cells into these survival states and their subsequent resuscitation might be a "bet-hedging" strategy allowing bacterial populations in general to withstand fluctuating environments (Grimbergen et al., 2015).

While the molecular mechanisms leading to spore formation are rather well characterized (Higgins and Dworkin, 2012; Paredes-Sabja et al., 2014; Setlow, 2014; Tan and Ramamurthi, 2014; Moir and Cooper, 2015), the underlying mechanisms causing bacterial persistence and the VBNC state are far from being fully understood (see below). The persistence of bacteria was first described in 1944 (Bigger, 1944) as a penicillin-insensitive, not inheritable state of a small subpopulation observed in an isogenic staphylococcal population. After antibiotic removal, the surviving bacterial cells started dividing again at normal growth rates generating again a similar fraction of persister cells upon renewed antibiotic treatment.

For bacterial pathogens, the persister state is most frequently observed after treatment with bactericidal antibiotics (Gollan et al., 2019; Bakkeren et al., 2020). It is of particular medical importance, apparently linked to persistent and chronic infections (Fauvart et al., 2011; Conlon, 2014; Kester and Fortune, 2014; Paredes-Sabja et al., 2014; Van Den Bergh et al., 2017; Jung et al., 2019). Recently, operational definitions and guidelines for *in vitro* studies of bacterial persistence have been reported (Balaban et al., 2019). According to these operational definitions, antibiotic tolerant bacteria are characterized by an exponentially slower killing rate in contrast to persisters which show a typical biphasic killing rate in comparison to that of antibiotic treated susceptible bacteria.

Generation of Bacterial Persister and VBNC Cells

Antibiotic-persistent subpopulations can be detected in bacterial populations even when growing under favorable culture conditions. Upon treatment with high doses of (especially bactericidal) antibiotics, the large majority of the population is killed and a small persister subpopulation already preformed stochastically is then selected by the antibiotic treatment. However, there is evidence that antibiotic stress as well as other stress conditions (discussed in more detail below) enhance persister formation (Ayrapetyan et al., 2015; Nierman et al., 2015; Harms et al., 2016; Mok and Brynildsen, 2018). Regardless of the persistence-triggering condition, in all cases the formed persisters can be resuscitated when the stressors are removed and normal growth conditions are restored.

The VBNC state, first reported by Xu and collegues (Xu et al., 1982), is also observed as stress-surviving bacterial subpopulations and seems to be closely related to the persister state. VBNC cells and persister cells share many features and appear to co-exist in the same bacterial population (Orman and Brynildsen, 2013; Ayrapetyan et al., 2015; Goncalves and De Carvalho, 2016; Zhao et al., 2017). Indeed, both persister and VBNC cells occur together in biofilms associated with infections (Spoering and Lewis, 2001; Li et al., 2014; Conlon et al., 2015). But unlike persister cells, VBNC cells have transiently lost the ability to grow in standard culture media. They can, however regain culturability in special complex media (Oliver, 2005). Some authors claim that VBNC cells may not represent a separate cell phenotype (Kim J. S. et al., 2018) and postulate that the term VBNC should be replaced with persister cells.

Persister Formation of Intracellular Bacterial Pathogens

Formation of persister cells has been observed for all bacterial pathogens, including IBPs (Mulcahy et al., 2010; Li et al., 2014). Persister formation appears to be largely responsible for the recalcitrant chronic infections caused by *Mycobacterium tuberculosis* and *Chlamydia* species (Fauvart et al., 2011; Cohen et al., 2013; Srinivas et al., 2020). However, the mechanism(s) causing persister formation of IBPs when replicating in mammalian host cells is poorly understood.

After internalization by mammalian phagocytic or non-phagocytic host cells, most IBPs normally replicate in specifically modified endo(phago)somal compartments, called pathogen-containing vacuoles (PCV) or inclusions (in case of *Chlamydia*). These IBPs are termed "vacuolar IBPs". Some IBPs termed "cytosolic IBPs" escape into the host cells' cytosol and proliferate there (Ray et al., 2009; Pizarro-Cerda et al., 2016; Martinez et al., 2018; Omotade and Roy, 2019).

The replication rate of "cytosolic IBPs" is normally considerable higher than that of the "vacuolar IBPs", presumably due to the better supply with essential nutrients in the cytosol. On the other hand, the vacuolar compartments may provide better protection against host innate immune attacks (Weiss and Schaible, 2015; Kim J. K. et al., 2018). In the context of persistence, it is interesting to note that persistent IBPs have been found exclusively in vacuolar compartments (see below), suggesting that these intracellular niches offer a better environment to reach the persistence state.

However, in both host cell compartments IBPs replicate in metabolically better defined environment than extracellular bacterial pathogens. The substantial progress over the last years in understanding the intracellular metabolism of IBPs and their adaptation to the metabolism of the host cells (Grubmüller et al., 2014; Eisenreich et al., 2017; Thompson et al., 2018; Best and Abu Kwaik, 2019; Eisenreich et al., 2019) may help to consider persistence of these bacterial pathogens from a metabolic point of view.

The focus of this review will be on persister formation of IBPs. First, we summarize the known bacterial factors and stress responses that have been associated with persister formation mainly in model bacteria (especially *Escherichia coli*). The possible impact of this knowledge on persistence of IBPs will be considered. Then, we will discuss the specific metabolic states of IBPs when growing within host cells which may favor persister formation. This discussion is based on the concept of bipartite metabolism which appears to be a rather common strategy for IBPs (Grubmüller et al., 2014; Eisenreich and Heuner, 2016; Häuslein et al., 2016; Häuslein et al., 2017a; Mehlitz et al., 2017; Best and Abu Kwaik, 2019). Indeed, this metabolic approach along with its impact on DNA replication initiation and cell division may explain the long-term survival states (persistence, VBNC) observed in subpopulations of IBPs mainly as a continuum of intracellular metabolic alterations. We are aware that this approach neglects possible immunological aspects, e.g., the ability of persistent bacterial pathogens to avoid their elimination by innate and/or adaptive immune responses (Ulrichs and Kaufmann, 2002; Sabbagh et al., 2018; Gupta et al., 2019).

IDENTIFIED BACTERIAL CELLULAR FACTORS AND PATHWAYS ENHANCING PERSISTER FORMATION

General Aspects

As mentioned above, bacterial persistence has been discovered 75 years ago, but the molecular mechanism(s) of persister formation is (are) still poorly understood (Kaldalu et al., 2016). Persisters can be apparently generated stochastically, probably due to the physiological heterogeneity of single cells in a bacterial population (Germain et al., 2015; Shan et al., 2017). The percentage of persisters in logarithmically growing cell cultures is small (<<1%), but significantly increased in the stationary-phase (Oliver, 2010). Persister formation is further triggered by various stress stimuli. A large number of molecular studies mainly performed with *E. coli* and a few other model bacteria, have identified different factors and stress response pathways that are apparently linked to persister formation. These include toxin-antitoxin (TA) systems (Maisonneuve et al., 2011; Balaban et al., 2013; Maisonneuve and Gerdes, 2014; Gerdes, 2016; Kedzierska and Hayes, 2016), oxidative stress response (Wu et al., 2012), RpoS-mediated general stress response (Mok et al., 2015; Liu et al., 2017; Mok and Brynildsen, 2018), stringent response together with the alarmone guanosine tetra-/pentaphosphate [(p)ppGpp] (Korch et al., 2003; Germain et al., 2015; Liu et al., 2017), DNA damage and SOS response (Dörr et al., 2009; Kreuzer, 2013; Völzing and Brynildsen, 2015), nutrient stress, and impaired energy production (Amato et al., 2013; Amato and Brynildsen, 2015; Shan et al., 2017; Mok and Brynildsen, 2018). All of these processes triggered by the corresponding stress stimuli lead to a significant increase of the persister fraction within bacterial populations.

It has been suggested that the toxin components of specific TA modules, more strongly expressed in some individual cells (possibly due to stochastic variation in ppGpp levels), are decisive factors for persister formation (Maisonneuve et al., 2013). However, strains with multiple deletions of genes encoding different TA modules (Maisonneuve et al., 2011) as well as ppGpp-negative strains (Maisonneuve et al., 2013) and *rpoS* deletion strains (Nguyen et al., 2011) still form persisters (though at reduced rates).

The discussed stress responses apparently involved in persister formation comprise complex regulatory networks that control the expression of multiple genes whose products are essential for coping with the stress conditions on the transcriptional, translational, and posttranslational level. In addition, many more genes (and in particular metabolic genes) than those directly involved in overcoming the stress situation are activated or repressed by the various stress regulators (Khil and Camerini-Otero, 2002; Jozefczuk et al., 2010; Yukihira et al., 2015; Christodoulou et al., 2018). Interestingly, transcriptome and metabolome studies indicate common metabolic pathways which are down- or up-regulated in a similar way by different stress conditions, such as temperature stress, oxidative stress, nutrient starvation, nutrient shifts or stationary phase (Jozefczuk et al., 2010). In addition, interactions and crosstalks exist between these regulons (Weber et al., 2005; Merrikh et al., 2009; Amato et al., 2013; Baharoglu et al., 2013; Leaden et al., 2018; Molina-Quiroz et al., 2018; Mitosch et al., 2019).

None of the described factors and pathways associated with persister formation converts the entire bacterial population into the persister state, although all (or at least most) cells in the population are subject to the respective stress stimulus and the subsequent response. This indicates that none of the above described stress conditions alone is the final cause of persistence and rather suggests that persistence is caused by the physiological heterogeneity triggered in bacterial populations under these conditions (Dhar and Mckinney, 2007; Gefen and Balaban, 2009). It is therefore more likely that a specific, yet unknown physiological state is responsible for persistence formation. This state appears to be reached stochastically in a few cells even of growing and unstressed bacterial populations (Amato et al., 2013; Radzikowski et al., 2017). The described stress conditions then stabilize and enhance this physiological state.

Specific Stress Conditions Enhancing Persister Formation in Model Bacteria and IBPs

Bacterial Toxin/Antitoxin Modules and Their Association With Persistence

Bacterial TA modules are composed of a toxin and an antitoxin component that neutralizes the toxin. At least four different types of TA modules have been identified based on the function of the antitoxin (Yang and Walsh, 2017; Harms et al., 2018). Type I and especially type II TA modules, most widely distributed among prokaryotes (Gerdes et al., 2005; Fozo et al., 2008; Leplae et al., 2011), have been shown to be involved in inducing persistence (Vazquez-Laslop et al., 2006; Rotem et al., 2010; Maisonneuve et al., 2011; Maisonneuve and Gerdes, 2014; Verstraeten et al., 2015; Gerdes, 2016; Page and Peti, 2016). In type I TA modules, the antitoxin is an antisense RNA which binds to the toxin-coding mRNA and blocks its translation (Fozo et al., 2008), whereas type II TA modules consist of toxin and antitoxin polypeptides that form an inactive complex (Gerdes et al., 2005; Gerdes, 2016). In type II TA modules, proteolytic degradation (usually by Lon or Clp proteases) of the antitoxin frees the active toxin. Degradation or depletion of the antitoxin may occur stochastically or in response to stresses and the unleashed toxin protein affects central cellular processes, including translation, DNA replication, cell division, and metabolism (Gerdes, 2016; Kedzierska and Hayes, 2016; Harms et al., 2018; Wilmaerts et al., 2019).

Notably, the contribution of type II TA modules to persister formation has been recently challenged (Harms et al., 2017a; Goormaghtigh et al., 2018). Goormaghtigh and colleagues (Goormaghtigh et al., 2018) provided evidence that an *E. coli* K-12 mutant strain lacking the 10 type II TA modules previously postulated to participate in persister formation produced similar levels of persisters as the wild-type bacteria in unstressed cultures and after exposure to antibiotics [but see also (Holden and Errington, 2018; Ronneau and Helaine, 2019)].

TA Modules in IBPs

Especially, type II TA modules have been found in most human bacterial pathogens including several vacuolar and cytosolic IBPs. An unusually large number of TA modules has been identified in *Salmonella enterica* serovars (24 TA modules including 4 type I and 19 type II) (Mcclelland et al., 2001; Di Cesare et al., 2016) and in *M. tuberculosis* (79 TA modules) (Slayden et al., 2018; Thakur et al., 2018). Type II TA modules are also present in *Bartonella* (Harms et al., 2017b), *Listeria monocytogenes* (Curtis et al., 2017; Kalani et al., 2018), *Shigella* serovars (Mcvicker and Tang, 2016), *Rickettsia* spp. (Socolovschi et al., 2013), and *Brucella* spp. (Heaton et al., 2012). Neither type I nor type II TA modules have been found in *Chlamydia* spp. and there is also no convincing evidence for the presence of such TA modules in *Coxiella*, *Francisella* and *Legionella* (Pandey and Gerdes, 2005; Leplae et al., 2011; Yamaguchi et al., 2011).

The contribution of TA modules to persistence of IBPs has been extensively studied in *S.* Typhimurium (Helaine et al., 2014; Di Cesare et al., 2016; Vandrisse et al., 2017; Rycroft et al., 2018) and *M. tuberculosis* (Korch et al., 2009; Albrethsen et al., 2013; Schuessler et al., 2013; Korch et al., 2015; Winther et al., 2016; Slayden et al., 2018). In *L. monocytogenes*, MazEF, a type II TA module which contributes to persister formation in many bacteria (Gerdes et al., 2005), obviously does not affect the persister formation upon treatment with antibiotics in lethal doses (Curtis et al., 2017).

The Stringent Response, the Alarmone (p)ppGpp, and the Association With Persister Formation

Next to TA modules, (p)ppGpp appears to play a major role in persister formation (Korch et al., 2003; Kim H. Y. et al., 2018). This alarmone is the molecular effector of the bacterial stringent response which leads to an extensive transcriptional reprogramming and to metabolic changes in response to nutrient deprivation (Potrykus and Cashel, 2008). In *E. coli* (and all members of the gamma-proteobacteria), RelA and SpoT synthesize (p)ppGpp (in the following termed only ppGpp) from GTP and GDP, whereas in *Bacillus subtilis* and many other bacteria, a single enzyme (Rel or Rsh) is responsible for this activity (Mittenhuber, 2001). Upon amino acid starvation, the uncharged tRNAs activate the ribosome-associated RelA to synthesize ppGpp, whereas carbohydrates and fatty acid starvation stimulate ppGpp synthesis by the cytoplasmic SpoT (Xiao et al., 1991; Seyfzadeh et al., 1993). In a complex with DksA, ppGpp binds to RNA polymerase and inhibits transcription initiated from stable RNA (i.e., rRNA and tRNA) promoters and upregulates transcription of amino acid biosynthesis and stress response genes (Potrykus and Cashel, 2008; Dalebroux and Swanson, 2012; Hauryliuk et al., 2015).

As mentioned above, a link between ppGpp and persister formation was first shown by Korch et al. (2003) and confirmed by other studies (Amato et al., 2013; Bokinsky et al., 2013; Germain et al., 2013; Maisonneuve et al., 2013; Kim H. Y. et al., 2018) demonstrating that increased ppGpp levels result in growth arrest and increased persistence.

Interestingly, ppGpp is also a regulator especially for type II, but also for type I TA modules (Maisonneuve et al., 2013; Maisonneuve and Gerdes, 2014; Verstraeten et al., 2015; Tian et al., 2017). For type II TA modules, increased production of ppGpp activates the Lon protease-dependent antitoxin degradation and the released toxin component appears to increase the generation of persisters by blocking central cell processes (see above). This ppGpp-mediated activation of type II TA modules has become a widely accepted model for persister formation that has, however, also come under critical debate recently (Chowdhury et al., 2016a; Shan et al., 2017; Maisonneuve et al., 2018). Transcriptional control by ppGpp has also been shown for the type I TA module HokB/SokB (Verstraeten et al., 2015; Wilmaerts et al., 2018) as outlined below in more detail.

Association of Stringent Response and ppGpp With Persister Formation in IBPs

According to Kegg Data Base, all IBPs except the obligate intracellular pathogens *Chlamydia* and *Rickettsia* spp. produce RelA or a RelA/SpoT-like enzyme (Rsh), and are able to synthesize ppGpp. SpoT is present as separate enzyme in IBPs belonging to the gamma-proteobacteria, i.e., *Shigella*, *Salmonella*, *Francisella*, *Legionella*, and *Coxiella*, but is absent in *Chlamydia* spp. and *Rickettsia* spp. (Mittenhuber, 2001; Clark et al., 2011). In *L. monocytogenes*, three genes code for ppGpp synthetases: one bi-functional RSH enzyme and two small synthases (Natori et al., 2009). *M. tuberculosis* carries a gene encoding RelMtb, a bifunctional RelA/SpoT homolog that modulates synthesis and hydrolysis of ppGpp during the stringent response (Avarbock et al., 1999; Hogg et al., 2004). A single RelA/SpoT-like Rsh enzyme has also been identified in *Brucella* (Dozot et al., 2006).

The involvement of ppGpp in persister formation of IBPs has been suggested for *S. enterica* und *M. tuberculosis* (Helaine et al., 2014). Helaine and coworkers reported that *Salmonella* living within macrophage (MP) vacuoles are exposed to potentially stressful conditions that induce the expression of 14 type II TA modules in a ppGpp/Lon-dependent manner, and this event apparently plays an important role in the formation of persister cells. It has also been shown that, in *S.* Typhimurium, persistence triggered by ShpAB (also a type II TA module with Lon-dependence) also occurs in a *relA* mutant, i.e., in the absence of ppGpp synthesis (Slattery et al., 2013). One should keep in mind, however, that the involvement of ppGpp in persister formation is clearly not restricted to its role in the activation of TA modules (Liu et al., 2017) (see also below).

In *M. tuberculosis*, the single Rsh enzyme (RelMtb), responsible for ppGpp production, is required for the long term survival under *in vitro* starvation conditions (Primm et al., 2000). Dahl and colleagues (Dahl et al., 2003) reported that RelMtb is critical for the successful establishment of persistent infection in mice by altering the expression of antigenic and enzymatic factors that may contribute to successful latent infection. Stringent response (involving ppGpp) mediates persistence in *M. tuberculosis* (Chuang et al., 2015). *M. tuberculosis* strains expressing a mutant ppGpp synthetase (RelMtb) are unable to persist in mice, also demonstrating that the RelMtb activity is required for maintaining mycobacterial titers during chronic infection (Weiss and Stallings, 2013). A RelMtb mutant does not slow down replication during nutrient starvation and carries out a similar metabolism as the wild-type strain in nutrient-rich media (Dutta et al., 2019). Furthermore, it has been reported (Wayne and Lin, 1982; Dutta and Karakousis, 2014) that in *M. tuberculosis* cultures growing under optimal conditions, translationally dormant cells pre-exist as a small subpopulation and that part of these pre-existing persisters are RelMtb-overexpressing cells (Srinivas et al., 2020). Interestingly, the involvement of ppGpp in persister formation seems to be restricted to pathogenic mycobacteria (Bhaskar et al., 2018).

Concerning the involvement of ppGpp in *L. monocytogenes*, a study by Taylor and colleagues (Taylor et al., 2002) showed that a *relA* mutant, which was unable to accumulate ppGpp in response to amino acid starvation, was avirulent in a murine infection model (in contrast to the wild-type strain), indicating an essential role of stringent response for survival and growth of *L. monocytogenes* in this host. A link to persistence is not apparent in this study. In *Francisella*, the involvement of ppGpp in the regulatory network governing virulence gene expression has been established (Charity et al., 2009; Dean et al., 2009; Cuthbert et al., 2017), but again the possible role in persistence remains unclear.

The presence of persistent, clinically relevant *Legionella pneumophila* strains in different natural environments, often in close association with free-living amoebae and multispecies biofilms, is well documented (Berjeaud et al., 2016; Abu Khweek and Amer, 2018), but virtually nothing is known on the mechanisms causing this persistence. *L. pneumophila* requires synthesis of ppGpp in response to amino acid starvation to reach a state which allows the bacteria to escape from infected amoeba (Hammer and Swanson, 1999). During its intracellular life cycle in host MPs, *L. pneumophila* switches between a replicative and a transmissive state (Swanson and Fernandez-Moreira, 2002). In these host cells, ppGpp seems to be required for the transmission, since a *relA*/*spoT* mutant is killed during entry to and exit from MPs. Further work showed, however, that RelA (which senses amino acid starvation) is dispensable in MPs, whereas the hydrolase activity of SpoT (and hence hydolysis of ppGpp) is essential for the conversion of the bacteria from the transmissive to the replicative phase in MPs (Dalebroux et al., 2009). The authors conclude that the SpoT-mediated ppGpp degradation (monitoring fatty acid biosynthesis; see above) is necessary for this alternation in MPs. However, the question whether ppGpp also plays a role in persistence formation of *L. pneumophila*, remains unanswered (Abu Khweek and Amer, 2018).

In *Brucella* spp., stringent response is induced by nutrient stress *via* ppGpp that is synthesized by a single, bi-functional Rsh enzyme. *Rsh* deletion mutants of *Brucella suis* and *B. melitensis* show an altered morphology and a reduced survival rate under starvation conditions in cellular and murine infection models (Dozot et al., 2006). ppGpp-Dependent cross-talk between nutrient, oxidative, and low-oxygen stress responses was demonstrated suggesting an important role of ppGpp in the adaptation of *Brucella* to the host (Hanna et al., 2013) and possibly in persistence and chronic infections (Ficht, 2003).

Coxiella and *Bartonella* spp. possess ppGpp synthases, but nothing is known on the possible participation of ppGpp in persister formation of these IBPs. *Chlamydia* spp. and *Rickettsia* spp. lack ppGpp synthases and hence are unable to produce ppGpp (Mittenhuber, 2001).

General Stress Response and Its Link to Persistence

In addition to the ppGpp-dependent stringent response, the general stress response (GSR) appears to be also linked to persistence (Boaretti et al., 2003; Hong et al., 2012; Schellhorn, 2014; Tkachenko et al., 2014; Harms et al., 2016; Trastoy et al., 2018). In *E. coli* (and related bacteria), GSR depends on the sigma factor S (RpoS). RpoS governs the expression of many stationary-phase-inducible genes in *E. coli* (Hengge-Aronis, 1996; Battesti et al., 2011) and the entry into the stationary phase is known to lead to increased persister formation (Wood et al., 2013). A variety of environmental stress conditions can also induce GSR, including nutrient deprivation, variations of temperature, biofilm production, high pH, oxidative stress, and hyperosmolarity. GSR is also connected with ppGpp and TA modules: ppGpp stimulates the accumulation of RpoS (Brown et al., 2002; Hirsch and Elliott, 2002; Dalebroux et al., 2010). Antitoxins of certain TA modules repress the expression of RpoS. However, upon stress, the antitoxins are degraded and RpoS expression is induced (Wang et al., 2011; Wang and Wood, 2011; Hu et al., 2012).

GSR in IBPs and Its Link to Persistence

Among the IBPs, GSR is controlled by RpoS in the gamma-proteobacteria *Shigella*, *Salmonella*, *Legionella*, and *Coxiella*, and the involvement of RpoS in formation of persistence has been suggested for salmonellae and shigellae (Trastoy et al., 2018). In *L. monocytogens*, GSR is regulated by the alternative sigma factor SigB (Mittenhuber, 2002), similar as in *B. subtilis* and a small group of other Gram-positive bacteria (Hecker and Volker, 2001). SigB is involved in the survival of both saprophytic and host-associated stresses by *L. monocytogenes* (Dorey et al., 2019). A SigB-related factor is also present in *M. tuberculosis* (Mittenhuber, 2002), where it plays a major role in determining the level of tolerance to several drugs and the amount of persisters surviving drug treatment (Pisu et al., 2017).

GSR in alpha-proteobacteria, such as *Rickettsia*, *Bartonella* and *Brucella*, is controlled by a cascade including the alternative sigma factor αEcfG (also termed σE1 or RpoE1), the anti-sigma factor NepR, and the anti-anti-sigma factor PhyR (Fiebig et al., 2015; Francez-Charlot et al., 2015). In *Brucella*, these factors control transcription of approximately 100 genes involved in persistence in a BALB/c mouse chronic infection model (Kim et al., 2013; Willett et al., 2016). To our knowledge, there are no reports on a possible involvement of GSR-mediated regulation cascade in persister formation of *Bartonella* and *Rickettsia*. The obligate intracellular *Chlamydia* lacks all of these GSR-mediating sigma factors.

Oxidative Stress, Reactive Oxygen Species, Oxygen Stress Response, and the Links to Persistence

The connection between persister formation and oxidative stress, the subsequent increased reactive oxygen species (ROS) production, and the oxidative stress response (OSR) thereby induced, has been extensively described (Dörr et al., 2009; Möker et al., 2010; Vega et al., 2012; Wu et al., 2012; Wang et al., 2017). Augmented ROS production alters the membrane potential and causes damage of proteins, lipids, and nucleic acids (in particular DNA) with a strong impact on persister formation (Wang et al., 2017).

In *E. coli*, an increased ROS level induces the transcription factors SoxRS and OxyR that are primarily involved in the expression of antioxidant activities. But SoxRS can also induce the expression of the AcrAB-TolC multidrug-resistant pump causing extrusion of antibiotics. As consequence, a larger fraction of cells become persisters in the presence of antibiotics (Wu et al., 2012).

Increased persister subpopulations are also observed upon treatment of bacterial populations with bactericidal antibiotics (Kohanski et al., 2007; Kreuzer, 2013; Belenky et al., 2015; Kawai et al., 2015). Bactericidal antibiotics—besides blocking their primary targets—lead to downstream effects including metabolic changes accompanied with increased production of ROS (especially hydroxyl radicals) which, as already mentioned above, damage essential cellular components, ultimately causing cell death. In line with this assumption is the finding that bacteriostatic antibiotics, which recognize the same primary targets as the bactericidal antibiotics, do not trigger ROS production (Kohanski et al., 2007).

The enhanced persister formation arising upon treatment of a bacterial population with bactericidal antibiotics is apparently connected with this oxidative stress and the subsequent OSR leading to several response reactions that favor persistence (Walawalkar et al., 2016; Tosic-Pajic et al., 2017).

IBPs frequently encounter oxidative stress during infection. MPs—host cells for most IBPs—generate ROS and reactive nitrogen species (RNS) upon activation. Thus, ROS and subsequent OSR could also contribute to persistence (and possibly) chronic infections of IBPs. However, convincing experimental data are missing to support this assumption.

DNA Damage-Induced SOS Response and Its Link to Persistence

The association of persister formation to SOS response induced by DNA damage has been primarily studied in *E. coli*. The SOS pathway is crucially involved in the repair of DNA damage in bacteria (Kreuzer, 2013). The key regulators controlling the SOS network are LexA and RecA. Mutants lacking the *lexA* or the *recA* gene are more susceptible to quinolones and exhibit significantly reduced persistence in presence of these antibiotics (Dörr et al., 2009; Fung et al., 2010; Wu et al., 2012), while the constitutive expression of these genes strongly enhances persistence under these conditions (Dörr et al., 2009). These results suggest that persistence triggered by quinolones is influenced by the ability of the bacterial cell to repair DNA damage.

A link between SOS response and specific TA modules has also been demonstrated. Deletion of the SOS-inducible TisAB pair causes high reduction of persister cells; however, deletion of other LexA-box-containing TA pairs has no effect on persister formation (Dörr et al., 2010; Lewis, 2010). TisB is a membrane peptide that causes a decrease of the proton motive force and ATP levels. The resulting ATP depletion could therefore be also responsible for the SOS-induced persister formation by the TisAB TA module (Unoson and Wagner, 2008; Lewis, 2010; Shan et al., 2017). These examples show the complex interactions of the various stress conditions and the resulting cellular responses ultimately causing persister formation.

Involvement of the GTPase Obg in Persistence

Obg (also known as ObgE and CgtA) is a highly conserved GTPase present in all bacteria. It appears to function as a regulator for fundamental cellular processes such as ribosome maturation, DNA replication and chromosome segregation (Sikora-Borgula et al., 2002; Sikora et al., 2006; Persky et al., 2009; Kint et al., 2014). Obg has also been found to be central in controlling persistence in *E. coli* and *Pseudomonas aeruginosa* (Verstraeten et al., 2015). In *E. coli*, Obg-mediated persistence depends on ppGpp and the type I toxin HokB. An elevated ppGpp level induced by Obg leads to enhanced expression of the type I TA module HokB/SokB. The increased expression of the pore-forming HokB toxin results in a collapse of the membrane potential causing ATP leakage associated with persistence (Wilmaerts et al., 2018). All IBPs possess Obg-like proteins (**Table 1**), but there are no reports showing their involvement in persistence of IBPs.

METABOLISM AND PERSISTENCE

Facts and Hypotheses

Undoubtedly, metabolism plays a central role in initiating, maintaining and ending the persister state (Amato et al., 2013; Orman and Brynildsen, 2013; Amato et al., 2014; Prax and Bertram, 2014; Hartman et al., 2017; Radzikowski et al., 2017; Cabral et al., 2018). It has been shown that antibiotics (and the other stressors mentioned above) cause specific changes of the bacterial metabolism which may favor persister formation (Yang et al., 2017; Zampieri et al., 2017). In order to survive antibiotic treatment (or any other of the above described stress conditions), the persister cells must (a) shut down or silence essential physiological cell functions which antibiotics or the other

TABLE 1 | Presence and absence of factors in intracellular bacterial pathogens (IBPs) that were previously found to be associated with persistence.

IBP	TA	Stringent response			General stress response		Oxidative response		SOS response		Obg/CgtA
		RelA	SpoT	Rel (Rsh)	RpoS	SigB	OxyR	SoxR/SoxS	LexA	RecA	
Vacuolar											
Salmonella Typhimurium	++	+	+		+		+	+/+	+	+	+
Mycobacterium tuberculosis	+++			+		+	+	WhiB3	+	+	+
Legionella pneumophila	(+)	+	+		+		+	–/–		+	+
Brucella mellitensis	+			+	(+)		+	–/–		+	+
Coxiella burnettii	(-)	+	+		+		+	–/–	+	+	+
Chlamydia trachomatis	-	-	-	-	-		-	–/–	-	+	+
Cytosolic											
Listeria monocytogenes	+	+	-			+	(-)	–/–	+	+	+
Shigella flexneri	++	+	+		+		+	+/+	+	+	+
Francisella tularensis	-	+	+		-	-	+	–/–	+	+	+
Rickettsia prowazekii	+	-	-	-	-	-	-	–/–	-	+	+

(+) or (-) indicate unclear findings, while more than one + indicates the presence of multiple TA modules in the respective IBP. For more details, see also text.

stressors would irreversibly damage, (b) maintain viability during stasis, and (c) resume growth once the stress is lifted.

However, it is difficult to determine the specific metabolism of persister cells mainly due to their abundance, their transient nature, and their similar morphology in comparison to normal cells (Orman et al., 2015; Rowe et al., 2016). Therefore, the knowledge on the metabolism of bacterial persister state(s) is still fragmentary and sometimes even contradictory [see, e.g., (Leszczynska et al., 2013; Orman and Brynildsen, 2013; Kim and Wood, 2017)], although it is of great importance since it may offer novel strategies for the elimination of persisters (Allison et al., 2011; Fauvart et al., 2011; Kim et al., 2011; Radzikowski et al., 2017).

We will now focus on the presently known facts concerning the metabolism of persister subpopulations. Most of these data have been obtained by studies with few prototrophic model bacteria (mainly *E. coli*) which may limit their general validity. The major conclusions are: (a) Persistence often described as a "dormant state" represents a specific metabolically active state (Spoering et al., 2006; Prax and Bertram, 2014; Radzikowski et al., 2016). Despite the large decline in metabolic activities, the persister cells continue to produce energy, energize their membranes and produce a special set of proteins (Babin et al., 2017; Ayrapetyan et al., 2018). (b) The persistence-promoting metabolic processes occur stochastically in a small fraction even of actively growing and unstressed bacterial populations and are significantly enhanced in the presence of antibiotics, by nutrient starvation, in the stationary growth phase, and under the above mentioned stress conditions (Dörr et al., 2010; Radzikowski et al., 2016; Gutierrez et al., 2017; Meylan et al., 2017; Yang et al., 2017). (c) The persister state requires a suitable carbon and energy source (other than glucose) which allows a low flux through specific core metabolic pathways (Spoering et al., 2006; Amato et al., 2014; Cabral et al., 2018). (d) The tricarboxylic acid (TCA) cycle activity has been shown to play an important role in aminoglycoside antibiotic (tobramycin) susceptibility and tolerance of *P. aeruginosa* (Meylan et al., 2017): stimulation of the TCA cycle by fumarate activates cellular respiration and proton motive force. It leads to tobramycin uptake, high susceptibility and cell death, whereas shunting of the TCA cycle by stimulation of the glyoxylate cycle enhances tolerance against this aminoglycoside. (e) The ATP level is an important factor in persisters formation (Conlon et al., 2016; Radzikowski et al., 2016; Shan et al., 2017; Cameron et al., 2018). It has been even concluded that "stochastic variation in ATP is the main mechanism of persister formation" and "the decrease in ATP provides a satisfactory explanation for the drug tolerance of persisters" (Shan et al., 2017); (f) Central cell processes, including transcription, translation, DNA replication and cell wall synthesis, the major targets for most antibiotics (and eventually other stressors), are greatly slowed down or even turned off (Lewis, 2007; Hurdle et al., 2011; Kwan et al., 2013; Wood, 2017). (g) The metabolism of persisters may depend on the type of selection pressure, i.e., metabolism of persisters selected by ß-lactam antibiotics (cell wall inhibitors) is different from that of persisters selected by ciprofloxacin or nalidixic acid (both are gyrase inhibitors) (Cabral et al., 2018; Barrett et al., 2019). (h) An interesting, metabolism-based model has been proposed in which a specific low metabolic flux is the basis for establishing persistence, while other factors (e.g., various TA modules, ppGpp, RpoS) have a modulating and/or stabilizing function (Radzikowski et al., 2016).

Metabolism of IBPs Within Host Cells and Possible Links to Persistence of IBPs

Regarding human IBPs, most work concerning the metabolism of persistent populations has been carried out with *M. tuberculosis*, *Chlamydia* spp. and *Salmonella* serovars where persistence appears to be a key factor for the often long lasting chronic infections. *M. tuberculosis* is a metabolically highly flexible pathogen able to adapt to the changing environments which this pathogen encounters during infection. Most studies dealing with metabolic aspects of persistent mycobacteria have been performed with various *in vitro* culture models and the murine *in vivo* model (Wayne and Lin, 1982; Dutta and Karakousis, 2014; Sohaskey and Voskuil, 2015). Persistence of *M. tuberculosis* in mice is promoted by the glyoxylate bypass (including isocitrate lyase, ICL) and deletion of the *icl* gene leads to reduced persistence. The *icl* mutant also shows decreased survival in activated but not in resting MPs (Mckinney et al., 2000). ICL, an enzyme involved in the mycobacterial glyoxylate and methylisocitrate cycle (Gould et al., 2006), is essential for the catabolism of lipids/fatty acids/cholesterol. Catabolism of host derived lipids including cholesterol is a major factor for the persistent state of *M. tuberculosis* (Bhusal et al., 2017; Bonds and Sampson, 2018).

The proper homeostasis of the oxido-reductive systems is likewise important for persistence and reactivation of *M. tuberculosis* (Kumar et al., 2011). *M. tuberculosis* possesses two succinate dehydrogenases (Sdh1 and Sdh2). Sdh1 but not Sdh2 is necessary for respiration through the electron transport chain under normoxic conditions. Sdh1 or Sdh2 appears to be essential for the respiratory adaptation to hypoxic environments leading to persistence (Hartman et al., 2014). Nitrate reduction also seems to serve a respiratory function upon a sudden shift of *M. tuberculosis* to hypoxia and the mycobacterial nitrate reductase activity is highly induced in the hypoxic state (Wayne and Hayes, 1998; Sohaskey, 2008). The genes for most of the functions important for the adaptation to hypoxic stress (and for persistence), including metabolic genes, such as those for nitrate reductase and enzymes involved in energy acquisition by alternative carbon substrates (e.g., fatty acids) and for synthesis of triacylglycerols, are under the control of the two-component system DosR-DosS (Park et al., 2003; Voskuil et al., 2003; Galagan et al., 2013). These and other studies, using *in vitro* and *in vivo* models of *M. tuberculosis* infection, show that the DosR regulon is an important control factor of persistence of *M. tuberculosis* under hypoxic conditions (Converse et al., 2009; Rustad et al., 2009; Leistikow et al., 2010; Liu et al., 2016).

Human-pathogenic *Chlamydia* species are obligate IBPs that cause a wide range of acute and chronic diseases. After having entered host cells (mainly non-phagocytic mucosal cells), these IBPs live in a vacuolar compartment, the "inclusion". During the intracellular infection cycle, *Chlamydia* exist in two different forms. The reticulate body (RB) is the intracellular non-infectious, but proliferating form, which converts into the the elementary body (EB), the non-replicative, but infectious form. Under special conditions, such as treatment with interferon-gamma (IFN-γ) or penicillin, nutrient deprivation, or co-infection with *Herpes* viruses (Deka et al., 2006), RBs convert into persistent, nonreplicative particles, termed aberrant reticulate bodies (ARBs), which may re-convert into RBs and infectious EBs when the unfavorable conditions subside (Elwell et al., 2016; Witkin et al., 2017; Xue et al., 2017; Panzetta et al., 2018). Both the *Chlamydia* cells and the host cells undergo massive metabolic changes during the different conversions (Käding et al., 2014; Shima et al., 2018; Yang et al., 2019). In the persistent ARB state, *Chlamydia trachomatis* ceases to produce its major structural and membrane components (Witkin et al., 2017), but the still ongoing basic metabolic reactions in the ARBs remain largely unknown. In the infected host cells, IFN-γ activates the expression of indoleamine-2,3-dioxygenase 1 (IDO1), an enzyme that degrades tryptophan to kynurenine suggesting that the depletion of tryptophan blocks the normal chlamydial developmental cycle. Human-pathogenic *Chlamydia* species are unable to synthesize tryptophan. Restoration of tryptophan supply reverts the ARBs to infectious EBs (Beatty et al., 1993; Beatty et al., 1995), suggesting that biosynthesis of proteins containing multiple tryptophan residues may be blocked in ARBs (see below).

The metabolic requirement of persistent *Salmonella* has been mainly determined in an *in vivo* model of persistence whereby BALB/c mice were infected intravenously with a *S.* Typhimurium derivative that survived but hardly proliferated in the systemically infected mice. Persistent subpopulations could be isolated from liver and spleen which survived treatment with enrofloxacin, a fluoroquinolone highly efficient against this strain *in vitro* (Barat et al., 2012). Almost all tested *Salmonella* metabolic activities, including ATP synthesis driven by aerobic respiration, were dispensable in this persistence model demonstrating extensive resilience of persistent *Salmonella* against metabolic perturbations. However, inactivation of the biosynthesis of unsaturated fatty acids and cyclopropane fatty acids resulted in clearance of persisters to non-detectable levels in these organs. These fatty acids probably modify the fluidity of the cell wall. Upregulation of *cfa* genes (encoding cyclopropane fatty acid synthesis) was also observed in surviving *S. enterica* of different serotypes (including Typhimurium and Enteritides) upon long-term inoculation of poultry feed (Andino et al., 2014). This finding also suggests that the continued synthesis of these fatty acids is essential for long-term survival of *Salmonella* under harsh conditions. Isocitrate lyase appears to be required for *S.* Typhimurium persistence during chronic infection in mice but not during the acute phase of salmonellosis (Fang et al., 2005). This indicates that the glyoxylate shunt may play a critical role in the ability of *Salmonella* to persist in mammalian hosts and suggests that persistent *Salmonella* may depend on the utilization of fatty acids similar to *M. tuberculosis* (see above) and other intracellular pathogens (Monack et al., 2004). This conclusion is in line with the observation that *Salmonella* fail to persist in Parδ null mice (Eisele et al., 2013). The peroxisome proliferator-activated receptor delta (PPARδ) is a eukaryotic transcription factor essential for sustaining fatty acid metabolism (Odegaard et al., 2008) which is upregulated in *Salmonella*-infected MPs (Eisele et al., 2013).

Persistent *L. monocytogenes* strains are a major problem in food-industry (Ferreira et al., 2014; Abee et al., 2016), since severe infections in humans can be caused by food-borne strains that resist food processing steps (Lianou and Sofos, 2007) and even treatment with antimicrobials (Fox et al., 2011). Transcriptome analysis of such persistent strains show—compared to non-persistent strains—enhanced expression of the *pdu*, *eut* and *cob-cbi* operons (encoding enzymes involved in propandiol utilization, ethanolamine utilization, and cobalamine synthesis, respectively) (Fox et al., 2011). Whether these metabolic capacities may also play a significant role in persistent human listeriosis (Kleemann et al., 2009) remains an open question (Garsin, 2010).

Interestingly, the normally cytosolically replicating *L. monocytogenes* persist after infection of immunodeficient (SCID) mice in large vacuoles of liver granuloma MPs (Birmingham et al., 2008). Prolonged infection of (non-phagocytic) human hepatocytes and trophoblast cells also leads to bacteria enclosed in vacuoles. The switch from active cyctosolic replication to the resting vacuolar phase is correlated with a decreased accumulation of ActA at the bacterial surface (Kortebi et al., 2017). The authors argue that the formation of these *Listeria*-containing vacuoles could potentially enable the persistence in epithelial tissues. Although none of these two reports address the metabolic conditions in these vacuolar compartments and the metabolic activities of the persistent vacuolar bacteria, it is reasonable to assume that the conditions within the vacuoles (especially the metabolic conditions) are better suited for a persistent state than those in the host cell cytosol.

Francisella tularensis, another cytosolic IBP, can also re-enter into membrane-surrounded vacuoles (FCVs) after its extensive replication in the cytosol (Checroun et al., 2006) possibly to survive the more stressful conditions in the cytosol. To the best of our knowledge, there are no reports on IBPs persisting in the cytosol of host cells.

Occurrence and Absence of Factors and Pathways Associated With Persistence in IBPs

As described above, several factors and pathways are associated with bacterial persistence. This correlation is mainly based on the observation that their expression enhances significantly the formation of persister cells whereas their deletion reduces persister formation. Most of these studies were performed with *E. coli* (see above). However, in contrast to *E. coli* and many extracellular bacterial pathogens, several human IBPs lack one or more of these factors and pathways, and *Chlamydia* spp. which belong to the most successful persisting bacterial pathogens, even lack all of them (**Table 1**). This fact suggests that other cellular processes must be responsible for persister formation in these IBPs. Common to all IBPs so far analyzed, including *Chlamydia*, is the (highly flexible) metabolic strategy that IBPs follow after internalization by mammalian cells which we termed "bipartite metabolism" (Grubmüller et al., 2014; Eisenreich et al., 2015).

The Essential Features of Bipartite Metabolism

"Bipartite metabolism" (BM) describes a mode of bacterial metabolism that is carried out by many (possibly all) IBPs replicating within mammalian cells (Grubmüller et al., 2014; Eisenreich and Heuner, 2016; Häuslein et al., 2016; Chen et al., 2017; Häuslein et al., 2017a; Mehlitz et al., 2017; Best and Abu Kwaik, 2019). BM uses as major energy source various host-derived energy-rich carbon compounds that are less essential for the host cell than glucose. The withdrawal of these substrates from the host cell does not lead to the same negative consequences as withdrawal of glucose (e.g., autophagy and apoptosis). The suitable carbon compounds include mainly C_3-metabolites such as pyruvate or glycerol, serine and cysteine that can be converted to pyruvate (Eylert et al., 2008; Alkhuder et al., 2009; Grubmüller et al., 2014; Puckett et al., 2014; Abu Kwaik and Bumann, 2015; Vanderven et al., 2015; Häuslein et al., 2016; Chen et al., 2017; Häuslein et al., 2017b; Mehlitz et al., 2017). Besides these C_3-substrates, malate (used, e.g., by *Chlamydia*), fatty acids or cholesterol (used, e.g., by *M. tuberculosis*) are also suitable energy-rich components. These carbon sources are finally oxidized to acetyl-CoA, which enters the TCA cycle yielding important intermediates, as well as the electron carriers NADH and $FADH_2$ which, as essential electron carriers, lead to ATP production by oxidative phosphorylation (OXPHOS). In the absence of a functional electron transfer chain (ETC), ATP can be also produced through substrate phosphorylation by conversion of acetyl-CoA *via* acetyl-phosphate to acetate. The latter ATP delivering step is present in all IBPs with the exception of *Chlamydia* spp. which can, however, import ATP from the host cell by ATP/ADP transporters (Tjaden et al., 1999).

De novo biosynthesis of IBP-specific cell components is normally restricted to those compounds that cannot be provided by the host cells. This concerns in particular cell wall components (e.g., meso-diaminopimelic acid, mDAP). For conducting these indispensable biosynthetic pathways, the IBPs import small amounts of host cell-derived glucose, glucose-6-phosphate or other carbohydrates that can be converted to glucose-6-phosphate with low energy consumption.

Thus, the concept of BM includes two distinct metabolic networks: a mainly catabolic part (P1) fed by the above mentioned C_3 substrates which leads to the production of energy and some essential metabolites produced in the TCA cycle (e.g., oxaloacetate, α-ketoglutarate, and succinate) and another, more anabolic part (P2), fed mainly by intermediates of the upper part of the glycolytic pathway and the pentose phosphate pathway (PPP), which is essential for the indispensable anabolic components (mainly essential components for cell wall biosynthesis) that cannot be delivered by the host cell. The P1 and P2 networks interact depending on the physiological state of the host cell (Eisenreich et al., 2017; Eisenreich et al., 2019). For example, most of the energy-delivering carbon substrates mentioned above are glucogenic, i.e., in principle they are able to produce glucose by entering the gluconeogenic pathway. Indeed, most IBPs possess all genes required for the gluconeogenesis enzymes (exceptions are *Rickettsia* spp.). However, this mode of glucose production requires a disproportionally large amount of energy (e.g., glucose production from pyruvate needs six molecules of ATP) and gluconeogenesis is usually too expensive for carbohydrate production under intracellular conditions.

Under nutrient-rich conditions of the host cell, some IBPs, like *C. trachomatis*, *L. pneumophila* and *M. tuberculosis*, may convert suitable carbon substrates to storage products such as glycogen (Gehre et al., 2016), polyhydroxybutyrate (PHB) (Gillmaier et al., 2016; Häuslein et al., 2017b) or triacylglycerols (Maurya et al., 2018). These polymers can then be degraded to useful carbon/energy substrates under starvation conditions.

All studies carried out so far (Eylert et al., 2008; Grubmüller et al., 2014; Häuslein et al., 2016; Chen et al., 2017; Häuslein et al., 2017a; Mehlitz et al., 2017) indicate that, in IBPs, most low-molecular precursors for macromolecules, e.g., amino acids, nucleotides, fatty acids and vitamins, are imported from the host cells, which in turn obtain most of these nutrients from the intercellular space of host tissue. Major exceptions are, however, the three non-essential amino acids Ala, Asp, and Glu, which are efficiently synthesized *de novo* by all IBPs tested so far. Interestingly, these amino acids (in their D-forms) are either directly needed in considerable quantities for the synthesis of cell wall components (peptidoglycan and (lipo) teichoic acids) or act, like Asp, as a precursor of meso-diaminopimelate (mDAP), which also represents an important building block for peptidoglycan and is synthesized *de novo* by all

IBPs except *Francisella*. The latter IBP probably uses Lys (which can again be obtained from the host cell) instead of mDAP for peptidoglycan synthesis. Thus, for carrying out intracellular replication, the IBPs import, in addition to the energy-delivering (e.g., glycerol) and anabolism-supporting (e.g., glucose-6P) carbon substrates, substantial amounts of low molecular metabolites from the host cell and restrict their anabolic activities to the biosynthesis of those components that cannot be provided by the host cells and, of course, to protein and nucleic acid biosynthesis.

Besides being an economic strategy for maintaining the metabolism of IBPs replicating in mammalian host cells, BM may also be advantageous for the expression of virulence factors that are essential for intracellular proliferation. Expression of these factors is often blocked (e.g by catabolite repression) when glucose is the major carbon source (Eisenreich et al., 2017). Furthermore, the metabolic flexibility provided by the BM strategy may lead to considerable metabolic heterogeneity in an IBP population during infection of mammalian host cells which may favor the formation of persister state(s). Indeed, metabolic heterogeneity in isogenic bacterial population has been extensively described (Ackermann, 2015; Sheik et al., 2016; Simsek and Kim, 2018) and its significance for persister formation discussed (Balaban et al., 2004; Jones and Lennon, 2010; Bald and Koul, 2013; Simsek and Kim, 2018).

Metabolic States of the Host Cells and the Intracellular Bacteria That May Lead to Persister Formation

Ignoring immunological aspects, one can easily visualize two well-defined metabolic borderline states of the host cells that will either lead to proliferation or killing of the IBPs. In the first state, the host cells provide to the IBP a well-balanced supply of the above mentioned necessary nutrients. Under these conditions, all metabolites and sufficient energy will be available for the IBP to successfully carry out BM. This metabolic state will enable the IBP to build up its macromolecular structures and to perform efficient proliferation.

In the second state, the host cell can neither provide the main energy-delivering carbon source (in the following termed ECS) nor the carbon substrate supporting the indispensable anabolic processes (termed ACS) of the IBP. Under these conditions, the internalized IBP will come under nutritional stress and the resulting ATP shortage will ultimately lead to the stop of all macromolecular biosynthesis and the breakdown of the ATP-dependent DNA repair functions—and finally to the bacterial cell death.

Most differentiated cells, such as non-activated MP or epithelial cells have a reduced basic metabolism (resembling more the second metabolic host cell condition). This metabolic state provides probably not enough ECS and ACS to support efficient replication of internalized IBPs. However, most IBPs apparently express specific fitness factors that can activate the metabolism of the resting host cells converting them either into hospitable or sometimes also hostile states [recently reviewed by (Eisenreich et al., 2019)].

A typical example are the different metabolic activation events of resting MPs by Gram-negative bacterial pathogens: LPS of Gram-negative pathogens leads to classic M1 polarization of MPs which is characterized by a metabolic switch from the OXPHOS-driven metabolism (occurring at low level in resting MPs) to a Warburg-like metabolism with induced glucose uptake, aerobic glycolysis, lactate production, enhanced PPP and decreased TCA cycle activities (Krawczyk et al., 2010; O'neill, 2014; Kelly and O'neill, 2015). This metabolic state of the host cell could provide an appropriate nutrient supply to internalized IBPs and allow in principle bacterial replication. However, the production of oxygen and nitrogen radicals associated with the M1 state is deleterious for most IBPs (Eisele et al., 2013; Xavier et al., 2013a; Price and Vance, 2014). On the other hand, the alternative activation of MPs, induced by IL-4 which may be triggered in MPs by other bacterial factors leads to the anti-inflammatory M2 phenotype. The metabolism of M2 polarized MPs is characterized by enhanced fatty acid oxidation, OXPHOS, and thus increased intracellular levels of unconsumed glucose. This host cell metabolism appears to be favorable for IBP replication and has led to the assumption that M2 MPs could in general represent comfortable host cells for IBPs (Xavier et al., 2013b; Buchacher et al., 2015). However, the metabolic programs that can be induced in resting MPs (e.g., by IBPs) appear to be more diverse than just the M1 or M2 states (Murray and Wynn, 2011; Guilliams et al., 2014; Murray et al., 2014). There is experimental evidence that the metabolism of IBP-infected MPs clearly differs from that of M1 or M2 MPs (Götz et al., 2010; Gillmaier et al., 2012; Mehlitz et al., 2017). This metabolic variability of MPs may thus lead to a considerable metabolic heterogeneity in the MP population when exposed to IBPs. The obvious variability in bacterial numbers ranging from single to many bacteria per host cell observed when primary MPs are infected with IBPs (Sheppard et al., 2003; Thöne et al., 2007; Gillmaier et al., 2012) could be caused—at least in part—by the metabolic flexibility of these host cells.

The BM strategy of the IBPs likely contributes to metabolic heterogeneity of the intracellular bacteria. In the following, we consider several metabolic scenarios of the host cell which an incoming IBP may face and how the bacterial cell may adapt its own metabolism to the different metabolic states of the host cell —also with regard to possible persister formation:

Scenario 1: The host cell is able to provide sufficient amounts of the energy-delivering carbon substrate (ECS) and the anabolism-supporting carbon substrate (ACS) to the intracellular IBP (**Figure 1A**). Under these conditions, the IBP can replicate at optimal rates as long as the host cell tolerates the increasing number of IBPs. Then, the IBPs will be released from the host cells by different mechanisms or spread into neighboring host cells. It is unlikely that bacterial persister cells are formed under these conditions.

Scenario 2: The host cell can provide ACS, but not ECS. This is an unlikely scenario, since the catabolism of ACS by the host cell will always lead to ECS production. Besides, this scenario would resemble the growth conditions which the IBP face in a rich *in vitro* culture media where excess of ACS (especially glucose) blocks the synthesis of virulence factors that are necessary for the intracellular IBP life cycle.

Scenario 3: The IBP may receive from the host cell sufficient supply of ECS, but no ACS (**Figure 1B**). This situation can occur when the whole host cell population runs out of glucose. It may also happen stochastically when either a single host cell is depleted of ACS (e.g., glucose) or unable to take up ACS. If ECS is still available in sufficient quantity, the TCA cycle of the IBP can function and enough ATP is produced by oxidative and/or substrate phosphorylation to maintain at least all repair functions in the IBP. In case ECS is a glucogenic substrate (which is true for most C_3-compounds serving as ECS), glucose (ACS substrate) may be produced by gluconeogenesis as long as there is enough ATP to keep this high energy-consuming pathway running. Under these conditions, the IBPs can still perform their own essential anabolic processes and can still multiply. Formation of persister cells is also less likely under these conditions.

Scenario 4: An additional shortage of ECS supply (this could again occur stochastically in single cells or in the entire host cell population) will lead to a continuous decrease of ATP production (**Figure 1C**). As a result, all high-energy-consuming anabolic processes, including protein, DNA, RNA and cell envelope synthesis, will gradually come to an halt. But even at low ECS concentration, low residual ATP synthesis is still possible by substrate phosphorylation through oxidation of the residual ECS to acetate. Indeed all IBPs (with the exception of *Chlamydia* spp.) have the capacity to convert ECS substrates *via* acetyl-CoA and acetyl-phosphate to acetate, thereby generating ATP. *Chlamydia*, missing this pathway, can however import ATP directly from the host cell (Schmitz-Esser et al., 2004). This residual ATP supply will still maintain necessary repair functions of macromolecules, especially of DNA (Dahan-Grobgeld et al., 1998), which is absolutely crucial for the viability of persister cells (Völzing and Brynildsen, 2015).

Under these ECS-limited conditions, various metabolic states can be anticipated which may lead to slow-growing and non-growing but surviving (persistent)? IBPs. Indeed, single cell analysis of IBP-infected host cells show a high heterogeneity of intracellular bacterial counts in *in vitro* and *in vivo* infection models, with most host cells containing few or often only one bacterial cell (Brown et al., 2006; Helaine et al., 2010; Watson and Holden, 2010; Gillmaier et al., 2012; Zuck et al., 2015; Dhar et al., 2016; Eisenreich et al., 2019).

DOES THE PREVENTION OF ATP-DNAA-DEPENDENT INITIATION OF DNA REPLICATION LEAD TO PERSISTER FORMATION?

DnaA Is the Essential Initiator Protein for DNA Replication in Bacteria Including All IBPs

The above described metabolic scenarios 1–4 can occur in all IBPs with varying probability and scenario 4 could lead to persister formation even in those IBPs that lack most functions claimed to enhance bacterial persister formation (**Table 1**). This is the case in particular for human-pathogenic *Chlamydia* species that can nevertheless successfully form persistence.

A common cellular process to which all of the discussed persister-enhancing conditions appear to converge is the initiation step of DNA replication. This critical step in the bacterial cell cycle depends in bacteria, including all human IBPs, on the initiator protein DnaA, more precisely on ATP-DnaA, the initiation-active form (Sekimizu et al., 1988). For recent reviews, see (Katayama et al., 2017; Dewachter et al., 2018; Hansen and Atlung, 2018). Inhibition of ATP-DnaA formation results in a reversible stop of DNA replication initiation and cell division will subsequently stop. The resulting replication-terminated DNA is in a closed circular conformation which is less sensitive to damage (e.g., by oxidative stress induced by bactericidal antibiotics) than DNA with stalled elongation which always yields two open replication forks that are highly susceptible to damage (Hanna and Carl, 1975; Zyskind et al., 1977; Ikeda et al., 2012; Molina-Quiroz et al., 2018). In the following, the most important aspects of the ATP-DnaA dependent initiation step of DNA replication is shortly summarized. For further details on this complex field, see expert reviews (Leonard and Grimwade, 2010; Katayama et al., 2017; Hansen and Atlung, 2018; Leonard et al., 2019).

Initiation of DNA replication needs a critical amount of ATP-DnaA per oriC (the unique origin of chromosomal DNA) (Fuller et al., 1984; Morigen et al., 2003; Skarstad and Katayama, 2013; Hansen and Atlung, 2018). The initiation-active ATP-DnaA complex is inactivated shortly after initiation of DNA replication at oriC by hydrolysis of ATP to ADP resulting in an initiation-inactive ADP-DnaA complex (Keyamura and Katayama, 2011; Kasho and Katayama, 2013; Kasho et al., 2017). This process prevents that more than one round of DNA replication is initiated per cell cycle—at least at the low replication rates which is the case for IBPs growing in host cells. Re-initiation of DNA replication requires a precisely regulated amount of DnaA protein and sufficient ATP to form again an active ATP-DnaA complex at the newly formed oriC. The critical level of DnaA protein in the bacterial cell is essentially obtained by the rejuvenation of ADP-DnaA to DnaA and by *de novo* synthesis of DnaA (Castuma et al., 1993; Fujimitsu et al., 2009; Dewachter et al., 2018). Renewal of DnaA is achieved by several (seemingly bacteria-specific) mechanisms. For details, see recent reviews (Dewachter et al., 2018; Hansen and Atlung, 2018). The amount of *de novo* synthesized DnaA appears to be regulated on the transcriptional, translational and post-translational level. Transcription of the *dnaA* gene proceeds from two promoters, p1 and p2, which is auto-regulated, i.e., both promoters are inhibited by ATP-DnaA (Braun et al., 1985). Earlier studies claimed that high levels of the alarmone ppGpp also repress *dnaA* transcription, especially from the major promoter p2 (Chiaramello and Zyskind, 1990). More recent data show that ppGpp rather prevents replication initiation by blocking the introduction of initiation-promoting negative supercoils through inactivation of RNA polymerase by binding of ppGpp and thus preventing gene transcription and negative supercoiling of DNA (Kraemer et al., 2019). The *dnaA* transcript is rather unstable and has a low translation frequency (Bernstein et al., 2002) further limiting the synthesis of DnaA protein. Although DnaA is a

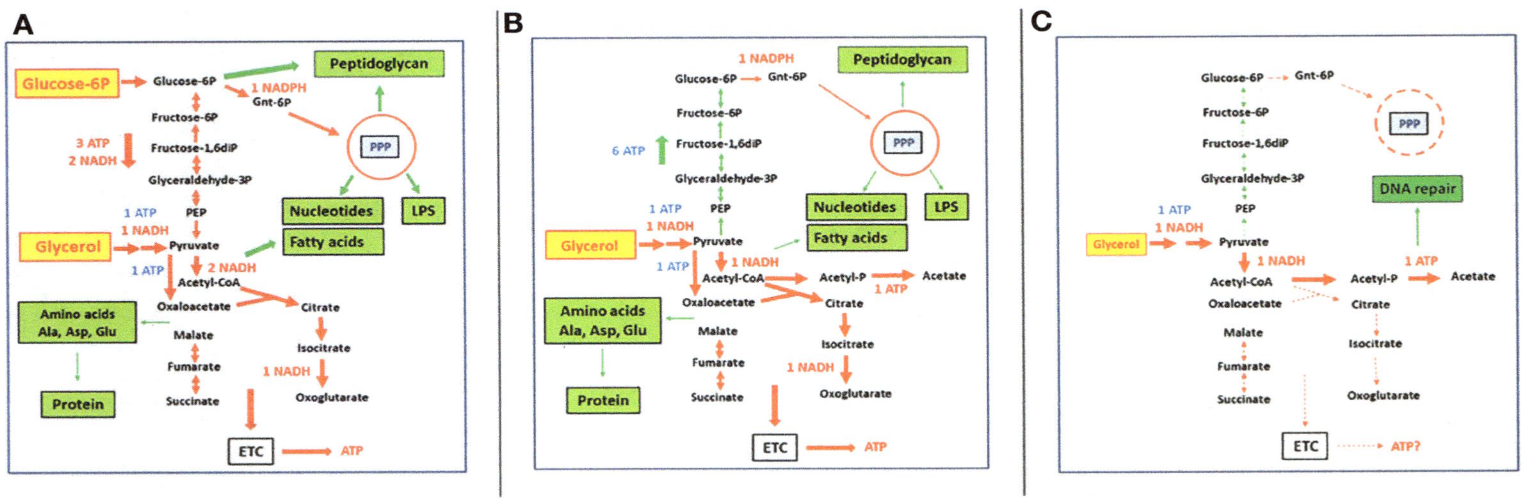

FIGURE 1 | Bipartite metabolism of IBPs on the example of intracellular *Listeria monocytogenes* (Grubmüller et al., 2014; Eisenreich et al., 2017) showing various metabolic states and their possible significance for formation of persistence. **(A)** Host cell provides to intracellular *L. monocytogenes* a sufficient amount of glycerol as an energy source and a limited amount of glucose-6-phosphate used mainly for anabolic processes. **(B)** Host cell is unable to provide glucose-6-phosphate but can still provide a sufficient amount of glycerol which has now to be used by *L. monocytogenes* as energy source and for maintaining anabolic processes. **(C)** Host cell can only provide a limited amount of glycerol which is now mainly used by the intracellular bacteria as energy source to provide sufficient ATP for repair functions (mainly for DNA repair). Red arrows indicate mainly catabolic processes and green arrows anabolic processes. The thickness of the arrows indicates the strength of the processes under the given nutritional conditions. Dashed thin arrows indicate reactions that are no longer occurring or severely restricted under the given condition. Red-labelled ATP, NADH, and NADPH mean formation and blue ones consumption by the corresponding processes. See text, for further details.

rather stable protein, excess DnaA can be acetylated at a highly conserved lysine (K178) by the acetyltransferase YfiQ (and non-enzymatically by acetylphosphate) which prevents DnaA from binding to ATP and hence to oriC (Li et al., 2017). In *E. coli*, the level of DnaA acetylation correlates with the frequency of replication initiation and reaches a peak at the stationary phase, leading to inhibition of initiation. This regulatory step of DnaA is reversible and deacetylation is catalyzed by the deacetylase CobB (Zhang et al., 2016). In *Caulobacter crescentus*, it has been shown that carbon starvation and nutritional depletion which take place, e.g., in the stationary phase lead to Lon-dependent proteolysis of DnaA and decreased translation of the *dnaA* transcript, thus reducing the accumulation of DnaA protein below the critical level necessary for initiation of replication (Leslie et al., 2015). Athough ATP binds to DnaA with high affinity, a critical cellular ATP/ADP ratio is required to charge the *de novo* synthesized and rejuvenated DnaA protein to form initiation-active ATP-DnaA. Finally, the membrane fluidity and especially its content of acidic phospholipids also play an essential role in the formation of a functional initiation complex between oriC and ATP-DnaA together with some other proteins (Norris, 1990; Castuma et al., 1993; Saxena et al., 2013; Katayama et al., 2017).

The central components involved in initiation of DNA replication, i.e., DnaA, ATP and oriC, are common to bacteria (Wolanski et al., 2014). The processes listed below (*Stochastic Occurrence* to *ATP Limitation*) leading to reversible stop of DNA replication initiation (due to the failure to form an active ATP-DnaA/oriC complex) are similar to the above discussed cellular events that favor bacterial persistence. It is therefore intriguing to hypothesize that persistence is linked to or even caused by stalling of DNA replication initiation due to insufficient cellular concentration of DnaA and/or ATP, or the failure to form an active ATP-DnaA/oriC complex. In the following, we point out that conditions known to lead to persister formation also negatively affect the formation of the ATP-DnaA complex and hence prohibit the initiation of DNA replication (see also **Figure 2**).

Stochastic Occurrence

In normal (i.e., unsynchronized) logarithmically growing *E. coli* cultures, DNA replication is in all possible stages. Most cells are in an ongoing replication process and their DNA will contain two replication forks due to the bidirectioal chromosomal replication (we will call this population of actively DNA replicating cells: a-cells). In a few cells, the replication is terminated but not yet re-initiated (this cell population will be called t-cells). The ratio of a-cells/t-cells depends on the growth rate and the growth phase, i.e., it will be higher in nutrient-rich culture media than in nutrient-poor ones and lower in the stationary than in the logarithmic growth phase. We postulate that the persister subpopulation observed in growing, unsynchronized bacterial populations always originates from the t-cell subpopulation.

Presence of Antibiotics

DNA replication inhibitors, like the gyrase-inhibiting quinolone antibiotics (e.g., nalidixic acid or ciprofloxacin), will quickly stop replication of a-cells. The stalled replication forks of a-cells are highly sensitive to damage and if not rapidly stabilized, these damages will lead to cell death by different processes, including double-strand breaks (Kuzminov, 1995; Kohanski et al., 2007; Masai et al., 2010; Liu and Imlay, 2013). The t-cell population, on the other hand, has a more stable, in most cases covalently closed circular DNA conformation (Nordström and Dasgupta, 2001) that can re-initiate replication once the antibiotic stress is released (Dewar and Walter, 2017). Antibiotics acting as inhibitors of translation or transcription (such as aminoglycosides, chloramphenicol or rifampicin) do not directly interfere with the DNA replication machinery, i.e., they do not disrupt the elongation step. These antibiotics can inhibit, however, initiation of DNA replication by blocking *de novo* synthesis of DnaA (Lark, 1972; Skarstad and Katayama, 2013), which is (at least in part) required for re-initiation of DNA replication. Besides, the bactericidal representatives among this group of antibiotics strongly interfere with DNA replication by formation of ROS and other processes that are lethal for the a-cell but less for the t-cell population (Yamaki et al., 1986; Hiraga et al., 1990; Miller et al., 2004; Kohanski et al., 2007; Liu and Imlay, 2013). This means that these antibiotics, which, similar to the gyrase-inhibiting antibiotics, cause enhanced persister formation, also lead to an increased t-cell subpopulation by inhibiting initiation of DNA replication. On the other hand, antimicrobials causing cross-linking of DNA, e.g., mitomycin, which lead to death even of persister cells (Keren et al., 2012; Chowdhury et al., 2016b) will also be lethal to t-cells, since re-initiation of cross-linked DNA can no longer occur.

TA Modules

Although most toxin components of TA modules target translation, some of them, e.g., ParE, CcdB, FicT, interfere with DNA replication by inhibiting the gyrase activity (Harms et al., 2018) and hence may exert similar inhibitory effects on DNA replication as the above described quinolone antibiotics. However, the majority of the toxins target translation in different ways (Harms et al., 2018). Thus, similar to translation-inhibiting antibiotics, TA modules may also exert inhibitory effects on *de novo* synthesis of DnaA and, hence, on initiation of DNA replication.

Increased Production of ppGpp

Amino acid starvation and other stress conditions lead to enhanced synthesis of ppGpp which increases persister formation and is also accompanied by inhibition of replication initiation due to different mechanisms including inhibition of RNA polymerase by ppGpp (Zyskind and Smith, 1992; Schreiber et al., 1995; Kraemer et al., 2019).

ATP Limitation

The formation of initiation-active ATP-DnaA requires not only a critical amount of DnaA protein but also a critical level of cellular ATP (Dewachter et al., 2018). Besides, DNA damage suffered by the t-cell population (e.g., by ROS) has to be repaired before proper reinitiation can occur. This repair process requires ATP and in the absence of sufficient ATP even t-cells cannot be revived.

Persistence Formation of IBPs and Its Possible Link to the DnaA-Dependent Initiation of DNA Replication

The formation of IBP persisters can also be explained by the above hypothesis. As outlined above, persister formation of all IBPs will be

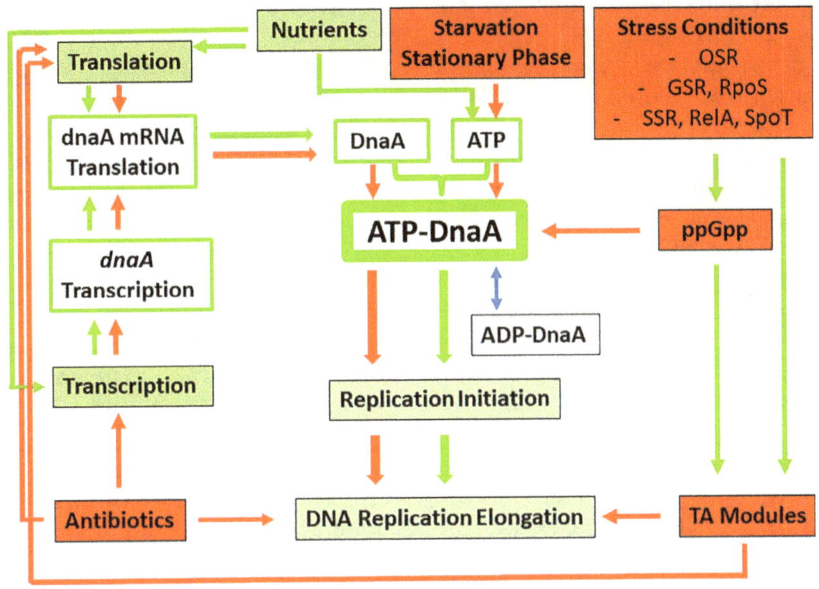

FIGURE 2 | Model showing the major processes necessary for the generation of the DNA initiation-active ATP-DnaA complex (green boxes and green arrows) and the factors, conditions and pathways (red boxes and red arrows) leading to inhibition of ATP-DnaA complex formation. Note that the latter situations are also identical to those leading to persister formation. OSR, oxidative stress response; GSR, general stress response; SSR, stringent stress response, leading to the generation of ppGpp *via* RelA and SpoT or Rsh. See text, for further details.

in general favored under intracellular (vacuolar) conditions which may reduce the metabolic activities and energy production of the IBPs (see **Figure 1C**) due to insufficient supply of essential nutrients from the host cell (Monack et al., 2004; Helaine et al., 2010; Helaine and Holden, 2013). It is likely that under these conditions the probability is increased that in some cells of an intracellular IBP population re-initiation of terminated DNA replication is prevented due to insufficient amount of ATP and/or DnaA protein, similar to cells in the stationary phase of *in vitro* growing bacterial cultures. We postulate that these IBP cells will become antibiotic-insensitive persisters.

Formation of persistence in IBPs on the basis of stalled replication initiation is outlined in the following more extensively on the example of *C. trachomatis*. This bacterium is an obligate IBP that lacks most factors and pathways that were previously associated with persister formation (**Table 1**). Yet, this pathogen is highly successful in persister formation.

As described above, *C. trachomatis* RBs enter the persister state when exposed to different conditions, such as treatment with IFN-γ or antibiotics including penicillin (Panzetta et al., 2018), azithromycin (Xue et al., 2017), and erythromycin (Clark et al., 1982), by amino acid starvation (Wyrick, 2010), or co-infection with *Herpes* viruses (Deka et al., 2006; Prusty et al., 2012). The persistent forms (ARBs) are viable, non-dividing and often enlarged cells that can revert to (RBs and) infectious EBs after removal of the persistence-inducing conditions (Panzetta et al., 2018).

ARBs generated in presence of IFN-γ carry out a generally reduced metabolism, slow down DNA replication, stop cell division, but continue to transcribe genes with different efficiencies (Ouellette et al., 2016), and show even enhanced synthesis of some proteins. The most striking ones among them are the chaperonins GroEL (Hsp60) and GroES (Witkin et al., 2017), and tryptophan synthase (TrpAB) (Belland et al., 2003). In the infected HeLa cells, IFN-γ reduces glycolysis, accompanied by the reduction of glucose transporter-1 (GLUT1) and of the hypoxia inducible factor-1a (HIF-1a) (Shima et al., 2018) leading to reduced ATP level in the host cells.

But in particular—as described above—IFN-γ induces indoleamine 2,3-dioxygenase (IDO). IDO strongly decreases the level of tryptophan in the host cell by degrading this amino acid to kynurenine thereby stopping the supply of tryptophan for intracellular *Chlamydia* which these bacteria are unable to synthesize. The ARBs formed under these conditions can readily be reactivated to normal infectious EBs when either IFN-γ is removed or tryptophan is added to the culture medium (Beatty et al., 1995; Muramatsu et al., 2016), suggesting that ARB formation is directly linked to the tryptophan deficiency.

Interestingly, proteins that continue to be synthesized or are even induced in the ARBs contain little or no tryptophan (e.g., GroES, GroEL, and TrpAB). Most other proteins contain on the average 1% tryptophan residues or more in their polypeptide chains (Ouellette et al., 2016) and, hence, their *de novo* synthesis will be reduced or even blocked. In this respect, the initiator protein DnaA is of particular interest, since it has to be (at least in part) *de novo* synthesized for initiating a new round of DNA replication. DnaA1 contains four Trp residues, three of which are within the first 50 amino acids of the 456 amino acid containing protein. It is therefore possible that—due to the IFN-γ induced tryptophan deficiency—*de novo* synthesis of DnaA may gradually cease, although the *dnaA* gene seems to be still transcribed (Belland et al., 2003). Ongoing DNA replication can be terminated, but not re-initiated due to the lack of DnaA. However, as transcription continues and is even increased for some Trp codon-rich genes (Ouellette et al., 2006; Ouellette et al., 2016), synthesis of some proteins containing no or little Trp is induced (see above) and production of other cell components, especially cell envelope components, can still occur and the ABRs will enlarge.

ATP is required for repair of possible DNA damage in the stalled DNA of the ARBs and it has been shown that both mitochondrial and chlamydial respiratory activities are necessary for ARB formation and maintenance (Liang et al., 2018). Thus, viability of the ARBs is probably maintained by self-produced and host cell-imported ATP (Schmitz-Esser et al., 2004).

Amino acid starvation of *Chlamydia*-infected host cells also leads to ARB formation. It has long been known that bacteria starved for amino acids can complete the ongoing DNA replication cycle, but cannot initiate a new one due to lack of *de novo* DnaA synthesis (Maaloe and Hanawalt, 1961; Lark et al., 1963; Abe and Tomizawa, 1967; Wolf et al., 1968).

An azithromycin-induced *in vitro* persistence model of *C. trachomatis* has been described (Xue et al., 2017) with the generation of ARBs that can revert after removal of azithromycin to infectious EBs. This macrolide antibiotic, as well as erythromycin which also leads to ARB formation (Clark et al., 1982), inhibit protein biosynthesis and hence will again block *de novo* synthesis of DnaA. Penicillin G, as well as other ß-lactam antibiotics, also induces formation of enlarged ARBs (Beatty et al., 1994; Skilton et al., 2009; Kintner et al., 2014). The penicillin-treated chlamydial cells do not divide and continue to enlarge without lysing. DNA replication continues resulting in ARBs with multiple genome copies (Lambden et al., 2006). Removal of penicillin leads to the regeneration of infectious EBs, although the large ARBs do not revert to the normal size RBs. Instead reversion to infectious EB particles occurs by budding RBs from the ARBs which develop to EBs (Skilton et al., 2009). Interestingly, the penicillin-induced ARBs are also resistant to azithromycin (Kintner et al., 2014), suggesting that the additional block of protein biosynthesis exerted by this antibiotic does not further affect the penicillin-induced ARB state. These observations are in accord with the assumption that the penicillin-induced ARBs carry several DNA replication-terminated genomes that are able to re-initiate once the stress condition is removed. This does not occur in the large ARB but rather in single RBs budding from the ARB. In these RBs, re-initiation of DNA replication may take place followed by the normal chlamydial development cycle.

C. trachomatis contains a homologue of the universal Obg/CtgA protein (**Table 1**) whose function in the chlamydial intracellular cycle is unknown. This essential GTPase has been shown to trigger persistence in *E. coli* by inducing type I TA modules (see above). A similar Obg-mediated persistence can be excluded for *C. trachomatis* since it lacks TA modules. However, Obg also plays a critical role in several basic cellular processes (Michel, 2005; Kint et al., 2014) including DNA replication by regulating the expression of *dnaA* in *E. coli* (Sikora et al., 2006).

CONCLUSIONS AND PERSPECTIVES

Persistence is a phenomenon common to most bacteria that probably evolved as a survival strategy against adverse environmental conditions. The formation of persistent human-pathogenic bacteria including IBPs, in particular *M. tubercuslosis*, *Salmonella* serovars, and *Chlamydia* spp., is a very important medical problem linked to chronic infections and to the development of antibiotic resistance. For recent reviews, see (Defraine et al., 2018; Thakur et al., 2019; Wilmaerts et al., 2019). A small subpopulation of persistent cells seems to be already present even in actively growing bacterial populations. These persisters appear to be generated spontaneously by yet unknown process(es). The proportion of persistent members in a bacterial population is increased in the stationary phase suggesting that the persistence trigger is increasingly present in this growth phase. Several stress conditions which the pathogens may encounter during infections, including nutrient starvation accompanied by reduced primary metabolism and energy production, release of the toxin components of various TA modules, stringent stress response with induction of the alarmone (p)ppGpp, general and oxidative stress responses, and especially treatment with various antibiotics, lead to a different but significant increase of the proportion of persistent subpopulations. The antibiotics not only select already existing (and apparently antibiotic-insensitive) persisters in a bacterial cell population, but directly enhance their formation. In fact, treatment with different antibiotics is the most common trigger for the increased incidence of persisters occurring during bacterial infections and also for the generation of persisters *in vitro*. However, none of the above mentioned conditions converts a bacterial population completely into the persistent state, but rather kills the majority of the bacterial population. This indicates that these various processes can stabilize and/or support specifically adapted cells to become peristers, but are not the direct cause of persister formation.

In this review, we cast a new hypothesis into the ongoing debate on the actual mechanism of bacterial persister formation which postulates that the persister state represents the cellular situation in which the DNA replication is terminated, but initiation of a new round of replication is prevented due to an insufficient amount of active ATP-DnaA initiator and possibly by an unaccessible OriC due to positive-supercoiled DNA. Re-initiation of DNA replication requires a precisely regulated amount of ATP-DnaA which depends in part on *de novo* synthesis of DnaA protein and a sufficiently high cellular ATP/ADP ratio. We provide circumstantial evidence that all physiological conditions leading to increased persister formation will also inhibit either DNA replication or protein biosynthesis (and hence *de novo* synthesis of DnaA), or decrease the ATP level thus preventing the formation of the critical cellular concentration of ATP-DnaA necessary for initiation of DNA replication at the newly formed origins of replication. The replication-terminated, but not yet re-initiated chromosomal DNA is in a positive-supercoiled closed circular conformation and hence less sensitive to DNA damage (especially by oxygen and nitrogen radicals) that occurs under virtually all stress conditions that lead to enhanced persister generation.

This mechanism of persister formation could be common to all human bacterial pathogens, even to those intracellular pathogens (e.g., *Chlamydia* and *Rickettsia* spp.) that lack most factors and stress pathways claimed to cause enhanced persistence (mainly identified in *E. coli*). But all bacteria studied so far, including these latter bacterial pathogens, require the ATP-DnaA dependent DNA replication initiation (Kaguni, 2006). As none of the bactericidal antibiotics used to screen for persisters target the ATP-DnaA complex and damage its function (Grimwade and Leonard, 2019), persister cells with terminated DNA replication but stalled initiation are able to re-initiate replication and cell division once the antibiotic is removed.

The correctness of this hypothesis can be experimentally tested and appropriate work is underway to confirm or to falsify it. If it holds true, it opens interesting new opportunities to combat bacterial persister formation.

AUTHOR CONTRIBUTIONS

All authors wrote the review. The overall concept and design is based on an idea of WG. All authors contributed to the article and approved the submitted version.

REFERENCES

Abe, M., and Tomizawa, J. (1967). Replication of the *Escherichia coli* K12 chromosome. *Proc. Natl. Acad. Sci. U.S.A.* 58, 1911–1918. doi: 10.1073/pnas.58.5.1911

Abee, T., Koomen, J., Metselaar, K. I., Zwietering, M. H., and Den Besten, H. M. (2016). Impact of pathogen population heterogeneity and stress-resistant variants on food safety. *Annu. Rev. Food Sci. Technol.* 7, 439–456. doi: 10.1146/annurev-food-041715-033128

Abu Khweek, A., and Amer, A. O. (2018). Factors mediating environmental biofilm formation by. *Legionella Pneumophila Front. Cell Infect. Microbiol.* 8, 38. doi: 10.3389/fcimb.2018.00038

Abu Kwaik, Y., and Bumann, D. (2015). Host delivery of favorite meals for intracellular pathogens. *PloS Pathog* 11, e1004866. doi: 10.1371/journal.ppat.1004866

Ackermann, M. (2015). A functional perspective on phenotypic heterogeneity in microorganisms. *Nat. Rev. Microbiol.* 13, 497–508. doi: 10.1038/nrmicro3491

Albrethsen, J., Agner, J., Piersma, S. R., Hojrup, P., Pham, T. V., Weldingh, K., et al. (2013). Proteomic profiling of *Mycobacterium tuberculosis* identifies nutrient-starvation-responsive toxin-antitoxin systems. *Mol. Cell Proteomics* 12, 1180–1191. doi: 10.1074/mcp.M112.018846

Alkhuder, K., Meibom, K. L., Dubail, I., Dupuis, M., and Charbit, A. (2009). Glutathione provides a source of cysteine essential for intracellular multiplication of *Francisella tularensis*. *PloS Pathog.* 5, e1000284. doi: 10.1371/journal.ppat.1000284

Allison, K. R., Brynildsen, M. P., and Collins, J. J. (2011). Metabolite-enabled eradication of bacterial persisters by aminoglycosides. *Nature* 473, 216–220. doi: 10.1038/nature10069

Amato, S. M., and Brynildsen, M. P. (2015). Persister heterogeneity arising from a single metabolic stress. *Curr. Biol.* 25, 2090–2098. doi: 10.1016/j.cub.2015.06.034

Amato, S. M., Orman, M. A., and Brynildsen, M. P. (2013). Metabolic control of persister formation in *Escherichia coli*. *Mol. Cell* 50, 475–487. doi: 10.1016/j.molcel.2013.04.002

Amato, S. M., Fazen, C. H., Henry, T. C., Mok, W. W., Orman, M. A., Sandvik, E. L., et al. (2014). The role of metabolism in bacterial persistence. *Front. Microbiol.* 5, 70. doi: 10.3389/fmicb.2014.00070

Andino, A., Pendleton, S., Zhang, N., Chen, W., Critzer, F., and Hanning, I. (2014). Survival of *Salmonella enterica* in poultry feed is strain dependent. *Poult Sci.* 93, 441–447. doi: 10.3382/ps.2013-03401

Avarbock, D., Salem, J., Li, L. S., Wang, Z. M., and Rubin, H. (1999). Cloning and characterization of a bifunctional RelA/SpoT homologue from *Mycobacterium tuberculosis*. *Gene* 233, 261–269. doi: 10.1016/S0378-1119(99)00114-6

Ayrapetyan, M., Williams, T. C., and Oliver, J. D. (2015). Bridging the gap between viable but non-culturable and antibiotic persistent bacteria. *Trends Microbiol.* 23, 7–13. doi: 10.1016/j.tim.2014.09.004

Ayrapetyan, M., Williams, T., and Oliver, J. D. (2018). Relationship between the viable but nonculturable state and antibiotic persister cells. *J. Bacteriol.* 200, e00249-18. doi: 10.1128/JB.00249-18

Babin, B. M., Atangcho, L., Van Eldijk, M. B., Sweredoski, M. J., Moradian, A., Hess, S., et al. (2017). Selective proteomic analysis of antibiotic-tolerant cellular subpopulations in *Pseudomonas aeruginosa* biofilms. *mBio* 8, e01593–17. doi: 10.1128/mBio.01593-17

Baharoglu, Z., Krin, E., and Mazel, D. (2013). RpoS plays a central role in the SOS induction by sub-lethal aminoglycoside concentrations in *Vibrio cholerae*. *PloS Genet.* 9, e1003421. doi: 10.1371/journal.pgen.1003421

Bakkeren, E., Diard, M., and Hardt, W. D. (2020). Evolutionary causes and consequences of bacterial antibiotic persistence. *Nat. Rev. Microbiol.* 18, 479–490. doi: 10.1038/s41579-020-0378-z

Balaban, N. Q., Merrin, J., Chait, R., Kowalik, L., and Leibler, S. (2004). Bacterial persistence as a phenotypic switch. *Science* 305, 1622–1625. doi: 10.1126/science.1099390

Balaban, N. Q., Gerdes, K., Lewis, K., and Mckinney, J. D. (2013). A problem of persistence: still more questions than answers? *Nat. Rev. Microbiol.* 11, 587–591. doi: 10.1038/nrmicro3076

Balaban, N. Q., Helaine, S., Lewis, K., Ackermann, M., Aldridge, B., Andersson, D. I., et al. (2019). Definitions and guidelines for research on antibiotic persistence. *Nat. Rev. Microbiol.* 17, 441–448. doi: 10.1038/s41579-019-0196-3

Bald, D., and Koul, A. (2013). Advances and strategies in discovery of new antibacterials for combating metabolically resting bacteria. *Drug Discovery Today* 18, 250–255. doi: 10.1016/j.drudis.2012.09.007

Barat, S., Steeb, B., Maze, A., and Bumann, D. (2012). Extensive *in vivo* resilience of persistent *Salmonella*. *PloS One* 7, e42007. doi: 10.1371/journal.pone.0042007

Barrett, T. C., Mok, W. W. K., Murawski, A. M., and Brynildsen, M. P. (2019). Enhanced antibiotic resistance development from fluoroquinolone persisters after a single exposure to antibiotic. *Nat. Commun.* 10, 1177. doi: 10.1038/s41467-019-09058-4

Battesti, A., Majdalani, N., and Gottesman, S. (2011). The RpoS-mediated general stress response in *Escherichia coli*. *Annu. Rev. Microbiol.* 65, 189–213. doi: 10.1146/annurev-micro-090110-102946

Beatty, W. L., Byrne, G. I., and Morrison, R. P. (1993). Morphologic and antigenic characterization of interferon gamma-mediated persistent *Chlamydia trachomatis* infection *in vitro*. *Proc. Natl. Acad. Sci. U. S. A.* 90, 3998–4002. doi: 10.1073/pnas.90.9.3998

Beatty, W. L., Morrison, R. P., and Byrne, G. I. (1994). Persistent chlamydiae: from cell culture to a paradigm for chlamydial pathogenesis. *Microbiol. Rev.* 58, 686–699. doi: 10.1128/MR.58.4.686-699.1994

Beatty, W. L., Morrison, R. P., and Byrne, G. I. (1995). Reactivation of persistent *Chlamydia trachomatis* infection in cell culture. *Infect. Immun.* 63, 199–205. doi: 10.1128/IAI.63.1.199-205.1995

Belenky, P., Ye, J. D., Porter, C. B., Cohen, N. R., Lobritz, M. A., Ferrante, T., et al. (2015). Bactericidal antibiotics induce toxic metabolic perturbations that lead to cellular damage. *Cell Rep.* 13, 968–980. doi: 10.1016/j.celrep.2015.09.059

Belland, R. J., Nelson, D. E., Virok, D., Crane, D. D., Hogan, D., Sturdevant, D., et al. (2003). Transcriptome analysis of chlamydial growth during IFN-gamma-mediated persistence and reactivation. *Proc. Natl. Acad. Sci. U. S. A.* 100, 15971–15976. doi: 10.1073/pnas.2535394100

Berjeaud, J. M., Chevalier, S., Schlusselhuber, M., Portier, E., Loiseau, C., Aucher, W., et al. (2016). *Legionella pneumophila*: the paradox of a highly sensitive opportunistic waterborne pathogen able to persist in the environment. *Front. Microbiol.* 7, 486. doi: 10.3389/fmicb.2016.00486

Bernstein, J. A., Khodursky, A. B., Lin, P. H., Lin-Chao, S., and Cohen, S. N. (2002). Global analysis of mRNA decay and abundance in *Escherichia coli* at single-gene resolution using two-color fluorescent DNA microarrays. *Proc. Natl. Acad. Sci. U.S.A.* 99, 9697–9702. doi: 10.1073/pnas.112318199

Best, A., and Abu Kwaik, Y. (2019). Nutrition and bipartite metabolism of intracellular pathogens. *Trends Microbiol.* 27, 550–561. doi: 10.1016/j.tim.2018.12.012

Bhaskar, A., De Piano, C., Gelman, E., Mckinney, J. D., and Dhar, N. (2018). Elucidating the role of (p)ppGpp in mycobacterial persistence against antibiotics. *IUBMB Life* 70, 836–844. doi: 10.1002/iub.1888

Bhusal, R. P., Bashiri, G., Kwai, B. X. C., Sperry, J., and Leung, I. K. H. (2017). Targeting isocitrate lyase for the treatment of latent tuberculosis. *Drug Discovery Today* 22, 1008–1016. doi: 10.1016/j.drudis.2017.04.012

Bigger, J. W. (1944). Treatment of staphylococcal infections with penicillin by intermittent sterilisation. *Lancet* 244, 497–500. doi: 10.1016/S0140-6736(00)74210-3

Birmingham, C. L., Canadien, V., Kaniuk, N. A., Steinberg, B. E., Higgins, D. E., and Brumell, J. H. (2008). Listeriolysin O allows *Listeria monocytogenes* replication in macrophage vacuoles. *Nature* 451, 350–354. doi: 10.1038/nature06479

Boaretti, M., Lleo, M. M., Bonato, B., Signoretto, C., and Canepari, P. (2003). Involvement of rpoS in the survival of *Escherichia coli* in the viable but non-culturable state. *Environ. Microbiol.* 5, 986–996. doi: 10.1046/j.1462-2920.2003.00497.x

Bokinsky, G., Baidoo, E. E., Akella, S., Burd, H., Weaver, D., Alonso-Gutierrez, J., et al. (2013). HipA-triggered growth arrest and beta-lactam tolerance in *Escherichia coli* are mediated by RelA-dependent ppGpp synthesis. *J. Bacteriol* 195, 3173–3182. doi: 10.1128/JB.02210-12

Bonds, A. C., and Sampson, N. S. (2018). More than cholesterol catabolism: regulatory vulnerabilities in *Mycobacterium tuberculosis*. *Curr. Opin. Chem. Biol.* 44, 39–46. doi: 10.1016/j.cbpa.2018.05.012

Braun, R. E., O'day, K., and Wright, A. (1985). Autoregulation of the DNA replication gene dnaA in *E. coli* K-12. *Cell* 40, 159–169. doi: 10.1016/0092-8674(85)90319-8

Brown, L., Gentry, D., Elliott, T., and Cashel, M. (2002). DksA affects ppGpp induction of RpoS at a translational level. *J. Bacteriol.* 184, 4455–4465. doi: 10.1128/JB.184.16.4455-4465.2002

Brown, S. P., Cornell, S. J., Sheppard, M., Grant, A. J., Maskell, D. J., Grenfell, B. T., et al. (2006). Intracellular demography and the dynamics of *Salmonella enterica* infections. *PloS Biol.* 4, e349. doi: 10.1371/journal.pbio.0040349

Buchacher, T., Ohradanova-Repic, A., Stockinger, H., Fischer, M. B., and Weber, V. (2015). M2 Polarization of human macrophages favors survival of the intracellular pathogen *Chlamydia pneumoniae*. *PloS One* 10, e0143593. doi: 10.1371/journal.pone.0143593

Cabral, D. J., Wurster, J. I., and Belenky, P. (2018). Antibiotic persistence as a metabolic adaptation: stress, metabolism, the host, and new directions. *Pharmaceuticals (Basel)* 11, 14. doi: 10.3390/ph11010014

Cameron, D. R., Shan, Y., Zalis, E. A., Isabella, V., and Lewis, K. (2018). A genetic determinant of persister cell formation in bacterial pathogens. *J. Bacteriol.* 200, e00303-18. doi: 10.1128/JB.00303-18

Castuma, C. E., Crooke, E., and Kornberg, A. (1993). Fluid membranes with acidic domains activate DnaA, the initiator protein of replication in *Escherichia coli*. *J. Biol. Chem.* 268, 24665–24668.

Charity, J. C., Blalock, L. T., Costante-Hamm, M. M., Kasper, D. L., and Dove, S. L. (2009). Small molecule control of virulence gene expression in *Francisella tularensis*. *PloS Pathog.* 5, e1000641. doi: 10.1371/journal.ppat.1000641

Checroun, C., Wehrly, T. D., Fischer, E. R., Hayes, S. F., and Celli, J. (2006). Autophagy-mediated reentry of *Francisella tularensis* into the endocytic compartment after cytoplasmic replication. *Proc. Natl. Acad. Sci. U.S.A.* 103, 14578–14583. doi: 10.1073/pnas.0601838103

Chen, F., Rydzewski, K., Kutzner, E., Häuslein, I., Schunder, E., Wang, X., et al. (2017). Differential substrate usage and metabolic fluxes in *Francisella tularensis* subspecies holarctica and *Francisella novicida*. *Front. Cell Infect. Microbiol.* 7, 275. doi: 10.3389/fcimb.2017.00275

Chiaramello, A. E., and Zyskind, J. W. (1990). Coupling of DNA replication to growth rate in *Escherichia coli*: a possible role for guanosine tetraphosphate. *J. Bacteriol.* 172, 2013–2019. doi: 10.1128/JB.172.4.2013-2019.1990

Chowdhury, N., Kwan, B. W., and Wood, T. K. (2016a). Persistence increases in the absence of the alarmone guanosine tetraphosphate by reducing cell growth. *Sci. Rep.* 6, 20519. doi: 10.1038/srep20519

Chowdhury, N., Wood, T. L., Martinez-Vazquez, M., Garcia-Contreras, R., and Wood, T. K. (2016b). DNA-crosslinker cisplatin eradicates bacterial persister cells. *Biotechnol. Bioeng.* 113, 1984–1992. doi: 10.1002/bit.25963

Christodoulou, D., Link, H., Fuhrer, T., Kochanowski, K., Gerosa, L., and Sauer, U. (2018). Reserve flux capacity in the pentose phosphate pathway enables *Escherichia coli*'s rapid response to oxidative stress. *Cell Syst.* 6569-578, e567. doi: 10.1016/j.cels.2018.04.009

Chuang, Y. M., Bandyopadhyay, N., Rifat, D., Rubin, H., Bader, J. S., and Karakousis, P. C. (2015). Deficiency of the novel exopolyphosphatase Rv1026/PPX2 leads to metabolic downshift and altered cell wall permeability in *Mycobacterium tuberculosis*. *mBio* 6, e02428. doi: 10.1128/mBio.02428-14

Clark, R. B., Schatzki, P. F., and Dalton, H. P. (1982). Ultrastructural analysis of the effects of erythromycin on the morphology and developmental cycle of *Chlamydia trachomatis* HAR-13. *Arch. Microbiol.* 133, 278–282. doi: 10.1007/BF00521290

Clark, T. R., Ellison, D. W., Kleba, B., and Hackstadt, T. (2011). Complementation of *Rickettsia rickettsii* RelA/SpoT restores a nonlytic plaque phenotype. *Infect. Immun.* 79, 1631–1637. doi: 10.1128/IAI.00048-11

Cohen, N. R., Lobritz, M. A., and Collins, J. J. (2013). Microbial persistence and the road to drug resistance. *Cell Host Microbe* 13, 632–642. doi: 10.1016/j.chom.2013.05.009

Conlon, B. P., Rowe, S. E., and Lewis, K. (2015). Persister cells in biofilm associated infections. *Adv. Exp. Med. Biol.* 831, 1–9. doi: 10.1007/978-3-319-09782-4_1

Conlon, B. P., Rowe, S. E., Gandt, A. B., Nuxoll, A. S., Donegan, N. P., Zalis, E. A., et al. (2016). Persister formation in *Staphylococcus aureus* is associated with ATP depletion. *Nat. Microbiol.* 1, 16051. doi: 10.1038/nmicrobiol.2016.51

Conlon, B. P. (2014). *Staphylococcus aureus* chronic and relapsing infections: Evidence of a role for persister cells: An investigation of persister cells, their formation and their role in *S. aureus* disease. *Bioessays* 36, 991–996. doi: 10.1002/bies.201400080

Converse, P. J., Karakousis, P. C., Klinkenberg, L. G., Kesavan, A. K., Ly, L. H., Allen, S. S., et al. (2009). Role of the dosR-dosS two-component regulatory system in *Mycobacterium tuberculosis* virulence in three animal models. *Infect. Immun.* 77, 1230–1237. doi: 10.1128/IAI.01117-08

Curtis, T. D., Takeuchi, I., Gram, L., and Knudsen, G. M. (2017). The influence of the toxin/antitoxin mazEF on growth and survival of *Listeria monocytogenes* under stress. *Toxins (Basel)* 9, 31. doi: 10.3390/toxins9010031

Cuthbert, B. J., Ross, W., Rohlfing, A. E., Dove, S. L., Gourse, R. L., Brennan, R. G., et al. (2017). Dissection of the molecular circuitry controlling virulence in *Francisella tularensis*. *Genes Dev.* 31, 1549–1560. doi: 10.1101/gad.303701.117

Dahan-Grobgeld, E., Livneh, Z., Maretzek, A. F., Polak-Charcon, S., Eichenbaum, Z., and Degani, H. (1998). Reversible induction of ATP synthesis by DNA damage and repair in *Escherichia coli*. *In vivo* NMR studies. *J. Biol. Chem.* 273, 30232–30238. doi: 10.1074/jbc.273.46.30232

Dahl, J. L., Kraus, C. N., Boshoff, H. I., Doan, B., Foley, K., Avarbock, D., et al. (2003). The role of RelMtb-mediated adaptation to stationary phase in long-term persistence of *Mycobacterium tuberculosis* in mice. *Proc. Natl. Acad. Sci. U. S. A.* 100, 10026–10031. doi: 10.1073/pnas.1631248100

Dalebroux, Z. D., and Swanson, M. S. (2012). ppGpp: magic beyond RNA polymerase. *Nat. Rev. Microbiol.* 10, 203–212. doi: 10.1038/nrmicro2720

Dalebroux, Z. D., Edwards, R. L., and Swanson, M. S. (2009). SpoT governs *Legionella pneumophila* differentiation in host macrophages. *Mol. Microbiol* 71, 640–658. doi: 10.1111/j.1365-2958.2008.06555.x

Dalebroux, Z. D., Svensson, S. L., Gaynor, E. C., and Swanson, M. S. (2010). ppGpp conjures bacterial virulence. *Microbiol. Mol. Biol. Rev.* 74, 171–199. doi: 10.1128/MMBR.00046-09

Dean, R. E., Ireland, P. M., Jordan, J. E., Titball, R. W., and Oyston, P. C. (2009). RelA regulates virulence and intracellular survival of *Francisella novicida*. *Microbiology* 155, 4104–4113. doi: 10.1099/mic.0.031021-0

Defraine, V., Fauvart, M., and Michiels, J. (2018). Fighting bacterial persistence: Current and emerging anti-persister strategies and therapeutics. *Drug Resist. Update* 38, 12–26. doi: 10.1016/j.drup.2018.03.002

Deka, S., Vanover, J., Dessus-Babus, S., Whittimore, J., Howett, M. K., Wyrick, P. B., et al. (2006). *Chlamydia trachomatis* enters a viable but non-cultivable (persistent) state within herpes simplex virus type 2 (HSV-2) co-infected host cells. *Cell Microbiol.* 8, 149–162. doi: 10.1111/j.1462-5822.2005.00608.x

Dewachter, L., Verstraeten, N., Fauvart, M., and Michiels, J. (2018). An integrative view of cell cycle control in *Escherichia coli*. *FEMS Microbiol. Rev.* 42, 116–136. doi: 10.1093/femsre/fuy005

Dewar, J. M., and Walter, J. C. (2017). Mechanisms of DNA replication termination. *Nat. Rev. Mol. Cell Biol.* 18, 507–516. doi: 10.1038/nrm.2017.42

Dhar, N., and Mckinney, J. D. (2007). Microbial phenotypic heterogeneity and antibiotic tolerance. *Curr. Opin. Microbiol.* 10, 30–38. doi: 10.1016/j.mib.2006.12.007

Dhar, N., Mckinney, J., and Manina, G. (2016). Phenotypic heterogeneity in *Mycobacterium tuberculosis*. *Microbiol. Spectr.* 4, 6. doi: 10.1128/microbiolspec.TBTB2-0021-2016

Di Cesare, A., Losasso, C., Barco, L., Eckert, E. M., Conficoni, D., Sarasini, G., et al. (2016). Diverse distribution of Toxin-Antitoxin II systems in *Salmonella enterica* serovars. *Sci. Rep.* 6, 28759. doi: 10.1038/srep28759

Dorey, A., Marinho, C., Piveteau, P., and O'byrne, C. (2019). Role and regulation of the stress activated sigma factor sigma B (sigma(B)) in the saprophytic and host-associated life stages of *Listeria monocytogenes*. *Adv. Appl. Microbiol.* 106, 1–48. doi: 10.1016/bs.aambs.2018.11.001

Dörr, T., Lewis, K., and Vulic, M. (2009). SOS response induces persistence to fluoroquinolones in *Escherichia coli*. *PloS Genet.* 5, e1000760. doi: 10.1371/journal.pgen.1000760

Dörr, T., Vulic, M., and Lewis, K. (2010). Ciprofloxacin causes persister formation by inducing the TisB toxin in *Escherichia coli*. *PloS Biol.* 8, e1000317. doi: 10.1371/journal.pbio.1000317

Dozot, M., Boigegrain, R. A., Delrue, R. M., Hallez, R., Ouahrani-Bettache, S., Danese, I., et al. (2006). The stringent response mediator Rsh is required for *Brucella melitensis* and *Brucella suis* virulence, and for expression of the type IV secretion system virB. *Cell Microbiol.* 8, 1791–1802. doi: 10.1111/j.1462-5822.2006.00749.x

Dutta, N. K., and Karakousis, P. C. (2014). Latent tuberculosis infection: myths, models, and molecular mechanisms. *Microbiol. Mol. Biol. Rev.* 78, 343–371. doi: 10.1128/MMBR.00010-14

Dutta, N. K., Klinkenberg, L. G., Vazquez, M. J., Segura-Carro, D., Colmenarejo, G., Ramon, F., et al. (2019). Inhibiting the stringent response blocks *Mycobacterium tuberculosis* entry into quiescence and reduces persistence. *Sci. Adv.* 5, eaav2104. doi: 10.1126/sciadv.aav2104

Eisele, N. A., Ruby, T., Jacobson, A., Manzanillo, P. S., Cox, J. S., Lam, L., et al. (2013). *Salmonella* require the fatty acid regulator PPARdelta for the establishment of a metabolic environment essential for long-term persistence. *Cell Host Microbe* 14, 171–182. doi: 10.1016/j.chom.2013.07.010

Eisenreich, W., and Heuner, K. (2016). The life stage-specific pathometabolism of. *Legionella Pneumophila FEBS Lett.* 590, 3868–3886. doi: 10.1002/1873-3468.12326

Eisenreich, W., Heesemann, J., Rudel, T., and Goebel, W. (2015). Metabolic adaptations of intracellullar bacterial pathogens and their mammalian host cells during infection ("pathometabolism"). *Microbiol. Spectr.* 3, 3. doi: 10.1128/9781555818883.ch3

Eisenreich, W., Rudel, T., Heesemann, J., and Goebel, W. (2017). To eat and to be eaten: mutual metabolic adaptations of immune cells and intracellular bacterial pathogens upon infection. *Front. Cell Infect. Microbiol.* 7, 316. doi: 10.3389/fcimb.2017.00316

Eisenreich, W., Rudel, T., Heesemann, J., and Goebel, W. (2019). How viral and intracellular bacterial pathogens reprogram the metabolism of host cells to allow their intracellular replication. *Front. Cell Infect. Microbiol.* 9, 42. doi: 10.3389/fcimb.2019.00042

Elwell, C., Mirrashidi, K., and Engel, J. (2016). *Chlamydia* cell biology and pathogenesis. *Nat. Rev. Microbiol.* 14, 385–400. doi: 10.1038/nrmicro.2016.30

Eylert, E., Schär, J., Mertins, S., Stoll, R., Bacher, A., Goebel, W., et al. (2008). Carbon metabolism of *Listeria monocytogenes* growing inside macrophages. *Mol. Microbiol.* 69, 1008–1017. doi: 10.1111/j.1365-2958.2008.06337.x

Fang, F. C., Libby, S. J., Castor, M. E., and Fung, A. M. (2005). Isocitrate lyase (AceA) is required for *Salmonella* persistence but not for acute lethal infection in mice. *Infect. Immun.* 73, 2547–2549. doi: 10.1128/IAI.73.4.2547-2549.2005

Fauvart, M., De Groote, V. N., and Michiels, J. (2011). Role of persister cells in chronic infections: clinical relevance and perspectives on anti-persister therapies. *J. Med. Microbiol.* 60, 699–709. doi: 10.1099/jmm.0.030932-0

Ferreira, V., Wiedmann, M., Teixeira, P., and Stasiewicz, M. J. (2014). *Listeria monocytogenes* persistence in food-associated environments: epidemiology, strain characteristics, and implications for public health. *J. Food Prot.* 77, 150–170. doi: 10.4315/0362-028X.JFP-13-150

Ficht, T. A. (2003). Intracellular survival of *Brucella*: defining the link with persistence. *Vet Microbiol.* 92, 213–223. doi: 10.1016/S0378-1135(02)00367-X

Fiebig, A., Herrou, J., Willett, J., and Crosson, S. (2015). General stress signaling in the Alphaproteobacteria. *Annu. Rev. Genet.* 49, 603–625. doi: 10.1146/annurev-genet-112414-054813

Fisher, R. A., Gollan, B., and Helaine, S. (2017). Persistent bacterial infections and persister cells. *Nat. Rev. Microbiol.* 15, 453–464. doi: 10.1038/nrmicro.2017.42

Fox, E. M., Leonard, N., and Jordan, K. (2011). Physiological and transcriptional characterization of persistent and nonpersistent *Listeria monocytogenes* isolates. *Appl. Environ. Microbiol.* 77, 6559–6569. doi: 10.1128/AEM.05529-11

Fozo, E. M., Hemm, M. R., and Storz, G. (2008). Small toxic proteins and the antisense RNAs that repress them. *Microbiol. Mol. Biol. Rev.* 72, 579–589, Table of Contents. doi: 10.1128/MMBR.00025-08

Francez-Charlot, A., Kaczmarczyk, A., Fischer, H. M., and Vorholt, J. A. (2015). The general stress response in Alphaproteobacteria. *Trends Microbiol.* 23, 164–171. doi: 10.1016/j.tim.2014.12.006

Fujimitsu, K., Senriuchi, T., and Katayama, T. (2009). Specific genomic sequences of *E. coli* promote replicational initiation by directly reactivating ADP-DnaA. *Genes Dev.* 23, 1221–1233. doi: 10.1101/gad.1775809

Fuller, R. S., Funnell, B. E., and Kornberg, A. (1984). The dnaA protein complex with the *E. coli* chromosomal replication origin (oriC) and other DNA sites. *Cell* 38, 889–900. doi: 10.1016/0092-8674(84)90284-8

Fung, D. K., Chan, E. W., Chin, M. L., and Chan, R. C. (2010). Delineation of a bacterial starvation stress response network which can mediate antibiotic tolerance development. *Antimicrob. Agents Chemother.* 54, 1082–1093. doi: 10.1128/AAC.01218-09

Galagan, J. E., Minch, K., Peterson, M., Lyubetskaya, A., Azizi, E., Sweet, L., et al. (2013). The *Mycobacterium tuberculosis* regulatory network and hypoxia. *Nature* 499, 178–183. doi: 10.1038/nature12337

Garsin, D. A. (2010). Ethanolamine utilization in bacterial pathogens: roles and regulation. *Nat. Rev. Microbiol.* 8, 290–295. doi: 10.1038/nrmicro2334

Gefen, O., and Balaban, N. Q. (2009). The importance of being persistent: heterogeneity of bacterial populations under antibiotic stress. *FEMS Microbiol. Rev.* 33, 704–717. doi: 10.1111/j.1574-6976.2008.00156.x

Gehre, L., Gorgette, O., Perrinet, S., Prevost, M. C., Ducatez, M., Giebel, A. M., et al. (2016). Sequestration of host metabolism by an intracellular pathogen. *Elife* 5, e12552. doi: 10.7554/eLife.12552.027

Gerdes, K., Christensen, S. K., and Lobner-Olesen, A. (2005). Prokaryotic toxin-antitoxin stress response loci. *Nat. Rev. Microbiol.* 3, 371–382. doi: 10.1038/nrmicro1147

Gerdes, K. (2016). Hypothesis: type I toxin-antitoxin genes enter the persistence field-a feedback mechanism explaining membrane homoeostasis. *Philos. Trans. R Soc. Lond B Biol. Sci.* 371, 20160189. doi: 10.1098/rstb.2016.0189

Germain, E., Castro-Roa, D., Zenkin, N., and Gerdes, K. (2013). Molecular mechanism of bacterial persistence by HipA. *Mol. Cell* 52, 248–254. doi: 10.1016/j.molcel.2013.08.045

Germain, E., Roghanian, M., Gerdes, K., and Maisonneuve, E. (2015). Stochastic induction of persister cells by HipA through (p)ppGpp-mediated activation of mRNA endonucleases. *Proc. Natl. Acad. Sci. U. S. A.* 112, 5171–5176. doi: 10.1073/pnas.1423536112

Gillmaier, N., Götz, A., Schulz, A., Eisenreich, W., and Goebel, W. (2012). Metabolic responses of primary and transformed cells to intracellular *Listeria monocytogenes*. *PloS One* 7, e52378. doi: 10.1371/journal.pone.0052378

Gillmaier, N., Schunder, E., Kutzner, E., Tlapak, H., Rydzewski, K., Herrmann, V., et al. (2016). Growth-related metabolism of the carbon storage poly-3-

hydroxybutyrate in *Legionella pneumophila*. *J. Biol. Chem.* 291, 6471–6482. doi: 10.1074/jbc.M115.693481

Gollan, B., Grabe, G., Michaux, C., and Helaine, S. (2019). Bacterial persisters and infection: past, present, and progressing. *Annu. Rev. Microbiol.* 73, 359–385. doi: 10.1146/annurev-micro-020518-115650

Goncalves, F. D. A., and De Carvalho, C. C. C. R. (2016). Phenotypic modifications in *Staphylococcus aureus* cells exposed to high concentrations of vancomycin and teicoplanin. *Front. Microbiol.* 7, 13. doi: 10.3389/fmicb.2016.00013

Goormaghtigh, F., Fraikin, N., Putrins, M., Hallaert, T., Hauryliuk, V., Garcia-Pino, A., et al. (2018). Reassessing the role of type II toxin-antitoxin systems in formation of *Escherichia coli* type II persister cells. *mBio* 9, e00640-18. doi: 10.1128/mBio.00640-18

Götz, A., Eylert, E., Eisenreich, W., and Goebel, W. (2010). Carbon metabolism of enterobacterial human pathogens growing in epithelial colorectal adenocarcinoma (Caco-2) cells. *PloS One* 5, e10586. doi: 10.1371/journal.pone.0010586

Gould, T. A., Van De Langemheen, H., Munoz-Elias, E. J., Mckinney, J. D., and Sacchettini, J. C. (2006). Dual role of isocitrate lyase 1 in the glyoxylate and methylcitrate cycles in *Mycobacterium tuberculosis*. *Mol. Microbiol.* 61, 940–947. doi: 10.1111/j.1365-2958.2006.05297.x

Grimbergen, A. J., Siebring, J., Solopova, A., and Kuipers, O. P. (2015). Microbial bet-hedging: the power of being different. *Curr. Opin. Microbiol.* 25, 67–72. doi: 10.1016/j.mib.2015.04.008

Grimwade, J. E., and Leonard, A. C. (2019). Blocking the trigger: Inhibition of the initiation of bacterial chromosome replication as an antimicrobial strategy. *Antibiotics (Basel)* 8, 111. doi: 10.3390/antibiotics8030111

Grubmüller, S., Schauer, K., Goebel, W., Fuchs, T. M., and Eisenreich, W. (2014). Analysis of carbon substrates used by *Listeria monocytogenes* during growth in J774A.1 macrophages suggests a bipartite intracellular metabolism. *Front. Cell Infect. Microbiol.* 4, 156. doi: 10.3389/fcimb.2014.00156

Guilliams, M., Ginhoux, F., Jakubzick, C., Naik, S. H., Onai, N., Schraml, B. U., et al. (2014). Dendritic cells, monocytes and macrophages: a unified nomenclature based on ontogeny. *Nat. Rev. Immunol.* 14, 571–578. doi: 10.1038/nri3712

Gupta, N., Maurya, S., Verma, H., and Verma, V. K. (2019). Unraveling the factors and mechanism involved in persistence: Host-pathogen interactions in *Helicobacter pylori*. *J. Cell Biochem.* 120, 18572–18587. doi: 10.1002/jcb.29201

Gutierrez, A., Jain, S., Bhargava, P., Hamblin, M., Lobritz, M. A., and Collins, J. J. (2017). Understanding and sensitizing density-dependent persistence to quinolone antibiotics. *Mol. Cell* 681147-1154.e1143. doi: 10.1016/j.molcel.2017.11.012

Hammer, B. K., and Swanson, M. S. (1999). Co-ordination of *Legionella pneumophila* virulence with entry into stationary phase by ppGpp. *Mol. Microbiol.* 33, 721–731. doi: 10.1046/j.1365-2958.1999.01519.x

Hanna, M. H., and Carl, P. L. (1975). Reinitiation of deoxyribonucleic acid synthesis by deoxyribonucleic acid initiation mutants of *Escherichia coli*: role of ribonucleic acid synthesis, protein synthesis, and cell division. *J. Bacteriol.* 121, 219–226. doi: 10.1128/JB.121.1.219-226.1975

Hanna, N., Ouahrani-Bettache, S., Drake, K. L., Adams, L. G., Kohler, S., and Occhialini, A. (2013). Global Rsh-dependent transcription profile of *Brucella suis* during stringent response unravels adaptation to nutrient starvation and cross-talk with other stress responses. *BMC Genomics* 14, 459. doi: 10.1186/1471-2164-14-459

Hansen, F. G., and Atlung, T. (2018). The DnaA Tale. *Front. Microbiol.* 9, 319. doi: 10.3389/fmicb.2018.00319

Harms, A., Maisonneuve, E., and Gerdes, K. (2016). Mechanisms of bacterial persistence during stress and antibiotic exposure. *Science* 354, aaf4268. doi: 10.1126/science.aaf4268

Harms, A., Fino, C., Sorensen, M. A., Semsey, S., and Gerdes, K. (2017a). Prophages and growth dynamics confound experimental results with antibiotic-tolerant persister cells. *mBio* 8, e01964-17. doi: 10.1128/mBio.01964-17

Harms, A., Liesch, M., Korner, J., Quebatte, M., Engel, P., and Dehio, C. (2017b). A bacterial toxin-antitoxin module is the origin of inter-bacterial and inter-kingdom effectors of *Bartonella*. *PloS Genet.* 13, e1007077. doi: 10.1371/journal.pgen.1007077

Harms, A., Brodersen, D. E., Mitarai, N., and Gerdes, K. (2018). Toxins, targets, and triggers: an overview of toxin-antitoxin biology. *Mol. Cell* 70, 768–784. doi: 10.1016/j.molcel.2018.01.003

Hartman, T., Weinrick, B., Vilcheze, C., Berney, M., Tufariello, J., Cook, G. M., et al. (2014). Succinate dehydrogenase is the regulator of respiration in *Mycobacterium tuberculosis*. *PloS Pathog.* 10, e1004510. doi: 10.1371/journal.ppat.1004510

Hartman, T. E., Wang, Z., Jansen, R. S., Gardete, S., and Rhee, K. Y. (2017). Metabolic perspectives on persistence. *Microbiol. Spectr.* 5, 10.1128. doi: 10.1128/microbiolspec.TBTB2-0026-2016

Hauryliuk, V., Atkinson, G. C., Murakami, K. S., Tenson, T., and Gerdes, K. (2015). Recent functional insights into the role of (p)ppGpp in bacterial physiology. *Nat. Rev. Microbiol.* 13, 298–309. doi: 10.1038/nrmicro3448

Häuslein, I., Manske, C., Goebel, W., Eisenreich, W., and Hilbi, H. (2016). Pathway analysis using ^{13}C-glycerol and other carbon tracers reveals a bipartite metabolism of *Legionella pneumophila*. *Mol. Microbiol.* 100, 229–246. doi: 10.1111/mmi.13313

Häuslein, I., Cantet, F., Reschke, S., Chen, F., Bonazzi, M., and Eisenreich, W. (2017a). Multiple substrate usage of *Coxiella burnetii* to feed a bipartite metabolic network. *Front. Cell Infect. Microbiol.* 7, 285. doi: 10.3389/fcimb.2017.00285

Häuslein, I., Sahr, T., Escoll, P., Klausner, N., Eisenreich, W., and Buchrieser, C. (2017b). *Legionella pneumophila* CsrA regulates a metabolic switch from amino acid to glycerolipid metabolism. *Open Biol.* 7, 170149. doi: 10.1098/rsob.170149

Heaton, B. E., Herrou, J., Blackwell, A. E., Wysocki, V. H., and Crosson, S. (2012). Molecular structure and function of the novel BrnT/BrnA toxin-antitoxin system of *Brucella abortus*. *J. Biol. Chem.* 287, 12098–12110. doi: 10.1074/jbc.M111.332163

Hecker, M., and Volker, U. (2001). General stress response of *Bacillus subtilis* and other bacteria. *Adv. Microb Physiol.* 44, 35–91. doi: 10.1016/S0065-2911(01)44011-2

Helaine, S., and Holden, D. W. (2013). Heterogeneity of intracellular replication of bacterial pathogens. *Curr. Opin. Microbiol.* 16, 184–191. doi: 10.1016/j.mib.2012.12.004

Helaine, S., Thompson, J. A., Watson, K. G., Liu, M., Boyle, C., and Holden, D. W. (2010). Dynamics of intracellular bacterial replication at the single cell level. *Proc. Natl. Acad. Sci. U. S. A.* 107, 3746–3751. doi: 10.1073/pnas.1000041107

Helaine, S., Cheverton, A. M., Watson, K. G., Faure, L. M., Matthews, S. A., and Holden, D. W. (2014). Internalization of *Salmonella* by macrophages induces formation of nonreplicating persisters. *Science* 343, 204–208. doi: 10.1126/science.1244705

Hengge-Aronis, R. (1996). Back to log phase: sigma S as a global regulator in the osmotic control of gene expression in *Escherichia coli*. *Mol. Microbiol.* 21, 887–893. doi: 10.1046/j.1365-2958.1996.511405.x

Higgins, D., and Dworkin, J. (2012). Recent progress in *Bacillus subtilis* sporulation. *FEMS Microbiol. Rev.* 36, 131–148. doi: 10.1111/j.1574-6976.2011.00310.x

Hiraga, S., Ogura, T., Niki, H., Ichinose, C., and Mori, H. (1990). Positioning of replicated chromosomes in *Escherichia coli*. *J. Bacteriol.* 172, 31–39. doi: 10.1128/JB.172.1.31-39.1990

Hirsch, M., and Elliott, T. (2002). Role of ppGpp in rpoS stationary-phase regulation in. *Escherichia coli J. Bacteriol.* 184, 5077–5087. doi: 10.1128/JB.184.18.5077-5087.2002

Hogg, T., Mechold, U., Malke, H., Cashel, M., and Hilgenfeld, R. (2004). Conformational antagonism between opposing active sites in a bifunctional RelA/SpoT homolog modulates (p)ppGpp metabolism during the stringent response [corrected]. *Cell* 117, 57–68. doi: 10.1016/S0092-8674(04)00260-0

Holden, D. W., and Errington, J. (2018). Type II toxin-antitoxin systems and persister cells. *mBio* 9, e01574-18. doi: 10.1128/mBio.01574-18

Hong, S. H., Wang, X., O'connor, H. F., Benedik, M. J., and Wood, T. K. (2012). Bacterial persistence increases as environmental fitness decreases. *Microb. Biotechnol.* 5, 509–522. doi: 10.1111/j.1751-7915.2011.00327.x

Hu, Y., Benedik, M. J., and Wood, T. K. (2012). Antitoxin DinJ influences the general stress response through transcript stabilizer CspE. *Environ. Microbiol.* 14, 669–679. doi: 10.1111/j.1462-2920.2011.02618.x

Hurdle, J. G., O'neill, A. J., Chopra, I., and Lee, R. E. (2011). Targeting bacterial membrane function: an underexploited mechanism for treating persistent infections. *Nat. Rev. Microbiol* 9, 62–75. doi: 10.1038/nrmicro2474

Hutchison, E. A., Miller, D. A., and Angert, E. R. (2014). Sporulation in bacteria: beyond the standard model. *Microbiol. Spectr.* 2, 5. doi: 10.1128/microbiolspec.TBS-0013-2012

Ikeda, M., Shinozaki, Y., Uchida, K., Ohshika, Y., Furukohri, A., Maki, H., et al. (2012). Quick replication fork stop by overproduction of *Escherichia coli* DinB produces non-proliferative cells with an aberrant chromosome. *Genes Genet. Syst.* 87, 221–231. doi: 10.1266/ggs.87.221

Jones, S. E., and Elliot, M. A. (2018). 'Exploring' the regulation of *Streptomyces* growth and development. *Curr. Opin. Microbiol.* 42, 25–30. doi: 10.1016/j.mib.2017.09.009

Jones, S. E., and Lennon, J. T. (2010). Dormancy contributes to the maintenance of microbial diversity. *Proc. Natl. Acad. Sci. U. S. A.* 107, 5881–5886. doi: 10.1073/pnas.0912765107

Jozefczuk, S., Klie, S., Catchpole, G., Szymanski, J., Cuadros-Inostroza, A., Steinhauser, D., et al. (2010). Metabolomic and transcriptomic stress response of *Escherichia coli*. *Mol. Syst. Biol.* 6, 364. doi: 10.1038/msb.2010.18

Jung, S. H., Ryu, C. M., and Kim, J. S. (2019). Bacterial persistence: fundamentals and clinical importance. *J. Microbiol.* 57, 829–835. doi: 10.1007/s12275-019-9218-0

Käding, N., Szaszak, M., and Rupp, J. (2014). Imaging of *Chlamydia* and host cell metabolism. *Future Microbiol.* 9, 509–521. doi: 10.2217/fmb.14.13

Kaguni, J. M. (2006). DnaA: controlling the initiation of bacterial DNA replication and more. *Annu. Rev. Microbiol.* 60, 351–375. doi: 10.1146/annurev.micro.60.080805.142111

Kalani, B. S., Irajian, G., Lotfollahi, L., Abdollahzadeh, E., and Razavi, S. (2018). Putative type II toxin-antitoxin systems in *Listeria monocytogenes* isolated from clinical, food, and animal samples in Iran. *Microb Pathog* 122, 19–24. doi: 10.1016/j.micpath.2018.06.003

Kaldalu, N., Hauryliuk, V., and Tenson, T. (2016). Persisters-as elusive as ever. *Appl. Microbiol. Biotechnol.* 100, 6545–6553. doi: 10.1007/s00253-016-7648-8

Kasho, K., and Katayama, T. (2013). DnaA binding locus datA promotes DnaA-ATP hydrolysis to enable cell cycle-coordinated replication initiation. *Proc. Natl. Acad. Sci. U.S.A.* 110, 936–941. doi: 10.1073/pnas.1212070110

Kasho, K., Tanaka, H., Sakai, R., and Katayama, T. (2017). Cooperative DnaA binding to the negatively supercoiled datA locus stimulates DnaA-ATP hydrolysis. *J. Biol. Chem.* 292, 1251–1266. doi: 10.1074/jbc.M116.762815

Katayama, T., Kasho, K., and Kawakami, H. (2017). The DnaA cycle in *Escherichia coli*: activation, function and inactivation of the initiator protein. *Front. Microbiol.* 8, 2496. doi: 10.3389/fmicb.2017.02496

Kawai, Y., Mercier, R., Wu, L. J., Dominguez-Cuevas, P., Oshima, T., and Errington, J. (2015). Cell growth of wall-free L-form bacteria is limited by oxidative damage. *Curr. Biol.* 25, 1613–1618. doi: 10.1016/j.cub.2015.04.031

Kedzierska, B., and Hayes, F. (2016). Emerging roles of toxin-antitoxin modules in bacterial pathogenesis. *Molecules* 21, 790. doi: 10.3390/molecules21060790

Kelly, B., and O'neill, L. A. (2015). Metabolic reprogramming in macrophages and dendritic cells in innate immunity. *Cell Res.* 25, 771–784. doi: 10.1038/cr.2015.68

Keren, I., Mulcahy, L. R., and Lewis, K. (2012). Persister eradication: lessons from the world of natural products. *Methods Enzymol.* 517, 387–406. doi: 10.1016/B978-0-12-404634-4.00019-X

Kester, J. C., and Fortune, S. M. (2014). Persisters and beyond: mechanisms of phenotypic drug resistance and drug tolerance in bacteria. *Crit. Rev. Biochem. Mol. Biol.* 49, 91–101. doi: 10.3109/10409238.2013.869543

Keyamura, K., and Katayama, T. (2011). DnaA protein DNA-binding domain binds to Hda protein to promote inter-AAA+ domain interaction involved in regulatory inactivation of DnaA. *J. Biol. Chem.* 286, 29336–29346. doi: 10.1074/jbc.M111.233403

Khil, P. P., and Camerini-Otero, R. D. (2002). Over 1000 genes are involved in the DNA damage response of. *Escherichia coli Mol. Microbiol.* 44, 89–105. doi: 10.1046/j.1365-2958.2002.02878.x

Kim, J. S., and Wood, T. K. (2017). Tolerant, growing cells from nutrient shifts are not persister cells. *mBio* 8, e00357-17. doi: 10.1128/mBio.00354-17

Kim, J. S., Heo, P., Yang, T. J., Lee, K. S., Cho, D. H., Kim, B. T., et al. (2011). Selective killing of bacterial persisters by a single chemical compound without affecting normal antibiotic-sensitive cells. *Antimicrob. Agents Chemother.* 55, 5380–5383. doi: 10.1128/AAC.00708-11

Kim, H. S., Caswell, C. C., Foreman, R., Roop, R. M.,. 2., and Crosson, S. (2013). The *Brucella abortus* general stress response system regulates chronic mammalian infection and is controlled by phosphorylation and proteolysis. *J. Biol. Chem.* 288, 13906–13916. doi: 10.1074/jbc.M113.459305

Kim, H. Y., Go, J., Lee, K. M., Oh, Y. T., and Yoon, S. S. (2018). Guanosine tetra- and pentaphosphate increase antibiotic tolerance by reducing reactive oxygen species production in *Vibrio cholerae*. *J. Biol. Chem.* 293, 5679–5694. doi: 10.1074/jbc.RA117.000383

Kim, J. K., Kim, Y. S., Lee, H. M., Jin, H. S., Neupane, C., Kim, S., et al. (2018). GABAergic signaling linked to autophagy enhances host protection against intracellular bacterial infections. *Nat. Commun.* 9, 4184. doi: 10.1038/s41467-018-06487-5

Kim, J. S., Chowdhury, N., Yamasaki, R., and Wood, T. K. (2018). Viable but non-culturable and persistence describe the same bacterial stress state. *Environ. Microbiol.* 20, 2038–2048. doi: 10.1111/1462-2920.14075

Kint, C., Verstraeten, N., Hofkens, J., Fauvart, M., and Michiels, J. (2014). Bacterial Obg proteins: GTPases at the nexus of protein and DNA synthesis. *Crit. Rev. Microbiol* 40, 207–224. doi: 10.3109/1040841X.2013.776510

Kintner, J., Lajoie, D., Hall, J., Whittimore, J., and Schoborg, R. V. (2014). Commonly prescribed beta-lactam antibiotics induce *C. trachomatis* persistence/stress in culture at physiologically relevant concentrations. *Front. Cell Infect. Microbiol.* 4, 44. doi: 10.3389/fcimb.2014.00044

Kleemann, P., Domann, E., Chakraborty, T., Bernstein, I., and Lohoff, M. (2009). Chronic prosthetic joint infection caused by *Listeria monocytogenes*. *J. Med. Microbiol.* 58, 138–141. doi: 10.1099/jmm.0.004234-0

Kohanski, M. A., Dwyer, D. J., Hayete, B., Lawrence, C. A., and Collins, J. J. (2007). A common mechanism of cellular death induced by bactericidal antibiotics. *Cell* 130, 797–810. doi: 10.1016/j.cell.2007.06.049

Korch, S. B., Henderson, T. A., and Hill, T. M. (2003). Characterization of the hipA7 allele of *Escherichia coli* and evidence that high persistence is governed by (p)ppGpp synthesis. *Mol. Microbiol.* 50, 1199–1213. doi: 10.1046/j.1365-2958.2003.03779.x

Korch, S. B., Contreras, H., and Clark-Curtiss, J. E. (2009). Three *Mycobacterium tuberculosis* Rel toxin-antitoxin modules inhibit mycobacterial growth and are expressed in infected human macrophages. *J. Bacteriol* 191, 1618–1630. doi: 10.1128/JB.01318-08

Korch, S. B., Malhotra, V., Contreras, H., and Clark-Curtiss, J. E. (2015). The *Mycobacterium tuberculosis* relBE toxin:antitoxin genes are stress-responsive modules that regulate growth through translation inhibition. *J. Microbiol.* 53, 783–795. doi: 10.1007/s12275-015-5333-8

Kortebi, M., Milohanic, E., Mitchell, G., Pechoux, C., Prevost, M. C., Cossart, P., et al. (2017). *Listeria monocytogenes* switches from dissemination to persistence by adopting a vacuolar lifestyle in epithelial cells. *PloS Pathog.* 13, e1006734. doi: 10.1371/journal.ppat.1006734

Kraemer, J. A., Sanderlin, A. G., and Laub, M. T. (2019). The stringent response inhibits DNA replication initiation in *E. coli* by modulating supercoiling of *oriC*. *mBio* 10, e01330-19. doi: 10.1128/mBio.01330-19

Krawczyk, C. M., Holowka, T., Sun, J., Blagih, J., Amiel, E., Deberardinis, R. J., et al. (2010). Toll-like receptor-induced changes in glycolytic metabolism regulate dendritic cell activation. *Blood* 115, 4742–4749. doi: 10.1182/blood-2009-10-249540

Kreuzer, K. N. (2013). DNA damage responses in prokaryotes: regulating gene expression, modulating growth patterns, and manipulating replication forks. *Cold Spring Harb Perspect. Biol.* 5, a012674. doi: 10.1101/cshperspect.a012674

Kumar, A., Farhana, A., Guidry, L., Saini, V., Hondalus, M., and Steyn, A. J. (2011). Redox homeostasis in mycobacteria: the key to tuberculosis control? *Expert Rev. Mol. Med.* 13, e39. doi: 10.1017/S1462399411002079

Kuzminov, A. (1995). Collapse and repair of replication forks in *Escherichia coli*. *Mol. Microbiol.* 16, 373–384. doi: 10.1111/j.1365-2958.1995.tb02403.x

Kwan, B. W., Valenta, J. A., Benedik, M. J., and Wood, T. K. (2013). Arrested protein synthesis increases persister-like cell formation. *Antimicrob. Agents Chemother.* 57, 1468–1473. doi: 10.1128/AAC.02135-12

Lambden, P. R., Pickett, M. A., and Clarke, I. N. (2006). The effect of penicillin on *Chlamydia trachomatis* DNA replication. *Microbiology* 152, 2573–2578. doi: 10.1099/mic.0.29032-0

Lark, K. G., Repko, T., and Hoffman, E. J. (1963). The effect of amino acid deprivation on subsequent deoxyribonucleic acid replication. *Biochim. Biophys. Acta* 76, 9–24. doi: 10.1016/0926-6550(63)90003-3

Lark, K. G. (1972). Genetic control over the initiation of the synthesis of the short deoxynucleotide chains in *E. coli*. *Nat. New Biol.* 240, 237–240. doi: 10.1038/newbio240237a0

Leaden, L., Silva, L. G., Ribeiro, R. A., Dos Santos, N. M., Lorenzetti, A. P. R., Alegria, T. G. P., et al. (2018). Iron deficiency generates oxidative stress and activation of the SOS response in *Caulobacter crescentus*. *Front. Microbiol.* 9, 2014. doi: 10.3389/fmicb.2018.02014

Leistikow, R. L., Morton, R. A., Bartek, I. L., Frimpong, I., Wagner, K., and Voskuil, M. I. (2010). The *Mycobacterium tuberculosis* DosR regulon assists in metabolic homeostasis and enables rapid recovery from nonrespiring dormancy. *J. Bacteriol* 192, 1662–1670. doi: 10.1128/JB.00926-09

Leonard, A. C., and Grimwade, J. E. (2010). Regulating DnaA complex assembly: it is time to fill the gaps. *Curr. Opin. Microbiol.* 13, 766–772. doi: 10.1016/j.mib.2010.10.001

Leonard, A. C., Rao, P., Kadam, R. P., and Grimwade, J. E. (2019). Changing perspectives on the role of DnaA-ATP in orisome function and timing regulation. *Front. Microbiol* 10, 2009. doi: 10.3389/fmicb.2019.02009

Leplae, R., Geeraerts, D., Hallez, R., Guglielmini, J., Dreze, P., and Van Melderen, L. (2011). Diversity of bacterial type II toxin-antitoxin systems: a comprehensive search and functional analysis of novel families. *Nucleic Acids Res.* 39, 5513–5525. doi: 10.1093/nar/gkr131

Leslie, D. J., Heinen, C., Schramm, F. D., Thuring, M., Aakre, C. D., Murray, S. M., et al. (2015). Nutritional control of DNA replication initiation through the proteolysis and regulated translation of DnaA. *PloS Genet.* 11, e1005342. doi: 10.1371/journal.pgen.1005342

Leszczynska, D., Matuszewska, E., Kuczynska-Wisnik, D., Furmanek-Blaszk, B., and Laskowska, E. (2013). The formation of persister cells in stationary-phase cultures of *Escherichia coli* is associated with the aggregation of endogenous proteins. *PloS One* 8, e54737. doi: 10.1371/journal.pone.0054737

Lewis, K. (2007). Persister cells, dormancy and infectious disease. *Nat. Rev. Microbiol.* 5, 48–56. doi: 10.1038/nrmicro1557

Lewis, K. (2010). Persister cells. *Annu. Rev. Microbiol.* 64, 357–372. doi: 10.1146/annurev.micro.112408.134306

Li, L., Mendis, N., Trigui, H., Oliver, J. D., and Faucher, S. P. (2014). The importance of the viable but non-culturable state in human bacterial pathogens. *Front. Microbiol.* 5, 258. doi: 10.3389/fmicb.2014.00258

Li, S., Zhang, Q., Xu, Z., and Yao, Y. F. (2017). Acetylation of lysine 243 inhibits the oriC binding ability of DnaA in *Escherichia coli*. *Front. Microbiol.* 8, 699. doi: 10.3389/fmicb.2017.00699

Liang, P., Rosas-Lemus, M., Patel, D., Fang, X., Tuz, K., and Juarez, O. (2018). Dynamic energy dependency of *Chlamydia trachomatis* on host cell metabolism during intracellular growth: Role of sodium-based energetics in chlamydial ATP generation. *J. Biol. Chem.* 293, 510–522. doi: 10.1074/jbc.M117.797209

Lianou, A., and Sofos, J. N. (2007). A review of the incidence and transmission of *Listeria monocytogenes* in ready-to-eat products in retail and food service environments. *J. Food Prot* 70, 2172–2198. doi: 10.4315/0362-028X-70.9.2172

Liu, Y., and Imlay, J. A. (2013). Cell death from antibiotics without the involvement of reactive oxygen species. *Science* 339, 1210–1213. doi: 10.1126/science.1232751

Liu, Y., Tan, S., Huang, L., Abramovitch, R. B., Rohde, K. H., Zimmerman, M. D., et al. (2016). Immune activation of the host cell induces drug tolerance in *Mycobacterium tuberculosis* both *in vitro* and *in vivo*. *J. Exp. Med.* 213, 809–825. doi: 10.1084/jem.20151248

Liu, S., Wu, N., Zhang, S., Yuan, Y., Zhang, W., and Zhang, Y. (2017). Variable persister gene interactions with (p)ppGpp for persister formation in *Escherichia coli*. *Front. Microbiol.* 8, 1795. doi: 10.3389/fmicb.2017.01795

Maaloe, O., and Hanawalt, P. C. (1961). Thymine deficiency and the normal DNA replication cycle. I. *J. Mol. Biol.* 3, 144–155. doi: 10.1016/S0022-2836(61)80041-7

Maisonneuve, E., and Gerdes, K. (2014). Molecular mechanisms underlying bacterial persisters. *Cell* 157, 539–548. doi: 10.1016/j.cell.2014.02.050

Maisonneuve, E., Shakespeare, L. J., Jorgensen, M. G., and Gerdes, K. (2011). Bacterial persistence by RNA endonucleases. *Proc. Natl. Acad. Sci. U. S. A.* 108, 13206–13211. doi: 10.1073/pnas.1100186108

Maisonneuve, E., Castro-Camargo, M., and Gerdes, K. (2013). (p)ppGpp controls bacterial persistence by stochastic induction of toxin-antitoxin activity. *Cell* 154, 1140–1150. doi: 10.1016/j.cell.2013.07.048

Maisonneuve, E., Castro-Camargo, M., and Gerdes, K. (2018). (p)ppGpp controls bacterial persistence by stochastic induction of toxin-antitoxin activity. *Cell* 172, 1135. doi: 10.1016/j.cell.2018.02.023

Martinez, E., Siadous, F. A., and Bonazzi, M. (2018). Tiny architects: biogenesis of intracellular replicative niches by bacterial pathogens. *FEMS Microbiol. Rev.* 42, 425–447. doi: 10.1093/femsre/fuy013

Masai, H., Tanaka, T., and Kohda, D. (2010). Stalled replication forks: making ends meet for recognition and stabilization. *Bioessays* 32, 687–697. doi: 10.1002/bies.200900196

Maurya, R. K., Bharti, S., and Krishnan, M. Y. (2018). Triacylglycerols: fuelling the hibernating *Mycobacterium tuberculosis*. *Front. Cell Infect. Microbiol.* 8, 450. doi: 10.3389/fcimb.2018.00450

Mcclelland, M., Sanderson, K. E., Spieth, J., Clifton, S. W., Latreille, P., Courtney, L., et al. (2001). Complete genome sequence of *Salmonella enterica* serovar Typhimurium LT2. *Nature* 413, 852–856. doi: 10.1038/35101614

Mckinney, J. D., Honer Zu Bentrup, K., Munoz-Elias, E. J., Miczak, A., Chen, B., Chan, W. T., et al. (2000). Persistence of *Mycobacterium tuberculosis* in macrophages and mice requires the glyoxylate shunt enzyme isocitrate lyase. *Nature* 406, 735–738. doi: 10.1038/35021074

Mcvicker, G., and Tang, C. M. (2016). Deletion of toxin-antitoxin systems in the evolution of *Shigella sonnei* as a host-adapted pathogen. *Nat. Microbiol.* 2, 16204. doi: 10.1038/nmicrobiol.2016.204

Mehlitz, A., Eylert, E., Huber, C., Lindner, B., Vollmuth, N., Karunakaran, K., et al. (2017). Metabolic adaptation of *Chlamydia trachomatis* to mammalian host cells. *Mol. Microbiol.* 103, 1004–1019. doi: 10.1111/mmi.13603

Merrikh, H., Ferrazzoli, A. E., and Lovett, S. T. (2009). Growth phase and (p) ppGpp control of IraD, a regulator of RpoS stability, in *Escherichia coli*. *J. Bacteriol.* 191, 7436–7446. doi: 10.1128/JB.00412-09

Meylan, S., Porter, C. B. M., Yang, J. H., Belenky, P., Gutierrez, A., Lobritz, M. A., et al. (2017). Carbon sources tune antibiotic susceptibility in *Pseudomonas aeruginosa* via tricarboxylic acid cycle control. *Cell Chem. Biol.* 24, 195–206. doi: 10.1016/j.chembiol.2016.12.015

Michel, B. (2005). Obg/CtgA, a signaling protein that controls replication, translation, and morphological development? *Dev. Cell* 8, 300–301. doi: 10.1016/j.devcel.2005.02.002

Miller, C., Thomsen, L. E., Gaggero, C., Mosseri, R., Ingmer, H., and Cohen, S. N. (2004). SOS response induction by beta-lactams and bacterial defense against antibiotic lethality. *Science* 305, 1629–1631. doi: 10.1126/science.1101630

Mitosch, K., Rieckh, G., and Bollenbach, T. (2019). Temporal order and precision of complex stress responses in individual bacteria. *Mol. Syst. Biol.* 15, e8470. doi: 10.15252/msb.20188470

Mittenhuber, G. (2001). Comparative genomics and evolution of genes encoding bacterial (p)ppGpp synthetases/hydrolases (the Rel, RelA and SpoT proteins). *J. Mol. Microbiol. Biotechnol.* 3, 585–600.

Mittenhuber, G. (2002). An inventory of genes encoding RNA polymerase sigma factors in 31 completely sequenced eubacterial genomes. *J. Mol. Microbiol. Biotechnol.* 4, 77–91.

Moir, A., and Cooper, G. (2015). Spore germination. *Microbiol. Spectr.* 3, 6. doi: 10.1128/microbiolspec.TBS-0014-2012

Mok, W. W. K., and Brynildsen, M. P. (2018). Timing of DNA damage responses impacts persistence to fluoroquinolones. *Proc. Natl. Acad. Sci. U. S. A.* 115, E6301–E6309. doi: 10.1073/pnas.1804218115

Mok, W. W., Orman, M. A., and Brynildsen, M. P. (2015). Impacts of global transcriptional regulators on persister metabolism. *Antimicrob. Agents Chemother.* 59, 2713–2719. doi: 10.1128/AAC.04908-14

Möker, N., Dean, C. R., and Tao, J. (2010). *Pseudomonas aeruginosa* increases formation of multidrug-tolerant persister cells in response to quorum-sensing signaling molecules. *J. Bacteriol.* 192, 1946–1955. doi: 10.1128/JB.01231-09

Molina-Quiroz, R. C., Silva-Valenzuela, C., Brewster, J., Castro-Nallar, E., Levy, S. B., and Camilli, A. (2018). Cyclic AMP regulates bacterial persistence through repression of the oxidative stress response and SOS-dependent DNA repair in uropathogenic *Escherichia coli*. *mBio* 9, e02144-17. doi: 10.1128/mBio.02144-17

Monack, D. M., Mueller, A., and Falkow, S. (2004). Persistent bacterial infections: the interface of the pathogen and the host immune system. *Nat. Rev. Microbiol.* 2, 747–765. doi: 10.1038/nrmicro955

Morigen,, Lobner-Olesen, A., and Skarstad, K. (2003). Titration of the *Escherichia coli* DnaA protein to excess datA sites causes destabilization of replication forks, delayed replication initiation and delayed cell division. *Mol. Microbiol.* 50, 349–362. doi: 10.1046/j.1365-2958.2003.03695.x

Mulcahy, L. R., Burns, J. L., Lory, S., and Lewis, K. (2010). Emergence of *Pseudomonas aeruginosa* strains producing high levels of persister cells in patients with cystic fibrosis. *J. Bacteriol* 192, 6191–6199. doi: 10.1128/JB.01651-09

Muramatsu, M. K., Brothwell, J. A., Stein, B. D., Putman, T. E., Rockey, D. D., and Nelson, D. E. (2016). Beyond tryptophan synthase: identification of genes that contribute to *Chlamydia trachomatis* survival during gamma interferon-induced persistence and reactivation. *Infect. Immun.* 84, 2791–2801. doi: 10.1128/IAI.00356-16

Murray, P. J., and Wynn, T. A. (2011). Obstacles and opportunities for understanding macrophage polarization. *J. Leukoc. Biol.* 89, 557–563. doi: 10.1189/jlb.0710409

Murray, P. J., Allen, J. E., Biswas, S. K., Fisher, E. A., Gilroy, D. W., Goerdt, S., et al. (2014). Macrophage activation and polarization: nomenclature and experimental guidelines. *Immunity* 41, 14–20. doi: 10.1016/j.immuni.2014.06.008

Natori, Y., Tagami, K., Murakami, K., Yoshida, S., Tanigawa, O., Moh, Y., et al. (2009). Transcription activity of individual rrn operons in *Bacillus subtilis* mutants deficient in (p)ppGpp synthetase genes, relA, yjbM, and ywaC. *J. Bacteriol* 191, 4555–4561. doi: 10.1128/JB.00263-09

Nguyen, D., Joshi-Datar, A., Lepine, F., Bauerle, E., Olakanmi, O., Beer, K., et al. (2011). Active starvation responses mediate antibiotic tolerance in biofilms and nutrient-limited bacteria. *Science* 334, 982–986. doi: 10.1126/science.1211037

Nierman, W. C., Yu, Y., and Losada, L. (2015). The *in vitro* antibiotic tolerant persister population in *Burkholderia pseudomallei* is altered by environmental factors. *Front. Microbiol.* 6, 1338. doi: 10.3389/fmicb.2015.01338

Nordström, K., and Dasgupta, S. (2001). Partitioning of the *Escherichia coli* chromosome: superhelicity and condensation. *Biochimie* 83, 41–48. doi: 10.1016/S0300-9084(00)01204-9

Norris, V. (1990). DNA replication in *Escherichia coli* is initiated by membrane detachment of oriC. A model. *J. Mol. Biol.* 215, 67–71. doi: 10.1016/S0022-2836(05)80095-6

Odegaard, J. I., Ricardo-Gonzalez, R. R., Red Eagle, A., Vats, D., Morel, C. R., Goforth, M. H., et al. (2008). Alternative M2 activation of Kupffer cells by PPARdelta ameliorates obesity-induced insulin resistance. *Cell Metab.* 7, 496–507. doi: 10.1016/j.cmet.2008.04.003

Ohnishi, Y., Seo, J. W., and Horinouchi, S. (2002). Deprogrammed sporulation in *Streptomyces*. *FEMS Microbiol. Lett.* 216, 1–7. doi: 10.1111/j.1574-6968.2002.tb11406.x

Oliver, J. D. (2005). The viable but nonculturable state in bacteria. *J. Microbiol.* 43 Spec No, 93-100.

Oliver, J. D. (2010). Recent findings on the viable but nonculturable state in pathogenic bacteria. *FEMS Microbiol. Rev.* 34, 415–425. doi: 10.1111/j.1574-6976.2009.00200.x

Omotade, T. O., and Roy, C. R. (2019). Manipulation of host cell organelles by intracellular pathogens. *Microbiol. Spectr.* 7, 2. doi: 10.1128/microbiolspec.BAI-0022-2019

Orman, M. A., and Brynildsen, M. P. (2013). Dormancy is not necessary or sufficient for bacterial persistence. *Antimicrob. Agents Chemother.* 57, 3230–3239. doi: 10.1128/AAC.00243-13

Orman, M. A., Mok, W. W., and Brynildsen, M. P. (2015). Aminoglycoside-enabled elucidation of bacterial persister metabolism. *Curr. Protoc. Microbiol.* 36, 17 19 11–14. doi: 10.1002/9780471729259.mc1709s36

O'neill, L. A. (2014). Glycolytic reprogramming by TLRs in dendritic cells. *Nat. Immunol.* 15, 314–315. doi: 10.1038/ni.2852

Ouellette, S. P., Hatch, T. P., Abdelrahman, Y. M., Rose, L. A., Belland, R. J., and Byrne, G. I. (2006). Global transcriptional upregulation in the absence of increased translation in *Chlamydia* during IFN-gamma-mediated host cell tryptophan starvation. *Mol. Microbiol.* 62, 1387–1401. doi: 10.1111/j.1365-2958.2006.05465.x

Ouellette, S. P., Rueden, K. J., and Rucks, E. A. (2016). Tryptophan codon-dependent transcription in *Chlamydia pneumoniae* during gamma interferon-mediated tryptophan limitation. *Infect. Immun.* 84, 2703–2713. doi: 10.1128/IAI.00377-16

Page, R., and Peti, W. (2016). Toxin-antitoxin systems in bacterial growth arrest and persistence. *Nat. Chem. Biol.* 12, 208–214. doi: 10.1038/nchembio.2044

Pandey, D. P., and Gerdes, K. (2005). Toxin-antitoxin loci are highly abundant in free-living but lost from host-associated prokaryotes. *Nucleic Acids Res.* 33, 966–976. doi: 10.1093/nar/gki201

Panzetta, M. E., Valdivia, R. H., and Saka, H. A. (2018). *Chlamydia* persistence: a survival strategy to evade antimicrobial effects *in-vitro* and *in-vivo*. *Front. Microbiol.* 9, 3101. doi: 10.3389/fmicb.2018.03101

Paredes-Sabja, D., Shen, A., and Sorg, J. A. (2014). *Clostridium difficile* spore biology: sporulation, germination, and spore structural proteins. *Trends Microbiol* 22, 406–416. doi: 10.1016/j.tim.2014.04.003

Park, H. D., Guinn, K. M., Harrell, M. I., Liao, R., Voskuil, M. I., Tompa, M., et al. (2003). Rv3133c/dosR is a transcription factor that mediates the hypoxic response of *Mycobacterium tuberculosis*. *Mol. Microbiol.* 48, 833–843. doi: 10.1046/j.1365-2958.2003.03474.x

Persky, N. S., Ferullo, D. J., Cooper, D. L., Moore, H. R., and Lovett, S. T. (2009). The ObgE/CgtA GTPase influences the stringent response to amino acid starvation in *Escherichia coli*. *Mol. Microbiol.* 73, 253–266. doi: 10.1111/j.1365-2958.2009.06767.x

Pisu, D., Provvedi, R., Espinosa, D. M., Payan, J. B., Boldrin, F., Palu, G., et al. (2017). The alternative Sigma Factors SigE and SigB are involved in tolerance and persistence to antitubercular drugs. *Antimicrob. Agents Chemother.* 61, e02267-17. doi: 10.1128/AAC.01596-17

Pizarro-Cerda, J., Charbit, A., Enninga, J., Lafont, F., and Cossart, P. (2016). Manipulation of host membranes by the bacterial pathogens *Listeria*, *Francisella*, *Shigella* and *Yersinia*. *Semin. Cell Dev. Biol.* 60, 155–167. doi: 10.1016/j.semcdb.2016.07.019

Potrykus, K., and Cashel, M. (2008). (p)ppGpp: still magical? *Annu. Rev. Microbiol.* 62, 35–51. doi: 10.1146/annurev.micro.62.081307.162903

Prax, M., and Bertram, R. (2014). Metabolic aspects of bacterial persisters. *Front. Cell Infect. Microbiol.* 4, 148. doi: 10.3389/fcimb.2014.00148

Price, J. V., and Vance, R. E. (2014). The macrophage paradox. *Immunity* 41, 685–693. doi: 10.1016/j.immuni.2014.10.015

Primm, T. P., Andersen, S. J., Mizrahi, V., Avarbock, D., Rubin, H., and Barry, C. E.,. 3. (2000). The stringent response of *Mycobacterium tuberculosis* is required for long-term survival. *J. Bacteriol* 182, 4889–4898. doi: 10.1128/JB.182.17.4889-4898.2000

Prusty, B. K., Böhme, L., Bergmann, B., Siegl, C., Krause, E., Mehlitz, A., et al. (2012). Imbalanced oxidative stress causes chlamydial persistence during non-productive human herpes virus co-infection. *PloS One* 7, e47427. doi: 10.1371/journal.pone.0047427

Puckett, S., Trujillo, C., Eoh, H., Marrero, J., Spencer, J., Jackson, M., et al. (2014). Inactivation of fructose-1,6-bisphosphate aldolase prevents optimal co-catabolism of glycolytic and gluconeogenic carbon substrates in *Mycobacterium tuberculosis*. *PloS Pathog.* 10, e1004144. doi: 10.1371/journal.ppat.1004144

Radzikowski, J. L., Vedelaar, S., Siegel, D., Ortega, A. D., Schmidt, A., and Heinemann, M. (2016). Bacterial persistence is an active sigmaS stress response to metabolic flux limitation. *Mol. Syst. Biol.* 12, 882. doi: 10.15252/msb.20166998

Radzikowski, J. L., Schramke, H., and Heinemann, M. (2017). Bacterial persistence from a system-level perspective. *Curr. Opin. Biotechnol.* 46, 98–105. doi: 10.1016/j.copbio.2017.02.012

Ramamurthy, T., Ghosh, A., Pazhani, G. P., and Shinoda, S. (2014). Current perspectives on viable but non-culturable (VBNC) pathogenic bacteria. *Front. Public Health* 2, 103. doi: 10.3389/fpubh.2014.00103

Ray, K., Marteyn, B., Sansonetti, P. J., and Tang, C. M. (2009). Life on the inside: the intracellular lifestyle of cytosolic bacteria. *Nat. Rev. Microbiol.* 7, 333–340. doi: 10.1038/nrmicro2112

Ronneau, S., and Helaine, S. (2019). Clarifying the link between toxin-antitoxin modules and bacterial persistence. *J. Mol. Biol.* 431, 3462–3471. doi: 10.1016/j.jmb.2019.03.019

Rotem, E., Loinger, A., Ronin, I., Levin-Reisman, I., Gabay, C., Shoresh, N., et al. (2010). Regulation of phenotypic variability by a threshold-based mechanism underlies bacterial persistence. *Proc. Natl. Acad. Sci. U. S. A.* 107, 12541–12546. doi: 10.1073/pnas.1004333107

Rowe, S. E., Conlon, B. P., Keren, I., and Lewis, K. (2016). Persisters: methods for isolation and identifying contributing factors - a review. *Methods Mol. Biol.* 1333, 17–28. doi: 10.1007/978-1-4939-2854-5_2

Rustad, T. R., Sherrid, A. M., Minch, K. J., and Sherman, D. R. (2009). Hypoxia: a window into *Mycobacterium tuberculosis* latency. *Cell Microbiol.* 11, 1151–1159. doi: 10.1111/j.1462-5822.2009.01325.x

Rycroft, J. A., Gollan, B., Grabe, G. J., Hall, A., Cheverton, A. M., Larrouy-Maumus, G., et al. (2018). Activity of acetyltransferase toxins involved in *Salmonella* persister formation during macrophage infection. *Nat. Commun.* 9, 1993. doi: 10.1038/s41467-018-04472-6

Sabbagh, P., Karkhah, A., Nouri, H. R., Javanian, M., and Ebrahimpour, S. (2018). The significance role of regulatory T cells in the persistence of infections by intracellular bacteria. *Infect. Genet. Evol.* 62, 270–274. doi: 10.1016/j.meegid.2018.05.001

Saxena, R., Fingland, N., Patil, D., Sharma, A. K., and Crooke, E. (2013). Crosstalk between DnaA protein, the initiator of *Escherichia coli* chromosomal replication, and acidic phospholipids present in bacterial membranes. *Int. J. Mol. Sci.* 14, 8517–8537. doi: 10.3390/ijms14048517

Schellhorn, H. E. (2014). Elucidating the function of the RpoS regulon. *Future Microbiol.* 9, 497–507. doi: 10.2217/fmb.14.9

Schmitz-Esser, S., Linka, N., Collingro, A., Beier, C. L., Neuhaus, H. E., Wagner, M., et al. (2004). ATP/ADP translocases: a common feature of obligate intracellular amoebal symbionts related to *Chlamydiae* and *Rickettsiae*. *J. Bacteriol* 186, 683–691. doi: 10.1128/JB.186.3.683-691.2004

Schreiber, G., Ron, E. Z., and Glaser, G. (1995). ppGpp-mediated regulation of DNA replication and cell division in *Escherichia coli*. *Curr. Microbiol* 30, 27–32. doi: 10.1007/BF00294520

Schuessler, D. L., Cortes, T., Fivian-Hughes, A. S., Lougheed, K. E., Harvey, E., Buxton, R. S., et al. (2013). Induced ectopic expression of HigB toxin in *Mycobacterium tuberculosis* results in growth inhibition, reduced abundance of a subset of mRNAs and cleavage of tmRNA. *Mol. Microbiol.* 90, 195–207. doi: 10.1111/mmi.12358

Sekimizu, K., Bramhill, D., and Kornberg, A. (1988). Sequential early stages in the in vitro initiation of replication at the origin of the *Escherichia coli* chromosome. *J. Biol. Chem.* 263, 7124–7130.

Setlow, P. (2014). Germination of spores of *Bacillus* species: what we know and do not know. *J. Bacteriol.* 196, 1297–1305. doi: 10.1128/JB.01455-13

Seyfzadeh, M., Keener, J., and Nomura, M. (1993). spoT-dependent accumulation of guanosine tetraphosphate in response to fatty acid starvation in *Escherichia coli*. *Proc. Natl. Acad. Sci. U. S. A.* 90, 11004–11008. doi: 10.1073/pnas.90.23.11004

Shan, Y., Brown Gandt, A., Rowe, S. E., Deisinger, J. P., Conlon, B. P., and Lewis, K. (2017). ATP-dependent persister formation in *Escherichia coli*. *mBio* 8, e02267-16. doi: 10.1128/mBio.02267-16

Sheik, A. R., Muller, E. E., Audinot, J. N., Lebrun, L. A., Grysan, P., Guignard, C., et al. (2016). *In situ* phenotypic heterogeneity among single cells of the filamentous bacterium *Candidatus* Microthrix parvicella. *ISME J.* 10, 1274–1279. doi: 10.1038/ismej.2015.181

Sheppard, M., Webb, C., Heath, F., Mallows, V., Emilianus, R., Maskell, D., et al. (2003). Dynamics of bacterial growth and distribution within the liver during *Salmonella* infection. *Cell Microbiol.* 5, 593–600. doi: 10.1046/j.1462-5822.2003.00296.x

Shima, K., Käding, N., Ogunsulire, I. M., Kaufhold, I., Klinger, M., and Rupp, J. (2018). Interferon-gamma interferes with host cell metabolism during intracellular *Chlamydia trachomatis* infection. *Cytokine* 112, 95–101. doi: 10.1016/j.cyto.2018.05.039

Sigle, S., Ladwig, N., Wohlleben, W., and Muth, G. (2015). Synthesis of the spore envelope in the developmental life cycle of *Streptomyces coelicolor*. *Int. J. Med. Microbiol.* 305, 183–189. doi: 10.1016/j.ijmm.2014.12.014

Sikora, A. E., Zielke, R., Wegrzyn, A., and Wegrzyn, G. (2006). DNA replication defect in the *Escherichia coli* cgtA(ts) mutant arising from reduced DnaA levels. *Arch. Microbiol.* 185, 340–347. doi: 10.1007/s00203-006-0099-3

Sikora-Borgula, A., Slominska, M., Trzonkowski, P., Zielke, R., Mysliwski, A., Wegrzyn, G., et al. (2002). A role for the common GTP-binding protein in coupling of chromosome replication to cell growth and cell division. *Biochem. Biophys. Res. Commun.* 292, 333–338. doi: 10.1006/bbrc.2002.6671

Simsek, E., and Kim, M. (2018). The emergence of metabolic heterogeneity and diverse growth responses in isogenic bacterial cells. *ISME J.* 12, 1199–1209. doi: 10.1038/s41396-017-0036-2

Skarstad, K., and Katayama, T. (2013). Regulating DNA replication in bacteria. *Cold Spring Harb. Perspect. Biol.* 5, a012922. doi: 10.1101/cshperspect.a012922

Skilton, R. J., Cutcliffen, L. T., Barlow, D., Wang, Y., Salim, O., Lambden, P. R., et al. (2009). Penicillin induced persistence in *Chlamydia trachomatis*: high quality time lapse video analysis of the developmental cycle. *PloS One* 4, e7723. doi: 10.1371/journal.pone.0007723

Slattery, A., Victorsen, A. H., Brown, A., Hillman, K., and Phillips, G. J. (2013). Isolation of highly persistent mutants of *Salmonella enterica* serovar typhimurium reveals a new toxin-antitoxin module. *J. Bacteriol.* 195, 647–657. doi: 10.1128/JB.01397-12

Slayden, R. A., Dawson, C. C., and Cummings, J. E. (2018). Toxin-antitoxin systems and regulatory mechanisms in *Mycobacterium tuberculosis*. *Pathog. Dis.* 76, 4 doi: 10.1093/femspd/fty039

Socolovschi, C., Audoly, G., and Raoult, D. (2013). Connection of toxin-antitoxin modules to inoculation eschar and arthropod vertical transmission in Rickettsiales. *Comp. Immunol. Microbiol. Infect. Dis.* 36, 199–209. doi: 10.1016/j.cimid.2013.01.001

Sohaskey, C. D., and Voskuil, M. I. (2015). *In vitro* models that utilize hypoxia to induce non-replicating persistence in Mycobacteria. *Methods Mol. Biol.* 1285, 201–213. doi: 10.1007/978-1-4939-2450-9_11

Sohaskey, C. D. (2008). Nitrate enhances the survival of *Mycobacterium tuberculosis* during inhibition of respiration. *J. Bacteriol* 190, 2981–2986. doi: 10.1128/JB.01857-07

Spoering, A. L., and Lewis, K. (2001). Biofilms and planktonic cells of *Pseudomonas aeruginosa* have similar resistance to killing by antimicrobials. *J. Bacteriol* 183, 6746–6751. doi: 10.1128/JB.183.23.6746-6751.2001

Spoering, A. L., Vulic, M., and Lewis, K. (2006). GlpD and PlsB participate in persister cell formation in *Escherichia coli*. *J. Bacteriol* 188, 5136–5144. doi: 10.1128/JB.00369-06

Srinivas, V., Arrieta-Ortiz, M. L., Kaur, A., Peterson, E. J. R., and Baliga, N. S. (2020). PerSort facilitates characterization and elimination of persister subpopulation in mycobacteria. *mSystems* 5, e01127–20. doi: 10.1128/mSystems.01127-20

Swanson, M. S., and Fernandez-Moreira, E. (2002). A microbial strategy to multiply in macrophages: the pregnant pause. *Traffic* 3, 170–177. doi: 10.1034/j.1600-0854.2002.030302.x

Tan, I. S., and Ramamurthi, K. S. (2014). Spore formation in *Bacillus subtilis*. *Environ. Microbiol. Rep.* 6, 212–225. doi: 10.1111/1758-2229.12130

Taylor, C. M., Beresford, M., Epton, H. A., Sigee, D. C., Shama, G., Andrew, P. W., et al. (2002). *Listeria monocytogenes* relA and hpt mutants are impaired in surface-attached growth and virulence. *J. Bacteriol.* 184, 621–628. doi: 10.1128/JB.184.3.621-628.2002

Thakur, Z., Saini, V., Arya, P., Kumar, A., and Mehta, P. K. (2018). Computational insights into promoter architecture of toxin-antitoxin systems of *Mycobacterium tuberculosis*. *Gene* 641, 161–171. doi: 10.1016/j.gene.2017.10.054

Thakur, A., Mikkelsen, H., and Jungersen, G. (2019). Intracellular pathogens: Host immunity and microbial persistence strategies. *J. Immunol. Res.* 2019, 1356540. doi: 10.1155/2019/1356540

Thompson, A., Fulde, M., and Tedin, K. (2018). The metabolic pathways utilized by *Salmonella Typhimurium* during infection of host cells. *Environ. Microbiol. Rep.* 10, 140–154. doi: 10.1111/1758-2229.12628

Thöne, F., Schwanhäusser, B., Becker, D., Ballmaier, M., and Bumann, D. (2007). FACS-isolation of *Salmonella*-infected cells with defined bacterial load from mouse spleen. *J. Microbiol. Methods* 71, 220–224. doi: 10.1016/j.mimet.2007.08.016

Tian, C., Semsey, S., and Mitarai, N. (2017). Synchronized switching of multiple toxin-antitoxin modules by (p)ppGpp fluctuation. *Nucleic Acids Res.* 45, 8180–8189. doi: 10.1093/nar/gkx552

Tjaden, J., Winkler, H. H., Schwöppe, C., Van Der Laan, M., Möhlmann, T., and Neuhaus, H. E. (1999). Two nucleotide transport proteins in *Chlamydia trachomatis*, one for net nucleoside triphosphate uptake and the other for transport of energy. *J. Bacteriol.* 181, 1196–1202. doi: 10.1128/JB.181.4.1196-1202.1999

Tkachenko, A. G., Kashevarova, N. M., Karavaeva, E. A., and Shumkov, M. S. (2014). Putrescine controls the formation of *Escherichia coli* persister cells tolerant to aminoglycoside netilmicin. *FEMS Microbiol. Lett.* 361, 25–33. doi: 10.1111/1574-6968.12613

Tosic-Pajic, J., Seklic, D., Radenkovic, J., Markovic, S., Cukic, J., Baskic, D., et al. (2017). Augmented oxidative stress in infertile women with persistent chlamydial infection. *Reprod. Biol.* 17, 120–125. doi: 10.1016/j.repbio.2017.03.001

Trastoy, R., Manso, T., Fernandez-Garcia, L., Blasco, L., Ambroa, A., Perez Del Molino, M. L., et al. (2018). Mechanisms of bacterial tolerance and persistence in the gastrointestinal and respiratory environments. *Clin. Microbiol. Rev.* 31, e00023-18. doi: 10.1128/CMR.00023-18

Ulrichs, T., and Kaufmann, S. H. (2002). Mycobacterial persistence and immunity. *Front. Biosci.* 7, d458–d469. doi: 10.2741/A788

Unoson, C., and Wagner, E. G. (2008). A small SOS-induced toxin is targeted against the inner membrane in. *Escherichia coli Mol. Microbiol.* 70, 258–270. doi: 10.1111/j.1365-2958.2008.06416.x

Van Den Bergh, B., Fauvart, M., and Michiels, J. (2017). Formation, physiology, ecology, evolution and clinical importance of bacterial persisters. *FEMS Microbiol. Rev.* 41, 219–251. doi: 10.1093/femsre/fux001

Vanderven, B. C., Fahey, R. J., Lee, W., Liu, Y., Abramovitch, R. B., Memmott, C., et al. (2015). Novel inhibitors of cholesterol degradation in *Mycobacterium tuberculosis* reveal how the bacterium's metabolism is constrained by the intracellular environment. *PloS Pathog.* 11, e1004679. doi: 10.1371/journal.ppat.1004679

Vandrisse, C. M., Parks, A. R., and Escalante-Semerena, J. C. (2017). A toxin involved in *Salmonella* persistence regulates its activity by acetylating its cognate antitoxin, a modification reversed by CobB sirtuin deacetylase. *mBio* 8, e00708-17. doi: 10.1128/mBio.00708-17

Vazquez-Laslop, N., Lee, H., and Neyfakh, A. A. (2006). Increased persistence in *Escherichia coli* caused by controlled expression of toxins or other unrelated proteins. *J. Bacteriol.* 188, 3494–3497. doi: 10.1128/JB.188.10.3494-3497.2006

Vega, N. M., Allison, K. R., Khalil, A. S., and Collins, J. J. (2012). Signaling-mediated bacterial persister formation. *Nat. Chem. Biol.* 8, 431–433. doi: 10.1038/nchembio.915

Verstraeten, N., Knapen, W. J., Kint, C. I., Liebens, V., Van Den Bergh, B., Dewachter, L., et al. (2015). Obg and membrane depolarization are part of a microbial bet-hedging strategy that leads to antibiotic tolerance. *Mol. Cell* 59, 9–21. doi: 10.1016/j.molcel.2015.05.011

Völzing, K. G., and Brynildsen, M. P. (2015). Stationary-phase persisters to ofloxacin sustain DNA damage and require repair systems only during recovery. *mBio* 6, e00731–e00715. doi: 10.1128/mBio.00731-15

Voskuil, M. I., Schnappinger, D., Visconti, K. C., Harrell, M. I., Dolganov, G. M., Sherman, D. R., et al. (2003). Inhibition of respiration by nitric oxide induces a *Mycobacterium tuberculosis* dormancy program. *J. Exp. Med.* 198, 705–713. doi: 10.1084/jem.20030205

Walawalkar, Y. D., Vaidya, Y., and Nayak, V. (2016). Response of *Salmonella* Typhi to bile-generated oxidative stress: implication of quorum sensing and persister cell populations. *Pathog. Dis.* 74, ftw090. doi: 10.1093/femspd/ftw090

Wang, X., and Wood, T. K. (2011). Toxin-antitoxin systems influence biofilm and persister cell formation and the general stress response. *Appl. Environ. Microbiol.* 77, 5577–5583. doi: 10.1128/AEM.05068-11

Wang, X., Kim, Y., Hong, S. H., Ma, Q., Brown, B. L., Pu, M., et al. (2011). Antitoxin MqsA helps mediate the bacterial general stress response. *Nat. Chem. Biol.* 7, 359–366. doi: 10.1038/nchembio.560

Wang, T., El Meouche, I., and Dunlop, M. J. (2017). Bacterial persistence induced by salicylate via reactive oxygen species. *Sci. Rep.* 7, 43839. doi: 10.1038/srep43839

Watson, K. G., and Holden, D. W. (2010). Dynamics of growth and dissemination of *Salmonella in vivo*. *Cell Microbiol.* 12, 1389–1397. doi: 10.1111/j.1462-5822.2010.01511.x

Wayne, L. G., and Hayes, L. G. (1998). Nitrate reduction as a marker for hypoxic shiftdown of *Mycobacterium tuberculosis*. *Tuber Lung Dis.* 79, 127–132. doi: 10.1054/tuld.1998.0015

Wayne, L. G., and Lin, K. Y. (1982). Glyoxylate metabolism and adaptation of *Mycobacterium tuberculosis* to survival under anaerobic conditions. *Infect. Immun.* 37, 1042–1049. doi: 10.1128/IAI.37.3.1042-1049.1982

Weber, H., Polen, T., Heuveling, J., Wendisch, V. F., and Hengge, R. (2005). Genome-wide analysis of the general stress response network in *Escherichia coli*: sigmaS-dependent genes, promoters, and sigma factor selectivity. *J. Bacteriol.* 187, 1591–1603. doi: 10.1128/JB.187.5.1591-1603.2005

Weiss, G., and Schaible, U. E. (2015). Macrophage defense mechanisms against intracellular bacteria. *Immunol. Rev.* 264, 182–203. doi: 10.1111/imr.12266

Weiss, L. A., and Stallings, C. L. (2013). Essential roles for *Mycobacterium tuberculosis* Rel beyond the production of (p)ppGpp. *J. Bacteriol.* 195, 5629–5638. doi: 10.1128/JB.00759-13

Willett, J. W., Herrou, J., Czyz, D. M., Cheng, J. X., and Crosson, S. (2016). *Brucella abortus* DeltarpoE1 confers protective immunity against wild type challenge in a mouse model of brucellosis. *Vaccine* 34, 5073–5081. doi: 10.1016/j.vaccine.2016.08.076

Wilmaerts, D., Bayoumi, M., Dewachter, L., Knapen, W., Mika, J. T., Hofkens, J., et al. (2018). The persistence-inducing toxin HokB forms dynamic pores that cause ATP leakage. *mBio* 9, e00744–18. doi: 10.1128/mBio.00744-18

Wilmaerts, D., Windels, E. M., Verstraeten, N., and Michiels, J. (2019). General mechanisms leading to persister formation and awakening. *Trends Genet.* 35, 401–411. doi: 10.1016/j.tig.2019.03.007

Winther, K., Tree, J. J., Tollervey, D., and Gerdes, K. (2016). VapCs of *Mycobacterium tuberculosis* cleave RNAs essential for translation. *Nucleic Acids Res.* 44, 9860–9871. doi: 10.1093/nar/gkw781

Witkin, S. S., Minis, E., Athanasiou, A., Leizer, J., and Linhares, I. M. (2017). *Chlamydia trachomatis*: the persistent pathogen. *Clin. Vaccine Immunol.* 24, e00203–17. doi: 10.1128/CVI.00203-17

Wolanski, M., Jakimowicz, D., and Zakrzewska-Czerwinska, J. (2014). Fifty years after the replicon hypothesis: cell-specific master regulators as new players in chromosome replication control. *J. Bacteriol.* 196, 2901–2911. doi: 10.1128/JB.01706-14

Wolf, B., Newman, A., and Glaser, D. A. (1968). On the origin and direction of replication of the *Escherichia coli* K12 chromosome. *J. Mol. Biol.* 32, 611–629. doi: 10.1016/0022-2836(68)90346-X

Wood, T. K., Knabel, S. J., and Kwan, B. W. (2013). Bacterial persister cell formation and dormancy. *Appl. Environ. Microbiol.* 79, 7116–7121. doi: 10.1128/AEM.02636-13

Wood, T. K. (2017). Strategies for combating persister cell and biofilm infections. *Microb. Biotechnol.* 10, 1054–1056. doi: 10.1111/1751-7915.12774

Wu, Y., Vulic, M., Keren, I., and Lewis, K. (2012). Role of oxidative stress in persister tolerance. *Antimicrob. Agents Chemother.* 56, 4922–4926. doi: 10.1128/AAC.00921-12

Wyrick, P. B. (2010). *Chlamydia trachomatis* persistence in vitro: an overview. *J. Infect. Dis.* 201 Suppl 2, S88–S95. doi: 10.1086/652394

Xavier, M. N., Winter, M. G., Spees, A. M., Den Hartigh, A. B., Nguyen, K., Roux, C. M., et al. (2013a). PPARgamma-mediated increase in glucose availability sustains chronic *Brucella abortus* infection in alternatively activated macrophages. *Cell Host Microbe* 14, 159–170. doi: 10.1016/j.chom.2013.07.009

Xavier, M. N., Winter, M. G., Spees, A. M., Nguyen, K., Atluri, V. L., Silva, T. M., et al. (2013b). CD4+ T cell-derived IL-10 promotes *Brucella abortus* persistence via modulation of macrophage function. *PloS Pathog.* 9, e1003454. doi: 10.1371/journal.ppat.1003454

Xiao, H., Kalman, M., Ikehara, K., Zemel, S., Glaser, G., and Cashel, M. (1991). Residual guanosine 3',5'-bispyrophosphate synthetic activity of relA null mutants can be eliminated by spoT null mutations. *J. Biol. Chem.* 266, 5980–5990.

Xu, H. S., Roberts, N., Singleton, F. L., Attwell, R. W., Grimes, D. J., and Colwell, R. R. (1982). Survival and viability of nonculturable *Escherichia coli* and *Vibrio cholerae* in the estuarine and marine environment. *Microb. Ecol.* 8, 313–323. doi: 10.1007/BF02010671

Xue, Y., Zheng, H., Mai, Z., Qin, X., Chen, W., Huang, T., et al. (2017). An *in vitro* model of azithromycin-induced persistent *Chlamydia trachomatis* infection. *FEMS Microbiol. Lett.* 364, 14. doi: 10.1093/femsle/fnx145

Yamaguchi, Y., Park, J. H., and Inouye, M. (2011). Toxin-antitoxin systems in bacteria and archaea. *Annu. Rev. Genet.* 45, 61–79. doi: 10.1146/annurev-genet-110410-132412

Yamaki, H., Ariga, H., and Tanaka, N. (1986). Inhibition of SV40 DNA replication *in vitro* by 1-N-acyl-3"-N-(trifluoroacetyl) kanamycin. *Biochem. Biophys. Res. Commun.* 136, 322–328. doi: 10.1016/0006-291X(86)90913-7

Yang, Q. E., and Walsh, T. R. (2017). Toxin-antitoxin systems and their role in disseminating and maintaining antimicrobial resistance. *FEMS Microbiol. Rev.* 41, 343–353. doi: 10.1093/femsre/fux006

Yang, J. H., Bhargava, P., Mccloskey, D., Mao, N., Palsson, B. O., and Collins, J. J. (2017). Antibiotic-induced changes to the host metabolic environment inhibit drug efficacy and alter immune function. *Cell Host Microbe* 22757-765.e753. doi: 10.1016/j.chom.2017.10.020

Yang, M., Rajeeve, K., Rudel, T., and Dandekar, T. (2019). Comprehensive flux modeling of *Chlamydia trachomatis* proteome and qRT-PCR data indicate biphasic metabolic differences between elementary bodies and reticulate bodies during infection. *Front. Microbiol.* 10, 2350. doi: 10.3389/fmicb.2019.02350

Yukihira, D., Fujimura, Y., Wariishi, H., and Miura, D. (2015). Bacterial metabolism in immediate response to nutritional perturbation with temporal and network view of metabolites. *Mol. Biosyst.* 11, 2473–2482. doi: 10.1039/C5MB00182J

Zampieri, M., Zimmermann, M., Claassen, M., and Sauer, U. (2017). Nontargeted metabolomics reveals the multilevel response to antibiotic perturbations. *Cell Rep.* 19, 1214–1228. doi: 10.1016/j.celrep.2017.04.002

Zhang, Q., Zhou, A., Li, S., Ni, J., Tao, J., Lu, J., et al. (2016). Reversible lysine acetylation is involved in DNA replication initiation by regulating activities of initiator DnaA in *Escherichia coli*. *Sci. Rep.* 6, 30837. doi: 10.1038/srep30837

Zhao, X., Zhong, J., Wei, C., Lin, C. W., and Ding, T. (2017). Current perspectives on viable but non-culturable state in foodborne pathogens. *Front. Microbiol.* 8, 580. doi: 10.3389/fmicb.2017.00580

Zuck, M., Feng, C., and Hybiske, K. (2015). Using fluorescent proteins to visualize and quantitate *Chlamydia* vacuole growth dynamics in living cells. *J. Vis. Exp.* 13, 51131. doi: 10.3791/51131

Zyskind, J. W., and Smith, D. W. (1992). DNA replication, the bacterial cell cycle, and cell growth. *Cell* 69, 5–8. doi: 10.1016/0092-8674(92)90112-P

Zyskind, J. W., Deen, L. T., and Smith, D. W. (1977). Temporal sequence of events during the initiation process in *Escherichia coli* deoxyribonucleic acid replication: roles of the dnaA and dnaC gene products and ribonucleic acid polymerase. *J. Bacteriol.* 129, 1466–1475. doi: 10.1128/JB.129.3.1466-1475.1977

Permissions

All chapters in this book were first published by Frontiers; hereby published with permission under the Creative Commons Attribution License or equivalent. Every chapter published in this book has been scrutinized by our experts. Their significance has been extensively debated. The topics covered herein carry significant findings which will fuel the growth of the discipline. They may even be implemented as practical applications or may be referred to as a beginning point for another development.

The contributors of this book come from diverse backgrounds, making this book a truly international effort. This book will bring forth new frontiers with its revolutionizing research information and detailed analysis of the nascent developments around the world.

We would like to thank all the contributing authors for lending their expertise to make the book truly unique. They have played a crucial role in the development of this book. Without their invaluable contributions this book wouldn't have been possible. They have made vital efforts to compile up to date information on the varied aspects of this subject to make this book a valuable addition to the collection of many professionals and students.

This book was conceptualized with the vision of imparting up-to-date information and advanced data in this field. To ensure the same, a matchless editorial board was set up. Every individual on the board went through rigorous rounds of assessment to prove their worth. After which they invested a large part of their time researching and compiling the most relevant data for our readers.

The editorial board has been involved in producing this book since its inception. They have spent rigorous hours researching and exploring the diverse topics which have resulted in the successful publishing of this book. They have passed on their knowledge of decades through this book. To expedite this challenging task, the publisher supported the team at every step. A small team of assistant editors was also appointed to further simplify the editing procedure and attain best results for the readers.

Apart from the editorial board, the designing team has also invested a significant amount of their time in understanding the subject and creating the most relevant covers. They scrutinized every image to scout for the most suitable representation of the subject and create an appropriate cover for the book.

The publishing team has been an ardent support to the editorial, designing and production team. Their endless efforts to recruit the best for this project, has resulted in the accomplishment of this book. They are a veteran in the field of academics and their pool of knowledge is as vast as their experience in printing. Their expertise and guidance has proved useful at every step. Their uncompromising quality standards have made this book an exceptional effort. Their encouragement from time to time has been an inspiration for everyone.

The publisher and the editorial board hope that this book will prove to be a valuable piece of knowledge for researchers, students, practitioners and scholars across the globe.

List of Contributors

Thomas E. Wood
Division of Infectious Diseases, Department of Medicine, Massachusetts General Hospital, Boston, MA, United States
Department of Microbiology, Harvard Medical School, Boston, MA, United States

Ezra Aksoy
Centre for Biochemical Pharmacology, William Harvey Research Institute, Queen Mary University of London, London, United Kingdom

Abderrahman Hachani
Department of Microbiology and Immunology, University of Melbourne at the Peter Doherty Institute for Infection and Immunity, Melbourne, VIC, Australia

Julian Pechstein, Jan Schulze-Luehrmann, Stephanie Bisle, Martha Ölke and Anja Lührmann
Mikrobiologisches Institut-Klinische Mikrobiologie, Immunologie und Hygiene, Universitätsklinikum Erlangen, Friedrich- Alexander-Universität Erlangen-Nürnberg, Erlangen, Germany

Franck Cantet and Matteo Bonazzi
Institut de Recherche en Infectiologie de Montpellier (IRIM), Centre National de la Recherche Scientifique (CNRS), Universite´ de Montpellier, Montpellier, France

Paul A. Beare
Coxiella Pathogenesis Section, Laboratory of Bacteriology, Rocky Mountain Laboratories, National Institute of Allergy and Infectious Diseases, National Institutes of Health, Hamilton, MT, United States

Christian Berens
Friedrich-Loeffler-Institut, Institut für Molekulare Pathogenese, Jena, Germany

Jacques Augenstreich and Volker Briken
Department of Cell Biology and Molecular Genetics, University of Maryland, College Park, MD, United States

Irshad Sharafutdinov, Delara Soltan Esmaeili, Aileen Harrer, Nicole Tegtmeyer and Steffen Backert
Department of Biology, Division of Microbiology, University of Erlangen-Nuremberg, Erlangen, Germany

Heinrich Sticht
Division of Bioinformatics, Institute of Biochemistry, University of Erlangen-Nuremberg, Erlangen, Germany

Sarah Wettstadt and Alain Filloux
MRC Centre for Molecular Bacteriology and Infection, Department of Life Sciences, Imperial College London, London, United Kingdom

Erh-Min Lai
Institute of Plant and Microbial Biology, Academia Sinica, Taipei, Taiwan

David R. Thomas, Patrice Newton, Nicole Lau and Hayley J. Newton
Department of Microbiology and Immunology, Peter Doherty Institute for Infection and Immunity, The University of Melbourne, Melbourne, VIC, Australia

Wenxin Mu, Xiaobing Chen, Haoyu Li, Zhi Wang and Bin Cheng
Hospital of Stomatology, Guanghua School of Stomatology, Guangdong Provincial Key Laboratory of Stomatology, Sun Yat-Sen University, Guangzhou, China

Yiqun Jia
Stomatology Center, Shenzhen People's Hospital, The Second Clinical Medical College of Jinan University, The First Affiliated Hospital of Southern University of Science and Technology, Shenzhen, China

Georgina L. Pollock
Centre for Innate Immunity and Infectious Diseases, Hudson Institute of Medical Research, Clayton, VIC, Australia

Jiyao Gan and Wibawa
Centre for Innate Immunity and Infectious Diseases, Hudson Institute of Medical Research, Clayton, VIC, Australia
Department of Microbiology and Immunology, University of Melbourne at the Peter Doherty Institute for Infection and Immunity, Melbourne, VIC, Australia

Nichollas E. Scott Joshua P. M. Newson and Rachelia R. Tania Wong Fok Lung
Department of Microbiology and Immunology, University of Melbourne at the Peter Doherty Institute for Infection and Immunity, Melbourne, VIC, Australia

Garrett Z. Ng and Ian van Driel
Department of Biochemistry and Molecular Biology, Bio21 Molecular Science and Biotechnology Institute, The University of Melbourne, Parkville, VIC, Australia

Jaclyn S. Pearson Elizabeth L. Hartland and Cristina Giogha
Centre for Innate Immunity and Infectious Diseases, Hudson Institute of Medical Research, Clayton, VIC, Australia
Department of Molecular and Translational Science, Monash University, Clayton, VIC, Australia

Tshegofatso Ngwaga, Deepika Chauhan and Stephanie R. Shames
Division of Biology, Kansas State University, Manhattan, KS, United States

Sabrina L. Slater and Gad Frankel
MRC Centre for Molecular Bacteriology and Infection, Department of Life Sciences, Imperial College London, London, United Kingdom

Jana Kamanova
Institute of Microbiology of the Czech Academy of Sciences, Prague, Czechia

Julia Monjarás Feria and Miguel A. Valvano
Wellcome-Wolfson Institute for Experimental Medicine, Queen's University Belfast, Belfast, United Kingdom

Tomoe Kitao
Department of Microbiology, Graduate School of Medicine, Gifu University, Gifu, Japan

Hiroki Nagai and Tomoko Kubori
Department of Microbiology, Graduate School of Medicine, Gifu University, Gifu, Japan
G-CHAIN, Gifu University, Gifu, Japan

Hazel Mak and Teresa L. M. Thurston
MRC Centre for Molecular Bacteriology and Infection, Imperial College London, London, United Kingdom

Ruth E. Dickenson and Charlotte Odendall
Department of Infectious Diseases, School of Immunology and Microbial Sciences, King's College London, London, United Kingdom

Noémie Alphonse
Department of Infectious Diseases, School of Immunology and Microbial Sciences, King's College London, London, United Kingdom
Immunoregulation Laboratory, Francis Crick Institute, London, United Kingdom

Wolfgang Eisenreich
Department of Chemistry, Chair of Biochemistry, Technische Universität München, Garching, Germany

Thomas Rudel
Chair of Microbiology, Biocenter, University of Würzburg, Würzburg, Germany

Jürgen Heesemann and Werner Goebel
Max von Pettenkofer-Institute, Ludwig Maximilian University of Munich, München, Germany

Index

www.ingramcontent.com/pod-product-compliance
Lightning Source LLC
LaVergne TN
LVHW081038210926
848485LV00004B/1049